AN INTRODUCTION TO

HUMAN GEOGRAPHY

ISSUES FOR THE 21ST CENTURY

Third Edition

Edited by

PETER DANIELS
School of Geography, Earth and Environmental Sciences,
University of Birmingham

MICHAEL BRADSHAW
Department of Geography, University of Leicester

DENIS SHAW
School of Geography, Earth and Environmental Sciences,
University of Birmingham

JAMES SIDAWAY
School of Geography, University of Plymouth

Harlow, England • London • New York • Boston • San Francisco • Toronto
Sydney • Tokyo • Singapore • Hong Kong • Seoul • Taipei • New Delhi
Cape Town • Madrid • Mexico City • Amsterdam • Munich • Paris • Milan

Pearson Education Limited

Edinburgh Gate
Harlow
Essex CM20 2JE
England

and Associated Companies throughout the world

Visit us on the World Wide Web at:
www.pearsoned.co.uk

First published 2001
Second edition 2005
Third edition 2008

ISBN: 978-0-13-205684-7

British Library Cataloguing-in-Publication Data
A catalogue record for this book is available from the British Library

Library of Congress Cataloging-in-Publication Data
An introduction to human geography : issues for the 21st century / edited by Peter
Daniels . . . [et al.]. — 3rd ed.
p. cm.
ISBN 978-0-13-205684-7 (paperback)
1. Human geography. I. Daniels, P. W.
GF41.I574 2008
304.2—dc22
2008008555

10 9 8 7 6 5 4 3 2 1
12 11 10 09 08

Typeset in 10/12.5pt Minion by 73
Printed and bound by Rotolito Lombarda, Italy

The publisher's policy is to use paper manufactured from sustainable forests.

AN INTRODUCTION TO

HUMAN GEOGRAPHY

Visit the *An Introduction to Human Geography*, *3rd edition* Companion Website at **www.pearsoned.co.uk/daniels** to find valuable **student** learning material including:

- Online tutorials offering valuable practical advice on how to improve your human geography skills
- Annotated weblinks to relevant internet resources to help with further research
- Online glossary of key terms
- Chapter-specific further reading, recommended by the authors
- Full list of chapter references for printing out
- Learning outcomes for each chapter

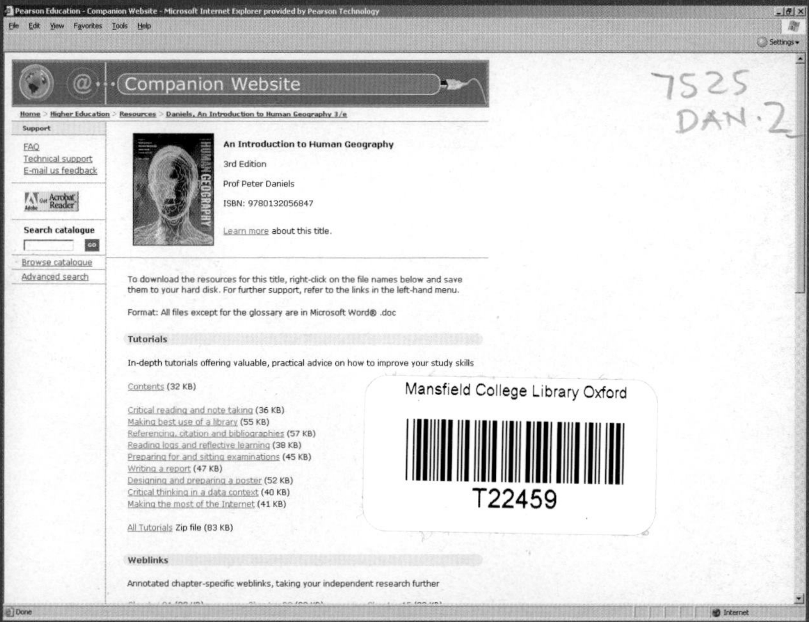

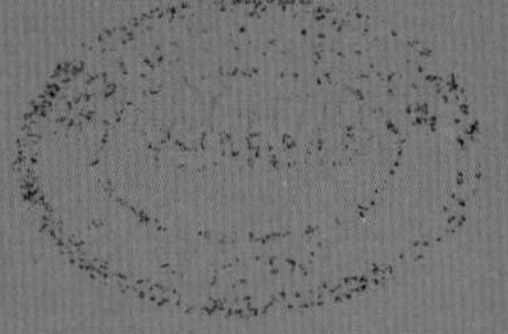

Dedicated to Jasmin Leila Sidaway

(see http://www.rgs.org/jasminleilaaward)

Brief contents

Contents in detail

SUPPORTING RESOURCES

Visit **www.pearsoned.co.uk/daniels** to find valuable online resources

Companion Website for students

- Online tutorials offering valuable practical advice on how to improve your human geography skills
- Annotated weblinks to relevant internet resources to help with further research
- Online glossary of key terms
- Chapter-specific further reading, recommended by the authors
- Full list of chapter references for printing out
- Learning outcomes for each chapter

For instructors

- PowerPoint slides, including useful figures and maps from the text
- Essay questions and exam questions with suggested answers for every chapter
- Suggested seminar activities based on the human geography study skills tutorials
- Learning outcomes for each chapter
- Online glossary

Also: The Companion Website provides the following features:

- Search tool to help locate specific items of content
- E-mail results and profile tools to send results of quizzes to instructors
- Online help and support to assist with website usage and troubleshooting

For more information please contact your local Pearson Education sales representative or visit **www.pearsoned.co.uk/daniels**

Guide to Spotlight boxes

		Page number	Historical perspectives	Models	Social and cultural themes	Globalization	Population and migration	Resource management	Geopolitics and the nation-state	Spatial inequalities	Economy and development	Cities and urban spaces	Agriculture and rural spaces	Money and space	Environmental issues
1.1	Bands	22	●		●										
1.2	Tribes	23	●		●								●		
1.3	Chiefdoms	24	●		●										
1.4	Regulated (pre-modern) states	24	●						●						
1.5	Market-based states	24	●												
2.1	Characteristics of states with a mature capitalist economy and society	40	●	●					●		●				
2.2	Characteristics of the period of merchant capitalism in Europe	42	●	●							●				
2.3	Proto-industrialization	51	●	●							●				
2.4	The factory system of production	51	●	●							●	●			
3.1	Organized capitalism	67		●	●						●				
3.2	Marxism	72		●	●										
3.3	Disorganized capitalism	78	●	●							●				
6.1	Sustainable development	140		●				●		●	●				●
6.2	Carbon offsetting	150				●		●							●
6.3	Environmental justice	153			●					●					●
7.1	Mobilizing the food chain	161		●		●		●			●		●		
7.2	Follow the thing: papaya	169				●		●			●		●		
7.3	Towards conventionalization? Fair trade and organic foods	171		●		●		●			●		●		
11.1	The hidden others of the countryside	250			●					●			●		
12.1	Climate porn	270													●
13.1	Cultures: a summary	274			●										
14.1	Perfect competition	297		●										●	
14.2	Transnational corporations	299		●		●				●				●	
14.3	Comparative advantage	308		●						●	●				
15.1	From Fordism to post-Fordism?	318	●	●		●					●				
15.2	Conceptualizing global divisions of labour?	323		●		●				●	●				
15.3	Types of clusters in the global economy?	328		●		●				●	●				
16.1	Services and the production process	341									●				
17.1	The functions of money	359												●	
17.2	The circulation of money as capital	367		●		●								●	
17.3	Rotating savings and credit associations (ROSCAs)	369		●	●									●	
17.4	Properties of monetary networks	370		●										●	
18.1	Consumerism as the end of culture?	382			●	●									
20.1	Irredentism and secessionism	429			●				●						
21.1	The Carter Doctrine: contexts and consequences	453	●						●						

Guide to Thematic Case studies

		Page number	Historical perspectives	Models	Social and cultural themes	Globalization	Population and migration	Resource management	Geopolitics and the nation-state	Spatial inequalities	Economy and development	Cities and urban spaces	Agriculture and rural spaces	Money and space	Environmental issues
1.1	The Roman Empire at its zenith (first to fourth centuries AD)	32	●						●	●					
1.2	European feudalism	33	●	●	●										
2.1	Summary of the *Laws of the Indies* by Philip II of Spain, 1573	48	●								●	●	●		
2.2	Tenement housing for industrial workers	59	●	●	●							●			
4.1	United Nations' Millennium Development Goal Number 4: reduce by two-thirds the mortality rate among children under five	92			●		●								
4.2	HIV/AIDS	93			●		●								
5.1	Nuclear power on the re-bound?	115						●		●					●
5.2	Energy security	126						●		●	●				●
5.3	Resource abundance: blessing or curse?	131			●			●		●	●				
6.1	Old-growth logging in Tasmania, Australia	137						●							●
6.2	Low impact development in Britain	141			●			●							●
6.3	Implementing a convention? The obstacles faced in tackling climate change	144				●		●		●	●				●
6.4	Consumption and the environment	149			●			●		●	●				●
6.5	Wilderness and conservation debates in Australia	154			●		●	●		●			●		●
6.6	What is your ecological footprint?	156		●		●		●		●					●
8.2	The 'strategic interests' of foreign aid	184							●		●				
8.3	Hurricane Katrina and the USA as 'Third World'	186			●					●	●				
8.5	States and the geopolitics of development	193	●			●			●	●	●				
8.6	Children and development	194			●				●	●	●				
8.7	The World Bank and the IMF	195			●	●			●	●	●			●	
8.8	Globalization, agriculture and development	200				●					●		●		
9.2	The creation of defended enclaves	214			●							●			
10.1	Geographies of institutional disinvestment	223			●					●	●	●			
10.2	Community opposition to asylum seekers	228			●		●					●			
12.1	The radical critique of limits to growth	259		●			●	●			●				●
12.2	Contesting wilderness	261						●					●		●
12.3	Desertification	265						●					●		●
12.4	Gay animals	266			●										
13.1	The globalization of culture: some examples	277			●	●					●				
13.2	Berlusconi breaks rank over Islam	281			●				●						
13.5	Bounded or hybrid national culture?	283			●	●			●						
13.6	Hybridity/Diaspora – some examples	285	●		●		●								

Guide to Thematic Case studies (cont.)

		Page number	Historical perspectives	Models	Social and cultural themes	Globalization	Population and migration	Resource management	Geopolitics and the nation-state	Spatial inequalities	Economy and development	Cities and urban spaces	Agriculture and rural spaces	Money and space	Environmental issues
14.1	Models of economic location	296	●	●						●	●				
14.2	An oligopoly: Nestlé	306				●					●				
16.2	ICT and the establishment of data-processing factories in the Caribbean and call centres in India	350			●						●				
16.3	Boeing's new spatial division of expertise	352				●				●	●				
18.5	Skateboarding and the consumption of urban space	390			●							●			
19.1	Marking territory	404			●				●			●			
19.2	'Ethnic' territorial cleansing	405	●		●				●						
19.4	Space, sport and gender	410			●							●			
21.1	The continuing Cold War?	457	●						●						

Guide to Regional Case studies

		Page number	Historical perspectives	Models	Social and cultural themes	Globalization	Population and migration	Resource management	Geopolitics and the nation-state	Spatial inequalities	Economy and development	Cities and urban spaces	Agriculture and rural spaces	Money and space	Environmental issues
3.1	Ireland in the twentieth century	75	●		●				●						
8.1	The United Kingdom and Uganda	182			●		●			●					
8.4	Bandung, non-alignment and the 'Third World'	187							●	●	●				
9.1	The Pearl River Delta, China	211										●	●		
10.3	Ethnic division in apartheid South Africa	230	●		●		●			●		●			
10.4	Gated communities in Uruguay	234			●					●		●			
11.1	Rural depopulation in the Pacific Islands: Niue	246			●		●			●			●		
11.2	Productivist agriculture in the Hereford cider industry	253									●		●		
13.3	Ubuntu and democracy in South Africa	282	●		●										
13.4	Zapatistas: democracy and indigenous rights in Chiapas, Mexico	282			●	●									
16.1	The changing economic structure of the West Midlands, United Kingdom	341								●	●				
17.1	From Bangkok to Whitley Bay . . .	365				●				●	●			●	
17.2	Grameen Bank	374			●						●			●	
18.1	Cultural geographies of McDonald's in East Asia	383			●	●									
18.2	Media products and cultural indigenization: the case of *Dallas*	385			●	●									
18.3	Western goods and cultural indigenization: the case of 'Les Sapeurs'	386			●	●									
18.4	The shopping mall experience: Woodfield Mall, Schaumberg, Illinois	388			●							●			
19.3	Religious-territorial segregation in Northern Ireland	407	●		●				●			●			
20.1	Nations without states: the Kurdish case	421	●						●						
20.2	Burma: a national geo-body or a mosaic of ethnic homelands?	423	●				●		●						

Contributors

Michael Bradshaw Professor of Human Geography, Department of Geography, University of Leicester. Economic geography of Russia; resource geography and globalization and energy security.

John R. Bryson Professor of Enterprise and Economic Geography, School of Geography, Earth and Environmental Sciences, University of Birmingham. Expertise-intensive industries; industrial design and competitiveness; spatial divisions of expertise; sustainability and innovation.

Neil Coe Senior Lecturer in Economic Geography, School of Environment and Development, University of Manchester. Global production networks and local economic development; geographies of local and transnational labour markets; geographies of innovation; institutional and network approaches to economic development.

Allan Cochrane Professor of Urban Studies, Faculty of Social Sciences, The Open University. Urban studies, particularly urban policy, the local state and restructuring.

Philip Crang Professor of Cultural Geography, Department of Geography, Royal Holloway, University of London. Cultural and economic geography, especially commodity culture.

Peter W. Daniels Professor of Geography, School of Geography, Earth and Environmental Sciences, University of Birmingham. Geography of service industries; services and globalization; trade, foreign direct investment and internationalization in services.

James Evans Lecturer in Geography, School of Environment and Development, University of Manchester. Ecological planning; local environmental governance; urban sustainability and the aesthetics of environmental science.

Carl Grundy-Warr Senior Lecturer in Geography, Department of Geography, National University of Singapore. Political geography; geopolitics; political geographies of forced displacement; political geographies and ecologies of environmental resource governance.

Phil Hubbard Professor of Urban Social Geography, Department of Geography, Loughborough University. Geographies of sexuality, gender and whiteness; urban consumption; geographies of higher education.

Brian Ilbery Professor of Rural Studies, Countryside and Community Research Institute, University of Gloucestershire. Agricultural change and policy; food supply chains and alternative food networks; property rights, tenancy and farm diversification.

Damian Maye Senior Research Fellow, Countryside and Community Research Institute, University of Gloucestershire. Agri-food restructuring; alternative food networks; farm tenancy; geographies of micro-brewing; biosecurity and disease.

Cheryl McEwan Senior Lecturer in Human Geography, Department of Geography, Durham University. Feminist and cultural geographies; post-colonial theory and development; geographies of transformation in South Africa.

Warwick E. Murray Associate Professor of Geography and Development Studies, School of Geography, Environment and Earth Sciences, Victoria University of Wellington, New Zealand. Development geography; globalization; rural change; Latin America and Oceania.

Jenny Pickerill Lecturer in Human Geography, Department of Geography, University of Leicester. Environmental and social justice; geographies of resistance and activism; Australian indigenous politics; green building solutions.

Jane Pollard Senior Lecturer in Urban and Regional Development Studies, School of Geography, Politics and Sociology, Newcastle University. Geographies of money and finance; political economy and regional economic development.

Marcus Power Lecturer in Human Geography, Department of Geography, Durham University. Geographies of development and post-colonialism; visuality and critical geopolitics; post-socialism and Lusophone Southern Africa.

John Round Lecturer in Human Geography, School of Geography, Earth and Environmental Sciences, University of Birmingham. Geographies of everyday life in post-Soviet regions; coping tactics as a response to economic marginalization; state–society relations.

Denis Shaw Reader in Russian Geography, School of Geography, Earth and Environmental Sciences, University of Birmingham. Historical and contemporary geographical change in Russia; history of geographical thought.

James Sidaway Professor of Human Geography, School of Geography, University of Plymouth. Development and political geography; history of geographic thought.

Terry Slater Reader in Historical Geography, School of Geography, Earth and Environmental Sciences, University of Birmingham. Historical geography; urban morphology; topographical development of medieval towns; urban conservation.

David Storey Senior Lecturer in Geography, Department of Applied Sciences, Geography and Archaeology, University of Worcester. Nationalism, territory and place; sport and national identity; rural change and development; heritage and rural place promotion.

Guided tour

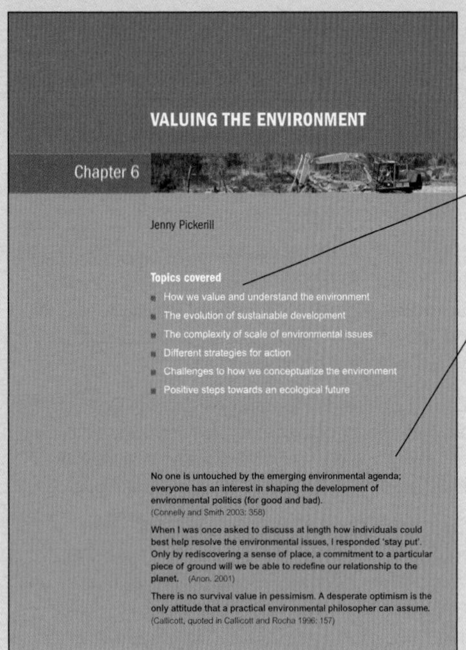

VALUING THE ENVIRONMENT

Chapter 6

Jenny Pickerill

Topics covered

- How we value and understand the environment
- The evolution of sustainable development
- The complexity of scale of environmental issues
- Different strategies for action
- Challenges to how we conceptualize the environment
- Positive steps towards an ecological future

No one is untouched by the emerging environmental agenda; everyone has an interest in shaping the development of environmental politics (for good and bad).
(Connelly and Smith 2003: 358)

When I was once asked to discuss at length how individuals could best help resolve the environmental issues, I responded 'stay put'. Only by rediscovering a sense of place, a commitment to a particular piece of ground will we be able to redefine our relationship to the planet. (Anon. 2001)

There is no survival value in pessimism. A desperate optimism is the only attitude that a practical environmental philosopher can assume.
(Callicott, quoted in Callicott and Rocha 1996: 157)

Topics covered introduce, at a glance, the key ideas and perspectives of the chapter.

Memorable **quotes** from important commentators past and present help contextualize and bring the subject to life.

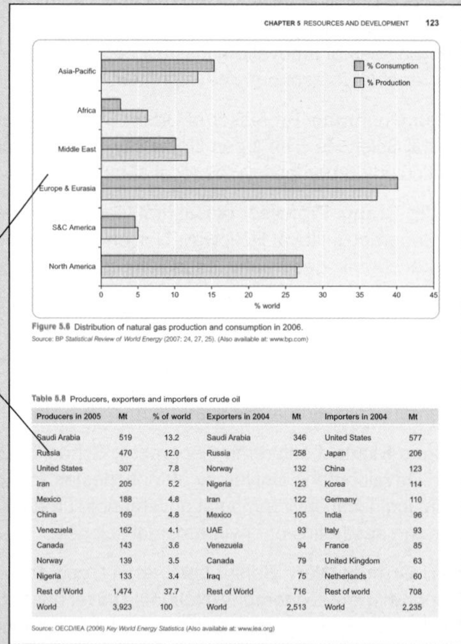

CHAPTER 5 RESOURCES AND DEVELOPMENT 123

Figure 5.8 Distribution of natural gas production and consumption in 2006.
Source: BP *Statistical Review of World Energy* (2007: 24, 27, 25). (Also available at: www.bp.com)

Table 5.8 Producers, exporters and importers of crude oil

Producers in 2005	Mt	% of world	Exporters in 2004	Mt	Importers in 2004	Mt
Saudi Arabia	519	13.2	Saudi Arabia	346	United States	577
Russia	470	12.0	Russia	258	Japan	206
United States	307	7.8	Norway	132	China	123
Iran	205	5.2	Nigeria	123	Korea	114
Mexico	188	4.8	Iran	122	Germany	110
China	169	4.8	Mexico	105	India	96
Venezuela	162	4.1	UAE	93	Italy	93
Canada	143	3.6	Venezuela	94	France	85
Norway	139	3.5	Canada	79	United Kingdom	63
Nigeria	133	3.4	Iraq	75	Netherlands	60
Rest of World	1,474	37.7	Rest of World	716	Rest of world	708
World	3,923	100	World	2,513	World	2,235

Source: OECD/IEA (2006) *Key World Energy Statistics* (Also available at: www.iea.org)

Figures and **tables** graphically illustrate the main points, concepts and processes in the text, to help your learning.

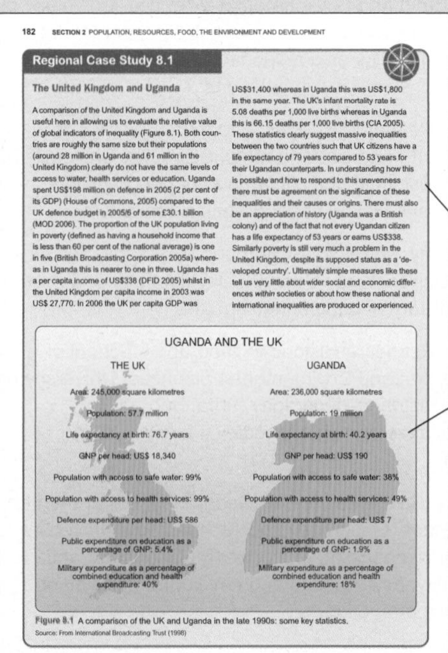

182 SECTION 2 POPULATION, RESOURCES, FOOD, THE ENVIRONMENT AND DEVELOPMENT

Regional Case Study 8.1

The United Kingdom and Uganda

A comparison of the United Kingdom and Uganda is useful here in allowing us to evaluate the relative value of global indicators of inequality (Figure 8.1). Both countries are roughly the same size but their populations (around 28 million in Uganda and 61 million in the United Kingdom) clearly do not have the same levels of access to water, health services or education. Uganda spent US$198 million on defence in 2005 (2 per cent of its GDP) (House of Commons, 2005) compared to the UK defence budget in 2005/6 of some £30.1 billion (MOD 2006). The proportion of the UK population living in poverty (defined as having a household income that is less than 60 per cent of the national average) is one in five (British Broadcasting Corporation 2005a) whereas in Uganda this is nearer to one in three. Uganda has a per capita income of US$338 (DFID 2005) whilst in the United Kingdom per capita income in 2003 was US$ 27,770. In 2006 the UK per capita GDP was US$31,400 whereas in Uganda this was US$1,800 in the same year. The UK's infant mortality rate is 5.08 deaths per 1,000 live births whereas in Uganda this is 66.15 deaths per 1,000 live births (CIA 2005). These statistics clearly suggest massive inequalities between the two countries such that UK citizens have a life expectancy of 79 years compared to 53 years for their Ugandan counterparts. In understanding how this is possible and how to respond to this unevenness there must be agreement on the significance of these inequalities and their causes or origins. There must also be an appreciation of history (Uganda was a British colony) and of the fact that not every Ugandan citizen has a life expectancy of 53 years or earns US$338. Similarly poverty is still very much a problem in the United Kingdom, despite its supposed status as a 'developed country'. Ultimately simple measures like these tell us very little about wider social and economic differences *within* societies or about how these national and international inequalities are produced or experienced.

Figure 8.1 A comparison of the UK and Uganda in the late 1990s: some key statistics.
Source: From International Broadcasting Trust (1998)

Regional case studies provide you with an insight into a real-life issue specific to a region.

A wide range of **maps**, both ancient and modern, create a clear visual representation of worldwide geographical data.

140 SECTION 2 POPULATION, RESOURCES, FOOD, THE ENVIRONMENT AND DEVELOPMENT

Spotlight Box 6.1

Sustainable development

The 1987 Brundtland Report adopted the position that it was possible to pursue economic growth without compromising the environment, and it is generally accepted as introducing the concept of sustainable development (the seeds of which had been sown some 15 years earlier). It also provided the first, and still widely used, definition of sustainable development: 'development which meets the needs of the present without compromising the ability of future generations to meet their own needs'.

However, its meaning has now become highly contested and vague, leading to weak interpretations. For example, how can present and future needs be determined, and whose needs does it refer to? Many attempts have been made to refine the concept, which have led to numerous, often contradictory, definitions. The result is that 'sustainable development' is often little more than a soundbite, signalling recognition that the environment matters but failing to provide substance.

The real success of the Brundtland Report was the emphasis it placed on the relationships between economic growth, social conditions and environmental degradation, an emphasis that placed sustainable development firmly on the global political agenda, and that over the next five years would see a concerted effort by UN agencies and many governments towards a global summit on environment and development. In 1992 – after five years of negotiations and preparatory meetings – the UN Conference on Environment and Development (also known as the Earth Summit) was held in Rio de Janeiro.

Q Is it morally justifiable and politically plausible for the developed world to try to dictate the terms of sustainable development and good environmental management to the rest of the world?

In essence it is a critique of everything about the way we currently live today, a critique of contemporary industrial society. Accepting that there are limits to growth means changing our notions of what makes a good life and learning to live more simply (thus slowing growth): for example, consuming less and more locally produced goods, producing less waste, travelling less and being conscious of our impact on the environment in our daily lives. The result is an emphasis on 'localization', living small-scale, perhaps more communally to share the load of everyday tasks, in energy-efficient buildings, and often with an emphasis on producing for your own needs, for example, through growing your own vegetables. See Case study 6.2 for a British example of low impact living.

The ecological modernization interpretation is far less radical and has been adopted by many more in mainstream politics. Supporters of ecological modernization criticize the deep green vision as naive and utopian with an unnecessary and ineffective focus on

Plate 6.2 Low impact house: Tony's Roundhouse, Brithdir Mawr, Wales.
(Jenny Pickerill)

Snappy **Spotlight boxes** improve your understanding by expanding on the central concepts of the chapter.

Key terms are highlighted throughout the text and are then collated to form a full Glossary at the back of the book. This helps you with unfamiliar concepts or jargon and is particularly helpful for revision purposes.

CHAPTER 21 GEOPOLITICAL TRADITIONS 457

Thematic Case Study 21.1

The continuing Cold War?

The Cold War in Asia was always a very complex mixture of ideological, territorial and national disputes. An extra dimension was added by the fact that China became a nuclear power in the early 1960s at the same time that the Chinese communists split with the USSR on the grounds of ideological differences.

Moreover, whilst it is now almost two decades since proclamations of the 'end of the Cold War' began to multiply in Europe, in the Asia-Pacific region remnants of the global Cold War confrontation remain in place. Despite substantial changes and accommodations, China, North Korea and Vietnam are still one-party 'communist' political systems. In addition, a series of divisions and territorial disputes whose history is deeply rooted in the Cold War remain unresolved. Most notably, the Korean peninsula remains divided between mutually antagonistic regimes and both the People's Republic of China and the offshore island of Taiwan have claimed to be the legitimate government of the whole of China and remain fundamentally opposed.

Arguably, therefore, there is a sense in which the narratives about 'the end of the Cold War' reflect the Euro-Atlantic experience rather than being globally valid. The 'end' may have happened in Europe and the Soviet Union may have ceased to exist years ago, but Cold War legacies and logics are active in East Asia (see Hara 1999) and beyond. However, as the main text details (see pages 000), the extent to which new geopolitical narratives, notably 'the war on terror,' have replaced the Cold War as organizing frames for Western strategy is much debated – and contested.

Plate 21.7 South Korean soldiers patrol their side of the demilitarized zone between North and South Korea.
(© Michel Setboun/Corbis)

of maps and scripts. A variety of interpretations and geopolitical 'models' now attempt to offer explanation and impose meaning on contemporary events.

Amongst the most influential of these has been the discourse of 'New World Order', articulated by then US President George Bush (senior) in the early 1990s. In Bush's vision, what is called the 'international community' (the member states of the United Nations, led by the USA, but with financial and military support from key allies such as Germany, the United Kingdom and Japan) would act as a kind of global police force, intervening where and when they felt necessary to maintain or restore 'order'. Both the 1990–1 Gulf War and the 1999 conflict with Yugoslavia were justified and conducted (in part) in the name of building such a 'New World Order'. But it was his son, President George W. Bush, who was left to articulate a new American strategy (which became the Bush Doctrine, of pre-emptive attack on states that may – it was claimed – threaten the United States or harbour terrorists) in the aftermath of 9/11. And even this was, arguably, partially within a long American geopolitical tradition of overthrowing overseas regimes during and before the Cold War, which has its roots in the rise of American power

Thematic case studies, often accompanied by thought-provoking images, investigate crucial geographical themes and illuminate how they can be applied to real life.

CHAPTER 9 CITIES: URBAN WORLDS 217

'spaces of utopia and dystopia'. This tension remains and cannot be wished away: cities are places of division, in which power relations are often expressed in built form and in social and political practices, as well as places within which it is possible to negotiate for alternative futures and to challenge what are sometimes presented as inevitable inequalities and hierarchies.

Learning outcomes

Having read this chapter you should be able to:

- Think geographically about cities as socio-spatial entities. Cities are the products of living social relationships, which have clear expressions both internally (as people come together within cities) and in terms of their changing position within wider economic and social networks of connection.
- Understand that the world is becoming increasingly urbanized – most people now live in cities. But, recognize that it is important to understand not only what links them together (what makes it possible to talk of 'cities' as recognizable phenomena) but also what helps to differentiate them.
- Systematically identify and critically explore key defining characteristics of cities, building on features drawn from Wirth (for whom cities are defined as being large, densely populated, permanent and socially heterogeneous).
- Recognize the implications of social heterogeneity and diversity for the ways in which people live their lives in cities, including the consequences of the existence of increasingly diverse and overlapping social worlds.

Further reading

Amin, A. and Thrift, N. (2002) *Cities: Reimagining the Urban*, Polity Press, Cambridge. This is an attempt to open up new ways of thinking about the city, reimagining their potential and reflecting the complexities of urban life.

Bridge, G. and Watson, S. (eds) (2000) *A Companion to the City*, Basil Blackwell, Oxford. A multidisciplinary collection of contemporary writings, which draws its examples from across the globe, and introduces a wide range of ideas about the city without seeking to impose any single approach on the reader.

Dear, M. (2000) *The Postmodern Urban Condition*, Blackwell, Malden, MA. A clear exposition of the LA school as a way of thinking about contemporary cities, which might usefully be read alongside Le Galès, P. (2002) *European Cities: Social Conflicts and Governance*, Oxford University Press, Oxford, which actively celebrates an alternative vision of European cities, with an emphasis on the importance of continuity and the role of embedded local elites.

Massey, D., Allen, J. and Pile, S. (eds) (1999) *City Worlds*, Routledge, London. This provides a relatively brief (185-page) introduction to geographical ways of thinking about cities. It highlights some of the tensions and ambiguities of urban life, exploring both the problems and potential of cities.

Robinson, J. (2005) *Ordinary Cities: Between Modernity and Development*, Routledge, London. This works well as a companion to the previous book. While sympathetic to approaches that seek to locate cities within a global system, it is also a powerful call to shift the focus of urban studies so that it is possible to incorporate the experience of a wider set of cities and to explore the ways in which networks of social relations overlap in cities.

Taylor, P. (2004) *World City Network: A Global Urban Analysis*, Routledge, London. A coherent and comprehensive approach to the study of cities which locates them as part of a world system.

Useful website

www.lboro.ac.uk/gawc/ A wealth of data and reflections on the external relations of world global cities.

For annotated, clickable weblinks and useful tutorials full of practical advice on how to improve your study skills, visit this book's website at **www.pearsoned.co.uk/daniels**

Learning outcomes provide a key checklist of learning milestones, to test your understanding of the chapter.

The **Further reading** section helps direct your independent study to the most appropriate texts, journals and websites.

Acknowledgements

As ever, the success of this project relies on the goodwill and enthusiasm of the contributors. We would like to thank the international panel of reviewers who assessed the second edition and suggested ways in which it might be improved. As a result, this edition includes a number of completely new chapters to produce a lively mix made up of those who have contributed from the outset, those who joined as new contributors to the second edition, and those who are new to the third edition. Whether revising previous contributions or starting from scratch, all have enthusiastically responded to the editors' edict that they should seek to adopt an accessible writing style that will engage readers and where possible enable them to relate issues to their daily lives using contemporary/everyday examples that makes the material more meaningful and less abstract. We have been keen to retain the global feel but have also encouraged the contributors to include some more specifically European examples or research. We have also sought to more closely integrate the case studies and artwork with the text so that the reader can see why a particular feature is situated at a particular point in a chapter, what it is there to illustrate, how can it provoke readers to think through the issue, and how it relates to the main narrative. We have also tried to ensure that the third edition has more of the cutting edge to offer readers; we were encouraged to finesse a more critical slant on human geography (by way of the latest debates, issues and controversies), by those who kindly fed back detailed comments on the second edition. The contributors have risen to this challenge and we would like to acknowledge their constructive response to the dialogue that this has necessitated along the way.

The editors and authors are especially indebted to Andrew Taylor at Pearson, for his unswerving support and encouragement during the evaluation and development of plans to follow up and improve on the successful Second Edition. His commitment to 'raising the bar' with respect to the coverage and the presentation of the third edition has spurred us on to build on previous strengths and rise to new challenges. Equally, we could not have succeeded without the help of Sarah Busby and Mary Lince, also at Pearson, and we thank them all for their calm demeanour when deadlines loomed and patience was tested with the excuses tabled when these had been stretched! The various members of the editorial and production staff at Pearson, who the editors never meet but who are critical to the design and completion of a book of this kind, also deserve mention.

There are also a number of individuals whose efforts deserve particular mention and thanks. Thanks are due to Kevin Burkhill, Drawing Office Technician, in School of Geography, Earth and Environmental Sciences at the University of Birmingham for his assistance with redrawing some of the charts/diagrams and converting some statistical material into appropriate charts. Sophie Hadfield-Hill in the Department of Geography at the University of Leicester took on the daunting task of trying to reconcile the revised chapters with the reference list at the end of the volume. This was achieved with great efficiency and saved us an enormous amount of time and energy. Denis Shaw wishes to acknowledge the help of John Bourne in the School of Historical Studies at the University of Birmingham. James Sidaway would like to thank Robina Mohammad and all of his colleagues, the support staff and the postgraduate students at the Department of Geography at the National University of Singapore and the School of Geography at the University of Plymouth; but in particular Veit Bachmann, Jon, Shaw, and Richard Yarwood, Geoff Wilson and Mark Wise at the latter and Carl Grundy-Warr, Shirlena Huang, Victor Savage and Henry Yeung at the former.

For all of the editors, production of this third edition has come at a time when we have been facing numerous challenges and demands on our time, both personal and professional. Most of all, the editorial team owes a particular debt of gratitude to James Sidaway for continuing to work energetically on this project through what have been especially difficult times. On a number of occasions it seemed that the third edition was proving one too many. However, we worked together and stuck to the task and thanks to all of the authors, the publisher, those mentioned above and many others, we have produced a text that we can all be proud of. Our hope is that it encourages a new generation of students to connect with human geography; after all, the vitality of the discipline depends on these students being enthused and critically and creatively engaging with human geography within and beyond the classroom.

PWD
MJB
DJBS
JDS

Birmingham, March 2008

The editors and publishers would like to thank the following reviewers of the second edition for their invaluable input into the shaping of ideas for this third edition.

Professor Cara Aitchison, University of the West of England
Dr Nicola Ansell, Brunel University
Dr Freerk J. Boedeltje, Radboud University Nijmegen, The Netherlands
Dr Luke Desforges, University of Sheffield
Professor Isabel Dyck, Queen Mary, University of London
Dr James Faulconbridge, Lancaster University
Dr Beth Greenhough, Queen Mary, University of London
Associate Professor Michael Haldrup, Roskilde University, Denmark
Associate Professor Brita Hermelin, Stockholm University, Sweden
Dr Simon Leonard, University of Portsmouth
Professor Anders Lundberg, University of Bergen, Norway
Associate Professor Per Gunnar Røe, University of Oslo, Norway
Dr Stephen A. Royle, Queen's University Belfast
Dr Sam Scott, University of Liverpool
Dr Bettina Van Hoven, University of Groningen, The Netherlands
Professor Neil Wrigley, University of Southampton

The editors and publishers would like gratefully to acknowledge the contribution of Ian Cook of the University of Exeter (with the assistance of Suzy Reimer, David Bell and Paul Cloke) to this third edition.

Publisher's acknowledgements

The publishers would like to thank all the contributors for their expert and generous input to this third edition. In particular, we would like to thank the four section editors – Peter Daniels, Mike Bradshaw, Denis Shaw, and James Sidaway – for their continuing dedication, skill and generosity with their time and experience, without which this project would not be possible.

We are grateful to the following for permission to reproduce copyright material:

Figure 1.1 from Simmons, I. G. (1996) *Changing the Face of the Earth: Culture, Environment, History*, 2nd Edition, Blackwell Publishing, Oxford; Figure 1.2 from Sherratt, A., *Cambridge Encyclopedia Archaeology* (1980), Cambridge University Press; Figure 1.3 Republished with permission of Taylor & Francis Informa UK Ltd. – Journals, from *Annals of the Association of American Geographers,* Whittlesey, D., Vol. 26, (4) (1936), permission conveyed through Copyright Clearance Center, Inc; Figure 1.4 The Greek colonisation of the ancient Mediterranean from Pounds, N.J.G. (1947) *An Historical and Political Geography of Europe*, Harrap, London p. 49; Figure 1.5 Map VI The road system of Roman Britain from *Roman Britain* (1984) edited by Salway, Peter. By permission of Oxford University Press; Figures 1.6, 1.7 and 3.2 based on *The Times Atlas of World History* (1989) © Collins Bartholomew Ltd., 1989, reproduced by permission of HarperCollins Publishers; Figure 2.1 Reprinted with the permission of the Free Press, a Division of Simon & Schuster Adult Publishing Group, from URBAN DEVELOPMENT IN CENTRAL EUROPE, Volume 1 of the International History of City Development Series, by E.A. Gutkind. Copyright © 1964 by The Free Press. Copyright renewed © 1992 by Peter C.W. Gutkind. All rights reserved; Figure 2.2 from *Descricao da Fortaleza de Sofala e das mais da India*, Fundacão Oriente, Lisbon (Carneiro, A de M 1990); Table 2.2 from R. I. Moore (ed), *The Hamlyn Historical Atlas*, Copyright © Octopus Publishing Ltd., 1981; Figure 2.3 from *Atlas of the British Empire*, Hamlyn, (Bayley, C. 1989), © Octopus Publishing Group; Figure 2.5 from *The Hamlyn Historical Atlas*, Hamlyn, London, (Moore, R.I. (ed) 1981); Figure 2.6 from *Yorkshire Textile Mills 1770–1930*, HMSO, National Monuments Record, © Crown copyright.NMR (Giles, C. and Goodall, I.H. 1992); Figure 2.7 from *Yorkshire Textile Mills 1770–1930*, HMSO, (Giles, C. and Goodall, I.H. 1992), courtesy of the Ironbridge Gorge Museum Trust; Figure 2.8 from Pounds, N.J.G., *An Historical Geography of Europe* (1990), Cambridge University Press; Figure 2.9A from Ward, D., *Poverty, Ethnicity and the American City, 1840–1925* (1989), Cambridge University Press; Figure 3.1 (left) from Mitchell, J.B., *Great Britain: Geographical Essays* (1962), Cambridge University Press; Figure 3.1 (right) from *The British Isles: A Geographic and Economic Survey*, Longman, London (Stamp, L.D. and Beaver, S. 1963); Figures 4.2, 4.3 and 4.4 from *World Population Prospects: The 1998 Revision*, United Nations, (1999); Figure 4.5 from Chung, R. (1970) Space-time diffusion of the demographic transition model: the twentieth century patterns (first published in *Geographical Journal*, Blackwell Publishing), in Demko, G.J., Rose, H.M. and Schnell, G.A. (eds) *Population Geography: A Reader* © 1970,

The McGraw-Hill companies Inc.; Figure 4.6 from Population Reference Bureau (2007) *Human Population: Fundamentals of Growth: Future Growth*, Available at http://www.prb.org/Educators/TeachersGuides/HumanPopulation/FutureGrowth/TeachersGuide.aspx?p=1; Figure 4.11 from UNAIDS (2006) *Report on the Global AIDS Epidemic 2006*, reproduced by kind permission of UNAIDS (www.unaids.org); Figure 4.12 from *Education and HIV/AIDS: A Window of Hope*, World Bank, (Bundy, D. and Gotur, M. 2002) © International Bank for Reconstruction and Development/World Bank; Figure 4.13 from *World Population Prospects: The 2004 Revision*, United Nations, (2005); Figures 5.1 and 5.3 from *Natural Resources: Allocation, Economics and Policy*, 2nd Edition, p. 15, p. 20 Routledge, (Rees, J. 1985); Table 5.1 and Figure 5.2 from Rees, J., Resources and environment: scarcity and sustainability, in *Global Change and Challenge: Geography in the 1990s*, p. 6, p. 9, Routledge, (Bennett, R. and Estall, R. eds 1991); Figure 5.4 Reproduced by permission of SAGE publications, London, Los Angeles, New Delhi and Singapore, from Jones, G. and Hollier, G., *Resources, Society and Environment Management*, Copyright (© Gareth Jones and Graham Hollier, 1997); Figures 5.7, 5.8 and 5.9 Source attribution: *BP Statistical Review of World Energy 2007*; Tables 5.8 and 5.9 from *Key World Energy Statistics* © OECD/IEA, 2006, p. 11, p. 13; Figure 5.11 from *Environmental Resources*, Mather, A.S. and Chapman, K., Pearson Education Limited © Prentice Hall (1995); Tables 6.1 and 6.3 from Carter, N., *The Politics of the Environment: Ideas, Activism, Policy* (2001), Cambridge University Press; Table 6.2 from The shallow and the deep, long-range ecology movement, Inquiry 16 pp. 95–100 (Oslo 1973), (Naess, Arne 1973); Table 6.4 from Garner, R., *Environmental Politics: Britain, Europe and the Global Environment* (2000), Macmillan, London. Reproduced with permission of Palgrave Macmillan; Table 6.6 *The Millenium Development Goals*, United Nations; Figure 7.1 from Hendrickson, M. K. and Heffernan, W. D. (2002) Opening spaces through relocalization: locating potential resistance in the weaknesses of the global food system, *Sociologia Ruralis*, 42, 347–69, Blackwell Publishing; Figure 7.2 from Phyne, J. and Mansilla, J. (2003) Forging linkages in the commodity chain: the case of the Chilean salmon farming industry, 1987–2001, *Sociologia Ruralis*, 43, 2, 108–27, Blackwell Publishing; Figure 7.3 First published in *Environment and Planning A*, 2005, Ilbery and Maye, 'Alternative (shorter) food supply chains and specialist livestock products in the Scottish-English borders'; Figure 8.1 from *The Bank, The President and the Pearl of Africa*, International Broadcasting Trust (IBT)/Oxfam, Oxford, (1998); Figure 8.3 The World Bank and the State: a Recipe for Change? By Nicholas Hildyard, © Bretton Woods Project, www.brettonwoodsproject.org; Table 9.1 Statistics © Philip's. Source: *Philip's Great World Atlas* (2003); Table 10.1 from *The Necessities of Life in Britain*, Working Paper, Poverty and Social Exclusion Survey, University of Bristol, (Pantazis, C., Gordon, D. and Townsend, P. 1999); Table 10.2 from *Dangerous Disorder: Riots and Violent Disturbances in Thirteen Areas of Britain, 1991–92* by Anne Power and Rebecca Tunstall, published in 1997 by the Joseph Rowntree Foundation. Reproduced by permission of the Joseph Rowntree Foundation; Table 10.3 from The changing dynamics of community opposition to human service facilities in *Journal of the American Planning Association*, Taylor & Francis Ltd., (Takahashi, L.M. and Dear, M.J. 1997), reprinted by permission of the publisher (Taylor & Francis Ltd., http://www.informaworld.com); Figure 14.1 from *CIA World Factbook* (2006) Geometry: UNEO (2006) The GEO Data Portal, as compiled from the FAO.http://geodata.grid.unep.ch Copyright © 2006, Maplecroft; Table 14.5 Z/Yen Limited, *The Competitive Position of London as a Global Financial Centre*, Corporation of London (November 2005); Figure 15.1 and Table 15.1 Reproduced by permission of SAGE publications, London, Los Angeles, New Delhi and Singapore, from Dicken, P., *Global Shift: Mapping the Changing Contours of the World Economy*, 5th Edition, Copyright (© Peter Dicken, 2007); Table 15.2 from Dedrick, J. and Kraemer, K.L., Is production pulling knowledge work to China? A study of the notebook PC industry, *Computer*, 39 © 2006 IEEE; Figure 15.3 from Foster, W., Cheng, Z., Dedrick, J. and Kraemer, K. L. (2006) *Technology and organizational factors in the notebook industry supply chain*, Figure 2, The Personal Computer Industry Center Publication, UC Irvine; Table 15.3 The growing power of retailers in producer-driven commodity chains: a 'retail revolution' in the US automobile industry?, unpublished manuscript, Department of Sociology, University of California at Santa Barbara, USA. (Kessler, J. 1998); Table 15.4 from *The Coffee Paradox: Global Markets, Commodity Trade and the Elusive Promise of Development*, Zed Books, London, (Daviron, B. and Ponte, S. 2005). Reproduced by permission of Zed Books Ltd., London and New York. Copyright © Benoit Daviron and Stefano Ponte 2005; Figure 15.6 from *Territories of Profit*, Stanford Business Books, Stanford, (Fields, G. 2004), with permission of Jeremy Shaw; Figure 15.7 from *Commodity Chains and Global Capitalism*, Gereffi, G., Copyright © (1994) by Gary Gereffi and Miguel Korzeniewicz. Reproduced with permission of Greenwood Publishing Group, Inc., Westport. CT; Figure 15.8 New models of regulating production networks adapted from Sum, N-L. and Ngai, P. (2005) Globalization and paradoxes of ethical transnational production: code of conduct in a Chinese workplace, *Competition and Change*, 9, 181–200, Figure 1 © Maney Publishing; Table 16.1 and Figure 16.3 from Mahajan, S. (2005) Input-Output Analysis: 2005, p. 27 and p. 23, Office for National Statistics, London; Figure 16.2 from *Service Worlds: People, Organizations, Technologies*, p. 52, Routledge, London, (Bryson, J. R., Daniels, P.W. and Warf, B. 2004); Table 16.2 from Bryson, J. R. (2007) A 'second' global shift? The offshoring or global sourcing of corporate services and the rise of distanciated emotional labour, forthcoming *Geografiska Annaler*, in press, Blackwell Publishing; Figure 16.4 from Transnational corporations and spatial divisions of 'service' expertise as a competitive strategy: the example of 3M and Boeing, Bryson, J. R. and Rusten, G., *The Service Industries Journal*, 28, (3) 1–17, (2008), Taylor & Francis Ltd., reprinted by permission of the publisher (Taylor & Francis Ltd. http://www.informaworld.com); Figure 17.1 from *Financial Geography*, Laulajainen, R., Copyright (© 1998) School of Economics and Commercial Law, Goteborg University. Reproduced by permission of Taylor & Francis Books; Figure 20.1 from *A People Without a Country: the Kurds and Kurdistan,* 2nd Edition, Zed Books, London, (Chailand, G. 1993); Figure 21.1 from *Caspian Oil and Gas* © OECD/IEA, 2000, Map 1, p. 21; Figure 21.2 The Monroe Doctrine by Charles Bartholomew, *Minneapolis Journal*, 1912; Figure 21.3 from *The Fate of the Forest: Developers, Destroyers and Defenders of the Amazon,* Verso, London and New York, (Hecht, S. and Cockburn, A. 1989); Figure 21.4 from *Political Geography*, 3rd Edition, Glassner, M.I., Copyright © (1993) John Wiley & Sons, Inc. Reprinted with permission of John Wiley & Sons, Inc.; Figure 21.5 from Mackinder, H. (1904) The geographical pivot of history, *Geographical Journal*, 23, 421–44, Blackwell Publishing; Figure 21.6 from Political geography and panregions in *Geographical Review*, America Geographical Society, (O'Loughlin, J. and van der Wusten, H. 1990).

Photographs (Key: b-bottom; c-centre; l-left; r-right; t-top): p. 18 STILL Pictures The Whole Earth Photo Library: Sebastian Bolesch/Das Fotoarchiv; p. 22 Michael Holford Photo Library; p. 28 STILL Pictures The Whole Earth Photo Library: blickwinkel/Luftbild; p. 29 Mary Evans Picture Library; p. 35 The Hereford Mappa Mundi Trust; p. 62 © Angelo Hornak/Corbis; p. 63 Mary Evans Picture Library; p. 66 © Gillian Darley/Edifice/Corbis; p. 68 Hulton Archive/Stringer/Getty Images; p. 71 © Alain Dejean/Sygma/Corbis; p. 73 Neil Emmerson/Robert Harding World Imagery; p. 74 © Reuters/Corbis; p. 75 Vintage Ads; p. 77 Paul Ridsdale/Photofusion Picture Library; p. 107 Ami Vitale/Panos Pictures; p. 110 ImageState/Alamy; p. 112 Evening Standard/Stringer/Getty Images; p. 113 Dita Alangkara/PA Photos; p. 114 Ria Novosti/Science Photo Library Ltd; p. 117 steven gillis hd9 imaging/Alamy; p. 129 Realimage/Alamy (t), FOX PHOTOS/Getty Images(b); p. 173 The Advertising Archives; p. 193 Martin Adler/Panos Pictures; p. 196 Oxfam GB; p. 199 PA WIRE/PA Photos; p. 207 © Bettmann Corbis (tl), © Hulton-Deutsch Collection/Corbis (tr) (br); © Underwood & Underwood/Corbis (bl); p. 209 © Bettmann/Corbis; p. 210 Amanda Hall/Robert Harding World Imagery (t), Thomas Schlegel/transit/STILL Pictures The Whole Earth Photo Library (b); p. 214 Dave Ellison/Alamy; p. 216 © Bohemian Nomad Picturemakers/Corbis; p. 225 Associated Press/PA Photos; p. 226 Janine Wiedel/Photofusion Picture Library; p. 229 Nottingham Evening Post; p. 258 Photodisc (tl), Paul Lunnon(tr) (bl) (br); p. 262 Bridgeman Art Library Ltd: Mr and Mrs Andrews, c.1748–9 (oil on canvas), Gainsborough, Thomas (1727-88)/National Gallery, London, UK; p. 263 Banksy; p. 268 NASA; p. 269 Ronald Grant Archive; p. 275 © Benjamin Rondel/Corbis; p. 279 Joanne O'Brien Photography; p. 286 © Dave G. Houser/Corbis; p. 300 © Walter Hodges/Corbis; p. 302 T. Paul Daniels; p. 312 Eugene Hoshiko/PA Photos; p. 316 Alex Segre/Alamy; p. 333 Sean Sprague/Panos Pictures; p. 334 Chris Stowers/Panos Pictures; p. 340 Dorothy Burrows/Photofusion Picture Library; p. 346 Topical Press Agency/Getty Images; p. 350 Fredrik Renander/Alamy; p. 352 Associated Press/PA Photos; p. 359 © Bettmann/Corbis; p. 363 AP Photo/Luke Palmisano/PA Photos; 377 © CLARO CORTES IV/Reuters/Corbis; p. 379 Vintage Ads: BMW UK Ltd; p. 380 Robert Stiggins/Express/Getty Images; p. 381 Stefan Boness/Panos Pictures; p. 384 National Geographic/Getty Images; p. 387 G.P. Dowling p. 389 Images of Birmingham/Alamy; p. 392 www.blackgoldmovie.com/dvd; p. 403 Photofusion Picture Library/Alamy; p. 418 Graham Grieves/Art Directors and TRIP photo Library; p. 425 Jim West/Alamy; p. 430 David R. Frazier Photolibrary, Inc./Alamy; p. 431 BAA Aviation Photo Library: In-Press: Anthony Charlton; p. 432 © Bill Ross/Corbis; p. 441 SPIA PRESS/Rex Features; p. 450 © Ira Nowinski/Corbis; p. 452 © Bettmann/Corbis; p. 453 Time Magazine; p. 456 2005 TopFoto; p. 457 © Michel Setbourn/Corbis.

Extract on page 66 from Poem: 'Slough' from *Collected Poems* by John Betjeman © 1955, 1958, 1962, 1964, 1968, 1970, 1979, 1981, 1982, 2001. Reproduced by permission of John Murray (Publishers) Limited; Extract on page 73 from We and They by Rudyard Kipling, with permission of A. P. Watt Ltd. on behalf of Gráinne Yeats; Extract on page 136 from Friends of the Earth (England, Wales and N. Ireland) 2006; Extract on page 224 from Running on empty in *The Guardian*, 14 May 1997 (Taylor, Ian) © Guardian News and Media Ltd 1997; Extract on page 364 from Your mortal enemy in *The Guardian*, 21 October 1998 (Korten, David C.) © Guardian News and Media Ltd 1997; Extract on page 386 from The world on a plate . . . by Caroline Stacey, *Time Out*, issue 1304, 16–23 August 1995.

In some instances we have been unable to trace the owners of copyright material, and we would appreciate any information that would enable us to do so.

GEOGRAPHY: A MIRROR TO THE WORLD

Introduction

Peter Daniels
Michael Bradshaw
Denis Shaw
James Sidaway

> Geography is indispensable to survival. All animals, including American students who consistently fail their geography tests, must be competent applied geographers. How else do they get around, find food and mate, avoid dangerous places? (Yi-Fu Tuan 2002: 123)

> We routinely make sense of places, spaces, and landscapes in our everyday lives – in different ways and for different purposes – and these 'popular geographies' are as important to the conduct of social life as are our understandings of (say) biography and history. (Gregory 1994: 11)

This book is the third edition of *An Introduction to Human Geography: Issues for the 21st Century.* The first edition was published in 2001. The task of compiling the first edition therefore dates back to 1997–98 so that it is now approaching a decade since we wrote the first introduction. While the third edition incorporates many revisions and further changes to the structure and contents, we have retained our original goal, which was to provide an introduction to human geography through a survey of the field, focusing upon contemporary *issues* and adopting contemporary approaches. What introductory textbooks in human geography choose to include and foreground (and what is excluded or neglected) has recently been the subject of heated debate in a leading disciplinary journal (*Transactions of the Institute of British Geographers*). According to one of the protagonists, new textbooks (in part via their influence on a prospective new generation of geographers) become part of what shapes the dominant themes for research and scholarship in a discipline. Textbooks are thereby implicated 'in strategies to mobilize support for a particular set of disciplinary practices.' (Johnston 2007: 437)

Textbooks also reflect *where* they are written and where they are read. For example, most textbooks written in North America devote a significant number of words to explaining what geography is and what constitutes a geographical approach. On the second page of one of the most widely used American human geography introductions, Knox and Marston (2004: 2) define human geography as: 'the study of the spatial organization of human activity and peoples' relationship with their environments'. The reason why introductory textbooks in the United States need to consider definitions of human geography and devote space to explaining what a geographical approach amounts to is that many of the students taking a module in introductory human geography are not geography majors, that is they will not go on to specialize in geography, and they will usually have experienced limited or no exposure to it as a discipline at high school. Geography is rarely taught in the American 'K–12' (Kindergarten to pre-University) school system. The institutional setting within which the present textbook has been put together is different. The editors and most of the contributors currently teach in British universities or in countries where the influence of a British style of educational system is more evident. The majority of their students have chosen to specialize in geography at a school or college and have made a choice to read for a degree in geography. A similar presence of pre-University geography is typically found in many other European and most Commonwealth countries. Consequently, many students from these states have ideas of what the subject matter of the discipline is (although this may turn out to be rather different from much of what they will subsequently encounter at university, see Bonnett 2003 and Stannard 2003). Allowing for this context, part of the rationale for many introductory human geography modules is to explore how university-level geography differs from students' experiences of the subject so far. In such modules students are challenged not to take the world for granted and to understand that human geography is not a direct reflection of a straightforward objective reality, but a 'social construction'. In other words, although certain experiences, beliefs and value systems are shared, these also vary and interpretations of the world differ from different vantage points in time and place. The very notion that there is a *single* notion of what human geography is about (that is valid for all times, places and perspectives) becomes problematic. Whose human geography, where and when?

Moreover, human geography and the world that it seeks to interpret and represent are dynamic. In the decade since we embarked on the first edition of this book a great deal has changed in the world; the spectacular and tragic events of 11 September 2001 in the USA, for example, have left many marks. Tracing these could begin with the landscape of Manhattan, where the 'twin towers' once stood, but it would include too the enlarged prison in Guantánamo, Cuba (where some of those accused of 'terrorism' have been de-tained by the US government) to Afghanistan and Iraq (whose governments were overthrown by US-led force and which are still under de facto foreign occupation as we write) and to the nature and extent of airport security procedures almost everywhere. At the same time, the human contribution to global climate change and other environmental transformations has moved more to the centre of political debates in many countries and in international organizations (see Chapters 5, 6 and 12). And dramatic natural disasters, such as the Indian Ocean Tsunami of 24 December 2004, the Kashmir earthquake of 8 October 2005, or Hurricane Katrina (New Orleans) on 25 August 2005, have not only caused enormous injury and immediate loss of life, but also have many longer-term social and political impacts. At the same time, ongoing processes of migration, urbanization and economic transformations (such as the fast pace of development in parts of China and India) are producing new spaces, connections and flows that require new maps and geographical narratives. New divisions are being created, old conflicts revived and on first appearances, the world might appear to be more fragmented and contested than it was at the start of the twentieth century. Equally, however, economic and cultural globalization continue: some music, fashions and commodities can be found almost anywhere – often merging and mixing – and economies across the world are deeply interconnected and so intertwined that we can even begin to wonder whether there is such a thing as a 'national' economy (see Chapters 13 and 14). And, for the first time ever, more than half of the world's population lives in cities, the largest of which are in what are known as 'developing (or sometimes 'less economically developed' or sometimes 'Third World') countries' (see Chapter 8 on these categories and Chapters 4 and 9 on cities and population). Yet, as some of the other chapters that follow will detail, these processes have many deep historical roots and antecedents.

Changing worlds: changing human geographies

Geographical knowledge is not – and should not attempt to be – static and detached from what is going on in the world, but is rather dynamic and profoundly influenced by events, struggles and politics beyond university life.

(Blunt and Wills 2000: x–xi)

After all, things do not happen outside of space and time, and always take *place*. There will always be a place for 'thinking geographically'.

(Hubbard et al. 2002: 239)

The root of the word geography combines *geo* (earth) and *graphy* (writing). To engage in geography is to write about the earth (which includes its lands and seas, resources, places and peoples) or, more widely perhaps, to represent the earth in text (which includes maps: some of the most complexly crafted of all texts). Of course, many other branches of knowledge such as history, anthropology, sociology, oceanography, politics and geology (from which geography draws) and many other sciences are also in some way or other about the world. Traditionally, what has been distinctive about human geography is that it puts an emphasis on people in places and spaces, on landscapes modified by human interventions and on complex spatial connections. Today, human geography embraces issues such as spaces of identity, consumption and sexuality, which are fundamental parts of the human geographies of our interconnected, contradictory, complex, conflict-ridden and fantastically diverse planet. Such topics coexist with (and complicate) longer-established geographical issues associated with, for example, resources, population, cities or the changing distribution of economic activity. Indeed, we would argue that some of the most interesting, but tricky, challenges for human geography arise from its ambition to incorporate human experiences, achievements and transformations in its efforts to understand and to influence the issues that it is addressing. In other words, human geography is not just about 'out there', it is also about 'in here'. Thus, you all create your own personal geographies derived, for example, from where you live, where you work or when you travel. Moreover, through such myriad social–cultural, political and economic geographies, we inhabit a world, as John Pickles (2004: 5) points out,

> that has, in large part, been made as a geo-coded world; a world where boundary objects have been inscribed, literally written on the surface of the earth and coded by layer upon layer of lines drawn on paper.

Such complexity and challenges (as well as diverse 'geo-codes') are evident when we consider 'globalization'. The latter is something that is often considered a defining characteristic of the times in which we live. As we edited the first edition, we were conscious of, and regularly argued amongst ourselves about what the term 'globalization' takes for granted, what kind of stances we should encourage the contributors to take towards it, and how far we could agree once we saw the arguments emerging in the chapters and subsections. Such discussions were repeated and finessed in preparing subsequent editions. Indeed, as many of the chapters have spelt out, there is certainly a significant sense in which a combination of technical, political, cultural and economic transformations throughout the twentieth century enhanced the sense of global interconnection and this is continuing in the twenty-first century. Take the case of this book. The copy that you are reading may well have been printed far from where you picked it up. It is published by a multinational publishing company, whose ownership and 'home' location may not be immediately evident. Its production involved technologies such as desktop and laptop computers (themselves usually manufactured in Asia, but whose components use raw materials more often from Africa), and software written and distributed by some of the world's largest US transnational companies

(Apple and Microsoft). From the very start of work on the first edition back in 1997, numerous drafts of the chapters and sections were emailed between authors and editors, sometimes between offices next door to each other, or sometimes across Europe. In turn, the second edition was prepared with one of the editors based in Singapore and the three others in the English Midlands, and contributors in a range of locations. And the most convenient place for the editors to meet to finalize outstanding matters for that edition turned out to be in Philadelphia, whilst we were all attending the centenary convention of the Association of American Geographers in March 2004. This edition has been finessed at meetings in Birmingham (where two of the editors are still based), but with follow-up communications and drafts circulated by email whilst some of us were at conferences or engaged in fieldwork (at various times) in India, Russia and Colombia. Just two decades ago the writing and production process would have been very different. There would have been much less dependence on electronic media, more extensive use of surface or air mail to exchange pieces of the manuscript, and the production of the book, copy-editing, proofreading and printing largely completed locally. Can we therefore see this book as both an artifact and representation of globalization?

To describe contemporary technological, economic, cultural or political tendencies as 'globalization' is, we agreed, to invoke a certain *geographical imagination*, a vision of the growing significance of a global scale of action, of the world as a single place. Yet, we know, many people and places remain relatively marginalized from these supposedly hypermobile ways of living and working. These experiences are made real and meaningful in the language that human geographers (and others) adopt. As an alternative to the term 'globalization', we might use other terms: 'imperialism', or 'capitalism', for example. Each carries particular connotations. In this way, 'globalization' serves as a particular concept that is used to make sense of the world and one on which a certain geographical imagination, of an increasingly connected and 'shrinking' world for example, is emphasized. Yet just because something is imagined and interpreted in particular ways and by reference to particular geographies, it does not make it any less real to those caught up in it. Globalization is seen by some as a broadly positive force, breaking down barriers, making capitalism more efficient and spreading its benefits throughout the world. For others it is a more negative process enabling another round of exploitation, often with the further destruction of local cultures and identities and further commodifying life and nature. Obviously, globalization is contested, in terms of both the meanings attributed to it and the evaluation of its consequences. Judge for yourself, but as you do so, do not make the mistake of assuming that everyone everywhere shares your vantage point and experiences of the world.

Some events and moments have had global coverage. The coming of the millennium (and perhaps '9/11') was represented to us (and lived and made around the world) as an enormous global collective likened by some to the moment when the American Apollo 8 spacecraft in 1968 captured the first image of the earth as a whole from space, or the reception of photographs of the earth from space by subsequent NASA missions (see Plate 12.4). These images have circulated widely and shape human perceptions of the planet (see Cosgrove 2001).

Yet the way that some events become 'significant' or 'global' reflects where they happen and who they affect. The spectacular losses of thousands of lives in Manhattan on 9/11, for example, became a global media event and subject of debate in the way that the death of more than a million from 1998 to 2003 in the Democratic Republic of Congo through war never did. Even the idea of a twenty-first century, which frames the title of this text, has its roots in a Western (Christian) timetable. According to the calendars of China, Islam and Judaism it is not the twenty-first century, and Islamic, Hindu, Jewish and Chinese 'New Years' do not coincide with the first of January. Furthermore, the calendar that produced 2000 as a millennium year was established first in the regions that we now refer to as Europe and the Middle East and then refined and disseminated over what we now call 'centuries'. In England the calendar now used was not established as official until 1751. And seemingly arcane controversies over whether 2000 or 2001 marked a 'new century' are in part rooted in the fact that European mathematics had no concept of zero, until it was imported, via the Arab–Islamic civilization, from India to Europe. So whilst the year 2000 is a Western construct, it turns out to rely on the idea of zero whose origins are elsewhere. One of the lessons here is that the West owes much of its identity and knowledge to ideas, people and artefacts originating outside what have been considered its geographical limits. Many observers subsequently claimed that a new age (and perhaps the geopolitics of a new century) began on a sunny New York morning in September 2001, when the drama of hijacked civilian aircraft being used as weapons rapidly became a global media event. Yet the many causes and consequences of '9/11' are far from new. As geographers you will be able to explore their links to flows of resources, power and geopolitics and to the technologies, cultures and economies of mobility that enabled the use of airliners as weapons to strike symbolic targets. Moreover, how the events and meanings of '9/11' are represented, experienced and understood in different places vary enormously.

Yet our academic geographies (particularly as researched and taught in Western Europe and North America) frequently still tend to be rooted in a Western vision, often taking this for granted and assuming that what is the norm is the Western 'developed country' experience. Human geography is only at the beginning of what seems likely to be a challenging but promising process of moving beyond such assumptions and outgrowing its modern roots as a Western science tied up with imperial knowledges of the 'rest' of the world. As it does so, assumptions, categories and taken-for-granted maps will come under scrutiny and be challenged. This book has set out some of the ways in which this is taking place as human geographers rethink what it means to represent the complexity, diversity and (uneven) interconnectedness of spaces, places and landscapes. As readers think more about this, it may be instructive to compare this book – whose contributors are mostly based in British universities – with three other introductory human geography texts (Le Heron et al. 1999; Wait et al. 2000; Knox and Marston 2001) written (and, like *An Introduction to Human Geography*, published by multinational information, media and publishing companies) from Australia, New Zealand and the United States respectively. Consider how the contents and style of this text, like those others, is marked by *where* it was written.

The origins of academic geography?

As a distinct subject (with students reading for a degree in it) geography has been present in European universities since the mid to late nineteenth century. Before then, however, geographical knowledge was studied and taught in many universities (not only those in Europe, but in the great centres of learning in the predominantly Islamic world, such as Baghdad and Cairo), as part of a variety of programmes of study – sometimes alongside mathematics and geometry for example, or as part of (or alongside) natural history, astronomy or cosmography (see Withers and Mayhew 2002). In the second half of the nineteenth century universities were reorganized around modern disciplines, increasingly with discrete departments and separate degree schemes. This happened as the old curriculum – based on classical learning in philosophy and sciences – began to break down with the rise of science, commerce and new capitalist rationalities. The fact that the late nineteenth century was also a time when many European states (chief amongst them Britain, France and the Netherlands) were engaged in overseas colonization and empire-building and all were keen to foster their sense of national identity and territorial coherence gave geography a new practical relevance. Children needed to be taught, it was argued, about their nation and its place in the world and their teachers thus required a degree in geography. At the same time, geographical knowledge had direct strategic and military relevance (this was the moment too of the birth of geopolitics, as detailed in Chapter 21) and commercial and imperial relevance, such as in schemes to exploit the perceived agricultural potential of colonies in Africa and Asia, for example. Nineteenth-century ideas about the relations between climate, environment and 'race', and (to use the language of that time) 'civilization' and progress, were caught up with the emergence of the discipline, but so too were the impacts of Darwin's ideas (which influenced physical geography too: in the conception of the way that landforms evolve).

The early years of the modern discipline were therefore inescapably tied up with nationalism and empires. This continued into the early twentieth century, with a growing number of geography departments being established at universities in the USA and Canada, in many Latin American countries, in the European colonies and dependencies (in places such as Singapore, Christchurch, Rangoon, Hong Kong and Jakarta) and in Japan. In some instances, such as Russia, the earliest departments were organized in the 1880s and this was tied to nation-building rather than overseas expansion. Such was also the case in Germany and in Scandinavia. By the time of the Second World War, geography was relatively well established – and the practical knowledges (of military 'intelligence' for example) it yielded in wartime helped to consolidate the discipline's place in universities in many countries. After 1945, it faced new challenges as the range of other disciplines expanded, but (with some exceptions such as the closure of a few departments in the United States), it benefited from the first big post-war expansion in the number both of universities and of students in the 1960s. Human geography increasingly reoriented itself to the technological and scientific spirit of the times, fed by a new phase of military competition in the Cold War (on this, see Barnes and Farish 2006, and see Johnston and Sidaway 2004 for a history of 'Anglo-American' human geography since 1945 and also Hubbard et al. 2002: 22–56 for a brief history of geographic thought). In an age of formal decolonization, where the old imperial disciplinary role was waning, human geography also found new fields of study and outlets for its graduates (such as urbanization, development and planning). There were, however, fierce debates about the appropriate focus (for example over the status of regional geography) and methods (such as the role of statistical analysis), which meant that what undergraduate students were exposed to (and thus had to learn) to pass a geography degree continued to change (though unevenly, depending on where they studied).

By the 1970s, the world was changing and so was human geography. This kindled an interest in the underlying economic causes of inequality in capitalist societies. In turn, capitalism entered a phase of heightened restructuring (involving recessions, new technologies and new forms and places of production and regulation, as detailed in Section 4). Reflecting the times, human geographers became more concerned with inequality, economic and political crises and contradictions. Feminist, ecological and other critiques also started to impact on human geography and feed into re-evaluations both of its history (the way that early twentieth-century imperial geography was shaped by racism and sexism, for example) and contemporary contents. Moreover, the boundaries between many of human geography's sub-disciplines, such as urban, political, historical or cultural geography, became more blurred. However, along the way, the modelling, data processing and visualizing capabilities of geographic information science continued to be refined (Fairbairn and Dorling 1997; Fotheringham, Brunsdon and Charlton 2000; Schuurman 2004) and the Internet, digitization and mobile telecommunications produced new capacities for communication and altered relations and perceptions of proximity and distance (see http://www.zooknic.com/ for work on the geographies of the Internet). For some, outside the discipline, the decline of the Cold War in the late 1980s brought the 'end of history'. For others, the development of information and telecommunications technology and the globalization of the economy were creating a 'borderless world'; some even proclaimed the 'end of geography'.

For the most part, the evolution of human geography has been driven by the adoption of ideas, concepts and theories from other disciplines. The challenge for human geographers has been to apply these to a geographical context. Economic geography, for example, initially took ideas from economics and business studies and applied them to help explain how economic activity is distributed across space. The geography of economic activities was seen as the outcome of this process with the job of the geographer being to map and describe that pattern of economic activity. This often seemed to relegate the economic geographer (and geographers more generally) to the role of suppliers of information for others to decipher.

More recently, however, all the talk of globalization has encouraged many other social scientists to reconsider geography. Of course, geographers have long claimed that *geography matters* for all the social sciences. But for some reason this is often acknowledged only when internationally renowned economists, political scientists and sociologists actually say that geography must be included in their analyses and interpretations. But as geographers we should beware; other disciplines tend to adopt their own 'take' on geography, one that is sometimes antiquated and simplistic – reinventing industrial location models in the name of 'geographical economics', for example, or attributing to tropical environments a key role in determining global patterns of development. Nevertheless, geographers can take some comfort and sense of possibility from the fact that other disciplines are recognizing the *spatiality of human society*.

Put simply, the fact that social and economic processes take place across space matters. Indeed the way they do so is vital to how they operate. Human geography is not just about *describing* the spatial manifestations of economy and society: it is about *explaining* how space is configured and shapes economies, societies and social processes. Thus, geography is not a passive outcome; it is a critical component of social and economic processes. More than that, geography in its broader definition provides an interface between the human and the natural worlds. We would argue that geography is a *key* subject for the twenty-first century, in part because many of the challenges that face humanity are at the interface between human societies and natural environments. One of the oldest themes of human geography (and indeed of geography as a whole, including its physical side), that of human–environment relations, has become an urgent agenda for the twenty-first century.

As this textbook has evolved so the environment and environmental issues have become more significant in our discussions. Initially, there was no chapter on the environment, though it was touched on in the chapter on resources and development. In this edition there are two chapters dealing explicitly with environmental issues and the social construction of nature. Furthermore many other chapters discuss issues such as global warming and competition over access to increasingly scare resources such as clean drinking water. In addition, the environmental costs of high levels of mass consumption and the consequences of economic development in the 'emerging economies' of the global South are also discussed. In short, this edition reflects the general environmental turn that we are currently experiencing as we come to accept that human society and the way we live is threatening the viability of the global ecosystem. This realization presents a major opportunity for geographers who are well equipped to deal with the interrelationships between environment and society.

Human geographers have also become more aware of the ways that knowledge is socially constructed. The way you see the world is partly a function of who you are and, hence, *where* you are from. Human geographers have come to realize that much of the knowledge and understanding that they claimed to be universal is in various ways 'Eurocentric'. Thus, as we have noted, 1 January 2000 was not the beginning of a new century or a new millennium for many of the world's populations; yet, an instant measured on the timescale of a particular religion that, until recently, dominated Western society, was widely taken to be a global event. Similarly, Eurocentrism has given a very 'white' view of a multi-ethnic world. This means that geographers must now reconsider the ways in which assumptions and value judgments shape the way they view the world. They must accept that *all* descriptions of the world are culturally determined, often politically motivated, and can always be contested.

A straw poll would quickly reveal that there remain differences between the authors, and among the editors, over the contents of the book, or how it should have been written. We have already sketched some of the ways that the nature and contents of introductory textbooks vary according to the location of their authors and targeted readers. What you see here is the result of compromise. The same type of book written by a group of geographers in, for example, Australia (see Wait et al. 2000), New Zealand (see Le Heron et al. 1999), India, Russia or South Africa would undoubtedly be different. At the same time, each reader of this book will interpret its contents in different ways. The position(s) that you occupy in your society and where you are on the globe when you read the book will tend to shape the way you look at the world and/or the way that you interpret its arguments. The majority of the readers will probably reside in the so-called 'Western' world. The first edition was reprinted in India in 2003 as a local edition (authorized for sale in India, Bangladesh, Pakistan, Nepal, Sri Lanka and the Maldives) and the second edition is to be published in Chinese. No doubt those readers will frequently have different worldviews from those of many students in the United Kingdom, for example. For all readers, a key factor shaping their life chances is where they were born (or at least what country they have citizenship of or the right of residence in, or are studying in). So, think about your position; realize that this book presents the views of human geographers working at a particular place and at a particular time. We have set out to challenge readers to think about the ways that human geography interprets the major social, cultural, economic, environmental and related issues that face the world in the early years of the twenty-first century. It is for you to use this book to inform your own insights and opinions as geographers; for some it could be part of a life-long engagement with and contribution to human geography.

Approach of the book

Writing an introductory textbook is a challenging task. No matter what its contents, or how they are organized, the finished product will not satisfy everyone. The book not only has to engage the attention and interest of the students who will learn with it, but also has to satisfy the instructors who will teach with it. Thus we are caught on the horns of a dilemma: how to make the text as accessible and as user-friendly as possible to students, while not appearing to be too naïve to the instructor assigning it. We also have to acknowledge that each geography programme is different, at least in detail if not in overall objective, and that a textbook fulfilling every need would be very large indeed.

Broadly speaking, however, most first-year geography programmes will have a course that offers an introduction to human geography. Such a course is nearly always compulsory. It is often supported with a 'world regional geography' course or a course devoted to topical 'issues in human geography'. Increasingly (though this was always the case in North America) such courses are required to meet the needs of non-geography students who are taking geography as an option or elective. This poses a major challenge for the instructor confronted by a class comprising those interested in human geography but coming to terms with a discipline whose contents, approach and language are very different from what they had expected, other students more interested in physical geography, and non-geographers who are there out of curiosity or requirement. Such a diverse audience also poses a challenge to the textbook writer.

If you browse in your university library (or at a bookshop or online bookseller), you will come across a variety of human geography textbooks. Broadly speaking these are of three types: systematic texts organized by major sub-discipline in human geography, topical texts organized by key topics or issues, and world regional texts using major regions (for example, 'Europe' or 'Southeast Asia') to introduce human geography. It will also be evident that these books are clearly aimed at different markets or audiences. Assigning a course text is almost a matter of fact in the United States (most students are aghast if there is no 'course text'), but elsewhere, certainly in the United Kingdom, assigned texts are a more recent phenomenon. Over the past couple of decades, however, an assigned text is more often seen as a necessary *starting point* for an introductory course in human geography almost everywhere. With student budgets often stretched, there is also a good case for purchasing a textbook that will cover more than one course and that will be of lasting use in later years. The benefit of regular updates in the form of new editions is that they will provide access to the most recent literature, although given the time it takes to publish and the limits to the scope that can be given to specific topics, textbooks should *always* be supplemented by other reading.

There is no doubt that introductory textbooks do not always travel well. This is because they are often designed for a particular educational system; but more because they also reflect a specific version of what is important in the world and what constitutes human geography. In addressing this problem, we have sought to produce a textbook that offers an *issue-oriented* introduction to human geography. The major sections focus on important issues facing the world at the beginning of the twenty-first century, but they also relate to major sub-disciplines in human geography. In revising the earlier editions we have added substantive chapters on social geography, the environment and environmentalism, and territoriality. In order to contain the length (and therefore cost) of the book, this has required hard choices and some themes have lost what in earlier editions were their own chapters, as we have worked them into other chapters. We believe that the issues we have selected are important and will continue to be so. They also reflect the interests and expertise of the authors and our sense of what captures the enthusiasm and imagination of the students we have taught over the years in a range of institutions on several continents. The individual chapters bring together ideas from a variety of fields in human geography. They also incorporate coverage of the globe, drawing ideas from a variety of places to reflect diversity and difference and inequality as well as the interdependent nature of the world. While this integrated approach may not sit well with the traditional ways of producing introductory human geography textbooks, we believe that it acknowledges the complexity of human society as well as the practice of contemporary human geography.

Structure of the book

This book is divided into five sections, supported by an Introduction. Section 1 provides an essential historical context for the four issue-oriented sections that follow. While it is natural for us to proclaim that geography matters, it is also the case that historical geography matters. A cursory examination of an atlas printed at the start of the last century will show you that over the past century the world has changed a great deal. Many of the contemporary issues discussed in this book, such as the geographies of development and of inequality, owe much to the way the world looked in 1900, or even earlier. This is not historical determinism, rather an acknowledgment that history is critical to an understanding of contemporary human geography. However, if we cast our minds even further back into the records of antiquity we are reminded that societies are transitory; nothing lasts forever. States and civilizations have come and gone and this should teach us that there is little permanent about the current world order. At the same time, what is past, and not always immediately visible, often has profound influence on the world we live in now. Section 1 highlights some of these issues and then focuses on the emergence of capitalism and its relationship with the making of the twentieth-century world that immediately preceded our own.

The remaining four sections comprise groups of chapters that focus on a related set of issues. It makes sense to work with Section 1 first; after that the sections can be utilized in whatever order you see fit. Each section starts with a brief summary of the important issues covered by each of the chapters and there is logic to the order in which the chapters are presented within each section. Section 2 examines the interrelationship between population change, resource production and consumption, global economic development and the environment. Section 3 focuses on social and cultural issues within an urban and rural context and at a global scale and includes a new chapter on the social construction of nature and what this means for the analysis and interpretation of how, for example, to manage the global environment in a sustainable way. Section 4 examines the globalization of economic activity, the emergence of a global financial system and means of exchange, and the importance of consumption to an understanding of the geography of economy. It incorporates two new chapters that, respectively, explore the geographies of global production networks and the significance of worlds of production and consumption that are centred around service industries. Section 5 considers a variety of political geographies at differing scales (from the local, to the national and global) and how they interface and intertwine.

How to use the book

Each chapter is designed to introduce a set of specific issues and particular fields in human geography. Each begins with a list of topics covered. The text is broken into manageable sections with supporting materials; important concepts and specific examples are presented in boxes. Each chapter concludes with a list of learning outcomes. These provide a means of assessing whether you have understood the content of the chapter. To assist you in reading beyond the text, each chapter has a set of initial recommended readings, at least some of which you should be able to find in your library (or via its website, in the case of access to journals). Where relevant, we have also included a list of websites that contain information you may find useful in your own research and writing. In addition, at the back of the book there is a glossary of important terms and an extensive bibliography. Such lengthy bibliographies are not always found in an introductory text; however, we want this book to remain an essential reference tool for the duration of your geography programme. Each chapter contains references to the bibliography giving you access to the specialist reading that is essential when you wish to research or otherwise follow through on specific topics in detail. One way of using this book is to read the chapters to support your lectures and seminars, the recommended readings to prepare for examinations, and the references, websites and bibliography to research and write assignments. In other words, reading this textbook is the beginning of your introduction to human geography, not the end.

So, where else do you go from here? As a geography student (or any other kind of student for that matter), you will be faced with choices, for example between human and physical geography or between more specialized subjects within one of those traditions. It is in the nature of academic disciplines to subdivide and to compartmentalize knowledge. Geography is unusual in that it sits at the intersections between the humanities, social sciences and natural sciences. This means that it internalizes both the disadvantages of disciplinarity and the potential advantages of multidisciplinarity. Much of the promise of geography lies on the margins between the sub-disciplines, between human and physical geography or, for example, between economic and cultural geography. In your studies you should make the most of this promise, seek out the space on the margins and look for the connections between the human and natural environments, the economic, the social and the cultural. Many of the challenges that face humanity in the new century also occupy these margins. The emphasis on places and spaces and how they are interconnected, imagined and represented, (the hallmarks of human geography), is itself promising. We hope that the chapters and arguments brought together in this book will (when supplemented by further reading, critical thought, writing and conversation) encourage our readers to continue to engage with this promise.

Learning resources

As we noted above, each chapter provides a set of recommended readings and, where relevant, useful websites. This chapter is no exception. Here we consider other resources that provide an introduction to human geography. The readings focus on other introductory texts and the study skills necessary to begin to conduct your own research. The website guide lists places to visit to find out more about geography as a discipline.

Further reading

This first set of readings provides a sample of other textbooks that you can use to supplement your reading. Individual chapters from them are often cited elsewhere as references in this book, but they are also worth browsing to get a sense of the different ways in which human geography is presented.

Cloke, P., Crang, P. and Goodwin, M. (eds) (2005) ***Introducing Human Geographies***, Arnold, London. The second edition of an innovative text: forty-three inviting short chapters that reflect the diversity of approaches in contemporary human geography. Designed more for the UK market, but well worth delving into wherever you may be.

Knox, P. and Marston, S.A. (2006) ***Human Geography: Places and Regions in Global Context***, 4th edition, Prentice Hall, Upper Saddle River, NJ. A comprehensive text written principally for the US market with lots of colour pictures and diagrams. Probably the best US textbook. Supported by a website, which is worth using, even if you do not have the book. See: http://wps.prenhall.com/esm_knox_humangeo_4/0,11860,3122929-,00.html

Le Heron, R., Murphy, L., Forer, P. and Goldstone, M. (1999) ***Explorations in Human Geography: Encountering Place***, Oxford University Press, Auckland. This is a stimulating textbook written by a group of New Zealand geographers specifically for New Zealand geography students. It may be difficult to find in the northern hemisphere, but reading it could turn upside down your own taken-for-granted view of the world.

Wait, G., McGuirk, P., Dunn, K. and Hartig, K. (2000) ***Introducing Human Geography: Globalization, Difference and Inequality***, Longman, French's Forest, NSW. Like the above, another rewarding Antipodean text; this one is published in New South Wales, Australia.

Research and study guides, readers and further insights into human geography

This section is dedicated to publications that will assist you in conducting geographical research and/or expand your knowledge of debates and approaches in the discipline. The list comprises dictionaries, and books on methodology, approaches and theory in human geography as well as a few (such as autobiographies of influential geographers) that are hard to categorize. You should also seek out the various 'readers' (comprising reprints of influential papers that usually first appeared in journals) that have been produced. Some bring together the most influential readings in a particular area of human geography, whilst others are collections of specially commissioned review essays that assess the status of particular areas of research.

Agnew, J., Livingstone, D. and Rogers, A. (eds) (1996) ***Human Geography: An Essential Anthology***, Blackwell, Oxford. A collection of classic geographical writings, and a sober reminder that there is very little that is wholly new.

Atkin, S. and Valentine, G. (eds) (2006) ***Approaches to Human Geography***, Sage, London, New Delhi and Thousand Oaks. A good way into the range of theoretical (and methodological) debates in human geography.

Benko, G. and Strohmayer, U. (eds) (2004) ***Human Geography: A History for the Twenty-First Century***, Arnold, London. This book seeks to address histories of continental European and anglophone geography and the relations between them.

Blunt, A. and Wills, J. (2000) ***Dissident Geographies: An Introduction to Radical Ideas and Practice***, Prentice Hall, Harlow. The geographies explored in the text all share political commitments to critique and challenge prevailing relations of power. From gay and lesbian geographies to geographies of anarchism and anti-racism, this book brings out some of the ways in which the production of geographical knowledge is tied to politics and struggles outside, as well as within, universities.

Bonnett, A. (2008) ***What is Geography?*** Sage, London, New Delhi and Thousand Oaks, CA. Not yet published as we go to press. It is evident from the title, however, that this starts with a complex and interesting question.

Castree, N. (2005) ***Nature***, Routledge, London and New York. A challenging introduction to how geographers have studied nature and what this tells us about the nature of academic geography.

Claval, P. (1998) ***An Introduction to Regional Geography***, Blackwell, Oxford. First published in French in 1993 as *Invitation à la Géographie Régionale*, this book draws on French (and some anglophone) traditions to review the past of and prospects for regional geography. An introduction to a significant tradition and work that overlaps with but is also beyond the mainstreams of English-language geography.

Clifford, N. and **Valentine, G.** (eds) (2003) ***Key Methods in Geography***, Sage, London. An introduction, covering methods used in both human and physical geography, and those they share.

Cloke, P. and **Johnston, R.** (eds) (2005) ***Spaces of Geographic Thought: Deconstructing Human Geography's Binaries***, Sage, London. A good collection to work through if you specialize in human geography beyond an introductory class.

Cloke, P., Crang, P. and **Goodwin, M.** (eds) (2004) ***Envisioning Human Geographies***, Arnold, London. This collection aims to sketch (and help to set) future agendas and directions for human geography. The plural 'geographies' in the title relates to the range of sub-disciplines considered in the chapters.

Cloke, P., Crang, P., Goodwin, M., Painter and **Philo** (2002) ***Practising Human Geography***, Sage, London. A more advanced (but very readable and inviting) guide to research methods and writing strategies in human geography.

Cresswell, T. (2004) ***Place: A Short Introduction***, Blackwell, Oxford. Explores how human geographers have studied, debated and complicated a concept that may seem self-evident and familiar, but which turns out to be a range of assumptions, ideologies and contests.

Daniels, S. and **Lee, R.** (eds) (1996) ***Exploring Human Geography: A Reader***, Arnold, London. The individual articles were not intended primarily for a student audience, but the collection provides a survey of human geography in the 1980s and 1990s.

Dorling, D. and **Fairbairn, D.** (1997) ***Mapping: Ways of Representing the World***, Prentice Hall, Harlow. Maps have always been associated with geography and geographers. This text will help you to understand how maps express the will to describe, understand and control.

Flowerdew, R. and **Martin, D.** (1997) ***Methods in Human Geography: A Guide for Students Doing a Research Project***, Longman, Harlow. An indispensable guide to a wide variety of the research techniques employed by contemporary human geography.

Gould, P. and **Pitts, F.R.** (eds) (2002) ***Geographical Voices: Fourteen Autobiographical Essays***, Syracuse University Press, Syracuse, NY. A collection of autobiographies reflecting the experiences and insights of some leading geographers of the past fifty years.

Hoggart, K., Lees, L. and **Davies, A.** (2002) ***Researching Human Geography***, Arnold, London. One of a number of books that provide an introduction to research methods in human geography, a good place to look when starting your own dissertation or research project.

Holloway, S.L., Rice, S.P. and **Valentine, G.** (eds) (2003) ***Key Concepts in Geography***, Sage, London. Worth reading alongside this text, especially so if you are taking introductory classes in human and physical geography, since this collection addresses both.

Holt-Jensen, A. (1999) ***Geography: History and Concepts***, (3rd edition, Sage, London, New Delhi and Thousand Oaks, CA. Translated from Norwegian, an introduction to the history, philosophy and methodology of human geography.

Hubbard, P., Kitchin, R., Bartley, B. and **Fuller, D.** (2002) ***Thinking Geographically: Space, Theory and Contemporary Human Geography***, Continuum, London. This book represents an attempt to demystify the nature and role of theory in contemporary human geography.

Johnston, R.J. and **Sidaway, J.D.** (2004) ***Geography and Geographers: Anglo-American Human Geography since 1945***, 6th edition, Arnold, London. First published in 1979, the sixth addition of this book provides an updated survey of the major trends in human geography since 1945 in the English-speaking world. Includes a discussion of geography as a discipline that will help you to better understand what academic geographers do and how the discipline is structured.

Johnston, R.J., Gregory, D., Pratt, G. and **Watts, M.** (eds) (2008) ***The Dictionary of Human Geography***, 5th edition, Blackwell, Oxford. Forthcoming as we go to press. Previous editions have proven to be an indispensable reference, and each one thicker and weightier than the last. This new edition is eagerly anticipated, but don't drop it on your foot when you get your copy!

Livingstone, D. (1992) ***The Geographical Tradition: Issues in the History of a Contested Enterprise***, Blackwell, Oxford. On publication this immediately became a landmark scholarly study of the long-term history of human geography.

Moseley, W., Lanegran, D. and **Pandit, K.** (eds) (2007) ***The Introductory Reader in Human Geography***, Blackwell, Oxford. The most recently published reader (collection of papers previously published in academic journals). A good way to get a feel for debates and approaches in a range of sub-disciplines (economic, political, urban and other parts of human geography).

Rogers, A. and **Viles, H.** (eds) (2003) ***The Student's Companion to Geography***, 2nd edition, Blackwell, Oxford. The second edition of a useful introduction; the sections contain information on the fields of geography, methods and data sources.

Thrift, N. and Kitchin, R. (2008) ***The International Encyclopedia of Human Geography***, Elsevier, Oxford. This A–Z will be the most comprehensive survey of human geography ever assembled. An online version will be available.

Tuan Yi-Fu (1999) ***Who am I? An Autobiography of Emotion, Mind and Spirit***, University of Wisconsin Press, Madison, WI. The autobiography of a Chinese-American who arrived in the United States as a 20-year-old graduate student in the 1950s and went on to become one of America's most original and respected geographers.

Women and Geography Study Group (1997) ***Feminist Geographies: Explorations in Diversity and Difference***, Longman, Harlow. A good way to begin to understand how feminist geographies have reshaped the study of human geography.

Wylie, J. (2007) ***Landscape***, Routledge, London. A text that examines how human geographers (and others) have conceptualized and narrated landscape.

Finally, as this introduction has sketched, human geography (understood as the subject, and those who study and teach it) has its own geography (understood as locations), in so far as it is unevenly developed (in some countries it is studied as part of the school curriculum, in other places – most significantly perhaps in the USA – it is not studied at schools). How and why this has come to be is in turn a complex reflection of wider histories and geographies, including the historical geography of empires (see Chapter 2), which meant that aspects of a British-style educational system were long imposed on many territories in Africa and Asia (and in Ireland). For more systematic reflections on the relative position of geography in the academy in a range of countries, see the set of papers in the *Journal of Geography in Higher Education,* **31** (1), compiled by and following a brief introduction by Kong (2007).

Useful websites

The Internet contains a wealth of information that is of relevance to an introductory course on human geography. It is impossible to cover all of the potential sources here. This section provides access to the sites of some of the major geographical societies in the United Kingdom and North America. All of these sites provide links to societies and organizations across the globe, as well as links to geography-related sites. In addition, many of these sites carry links to electronic versions of their journals (although to access these, you will probably have to go through the website of your university library) and other publications as well as information and links to resources such as maps, statistics, atlases or photographic collections. Your own department will have a website; visit it and find out what the academics who teach you are also saying about what they do. Visit the sites of other departments across the world and discover how geography is taught in different places. Thus, the Internet provides you with your own virtual guide to human geography. With the partial exception of the Canadian Association of Geographers, the links below focus on the anglophone world. Use the International Geographical Union website to seek out geographers across the globe.

www.aag.org The website of the Association of American Geographers. Founded in 1904, the AAG is the primary academic geography organization in the United States. This site provides a wealth of information on the Association's activities, as well as access to its two scholarly journals: the *Annals of the Association of American Geographers* and *Professional Geographer*.

www.amergeog.org The website of the American Geographical Society. The Society was established in 1851 and is the oldest professional geographical organization in the United States. The society publishes the *Geographical Review* and *Focus*.

www.cag-acg.ca The website of the Canadian Association of Geographers. Founded in 1951, the CAG-ACG is the primary academic organization in Canada. The association publishes *The Canadian Geographer* and organizes national and regional meetings.

www.geography.org.uk The website of the Geographical Association, which was founded in 1893 and is the national subject teaching organization for all geographers in the United Kingdom. The Association publishes three journals: *Primary Geographer*, *Teaching Geography* and *Geography*.

www.iag.org.au The website of the Institute of Australian Geographers. The IAG was founded in Adelaide in 1958 and is the principal body representing geographers and promoting geography in Australia. The institute publishes the journal *Geographical Studies*.

www.igu.net.org The website of the International Geographical Union, the international umbrella organization for academic geography. This website contains a comprehensive list of geography departments throughout the world.

www.nationalgeographic.org A public face of geography in the English-speaking world. Home to the *National Geographic* magazine, the site also provides access to the maps and photographs that appear in its magazine, as well as a searchable index.

www.rgs.org The home of the Royal Geographical Society (with the Institute of British Geographers). The RGS was originally founded in 1830 and the RGS-IBG is the primary academic organization in UK geography. The website provides information on the societies as well as electronic access to its three key journals: *Area*, the *Geographical Journal* and *Transactions*.

In addition to the websites of the various learned societies, there are dedicated 'web portals' that provide access to a wealth of information to support your geographical studies. The two examples below are UK-based, but their outreach is international and their links may take you to portals or subject gateways based elsewhere.

www.jisc.ac.uk/resourceguides Resource Guide for Geography and the Environment. This website professes to be the essential guide for exam revision and dissertations and theses. It provides access to online catalogues, publications, and subject gateways and data services. It requires an Athens username and password and therefore may be limited to UK users.

www.geosource.ac.uk Geosource: Geography and Environment Information on the Internet. This is a free service led by the University of Manchester. It currently contains over 5,000 individually selected information-rich Internet resources covering five subject areas: the environment, general geography and environmental science, geographical techniques and approaches, human geography and physical geography.

As you will probably already know, there are dozens of commercial websites that will help you to try to short-cut and circumvent learning and scholarship, by buying putatively top-grade custom-written essays for almost any assignment. Like other even seedier sides of the Net, these have their own geography (in terms of by whom and *where* they are produced, *where* the money goes, what kind of economies and technologies enable them and so on). They are, of course, a specialist form of fraud known as plagiarism and severely sanctioned in all Universities when detected.

For annotated, clickable weblinks and useful tutorials full of practical advice on how to improve your study skills, visit this book's website at **www.pearsoned.co.uk/daniels**

WORLDS IN THE PAST: CHANGING SCALES OF EXPERIENCE AND PAST WORLDS IN THE PRESENT

Section 1

Edited by Denis Shaw

The world in which we live today has been moulded and is constantly being remoulded by the forces of capitalism (defined in Chapter 2, pp. 40–41). As modern capitalism spread across the globe it inevitably changed the societies which preceded it. But the pre-capitalist societies, while adapting to capitalism, did not necessarily lose all their distinctive features, and capitalism itself was changed in the process of interacting with such societies. The fact that human societies in the modern world differ among themselves in all kinds of ways is only partly the result of the way global capitalism works and of the ways in which societies are now adapting to the various forces affecting them. Societies also differ because of their past histories, histories that stretch back into pre-capitalist times. The world into which modern capitalism spread was itself already enormously varied and had been changing in innumerable ways over thousands of years. This section presents a brief survey of that long history and argues that the human geographer cannot understand the present-day world without knowledge of the worlds which have existed in the past.

Chapter 1 of this section considers what the world was like before capitalism and what if any generalizations can be made about this long period of human history. Chapter 2 analyses the rise and spread of modern capitalism down to the end of the nineteenth century. Its history during this period is intimately linked to the rise of the European powers and to the worldwide spread of their influence. This, of course, is not to deny that capitalism probably sprouted, and then died, in other societies in other periods, or that it might have gained worldwide significance on the basis of a non-European core had circumstances been different. During the twentieth century, which is the subject of Chapter 3, capitalism became a truly global phenomenon and radically changed in the process. This chapter thus provides a more immediate background for the sections that follow.

In summary, this part of the book emphasizes the variety and complexity of the world before globalization. And it demonstrates that only by knowing something about that world can we understand globalization itself.

PRE-CAPITALIST WORLDS

Chapter 1

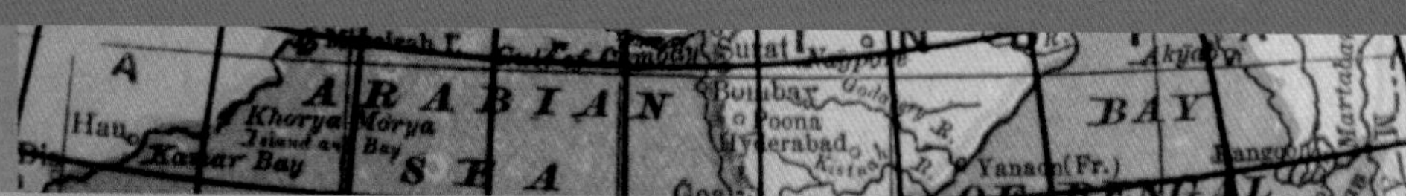

Denis Shaw

Topics covered

- Why geographers need to know about the past
- Ways of classifying past and present societies
- Bands, tribes, chiefdoms, regulated states, market-based states
- Hunting and gathering
- The invention and spread of agriculture
- The rise of cities, states and civilizations
- Medieval feudalism
- Problems of studying past societies

Much of this book is about how capitalism has spread across the globe and even now is remoulding human societies in new and unexpected ways. This chapter is about what the world was like before capitalism. In other words it is about how human beings have lived during most of their existence on this planet. It will also say something about why it is important for human geographers to be aware of that past.

Capitalism has changed the world profoundly and often very quickly. In the first half of the twentieth century it was still possible for geographers and others to travel to regions where its influence was minimal and where people still lived in pre-capitalist societies. Today this is much less true. Even so, the geographer can still find numerous areas where the way of life has many pre-capitalist traits, though perhaps increasingly influenced by the modern world. Let us consider some examples.

About 2 per cent of the population of Australia is composed of Aborigines whose ancestors first settled the continent up to 50,000 years before white Europeans (Broome 1994). Today their way of life ranges from that of suburban professionals to remote outstations and homelands where people live partly off the land in traditional style. When the British established their first penal colony on Australian soil in 1788, more Aborigines lived on the continent than today, and they were divided into some 500 groups displaying a wide variety of languages, cultures, economies and technologies. All were hunters and gatherers leading a semi-nomadic existence and using some similar tools, like stone core hammers, knives, scrapers and axe heads as well as wooden implements like spears, digging sticks and drinking vessels. However, there was also a variety in technology and economy that reflected the diverse environments in which Aboriginal peoples lived. Whether living on coasts, riverbanks, in woodland or desert regions, they were efficient and skilled at exploiting the flora and fauna of their surroundings. Aboriginal peoples who hunt and fish today are often using skills honed by their ancestors over thousands of years. Their languages and cultures similarly reflect the experiences of many hundreds of generations.

The Inuit peoples living in the Arctic, mainly coastal, territories of Greenland, northern Canada, Alaska and the extreme eastern tip of Russia, are another group who in the past lived almost entirely by hunting and gathering (Sugden 1982). Hunting and gathering still plays an important role in many of the remoter communities. Like the Australian Aborigines, the Inuit showed a remarkable ability to exploit the resources of what, in their case, is an extremely harsh environment. Because of their environment, Inuit communities demonstrated less economic variation than the Aborigines but they were far from uniform. There were marked seasonal movements depending on the character of the local resources as groups moved between exploiting the fish and mammals of the open ocean, hunting seals on the ice, and seeking caribou, musk ox and similar fauna inland. The rich Inuit culture is also varied but by no means as varied as the Aboriginal (there are only two interrelated languages, for example). Considering the vast geographical spaces that the Inuit occupy, the latter fact is remarkable and may reflect a relative ease of communication by sledge and boat as well as the frequency of migrations. Like the Aborigines, the Inuit have had to adapt in the recent period to an intrusive,

Plate 1.1 Pastoral farmers with their cattle, West Africa.

(© Sebastian Bolesch/Das Fotoarchiv/ Still Pictures)

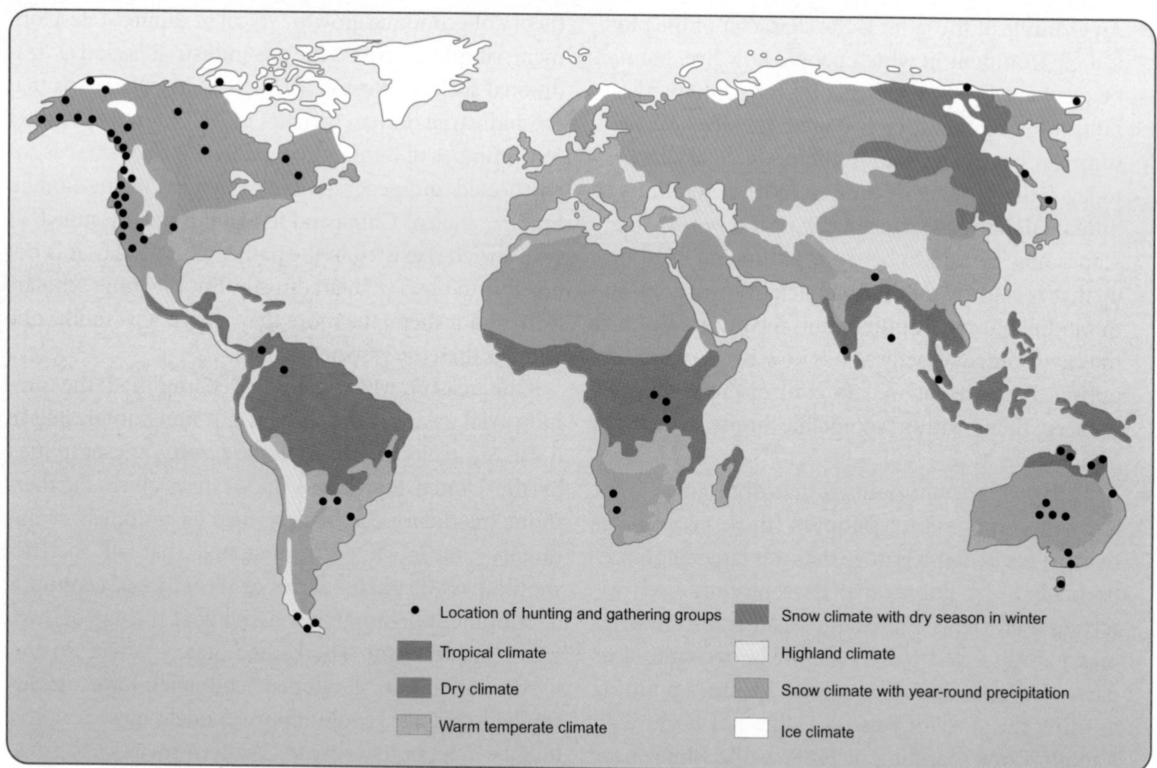

Figure 1.1 World distribution of hunter–gatherers today and in the recent past, with indication of their habitats (based on a classification of climate).

Source: Adapted from Simmons (1996: 48)

mainly white culture, but are still sufficiently conscious of their distinctive histories and life-styles to have a strong sense of identity.

The Australian Aborigines and the Inuit are but two of many human groups existing either now or in the recent past who have lived by hunting and gathering (see Figure 1.1). The present-day world also contains other groups whose way of life to a greater or lesser extent reflects pre-capitalist characteristics. Pastoral nomads, for example, live mainly by raising and herding domestic animals (cattle, sheep, goats, camels, yak, reindeer and others), that provide them with food, clothing and other necessities. Pastoralists are particularly found in marginal (semi-arid, sub-arctic, sub-alpine) lands in parts of Eurasia and Africa (see Plate 1.1). Another example is the many peasant farmers still to be found in parts of sub-Saharan Africa, southern Asia and elsewhere. Some still grow much of their own food and market relationships have yet to become of central importance to their lives (Wolf 1966). To the extent that such people eat and otherwise depend on what they grow and raise rather than on what they can sell or earn outside the farm, their way of life is similar to that of peasants over the centuries. Of course, nowadays the number of peasants is declining and they are increasingly exposed to the influences of the outside world. But again they remind us of a world before capitalism.

The point, then, is that today's world still contains many societies whose ways of life differ from those lived by most of this book's readers and which recall earlier periods in human history. In order to understand those ways of life and the world as it is now, the geographer must know something about the world before capitalism. Learning about that world helps us to see how capitalism has changed the world and the pluses and minuses of that process.

A number of arguments can be advanced to suggest why it is important to know something about the world before capitalism:

- Understanding the past helps geographers and others to understand themselves and their societies. Societies and individuals are products of the past, not just of the present. The present cannot be understood in ignorance of the past. Studying the past provides answers not merely to questions about how things were but also about how things are now (and, more tentatively, about how things will be in the future).

- An example of the latter is the character of the physical environment in which people now live. Human beings have lived on this planet for so many thousands of years that over vast areas the physical environment has been profoundly modified. The world today is the product of thousands of generations of human activity. In fact, can any of the world's landscapes now be said to be truly 'natural'? This shows us that it is likely that human activity will continue to modify our environment into the future. Keeping the environment exactly as it is now is a very unlikely option. But how far we can control those changes, or keep them within acceptable limits, is a more difficult question.
- A further important point is that Aborigines, Inuit and other **indigenous peoples** (those peoples native to a particular territory that was later colonized, particularly by Europeans) have recently been asserting their rights. For example, Australian Aborigines have asserted their right to be recognized as part of the Australian national identity, pointing out that the story of Australia does not begin with Captain Cook's landing in 1788, as the whites have so often assumed, since the Aborigines were in Australia long before. Indigenous peoples have also been claiming rights to local resources long ago usurped by outsiders and are trying to protect their distinctive cultures. Such peoples often feel they are the victims of the past. By studying that past and trying to understand the variety of cultures and ways of life that exist in the world and how they came to be, geographers are more likely to understand and respect such feelings.

Thus, while human geography is primarily about the present, it cannot afford to ignore the past. Some of the above points will be illustrated in the following pages.

1.1 Making sense of the past

Despite the spread of capitalism and globalization, the world of today remains immensely varied and complex. The same is true of the world in the past. The question for the geographer is how to make sense of this complex past; how to make it amenable to geographical analysis and understanding.

Different scholars have tried different ways of answering this question, but all suffer from shortcomings. For example, the economist W.W. Rostow, in his famous theory of economic growth typical of capitalist development, simply described the pre-industrial period as 'traditional society' (Rostow 1971: 4–6). Yet the fact is that pre-industrial human societies ranged from the smallest communities of hunters and gatherers to societies as sophisticated and geographically extended as the Roman Empire, ancient China and feudal Europe. Inasmuch as such societies existed in the past, and not today, it is not possible to observe them directly. But the more scholars learn about them, the more they realize how unlike one another these early societies were.

One reason why Rostow oversimplified the pre-industrial past was that he was not much interested in it. He was really interested in modernity, as exemplified by the United States and its Western allies. Furthermore, his theory can be described as 'unilinear evolutionary' inasmuch as it suggested that all societies should pass through a series of set stages of economic development before finally arriving at the age of 'high mass consumption'. The United States (where Rostow lived) and other 'developed' countries have already reached this age. Evolutionary models have certainly long been a popular way for Western societies to try to make sense of the past. One of the attractions of such models has no doubt been their tendency to suggest that the West is the most 'developed' and thus most 'progressive' part of the world and, by implication, the best. But quite apart from the questionable assumption that all societies are seeking to imitate the West, this raises important issues about the meaning of 'progress' and 'development', two terms that certainly carry very positive connotations in the West. For example, while those parts of the world usually deemed most developed and wealthiest certainly use most energy and enjoy access to more material goods, they also make huge demands on the environment, perhaps ultimately to everyone's undoing. Thus, what is progress in this context? Similarly, it has often been noted that the price people tend to pay for more 'development' and wealth is that they have to work harder and for more hours. Unless hard work is regarded as a virtue in itself, this again raises questions about the nature of progress. The point here is not to disparage all forms of evolutionary theory or idea (for example, to deny that human societies have, by and large, become more complex and spatially extended through time; see Dodgshon 1987). Rather, it is to suggest that what is most recent or new or complex is not necessarily best. Thus, in what follows, words like 'modern' do not imply 'better', nor do 'ancient' or 'primitive' imply 'worse'.

An alternative method of trying to make sense of the past, and one that does not have the unilinear evolutionary structure of, say, Rostow's theory, is to classify human societies into a series of types. Classification can be described as a way of simplifying a complex world by grouping together phenomena that are regarded as having some common feature or property, particularly where the latter is deemed especially significant. Thus Marxists have commonly grouped societies in accordance with their prevailing mode of production, or in other words with the way in which material production is related to social structure (Hindess and Hirst 1975). As a geographer, primarily interested in spatial structure, Robert Sack classified societies in accordance with their use of territoriality (Sack 1986). The point is that there is no right or wrong mode of classification. It all depends on what the purpose of the classification is. Most social theorists have wanted to claim that their mode of classifying societies is particularly significant for understanding how and why societies differ. The problem is that there is a large measure of disagreement about what the best mode is, and each has its shortcomings.

1.2 A classification of human societies

As an example of the kinds of difficulties that face any attempt to classify past and present human societies, this chapter will focus on one mode of classification that has had widespread appeal not only for human geographers but also for anthropologists and other scholars. This is the mode of classifying human societies into bands, tribes, chiefdoms and states (regulated and market-based) (see Bobek 1962; Service 1971; Dodgshon 1987). Spotlight boxes 1.1 to 1.5 describe, in simplified form, the major characteristics of each type of human society according to this classification. It will quickly be seen that the classification describes a range of societies from the simplest and most primitive to the capitalist societies of today. What it does not suggest is the circumstances under which one kind of society may change into, or be succeeded by another.

One reason why this mode of classification has had such appeal is because it suggests that over the course of human history larger and more complex societies have tended to appear. It also relates size and complexity to the way societies occupy space and their relationships to their physical environment. However, it is worth reiterating that this is only one way of classifying human societies. Indeed some scholars reject the whole idea of classifying human societies because they feel it leads to misapprehensions about how societies are formed and survive, and even to dangerous notions of unilinear evolution. It thus needs to be remembered that this scheme is suggestive only, and does not necessarily describe the way any given society has actually changed through time. In particular, it does not stipulate that societies must develop through a series of set stages, or that one type of society necessarily leads to the next.

Subsequent sections will outline some of the social processes that might have helped to produce, change and perhaps destroy the different kinds of human society described in Spotlight boxes 1.1 to 1.5. In so doing, some of the limitations of the classification system used will become apparent. Before proceeding, however, it is worth making one important point. Our knowledge of prehistoric societies (societies which have left no written records) derives mainly from archaeological data, or material left by such societies and the traces of their environmental impacts. Methods for analysing such data have become ever more sophisticated, but the data themselves obviously become scarcer the older they are. Moreover, there are many kinds of questions, for example about the way early people viewed the world around them, which cannot be answered directly from the archaeological evidence. New discoveries are always liable to change our understanding of past societies, and especially of the earliest ones. Our knowledge of the latter societies, and of the dates attaching to them, must therefore be regarded as especially tentative.

1.3 Hunting and gathering

The time when the first human-type species (hominids) first appeared on earth is very uncertain, but a date of at least two and a half million years ago is often given. Homo erectus, forerunner of the modern human species, is believed to have appeared around 1.9 million years ago, and the modern species (Homo sapiens sapiens) at least 40,000 years ago (but see Foley 1995, for a contrary view). Since agriculture appears to have arisen less than 10,000 years ago, hunting and

Spotlight Box 1.1

Bands

- Bands are societies of hunter–gatherers. Most of human history has been lived in this way.
- Bands live by hunting wild animals and gathering food from the surrounding flora.
- Because of the pressures they place on the surrounding environment, bands are almost always nomadic.
- Bands are the smallest and simplest of human groups.
- Bands rarely exceed 500 people.
- For most of the time band members generally live in smaller groups of 25–100 people, which allows for efficient exploitation of the environment.
- Bands usually have a minimum of social differentiation and functional specialization.

gathering in bands have been the basic occupations for much of humanity's existence (Spotlight box 1.1).

As noted already, only a limited amount is known about early human societies (see Plate 1.2). In addition to archaeological evidence, much has to be inferred, for example from hunter–gatherers who are still in existence or who existed until quite recently. Needless to say, such inference can be dangerous because modern hunter–gatherers have probably been influenced by other, more 'advanced', societies that now share their world. Their way of life is thus unlikely to be identical with those who existed in the remote past.

The distribution of hunter–gatherers in the recent past shows that they have existed in every major climatic zone and have practised enormously varied economies, depending on the nature and potentials of the local environment (Figure 1.1). This fact alone suggests that a simplistic evolutionary model of human societies will fail to do justice to their complex histories. Some generalizations do seem possible, however – for example, that in recent times groups living in the highest latitudes (above 60 degrees) have relied largely on hunting, and those below 39 degrees on plant collecting. Fishing seems to have played an important role for many groups living in intermediate latitudes (Lee 1968).

Hunter–gatherers have lived in this world for such a long period that today's environment is the way it is partly as a result of their activities. For example, any notion that early human beings lived in complete harmony with their environments or always had 'sustainable' economies is simply untrue. Thus hunter–gatherers are credited with the extinction of many animals. An outstanding case is North America where perhaps two-thirds of the large mammal fauna living there just before the arrival of the ancestors of modern native Americans (via the Bering land bridge, possibly about 12,000 years ago) subsequently disappeared. The most likely explanation is the effects of hunting. The discovery of fire, its use by hominids

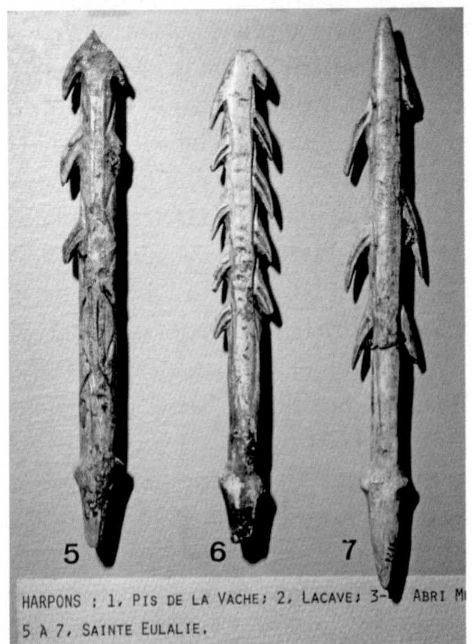

Plate 1.2 Artefacts from early human sites: flints and bone harpoons.

(Michael Holford Photo Library)

long predating the appearance of Homo sapiens sapiens, was also an important instrument whereby human beings changed their environments. Some scholars have argued that the present-day appearance of such major biomes as the African savannas and the prairies of North America is the result of the human use of fire over thousands of years (Simmons 1996). More recently, contacts between hunter–gatherers and agricultural and industrial societies frequently had even more far-reaching environmental effects. At the same time, other scholars have asserted that the relations between such groups and their physical environments were rarely if ever exploitative in the modern sense. Hunter–gatherers lived in an intimate relationship with their environment and frequently regarded it, or aspects of it, as sacred. Some scholars have thus described their relationship as 'organic' in a way which has been forgotten by our modern societies, much to the detriment of today's global environment (Merchant 1980).

Scholars have wondered about processes whereby bands evolved into tribes, or hunter–gathering economies into agricultural ones (Harris 1996a, b). It has been pointed out, for example, that where the natural environment was especially favourable, the packing of bands tended to be denser, leading to less mobility and perhaps to semi-permanent settlement and closer interaction between bands. There may also have been a much greater degree of manipulation of the environment and its resources than might be expected in a pure hunter–gatherer economy. Altogether the life of bands was much more variable and a good deal less stable than the above classification scheme suggests. The boundary between band and tribe is thus blurred.

1.4 Human settlement and agriculture

Just as the boundary line between band and tribe may not always be easy to define in practice, the same can be said of the boundary between tribe and chiefdom. Some scholars, for example, doubt whether the category 'tribe' (Spotlight box 1.2) is particularly helpful, implying as it does the lack of a hierarchical social structure (Friedman and Rowlands 1977). Such scholars tend to believe that the appearance of agriculture will have sparked off competition for access to the best land, or at least rules whereby such land was allocated and inherited, and that some people and groups will inevitably have lost out. They therefore see a social hierarchy (characteristic of chiefdoms and states – see Spotlight boxes 1.3 and 1.4) beginning to emerge even as agriculture and the process of permanent settlement began.

Be that as it may, it is now widely accepted that the traditional picture of agriculture being invented in a small number of 'hearths' and then spreading across the globe is far too simple. Bands of hunters and gatherers are known to develop an often intimate knowledge of their local environments, and no doubt human beings understood much about the factors influencing plant and animal development long before they began to practise full-blown agriculture. David Harris (1989, 1996a, b) has argued that the development of agriculture was a long drawn-out affair and that there must have been much trial and error before it finally emerged in some places in a fully recognizable form. Some prehistoric northern Australian Aborigines, for example, knew about agriculture, but never adopted it.

Spotlight Box 1.2

Tribes

- Tribes appeared with the invention and spread of agriculture.
- Agriculture usually demands considerable investment of human effort into a relatively small area. This reduces the propensity to migrate (but pastoral nomads are an exception).
- The appearance of agriculture usually allows more people to live in a smaller space. Agricultural societies thus tend to be bigger than bands, with higher population densities.

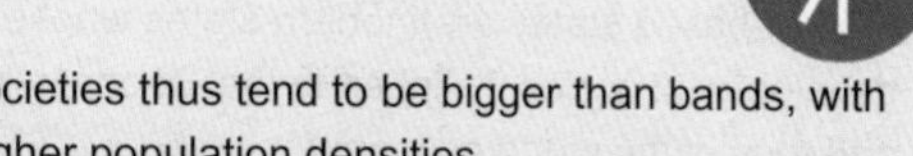

- Settlements tend to become more fixed than in bands.
- The greater spatial fixity of society encourages intermarriage and the development of extended family networks. Tribes thus develop a sense of kinship and of common descent.
- Socially, tribes are relatively egalitarian, at least as between kinship groups.

Spotlight Box 1.3

Chiefdoms

- Chiefdoms arose with the emergence of 'ranked and stratified' societies.
- 'Ranked' societies are societies where groups and individuals have, on a relatively permanent basis, different degrees of status and power.
- 'Stratified' societies are societies where groups and individuals have, on a relatively permanent basis, different degrees of material wealth.
- A chiefdom implies the presence of a permanent ruling group and/or individual, though kinship remains the chief bond between ruler and subject.
- Chiefdoms imply a greater degree of centralization and control within society than in a tribe (shown, for example, in the imposition of taxes and tribute on the ordinary subjects). Spatially this might be reflected in the greater importance accorded to a central settlement (proto-city) where the ruler resides.
- The first chiefdoms began to appear about 3000 BC in Europe, but earlier elsewhere.

Spotlight Box 1.4

Regulated (pre-modern) states

- Whereas the chiefdom is organized around the principle of kinship, the state is organized on the basis of territory (viz. one is subject to the state if one resides in the territory of that state).
- States, whilst also based on social inequality, tend to be larger and administratively more complex than chiefdoms. This implies greater functional specialization and greater probability of the rise of urban forms.
- Pre-modern states existed in a world before modern capitalism. The market was not central to their functioning and was often controlled ('regulated') in various ways. Thus economic relationships were generally subordinated to political, social and religious considerations.
- Other forms of social relationship, for instance landholding, also often tended to be controlled (regulated) rather than being determined by the market.

Spotlight Box 1.5

Market-based states

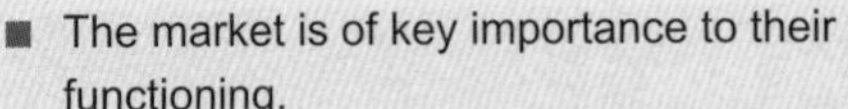

- Market-based states are modern states whose development is closely linked to that of capitalism and thus of the world economy.
- The market is of key importance to their functioning.

Something can, however, be said about when, and perhaps why, agriculture appeared. The archaeological evidence makes it possible to detect and date the remains of domesticated varieties of plants and animals. According to Simmons (1996: 93) there were probably three foci for the initial surges of domestication: around 7000 BC (to use the Christian chronology) in south-west Asia, around 6000 BC in South-east Asia, and around 5000 BC in Meso-America. Different species of plants were associated with each, for example: wheat, barley and oats in south-west Asia; rice in south-east Asia; and maize, squash and gourds in

Meso-America. Domesticated animals also began to appear about the same time (though the dog, domesticated from the wolf, appeared before agriculture). Sheep, goats, pigs and cattle are all associated with south-west Asia (Simmons 1996: 87–134), though pigs may also have been domesticated independently in eastern Asia.

As to why agriculture was adopted, scholars differ. The simple answer, that it was more efficient than hunting and gathering, hardly applies to naturally productive environments where it clearly involved much greater effort relative to the return achieved. Some scholars favour population pressure. Others, however, argue that this view is over-deterministic and point out that even hunter–gatherers had the means of controlling population growth. Such scholars tend to favour more complex explanations that embrace cultural preferences and choices as well as environmental and population pressures (Maisels 1993: 25–31).

Just as the factors that led up to the appearance and spread of agriculture are by no means straightforward, the same is also true of its further development. No simple evolutionary model can account for how agriculture has changed and developed throughout the world. Everywhere traditional agriculturalists showed a remarkable ability to adapt their practices to local environmental conditions. Eventually they largely displaced the hunter–gatherers (Figure 1.2) and gave rise to an enormous variety of agricultural systems (Figure 1.3).

Agriculture's effects on the landscape were much greater than the effects of hunting and gathering. It changed the vegetation cover of wide areas as forests and grasslands disappeared under fields, practices like terracing and irrigation were introduced, and the effects on soil composition abetted erosion in some places. The grazing of animals and other forms of resource use frequently changed the species composition of forests, grasslands and other areas. Agricultural systems, crops and domesticated animals spread way beyond their initial locations. An outstanding example is the cultivation of rice, which may have started in east Asia around 5000 BC, if not before, and spread as far as Egypt and Sicily by Roman times. Such developments long preceded the European overseas expansion beginning in the fifteenth century AD, which was to have even more far-reaching consequences both for the geography of agriculture and for the environment (see Chapter 2, pp. 44–8).

There is no doubt that the spread of agriculture (whether as a result of migration or through cultural diffusion) had far-reaching effects on human society, encouraging permanent settlement and the emergence of tribal systems (though Crone, 1986, questions the nature of the link between agriculture and tribes). Tribal members, living permanently side by side, almost inevitably developed similar cultural traits such as common languages. As suggested in Spotlight box 1.2, tribes also typically develop myths of common ancestry (indeed, this seems to be part of their definition). In the late nineteenth and early twentieth centuries, a period much influenced by ideas based on Darwinian evolution as well as by myths of European racial and cultural superiority bolstered by European overseas imperial expansion, it became fashionable, on very limited evidence, for scholars to imagine that modern ethnic differences have a biological basis and that certain 'races' or ethnic groups are inherently superior to others. In Britain, for example, many historians and archaeologists regarded the modern English as deriving almost entirely from Anglo-Saxon invaders with both racial and cultural characteristics which equipped them to dominate the 'inferior' Celtic peoples. Similar ideas, of course, reached their ultimate absurdity in the racist theories propounded by Adolf Hitler and the Nazis. Nowadays, with the availability of new scientific approaches like molecular archaeology and DNA studies, we know that the origins of the English and of other peoples are much more complex than was once thought and that different peoples are in fact remarkably similar in their capabilities (Miles 2006: 18–31). Different human groups, settling in the same region, may gradually have adopted similar languages and cultural patterns, thus slowly fusing into one people or ethnic group even though they were not originally closely interrelated. In this way the study of even the remote past can cast light on issues which cause contention today and, as in this case, help to discredit theories based on crude racist prejudices and a misreading of history.

1.5 Cities and civilization

The invention of agriculture was only one of the events that moved human societies along a road that ultimately led to the kinds of societies which predominate in the world today. Others included the appearance of cities, states and civilizations. Cities and states usually go together. Maisels (1993: 302), for example, argues that the functional specialization associated with cities allowed rulers to rule more effectively and also to put a

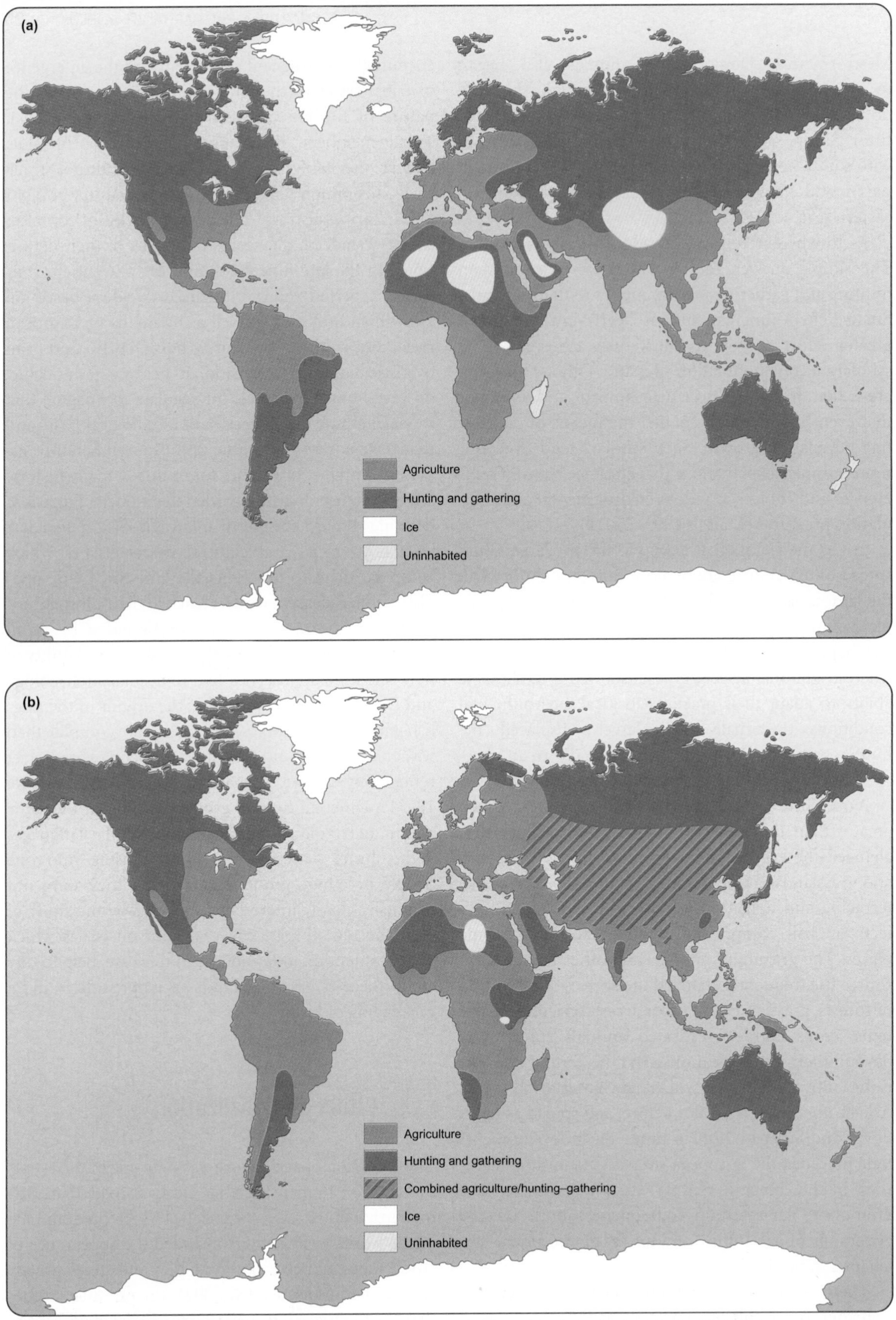

Figure 1.2 World distribution of agriculture and hunter–gathering about AD 1 (a) and AD 1500 (b).

Source: Sherratt (1980: 97, 117)

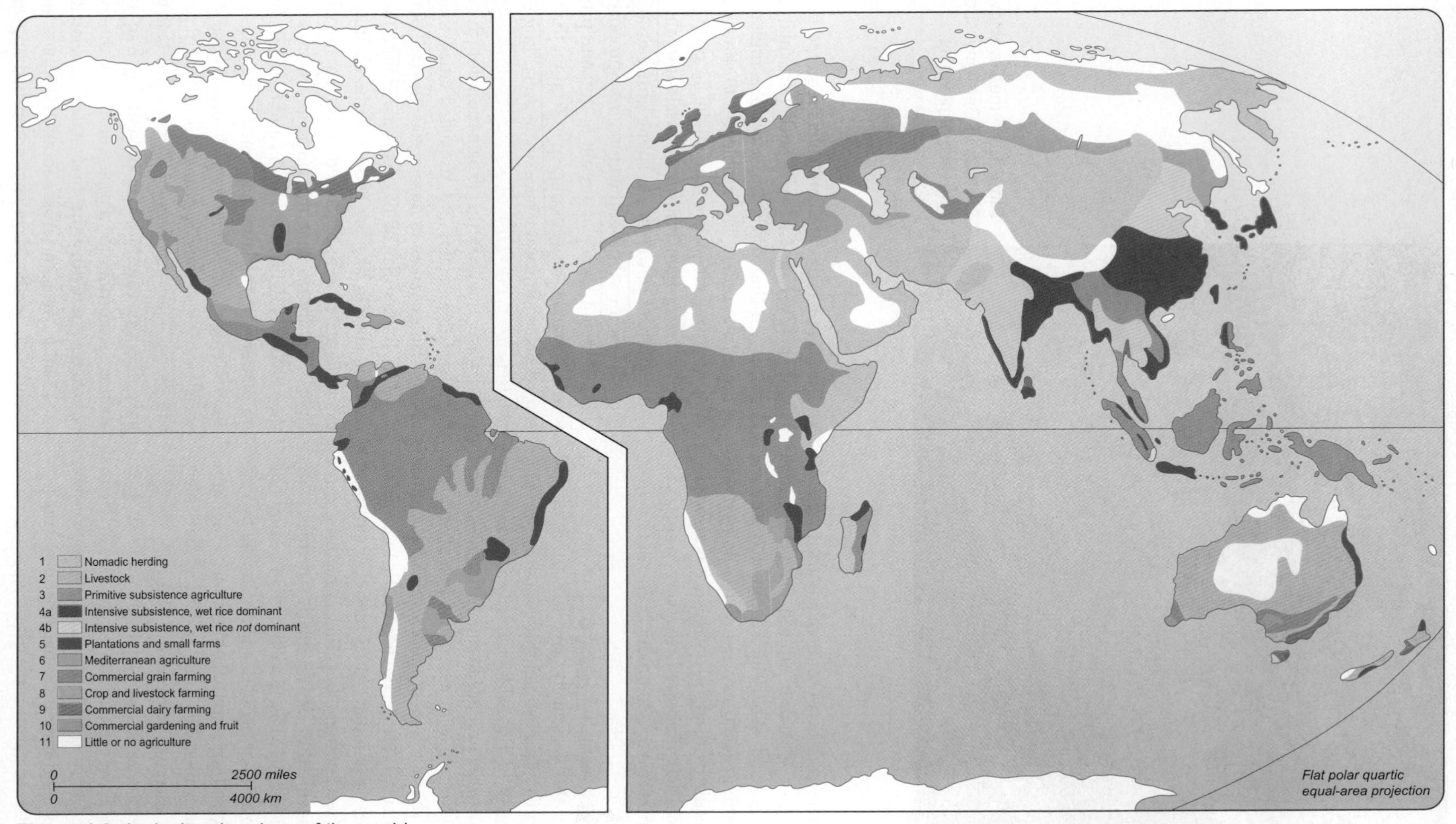

Figure 1.3 Agricultural regions of the world.

Source: After Whittlesey (1936)

Plate 1.3 A medieval town, Altstadt, in Germany, showing evidence of planning.
(Blickwinkel/Luftbild)

greater social distance between them and their subjects. The gathering of specialists around the person of a state's ruler fostered urban life, commerce and the many other things (such as writing, technology, religion, and speculative thought) with which is associated the word 'civilization'. Indeed, the very words 'city' and 'civilization' have a common Latin root (civitas, meaning 'state').

As is the case with the appearance of agriculture, the processes that eventually gave rise to cities and civilizations are far from straightforward. It is clear that the development of a social hierarchy was a necessary precursor, but what produced that? Different scholars have debated the merits of alternative potential causes such as land shortages, inheritance rules, or the emergence of particularly strong or charismatic individuals. But while these and other factors may have produced **chiefdoms** (see Spotlight box 1.3), something more seems to be required to explain the rise of cities, states and civilizations. As Maisels writes, chiefs enjoy only 'hegemony' over their peoples, and are bound by rules of kinship which imply mutual obligations. Chiefdoms were notoriously unstable. Kings and other state rulers, by contrast, exercise 'sovereignty' (ultimate power) over a **territory** and its peoples (Maisels 1993: 199). Thus scholars have cast around for particular factors which enabled individual leaders to have their power recognized as legitimate by subject peoples and persuaded the latter to pay over the tribute, food surpluses and services which cities and rulers required to maintain themselves in existence. Such particular factors might have included religious sanctions, military skills or organizing power. The problem is to show how the kind of social inequality that might be produced by such processes then transforms itself into the stable and legitimate power that a ruling group exercises in the state. Some scholars, therefore, have proposed that states came into being as a result of different processes working together in a contingent way rather than being caused deterministically by a single overriding process (Maisels 1993: 199–220).

Whatever may be the explanation for the initial development of states and associated cities, they gave rise to major changes in human societies (see Spotlight box 1.4). Rather than being organized around the principle of kinship like the tribe and the chiefdom, for example, states were organized on the territorial principle. This means that the rulers of states exercised their rule over defined territories and their inhabitants, no matter who the latter happened to be. This proved an extremely powerful way of organizing human societies. States thus appeared in different parts of the globe. The first organized states are believed to have developed in Sumer in present-day Iraq around 3000 BC. According to Maisels, these took the form of city-states. In China the first state, associated with the Shang dynasty, arose about 1500 BC, and by about 1000 BC the Maya civilization had appeared in

Meso-America. In Europe the first states are commonly taken to be the Greek city-states that began to develop in eastern Greece and the neighbouring islands about the eighth century BC. Whether the state arose independently in different places across the globe, or whether the idea of the state was somehow diffused from one or two initial foci, is uncertain.

Early states took many different forms. All faced difficulty in enforcing and maintaining their territoriality in the context of poor communications, and different states approached this problem in different ways. The Greek city-states, for example, typically occupied only small areas and restricted their population sizes. When population in any one state exceeded a certain threshold, that state would found a self-governing colony elsewhere on the sea coast. Travel by sea was generally easier than overland travel and thus settlements could keep in touch (for an analogous example, see Plate 1.4). In this way the Greeks eventually colonized much of the Mediterranean coast (Figure 1.4). The huge Roman Empire, by contrast, could not rely solely on sea travel. This was a much more centralized polity and only by having superb military organization, an elaborate system of roads and fortifications, good administration, and a state-controlled religion centred on the person of the emperor, could it survive for several centuries (Figure 1.5 and Case study 1.1). The Mongol empire was also very extensive, controlling much of Asia and a good slice of central and eastern Europe in the thirteenth and fourteenth centuries AD. The Mongols were pastoral nomads whose formidable warrior skills were based on their horsemanship and their ability to outmanoeuvre their enemies. But nomadic empires had a tendency to contract as quickly as they arose. It was difficult for the emperor to maintain control over his swift-moving armies of mounted warriors. Finally, medieval Europe's answer to the problem of territoriality was to decentralize political power through kings to territorial lords and minor nobles (see Case study 1.2). In return

Plate 1.4 A nineteenth-century Polynesian canoe. Travel by water was the most efficient form in pre-industrial times, allowing vast distances to be covered in remarkably short periods of time.

(Mary Evans Picture Library)

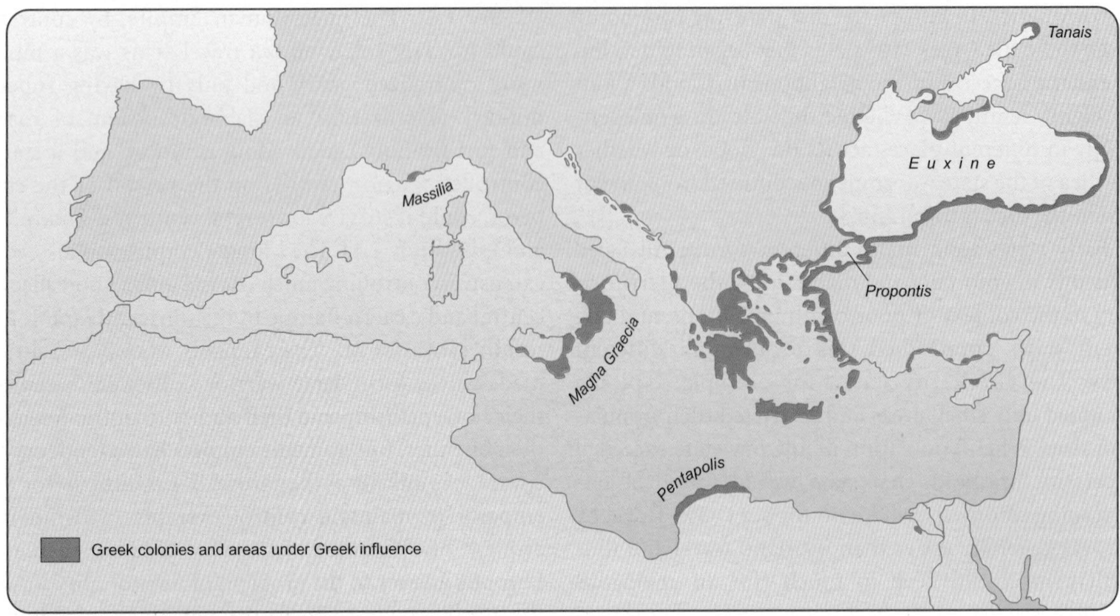

Figure 1.4 The Greek colonization of the ancient Mediterranean c. 550 BC.

Source: After Pounds (1947: 49)

for their rights to their land and other privileges, lesser nobles owed their lords a duty of military support in time of need, which meant having to appear with an army of knights, retainers and other soldiers when ordered to do so. The lords in turn owed loyalty to their overlords, and so on in a hierarchy that ended with kings or other rulers. Medieval European states were thus quite decentralized and unstable, as regional lords and their underlings were often tempted to rebel (Figure 1.6). Only after centuries of struggle between rulers, lords and ordinary subjects did the European states of today finally appear.

What all this tells us is that the present-day pattern of states is contingent on and the outcome of quite random historical processes. Had other processes taken place, the outcomes might have been totally different. In other words, there was no historical inevitability about the emergence of states like the United Kingdom, France, China or Nigeria, whatever present-day nationalists might like to think. Often enough state-building has been a murky and messy business in which many innocent people have suffered, and it usually lacked the glorious victories and heroic deeds so often celebrated in modern national mythologies. And if the existence of modern states is the product of historical chance, why should things be any different in the future?

1.6 Pre-capitalist societies

It should be clear by now that pre-capitalist societies were by no means simple. They differed among themselves in all kinds of ways. At the same time, when compared with present-day capitalist societies, they appear to have had a number of distinguishing features. This section will consider some of the more obvious ones.

Geographers and others have sometimes been tempted to think that past societies, lacking modern communications, were essentially small in scale and localized, with life based around the community. It is hoped that enough has been said already to suggest how over-simplified this view is. Huge empires had to be administered and defended, great cities like classical Rome had to be fed. Classical Rome imported some 17 million bushels of wheat each year from Egypt, North Africa and Sicily (Simmons 1996: 109). In the Roman Empire and other societies, high-value goods often travelled much greater distances. Even the Inuit, living the lives of hunters and gatherers, sometimes travelled hundreds of miles each year on hunting trips, and their geographical knowledge frequently covered an even wider span. Enough has been said above about migrations and the diffusion of artefacts and practices across great distances to suggest how space could be

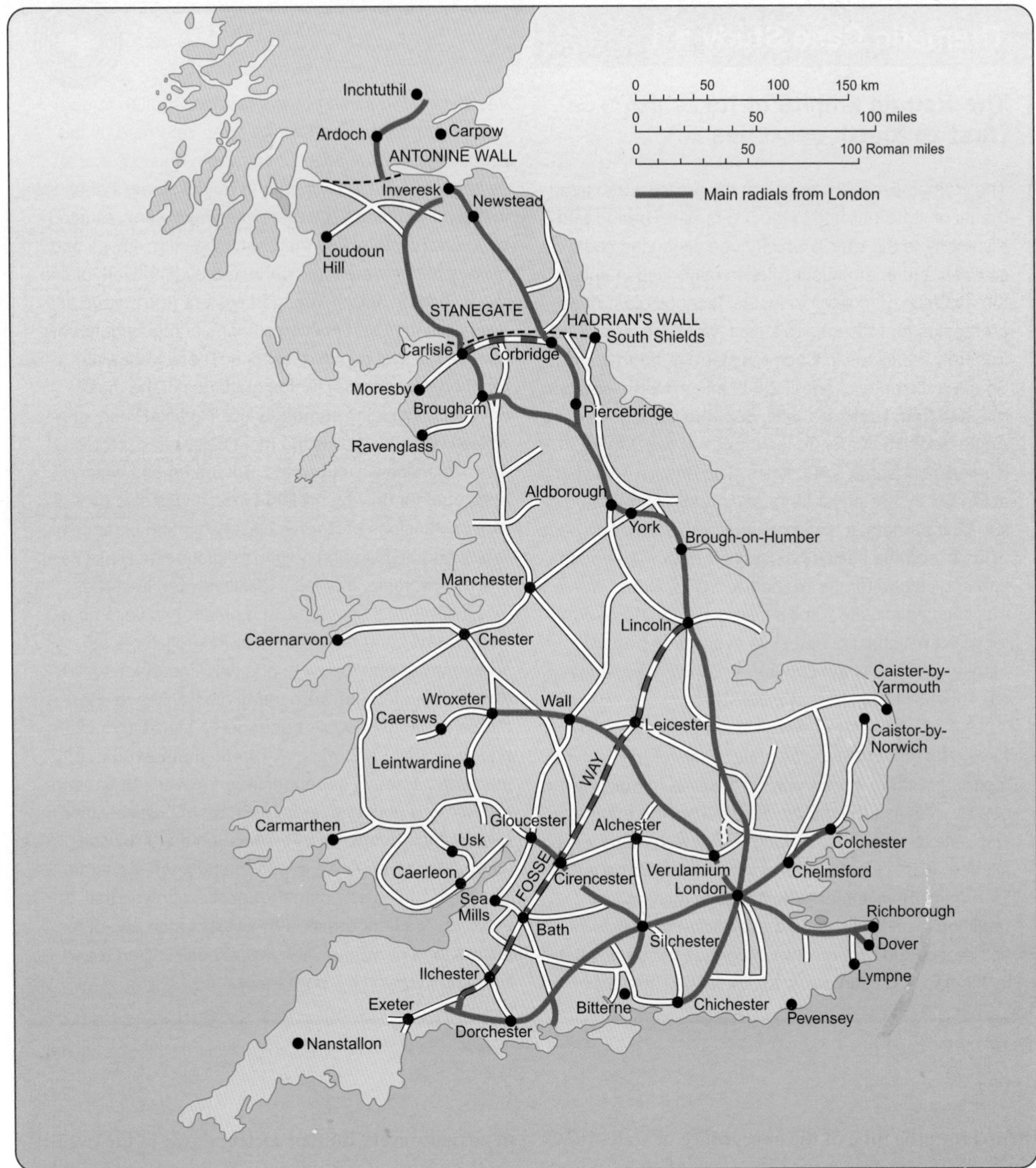

Figure 1.5 Roman Britain.

Source: © Oxford University Press 1984. Redrawn from map VI from Roman Britain in *Oxford History of England*, edited by Peter Salway (1984) by permission of Oxford University Press

overcome even if individuals were immobile, at least by today's standards.

There was, however, no avoiding the problems of communication in pre-modern societies. Most people in agricultural societies lived lives which were bound to their villages and the surrounding regions. Their work consisted essentially in winning the means of subsistence for themselves and their families from their environment, as well as paying the taxes and meeting the other demands which were made upon

Thematic Case Study 1.1

The Roman Empire at its zenith (first to fourth centuries AD)

The Roman Empire was one of the greatest achievements of state-building in the pre-modern period and is testimony to the way in which huge distances could be crossed and enormous territories controlled in spite of the absence of modern industrial technologies. At its greatest extent the empire's east–west axis stretched from the Caucasus to Cape Finisterre in north-west Spain, a distance of about 2,800 miles. North–south it reached from Hadrian's Wall, near the present-day Anglo-Scottish border, to the fringes of the Sahara in North Africa, some 1,600 miles. The empire was characterized by two major languages (Latin and Greek), low tariff barriers, a common currency, a common code of laws, and the basis of a common system of education and culture for the Romanized elite. Rome was a truly 'universal state', which lasted in its mature form for half a millennium, and in its eastern (Byzantine) manifestation until the capture of Constantinople by the Turks in 1453.

Like all early states, Rome had problems in enforcing its territoriality – the sheer scale of the empire meant that there were always difficulties in maintaining control over its far-flung provinces. That the empire was able to endure for so long (albeit with many vicissitudes) was a tribute to its highly developed capacity for organization and an astuteness in gaining and retaining local loyalties. The acquisition of new territories proceeded by a mixture of conquest, colonization by Roman and Latin colonists, the establishment of vassal kingdoms and powers, alliances (often forced) and other means. The Roman army, reformed by the Emperor Augustus after 13 BC, was a masterpiece of military organization superior to any of Rome's enemies in this period. The seas and waterways were patrolled by the navy, protecting merchant shipping. Towns and military bases were interconnected by a network of fine roads, and an official transport system (the *cursus publicus*) with inns and posting stations at regular intervals speeded official communications. There were even roadbooks (*itineraria*) for the guidance of the many travellers, including some tourists. Fortified lines, of which Hadrian's Wall is the most celebrated, protected the most vulnerable frontiers. But the empire was not held together only by military force. It was also by judicious extensions of Roman citizenship and other privileges, and by fostering helpful religious practices like emperor worship, that local loyalties were ensured.

The empire's economic and cultural achievements were many. Long-distance trade in such items as grain, metals and luxury goods was important, for example, though there were many hindrances to commerce and most people depended on local agriculture and manufacture. It must also be remembered that there was a heavy dependence on slavery. In the end the Romans could never be assured of political stability, even in their heartlands. Reliance on the army contributed to the empire's eventual undoing – generals and their armies competed for political power, the economy was undermined by overtaxation and other factors, and growing military weakness invited rebellion, and invasion by 'barbarians'.

them. From the time of the appearance of states down until the twentieth century the lives of most people were lived in the context of such **peasant** communities. As mentioned earlier, such communities are still characteristic of parts of the world today, though increasingly eroded by commercialism and growing contacts with modernity.

Scholars researching traditional peasant societies have frequently pointed to the many cultural differences between such rural dwellers and the cities that ruled over them and which they might occasionally visit. A case in point is religion. Religion has been a fundamental factor in social organization and in outlook in virtually every human society down to the twentieth century. Early states were generally associated with an official religion that legitimized the established order and which was often practised in cities but was rarely fully understood by the peasants, most of whom were probably illiterate (see Plate 1.5). Even where the peasants officially followed the same religion as the elite, as in medieval Europe, they almost always interpreted its teachings in their own ways, mixing them with their own superstitions and 'pagan' beliefs. The life and outlook of the ruler and the elite, if not of all cities, were entirely alien to them. And equally their way of life was alien to the cities.

Thematic Case Study 1.2

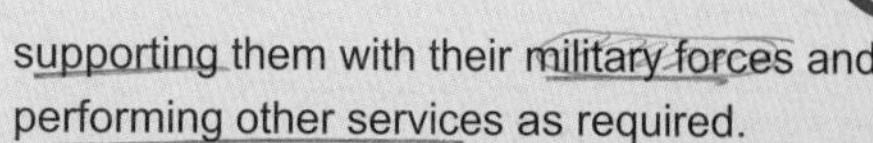

European feudalism

- Feudal society was hierarchical – ranked and stratified.
- The majority of the population were subsistence farmers, engaged in extensive forms of agriculture (for example, by cultivating strips in open fields shared with others, raising some livestock) and exploiting local resources like pastureland, woodland and fish.
- Many of the rural population were obligated to their lords and landholders, for example not to move away from their villages, to work the lord's land, to pay the tribute and taxes the lord and the state required, to serve the lord militarily and to perform other services.
- Lords held their land conditionally from their overlord and ultimately from the king or other ruler. Land could not easily be sold. Lords exercised jurisdiction over those living on their estates.
- Lords owed allegiance to their overlords (and they in turn to their overlords and so on), for example in supporting them with their military forces and performing other services as required.
- Feudal states were quite decentralized, parcelled out into lordships and they in turn into jurisdictions of various kinds.
- Towns were generally small and few and far between (see Plate 1.3). Most were centres of trade and crafts. Their residents (merchants, burgesses) often enjoyed freedoms and privileges ('liberties') denied to most rural dwellers.
- Merchants and itinerant traders joined together networks of local and long-distance trade, linking local fairs and markets with towns, ports and so on. Craft production and trade were generally subject to a variety of restrictions and controls, and often played only a minor role in the lives of rural dwellers.
- The medieval world-view was strongly influenced by the Church, whose priests and officials were among the few educated people and were universally present.

For this reason, it would be completely misleading to think of the subjects of pre-capitalist states as 'citizens' in the modern sense. Traditional rulers knew little of the countryside, where most of their subjects lived, and cared even less (except, perhaps, where they had landed estates, and even then the running of the estate could be left to officials). What mattered to rulers was law and order, and extracting the taxes and tribute the state needed to maintain itself in existence. The idea that rulers should care for the welfare of their ordinary subjects, let alone consult them about their policies, is very modern indeed.

It is difficult for most people living in a world dominated by capitalist relationships to conceive of societies where this was not the case. Before capitalism, most people were engaged in subsistence activities: hunting, gathering, farming, fishing or whatever. They might have to pay taxes or tribute, and they might trade some of what they gathered or produced, but market trade, generally organized by a small minority of merchants and traders, was often quite marginal to them. Where market trade did exist in early societies, it was frequently hedged about by laws and restrictions of various kinds. There might be laws against the taking of interest on loans, for example, regulations on price, or restrictions on where trade might take place or who might engage in it. Only with the rise of modern capitalism did the market become central to the way societies functioned.

It is very difficult to generalize about pre-capitalist societies and almost any generalization is open to objection. Case study 1.2 describes the main features of the feudal society that existed in medieval Europe, the society that was the seedbed for the development of world capitalism (see Chapter 2). Even here, however, one must be careful – the features described would be more or less true, depending on the date and the place being considered.

1.7 The heritage of the past

Nowadays the past, or what is commonly referred to as **heritage**, is big business. Medieval cities and cathedrals, historic villages and battlefields, monuments,

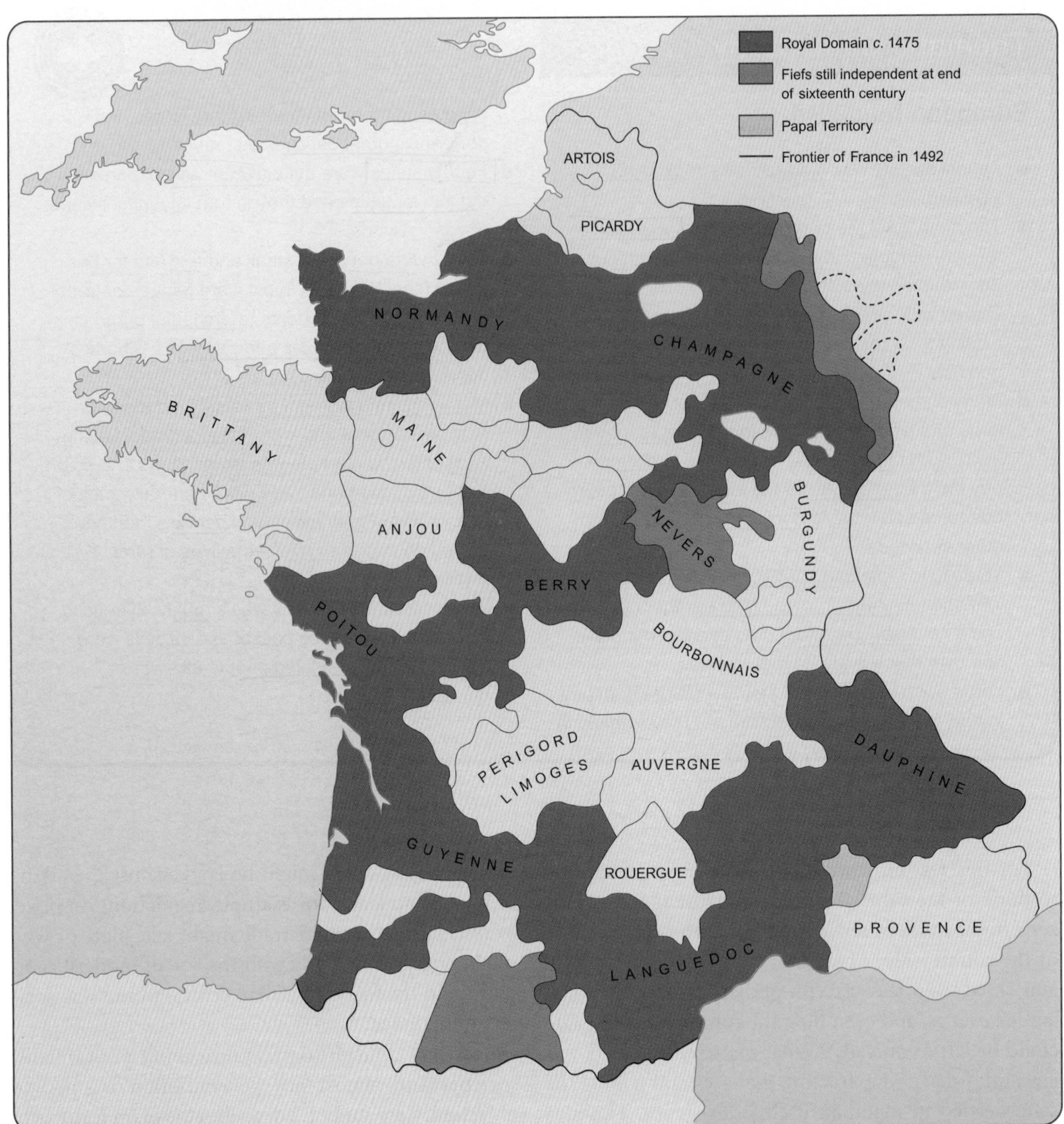

Figure 1.6 France towards the end of the feudal period, showing the political fragmentation typical of many parts of Europe prior to the appearance of modern states.

Source: Based on *The Times Atlas of World History* (1989: 15.1). © Collins Bartholomew Ltd 1989, reproduced by permission of HarperCollins Publishers

gardens and ruined temples are objects sought out by international tourists and cultural visitors, whilst tourist providers and also those charged with the conservation of the past compete to sell the past to visitors. What cannot be viewed out of doors is gathered together into museums which become ever more elaborate and ever more tied to the leisure industry rather than to the educational role which defined them in the past. Cities which are in the business of selling themselves to international investors and global companies use their cultural resources, including heritage, to do so. Altogether, it seems, our age is obsessed with the past, and nostalgia has become part of almost everyone's life.

Not surprisingly, perhaps, the present-day significance of heritage has not escaped the notice of human

Plate 1.5 The celebrated Mappa Mundi or world map, dating from c. 1300 AD, in Hereford Cathedral, England. Constructed to show Jerusalem in the centre of the world, this map typifies the medieval Christian view of a world explicable only in religious terms and symbols.

(By permission of the Hereford Mappa Mundi Trust)

geographers and other students of society (Graham et al. 2000; Johnson 2003). Geographers have been concerned to know what it is about the past which appears to attract so much public interest and how people understand those facets of the past which they encounter. More specifically, since the past is no longer with us and cannot be encountered directly, scholars have begun to ask about which particular past is being viewed under the guise of heritage, and about the ways in which the past is manipulated for public consumption. For the fact is that there are many pasts, depending on the viewpoints of those who wish to describe them and on the messages which those in control of heritage wish to purvey. Whether it be national governments, wishing to convey some lessons about supposed national 'glories' in the past, museum curators trying to explain particular historical events or epochs, or the guardians of historic sites aiming to entertain and titillate tourists, the scope for historical distortion (even if it means only telling one side of the story) is immense. History, in other words, is all around us, and history is controversial. Only by knowing something about the past can the geographer hope to avoid ignorance about the present.

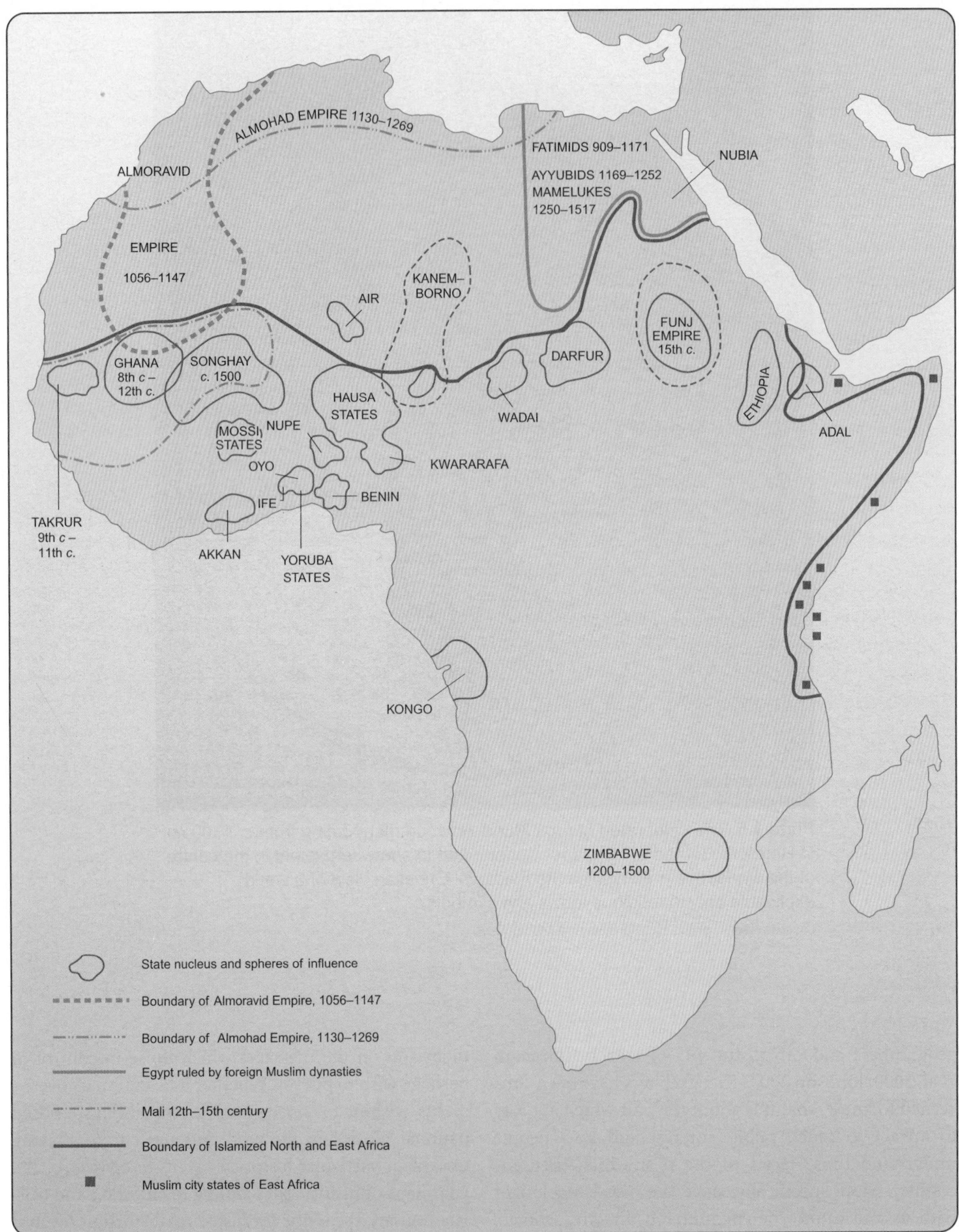

Figure 1.7 The great empires of Africa (AD 900–1500) before European colonization.

Source: Based on *The Times Atlas of World History* (1989: 136–7). © Collins Bartholomew Ltd 1989, reproduced by permission of HarperCollins Publishers

Learning outcomes

Having read this chapter, you should begin to appreciate that:

- The surviving evidence necessarily limits knowledge of past societies, since they cannot be observed directly. In general, the further back in time past societies existed, the more uncertain our knowledge about them becomes.
- People in the past lived very different lives from those of most people alive today. But it is important not to oversimplify or overgeneralize about past societies that varied enormously in their structures and ways of life and in their degree of complexity.
- Classifying societies into types is a way of making sense of the past and of trying to understand the geographical characteristics of varied societies. But all classifications have their shortcomings.
- Overcoming the friction of space was a severe problem in pre-capitalist societies. But it is important not to exaggerate its effects – people often travelled, and often migrated, long distances even in 'primitive' societies, ideas and artefacts travelled vast distances, and empires were successfully established (see Figure 1.7).
- The landscapes that can be seen today have been profoundly influenced by past societies, including hunter–gatherer ones. Our world is the product of generations of human activity.
- Change and progress are two different things. Social change almost always means winners and losers. The human record suggests that the latter have often been the majority.
- We must not assume that our present-day society is the world's most successful or progressive. We should be prepared to learn from others and to allow for the possibility that other ways of organizing society might be better at coping with some problems (for example, that of environmental degradation) than our own.
- There are many pasts depending on the viewpoint of the observer. Many assumptions adopted by people alive today are based on misunderstandings or distortions of the past. We should always be prepared to test our ideas and assumptions against the historical evidence. We should be prepared to do the same with the ideas and assumptions of others.

Further reading

Crone, P. (1989) ***Pre-Industrial Societies***, Blackwell, Oxford. An excellent introduction to the character of pre-industrial societies, from the invention of cities and civilization. Written for undergraduates.

Dodgshon, R.A. (1987) ***The European Past: Social Evolution and Spatial Order***, Macmillan, Basingstoke. An original examination of European prehistory and history, taken from a geographical perspective. Based on the social classification system highlighted in this chapter. Very detailed, but repays careful study.

Dodgshon, R.A. and Butlin, R.A. (eds) (1990) ***An Historical Geography of England and Wales***, 2nd edition, Academic Press, London. An historical geography of the region from prehistoric times to 1939. Chapters 4 and 5 illustrate the issue of feudalism.

Simmons, I.G. (1996) ***Changing the Face of the Earth: Culture, Environment, History***, 2nd edition, Blackwell, Oxford. A history of the impact of humans on the natural environment from the earliest times to the present. Contains excellent chapters on pre-capitalist societies, especially from the environmental point of view, but tends to be a little technical in places.

The Times Archaeology of the World (1999). New edition, Times Books, London. A lavishly illustrated survey of archaeology.

The Times Atlas of World History (1999). New edition, Times Books, London. A highly acclaimed survey of world history beginning with human origins. This is in fact more a history book with maps than a true atlas and fails to provide detailed maps despite its large format. But it is lavishly illustrated and contains a wealth of factual and explanatory material.

Useful websites

www.thebritishmuseum.ac.uk The British Museum, London. The museum holds a unique collection of art and antiquities from ancient and living societies across the globe. The website gives details of the museum's holdings and current exhibitions, hosts discussions on prehistory and past civilizations, and gives guidance on how to find further information.

www.si.edu The Smithsonian Institution, Washington, DC. The Smithsonian is a focus for many kinds of scientific and cultural endeavour in the United States. The website contains much that is useful for readers of this chapter, especially in relation to the history and prehistory of North America.

www.jorvik-viking-centre.co.uk The website of the Jorvik Viking Centre in York, England, famed for its reconstruction of life in York in the Viking period.

www.channel4.com/history/timeteam Channel 4 (UK)'s popular 'Time Team' programme is devoted to the archaeological investigation of a variety of sites, particularly in Britain.

http://english-heritage.org.uk The website of English Heritage, which is responsible for the upkeep of many prehistoric and historic sites in England, many of which are open to the public.

www.besthistorysites.net Entitled 'Best of History Websites', a comprehensive guide to history-oriented resources online, starting with prehistory. For teachers, students and others.

http://whc.unesco.org/ The official world heritage site of UNESCO which designates and seeks to protect 'World Heritage', defined as 'cultural and natural heritage around the world considered to be of outstanding value to humanity'.

For annotated, clickable weblinks and useful tutorials full of practical advice on how to improve your study skills, visit this book's website at **www.pearsoned.co.uk/daniels**

THE RISE AND SPREAD OF CAPITALISM

Chapter 2

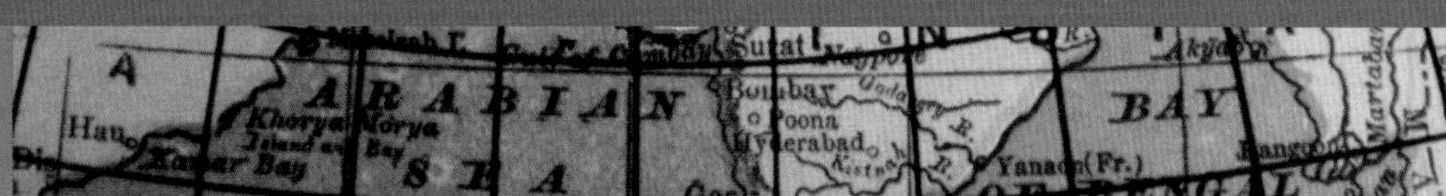

Terry Slater

Topics covered

- Definitions of capitalism
- The cyclical nature of capitalism
- The transition from feudalism to capitalism
- The beginnings of European imperialism
- Colonial commerce
- Transatlantic migrations and the slave trade
- European colonial empires and racism
- Industrial and agricultural transformations
- New modes of transportation
- Industrial urbanism

Between 1500 and 1900 there was a fundamental change in the way an increasingly large part of the world was organized. Beginning in particular parts of England, and spreading to other parts of Europe, the capitalist economic system was to change, or influence substantially, not only the economies, but also the political, social and cultural dimensions of newly powerful nation-states. By 1900, capitalism was the dominant socio-economic system over a large part of the world, a by-product of the colonial empires of those nation-states.

2.1 What is capitalism?

A number of writers have provided theoretical frameworks for understanding the workings and ramifications of the capitalist system. The first was devised by the Scottish philosopher, historian and father of classical economics, Adam Smith, in his book *An Inquiry into the Nature and Causes of the Wealth of Nations* (1776). Like all theoretical models, Smith's is a simplification of reality, but it introduced a terminology and series of conceptualizations that are still familiar to us today. Smith assumed the model was driven by people's selfish desires for gain and self-interest. Thus, production takes place to generate profit; surplus profits are accumulated as capital; and the basic rule of the system is 'accumulate or perish'. Integral to the system is the determination of prices which, in a free market, said Smith, are determined by the supply of, and demand for, the factors of production. In order to maximize their profits, industrialists will always seek to minimize their costs of production, including the wages that they pay their workforce, so as to outcompete other producers by having lower prices. One way of doing this was through the **division of labour**, dividing manufacturing tasks into simple, repetitive operations that could be performed by unskilled, and therefore cheap, labour (see Spotlight box 15.1). Another important way in which manufacturers could seek **competitive advantage** over their fellows was to be first in the use of new technology. The significance of the invention and adoption of new ways of doing things, whether through science and technology, or ideas and organization, has often been crucial to firms, industries and regions in getting ahead of competitors. This is still very clearly so today in technological 'hotspots' like the Milano and Stuttgart regions of the European space economy. Most scholars agree that the capitalist system began to cohere into an integrated whole in a particular place (England), at a particular time (the seventeenth century), and that it then took another long period (until the mid-nineteenth century) before it matured in the states of north-west Europe and in North America (Green and Sutcliffe 1987: 6–7) (Spotlight box 2.1).

2.1.1 Cyclical characteristics of economic development

Smith went on to model the workings of a national economy over a year (macroeconomics) to elucidate the way in which capitalism works in a series of interconnected cycles. These cycles have been studied

Spotlight Box 2.1

Characteristics of states with a mature capitalist economy and society

- The majority of the population in these countries were wage labourers forming a 'working class' who sold their labour power for wages in cash or kind so as to purchase food and other commodities to survive.
- The majority of these wage labourers were male and worked outside the home; females increasingly 'worked' in the home, both in the domestic care of their families and by 'selling' surplus labour time for minimal wages.
- Wage labourers were increasingly closely supervised by managers of the production processes.
- The means of production of wage labourers was owned by capitalists whose aim was to make a profit on their investment.
- There was an increasing disparity between the 'profit' of capitalists and the 'wages' of workers with an increasing propensity for conflict between the two.
- The vast majority of goods and services, including fixed property like land, were distributed through monetary exchange.

Table 2.1 Logistic and Kondratieff cycles

Cycle	Period	Growth phase	Stagnation phase
Logistic II	c.1450–c.1750	c.1450–c.1600	c.1600–c.1750
KI	1770/80–1845/50	1770–1815	1815–1845/50
KII	1845/50–1890/96	1845/50–1875	1875–1890/96

subsequently by other economists, most notably the Russian Kondratieff (1925), whose name is now used for the roughly fifty-year-long cycles of boom and depression that have characterized the capitalist world since the mid-eighteenth century (Table 2.1).

In seeking the explanation for these cycles of growth and stagnation, Schumpeter (1939) argued that technical innovation leading to the development of new industries was the key to understanding the growth phase of the cycle. Because technical innovation is spatially uneven, then so too has been the geography of economic development under capitalism. The question then arises as to why the geography of innovation is uneven. Recent research has suggested that the socio-institutional structures of regions of innovation, or of those lacking innovation, are the key to explanation: in other words whether educational, governmental and social organizations encourage or discourage enterprise and change. Others have criticized the technological determinism of these theories (Mahon 1987).

2.2 Other perspectives, other stories

2.2.1 Marxism

The viewpoints of orthodox economics are not the only interpretation of the transformation of large parts of the world over the past three hundred years. Karl Marx (1867) took a very different perspective in his analysis of capitalism in the mid-nineteenth century, *Das Kapital* (see Spotlight box 3.2). He proposed that profit arises out of the way in which capitalists (the bourgeoisie) dominate labour (the workers) in an unequal class-based relationship. This unequal class relationship was perceived by Marx to lead inevitably to class conflict. In his later writing he sketched out ways in which labour could gain control of the means of production and thereby 'throw off their chains'. This was to have dramatic long-term consequences for the socio-political organization of the world through most of the twentieth century between the Russian Revolution and the upheavals of 1989 (see Chapter 3, pp. 71–3). We need to note, however, that Marx's writing is dominated by an historical, but not a geographical, perspective: it traces class relationships in time but not in space. Marxist interpretations of capitalist development have therefore always had difficulty in conceptualizing uneven spatial development.

2.2.2 World systems theory

Another way of conceptualizing the nature of these changes is to be found in the writings of Immanuel Wallerstein (1979), whose 'world-systems theory' provides a threefold categorization of historical socio-economic systems. Wallerstein proceeded to delimit the spatial characteristics of the capitalist world economy into core, periphery and semi-periphery. These terms are not used in the everyday sense but signify areas in which particular processes operate. For Wallerstein, core processes are those characterized by relatively high wages, advanced technology and diversified production; periphery processes are characterized by low wages, simple technology and limited production. Between the two is the semi-periphery: areas that exploit the periphery but which are exploited by the core, so that they exhibit a mixture of both core and peripheral processes (Taylor 1989: 16–17). In the period between the seventeenth and the end of the nineteenth century some regions of the world were in the core (north-west Europe, for example) or periphery (central Africa) throughout the period. Other regions moved from the semi-periphery into either the core (the southern states of the USA, for example) or the periphery (the states emerging from the ruins of the Ottoman/Turkish empire). Wallerstein's theory has been very influential in explanations of the growth of globalization, but it is an explanation only at the 'structural' level. It says little about the complexities of the social and economic networks that enable the system to work, or of the resistance by groups or individuals trying to change it (Ogborn 1999).

2.3 The transition from feudalism to capitalism

The transformation of the European economy into a capitalist one was a long drawn-out process and historians continue to argue the precise causes of the changes that took place. Writers using an avowedly Marxist frame of reference (Dobb 1946; Kaye 1984) have made many of the most significant contributions in this debate. To them, the development of a class of wage labourers is crucial, together with the assumption of political power by the new class of capitalists. Geographers have been particularly concerned to trace the outworking of these historical processes of social and economic transformation in the particular space economies of local regions (Gregory 1982; Langton 1984; Stobart and Raven 2004).

The disintegration of feudalism (see Case study 1.2), with its carefully controlled market system, was considerably speeded by the after-effects of the Black Death in mid-fourteenth-century Europe. The European population was reduced by between one-third and a half, and towns were especially hard hit. The resulting labour shortage meant that enterprise and new ways of doing things were more likely to be rewarded, since the feudal elite required more revenue to maintain their power. At the same time, the slackening of social and cultural controls meant that new ways of thinking could flourish, especially in the growing towns.

By the later fifteenth century, larger European towns were dominated by what has come to be called **merchant capitalism**. Merchants were both the providers of capital and principal traders in a regionally specialized, complex, Europe-wide trading nexus based not on luxury goods as in the medieval period, but on bulky staples like grain and timber, and on an increasing array of manufactured products. Towns that flourished economically in fifteenth- and sixteenth-century Europe were those whose inhabitants were able to copy and manufacture imported products more cheaply than the exporting region, and develop a constant stream of new, innovative, marketable products (Knox and Agnew 1994: 154–61) (Spotlight box 2.2). For most commentators, the key question then becomes: why did these processes of transformation coalesce first in sixteenth- and seventeenth-century England?

We can begin to answer this question by suggesting that, since land was one of the key factors of production, critical in the sixteenth century was the enormous transfer of land in England from conservative ecclesiastical ownership to secular ownership. This was the consequence of the dissolution of the monasteries by Henry VIII in the later 1530s. Even the monastic buildings could be adapted by capitalist manufacturers, as the Wiltshire clothier William Stumpe showed at Malmesbury Abbey where he installed 300 weavers in the 1540s (Chandler 1993: 487–9). Land was also being transformed in the sixteenth and seventeenth centuries

Spotlight Box 2.2

Characteristics of the period of merchant capitalism in Europe

- Increasing numbers of people sold their labour for money wages. They ceased to work on their own land. This led to increased consumer demand for food, clothing and household goods; in many places there was a notable rise in living standards.
- More producers of both agricultural and craft-manufactured products began to accumulate capital as they produced for this growing market. These prosperous yeoman farmers and manufacturers were the foundation of a new class of capitalists.
- The removal of feudal market restrictions and controls by guilds on production led merchants to invest in the reorganization of production on a capitalist basis.
- Technical innovations transformed industries; the most significant was probably the development of the printing press by Gutenberg, in Mainz, enabling knowledge to be diffused cheaply and rapidly.
- The rediscovery of Classical knowledge led to new ways of seeing the world and, consequently, to its rebuilding to reflect these new images, especially in towns.
- Ultimately, these changes led to religious upheaval and political revolution whereby the new capitalist class became dominant in the governance of nation-states.

Figure 2.1 The palace, gardens, park and town of Karlsruhe, Germany, founded in 1715 by the Margrave Karl Wilhelm of Baden.

Source: Gutkind (1964: 301). Reprinted with the permission of the Free Press, a Division of Simon & Schuster Adult Publishing Group, from *Urban Development in Central Europe*, Volume 1 of the International History of City Development Series, by E.A. Gutkind. Copyright © 1964 by The Free Press. Copyright renewed © 1992 by Peter C.W. Gutkind. All rights reserved.

through the process of enclosure. This enabled livestock to be raised more efficiently for the urban meat markets and experimentation in new agricultural techniques to be undertaken by yeoman farmers (Butlin 1993: 178–9).

In the seventeenth century it was the social, religious and political transformation of England that was critical, according to Dobb (1946). The struggles between Crown and Parliament during the Civil War and Commonwealth (1642–60) are seen as a conflict between landowners, and capitalist yeomen and manufacturers. Though this struggle was continued through the Restoration (1660) and Glorious Revolution (1688), by the end of the seventeenth century the capitalist bourgeoisie were politically predominant and were able to transform the state to their own advantage.

Elsewhere in Europe landowners remained preeminent over a predominantly peasant agricultural workforce and this period is known as the 'Age of Absolutism'. Landowners, especially the rulers of small states, amassed enormous wealth from their control of land. This wealth was expended on increasingly spectacular landscapes of display: huge palaces, filled with works of art and every luxury, set in carefully designed, and often very intricate, landscapes reliant on vast inputs of labour for their maintenance (Figure 2.1). King Louis XIV of France had set the model at his palace of Versailles, outside Paris, using the resources of a much larger state. Such conspicuous displays of wealth and privilege by the few led, ultimately, to a much bloodier revolution than in England. The French Revolution saw the slaughter of not simply the royal house, but of the landowners, intellectuals and the bourgeoisie, since this was the first of the workers' revolutions, with its cry of '*Liberté, egalité, fraternité*'. However, it led rapidly to the totalitarian militarism of Napoleon.

2.4 An expanding world

As the European economic and cultural world was gradually transformed, European travellers began to voyage beyond the shores of Europe. They travelled overland into Asia, drawn both by a thirst for greater knowledge and, more significantly, by the rewards that could be reaped from direct exploitation of scarce commodities for European consumption. These included spices, sugar, silk, muslins, porcelain and the like, which had previously been traded through the eastern Mediterranean. In the late fifteenth century, however, this region was coming under Turkish (Ottoman) control. The Portuguese, funded by Italian bankers, were the first to navigate round Africa into the Indian Ocean, with its Arab-dominated sea-based trading networks, whilst soon after (in 1492), Christopher Columbus, again funded by Italian (Genoese) merchant capital, sailed westwards across the Atlantic to open the Americas to European exploitation.

For Spain and Portugal, this is the heroic Age of the Navigators, as they moved rapidly to take control of these new maritime trade routes looking out to the Atlantic. The merchants of Lisbon and Seville grew rich on this trade. For the rest of the world it marks the beginning of the European colonial empires that were to dominate and control almost every aspect of life for the next four hundred years. The Portuguese established a far-flung network of fortified port bases around Africa and the Indian Ocean through which trade in gold, spices and textiles was conducted (Figure 2.2). Equally significant was their use of African slaves to produce sugar on plantation farms in their Brazilian territory. The Spanish founded a more militaristic, oppressive and exploitative empire in central and south America which saw the decimation of local populations through warfare and the catastrophic effects of European diseases, especially smallpox and measles. It also brought enormous inflows of gold and silver bullion into Spain and, ultimately, onwards into the European economy. Blaut (1993) sees this as the event that more than any other kick-started the European capitalist world system.

Fairly quickly, control of the products of this new maritime world economy moved from south-west to north-west Europe. The Dutch began to compete successfully with the Portuguese in the Indian Ocean and the centres of European capital moved from Lisbon and Seville first to Antwerp by the mid-sixteenth century,

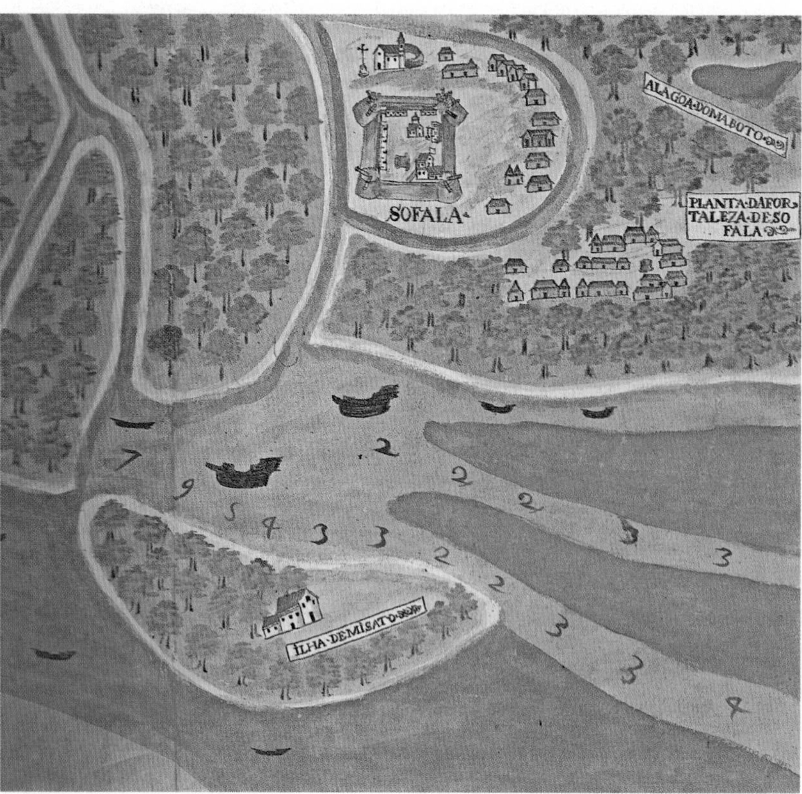

Figure 2.2 The Portuguese fortress and trading station of Sofala, in modern Mozambique, in 1558. For a time it was the administrative centre for all Portuguese trading with India.
Source: Carneiro (1990)

and then to Amsterdam at the end of the century. The merchants of these ports successfully integrated the Atlantic trade with that flowing from the Baltic and from central Europe via the Rhine valley. It was at Antwerp that the first stock exchange was established, enabling capital raised in one business to fund investment in another. Amsterdam's rise was based on having greater military and naval power than its neighbour and using it to enforce the privileges of a distinctive trading monopoly: the state-licensed joint-stock charter company (the Dutch East India Company) (Dodgshon 1998: 74–83).

At the same time, French and English fishermen had followed John Cabot's exploratory voyage across the North Atlantic to begin to exploit the rich fishing grounds off Newfoundland and, subsequently, to trade with native Americans for fur pelts. From these small beginnings French and English merchant influence began to expand through the seventeenth century in the Indian Ocean (using similar company trading monopolies to the Dutch); in the Caribbean, where sugar plantations and slave labour were copied from the Portuguese; and in North America. They superseded the Dutch in the early eighteenth century through superiority of arms, but it was not until the later eighteenth century that the British emerged dominant in the developing world economy. Consequently the primary city of the European capitalist system moved once more, from Amsterdam to London (Dodgshon 1998).

The colonialist trading systems that developed in the period up to 1770 have often been described in simple dualist, oppositional, terms (core–periphery; dominant–subordinate; metropolitan–colonial, for example). Blaut (1993) is especially critical of the Eurocentric character of this type of argument. More recently, a more complex, multi-layered interpretation has been posited. We need to note, first, that different parts of the developing world economy operated in somewhat different ways, were based on different product interactions between raw materials, manufacture and consumption, involved complex transportation flows, and required different politico-military frameworks to make them function properly. Meinig's (1986) model of the Anglo-French mid-eighteenth-century North Atlantic economy uncovers some of these complexities in terms of the commercial, political and social systems that were required for the geography of colonial capitalism, and Ogborn (1999) has added details of networks of individuals and patterns of resistance (Figure 2.3). The North Atlantic 'triangular trade' is certainly the best-known of these colonial trading systems.

2.4.1 Colonial commerce

In Meinig's North Atlantic commercial system, London and Paris acted as the source of finance, commercial intelligence and marketing (London's Royal Exchange was founded in 1566). The expanding industries of particular regions of western Europe were growing, in part because they supplied a developing colonial market with manufactured goods such as textiles, tools, armaments and household necessities. These goods were assembled in Atlantic ports such as Bristol, Liverpool, Bordeaux and Le Havre. After crossing the Atlantic they were stored, disassembled and distributed from colonial ports such as New York, Charleston and New Orleans (Figure 2.4, Plate 2.1) by major traders. They sold on to local traders towards the colonial frontier. At the frontier a barter system of exchange was as likely to be found as a money economy. Nonetheless, fur, fish, timber and agricultural produce flowed in the opposite direction for storage and transhipment back to Europe (Figure 2.3).

In North America, tobacco, rice, sugar, rum and cotton were the major export products. All were produced on the plantation system using slave labour. By 1750, some 50,000–60,000 people were being forcibly transported across the Atlantic each year from all parts of the western African seaboard. Many died en route or shortly afterwards but, altogether, some 3.8 million Africans had been transported by 1750, just over half to Latin America, the remainder to English, French and Dutch North America and the West Indies. These staggering figures meant that African-Americans were easily the largest new culture group to be established in both North and South America in the colonial period (Meinig 1986: 226–31).

The African abolitionist Olaudah Equiano, who gained his freedom in England, wrote forcefully from his own experience about the horrors of the 'middle passage' voyage from West Africa to the West Indies. He said:

> The stench of the hold while we were on the coast was intolerably loathsome . . . but now that the whole ship's cargo were confined together, it became absolutely pestilential . . . The air soon became unfit for respiration . . . and brought on a sickness among the slaves, of which many died . . . This wretched situation was aggravated by the galling of the chains . . . and the filth of the necessary tubs (toilets) into which the children often fell and were almost suffocated. The shrieks of the women and the groans of the dying, rendered the whole scene of horror almost inconceivable.
>
> (1789)

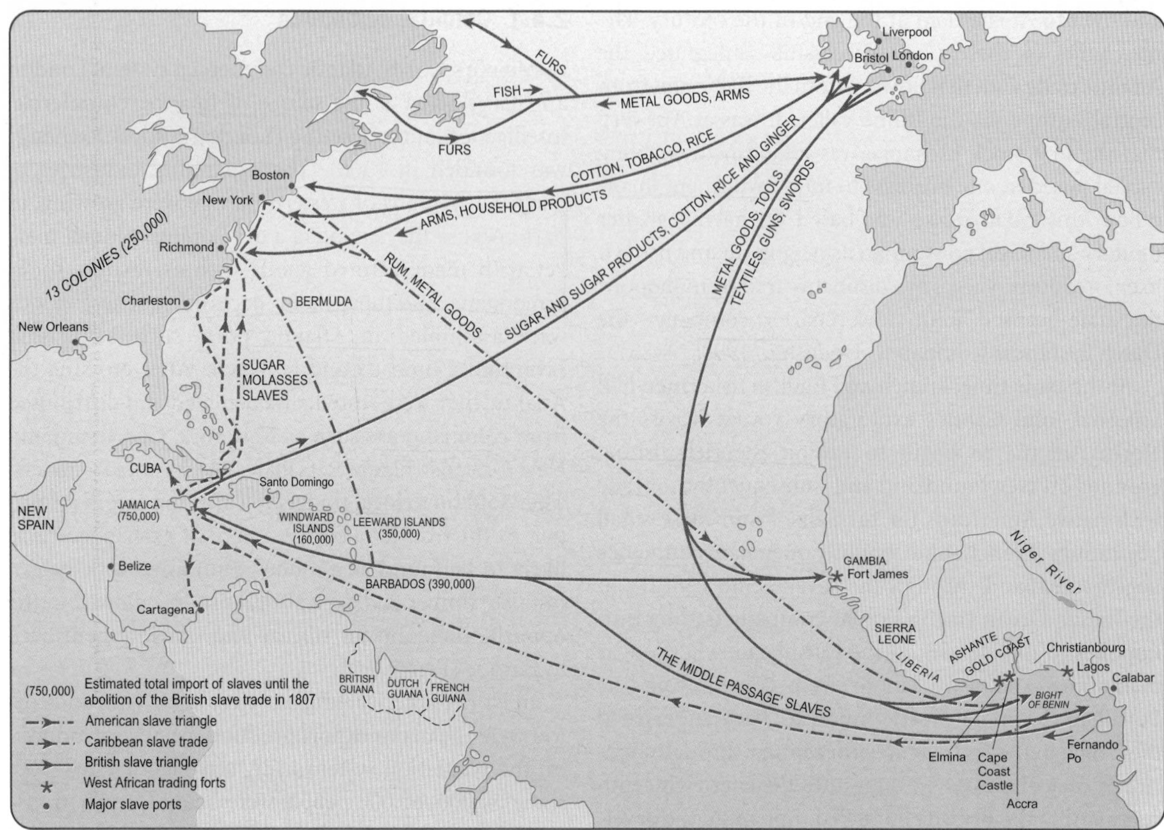

Figure 2.3 The eighteenth-century North Atlantic trading system between Britain, Africa and North and Central America.

Source: Bayley (1989: 46–7)

By 1750, the trade in slaves was dominated by large European companies using specially constructed and fitted ships, but many smaller traders from the Atlantic ports would sail first to African trading stations to exchange manufactured products, including guns, tools and chains, for slaves. The slaves would then be taken to the West Indies or Charleston, with ships returning to Liverpool or Bristol with a cargo of sugar, tobacco, rum or rice, and sometimes slaves such as Equiano, too. Most of the latter were destined for domestic service in aristocratic households (Figure 2.3).

2.4.2 Colonial society

The slave trade was perhaps the most significant aspect of the cultural transformation effected by the eighteenth-century capitalist world economy in the Americas and, it should not be forgotten, in Africa. One of the distinctive features of this transformation in the Americas was the 'othering' of Africans by Europeans. They were regarded as in every way inferior to Europeans; they were legally and socially defined as different; and they therefore lived their lives separately from white settlers. The social consequences of this were to reverberate through the next 250 years, into our own times, where spatially segregated city neighbourhoods based on ethnicity are still a commonplace in North American cities, whilst from the 1950s West Indian migration to Britain saw similar ethnic areas develop in many British cities. Where Africans were in the majority in the eighteenth-century colonies, as they were in the West Indies, and in the Carolinas and Louisiana, this enabled distinctive African-American cultures to develop within the confines of political and economic enslavement.

There were, of course, other trans-Atlantic migrations throughout the time period covered in this chapter. The earliest included those seeking to escape religious persecution in Europe, or to establish religious utopias in the new continent. The Puritan 'Founding Fathers' of New England were in this category and, right through to the Mormon migration in

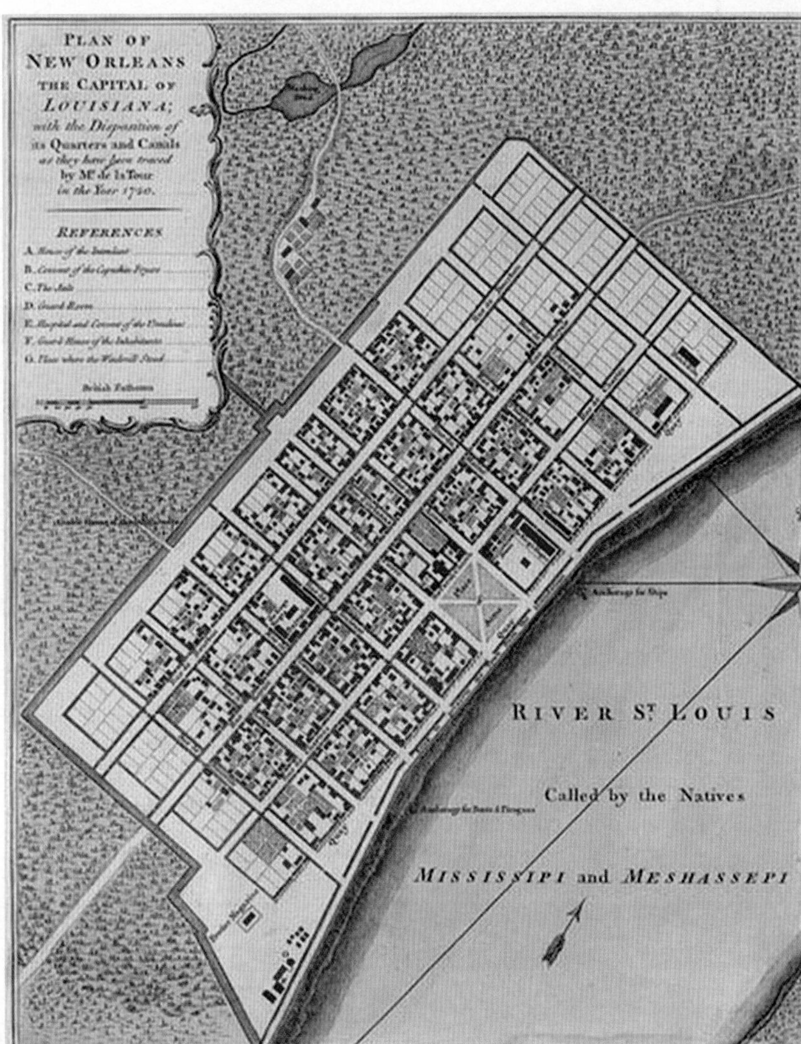

Figure 2.4 Plan of the French colonial entrepôt city of New Orleans at the head of the Mississippi delta in 1720, published in London in 1759 by Thomas Jefferys.

Plate 2.1 Madame John's Legacy, c.1788. One of the oldest houses in New Orleans, it was built during the Spanish colonial period (1765–1803). Commercial functions occupy the brick-built ground floor, residence the balcony floor.
(Author)

the early nineteenth century, such groups continued to leave Europe's shores for what was perceived to be a better new world. Religious faith also inspired another group of settlers who came to convert the native people of the Americas to Christianity. The competition in Europe between Catholic and Protestant versions of the faith – between Reformation and Counter-Reformation – was exported to the colonies since both native Americans and African slaves were perceived as being in need of conversion. Consequently, Jesuit monasteries or Franciscan friaries often stand at the heart of Spanish colonial towns and cities, whilst their dedicatory saints gave name to many of today's largest cities: San Francisco, Los Angeles and San Diego, for example (Conzen 1990).

By the beginning of the eighteenth century, significant migrations of Scots, Irish, French, Germans, Dutch, Swiss and Moravians had taken place, but the dominant migrant group were the English. Between 1700 and 1775 the population of the eastern colonies grew tenfold to some 2.5 million and, by 1820, the population of the United States surpassed that of Britain (Lemon 1987: 121). The losers were native Americans who were dispossessed of their lands east of the Appalachians and often taken into slavery.

2.4.3 Colonial politics

There were major differences in the way European states governed their colonial empires. The Spanish developed a highly centralized system in which all aspects of policy were laid down in Madrid. The Laws of the Indies, which laid down in meticulous detail how colonization should take place and how new towns should be planned, are symptomatic of this centralization (Nostrand 1987) (Case study 2.1). The French governed through a similar centralized structure of military, administrative and ecclesiastical strands of authority linked to government departments and the Crown in Paris. The English, in contrast, allowed each colony to be virtually self-governing and the links to London were many and various. English colonial plantation owners and merchants were not averse to threatening a governor if their capitalist trading relations were threatened. It should occasion no surprise that it was the English colonies that eventually first rebelled and fought for their independence. Overlapping these colonial territories were the older territorial patterns of native groupings, those of the Iroquois and Creeks remaining well into the eighteenth century (Meinig 1986: 262).

Thematic Case Study 2.1

Summary of the *Laws of the Indies* by Philip II of Spain, 1573

Laws 1–14
Establish that exploration cannot take place without royal permission; that governors should learn all they can about their territory; they should consult with and negotiate with local 'Indians'. 'Discoverers' should take possession of land and should name rivers, hills and settlements.

Laws 15–31
Lay down that Spaniards should treat 'Indians' in a friendly manner; they should help priests in their work of conversion.

Laws 32–42
Concern the type of region that should be explored, pacified, settled and brought under Spanish mandate; the siting of towns in such regions, and the establishment of local government.

Laws 43–109
Establish the legal and taxation regimes for new colonies and lay down that each town should have at least thirty families. The ideal size of house plots, farms, herds, commons is established.

Laws 110–128
Provide details of the plans of new towns, including the size of the plaza, the orientation and breadth of the streets, the location of churches and public buildings, and the allocation of house plots by lot.

Laws 129–135
Describe provision for common land and farms and the character of domestic buildings.

Laws 136–148
Lay down relationships with local 'Indians' and the way in which they should be converted to Christianity so that they 'can live civilly'.

Source: Crouch et al. (1982: 6–19)

2.5 Imperialism and racism

A second phase of colonial expansion, through the nineteenth century, was geographically distinct in that it was focused on Asia and, especially after 1880, on Africa. It is also marked by the development of a much more virulent racism in the 'othering' of all non-Europeans. The consolidation of European states in the nineteenth century, especially of Germany and Italy, and the increasing economic competition they offered to the eighteenth-century leaders of the world economy, Britain and France, led them to seek new opportunities through the colonial exploitation of hitherto economically peripheral lands. Britain seized the Dutch Cape Colony at the tip of southern Africa in 1806 to prevent its use by Napoleon's navy. By 1820, 4,000 British settler graziers were in conflict with African pastoralists, despite the fact that the low-wage labour of those same Africans was essential to maintain the farms (Lester 1999). At much the same time, the first British farmers were beginning to establish themselves in Australia as its period as a convict colony began to draw to a close. The recent novel by Kate Grenville, *The Secret River* (2006), gives a perceptive imaginative account of the struggle for land between farmers and Aborigines at this time and the consequences for both cultures of that struggle. Meanwhile, in Britain itself, Christian-inspired reformers succeeded in bringing an end to the slave trade in 1807, and then to slavery in 1833. Sadly, the invective of both populist and bourgeois writing opposed to abolition characterized Africans, Australian Aborigines, Indians and native Americans as not only unchristian and therefore 'uncivilized' but also as less than human; inferior to Europeans in every way; both savage and yet child-like, and therefore needing to be controlled and disciplined. By the middle of the nineteenth century, these attitudes were widespread amongst all classes in Britain and were fuelled further by popular antagonism to events such as the Indian Mutiny (1857), the Zulu rebellion (1879) and the Maori wars (1860s). By the late nineteenth century, Darwin's ideas on evolution were being misused to give a scientific veneer of respectability to this 'othering' of non-Europeans, to the extent that Australian Aborigines were being shot as 'vermin' rather than seen as fellow humans.

The mid-nineteenth century marked both the deliverance of one continent from colonialism to self-government, as South America broke free from Spanish and Portuguese control, and the colonizing of another, as European powers raced to divide Africa amongst themselves. The interior of Africa remained largely unknown to Europeans until the 1860s except for Christian missionaries. But by the end of the century, borders had been delineated and fought over, natural resources were being rapaciously exploited, and the polities, cultures and economies of native peoples had been disrupted or destroyed (Figure 2.5). Again, this period of history reverberates down to the beginning of the twenty-first century because the boundaries drawn to the satisfaction of European soldiers, administrators and traders did not correspond to African ethnic divisions. In many respects the origins of the genocides of Biafra (1960s' Nigeria) and Rwanda (1990s), and the long civil war in Sudan, which is still ongoing in Darfur, can be laid at the door of the 'scramble for Africa' in the later nineteenth century.

2.6 Industrialization

2.6.1 Proto-industrialization

The early phases of capitalist industrial development in Europe, up to about 1770, are categorized in various ways. To some writers this is the period of merchant capitalism (Spotlight box 2.2), to others this is the period of **proto-industrialization** (Spotlight box 2.3). Both terms have generated a substantive academic debate as to precisely what processes of transformation were involved.

Textile manufacture is the most studied production process of this period. Its transformation began in several parts of Europe as early as the fifteenth century. By the early eighteenth century, more and more of the production processes were being mechanized and industrialists were concentrating their machinery into single, multi-floor buildings. By the 1770s, the mechanization of spinning had been completed and weaving followed in the 1820s. The logic of this transformation was completed by steam power and the factory was established as both a building (Figure 2.6) and a system of production (Spotlight box 2.4). Different sectors of the textile industry went through this transformation at different times, and some sectors declined absolutely as capital was diverted to the most prosperous sectors. The textile industry, more than any other, had a relentless tendency to geographical specialization and concentration with the result that many towns and regions were dangerously dependent on a single industry (Laxton 1986). By the 1770s in Britain, west Yorkshire

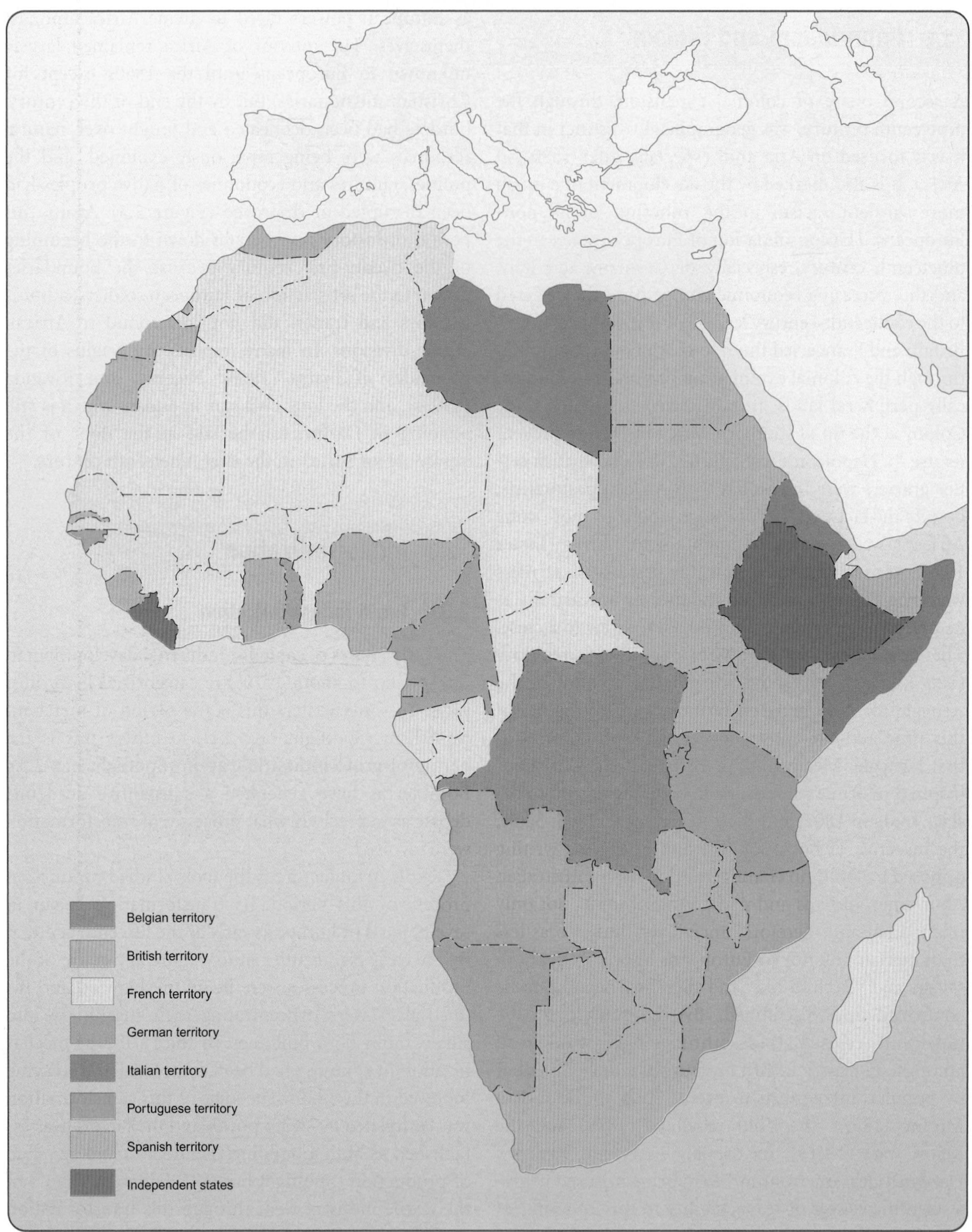

Figure 2.5 Africa in 1914 after the European imperial contest to divide the continent.

Source: Moore (1981: 74)

Spotlight Box 2.3

Proto-industrialization

- Many early industrial processes were located in the countryside.
- Most industries were characterized by some type of 'domestic' production system whereby the capitalist/merchant provided raw materials/machinery for producers who worked in their own homes.
- Producers generally worked in family units and were paid piece rates according to their production each week or month.
- Machinery was frequently water-powered.

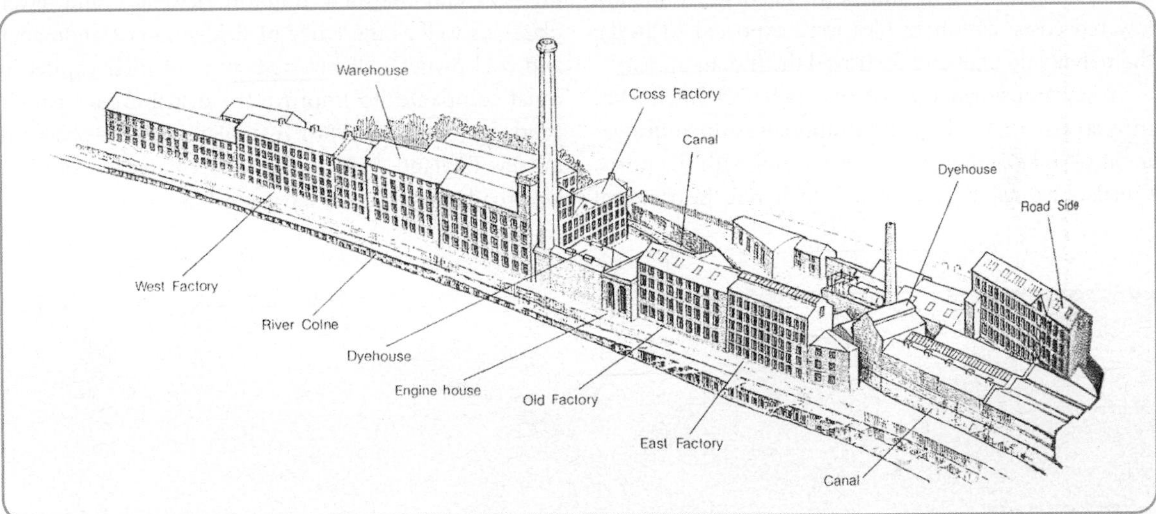

Figure 2.6 Starkey's woollen textile mill, Huddersfield, c.1850. Though this mill was steam-powered, by mid-century its powerlooms were still housed in multi-storey buildings rather than a single-storey shed, since the site was a restricted one.

Source: Giles and Goodall (1992: 102) © Crown Copyright. NMR

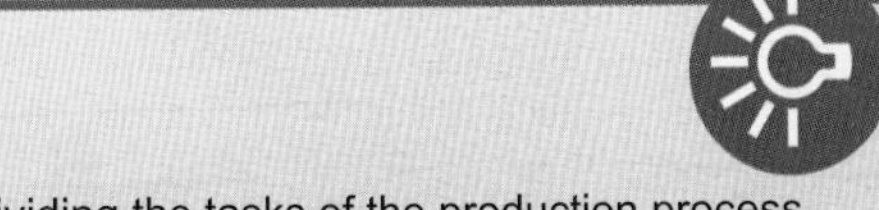

Spotlight Box 2.4

The factory system of production

Under the factory system the capitalist:

- Had complete control of the production process from receipt of raw materials to finished product.
- Had control of the labour force on whom a new, disciplined time geography could be imposed (early factories often worked day and night).
- Could apply capital to the development of new machinery to simplify processes and reduce labour costs through mechanization; and,
- By dividing the tasks of the production process, could reduce labour costs further by employing women and children.
- 'I began work at the mill in Bradford when I was nine years old . . . we began at six in the morning and worked until nine at night. When business was brisk, we began at five and worked until ten in the evening'. (Hannah Brown, interviewed in 1832).
- Legislation dating from this time remains effective in controlling the age at which children and young people can work and their hours of employment.

was beginning to dominate light worsted woollen cloth, Lancashire and the north Midlands cotton cloth production, the Welsh borders flannel, and the east Midlands hosiery.

The complex interaction of resources, labour skills, technological innovation, and capital circulation in particular regional economies, which increased production and lowered prices, meant that new markets had to be developed and transport improved. Market opportunities were to be found both in the growing urban markets at home, and among the colonial populations overseas. Thus, cotton textiles were first imported to Britain from India; the techniques of production were learnt; manufacture commenced; factory production lowered costs; cotton textiles were exported to India; there they undercut and destroyed the Indian industry.

A key transformation of the transport infrastructure, so essential to industrialization, was the improvement of links with growing towns, and with the ports. Canals and turnpike (toll) roads, which were the principal innovations before the 1820s, also lowered the costs of raw materials, especially of coal. This is graphically illustrated by the way in which urban populations celebrated the opening of new canals, not for their industrial potential, but because they dramatically reduced the cost of domestic heating. In Britain, the construction of better roads, and of the canal network, mostly between 1760 and 1815 (the growth phase of the first Kondratieff cycle, Table 2.1) was financed by private capital. The state was involved only in providing the necessary legislation to enable construction to take place and fees or tolls to be levied. Thus Josiah Wedgwood (china and pottery), Abraham Darby (iron manufacturer) (Figure 2.7) and Matthew Boulton (jewellery and silver plate), as well as the Duke of Bridgewater (landowner and coal owner), all invested some of their capital in canal companies to improve the distribution of their products (Freeman 1986). A similar pattern is observed in New England. In France and Germany, by contrast, new roads were seen to have a military function in the

Figure 2.7 The world's first iron bridge, constructed across the River Severn, east Shropshire, in 1788 by Abraham Darby III's Coalbrookedale Company, was part of the transport infrastructure improvements financed by local industrialists.

Source: The Ironbridge Gorge Museum

centralization of state control and were therefore largely financed by the state through taxation.

2.6.2 Agricultural change

It was not just industries such as textiles, metal manufacturing and pottery that were transformed by early capitalist production: society needed food production to keep pace with a rising population that was increasingly employed full-time in manufacturing. Consequently, the eighteenth century saw the transformation of agriculture into increasingly capitalistic modes. In parts of Europe, especially in Britain, self-sufficient peasant farming began to come under pressure from landowners who used state legislation to remove common rights to land in favour of private ownership through enclosure. Ownership was increasingly concentrated into fewer hands, and farms were consolidated from scattered strips to single-block holdings, again especially in Britain and, rather later, in Scandinavia and northern Germany. Elsewhere in Europe this process was delayed until the twentieth century. In northern and western Europe, landowners invested in transforming production for the new urban markets, using technology and new techniques and crops first developed in the Low Countries in the seventeenth century. In eastern Europe, the semi-periphery, this period saw a reversion to serfdom and near-feudal relations of production. Capital was also deployed to increase the area of land under intensive, rather than extensive, production. Fenland and coastal marshes were drained; heathland soils improved for cultivation, and the moorland edge of improved pastures pushed higher up hill and mountain slopes.

In the nineteenth century, European agriculture continued to increase production but was unable to keep pace with the demand from the growing urban industrial population. In 1840, Britain repealed the Corn Laws, which had protected local grain producers, and opened its markets to colonial and American producers. New methods of extensive grain farming were used in mid-western North America, and in the new British colonies in Australasia and South Africa. The invention of the steam ship, allowing more rapid transport, and of refrigeration and meat canning from the 1870s, led to similarly extensive ranch grazing of sheep and cattle in these countries. Together with the Pampas grasslands of Argentina, all became an important part of the semi-periphery of Britain's global economy. This new export-dominated agriculture was reliant on overseas investment by British capital in new railway networks to transport these products to slaughterhouses and industrial packing plants at the port cities whilst, in the USA, Chicago was growing rapidly on the same economic foundations (Cronon 1991; D.L. Miller 1997). It was accompanied, too, by new waves of migration to the colonies and the mid-west of the USA (Table 2.2), especially from Ireland and Scotland,

Table 2.2 European emigration 1881–1910

Source countries	Destination countries	Numbers of migrants
Great Britain	N. America, S. Africa, Australasia	7,144,000
Italy	USA and N. Africa	6,187,000
Germany	North and South America	2,143,000
Austria/Hungary	North and South America	1,799,000
Russia	USA	1,680,000
Scandinavia	North America	1,535,000
Spain	Central and S. America; N. Africa	1,472,000
Ireland	USA	1,414,000
Portugal	South America	775,000
S.E. Europe	USA	*c.*465,000
France	Central and S. America	223,000
Low Countries	USA	171,000

Source: Moore (1981: 57)

where the Great Famine and the Highland Clearances, respectively, were other manifestations of capitalist agricultural change in the British Isles. The United States absorbed enormous numbers of European migrants throughout the nineteenth century. Irish migrants predominated in the 1840s and 1850s, followed by Scots and English, Scandinavians and Germans through the second half of the century, with Russians, Italians and south-central Europeans from the 1880s onwards (Table 2.2) (Ward 1987).

This later colonization, and the internal colonization of the United States mid-west in particular, were grandly characterized as a moving frontier of settlement by the American historian, Turner, writing in 1894, who suggested that it also transformed the 'character' of frontier settlers, making them self-reliant, opportunistic, individualistic and democratic. Recent commentators have noted that Turner's hypothesis says little about the continued sweeping aside of the rights of the native populations of these lands who fought a long and ultimately unsuccessful attritional battle to retain it for their own use (the land was regarded as 'open' or 'free' for white settlement); nor does it give any credit for the settlement process to women, who provided both the domestic labour and, often, especially in the initial phase of settlement, much of the farm labour too. The advancing frontier of capitalist agriculture was as much small-scale, incremental and domestic as it was wide-sweeping and large-scale.

2.6.3 Factories and industrial production

Some commentators have seen the 'Industrial Revolution' in Britain as a short period, between 1770 and 1830, of rapid transformation whereby the country's economy moved from an agricultural to an industrial manufacturing basis. More recent writers see the transformation as much more drawn out, extending from the seventeenth century and into the early twentieth century. They also see it as much more regionally diversified and geographically uneven, with particular regions specializing in particular products which transformed their production systems at different times. Most attention has been given to the textile and iron and steel industries, both of which are characterized by large factories requiring massive capital investment in buildings and machinery, driven by steam power, with manufacturing processes increasingly vertically integrated, and employing a large, and increasingly disciplined labour force (Figure 2.6). The same forces were at work in New England. By 1855, there were 52 cotton mills in Lowell, Massachusetts, employing more than 13,000 people, two-thirds of them women (Groves 1987). In Europe, Lille (France), Ghent (Belgium) and the Wupper valley of Germany developed as centres of cotton textile manufacture (Pounds 1990: 402).

Many other industries were developed in the same way as new markets arose on a national and international basis and production expanded. Food and drink, for example, is rarely thought of as a factory-based industry, but brewing, once the prerogative of almost every village inn, became increasingly an urban, large-scale industry with regional markets for its beers. In England, Burton-upon-Trent developed rapidly as the 'brewing capital' of Britain, with huge factory-scale breweries, on the basis of its colonial contracts to supply bottled 'India Pale Ale' to troops and civilians stationed in the Indian sub-continent as a substitute for local water. For other industries, however, including many manufacturing industries, 'factories' remained small-scale workshops employing fewer than 50 people (the Sheffield cutlery industry in northern England, and wire-drawing in the eastern Ruhr in Germany are good examples). Indeed, many industries remained almost domestic in scale well into the twentieth century (for example, the Birmingham jewellery and Coventry watch industries, both in central England).

The nineteenth century is characterized not only by this enormous variety in the scale of production, but also by an equally enormous variety of manufactured products. The tag 'The Workshop of the World' is as true of nineteenth-century Britain in general as it is of several of the country's manufacturing cities. In other British, European and American regions there was an increasing specialization so that local economies were dependent on maintaining market advantage on a single industry: the Polish textile city of Łódź and Lancashire's cotton mill towns were of this kind. Such places also produced distinctive working-class cultures, which, in the case of Łódź, was also predominantly Jewish (Koter 1990). Again, in those same mill towns much factory employment was female (see the example of Lowell above), whereas in heavy industry centres, such as ship-building in Newcastle and Sunderland, in the north-east of England, waged employment was male-dominated and women's roles were primarily domestic. E.P. Thompson, in his classic Marxist interpretation of *The Making of the English Working Class* (1963), is especially sensitive to the variety of experiences of working people in this period of aggressive industrial capitalism. There are, however, some common features of nineteenth-century industrial development in Britain and, subsequently, in Europe and North America. First was the fact that the power supply for the majority of industries came to be dominated by steam. There was therefore a move by many industries to the coalfield regions to reduce the cost of fuel (Figure 2.8). The development of the Ruhr region of Germany (Pounds 1990: 412–29), and of western Pennsylvania in the USA (Meyer 1990: 256–8), are good examples of this process. The reason for this change in fuel supply was the invention and commercial development of the rotative steam engine by James Watt and Matthew Boulton in 1769–75.

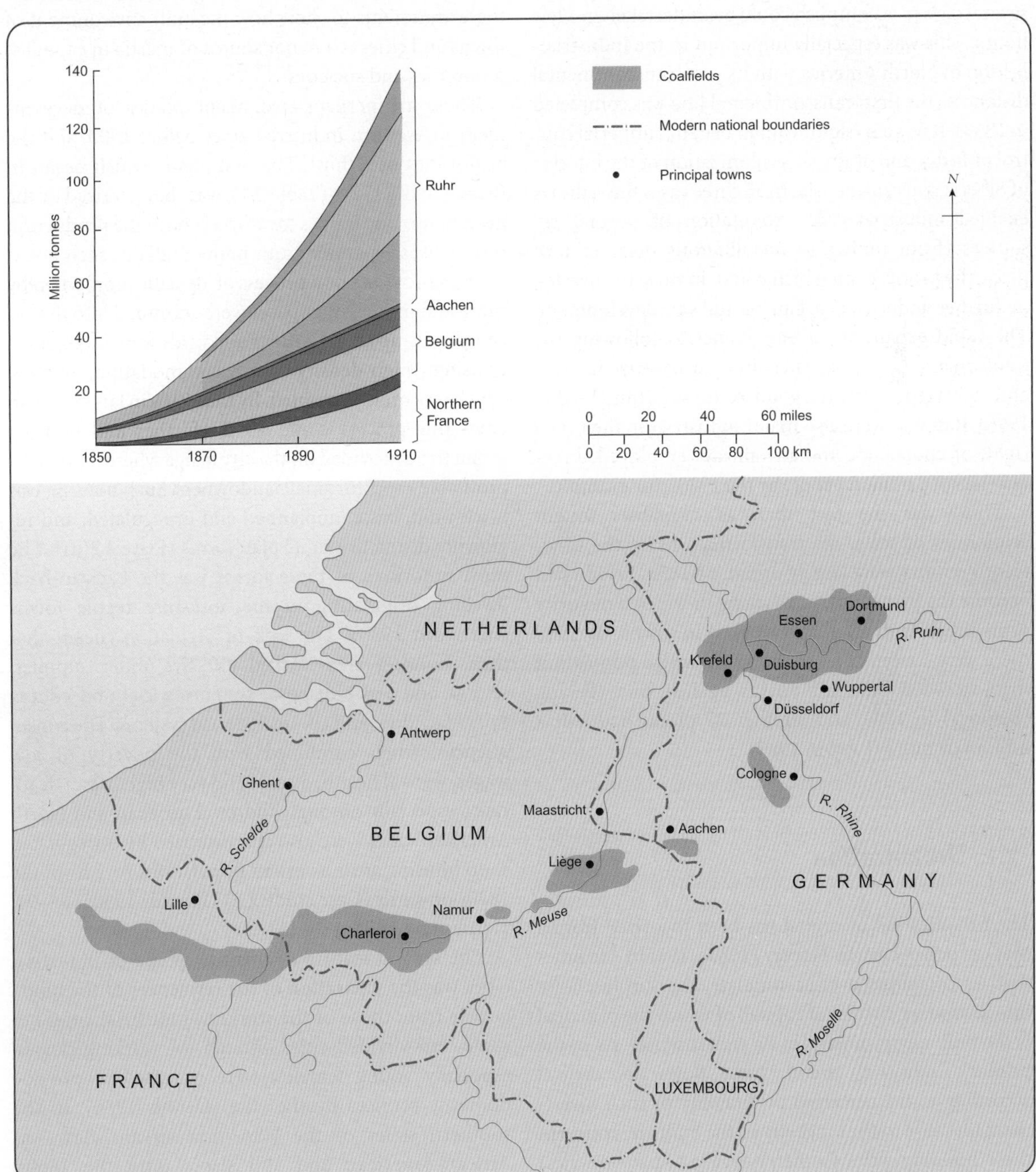

Figure 2.8 The coalfields of north-west Europe saw a concentration of new industrial enterprises after the 1780s as steam-powered machinery became normal in textile manufacturing, iron and steel making and metal goods manufacturing.

Source: Pounds (1990: 44)

Secondly this, in its turn, was fundamental to the invention and development of railways that transformed the transport costs of both raw materials and finished products almost as dramatically as it reduced the friction of distance. The Western and, increasingly, the colonial worlds saw a new time geography from the 1840s onwards, whereby journeys previously requiring several days to accomplish could be undertaken in a few hours. This was especially important in the industrialization of North America with its vast transcontinental distances (the first transcontinental line was completed in 1869). It was also significant in Britain's imperial control of India, and of Russia's colonization of the interior of Siberia and central Asia. In all three cases the railways enabled industrial-scale exploitation of natural resources (from timber to metalliferous ores) to take place, the profits from which could, in turn, be invested in further industrial, urban or railway development. The rapid expansion of San Francisco following the California gold and Nevada silver discoveries in 1849 and 1859 respectively is a good example of this (Walker 1996). Railways were also major industries in their own right, of course; the Indian railways employed three-quarters of a million people by the 1920s, for example.

Third was the fact that, as capitalists sought economies of scale, the transformations of the nineteenth century were largely urban. By 1851, Britain had become the first country anywhere in which a majority of its inhabitants lived and worked in towns. By 1900 there were sixteen cities in the world with a population that exceeded one million, as against one (Peking [Beijing]) in 1800, and another 27 places had half a million or more (Lawton 1989).

2.7 Urbanization

The urbanization of capital has been theorized from a Marxist perspective by Harvey (1985a, 1985b). He notes that, as capitalists over-accumulate, the surplus flows into secondary circuits of capital, of which the principal is the built environment (in its widest sense). As in the primary industrial circuit, these flows are cyclical according to the perceived profitability of such investment. Because of the longevity of the built environment these 'building cycles' (sometimes called Kuznets cycles) are much longer than the business cycles, somewhere between 15 and 25 years. Crises may lead to the loss of profitability of an investment, but they do not often lead to the loss of building fabric. Western cities are therefore made in the image of past capitalist decisions and subsequent adaptations to fit new circumstances. Harvey goes on to suggest ways in which class struggle is written into the landscape of the Western city (1985a: 27–31), which he explores in more detail in an analysis of Paris in the third quarter of the nineteenth century. More recently, Dodgshon (1998: 148–61) has built on these arguments to show how the built environment of towns and cities is a major source of inertia in capitalist economies and societies.

There are perhaps two major phases of development in Western industrial cities before 1900, and the beginnings of a third. The first phase, which began in Britain in the 1770s (Table 2.1), was characterized by the need to provide homes for workers once the production process was separated from home. Initially, such housing was provided by a process of 'densification' whereby more and more living spaces were crammed into the existing built-up area. Gardens and yards were used for increasingly high-density basic accommodation, much of which was multi-occupied. By the 1770s in larger British cities this process could go no further and housing began to be provided on the city fringe where it proved a profitable 'crop' for small landowners. Such housing was brick-built, basic, unplanned and unregulated, and regionally distinctive in its plan forms (Plate 2.2(a)). The most notorious of these forms was the back-to-back dwelling originating in the Yorkshire textile towns (Beresford 1988). The lack of regulation meant that these housing areas lacked effective water supplies, sewage and waste disposal systems, social and educational facilities, and connected road systems. The consequences, when combined with the poverty of low wages, were ill health, disease and very high urban death rates, especially among children. European and North American cities were also characterized by unregulated slum housing areas as urban populations rose from the 1820s onwards (Case study 2.2) (Pounds 1990: 368–91; Homburger 1994: 110–11).

The second feature of this first phase of industrial cities was the separation of the residences of the bourgeoisie from those of the workers. Industrial cities became class-divided cities, though the working classes, especially young females, were recruited to provide domestic services by the elite (Dennis 1984). In the southern states of the USA that servant class was African-American and still slaves until the 1860s. Thirdly, the capitalist land market began to value more highly the accessibility of the city centre and so the central business district began to emerge, as did districts in which industry and warehousing were the predominant

(a)

(b)

(c)

(d)

Plate 2.2 English nineteenth-century industrial housing. **(a)** Back-to-back courtyard house, Birmingham c. 1840. The door on the right leads on to a court of 12 identical houses fronting a narrow courtyard. Such houses were found in the Midlands in the period 1770–1860. **(b)** Tunnel-back bye-law houses, Teignmouth, Devon c.1880. The distinctive rear elevation resulted from bye-laws derived from the 1875 Public Health Act. These houses are found in all cities in the period 1875–1914. **(c)** Harborne Tenants Association, Birmingham, 1908. These houses derive from the garden city movement. Experimental estates of the Edwardian period provided the model for 20th-century municipal housing. **(d)** Improved industrial dwellings, east London c. 1885. Industrial Dwellings Associations experimented with tenement blocks with open galleried staircases.

(All photographs, the author)

Thematic Case Study 2.2

Tenement housing for industrial workers

- The 1865 New York Report on the Sanitary Condition of the City illustrated tenements (Figure 2.9, Plan A) on what had originally been four lots facing First Avenue. The apartments in the five-storey front block were of two or three rooms, only one of which was heated.
- In the southernmost lots dark tunnels provided access to the yards where smaller, cheaper two-room apartments occupied four-storey buildings to the rear of the lots.

SKETCH PLAN A

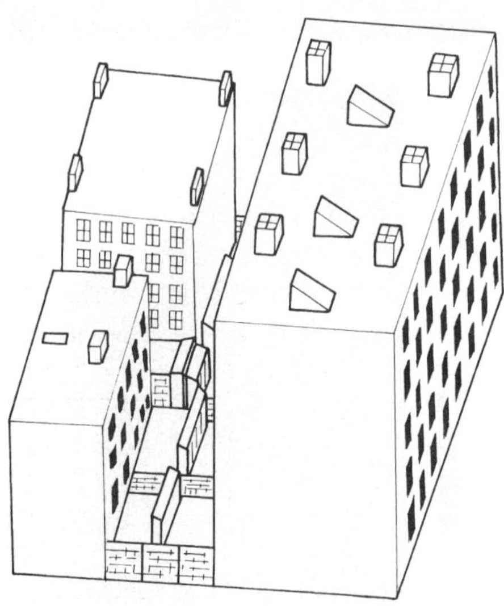

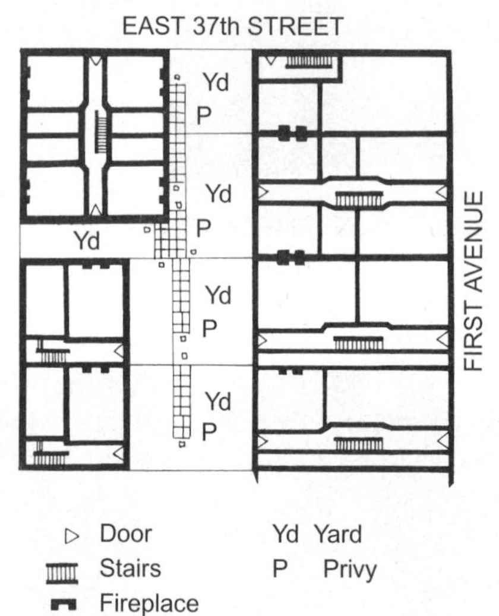

SKETCH PLAN B

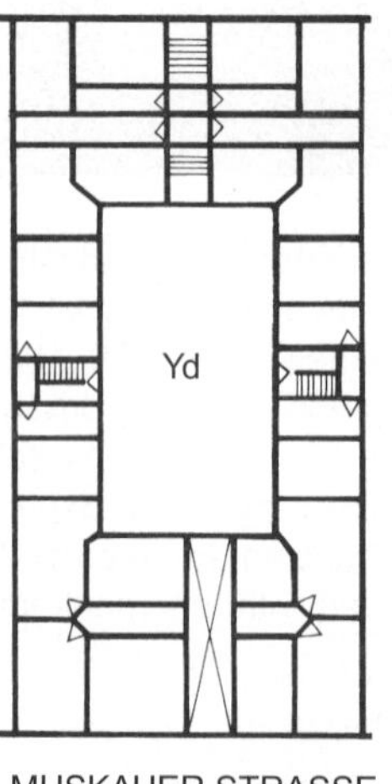

Figure 2.9 Plans of tenements in New York (A) and Berlin (B).

Sources: Ward (1989: 34); Borgelt et al. (1987: 11)

- On the northern lots, East 37th Street allowed for crossways sub-division of the property and a block of small two-room apartments had been built.
- Privy blocks occupy the yards which would also normally have been strung with drying washing.
- Berlin was expanding rapidly at the end of the nineteenth century as the capital of newly unified Germany. Workers lived in 'Mietskasernen' (single-room barrack blocks) earlier in the century.
- In the 1890s 'Mietshaüser' (tenement blocks) were the normal housing of workers and artisans.
- Some of these were 'L'-shaped on the lot, some 'C'-shaped, but many were built around a central court-yard, as in Muskauer Strasse (Plan B).
- Shops often occupied the ground floor on the street frontages of these blocks; a central passageway gave access to the yard where four stairwells accessed the apartments. Most were five or six storeys, so that the yards were dark and gloomy.
- Apartments were of between one and three rooms, but by the 1890s usually had integral sanitation.

land use, often close to the port facilities of canal, river frontage or harbour. One of the classic portrayals of this phase of urban development in Britain is Engels's (1845) *The Condition of the Working Class in England.* Warehouse districts, of course, have recently been 'revalued' in many cities to become today's loft apartments or high-tech offices (Zukin 1989).

The second phase of development, from 1840 to the 1890s (Table 2.1), was one of increasing regulation in towns and cities. New forms of local government and the collection of statistical information led to building regulations; the provision of sewerage and better water supplies; the provision of new cemeteries and parks; and better urban transport, especially tramways. All this led in turn to the increasing suburbanization of the better-paid elements of the urban workforce in the last quarter of the nineteenth century. These transformations occurred in most European core economies, and in North American cities, but with temporal variations. Thus, in England, the characteristic late-nineteenth-century dwelling was a suburban, brick-built, tunnel-back, terraced house (Plate 2.2(b)). In North America, excepting only the largest cities, the vast majority of urban housing was timber-built; in much of the rest of Europe, from Stockholm to Naples (as well as in Scotland), the apartment or tenement block was the norm (Plate 2.2(d)) (Case study 2.2). The North American city, from the 1840s, was also characterized by the ghettoization of immigrant populations: first Irish and Germans, then freed slaves from the southern states after the Civil War, then Italians, and finally, from the 1890s, Jewish migrants from the Russian empire of eastern Europe (Ward 1987).

Central business districts were modified by two other major transformations in the capitalist system in this second phase. First, banking services were increasingly required to enable surplus capital to be safely stored, efficiently invested and recycled towards new opportunities; such banks were but one aspect of developing office quarters, including, for example, insurance offices, land agents and other property services, and legal services. Many of these well-constructed buildings will have been found alternative uses today, as cafes, pubs or homes, because they have become valued parts of city conservation areas. This tertiary employment sector grew rapidly in the second half of the nineteenth century and 'the office' developed as both a specialized building type and a means of production (Daniels 1975). It was increasingly characterized by female employees using typewriters (the key technological invention) supervised by male managers. In North America the building type showed an increasing propensity to both large floor areas and height at the end of the century. The skyscraper was born in Chicago and relied on the development of steel-frame construction methods and the invention of the elevator (Gad and Holdsworth 1987). Second was the growth of consumption. This led to the development of a variety of new retail spaces in city centres, including department stores and arcades, as well as an enormous growth in the variety of specialized shops.

The incipient third phase of urban development reflected a minority interest in improving the living conditions of workers by industrialists. Model settlements have a history that dates to the beginning of the factory system. Modern commentators note both the generally higher standard of accommodation in these places, and the capitalist control of the home life as well as the working life of their workers. At the century's end the experimental garden suburb settlements at Bournville and Port Sunlight in England (Plate 2.2(c)) were widely admired and imitated in Germany (Margarethenhöhe in Essen, for example) and the USA (Radburn, New Jersey). At the same time, the first experiments in the large-scale provision of social housing were underway in London using apartment-block housing. These two themes, the garden suburb and social housing, were to combine powerfully in the development of modern planning to shape the urban environment in the first half of the twentieth century in Europe and its empires, but not in North America.

2.8 Conclusion

Within three centuries, Western capitalism had utilized its ill-gotten gains from its first colonial adventures to develop a series of specialized industrial regions supplying worldwide markets. Food production, industrial manufacture, service provision, consumption and transportation were all radically transformed by the capitalist enterprise. Cultures, societies and governments were necessarily impelled to change, too. States became more centralized and powerful. The difference between rich and poor both at the level of individuals and between countries became more marked. For some individuals and groups capitalism brought prosperity, improved living conditions and greater freedom; for others it brought destruction of local cultures, impoverishment, degradation and slavery. At the end of the nineteenth century the South African Boer War gave the first glimpse of the industrialization of warfare which was to scar the twentieth century so deeply.

Learning outcomes

Having read this chapter, you should know that:

- Scholars have interpreted and theorized the development of capitalism in different ways.
- The European colonialist enterprise was critical in the evolution of capitalism.
- Industrial capitalism transformed all sectors of the economy, including the built environment and social relations between individuals and classes.
- Industrial capitalism was essentially urban.

Further reading

Blaut, J.M. (1993) ***The Colonizer's Model of the World: Geographical Diffusionism and Eurocentric History***, Guilford Press, New York and London. This is a very readable and thought-provoking polemic of post-colonial writing, providing an alternative explanation of the 'success' of European capitalism.

Dodgshon, R.A. (1998) ***Society in Time and Space: A Geographical Perspective on Change***, Cambridge University Press, Cambridge. A more advanced text which provides much more detailed arguments and evidence for the themes of this chapter.

Knox, P. and Agnew, J. (2004) ***The Geography of the World Economy***, 4th edition, Edward Arnold, London. A well-written and popular textbook which will give readers another perspective on this period and much else besides.

Langton, J. and Morris, R.J. (eds) (1986) ***Atlas of Industrializing Britain***, 1780–1914, Methuen, London and New York. Historical atlases usually treat this period very well with innovative cartography and thought-provoking texts. This one deals with Britain in considerable detail.

Mitchell, R.D. and Groves, P.A. (eds) (1987) ***North America: The Historical Geography of a Changing Continent***, Hutchinson, London. There are a number of good historical geographies of North America. This one is well written, thoughtful and copiously illustrated.

Ogborn, M. (2007) ***Indian Ink: Script and Print in the Making of the English East India Company***, University of Chicago Press, Chicago. This book explores the geographies of power and knowledge in the rise of the British Empire in India.

Useful websites

www.besthistorysites.net/ Entitled 'Best of History Websites', a comprehensive guide to history-oriented resources online. For teachers, students and others.

For annotated, clickable weblinks and useful tutorials full of practical advice on how to improve your study skills, visit this book's website at **www.pearsoned.co.uk/daniels**

THE MAKING OF THE TWENTIETH-CENTURY WORLD

Chapter 3

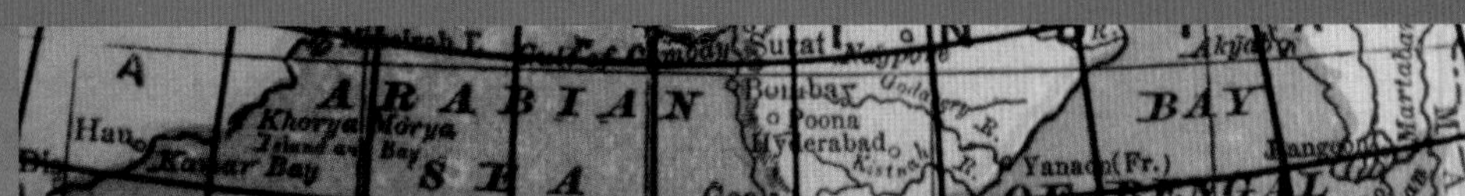

Denis Shaw

Topics covered

- The second industrial revolution
- Fordism – new patterns of production and consumption in the twentieth century
- Organized capitalism
- Challenges to Western-style capitalism: Nazism, the Islamic revival
- Challenges to capitalist-type development: Marxism, communism
- The end of European imperialism; informal imperialism
- The Cold War and the collapse of communism
- Disorganized capitalism

The twentieth century could be said to have been the period when capitalism finally triumphed over most of the globe. But it was neither a straightforward triumph nor an unchallenged one, and capitalism itself was changed in the process. This chapter is concerned with the various spaces created by and in response to twentieth-century capitalism – spaces of resistance and reinterpretation as well as spaces of adaptation and acceptance. The patchy and unequal world in which we now live reflects the erratic and conflict-laden nature of the processes which have produced it, and we need to know something about those processes to understand the world as it is now.

3.1 The changing capitalism of the early twentieth century

In the autumn of the year 1933 the writer and journalist J.B. Priestley set out by bus on a trip that was to take him the length and breadth of England and which he later described in his *English Journey* (Priestley 1937). As he left London by the Great West Road, Priestley noted how the road 'looked odd. Being new, it did not look English. We might have suddenly rolled into California.' What struck Priestley as particularly odd was 'the line of new factories on each side' of the road. 'Years of the West Riding', he explained (he was born and raised in Bradford), 'have fixed forever my idea of what a proper factory looks like: a grim, blackened rectangle with a tall chimney at one corner. These decorative little buildings, all glass and concrete and chromium plate, seem to my barbaric mind to be merely playing at being factories.' Armed with a copy of the now celebrated textbook, *The British Isles: Geographic and Economic Survey* by Stamp and Beaver (1933), which was later to be used by generations of geography undergraduates, Priestley went on to make some astute geographical and social points about these factories: 'Actually, I know, they are tangible evidence, most cunningly arranged to take the eye, to prove that the new industries have moved south. You notice them decorating all the western borders of London. At night they look as exciting as Blackpool. But while these new industries look so much prettier than the old, which I remember only too well, they also look far less substantial. Potato crisps, scent, tooth pastes, bathing costumes, fire extinguishers; those are the concerns behind these pleasing facades' (Priestley 1937: 3–5).

In these few words, Priestley summarized some of the major ways in which the industrial world of the twentieth century was to differ from its nineteenth-century predecessor. The fact that he was making his journey by road was itself significant; thirty years before he would have had to go by rail. The new factories he observed were the products of the technological changes that had been transforming industrial capitalism since the late nineteenth century, and many had clearly developed to serve an expanding consumer market (see Plates 3.1 and 3.2). And the location of the factories by the new arterial highway leading westwards out of London was the result not only of a revolution in transport and communications but also of

Plate 3.1 The Hoover factory, Perivale, west London. This splendid example of Art Deco architecture, design by Gil Wallace in 1932, reflects the new consumer industries which were being established in the years after the First World War. (© Angelo Hornak/CORBIS)

Plate 3.2 Using a vacuum cleaner or Hoover, one of the new consumer appliances which were becoming widely available by the 1930s. Photos and advertisements of the period almost always depicted women performing such domestic tasks.

(Mary Evans Picture Library)

the locational freedom deriving from the availability of electricity and other fuels. The textile industries that Priestley remembered from his childhood in Bradford were tied to the coalfields; the newer industries that were appearing in London by the 1930s no longer needed coal, and were much cleaner and brighter in consequence.

Of course, what Priestley saw along the Great West Road was by no means representative of all the technological changes that had been affecting the industrial economies of Britain and other countries for the previous few decades. What had been happening in these countries was that a whole series of new industrial branches had been developing to supplement, and eventually to eclipse, the traditional activities based on coal, iron and textiles. Not all of these were as pleasing to the eye as those observed by Priestley. In the second half of the nineteenth century, for example, the metallurgical industries had been transformed as a result of a series of inventions allowing the production of cheap steel. Next came the rapid development and proliferation of different branches of the chemical industry (alkalis, dyestuffs, pharmaceuticals, explosives, lacquers, photographic plates and film, celluloid, artificial fibres, plastics). The electricity industry, which began to flourish by the end of the nineteenth century, was dependent on earlier inventions, like the steam turbine. About the same time came the rise of the motor industry, which was in turn associated with other industries like oil and rubber. When Priestley set out on his journey, society was already beginning to adjust to the impact of the many new activities catering to the consumer (most notably, domestic appliances – see Plate 3.2) and to new means of transportation (car, bus, aircraft). Of course the full impact of such developments was to come later, after the Second World War, while some technologies, like regular TV broadcasting, the jet engine, nuclear power and the microchip, still lay in the future.

So profound were the technological and accompanying social changes that affected industrial capitalism from the late nineteenth century that some historians have described them as a 'second industrial revolution' (Landes 1969: 4). But it is important to remember that the older industries – coal, textiles, railways, some forms of engineering – did not die immediately or indeed quickly. One of the features of the changing industrial geography of the late nineteenth and early twentieth centuries was that countries like the USA and Germany, whose industrialization came later than Britain's, now began to forge ahead on the basis of the newer industries described above. Britain remained overdependent on the older and less dynamic branches (Figure 3.1).

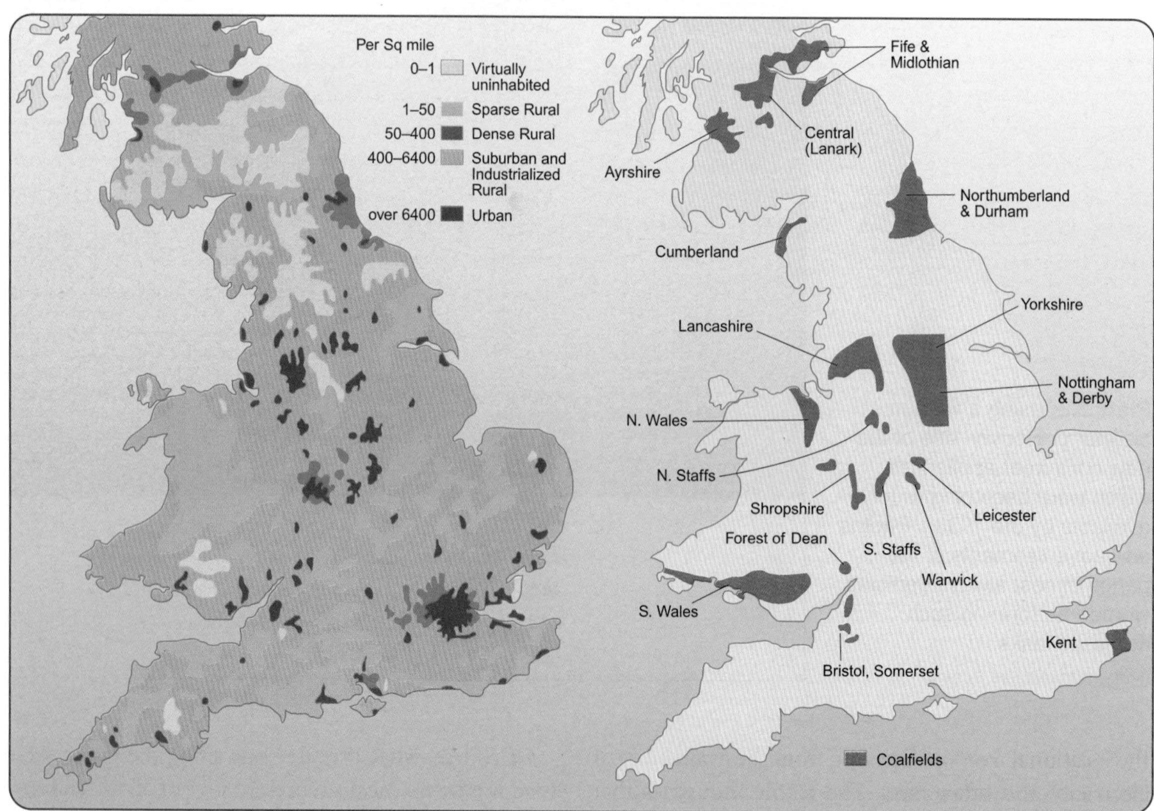

Figure 3.1 The geography of the UK population and coalfields compared. With the significant exception of London, there is a strong correlation between the geography of population and that of coalfields, reflecting both the early stage at which Britain industrialized and urbanized, and the country's long dependence on nineteenth-century industries.

Sources: Population, based on Mitchell (1962); coalfields, Stamp and Beaver (1963: 286)

For the first half of the twentieth century, the industrial changes described above only directly affected certain parts of the world, notably Western Europe, North America, Japan, and by the 1930s the Soviet Union and some other areas. Much of what was later to become known as the Third World, or the developing world, was still agricultural. Yet, in continuation of earlier processes (see Chapter 2, pp. 45–46, 53–54) many colonies and other regions were now being organized commercially to supply the industrial countries with raw materials and tropical products – for example, bananas and sugar from Central America and the Caribbean, Brazilian coffee, Indian tea, Malaysian rubber. They were thus being tied in to the capitalist world economy. Gradually certain of these countries began to adopt the technologies of the industrialized world – the rise of the Indian textile industry is one example – but only later in the twentieth century did industrialization become more widespread.

Thus the foundations of what is now known as a global economy were already being laid in Priestley's day, or even earlier. By the beginning of the twentieth century capitalism had become a world phenomenon, tying far-flung countries together by means of international trade and fostering international capital flows through major financial centres like London. The first multinational corporations were already appearing. All this was aided and abetted by the new systems of communication and transport – telegraph, telephone, radio (from the 1900s), fast steamships, aircraft – which were beginning to span the globe. Of course, none of this bears comparison with the forces of globalization which were to become so significant later in the century. Yet the world was already becoming a smaller place (see Plate 3.8). J.B. Priestley himself suggested this when he compared the Great West Road to 'California'. What might this mean to the average English reader in the 1930s? The answer is – a great deal. The Hollywood film industry was in its heyday and the people, homes and landscapes it portrayed were being viewed, and copied, the world over.

3.2 Organized capitalism

It would be a mistake to suppose that the advance of capitalism in the twentieth century was a story of unmitigated triumph. On the contrary, its fluctuations and misfortunes were such that one historian felt constrained to call the period the 'Age of Extremes' (Hobsbawm 1995). The first half of the century – Hobsbawm's 'Age of Catastrophe' – was particularly disturbed, with two world wars (1914–18 and 1939–45) and a deep world economic depression (1929–33). By contrast, the years between 1945 and 1973 were ones of growing prosperity across much of the globe (Hobsbawm's 'Golden Age') to be followed once more by a disturbed period in the wake of the shock oil price rises of the 1970s.

It was Karl Marx who originally emphasized the unplanned, competitive and even chaotic nature of capitalism's development. But one of the features of the twentieth century has been the attempt, by both government and private agency, to regulate and even to control it. The reasons for this phenomenon are many, but they are no doubt linked both to the precipitate nature of technological change during this period and to the severe fluctuations mentioned above. Attempts to 'organize' capitalism have taken a number of different forms. For example, already in the late nineteenth century, and especially in the USA and Germany, there were moves towards the formation of inter-firm agreements, cartels and larger companies and corporations. Large corporations could more easily marshal the huge capital resources which modern industry requires, and also influence their markets more effectively. As noted already, the first multinational corporations appeared at this time, those based in the USA being most notable. Nineteenth-century examples include the German electrical firm, Siemens, and the US Singer sewing machine company. Twentieth-century examples include US-based Hoover, Ford, Coca-Cola, Pepsi Cola, Nabisco (Shredded Wheat), and Kellogg's, all of which invested in Britain in the 1930s. Most of the major international oil firms also date from this period.

A further way in which capitalism became more 'organized' was the phenomenon of mass production, linked especially to the growing consumer market. Here two Americans are regarded as particularly significant. F.W. Taylor (1856–1915) is especially associated with time-and-motion studies, whereby complex tasks on the factory floor could be completed more efficiently and productivity increased. The other is Henry Ford (1863–1947) who organized car production in his Dearborn, Michigan, plant using modern methods like the assembly line and interchangeable parts. The result of his centralized approach was a significant reduction in the cost of producing cars, which could now be manufactured on a mass basis. The ensuing mass production methods and associated patterns of mass consumption are frequently referred to as **Fordism**.

Plate 3.3 A suburban house built in the Modern style in the 1930s, Essex, England. The suburbanization of the inter-war period was a product of the affluence and increased mobility of some middle-class groups during this period. Suburbanization was to become even more prominent after the Second World War.
(© Gillian Darley; Edifice/CORBIS)

Governments were also affected by the desire to 'organize' capitalism and to tackle the many problems to which it seemed to give rise. At the international level, the USA, which emerged after the Second World War as the undisputed leader of world capitalism, took the lead in establishing a series of institutions like the World Bank, the IMF and the General Agreement on Tariffs and Trade (GATT) to ease international monetary payments, promote trade and encourage economic development. At the national level, numerous countries, especially some of the west European ones, pursued democratic agenda of various kinds, such as attempts to construct 'welfare states' to tackle such social problems as unemployment, ill health, old age and social inequity. Arguably the experience of government planning and controls in wartime (and possibly also a fear of communism – see below) helped pave the way for the optimistic belief in planning and large-scale social engineering that characterized the post-war period. In western Europe especially this was the era of bold experiments in new town and city development, slum clearance and ambitious social housing schemes, regional planning, and extensive controls over land use (Hall 2002). No doubt the success of these schemes was dependent on the spreading affluence that accompanied Hobsbawm's 'Golden Age'. Across western Europe and North America a tide of suburbanization signalled not only a growing ability to own one's own home in a desirable location but also the availability of the social and physical infrastructure, the private cars and the many new consumer products which now made such a goal possible for many (see Plate 3.3).

The writers Scott Lash and John Urry have described the era ushered in by the methods of Ford and Taylor but reaching its apogee during the period 1945–73 as 'organized capitalism' (Lash and Urry 1987). Some of its principal features (which were especially characteristic of the developed world) are described in accordance with their views in Spotlight box 3.1.

Needless to say, such generalizations would be more or less true, depending on the time and location being considered. Capitalism had different histories in different places, and the exact form it took had much to do with the long-term evolution of each society affected by it.

3.3 Challenges to liberal capitalism

Come, bombs, and blow to smithereens,
Those air-conditioned, bright canteens,
Tinned fruit, tinned meat, tinned milk, tinned beans,
Tinned minds, tinned breath.
Mess up the mess they call a town –
A house for ninety-seven down
And once a week a half-a-crown
For twenty years

From 'Slough' by John Betjeman (1937)[1]

John Betjeman's famous fulminations against the town of Slough, situated just west of London and experiencing developments similar to those observed by J.B. Priestley a few years earlier, are in fact a hymn against

[1]Poem: 'Slough' from *Collected Poems* by John Betjeman © 1955, 1958, 1962, 1964, 1968, 1970, 1979, 1981, 1982, 2001. Reproduced by permission of John Murray (Publishers) Limited

Spotlight Box 3.1

Organized capitalism

- Extractive and manufacturing industries are the dominant economic sectors.
- There is an accent on economies of scale, leading to the importance of large industrial plants. Such plants may structure entire regional economies around themselves. Examples might include (in the United Kingdom): the West Midlands, based on cars and engineering; Lancashire, based on cotton textiles; the North-East, based on mining, shipbuilding and heavy engineering; and (in the USA): Detroit, based on cars and engineering; and Philadelphia, based on textiles and port-related activities.
- Manufacturing plants are controlled centrally by big industrial corporations – there is an emphasis on mass production and standardization.
- There is state regulation of the economy to overcome problems generated by the market, for example regional unemployment problems.
- Big industrial cities are the spatial expression of large-scale manufacture.
- There is state-controlled welfare provision to even out social inequalities, address unemployment problems and raise health standards.
- In culture and social provision, there is an accent on mass provision, for example in housing, consumer goods, TV programming and newspapers. The emphasis on mass coverage and standardization leaves relatively little choice, reflecting a modernist perspective.

Source: after Lash and Urry (1987)

modernity. Betjeman was railing against many of the social repercussions of the profound twentieth-century changes noted earlier in this chapter. In this he was by no means alone. Indeed, according to some commentators (e.g. Wiener 1985), Britain has been characterized by an anti-urban and anti-industrial culture that made it very difficult for the country to adapt to twentieth-century change. However, rather similar problems have also occurred in other countries where modernity (here defined as the spectrum of economic, social, political and cultural changes associated with twentieth-century capitalism) has brought problems of adaptation.

It has already been noted that the twentieth century cannot be described as a century of uninterrupted progress for capitalism, particularly in its Western 'liberal' form. Capitalism has been subjected to a series of challenges and political struggles that greatly affected the course of twentieth-century history, and in various parts of the world there have been attempts to create spaces in which alternatives to liberal capitalism can flourish. Interestingly enough, in terms of Wallerstein's world systems theory discussed in Chapter 2 (see p. 41) most of these attempts have been associated with countries outside the core, or with those like Germany after its defeat in the First World War struggling to rejoin the core states. Particularly important for the political geography of the twentieth century were the attempts by Marxists and others to reject the capitalist development model entirely and to reconstruct society on a new basis. This issue will be discussed in the next section. This section will focus on two very different movements that have also challenged Western-style capitalism in profound ways, but without discarding it entirely: Nazism (National Socialism) with its close relative, Fascism; and the Islamic revival.

Although they had nineteenth-century antecedents, both Fascism and Nazism were essentially products of the inter-war years. Fascism, under its leader Benito Mussolini, ruled in Italy from 1922 until its final defeat in the Second World War in 1945. Nazism under Adolf Hitler ruled Germany from 1933 until it too was defeated in 1945. Various Fascist or neo-Fascist groups ruled or were active elsewhere in Europe, and in some other regions, during this period, and to a lesser extent since.

There is no doubt that Nazism was by far the most influential Fascist movement after 1933, and so the following remarks will be devoted to it (Kershaw 2000). One of the problems of discussing Nazism or other forms of Fascism is their lack of a consistent ideology or philosophy. However, certain general points can be made. Like certain Western intellectuals and others, the Nazis were moved by a dislike of facets of capitalist modernity, such as commercialism, materialism, individualism, threats to the traditional family, like the rise of female employment (see Plate 3.4), and similar tendencies that they associated with the 'decadent'

Plate 3.4 Women workers in armaments plants during the First World War. Although Nazis and some other political groups were later to oppose female employment, the labour shortage during the First World War gave women job opportunities which were increasingly taken as the twentieth century advanced.
(Hulton Archive/Stringer/Getty Images)

Western democracies. They also despised Western-style parliamentary democracy, with its plurality of political parties and class divisions. In its place, they advocated the concept of a single national community, a *Volksgemeinschaft*, headed by a single Leader or Führer (Hitler), who was regarded as representative of, and chosen by, the people (this was the Nazi concept of 'democracy'). The Leader's power was absolute. Such a creed, however, seemed exceedingly unlikely to come to power in Germany in the early 1930s had it not been for the extreme circumstances reigning there. One was a general sense of resentment at Germany's defeat in the First World War and subsequent national humiliations. As extreme nationalists, the Nazis promised to avenge this defeat. Another was the dire straits to which many of the middle class had been reduced by post-First World War inflation and the Depression that began in 1929. There was also the fear of the many strikes and disorders perpetrated by communists and other left-wing groups (as well as by the Nazis), that were, of course, encouraged by the selfsame economic difficulties.

One of the oddities of Nazism was that it was, at one and the same time, both reactionary and modernizing (Herf 1984). On the one hand, the Nazis looked backwards to an imagined heroic and rustic Germany of the past, to Nordic myths, Germanic towns and landscapes (see Rollins 1995; Hagen 2004) and happy peasants tilling the fields in traditional costume. They tried to bolster the 'traditional' family: women were to remain at home and raise children for the fatherland. There were even attempts to build villages and garden settlements to reflect such ideals. On the other hand, the Nazis were also modernizers who built the autobahns, fostered industry and spent vast sums on the military. Their aim was to turn Germany into a superpower, able to dominate the European continent and regions beyond (see Chapter 21, pp. 477–50).

One of the things that distinguished Nazism and Fascism from other right-wing tendencies was that they were mass movements. The masses were to be organized and captivated by ceremonial and display, by mass rallies and triumphal processions (rather than, for example, by rational persuasion) (Mosse 1975; see also Atkinson and Cosgrove 1998). The landscape of Germany was dotted with places where such ceremonials could be performed, the Nuremberg Party Rally grounds being perhaps the most notorious. Had Nazism survived, the whole of central Berlin would have been reconstructed on a grandiose scale to reflect the supposed glories of the Third Reich (Speer 1995: 195–223). Recently geographers have become fascinated by the ways in which images, symbols and other forms of representation are now used to market commodities and to sell ideas (see Chapters 13 and 18). The Nazis were among the first to see the significance of image in a mass society and to realize the many ways in which the manipulation of images and other forms of representation can be used to attain and to hold on to power.

Nazism's central and most notorious feature was its racism. Building on common European assumptions of racial and cultural superiority (prejudices which have by no means disappeared, even today), they taught that the Germans and related Aryan **races** were equipped by nature to dominate the globe. All non-Aryan peoples were regarded as inferior, particularly the Jews who, because of their culture, religion and cosmopolitan ways, seemed to represent all that the Nazis feared and hated. As time went on, it became clear that the Nazis meant to exterminate the Jews (they killed six million of them) as well as others (gypsies, homosexuals, the insane, certain religious groups) who could have no place in the world they intended to reconstruct. The Nazi death camp at Auschwitz in Poland, which has recently been researched by geographers (Charlesworth et al. 2006), is kept as a memorial to the huge numbers who were murdered and as a solemn warning to today's world of the horrific consequences of racial prejudice.

Dislike of some of the features associated with Western-style capitalism has also characterized the other movement to be considered here: the Islamic revival. Islam, of course, is quite different from twentieth-century secular (and essentially 'Western') movements like Fascism. Islam is one of the world's ancient religions, with many millions of adherents in a swathe of territory stretching from North Africa, across southern Asia and down to Indonesia, with many others living beyond (Figure 3.2) (Park 1994; Esposito 1995). It has a rich cultural heritage, and indeed for many centuries in the medieval period the Islamic (or Muslim) world easily outstripped Europe in terms of intellectual, scientific and cultural achievement. The Islamic revival is essentially a product of the twentieth century, particularly of its second half, and, in the opinion of many, represents a reaction against the inroads of secularism, Western colonialism, and some forms of modernization that have affected the Islamic world during this period. Parts of the Muslim world, for example, were colonized by one or more of the European powers either several centuries ago or during the twentieth century. Furthermore, beginning with Kemal Ataturk's founding of the modern, secular state of Turkey in the early 1920s, many Muslim political leaders have embraced Western-style political, social and economic ideas in the hope of modernizing their countries.

In the opinion of some, the Muslim faith recognizes no divisions between the realms of the sacred and the secular and exhorts its followers to obey its injunctions in every aspect of their lives. The whole of life and society, in other words, is to be regulated in accordance with Islamic teaching (though there is room for debate and disagreement over how far modern social changes can be reconciled with those teachings). Some devout Muslims were particularly offended by the programmes of Westernization and technological development adopted by several Muslim states in the 1960s and 1970s (Scott and Simpson 1991; Park 1994). Secular education, 'immodest' dress and liberal sexual attitudes, the use of alcohol, Western-style entertainment, growing social inequalities and profligate spending by

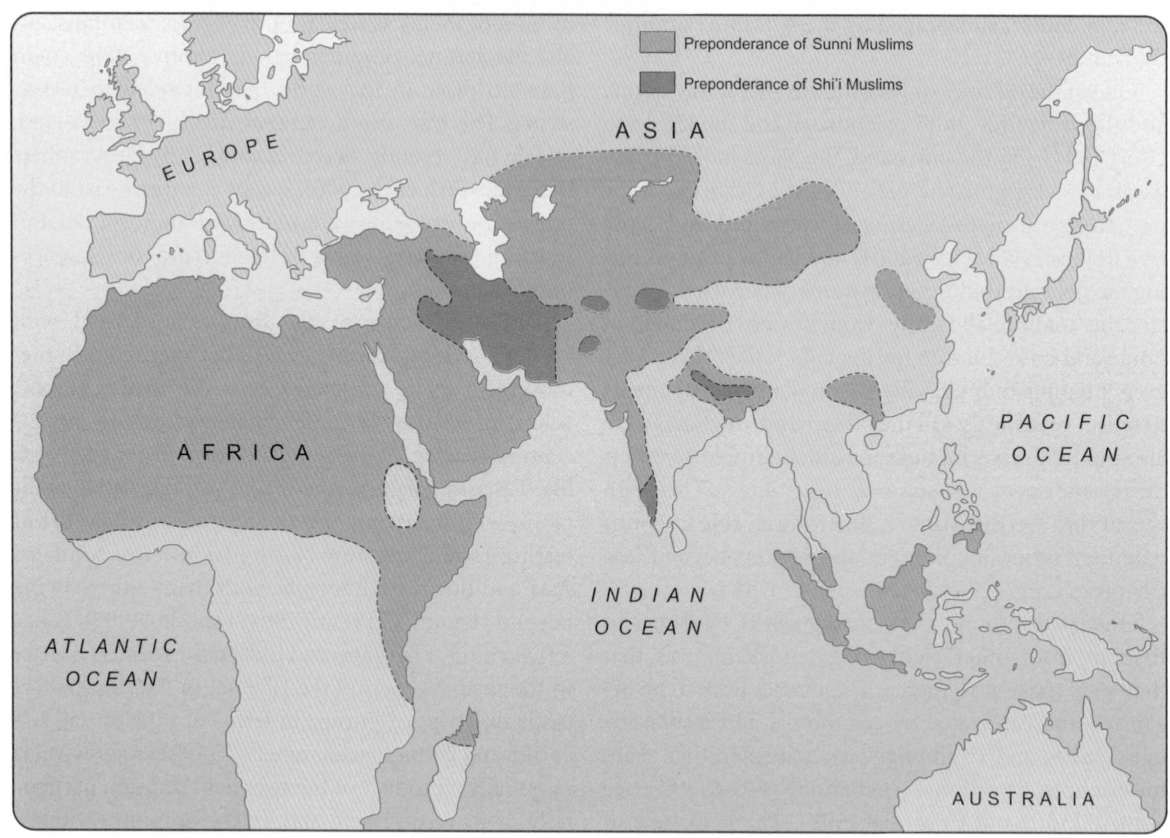

Figure 3.2 The geography of Islam.

Source: Based on *The Times Atlas of World History* (1989: 105). © Collins Bartholomew Ltd 1989, reproduced by permission of HarperCollins Publishers

new social elites (offending against Muslim teachings on charity and social justice) have all been subject to criticism. In the minds of many Muslims, such tendencies are associated with the West and its 'degenerate' society.

Across the Muslim world, there have thus been movements for Islamic revival and renewal, often taking a political form and demanding a return to the original purity of Islam. Sometimes these movements take a particularly militant and anti-Western stance, though this is by no means always the case (Saudi Arabia is an interesting case of a pro-Western state whose society is nevertheless built on strict Islamic principles). Such movements may sometimes threaten existing regimes that are supported by the West. However, the Islamic world is so variable that the political consequences of the revival are very difficult to predict. Perhaps the best-known case of a country that experienced an Islamic revolution is Iran, whose previous pro-Western ruler, the Shah, was overthrown in 1979. Here the Islamic clergy took direct control of the government in order to build an Islamic republic (see Plate 3.5). This entailed, among other things, the implementation of strict Islamic (Sharia) law that was intended to nullify some of the less acceptable Western cultural influences (Stempel 1981).

One feature of the Islamic revival that has attracted much attention in the West is the veiling of women. This has often been regarded in the West, and in the Islamic world also, as a return to the past when women were exploited and confined to a life of domesticity. An interesting point, however, is that in parts of the Muslim world some educated women have taken the veil voluntarily, seeing it as a protest against the sexual and commercial exploitation of women that is sometimes regarded as a product of Western-style capitalism. Another interesting but less publicized feature of the revival has been the attempt to move to an Islamic system of (zero-interest) banking. This is regarded as not only in accordance with Islamic ethical norms but also much fairer than Western-style banking in spreading risk. It suggests an attempt to construct an

Plate 3.5 Street demonstration with a poster of Ayatollah Khomeini, figurehead of the Iranian Islamic revolution.
(Alain Dejean/Sygma/Corbis)

economic system based on Islamic rather than Western-style commercial principles.

Nowadays many Western countries contain significant Muslim minorities and international migration continues to add to their numbers. Most unfortunately, migration pressures and fears over terrorism which grew from the 1990s have been used by some groups and even media to advocate discriminatory policies directed at Muslims. It is only to be hoped that the twenty-first century will not witness a return to the bad old days of intolerance and racism which so marred the first half of the twentieth.

3.4 Communism and the command economy

Perhaps the most significant challenge to capitalism in the twentieth century came from Marxist-style communism (Calvocoressi 1991). By the 1960s and 1970s up to one-third of humanity was living under communist governments which explicitly rejected capitalism as an acceptable way of organizing society. The reasons for that rejection and why it largely failed must now be considered.

An outline of some of the principal features of Marxism is given in Spotlight box 3.2. An important point is that Marx's teachings failed to change those societies at which they were initially aimed – the industrial societies of Western Europe and North America. Marx himself had expected that communism would find support among the growing industrial working classes of countries like Germany and Britain, where factories were bringing such groups together in increasing numbers. However, it was by no means obvious, as the years passed, that the workers of those countries were necessarily being increasingly exploited, as Marx seems to have expected (later, these countries were accused by Marxists of exporting exploitation to colonies and other less developed regions). Instead, Marxism triumphed in Russia (in 1917), in what was in fact the least industrialized of Europe's great powers. Thus, whereas in terms of Wallerstein's world systems theory Marxism was expected to find favour in the core countries, in fact it initially triumphed in a semi-peripheral one. That it did so changed the character of Marxism, which was now faced with the challenge of building socialism in a peasant society, and in virtual isolation from the rest of the world.

What happened after 1917 in Russia (or the Soviet Union as it was now to be called) was of profound importance for the other countries which later adopted communist systems, if only because Russia was the pioneer. What happened there began to assume something of the character of orthodoxy (Hosking 1992). In view of the difficulties they faced, and in all likelihood because of their own inclinations, the Bolsheviks (as the Russian or Soviet communists were called) adopted a highly centralized political system that brooked no opposition. Eventually, from the late 1920s, they implemented a fully centrally planned economic system. This involved the abolition of private enterprise and virtually all forms of market relations, and the collectivization of agriculture. The purpose of this extraordinary economic system was both to speed the process of economic development and to build up the country's military resources. As the 1930s advanced, it became clear that what had earlier seemed a rather vague threat from the outside capitalist world was beginning to take a concrete and menacing form in the guise of Nazi Germany. The centrally planned or command economy (which involved much suffering on the part of the Soviet people) proved equal to this challenge. In the ensuing war with Germany (1941–5), the Soviet Union emerged victorious, but only after sustaining enormous losses.

Soviet victory in the Second World War greatly enhanced that country's prestige. Moreover, because the Soviet armies were now in occupation of much of central and eastern Europe, they were able to ensure that regimes friendly to the Soviet Union (that is, communist regimes) would assume power in those regions. Communism soon spread into other countries, notably

Spotlight Box 3.2

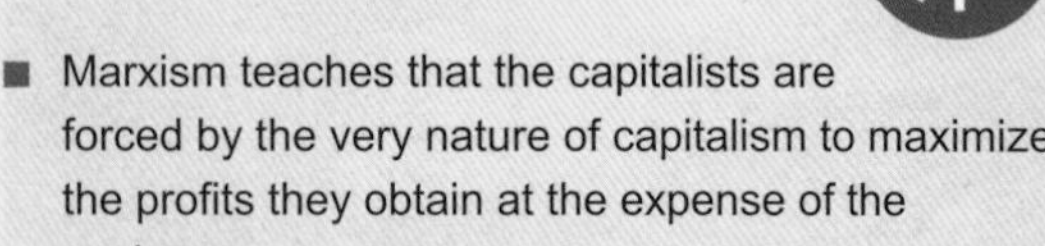

Marxism

- Marxism, which derives from the teachings of Karl Marx (1818–83), is related to other forms of socialism which seek to moderate or reform the injustices of capitalism.
- Unlike some other forms of socialism, Marxism regards capitalism as an innately unjust and exploitative system.
- According to Marxism, capitalism divides society into antagonistic classes: those who own the main sources of wealth (the capitalists) and those who must live by selling their labour to the capitalists (the proletariat).
- Marxism teaches that the capitalists are forced by the very nature of capitalism to maximize the profits they obtain at the expense of the workers.
- Marx thought that eventually capitalism would become so exploitative and prone to crises that its downfall was inevitable.
- It would then be replaced by a much more just, classless society (socialism, gradually maturing into communism).
- In the meantime Marx exhorted the world's workers (especially the industrial workers) to organize politically to hasten capitalism's downfall.

China (in 1949), south-east Asia and beyond: in other words, into Wallerstein's 'periphery'. All these countries initially followed the Soviet development model, but soon found that it was necessary to adapt it to their own needs. In the meantime the spread of communism, and Soviet ambitions, excited the suspicions of the capitalist West. From the 1950s, therefore, the world was split into two armed camps, both equipped with nuclear weapons. The ensuing confrontation, known as the Cold War, profoundly influenced both sides and encouraged their militarization. However, many countries, especially in the developing world, tried to avoid taking sides, while China, though communist, began to pursue its own version of communism outside the Soviet sphere.

While the actual form that communism took in the Soviet Union and other countries may have been a modification of Marx's own ideas, it did represent a radical departure from the capitalist development model. Not only was the command economy a very different, state-centred approach to economic development (and one which was copied in many parts of the developing world after the Second World War, with varying degrees of success), it regarded itself, and was regarded, as a threat to the whole idea of capitalism (Nove 1987). Internally, whilst it led to attempts to reconstruct society on a different basis, it did in fact give rise to new forms of spatial inequality (Bater 1986). Interestingly enough it also had certain spatial features in common with Nazism and Fascism, such as the emphasis on creating urban spaces specifically for the purpose of mass ceremonial and display (Plate 3.6). The creation of 'spaces of terror', such as concentration camps for the incarceration or elimination of those deemed unacceptable to the regime, was also a feature of the two systems (Moran 2004; Pallot 2005).

In the end, communism failed to prove itself a successful viable challenger to capitalism. Especially from the 1970s, the Soviet Union and its eastern European allies fell behind their capitalist rivals in terms of productivity, flexibility and innovation. Whether this was because of problems inherent to command economies as such, or whether it has more to do with mistakes made by the various political leaderships, is hard to say. Whatever the reasons, by the end of the 1980s practically the whole of the communist world was in a state of economic and political crisis. The subsequent fall of communism in eastern Europe, and the splitting of the Soviet Union into fifteen separate states in 1991, signalled the end of the Cold War. Since then the post-communist states, plus China which continues to pursue its own version of socialism, have been struggling to adapt to the market economy and in other ways to cope with the problems of post-communist transformation (Gwynne, Klak and Shaw 2003: 59–72, 101–8).

It is worth asking ourselves how different the world might have been today had the Soviet Union survived. Would the United States and its allies have invaded Iraq in 2003 and intervened elsewhere had there still been a Cold War? Would growing social inequalities and the conspicuous consumption by wealthy elites which are so characteristic of today's capitalism have been restrained had there still been a threat from communist

Plate 3.6 Central Moscow, showing one of the open spaces created or enhanced by the Soviet dictator Joseph Stalin for official communist demonstrations and ceremonial display.
(Neil Emmerson/Robert Harding)

and socialist ideas and labour movements? Altogether it is difficult to believe that the apparently triumphant capitalism of the late twentieth century, and the Western countries which so benefit from it, will not be faced by further significant challenges as the new century unfolds.

3.5 The end of imperialism?

Father, Mother, and Me
Sister and Auntie say
All the people like us are We,
And everyone else is They.
And They live over the sea,
While We live over the way,
But – would you believe it? – They look upon We
As only a sort of They! –

From 'We and They' by Rudyard Kipling (1912: 763–4)[1]

At the beginning of the twentieth century Britain and several other European powers sat proudly at the centre of a series of empires that spanned the globe (Figure 3.3). As noted in Chapter 2, these empires were the products of a long period of European exploration, settlement, economic exploitation and imperial rivalry. Something of the complacency and condescension with which Europeans commonly regarded their empires at this period is nicely captured by Rudyard Kipling in his comic poem 'We and They', quoted above.

The early years of the twentieth century were not in fact the high points of empire. After the First World War (1914–18), Britain, France and other victorious powers helped themselves to the former German colonies, while Italy invaded the independent African state of Abyssinia (Ethiopia) as late as 1935–6. However, the future of European imperialism was already being questioned even before the war. The English liberal J.A. Hobson, and later the Russian revolutionary V.I. Lenin, popularized the idea that Europe's overseas colonies were being economically organized and exploited mainly for the benefit of the European 'mother countries', forming an undeveloped periphery to the European core. Lenin taught that imperialism was an inevitable consequence of capitalism – its 'highest stage'. In the meantime various rumblings of discontent were being felt in various parts of the European

[1]We and They by Rudyard Kipling, with permission of A. P. Watt Ltd. on behalf of Gráinne Yeats

Plate 3.7 The fall of the Berlin wall in 1989 signalled the end of communism in the Soviet bloc and the removal of a major twentieth-century challenge to world capitalism.
(© Reuters/Corbis)

empires. Britain, for example, felt constrained to grant home rule to a number of its colonies where white settlement had been significant – Australia, New Zealand, Canada and South Africa. It was also considering doing something similar in Ireland (see Case study 3.1), Egypt and India by 1914. Russia's defeat by Japan in the war of 1904–5, almost the first time that a 'non-white' people had humbled a major European power, suggested that European imperialism might be challenged successfully.

It was, however, the three great episodes of the first half of the twentieth century – the First World War, the Great Depression (1929–33) and the Second World War – that gave conclusive evidence of European weakness and fatally undermined European imperialism. Nationalism, which had had such an impact on the political geography of Europe (see Chapter 20), had also influenced the colonial world where 'national liberation movements' began to demand independence for their countries. Starting with the independence of India in 1947, the next quarter of a century witnessed the break-up of all the European empires (though the Soviet Union, successor to the old Russian empire, finally disappeared only in 1991). Many new, independent states appeared on the map of Africa, Asia and other regions, though not, unfortunately, without considerable turmoil in some cases. The world's political geography was transformed.

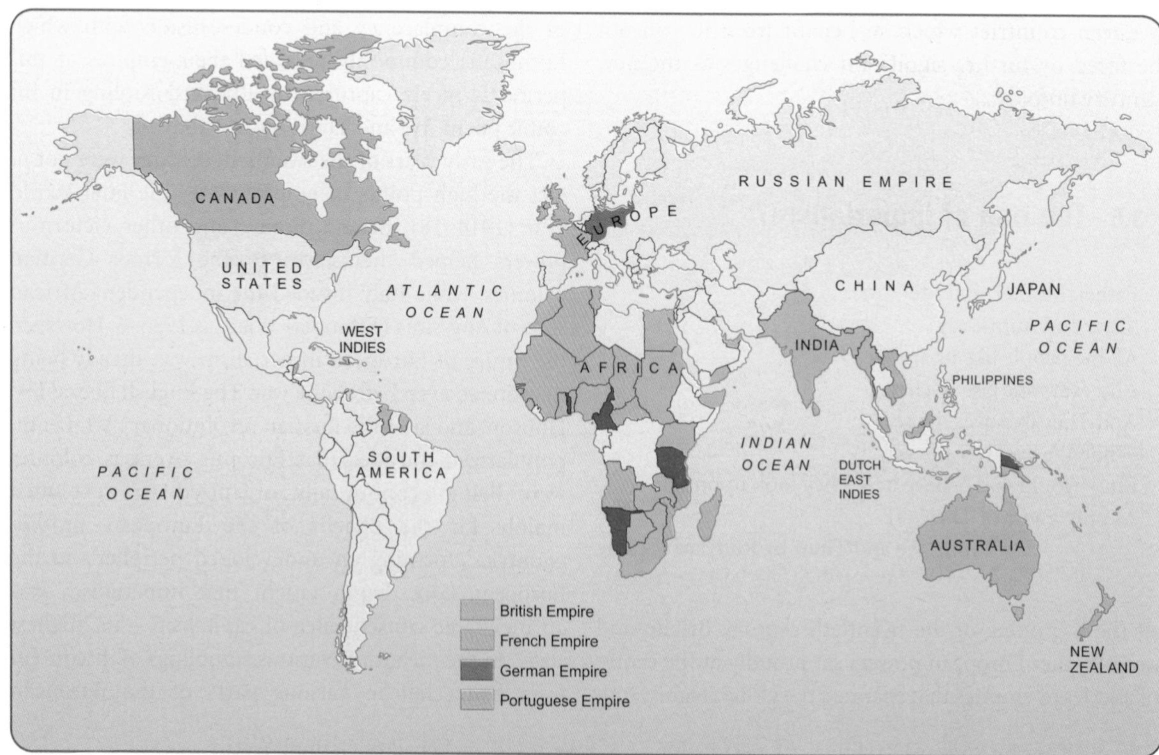

Figure 3.3 The world in 1914 showing British, French and German empires.

Regional Case Study 3.1

Ireland in the twentieth century

Ireland's history in the twentieth century illustrates many of the problems bequeathed by imperialism to newly independent countries.

Although the English had begun to penetrate Ireland as early as the twelfth century, they were never able to assimilate the territory into the English state. Differences between the two countries were exacerbated at the time of the Reformation when England and Scotland became Protestant but most Irish remained Catholic. Sporadic resistance to English/British rule was met by repression. In the nineteenth century the rise of Irish nationalism stimulated several unsuccessful attempts to secure autonomy (home rule). Only after the violence of the Easter Rising in Dublin in 1916 and revolution in 1919–22 did most of Ireland achieve dominion status under the British Commonwealth, becoming the Irish Free State in 1922.

Irish autonomy, however, did not appeal to the Protestants who formed a majority population in the north-eastern part of the island. Most of these people were descendants of settlers who had been deliberately introduced to the region by the English government in the seventeenth century in an attempt to subdue the Irish. Protestants, who had traditionally had a largely privileged status in Ireland, feared incorporation into a Catholic state and demanded continued union with the British Crown. As a result of various manoeuvres the northern Protestants were enabled to establish their own statelet of Northern Ireland as a continuing part of the United Kingdom in 1921. However, the six counties of Northern Ireland contained a substantial minority of Catholics, and the partition of Ireland was widely resented by nationalists on both sides of the new border.

After independence, governments of the Irish Free State became concerned to secure a greater measure of freedom from Britain and also to assert a post-colonial Irish national identity. Policies such as economic protectionism, attempts to revive the Irish language and Gaelic culture, and imposition of a Catholic moral code and censorship, were thus designed to build a new Ireland, free from the influence of the former imperial master. These policies culminated in Irish neutrality during the Second World War, and declaration of an Irish Republic and secession from the Commonwealth in 1949. However, this also had the effect of strengthening partition, as northern Protestants clung ever more tenaciously to the British connection whilst Northern Irish Catholics often suffered discrimination.

The violence and terrorism that partition has abetted has cast doubt on the extent to which it is a suitable answer to Irish problems. Fortunately recent developments on both sides of the border provide hope of better times in the future. Thus the Irish Republic now seems to be moving away from the narrow nationalism of the past and is evidently seeking a broader and more inclusive identity within the ambit of the European Union and the wider world. Meanwhile Protestant unionists and Catholic nationalists north of the border, divided and segregated for so long, have been building a long-term agreement.

(See also Case study 19.3, Spotlight box 20.1 and Foster 2001)

In giving (or being forced to give) independence to their colonies, the former imperial powers hoped that they would adopt European-type political systems and capitalist economic systems, partly because these appeared the best basis for future development, and partly because they seemed a reasonable way of upholding European influence. The elites who were now to hold power in the new states were often sympathetic to these aims, since they had frequently been educated by Europeans and wished to see their countries modernized along European or Western lines. However, the adoption of Western models was not always a suitable response to their problems. For example, many of the new states attempted to copy the European idea of the nation-state, hoping to unite their peoples around a common sense of national identity. But the old colonial boundaries had generally been drawn up to suit imperial convenience rather than that of local communities. Thus these boundaries, now the boundaries of independent states, frequently grouped peoples into one state who had no common culture or history while dividing others who did so. This contributed little to the political stability or unity of the new states. Another problem was the feeling among many citizens

Plate 3.8 Imperial Airways flies to Australia. Imperial Airways was the major British airline company which operated at ever more ambitious international and intercontinental scales during the 1930s.
(Vintage Ad Gallery)

of the new states that the capitalist economy was responsible for the underdevelopment of their countries. This set off a search for socialist or communist alternatives, much to the annoyance of the West, and encouraged debates about the meaning of 'development' (see Chapter 8, pp. 186–9). Finally, fundamental questions were frequently posed about how far modernization, as generally understood, was compatible with the traditions of the former colonial peoples. The Islamic revival, discussed above, can be seen as one response to this dilemma.

Just as imperialism had a profound impact on the political and social geographies of enormous areas in Africa, Asia and the Americas, it equally affected the imperial countries themselves. One of the most important manifestations of this in the twentieth century was the flow of migrants from colonies and former colonies, especially in the tropics and sub-tropics, to take up jobs in the former imperial states. This naturally had a far-reaching cultural impact in cities and regions in Western Europe. In fact the whole experience of imperialism led to the mixing of peoples and cultures on a grand scale. It also led to widespread questioning of long-held assumptions about European (and often male) cultural superiority (see Chapter 2, p. 49).

Many scholars have argued that the end of European colonialism did not mean the end of exploitative relationships between the core countries of the world economy (including the former imperial powers) and what was now increasingly referred to as the 'Third World'. According to such thinkers, the formal imperialism of the colonial era had merely been replaced by a more 'informal' alternative, but the basic situation of the core exercising hegemony over the periphery had not really changed (Frank 1969; Wallerstein 1980). From the 1970s, however, some fundamental changes seemed to affect the world economy which, in the opinion of certain scholars, demanded that international relationships be viewed in a new way. Lash and Urry have described these changes as 'the end of organized capitalism' (Lash and Urry 1987).

3.6 The disorganization of capitalism

In 1960 the industrialized areas of Western Europe and North America produced almost 80 per cent of the world's industrial output. Even Japan accounted for only around 4 per cent. Much of the Third World remained agricultural. Only after this time did industrialization spread beyond its traditional centres (which since the 1930s had included the Soviet Union). Meanwhile many of the older industrial countries began to lose industries, even some of those which had arrived with the twentieth century.

Of course, some areas of the 'Third World' benefited far more from this industrial spread than did others. Most spectacularly, the Newly Industrializing Countries of east Asia, Brazil, Mexico and certain others soon seemed set to join the industrial core. Yet others, like certain Middle Eastern states, earned huge revenues from their energy exports. But there remained many areas which missed out on the new developments (even so the latter part of the twentieth century was a period of unprecedented population growth and urbanization across much of the developing world – see Chapters 4 and 9). The term 'Third World', used to group together countries with such disparate economic characters, seemed increasingly redundant, and the world as a whole seemed as unequal as ever.

How is one to explain the changes affecting the world's economic geography in the last third of the twentieth century? Geographers and others argued that such changes are part of the process of globalization. Speedier communications meant that the world was becoming a much smaller place as the twentieth century drew towards its close. Capitalism itself was now a truly global phenomenon as markets were internationalized and finance became fully mobile. Before 1960, despite the importance of international trade, the world economy was structured around individual states. After 1960, the world economy became in effect transnational as the boundaries of individual states became ever less important to its functioning. Thus this period witnessed the rise to global importance of the transnational corporations, commercial conglomerates which became major players on a world scale. Because of the wealth and political influence they wielded, such huge companies became increasingly free to switch their operations from country to country as economic circumstances dictated. States, which had previously seemed unchallenged within their own frontiers, found it ever more difficult to control their own economies and began to bid against one another to attract footloose investment and the favours of the transnational corporations. Many industries (and not only industries) now began to locate in parts of the Third World, where costs were cheaper, whilst core industrial countries began to experience deindustrialization and a switch into services and 'control' functions (the headquarters of the transnational corporations still tended to be located in the traditional core countries). A further important result of the development of information technology was that production became much more flexible than before and more geared up to highly specialized markets, changing fashion and the whim of the individual consumer. This new, more flexible approach to production is sometimes known as **post-Fordism**.

Later chapters of this book will explore some of the economic and social implications of these changes. However, it is worth stressing here how unsettling such developments have been, especially for the core countries. The years of growing affluence after 1945 were succeeded, from the mid-1970s, by a period of greater uncertainty as deindustrialization set in, unemployment rose, insecurity became more widespread, and social inequality became more apparent. By the century's end it seemed that Western-style capitalism and democracy were increasingly threatened, both at international and national scales, by such phenomena and

Plate 3.9 Deindustrialization in the 1980s afflicted many parts of the world in the wake of the 1970s' oil price shocks and of major changes in the world economy.

(Paul Ridsdale/Photofusion)

Spotlight Box 3.3

Disorganized capitalism

- The onset of 'disorganized capitalism' is marked by a decline in the relative importance of extractive and manufacturing industries.
- There is a relative increase in the importance of service and consumer industries, especially in employment.
- The use of flexible technologies encourages a reduction in the average size of manufacturing plants with more accent on labour-saving investments and more flexible employment processes, all induced by competition.
- Because of the need for flexibility and cost-cutting, industrial firms tend to 'hive off' many of the services and supporting activities they need to other firms and organizations. There are thus more opportunities for small firms, changing the traditionally specialized nature of the regional economy.
- Regional economies are also affected by the greater emphasis on non-standardized production – traditional regional specializations become less marked.
- The global economy reduces the effectiveness of state attempts at economic regulation – from the state's point of view, the economy becomes less predictable.
- Rising costs, demands for reduced taxation, and growing social inequality challenge the idea of a centralized welfare state.
- Smaller, more footloose industries, the rise of services, better communications and other factors reduce the traditional importance of big, industrial cities by comparison with small towns and rural areas.
- There is a rise in importance of the educated social strata needed to work in the new administrative, control, service and related activities – the so-called 'service class' – with more sophisticated and individualized tastes in consumption and other areas. The age of mass cultural provision is replaced by greater cultural fragmentation and pluralism (sometimes referred to as post-modernism). There is a commensurate decline in faith in large-scale planning and similar activities associated with modernism.

Source: after Lash and Urry (1987)

also by international terrorism. Furthermore, ever more searching questions were being asked about global environmental threats and about capitalism's responsibility for climate change and similar problems. The world was no longer the self-confident, cosy, Eurocentric place it had been at the beginning of the century. Certain writers, notably Francis Fukuyama, celebrated liberal capitalism's victories over several of its twentieth-century rivals, particularly Fascism and communism (Fukuyama 1992). But this was a mood shared by few others. In the wake of 9/11, the world has never seemed more unstable nor the future more uncertain.

Just as Lash and Urry have used the term 'organized capitalism' to describe the years down to about 1973 when Fordism reached its apogee, so they have described the last years of the century as those of 'disorganized capitalism'. Some of its more prominent features are listed in Spotlight box 3.3, once again paying particular attention to how the changes have affected the core countries of the world economy.

3.7 Conclusion

The twentieth century was a period of rapid economic and social change over most parts of the world. It was also a period when the world seemed to become smaller and most regions were gradually drawn into an ever more embracing global system. Yet it would be a mistake to imagine that this was an uncontested process, or one that threatens to bring about a global uniformity. The legacy of the twentieth century is a world that is both dynamic and uneven, and therefore very uncertain. The rest of this book tries to grapple with this uncertainty.

Learning outcomes

Having read this chapter, you should know that:

- Capitalism is inherently dynamic and unstable. It has been so throughout the twentieth century, and is likely to continue to be so in the future.
- The concepts of 'organized' and 'disorganized' capitalism are ways of trying to make sense of the changes which affected capitalist societies during the twentieth century.
- 'Fordism' and 'post-Fordism' (concerning production and consumption) and 'modernism' and 'postmodernism' (concerning culture) similarly try to make sense of twentieth-century change. No concepts, however, can do justice to the complexity of change during this period.
- Western-style liberal capitalism is only one variant of capitalism. It has been challenged in various ways in the twentieth century, some of which have had long-term consequences for different parts of the globe. Future challenges might prove more successful than past ones.
- The twentieth century has been an era of nation-states and of nationalism. Towards the end of the twentieth century the role of the nation-state seemed increasingly challenged by globalization. Nationalism, however, may yet flourish as a response to globalization.
- Modernity may not lead to the disappearance of traditional cultural practices. The Islamic revival, and similar religious revivals across the world, are a case in point.
- Despite the view of some that globalization is leading to the emergence of a global culture, it might actually increase the differences between places and hence the importance of geography.

Further reading

Fieldhouse, D.K. (1999) *The West and the Third World*, Blackwell, Oxford. Provides a comprehensive survey of the relationship between the West and the Third World, and the debate over its effects, during the twentieth century. Combines theoretical discussion with empirical evidence.

Godlewska, A. and Smith, N. (eds) (1994) *Geography and Empire*, Blackwell, Oxford. Discussions of the interplay between the rise of geography as an intellectual discipline and the development of the European empires culminating in the decolonization of the post-war period and the post-colonial experience.

Gwynne, R.N, Klak, T. and Shaw, D.J.B. (2003) *Alternative Capitalisms: Geographies of Emerging Regions*, Arnold, London. Examines the effects of globalization and recent economic and political transformations in the world's 'emerging regions', with particular reference to Latin America and the Caribbean, East Central Europe and the former Soviet Union, and East Asia.

Hall, P. (2002) *Cities of Tomorrow*, 3rd edition, Blackwell, Oxford. A splendid account of urban development in the late nineteenth and twentieth centuries in different parts of the world. The accent is on urban planning and design, but there are many social insights.

Hobsbawm, E. (1995) *Age of Extremes: The Short Twentieth Century 1914–1991*, Abacus, London. What quickly became a classic account of twentieth-century history, written by a doyen of British Marxist historians. Any chapter is worth reading, but geography students will find those dealing with economic, social and cultural change especially revealing.

Knox, P. and Agnew, J. (2003) *The Geography of the World Economy*, 4th edition, Edward Arnold, London. A survey of geographical change in the world economy which includes an account of twentieth-century developments.

Useful websites

www.si.edu/ The Smithsonian Institution, Washington, DC. The Smithsonian is a focus for many kinds of scientific and cultural endeavour in the United States. The website is a very useful source for twentieth-century history and developments, with particular emphasis on the United States.

www.besthistorysites.net/ Entitled 'Best of History Websites', a comprehensive guide to history-oriented resources online. For teachers, students and others.

For annotated, clickable weblinks and useful tutorials full of practical advice on how to improve your study skills, visit this book's website at **www.pearsoned.co.uk/daniels**

POPULATION, RESOURCES, FOOD, THE ENVIRONMENT AND DEVELOPMENT

Section 2

Edited by Michael Bradshaw

This section encapsulates the major challenges that face the world towards the end of the first decade of the twenty-first century, namely the need to strike a balance between population growth, resource consumption, the drive for economic development and sustaining the planet's ecosystem. We are now coming to realise that the processes of economic development that dominated the twentieth century are no longer sustainable, economically or ecologically. Somehow we have to find a way of improving standards of living and reducing levels of inequality without increasing the stress we place on the environment. If the last quarter of the twentieth century was dominated by the rhetoric of sustainable development, we now have to make the difficult choices needed to turn that rhetoric into reality.

This section makes clear both the scale and the complexity of delivering a more sustainable approach to development. The chapters highlight the diversity of problems that exist at the global, interstate and local level. In this era of heightened globalization, developments in one locale, region or state can have global impact. For example, the so-called 'developed' world has achieved high levels of consumption, but has exhausted indigenous sources of energy and food supply, making consumers increasingly reliant upon global chains of supply. The same regions now face a shortage of skilled workers and an increasing elderly population. Elsewhere in the world there are regions and states that suffer from an absence of economic development and an inability to compete in the global marketplace for much-needed resources: a problem often compounded by the fact that they have to export food and natural resources to generate income to support the trappings of statehood. The latter often leads to the misappropriation of resource rents and conflict. While the chapters highlight differences, they also stress the interconnected nature of the processes that generate global problems. The emergence of new centres of growth, such as India and China, highlight the dynamic nature of the global system and the fact that new sources of demand for resources are placing additional stress on the planet's ability to support global capitalism. Perhaps the only solution is to change the relationship between environment, society and development?

DEMOGRAPHIC TRANSFORMATIONS

Chapter 4

John Round

Topics covered

- The geographies of population growth
- Changes in life expectancy and fertility
- The impact of disease and economic change on demographic futures
- Population ageing and the role of migration in ageing societies

4.1 Introduction

Demographic transformations influence almost every topic discussed in this book. Changes in population structure impact on the environment, influence economic development, alter government strategies, stimulate migration and, among many other examples, can lead to social upheaval. While it is clear that the world's population is growing rapidly, within this growth there are many geographies for this chapter to explore. To begin we first look at global population growth over the past few centuries, highlighting how the world's population has risen rapidly since the end of the Second World War and the variations in where this has occurred. After detailing some future population predictions, and showing how tiny variations in **birth/death rates** can alter the projections, the chapter then investigates case study examples where factors such as disease, economic upheaval and, conversely, economic stability have rapidly altered population structures. The illustrations include the impact of HIV/AIDS on sub-Saharan Africa, Russia's post-Soviet demographic crisis and Western Europe's ageing population and the subsequent effects of inwards migration. These examples have seen demographers reassess the traditional **demographic transition model**, which suggested that societies reach a plateau where death rates are slightly below birth rates ensuring gradual population growth, highlighting how difficult it is to predict future population growth rates. Of course, demography is not just concerned with overall population numbers. Life expectancy, family size, fertility levels, migration and ageing are some of the other demographic challenges that countries must face. Such issues are explored through various case studies but, by way of introduction, we examine the geographies of population growth over the last two millennia.

4.2 Geographies of population growth

It is impossible to state accurately what the world's population is. Although most countries conduct regular population censuses these are not an exact science. The data generated by these surveys might be unreliable because, for example, statistical methods prove to be imperfect, households might wish to hide the number of people living there for tax reasons, or governments might inflate numbers for political reasons. For example, the 2001 census in Great Britain showed that there were 800,000 fewer young men living in the country than was previously thought (www.statistics.gov.uk). Initially migration was thought to be the cause but eventually 190,000 men were added to the total figure. Such variations have serious implications for government policy. Many local authorities argued that the census underestimated their populations, which would lead to them losing government funding. Nottingham city council argued that the census failed to include 16,000 people, which would cost them over seven million pounds in funding per year (www.news.bbc.co.uk). The Russian census of 2002 declared that almost 1.1 million people lived in the region of Chechnya. Many observers argued that after over a decade of war and outward migration the Russian government was inflating the figure to try and show that life was returning to normal there. The Danish Refugee Council, having surveyed the region in the same month as the official census, suggested the figure was nearer to 700,000 (www.drc.dk). Similarly, it was more symbolic than based on fact that a baby born in Sarajevo in October 1999 was declared by the United Nations as the sixth billionth person on earth. What is irrefutable, however, is that over the past fifty years a population explosion has taken place. It is estimated that at the start of the twentieth century there were approximately 1.3 billion people on earth; by the century's end the world's total population had grown by approximately 4.7 billion people. Even more amazingly, the vast majority of this increase has taken place in the past fifty years as the estimated population in 1950 was just over two billion (see Figure 4.1)

Between 1900 and 1920 the world's average population growth rate was approximately 0.5 per cent per year. This growth rate doubled during the 1920s, remaining constant until the end of the 1940s. During the 1950s the rate rose rapidly to 1.8 per cent, rising again to 2 per cent the following decade. Obviously such increases are compound (the yearly increases are in addition to previous growth), meaning that a growth rate of 2 per cent allows the world's population to double every 35 years. From this peak, growth rates fell to 1.7 per cent in the 1980s to the current average of approximately 1.2 per cent increase per year. This, estimates the United Nations population division (www.un.org/popin), sees an annual population growth of 77 million people per year down from a peak of 83 million a year in the 1980s.

One of the most challenging problems that demographers face is trying to project what future population

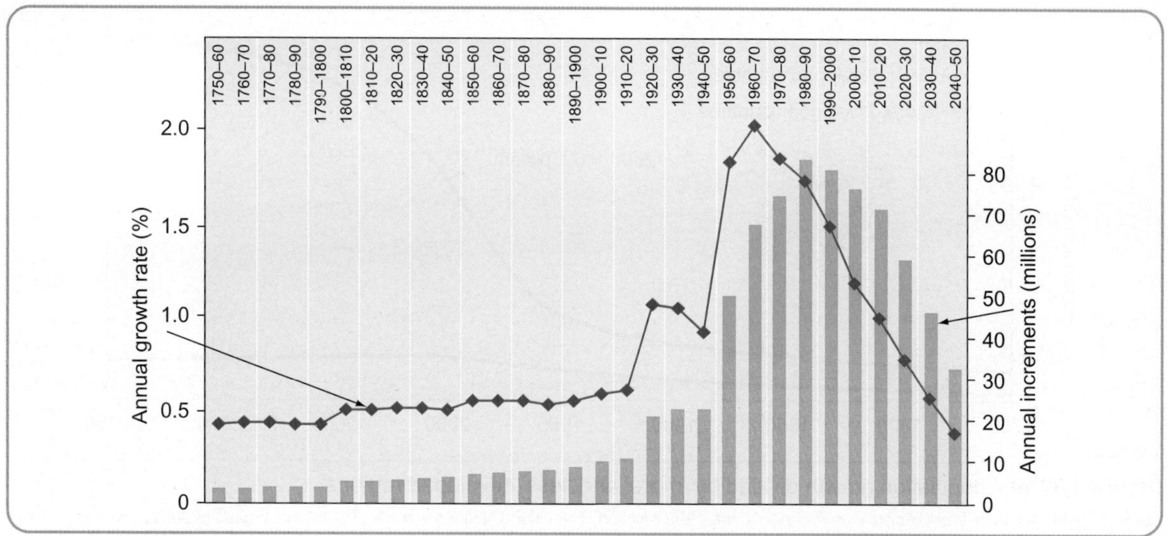

Figure 4.1 World population growth, 1750–2050.

Source: UN (2003) and author's calculations. Data from 2010 onwards are based on medium-variant projections

growth rates will be. It is now expected that growth rates will continue to decline, with the United Nations predicting a fall below 1 per cent by around 2020, continuing to decline to 0.3 per cent by 2050 (which will be the lowest levels since the mid-eighteenth century). However, given the current population expansion, even these more modest growth rates will see the world's population grow by over 35 million a year until 2050. This would then give an overall population of 8.9 billion, 3 billion more than present. However, given the compound nature of population growth, as we shall see below, it takes only small changes in fertility/death rates to produce large changes in population totals over a long period. Thus, as Figure 4.2 shows, the United Nations produces a range of predictions based on different **fertility rates** for 2050, ranging from 7.3 billion, and already declining, to 10.7 billion.

While the world's population is growing rapidly this overall figure tells us little of the geographies that exist within it. The most striking aspect is how population size in the '**more developed countries**' has remained relatively stable since the late 1950s, at around 1 billion. In contrast the total population of the '**less developed countries**' has increased rapidly from approximately 1 billion to 5 billion since 1950. It must be noted here that terms such as 'more developed' or 'less developed' are extremely problematic, as is discussed in Chapter 8, but they are used reservedly here as they are the terms employed by the United Nations when discussing demographic issues. This difference in growth rates is shown starkly by Figure 4.3.

Even within the category of 'less developed' there are still further variations, the reasons for which will become more apparent below. The 49 least developed

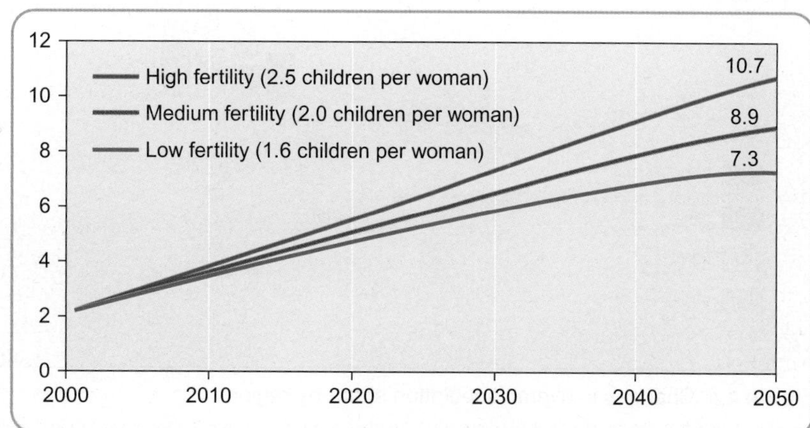

Figure 4.2 How changes in fertility rates might impact upon future population growth.

Source: United Nations, *World Population Prospects*, the 1998 revision. The United Nations is the author of the original material.

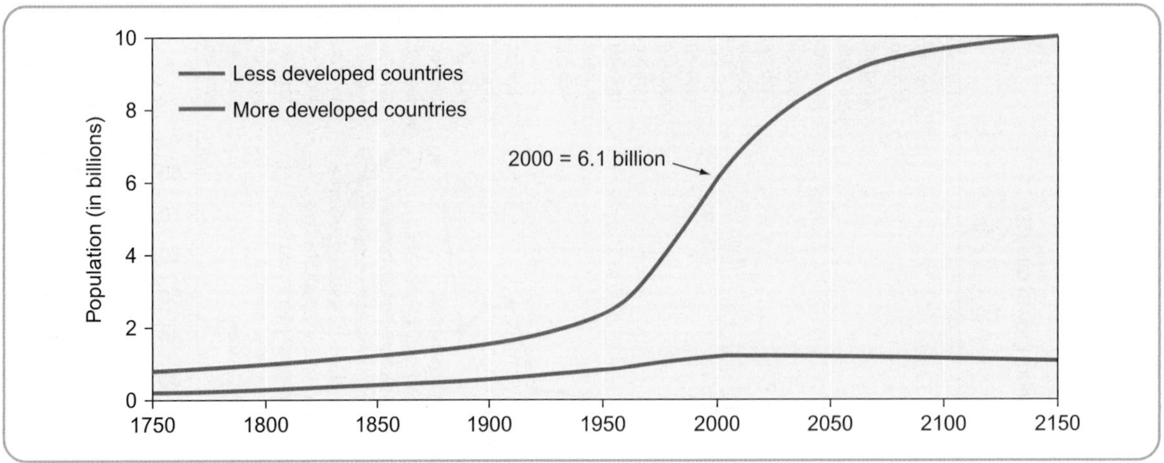

Figure 4.3 Total population growth divided by 'more' and 'less' developed regions.

Source: United Nations, *World Population Prospects*, the 1998 revision. The United Nations is the author of the original material.

countries, according to UN data, experienced the greatest increases in population growth. This group only held 8 per cent of the world's population in 1950 but in the subsequent population explosion they have contributed around 14 per cent of the growth. Such differences in population growth rates has, unsurprisingly, altered the relative distribution of the world's population. For example, the 'more developed' countries in 1950 combined to provide approximately a third of the world's population; by 2003 this had fallen to under a fifth. As Figure 4.4 shows, this has resulted in significant changes in overall population share by region.

As the map demonstrates, Europe's share of the world's population fell from approximately a quarter to around an eighth between 1900 and 2000, with Africa the biggest gainer in percentage terms. What is most significant here is how quickly these changes have taken place with most of the changes taking place since 1950. Table 4.1 details these changes in numerical terms.

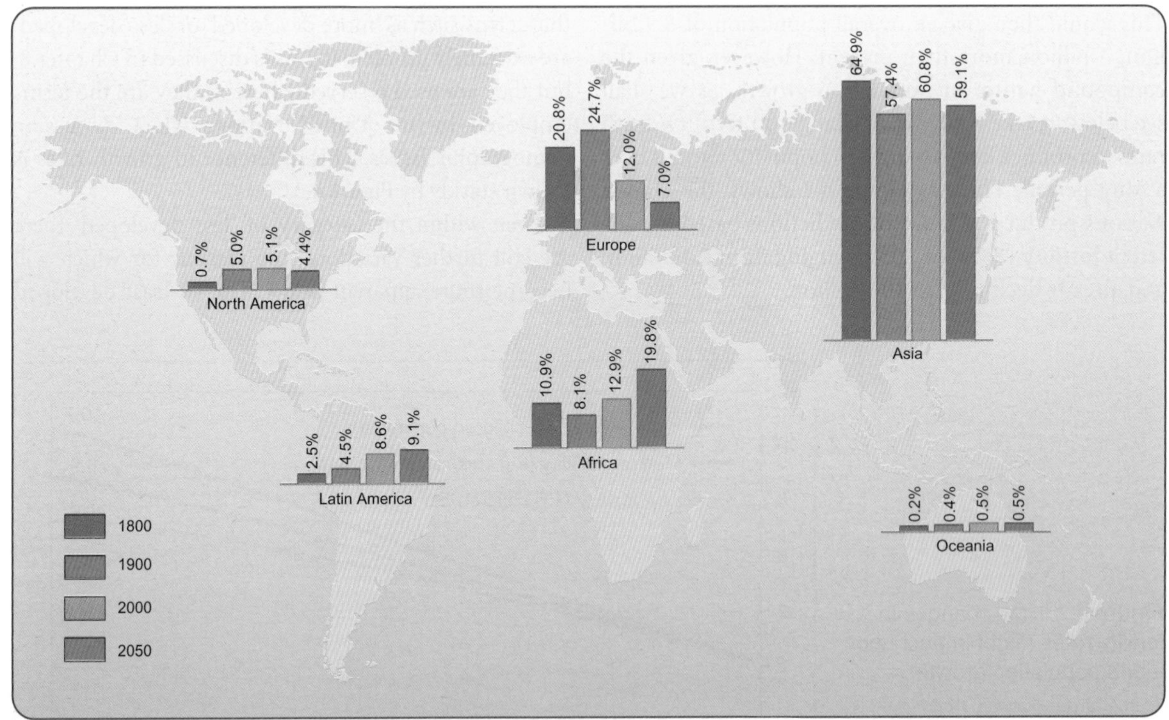

Figure 4.4 Changes in overall population share by region.

Source: United Nations Population Division (1998). The United Nations is the author of the original material.

Table 4.1 Population growth, 1950–2050, for development categories and major areas

Category and major area	Population (millions)			Share of world population (%)			Change (millions)		Share of world change (%)	
	1950	2003	2050	1950	2003	2050	1950–2003	2003–2050	1950–2003	2003–2050
World	2519	6302	8919	100.0	100.0	100.0	3783	2617	100.0	100.0
MDR	813	1203	1220	32.3	19.1	13.7	391	16	10.3	0.6
LDR	1706	5098	7699	67.7	80.9	86.3	3392	2601	89.7	99.4
LDC (within LDR)	*200*	*718*	*1675*	*8.0*	*11.4*	*18.8*	*518*	*956*	*13.7*	*36.5*
Africa	221	851	1803	8.8	13.5	20.2	629	953	16.6	36.4
Asia	1399	3823	5222	55.5	60.7	58.6	2425	1399	64.1	53.4
Europe	547	728	632	21.7	11.5	7.1	180	−96	4.8	−3.7
L. America	167	528	768	6.6	8.4	8.6	361	240	9.5	9.2
N. America	172	319	448	6.8	5.1	5.0	148	129	3.9	4.9
Oceania	13	31	46	0.5	0.5	0.5	19	14	0.5	0.6

Note: data for 2050 are medium-variant projections. MDR More Developed Regions, LDR Less Developed Regions, LDC Least Developed Countries (subset of LDR), L. America Latin America and the Caribbean, N. America Northern America.

Source: UN (2003) and author's calculations

By 2050 Europe's share of the world's population is likely to have fallen to 7 per cent, while Africa's is expected to be nearly three times this. Over the period 2003–2050 Africa's population is projected to grow by just under 1 billion, representing over one-third of global growth (36.4 per cent) compared to only one-sixth (16.6 per cent) for 1950–2003. Meanwhile, Asia seems destined to remain the main contributor in absolute terms, adding a further 1.4 billion up to 2050, but this is a marked reduction in the pace of growth compared to its 2.4 billion gain between 1950 and 2003.

Rapid changes can also be seen at a country level. In 1950 only 8 countries, using current boundaries, had a population of over 50 million. By 2006 this figure has grown to 22. China has seen its population rise from approximately 550 million in 1950 to over 1.3 billion in 2001, whereas Europe's population has risen from around the same figure to 735 million in the same period. This is despite attempts to dampen population growth in China through government policies such as the 'one child per family' campaign. This policy made it very difficult for families to have more than one child as any extra would incur heavy education, health care and taxation costs. Since the mid-1990s, birth rates have fallen steadily but they are still more than double death rates, producing a significant growth rate. Given the compound effects of previous population growth, this will mean that China's population will continue to grow for the foreseeable future. While it is currently the world's most populated country, the declining growth rate means that it is likely that India will overtake it in the next forty years. In its 2001 census India was said to have a population of just over a billion people. This is a staggering rise of over 21 per cent from the 1991 census. The Indian government predicts that by 2025 the population will rise to 1.4 billion, a further increase of 36 percent. It is estimated the total population will increase by a further 400 million by 2050.

4.3 Geographies of changing birth and death rates

To explain changes in population growth rates demographers often use the demographic transition model. As Figure 4.5 shows, this model has three stages to differentiate between the stages which most countries have passed through in their demographic history (see Ronald, 2003, for an overview of its development).

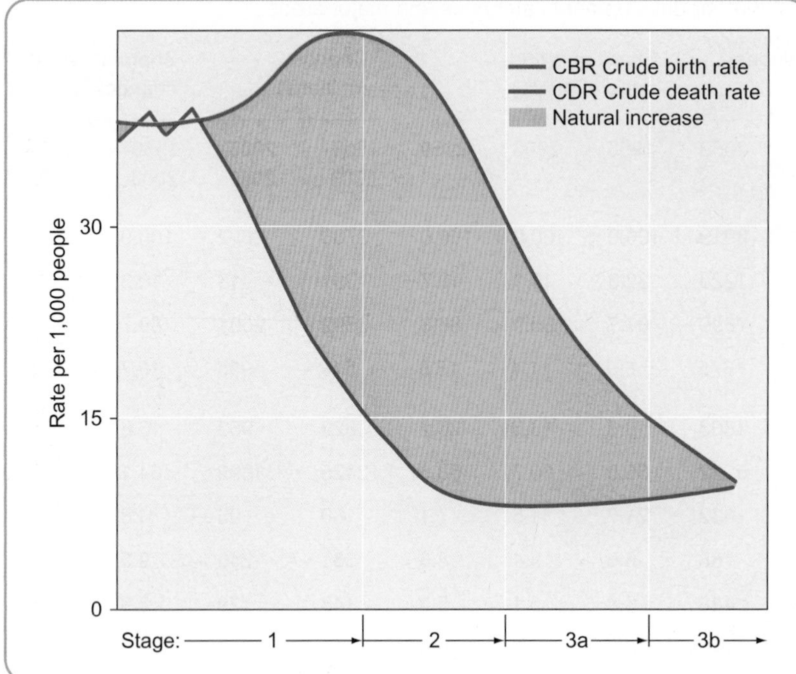

Figure 4.5 The demographic transition model.

Source: Modified from Chung, R., Space–time diffusion of the demographic transition model: the twentieth-century patterns, in Demko, G.J., Rose, H.M. and Schnell, G.A. (eds) *Population Geography: A Reader* © 1970 The McGraw-Hill Companies Inc. Reprinted by permission of The McGraw-Hill Companies Inc.

Stage one was a period characterised by rudimentary health care, with death from what are now easily treatable diseases commonplace, leading to very high crude death rates (CDR – deaths per 1000 people). Birth rates (CBR – births per 1000 people) were also high as large families were the norm, partly to ensure that parents would have support when they could no longer work. With birth and death rates relatively equal, overall population size does not increase dramatically during this stage.

As health care and sanitation systems develop, death rates fall as fewer people die from easily preventable diseases and countries move into stage two of the model. However, there is usually a period during which birth rates remain high. This might be for cultural reasons or because economic growth sees an increasing demand for family labour. This gap between death and birth rates sees an increasing population. As death rates can fall rapidly this can often cause a rapid rise in population growth. Figure 4.6 demonstrates how this occurs.

From these figures we can see that in less developed countries, from the middle of the previous century death rates fell dramatically as health care began to improve. However, birth rates, although declining, did not converge with death rates, resulting in a rapid increase in the rate of natural population growth. More developed countries have entered the third stage of the demographic transition model where birth and death rates converge. Birth rates have fallen rapidly for a combination of reasons:

- Increased economic prosperity and the development of pension systems means that many parents no longer have to rely on their children for support during old age.
- Increased urbanization means that families do not have the space to have many children.
- Increased social and economic choices for women include improving opportunities in the workplace, much greater levels of participation in higher education than in previous generations and the availability of reliable contraceptives.
- Cultural attitudes towards large families have changed so that rather than being the norm, having many children is seen as unusual and an increasing number of women are deciding not to have children at all.
- Increasing levels of divorce and single-parent families means that large families are less likely.
- A rapid increase in the average age of first birth (see Figure 4.7). If the first birth is delayed, then a large family is less likely.

All of these factors have combined to bring about a dramatic fall in fertility rates, as shown in Figure 4.8. As the graph demonstrates, the **EU** average fertility rate is

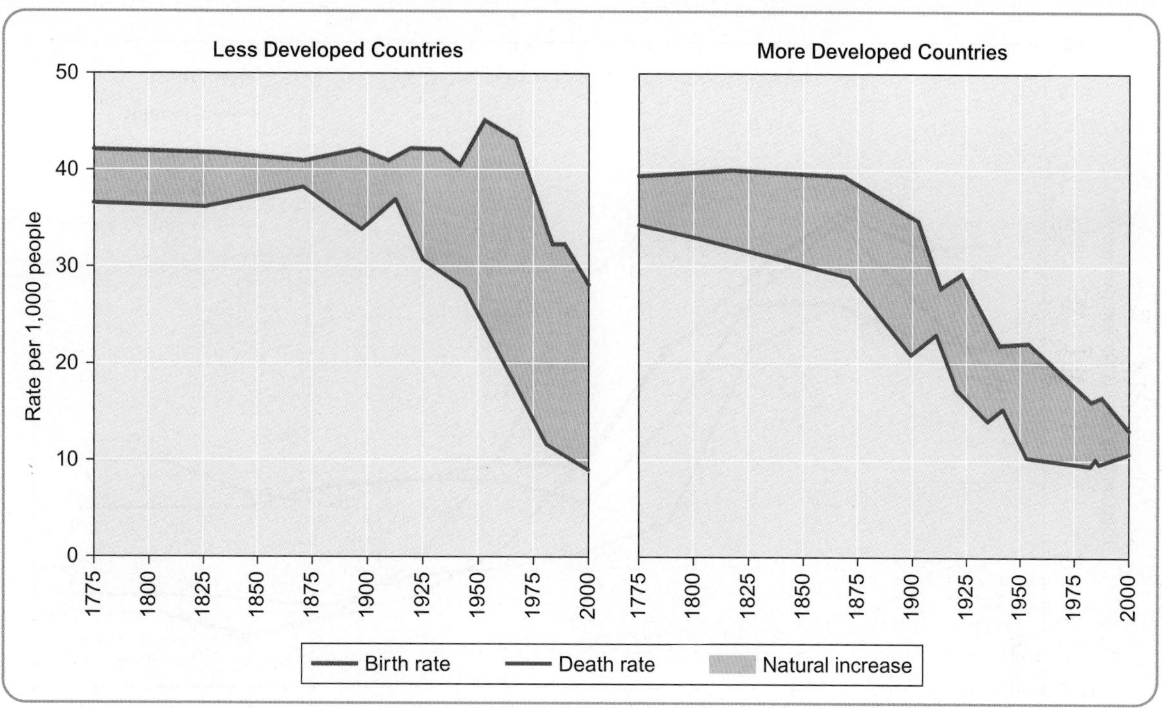

Figure 4.6 How changes to birth and death rates can lead to changes in natural population growth.

Source: Population Reference Bureau (2007)

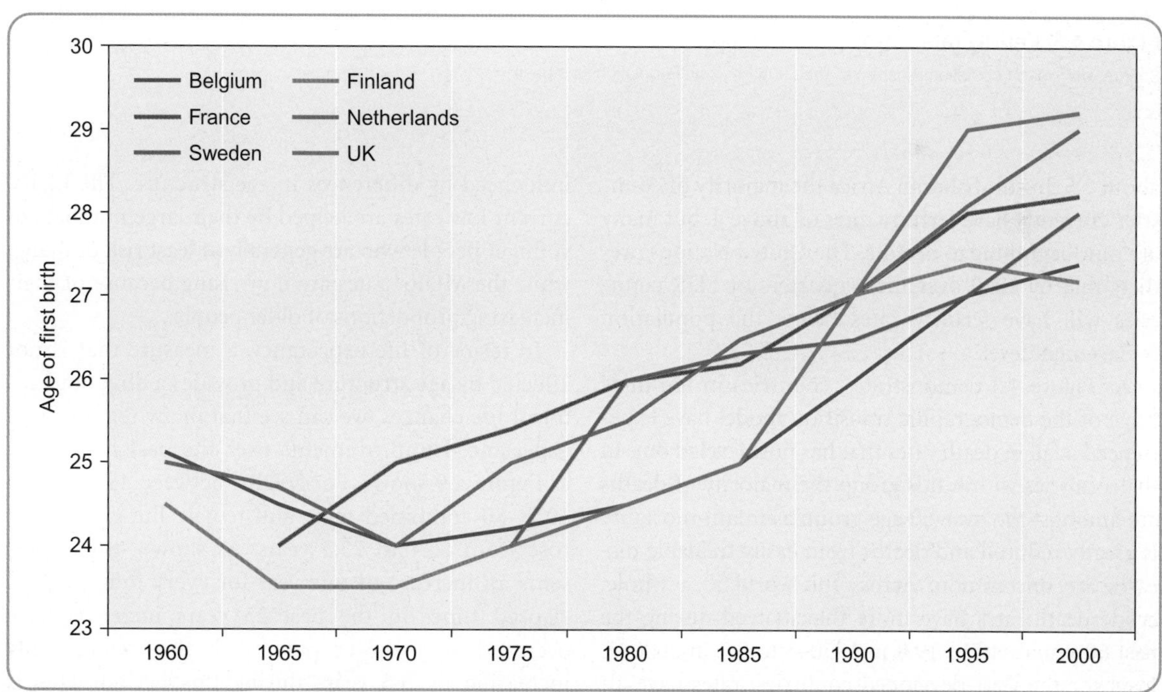

Figure 4.7 Increase in age of first birth.

Source: data from Eurostat.org

now well below the population replacement level of 2.1. This figure is the minimum needed for a population not to decline in over the long term. While fertility rates have fallen across the world, in many regions they are still well above the population replacement level. In the 1960s fertility rates averaged around 6, but since then it has fallen to below half that figure. In Asia and South America the average fertility rates are currently

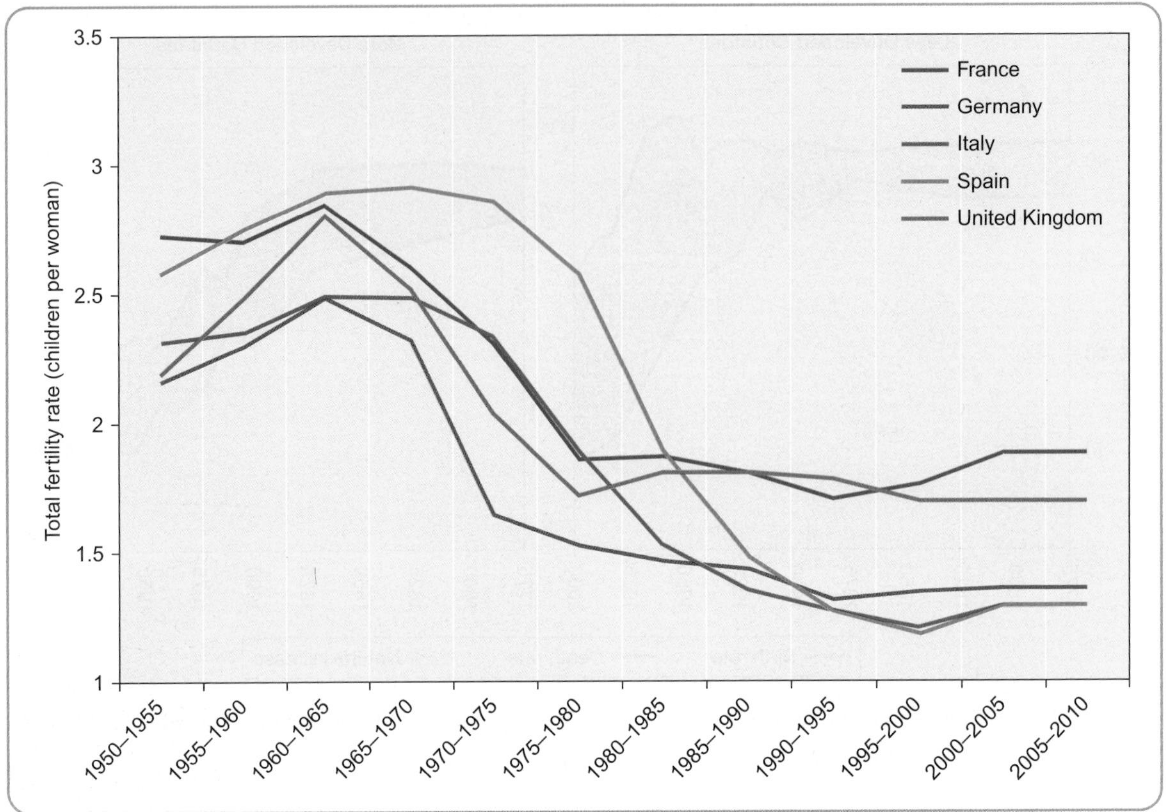

Figure 4.8 Fertility rates in Western Europe.

Source: UN World Population Prospects: The 2006 Revision Population Database

about 2.5. In sub-Saharan Africa the majority of countries currently have fertility rates of above 4, but many are now beginning to decline. The United Nations predicts that by 2050 over three-quarters of LDR countries will have fertility rates below the population replacement level.

As Figure 4.5 demonstrates, countries in the third stage of the demographic transition model have experienced a fall in death rates that has now leveled out. In the countries within this group the majority of deaths are amongst the over-60 age group as infant mortality is greatly reduced and deaths from easily treatable diseases are uncommon. Across the world as a whole, crude death rates have more than halved during the past 50 years – from 19.8 in 1950–5 to 9.1 in 2000–5. Even for the least developed countries, rates have almost halved, dropping from 27.9 to 15.1. The greatest change has occurred in the more developed parts of the LDRs, because the LDRs in aggregate have seen their death rate plummet by two-thirds, ending up at only 8.8 per 1,000, lower than the average for the MDRs. These trends in crude death rate are, however, partly influenced by differences in age structure. The LDRs' current low rates are helped by their large numbers of younger people who are generally at least risk of dying, while the MDRs' rates are now rising because of their increasing proportions of older people.

In terms of life expectancy, a measure that is not affected by age structure and provides a direct indicator of life chances, we can see that many regions have experienced improvements over the past fifty years, as Figure 4.9 shows. For MDRs, between 1950–5 and 1975–80 combined male and female life expectancy rose from 66.1 to 72.3 years. As shown, this represents an increase of one year for every four years of elapsed time. In the past 25 years increases have occurred at a slower pace, with the average only increasing by 3.5 years during this period. This is partly explained by the decline in life expectancy in former communist countries, the reasons for which are explored below, where men can expect to live for only 64 years compared to 74 years in Western Europe. In LDRs, life expectancy has increased rapidly but again the rate of growth is slowing. Here, in the 25 years

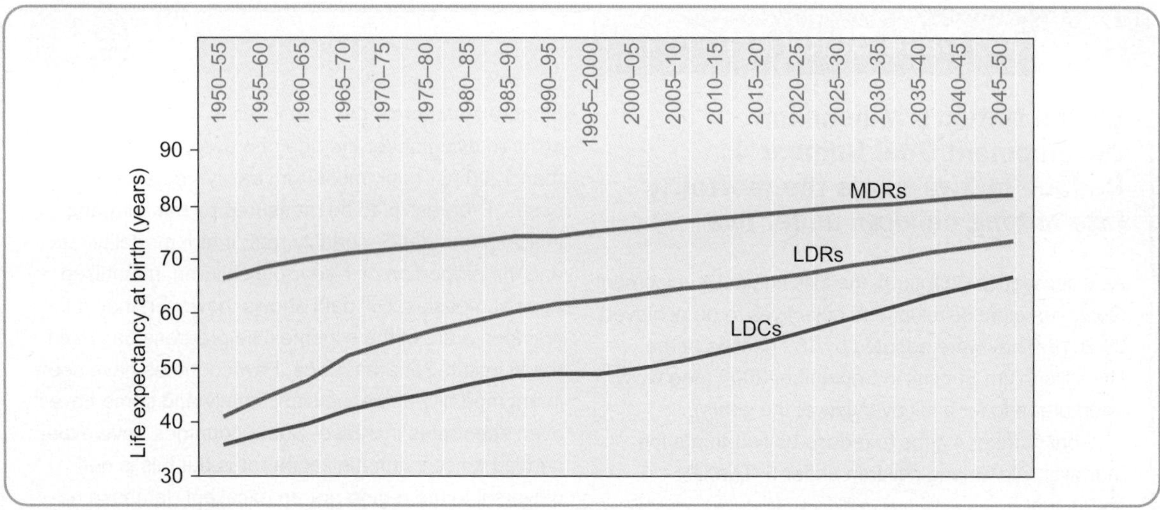

Figure 4.9 Life expectancy at birth (both sexes) 1950–2050 by development category.

Source: Data from UN (2003). Note: data from 2000 are projections. Least Developed Countries are a subset of Less Developed Regions

from 1950, life expectancy rocketed from 41 to almost 57 years, but over the next quarter of a century average life expectancy advanced by a further 7 years only. The earlier progress can be attributed to the rapid reduction in malaria, principally because of the use of techniques developed to protect soldiers fighting in south-east Asia in the early 1940s and to sustained programmes of curbing infectious and water-borne diseases like smallpox, tuberculosis and cholera. Within the space of 20 years, thanks to the importing of mortality control measures from the MDRs, many countries were able to pass through an **epidemiological transition** that had taken over a century for many European countries. Even so, life expectancy remains stubbornly low in the least developed countries in general and in sub-Saharan Africa in

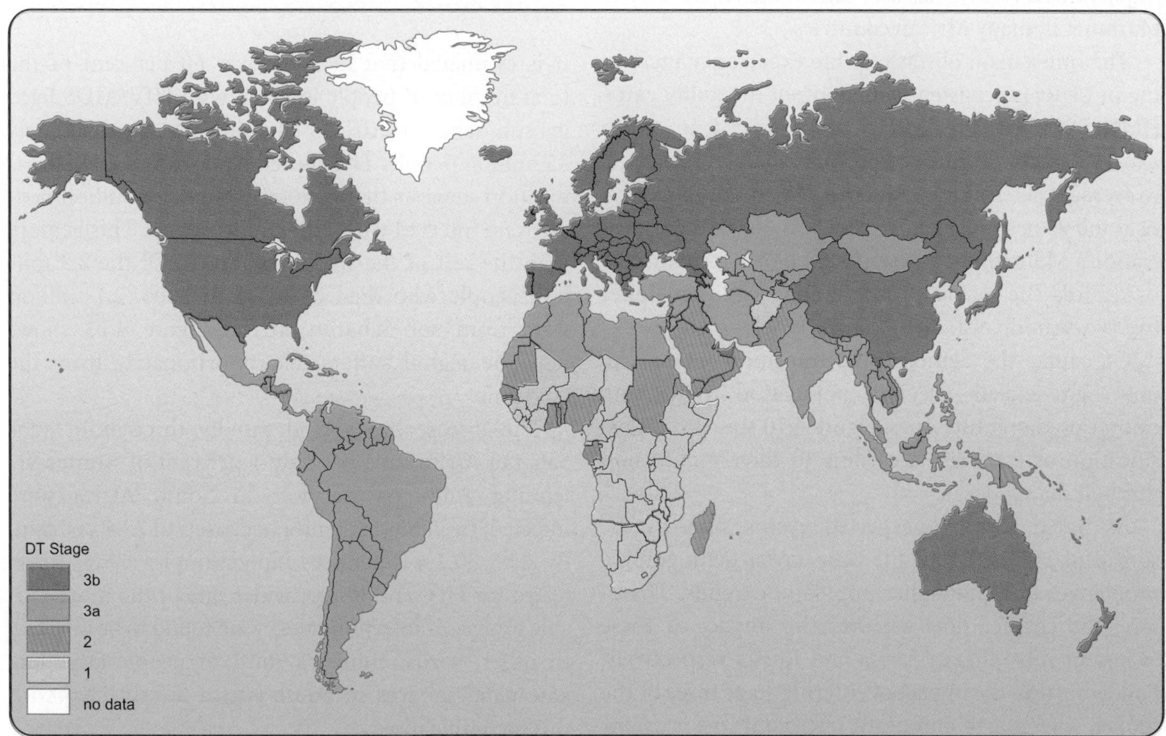

Figure 4.10 Countries classified on the basis of stage reached in the demographic transition by 2000–5.

Source: Data from UN (2003)

Thematic Case Study 4.1

United Nation's Millennium Development Goal Number 4: Reduce by two-thirds the mortality rate among children under five

As discussed in Chapter 8, the Millennium Development Goals are eight development objectives to be achieved by 2015. They were adopted by 189 nations at the UN Millennium Summit in September 2000 (see www.undp.org/mdg for a full overview of the goals).

Goal number 4 aims to reduce by two-thirds the mortality rate among children under 5. The UN states that:

> One of the darkest characteristics of poverty is that it seems to prey on the vulnerable and defenceless. In low-income countries, one out of every 10 children dies before the age of five. In wealthier nations, this number is only one out of 143.

In more stark terms this translates to 11 million children dying under the age of 5 every year, more than 1200 per hour, most from easily treatable diseases. Progress is to be measured by the following indicators: under-5 mortality rate; infant mortality rate and the proportion of 1-year-old children immunised against measles. UN data shows, however, that in the first four years of the scheme little progress has been made in sub-Saharan Africa. Few countries have seen infant mortality decrease significantly and some have even seen rates increase. Some countries have experienced a rise in immunization rates but this is not universal to the region (for an excellent database on Millennium Development Goal data go to www.un.org/millenniumgoals). Without rapid increases in funding to prevent easily preventable child deaths the United Nations estimates that in some parts of sub-Saharan Africa the goals will not be reached until 2115.

particular. This can often be attributed to armed conflict, such as the recent genocide in Rwanda, and the impact of HIV/AIDs has also caused life expectancy to plummet in many African countries.

The other main obstacle to life expectancy increasing in LDRs is persistently high **infant mortality rates** (IMR). Whereas the IMR is now under 8 in North America, most of Europe and Australia/New Zealand, it averages 32 in Latin America, 68 in south-central Asia and 95 in sub-Saharan Africa. It is one of the United Nation's Millennium Development Goals to reduce by two-thirds the mortality rate among children under five (www.undp.org). See Case study 4.1.

Of course the demographic transition model can only help explain previous population trends and cannot predict future ones. Figure 4.10 shows the classification of countries according to their progression through the model.

As noted above, unexpected events such as the spread of HIV/AIDS or the post-Soviet demographic problems can rapidly alter population trends. Therefore, the chapter now explores the impact of these events in sub-Saharan Africa and Russia respectively. Following this, the impact of entering stage three of the model, where birth and death rates converge is examined through a case study of the problems facing an ageing Western Europe.

4.4 The geographies of HIV/AIDS in sub-Saharan Africa

It is estimated that in 2006 over 60 per cent of the total number of people infected with HIV/AIDS lived in sub-Saharan Africa, equating to approximately 25 million people. The United Nations UNAIDS organization suggests that in 2006 around 2.8 million people were infected with HIV in the region. This is more than the rest of the world combined. Of the 2.9 million people who died of AIDS in 2006, 2.1 million were from sub-Saharan Africa. Figure 4.11 shows how the region suffers disproportionately from the epidemic.

The disease has spread rapidly throughout sub-Saharan Africa. In 1990 only 1 per cent of women attending ante-natal services in South Africa were infected; by 2000 this figure had risen to 22.4 per cent. By 2005, 30.2% (attendees' blood samples are routinely tested for HIV and this provides one of the more reliable sources of infection rates) were found to be infected. In other words, almost a third of people attending ante-natal services in South Africa are infected with this incurable disease.

Why are infection rates so high in the region when compared to the rest of the world? There are many

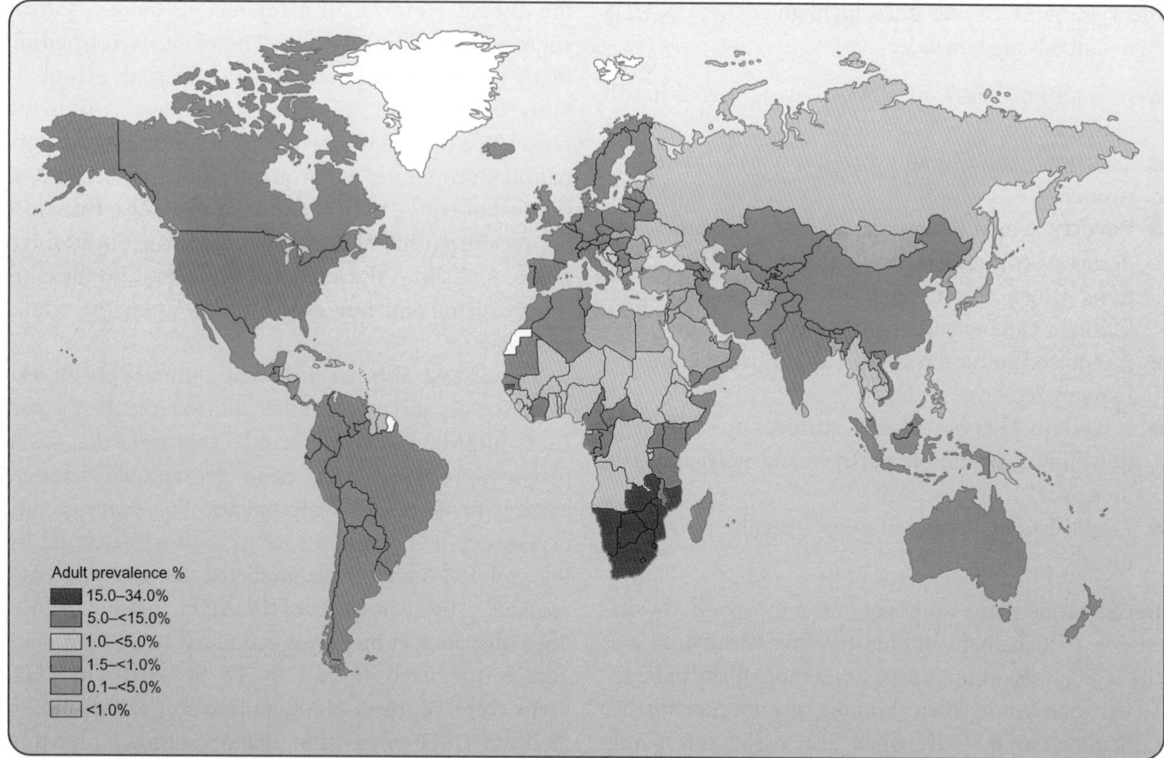

Figure 4.11 Geographies of HIV infection.
Source: UNAIDS (2006)

Thematic Case Study 4.2

HIV/AIDS

HIV (Human Immunodeficiency Virus) is a virus that can develop into AIDS (Acquired Immunodeficiency Syndrome). This syndrome leads to the failure of the immune system making minor infections life-threatening. There is no cure for this disease. The virus is most commonly passed on by sexual contact and the sharing of infected needles and can also be transmitted during pregnancy and through breastfeeding. While the scale of the pandemic is clear the origins of HIV/AIDS are not easily identifiable. There are many competing theories as to why and where the disease began but it first came to the attention of the medical profession in the early 1980s when groups of patients were suffering from diseases rarely seen within their age group and which were extremely resistant to treatment (see www.avert.org or www.unaids.org for further details). By the mid-1980s the disease was named, tests became available and public awareness campaigns on how to avoid infection began. Despite this the number of people becoming infected grew rapidly. One of the main causes of this is the 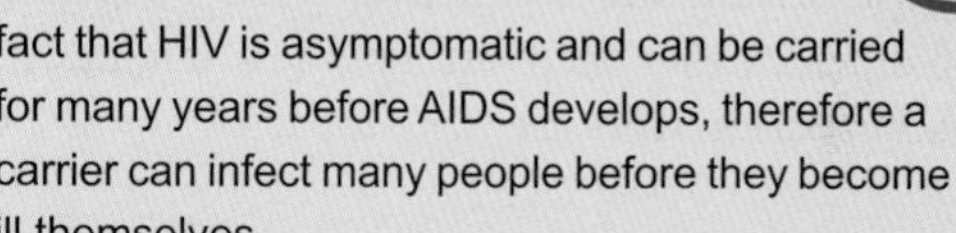fact that HIV is asymptomatic and can be carried for many years before AIDS develops, therefore a carrier can infect many people before they become ill themselves.

While there is no cure, drugs, known as antiretrovirals, can slow down the progression of the disease. This can greatly improve the health and life expectancy of people living with HIV/AIDS and they can block the transmission of the virus from mother to child. They are not a cure and if treatment is missed or patients develop resistance to the drug then the disease reverts to its normal course. The high cost of the drugs means that many people in low and middle income countries are not able to access them. In late 2006, UNAIDS estimated that out of over 7 million people living with HIV in these countries less than 2 million had access to antiretroviral medication. In 2005 the G8 countries pledged that by 2010 universal access would be reached. Whether this is feasible is open to debate but such proposals indicate that the scale of the issue is now fully acknowledged and that large-scale funding must be made available to tackle it.

theories, some of the ones highlighted by UNAIDS (www.unaids.org) include:

- A traditional lack of discussion of sexual-health matters.
- The low social and socio-economic status of women.
- **Poverty**, which in some cases leads to women undertaking commercial sex work, migrant labour patterns which means that family life is disrupted leading to an increase in partners.
- A reported cultural resistance to the use of condoms in many regions.
- In certain regions cultural attitudes towards relationships, i.e. multiple partners and polygamy, can increase risk of infection.
- Established epidemics of other sexually transmitted diseases.

In contrast to many other regions of the world, the disease is predominant amongst young women. Across the region this group is four times more likely to be infected than young men. Women also endure further difficulties, as they are often also caring for family members who are ill from the disease. With infection rates reaching 50 per cent in some urban areas, it is no surprise that it is estimated that by 2010, 40 million children will have been orphaned by the disease. Childcare is then often passed on to elderly family members who themselves are often suffering from ill health. UNAIDS suggests that there are now few families in sub-Saharan Africa who are not impacted upon by the disease. However, in many regions there is a great stigma attached to infection. Therefore, even when ill, many people are still reluctant to be tested: this obviously denies them appropriate treatment. Given the prevalence of infection within the working-age group, national economies suffer greatly, not just because of the cost of trying to treat the disease but also from lost tax revenues, skill shortages and general uncertainty. Figure 4.12 shows the impact of the disease on the education sector and how this impacts upon the whole economy.

The disease also has a dramatic impact on life expectancy. As discussed earlier, during the 1970s and 1980s improving levels of health care meant that fewer people were dying from easily preventable diseases, leading to rising life expectancies. For example, life expectancy in Botswana rose to approximately 65 by the mid-1980s and it was predicted that it would have reached 70 by 2010, but for HIV/AIDS. Instead, in 2006 life expectancy at birth was estimated to be 33.9 years and is predicted to fall to 27 by 2010. In 2006 there were 23 births/1000 population compared to 29 deaths/1000 population, demonstrating a return to the early stages of the demographic transition model. Figure 4.13 highlights this dramatic decline in life expectancies. UNAIDS has recently predicted that a fifteen-year-old boy in Botswana has an 86 per cent chance of dying of AIDS.

There is also a clear political economy to the disease in sub-Saharan Africa. Many governments still deny the severity of the disease, and many dispute the causes

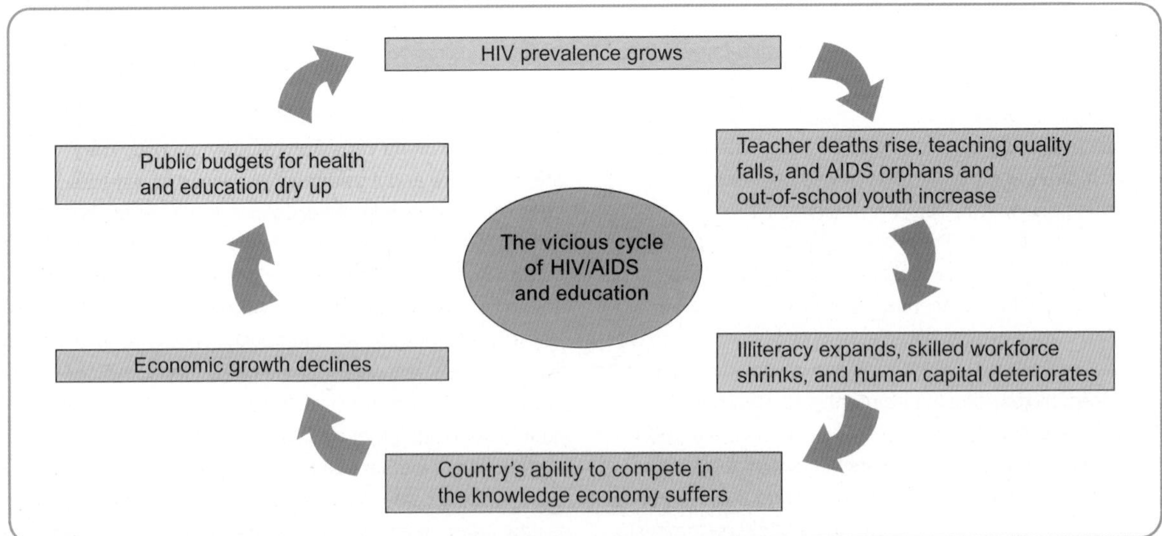

Figure 4.12 The impact of HIV/AIDS on education systems and the broader economy.

Source: Bundy and Gotur (2002)

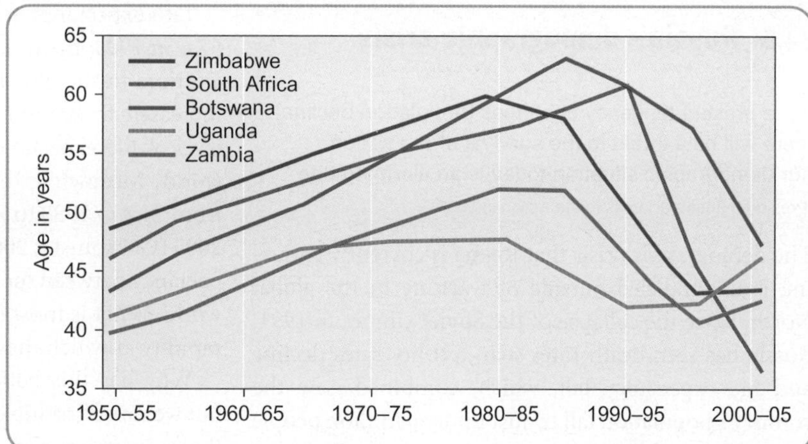

Figure 4.13 Declining life expectancy in five sub-Saharan countries.

Source: United Nations Population Division (2004). The United Nations is the author of the original material.

of the epidemic. The South African President, Thabo Mbeki, has publicly disputed whether the HIV virus leads to the development of AIDS (McGreal 2002). He has also argued that in South Africa far fewer people have died of the disease than is officially reported. Such statements lead to confusion amongst the general public and hamper attempts to educate people in how to avoid contracting the disease. Furthermore, within the region there is a lack of a coordinated response to the disease. However, it is argued that in many areas Western ideals are imposed as a solution to the problem. For example, some religious groups suggest abstinence rather than condom use as a solution, which does not reflect the realities of the situation. On a global scale, there is much debate about the cost of anti-viral drugs which can increase dramatically the life expectancy of those living with AIDS. Many sub-Saharan states cannot afford the high cost of these drugs, partially because of the economic impact of the disease as discussed above. Therefore, pressure is put upon Western drug companies to make treatment more freely available. Yet there is resistance to providing cheap drugs, given the cost in developing them and concerns as to whether they will be distributed correctly. In order to develop approaches which have some chance of success, UNAIDS suggests the following areas must be focused upon:

- HIV/AIDS can no longer be seen as an 'individual' disease and needs to be tackled on a national level.
- Educational and preventative measures must take into account social and cultural norms.
- Women's double burden of looking after the family while sick themselves must be considered.
- The socio-economic structure of the disease must be considered – young women must be targeted for education and empowerment.
- Interested parties must increase levels of cooperation.
- Western states need to be prepared to support, both financially and politically, the restructuring of health and education systems.

When looking at the geographies of infection within sub-Saharan Africa, it is clear that some countries implemented the above suggestions at an early stage of the disease. For example, Senegal has stopped infection rates at between 1 and 2 per cent and Uganda has cut HIV/AIDS prevalence from double-digit figures to around 7 per cent. This demonstrates that countries with the political will and the careful use of resources can contain the spread of HIV/AIDS. The first case of HIV/AIDS in Uganda was detected in 1984 and by 1986 a national committee was set up, which the President personally led, to examine responses to the disease. Remember that in 2006 the South African president was still disputing the cause of the disease. In 1991 a public awareness campaign was launched across all sectors of the media with pop songs, billboards and radio messages all popularizing HIV/AIDS protection messages. As a result HIV/AIDS rates peaked in 1993 and now 98 per cent of Ugandans know that the disease is mainly transmitted through unprotected sex. As the United Nations points out, if such measures are not adopted by all sub-Saharan countries, then the epidemic will continue to destroy generations and will provide a further barrier to sub-Saharan Africa's political and economic development.

4.5 Russia's demographic crisis

if the present tendency continues [population decline] there will be a threat to the survival of the nation . . . our demographic situation today is an alarming one

(President Putin's address to the nation in 2000)

The demographic crisis that Russia is currently enduring is unparalleled outside of wartime in the global North. Since the collapse of the Soviet Union, in 1991, Russia has seen death rates soar, fertility rates decline and life expectancy fall, which, combined, saw the country's population fall by just under a million people in 2005 alone. In the period since the collapse of the Soviet Union, Russia's population has fallen by approximately 2.5 million people. However, high levels of in-migration masked the true scale of the problem as Russians returned to their homeland from the now former Soviet states. Over 5.5 million people made this journey suggesting that the natural decrease is over 8 million people. As Figure 4.14 shows, Russia's population was growing in the late 1980s and while migration partially compensated for the rapid decline in natural population growth in the 1990s this has now slowed. As the figure shows, the overall population decrease at the turn of the century reached over 800,000 people per year (data from Marquez 2005).

Life expectancy fell for men from 63.8 in 1990 to 57.6 in 1994, the quickest fall ever recorded outside of wartime in the global North, and this figure has only increased to around 59 by 2005. This extremely low level of life expectancy is the lowest in the 'developed world'. Meanwhile the life expectancy for women fell from 74.3 in 1990 to 71.2 in 1994 rising to around 73 in 2005 (Goskomstat 2006). The disparity between life expectancy between men and women is the highest in the world. What is most startling about these figures is the rapidity at which these falls have occurred.

Why has this happened? There are many theories but we must first look at the many demographic shocks the country has experienced over the past 100 years as these laid the foundations for the current problems. The Russian revolution in 1917 led to many years of civil war and famine, resulting in many millions dying prematurely. Not long after this situation had stabilized, the Second World War broke out. Russia experienced some of the heaviest losses, with recent estimates suggesting that over 14 million Russians were killed during this period. The end of the war did not bring stability to the country as Stalin ushered in a period of 'great terror' with up to 2.5 million people exiled to prisoner camps in the Russian far north. It is estimated that during Stalin's reign, between 1927 and 1953, at least 2.8 million people died in these camps (Applebaum 2003). It was not until Stalin's death that

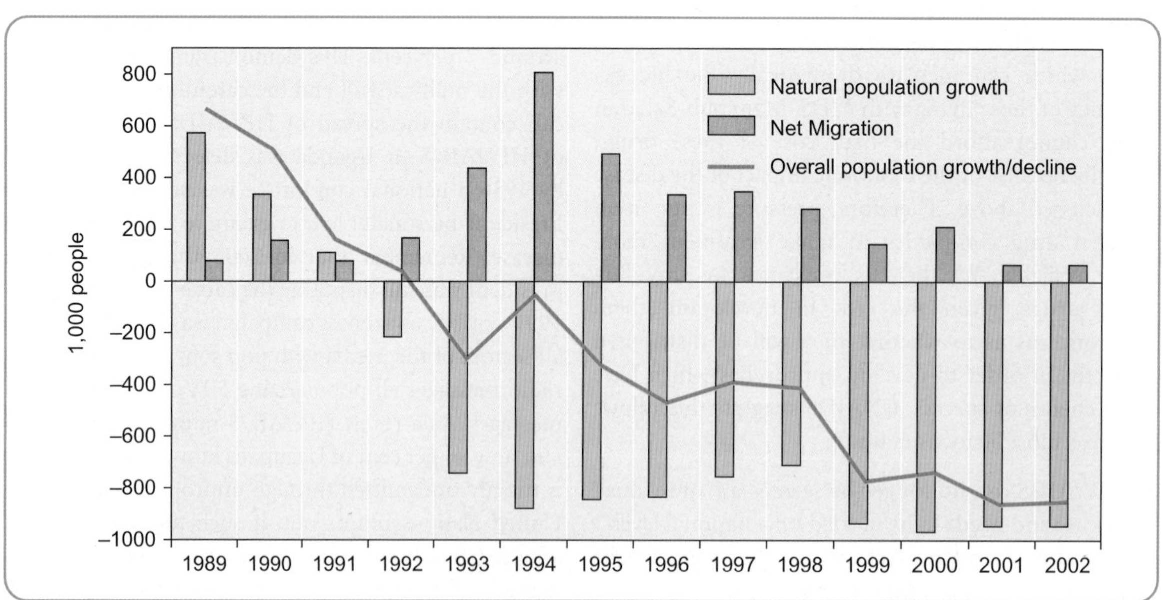

Figure 4.14 The decline of Russia's population 1989–2002.

Source: Goskomstat (2003)

Russia's demographic situation began to compare to that in Western Europe. Russia soon completed the first stage of the **epidemiological transition model**, the 'sanitary revolution' meaning that people should no longer die of easily preventable causes, and the creation of a countrywide health system. Even then, there were still great demographic imbalances, as more men died in the above events than women. The legacy of this is that in 2002, there were 67 million men compared to 77 million women in Russia. However, Russia was, at this point, beginning to move through the demographic transition model as birth rates began to rise while death rates fell.

In the 1960s and 1970s, life expectancy did not converge with levels in Western Europe, mainly for cultural reasons. Traditionally Russia has had some of the world's highest alcohol consumption rates, and a limited range of food products meant that many people led unhealthy lifestyles. In mid-1985 President Gorbachev tried to tackle this by introducing limits on when, where and how much alcohol could be sold. Whilst very unpopular, these measures did see some increase in life expectancy. However, these gains were short-lived as the collapse of the Soviet Union in 1991 meant such reforms were abandoned. As noted above, this period ushered in great demographic uncertainty and Russia diverted from the path suggested by the demographic transition model. As with other countries Russia had seen birth and death rates converge which would have led to a stable population. As Figure 4.15 shows, this has changed dramatically from 1992. Note how this differs from Figure 4.5 above.

Until 1991 birth and death rates were converging and Russia's population should have stabilized. Since 1992 the gap between birth and death rates led to the rapid decline in the country's population. There are many theories as to why this demographic crisis has developed and rather than one single cause it is arguably a mixture of several factors.

Historical lag This argues that the current situation is a continuation of the decline in health of the Russian population since the 1960s and was so rapid because of the suppressed deaths delayed by Gorbachev's anti-alcohol campaigns of the mid-1980s. In other words, those who would have died from alcohol abuse in the 1980s but were not able to do so because of the restrictions were now able to do so. Furthermore, while impressive in terms of the number of doctors and hospitable beds, the Russian health-care system had not adopted all of the most recent medical advances. While there is no doubt that these issues did contribute to the problem it does not really explain why the crisis has lasted for so long.

Poverty Current estimates suggest that 20 to 30 million Russians can be classed as poor, as for many the real worth of their incomes has declined rapidly after the collapse of the Soviet Union. However, the collapse did not bring starvation as people/households developed alternative survival methods, such as undertaking

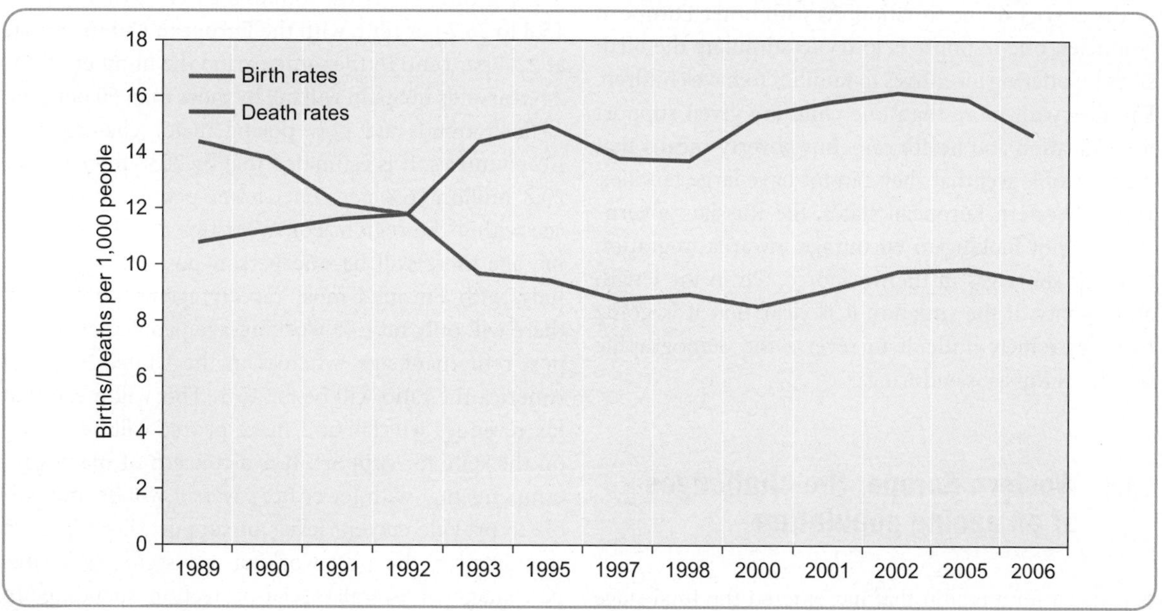

Figure 4.15 Changes in birth rates and death rates in post-Soviet Russia.

Source: Goskomstat (2006)

informal work or growing their own food, for example. Given that most people can earn enough money to buy enough food for their household, this can only be part of the problem.

Access to services While Russia still has a free health-care system that has in theory enough doctors and nurses, the system is starved of funding. This means that the wages for many medical staff are only just above poverty levels and equipment is often poor. Many people say that they have to pay to obtain free medicines and high poverty levels mean that many cannot afford all of the medicines that they need.

Cultural issues Many middle-aged men have found it difficult to adapt in the post-Soviet era. They often have had to find work outside of their professions and they feel that the skills they developed in the Soviet era are no longer required. Often this leads to frustration and increasing stress levels, which may often manifest in unhealthy behaviour, such as binge drinking.

It is a mix of the above factors that sees Russia population currently falling by almost a million people per year. It was assumed during the 1990s that as the Russian economy developed, the situation would rectify itself. The fact that by 2007 the problems are more serious than ever demonstrates that this is not going to happen. The United Nations has forecast that, given current trends, by 2050 Russia will fall from the 7th to the 14th most populated country. As the opening quote shows, the Russian government is now aware of the gravity of the situation. As with other European countries, one response is to try to stimulate the birth rates by offering incentives to families to have children. Families with more than one child are given support for education and health care, but poverty means that many people feel that they cannot have large families. Unlike Western European states, the Russian government is not looking to encourage inwards migration to cover shortages in the workforce. Given the length and gravity of the situation it is clear that it is going to be extremely difficult to reverse the demographic problems Russia is enduring.

4.6 Western Europe: the challenges of an ageing population

It is usual for a region that has entered the final stage of the demographic transition model, where birth and death rates converge, to experience population ageing, which is when the average age of a population rises as people live longer and fewer people are born. This section examines why populations age and considers the long-term challenges this presents. According to UN calculations, between 1950 and 2000 the average age of the world's population rose from 23.6 to 26.4 years and is expected to rise steeply by 2050 to 36.8. While less developed regions are experiencing the fastest rates of growth, the 'oldest' countries are still to be found in the more developed regions (Figure 4.16). Europe will experience the highest percentage share of people over 65. This figure is significant as it is assumed that people over this age will not be working and that many will be dependent on state pensions for their income. Also this group will be less likely to be paying tax, will use health-care systems more frequently and will, obviously, not be having children. With fertility rates falling, there will be less young children and then over time fewer working-age adults to support those that have retired.

For a country's population to grow, its fertility rate must be above what is known as the population **replacement rate**. This replacement rate is based on the idea that if a woman has less than two children, then she is not 'replacing' herself and her partner. Therefore for a population to grow, fertility must be above 2.1. In Italy, for example, fertility rates are currently 1.2: thus its population cannot naturally replicate itself. In 2000, 18 per cent of the Italian population were aged over 65. By 2050 this figure will increase to 35.9 per cent. For Great Britain the figure will rise from 15.8 to 26.7 per cent, with the European Union average at 22.4 per cent. In the same period the number of 20–24-year-olds in Spain will fall by more than 60 per cent.

Such trends are now posing major challenges to EU countries. It is estimated that by 2030 there will be 20.8 million (6.8 per cent) fewer people of working age within the region. For every two people of working age there will be one person post-retirement. In Italy, with Europe's most rapidly ageing population, there will only be 1.45 working-age people per person post-retirement age, whereas in the United States of America the ratio will be 3.2 to 1. This will mean that tax revenues will fall and more people will be relying on the state for support. It is a concern of many governments that with fewer tax payers it will be impossible to provide current levels of support. Even in 2005 spending on the post-working age sector is a large percentage of overall social-protection spending. In Italy and Greece the figure is over 60 per cent. As the European Commission has stated, current levels of

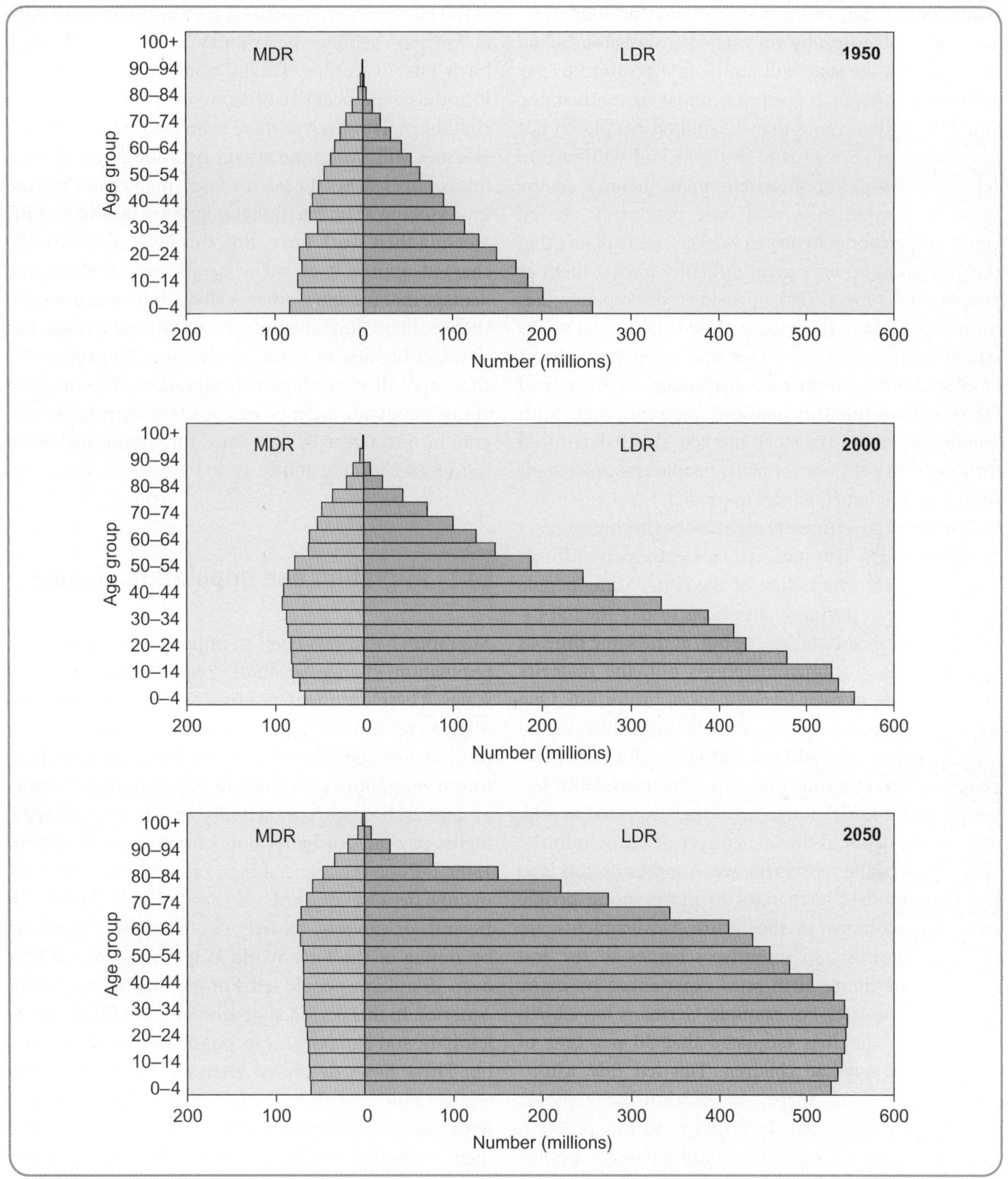

Figure 4.16 Age structure of More and Less Developed Regions, 1950, 2000, and 2050.

Source: data from the UN (2003). Note: the 2050 data are based on medium-variant projections

social provision are not sustainable, as populations age, unless taxes are raised to prohibitive levels (EU 2005).

Such trends increase individual vulnerability as societies age. The creation of a post-war welfare state in Britain was based on a 'cradle to the grave' principle. People worked in the knowledge that their state pension would provide enough income for a satisfactory standard of living and that they would be able to turn to the health-care system for free assistance in times of need. Given the growing disparity between the tax base and those reliant on state incomes such a framework is no longer viable. The British government has recently made it clear that young people should no longer expect to be able to 'get by' on a state pension alone when they retire as the state will not be in a position to pay such pension levels. It is estimated that currently in the United Kingdom more than 12 million people do not have personal or company pensions and will have to rely on the state for their retirement income. At the same time, companies who have previously offered generous pension schemes to workers are realizing that they are going to have great difficulty paying them as people live longer. This pensions discrepancy has meant that many firms have closed their final salary schemes (whereby paying a set amount each month an employee was guaranteed a percentage of their final salary as their monthly pension), replacing them with schemes linked to the stock market. This has reduced the projected pensions for many people and made their future income much harder to predict.

European governments are now beginning to react to this problem. The task is made extremely difficult by the very emotive nature of the issue. Many people would baulk at paying higher taxes to pay for the increasing cost of societal ageing but, at the same time, as the European Commission points out, the majority wish to maintain their levels of social protection. One solution is to increase the retirement age, which would increase tax revenue and reduce the length of time that people will receive state pensions. Understandably, few people wish to work longer than they expected to and, often, strike action is threatened even at the mention of an increase of the retirement age. Another option is to try and stimulate birth rates so more young people enter the economy in the future. Governments are therefore resorting to incentive schemes to try and arrest the declining birth rates experienced by many European countries. For example, Germany has a birth rate of 1.37 and it is estimated that 30 per cent of women have not had children. This led *Bild*, a top-selling German newspaper, to exclaim: 'Baby shock: Germans are dying out'. In response to this problem the German government introduced a 'parents money' law which means from January 2007 parents who remain at home to look after their child will receive 67 per cent of their previous income for a year (up to €1,800 per month). Those on a low income will receive 12 monthly payments of €450. This is extremely generous when compared to the British system wherein a new mother receives 90 per cent of previous pay for 6 weeks and then a flat €150 (£102) for 26 weeks only. Men can take two weeks paid paternity leave at €150 per week. In Norway mothers receive 10 months at full pay and men have to take 4 weeks paternity leave at full pay, helping the country to the third highest birth rate in Europe. Poland and Italy have recently introduced an incentive of a one-off payment for each child born. However, as there is no guarantee that such schemes will work, and it will be a long period until this cohort reaches the labour force, many countries are now looking at international migration as a way of increasing their work force. But, this is a very politically charged approach as many people worry about the damage that foreign workers will do to their economy – such as, increasing unemployment rates or a reduction in wages because of extra competition. To explore the challenges that international migration poses to individual countries, we now examine the increase in migration into Great Britain since the expansion of the European Union in 2004.

4.7 Migration and population change

Migration has always been an important component of population change at local, regional and national scales. There are many reasons why people decide to migrate to another region or country. These can be push factors such as political instability, civil conflict, forced migration, a lack of job opportunities, famine, or pull factors such as stability in the new country, higher pay rates or better opportunities. From around 1800, Europe experienced high rates of out migration, mainly because of push factors such as famine, in Ireland, or poverty, in Italy, for example. Up to the beginning of the First World War it is estimated that over 55 million people left Europe, mainly for North America. In this period approximately 9 million people left Italy, and Norway saw its population fall by a quarter. These flows decreased after the First World War as economic growth in Europe meant that there were fewer push factors and recipient countries tightened their migration criteria. For example, in 1924 the United States of America introduced migration quotas and then the Great Depression of the early 1930s diminished previous pull factors. While after the Second World War there was a significant baby boom in Western Europe, this did not provide a rapid enough response to the need for labour as the region's economies grew. Thus, rather than seeing mass emigration the region proactively encouraged immigration.

This was usually along colonial lines as infrastructures were in place to facilitate migration. For example, the British government advertised for workers in the West Indies to migrate to fill labour shortages in transport and hospitals. The further tightening of migration controls in the USA made Britain a more attractive destination and by the late 1950s up to 50,000 people per year were making the journey. Inwards migration was also stimulated by India gaining independence from colonial rule as religious tension meant that many people were displaced from their home region. The majority of migrants who came to Britain came from three regions: Gujarat, Punjab and Mirpur (in Pakistan after India was partitioned into two states after independence). None of these regions were particularly poor but they did have a tradition of international trade and migrants had previously settled in Britain. The formation of **diasporas** (communities of migrants) enabled further migration as new arrivals had a starting point from which to build their new lives, such as family or community support. Migration streams into Great Britain peaked in 1960–62, as legislation was subsequently introduced that made it harder to enter the country. This was passed against a backdrop of fear that increasing immigration would erode British values and would lead to violence between ethnic groups. The epitome of this was Enoch Powell's 'Rivers of Blood' speech in which he claimed the country would be 'overrun' if migration was not halted. Although this speech saw him expelled from his Shadow Cabinet post, he did receive popular support. Therefore, it was no surprise that migrants often suffered from racism and struggled to find well-paying work. When Ugandan Indians were expelled en masse by the dictator Idi Imin many wished to move to Britain because the diasporas that already existed. Fearing this, Leicester City Council took out adverts in the Ugandan press telling migrants not to move there stating that they would be unable to help them if they did. Despite this, many did move to the region and the city embraced the challenge and is now recognised across the world as a model for multicultural relations.

With birth rates falling in many Western countries, cross-border migration is now again a topic of much debate, as it constitutes a higher percentage of population growth than previously. Furthermore, the overall number of international migrants is steadily growing. In 2002, the United Nations estimated that the number of international migrants was over 175 million, double the total in 1975. In MDR countries, on average 10 per cent of the population are classed as migrants and there is a net flow of over two million people between LDRs and MDRs. Castles and Miller (2003) argue that the forms of migration are changing and that we are entering an 'age of migration'. They identify four main areas where migration is changing:

1. Migration is becoming more 'global', in the sense that more and more countries are being affected at the same time and the diversity of areas of origin is also increasing.
2. Migration is accelerating, with the number of movements growing in volume in all major regions at the present time.
3. Migration is becoming more differentiated, with no one type of movement dominating a country's flows but instead with combinations of permanent settlers, refugees, skilled labour, economic migrants, students, retirees, arranged brides, and so on.
4. Migration is increasingly feminized, with women not only moving to join earlier male migrants but now playing a much fuller part in their own right, notably being labour migrants themselves as well as often being dominant in refugee flows.

In Europe the biggest impact on international migration was the collapse of the Soviet Union and the subsequent accession of many east European countries to the European Union in 2004 and in 2007. During the Soviet period it was extremely difficult for east Europeans to travel to Western Europe. When the Soviet system began to collapse, signified by the fall of the Berlin Wall in 1989, people in the region became free to leave their countries. However, many west European countries imposed strict visa regimes to stop mass migration from the region. This meant that people could visit Western Europe for a vacation but were unable to migrate to look for work. This situation changed when the European Union was expanded eastwards in May 2004. This meant that over 75 million people joined the European Union, many of them from counties where incomes are much lower than in Western Europe. Prior to this expansion EU members had the right to move freely between countries to look for work. However, existing member states were worried that if the new east European members were given the same migration rights, then too many people would come to Western Europe looking for work. As a result, many countries enforced legislation that meant they did not fully open their borders to new member states. This caused great anger in the accession states as such restrictions were not included in the negotiations that took place prior to entry into the European Union. Britain, Ireland and Sweden were the only countries not to impose numerical restrictions.

Prior to the expansion taking place, the opening of Britain's borders caused much concern. There were fears that too many people would move in looking for work putting social welfare systems under great strain and that unemployment rates would rise rapidly. There was also a great underestimation by the government of the number of people that would move from Eastern Europe. Initial estimates suggested that approximately 15,000 people per year would migrate to look for work. However, between May 2004 (when accession took place) and June 2006, almost 450,000 people applied to Worker Registration Scheme (WRS), which the British government put in place to monitor labour migration. Of course, it is possible that the true migration figure is much higher as people can work illegally in order not to pay taxes and the scheme does not include self-employed workers. It is extremely difficult to calculate an accurate figure as migration figures are taken for a year as a whole. For example, if a migrant makes a short trip home then they are counted twice in the migration figures for that period. However, it is clear that far more migration has taken place than was expected, leading to much controversy over whether it benefits the country.

In Britain the centre-right press continually publishes stories criticizing government migration policy, arguing that unfair pressure has been put on health care, education and other welfare systems and that migrants are forcing down wage rates and keeping unemployment figures high. However, others argue that in fact migrants are predominantly undertaking low-paid, low-skilled work, which employers were previously finding difficult to recruit to. Government data suggests that between May 2004 and June 2006, 82 per cent of migrants were aged between 18 and 24 and that only 3 per cent had dependants under the age of 17 with them. Just under 6,000 people had applied for Income Support of which only 768 applications were given further consideration. This suggests that the majority of migrants have moved to work rather than to look for state benefits. This migration was helped by existing migrant diasporas but the majority of migration has occurred to places where work is available. This has seen the creation of new communities, which is evidenced by the opening of shops selling Polish food and goods, or adverts for banking in Lithuanian, for example. This is not to deny that migration is causing pressures in some areas. For example, it is estimated that in Southampton up to 10 per cent of the current population are migrants from the new accession states. This inevitably means that demands on state services will increase. At the same time this does not mean that such issues are insurmountable. However, despite the fact that increased migration has clearly not led to a rapid rise in unemployment rates, migration is still an extremely emotive issue. In response to public pressure the British government put restrictions on the number of Romanian and Bulgarian workers allowed to move to Great Britain after their countries joined the European Union in 2007. This is despite the fact that for 2005 the government estimated that the overall economic benefit to the country arising from increased migration was over £500 million.

4.8 Conclusions

It is clear that the world has undergone many different demographic transformations, but the past fifty years have proved to be quite dramatic. The world's population has grown by over 5 billion during this period with the vast majority of the growth occurring within less developed regions. This obviously presents challenges and in many instances has put great strain on environmental, food, housing and welfare infrastructures. Conversely, many countries in the more developed regions are now experiencing natural population decline as fertility rates fall. This has posed many new challenges to governments, as ageing populations will make the provision of suitable levels of welfare increasingly difficult. As fertility rates fall below replacement rates in many Western European countries, populations would fall if it was not for international migration. Again increasing migration brings its own set of questions, and pressures on infrastructures.

As noted above, predicting future population levels is extremely difficult. Even small changes in fertility rates can cause large differences in the medium term. Furthermore, as our case studies have shown, unexpected events can lead to very rapid changes in population levels. HIV/AIDS, while not causing sub-Saharan Africa's population to fall, is extremely damaging to the future of the region and has caused life expectancy to plummet. Russia's post-Soviet demographic crisis is another example of unexpected population change that will have a long-lasting impact on the country. Without knowing where such impacts will occur next, or their scale, it is impossible to state with any confidence the rate of population growth over the next fifty years.

Notwithstanding these caveats, it is clear that the world's population will continue to grow and it is likely that the overall population will grow to over 7.5 billion, barring a catastrophe such as global warfare or disease pandemic, which will pose many questions for governments. Given the difficulties in trying to ensure an adequate diet and housing for the current population, further growth will only exacerbate these pressures.

Learning outcomes

Having read this chapter, you should be able to:

- Recognize the dynamic nature of populations, not only in terms of size but also with respect to the geographies which exist within the overall growth.
- Explore how sudden economic changes or disease alters the course of countries through the demographic transition process.
- Appreciate that the completion of the demographic transition no longer appears to herald the end of transformations in the population.
- Acknowledge the increasing importance of migration as a component of national population change alongside fertility and mortality.
- Assess the significant problems that ageing populations will have in terms of social and economic development.

Further reading

Castles, S. and Miller, M.J. (2003) ***The Age of Migration: International Population Movements in the Modern World***, Palgrave Macmillan, Basingstoke. This book, now in its third edition, provides a comprehensive survey of past trends and current developments in international migration. Look especially at Chapter 6 on the globalization of international migration and Chapter 12 on the post-Cold War era.

Harper, S. (2005) ***Ageing Societies***, Hodder Arnold, London. This book explores the issues of population in both developing and mature countries. It examines the necessity for people to extend their working lives, the changes ageing societies have on families, the challenges it poses for state social provision and the impact these issues will have on developing regions.

Hugo, G. (2007) **Population geography, *Progress in Human Geography***, 31(1): 77–88. One of a series of articles looking at developments in population geography in the Southern hemisphere. This article looks at population vulnerability and migration.

Jones, H. (1990) ***Population Geography***, Paul Chapman, London. This is a clear introduction to geographical perspectives on population. Look especially at the chapters on population growth and regulation, international variations in mortality, fertility in developed countries, fertility in less developed countries, and international migration.

National Research Council (2000) ***Beyond Six Billion: Forecasting the World's Population***, National Academy Press, Washington, DC. This presents the findings of a US National Academy of Sciences panel on population projections. It contains detailed examinations of transitional and post-transitional fertility, mortality and life expectancy, and international migration, together with assessments of the accuracy of past projections and of the uncertainties in current population forecasts.

UNAIDS (2006) ***Report on the global AIDS epidemic 2006.*** Available to download at http://www.unaids.org/en/HIV_data/2006GlobalReport/default.asp. This report gives a comprehensive overview of the challenges that countries and international organizations face when confronting the epidemic. It also details many success stories and the development of good practice, and provides country and regional statistics.

United Nations (2006a) ***Population challenges and development goals*** – available to download at http://www.un.org/esa/population/publications/pop_challenges/Population_Challenges.pdf. The report provides an overview of global demographic trends and explores in more detail population size and growth, urbanization and city growth, population ageing, fertility and contraception, mortality and international migration.

Useful websites

http://www.un.org/esa/population/unpop.htm Provides access to extracts from the United Nations' population-related publication and statistics.

www.census.gov/ipc/www/idbnew.html US Census Bureau (online) International Data Base, United States Bureau of the Census, Washington, DC. This is a computerized source of demographic and socio-economic statistics for 227 countries and areas of the world from 1950 to the present, and projected demographic data to 2050, freely available on the World Wide Web.

www.prb.org Home page of the Population Reference Bureau, based in Washington, DC. This provides news updates on all aspects of world and US population, plus information about its own publications (including its monthly *Population Today* and *Population Bulletin*) and links to other websites.

www.populationaction.org Home of Population Action International, an organization dedicated to advancing policies and programmes that slow population growth in order to improve quality of life for all people.

http://popindex.princeton.edu/ The primary reference tool to the world's population literature, maintained by the Office of Population Research at Princeton University, with a searchable and browsable database containing over 46,000 abstracts as published in *Population Index* 1986–2000.

http://www.unaids.org Provides a great deal of statistical data on the epidemic and provides access to publications.

www.un.org/millenniumgoals/ Provides background to the Millennium Development Goals and has numerous publications and reports.

For annotated, clickable weblinks and useful tutorials full of practical advice on how to improve your study skills, visit this book's website at **www.pearsoned.co.uk/daniels**

RESOURCES AND DEVELOPMENT

Chapter 5

Michael Bradshaw

Topics covered

- The nature of natural resources
- Ways of defining and classifying resources
- The factors that determine resource availability
- The changing geography of energy production and consumption
- The relationship between energy consumption and economic growth
- Global energy dilemmas

Resources are defined by society, not by nature.
(Adapted from Rees 1985:11)

Resources are not, they become; they are not static but expand and contract in response to human wants and actions.
Zimmerman (Peach and Constantin 1972:16)

The large-scale exploitation of the planet's resource base may well go down in history as a defining characteristic of the twentieth century. Today the scale of resource consumption threatens the entire global ecosystem (see Chapter 6). Throughout the last century access to, and control over, natural resources was a source of conflict. Equally, patterns of resource consumption remain a major marker between the so-called 'developed' and 'developing' worlds (see Chapter 8 for further discussion). According to the World Bank (2007), in 2005–6 the so-called 'high-income countries' accounted for 15.7 per cent of the world's population, but produced 77.7 per cent of global Gross Domestic Product (GDP) and accounted for 50 per cent of the world's CO_2 emissions. In the early twenty-first century, society is increasingly aware of the finite nature of the planet's resource base. Yet, we still seem to have a blind belief that technological progress will solve the inevitable shortage of non-renewable resources. But more recently we have come to understand that before the resources physically run out we will have to find alternative ways of doing things to avoid destroying the planet's ecosystem (see Chapter 6). This chapter explores the nature of natural resources and considers the relationship between resource production and consumption, and economic development. The chapter is divided into three sections, each with a distinct task. The first section examines the meaning of the term 'natural resource', it evaluates the various ways of classifying resources and analyses the diverse factors that influence their availability. The second section analyses the specific case of energy resources and the changing geographies of production and consumption. The third, and final, section considers the interrelationship between energy consumption and economic development. The chapter concludes by discussing the resource dilemmas that currently face the different regions of the world.

5.1 Natural resources

At any moment in time, the planet Earth holds a finite stock of resources, the **resource base**. However, what human societies have considered a resource has varied through time and across space. Thus, the notion of what constitutes the earth's resource base also changes. As both Zimmerman and Rees acknowledge above, something is a resource because human society attaches value to it. Many of the things we value today as 'resources' held no value in the past and may hold no value in the future. For example, petroleum was not an important source of energy until relatively recently in human history. Followers of science fiction programmes, such as *Star Trek*, see a future based on resources and technologies that have yet to be created.

5.1.1 Defining resources

In his seminal work *World Resources and Industries*, Zimmerman (Peach and Constantin 1972: 9) states: 'The word "resource" does not refer to a substance, but a function that a thing or a substance may perform, or to an operation in which it might take place.' He goes on to note that 'resources are . . . as dynamic as civilisation itself.' In fact, one can suggest that each major human civilization was sustained by a particular set of resources and technologies for their exploitation (Simmons 1996). Thus, the archaeological record talks of the *Stone Age*, the *Bronze Age* and the *Iron Age*. In reference to the Industrial Revolution of the nineteenth century, Simmons (1996: 208) states that 'industrialization based on fossil fuel energy represents a turning point in the history of human–nature relations.' If the nineteenth century was based on the exploitation of coal, then the twentieth century will been seen as the era of oil and gas (Hall et al. 2003). Thus, the changing notion of what constitutes a 'resource' is an important factor in shaping the relationship between human societies and the natural environment. Undoubtedly, the hunter–gatherer communities living in the Palaeolithic period had a very different relationship with nature from the industrialists of the nineteenth century (see Section 1). As Blunden (1995: 164) observes, 'Because definition as a resource depends on usefulness to human society, natural materials may be required as resources by societies in some times and places but not in others.' At the time that the industrial revolution was taking place in Europe there were societies located elsewhere on the planet that were still at the hunter–gatherer stage in terms of their concept of what constitutes a resource and in terms of their relationship with nature. This example adds a further dimension to the complexity of the term 'resource'. Its definition can vary across space at the same time. While it is still the case today that people in many parts of the world live in what Mather and Chapman (1995: 139) call 'low-energy societies' (dependent upon plants, animals and human labour), one of the defining characteristics of the early twenty-first century, and a consequence of the processes of globalization, is that the majority of societies now have at least some shared notion of what sort of things constitute resources. The Inuit living in northern Canada are as dependent on gasoline for their skidoos as the commuters in their cars in Los Angeles or London are. Furthermore, echoing Lenin's famous edict that 'Communism is electrification plus Soviet labour power', the provision of electricity is seen as a central component of the modernization process in all types of societies. Witness the chaos that ensued when large parts of the US east coast lost all electricity in late 2003. A similar power outage occurred in Italy and there are growing concerns about the ability of national electricity systems to handle ever-increasing demand.

If the notion of resource is dynamic and intimately linked to the evolution of human society, it follows that so-called 'technological progress' both creates and destroys resources. As new technologies emerge, dependent on particular resources, so old technologies and their associated resources become redundant. For example, today we attach little value to flint (it has some value as a building material), yet in the Stone Age it was an essential resource for making tools. Because these resources had 'use value' they were also the subject of trade. On the other hand, many of the resources that were valued in the Bronze Age, such as copper and zinc, are still valued today but they are put to different uses. Thus, some resources retain their value as new ways of using them are discovered. The use value of resources also means that control over their supply is an important part of political and economic power. In theory at least, resource-rich regions are able to exploit this natural advantage in their dealings with resource-poor regions. Consequently, gaining control over particular resources has been at the heart of many wars and much conflict (see LeBillion 2007 for a review of the recent literature on resources and conflict). The European colonization of what we now call the 'developing world' was in part motivated by a desire to discover and control new sources of resources. It is also no surprise that many of the world's largest corporations, until recently at least, were involved in resource development. Our discussion so far has implied that all resources are homogeneous and that they share the same characteristics. However, we know that there are many different types of natural resource and the various ways of classifying them is the subject of the next section.

Plate 5.1 The use of firewood for cooking in a 'low-energy' society.

(Ami Vitale/Panos Pictures)

5.1.2 Classifying resources

Natural resources are commonly divided into two types: **non-renewable** or **stock resources** and **renewable** or **flow resources**. Figure 5.1 presents a classification of resources on this basis. Stock resources are those, mainly mineral, that have taken millions of years to form and so their availability is finite. Hence, we also refer to them as non-renewable as there is no possibility of their being replenished on a timescale of relevance to human society. For example, on the basis of the geological timescale, it is possible for new deposits of coal, oil and gas to be created; however, the time this will take means that we only have available to use the stock of already created fuel minerals. Within the category of stock resources, it is also useful to distinguish between those that are consumed by use, such as fuel minerals, those that are theoretically recoverable and those that are recyclable, such as aluminium. A further characteristic of stock resources is that they tend to be highly localized, that is they are found in

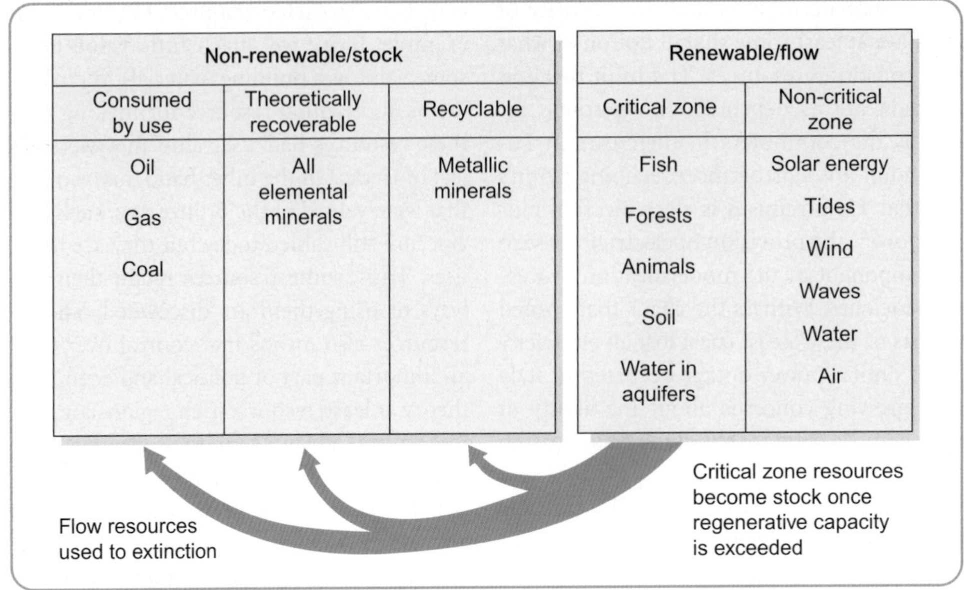

Figure 5.1 A classification of resource types.

Source: J. Rees, *Natural Resources: Allocation, Economics and Policy*, 2nd edn, Routledge, 1985, p. 15

relative abundance only in specific places, such as ore deposits and coalfields. Some stock resources are more abundant than others and their relative scarcity affects their value. For example, aggregate minerals, like sand and gravel, are relatively abundant, while precious metals, such as gold and silver, are relatively rare.

Renewable or flow resources are those that are naturally renewed within a sufficiently short time-span to be of use to human society. Again we can distinguish between different types of stock resource. Figure 5.1 divides flow resources into 'critical zone' and 'non-critical zone' resources. The distinction here is between those flow resources whose continued availability is dependent upon management by society (critical zone) and those that will continue to be available independent of the actions of society (non-critical zone). As indicated by the arrows, it is possible for critical zone flow resources to become stock resources if they are mismanaged and their regenerative capacity is exceeded. Thus, the aim of resource management is to ensure that exploitation of a particular renewable resource does not damage its capacity to replace itself. In recognition of the increasing challenge, and the numerous failures, to manage renewable resources, Rees (1991: 8) has developed an alternative to the conventional two-part typology of natural resources: 'All resources are renewable on some timescale . . . what matters for the sustainability of future supplies is the relative rates of replenishment and use . . . it seems better . . . to think in terms of a "resource continuum" than the conventional two-part typology' (Figure 5.2). The combined impact of population growth and industrialization is placing increasing stress on the planet's finite supply of fresh water and clean air. According to UNEP (1999: xxii), about 20 per cent of the world's population currently lacks safe drinking water, while 50 per cent lacks access to a safe sanitation system. In 2000 the international community committed to the so-called 'Millennium Goals' (see www.developmentgoals.org); Goal 7, to ensure environmental sustainability, set a target to halve by 2015 the proportion of people without sustainable access to safe drinking water and basic sanitation. In sub-Saharan Africa alone 300 million people lack access to improved water sources and 450 lack adequate sanitation services. Meeting the Millennium Development Goals will require providing about 1.5 billion people with access to safe water between 2000 and 2015. The resource continuum recognizes that (clean) air and water supply now have scarcity value and should be considered as finite.

In our discussion so far we have highlighted the complexity of the notion of natural resource and considered the various ways of classifying resources, but what factors actually affect the availability of particular resources?

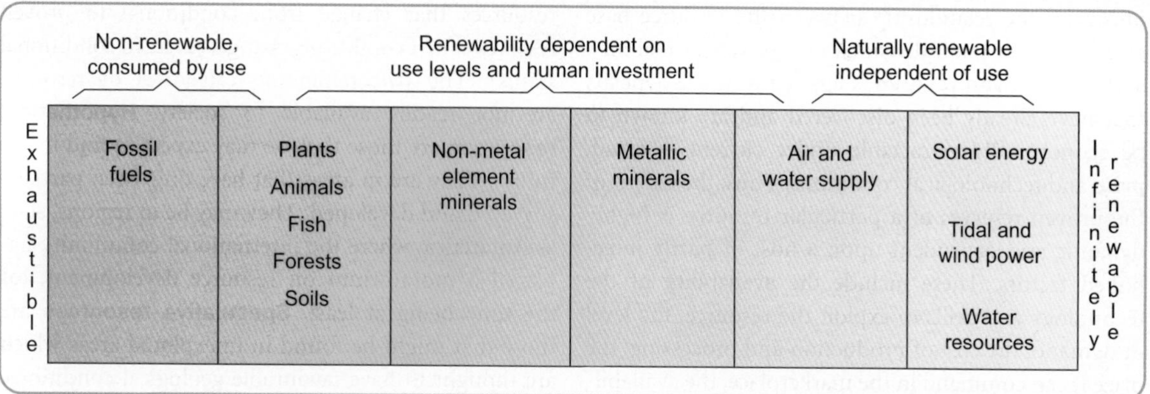

Figure 5.2 The resource continuum.

Source: J. Rees, Resources and environment: scarcity and sustainability, in Bennett and Estall (eds) *Global and Challenge: Geography in the 1990s*, Routledge, 1991, p. 9

5.1.3 Resource availability

As was noted earlier, at any moment in time there is a finite stock of natural resources on the planet, the resource base. Each resource has in turn its own resource base, the total quantity of a substance or property on the planet, for example, the total amount of oil in existence today. However, that total resource base is not the amount available for human exploitation. Figure 5.3

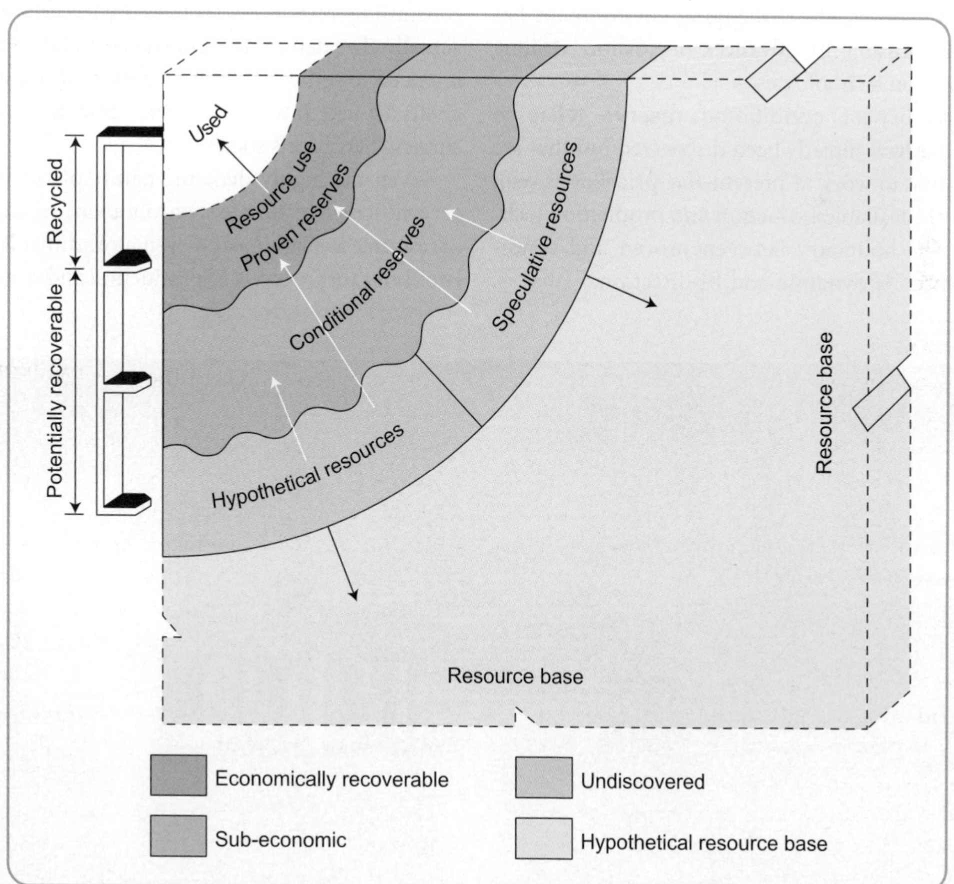

Figure 5.3 Resource availability.

Source: J. Rees, *Natural Resources: Allocation, Economics and Policy*, 2nd edn, Routledge, 1985, p. 20

illustrates the relationship between the resource base and the various sub-divisions of resource availability. The term **proven reserve** is applied to those deposits that have already been discovered and are known to be economically extractable under current demand, price and technological conditions. Thus, the extent of the proven reserves of a particular resource is highly dynamic and dependent upon a host of partly interlinked factors. These include the availability of the technology and skills to exploit the resource, the level of demand, the cost of production and processing, the price it can command in the marketplace, the availability and price of substitutes, and the environmental and social costs of developing the resource. These factors determine whether or not a particular resource will be exploited in a particular place and the global level of supply of that resource. The extent to which each of these factors influences resource development also varies across space and time. Today we have the technological ability to recover resources in geological and environmental conditions that were previously uneconomic. For example, the exploitation of North Sea oil and gas in a physically challenging environment has been made possible by advances in offshore drilling and production technologies.

The category of **conditional reserve** refers to deposits that have already been discovered but that are not economic to work at present-day price levels with the currently available extraction and production technologies. The boundary between proven and conditional reserves is dynamic and bi-directional (that is, resources that change from conditional to proven reserves can, if conditions change, revert to conditional status). The two remaining categories of resource are not readily available to society. **Hypothetical resources** are those that we may expect to find in the future. They are in areas that have only been partially surveyed and developed. They may be in regions, such as Antarctica, where the international community has placed a moratorium on resource development, for the time being at least. **Speculative resources** are those that might be found in unexplored areas which are thought to have favourable geological conditions. Finally, there remains a large part of the earth about which we have no information on its potential resource base. The strength of this classification is that it stresses the highly dynamic nature of the concept of resource reserve. The danger is that it leads to the view that society will never run out of resources because there are always more to discover and technological progress will continue to make new resources available for exploitation. Even if the latter were the case, and clearly it is not, the planet now faces the additional problem that consuming resources, such as hydrocarbon fuels, is actually threatening the global ecosystem. Thus, there is a need to rethink the whole notion of resource availability to take into account the ecological cost of our current 'fossil fuels society'.

Given the highly dynamic nature of resources, it is very difficult to estimate at any moment in time the level of resource availability. It is even more difficult to speculate about future levels of production and consumption

Plate 5.2 As offshore technology advances, so more remote resources can be exploited.
(ImageState/Alamy)

Table 5.1 The dimensions of resource scarcity

Type of scarcity	Concern
Physical scarcity	• Exhaustion of minerals and energy. • Human populations exceed the food production capacity of the land. • Depletion of renewable resources such as fish, soils or timber. • Growing demand for water for human use threatens aquatic ecosystems and the ability of river systems to replenish themselves.
Geopolitical scarcity	• Use of minerals exports as a political weapon (e.g. sales embargoes). • Shift in the location of low-cost minerals sources to 'hostile' or unstable blocs of nations (e.g. growing resource nationalism)
Economic scarcity	• Demand at current price levels exceeds the quantity supplied (therefore shortages). • Needs exceed the ability of individuals or countries to pay for resource supplies. • Rich economies can always outbid the poor for essential resources, creating unequal patterns of resource use. • Economic exhaustion of specific minerals or renewable resources causes economic and social disruption in producer regions or in nations dependent on them.
Renewable and environmental resource scarcity	• Distribution of essential biogeographical cycles (e.g. the carbon dioxide cycle and the greenhouse effect) threatening sustainability of life on earth. • Pollution loads exceeding the 'absorptive' capacity, causing economic health and amenity problems. • Loss of plant and animal species (biodiversity) and landscape values, with wide, but poorly understood, long-term consequences.

Source: Adapted from J. Rees, Resources and environment: scarcity and sustainability, in Bennett and Estall (eds) 1991 *Global Change and Challenge: Geography in the 1990s*, Routledge, p. 6

and, thus, the possibility of resource scarcity. This is not just because extrapolation on the basis of current trends is often misleading, but also because there are a whole variety of factors that can promote resource scarcity. Table 5.1 identifies a variety of different types of scarcity. The resource crisis of the 1970s was motivated by geopolitics, but by increasing the price of energy, it brought about other forms of scarcity. For example, the increased price of oil had a major negative impact upon those countries in the 'developing world' that had embarked upon industrialization and had become increasingly dependent upon imported oil. However, it also shocked the 'developed world' into the realization that energy resources were finite, energy conservation was worthwhile and that alternative sources of energy were required. The 'developed world', and particularly the United States, came to realize the strategic importance of securing access to energy supplies. Critics of the war in Iraq say that one of the primary motivations for toppling the regime of Saddam Hussein was the desire on the part of the United States and her allies to secure control of Iraq's substantial oil reserves. The rejuvenation of Iraq's oil industry and subsequent exports are a key component of the country's path to recovery, but that is still some way off and it is too soon to tell what impact increased Iraqi oil exports will have on the world price of oil. Today, despite continued geopolitical tensions in the Middle East and growing resource nationalism elsewhere, we still have a relative abundance of oil, but the rapid growth of energy demand in the emerging world, especially China and India, is causing concern that we are entering a period of sustained high energy prices. At the same time, as noted above, society is increasingly concerned about the ecological consequences of such increased energy consumption. The IPCC Fourth Assessment Report, Working Group Three (2007: 3) suggests that between 1970 and 2004 the largest growth of greenhouse gases came from the energy supply sector, with an increase of 145 per cent. The combination of increasing demand as a result of economic development in the 'global south' and the challenge of climate change has led some commentators to talk about a 'New Energy Paradigm' that combines traditional concerns about security of supply at reasonable prices, with the need to devise energy policies that address climate change (Helm 2007). The remainder of this chapter focuses on contemporary trends in the geography of energy production and consumption and the relationship between energy consumption and economic development.

Plate 5.3 The energy crisis in 1973 brought queues at the petrol pumps in many western countries and a three-day week in the UK.
(Evening Standard/Stringer/Getty Images)

5.2 Fuelling the planet

Of all the different types of resource that we have discussed so far, in recent history it is those that provide a source of energy that have been the most sought after. Figure 5.4 provides a simple classification of the different types of energy resource. For most of human history societies have utilized renewable sources of energy; flow resources such as wood that can be depleted, sustained or increased by human activity; and continuous resources, such as water to drive watermills or the wind to turn windmills, that are available irrespective of human activity. They have also used draught animals, horses, buffalo, etc., but these need to be fed and housed. The industrial revolution changed the way in which in certain parts of the world society powered their economies, while the rest of the world remained dependent upon renewable resources.

5.2.1 The dominance of fossil fuels

Since the invention of the steam engine about 200 years ago, much of human society has become ever more dependent upon the exploitation of non-renewable energy resources. Today three fossil fuels account for over 85 per cent of the annual sale of the world's most important minerals. These resources were formed from the decomposition of organic materials millions of years ago and have been transformed by heat and pressure into coal, oil and natural gas. They represent the classic, non-renewable resource found in specific locations and in a variety of forms and conditions. For example, commercial coal deposits are of three types: lignite (brown coal), bituminous coal and anthracite. There are different grades of oil along a scale from light to heavy and individual refineries are designed to process particular grades of crude oil. Natural gas is often found in association with oil, but is also found on its own. Until recently, the natural gas associated with oil was simply flared off. Today it is often re-injected into the oilfield to enhance oil recovery or used as a hydrocarbon resource in its own right. The mineral content of natural gas also varies: for example, so-called 'sour gas' has high sulphur content. This sulphur has to be extracted before the gas can be transported by pipeline. The advantage of fossil fuels as a source of energy is that they are readily accessible, easy to convert

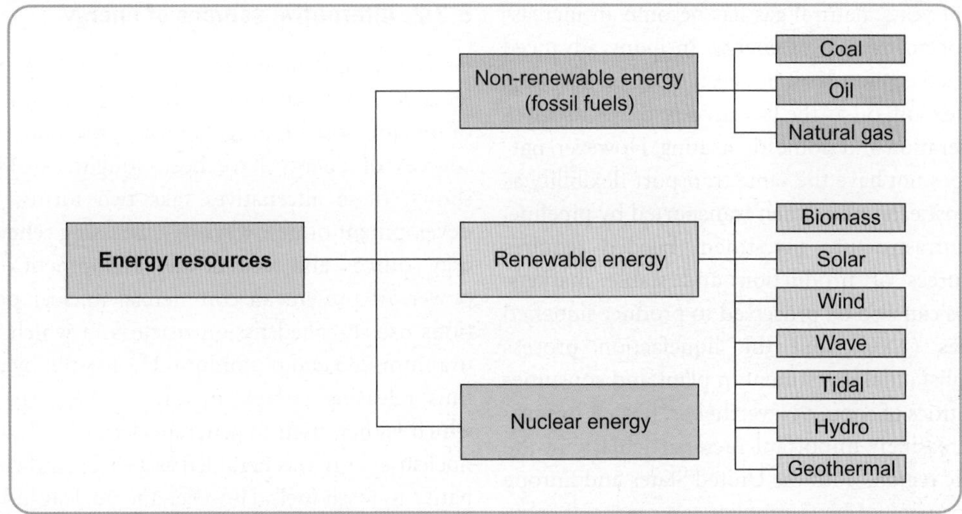

Figure 5.4 Categories of energy resources.
Source: Jones and Hollier (1997: 173)

using proven technologies, and cost-efficient in production and use.

Of the three types of fossil fuel, oil is the most versatile. Today coal is used in many industrial processes and as a source of heat and remains the most important fuel for electricity generation. However, it is a relatively bulky resource and in recent years, particularly in Western Europe, it has been unable to compete with oil that is more versatile, or with cleaner natural gas. The great advantage of oil is its transportability. It can be transported over long distances in large volumes by pipelines, it can cross oceans in tankers, or it can be moved in smaller quantities by rail and road tankers. It has spawned a huge petrochemical industry that now produces a vast range of products. Many of these products, such as plastics and textiles, have substituted products based on renewable resources, such as wood and cotton, for products based on non-renewable resources. At the same time, oil has created many resources for which we have yet to develop commercially viable substitutes, such as jet engine fuel. This means that advanced industrial societies are totally dependent upon the continued supply of oil.

Plate 5.4 Special LNG tankers are used to distribute liquefied natural gas to markets worldwide.
(AP Photo/Dita Alangkara)

In recent years, natural gas has become an increasingly important source of energy. In many advanced industrialized economies it has even replaced coal and, to a lesser extent, oil as the favoured resource for electricity generation and domestic heating. However, natural gas does not have the same transport flexibility as oil. It is most economic when transported by pipeline, but the infrastructure investment needed requires secure sources of production and stable markets. Natural gas can also be processed to produce liquefied natural gas (LNG), but the liquefaction process requires substantial investment in plant and consumes large quantities of energy. Nevertheless, LNG is becoming an increasingly important fuel, particularly in the Asia-Pacific region. Both the United States and Europe see the expansion of LNG supplies as a means of compensating for declines in domestic production and diversifying their sources of supply.

5.2.2 Alternative sources of energy

Given the finite nature of hydrocarbon fuels, the problems of scarcity and the environmental consequences of burning fossil fuels, it is no surprise that alternative sources of energy have been sought. As Figure 5.4 shows, these alternatives take two forms: first, the development of new ways of harnessing renewable energy sources, and second, the development of nuclear power. At present all commercial nuclear power stations use so-called fission reactors in which atoms of uranium-235 and plutonium-239 are split by neutrons, thus releasing energy, mainly in the form of heat, which is then used to generate electricity. In the 1950s nuclear energy was heralded as a cheap and clean alternative to fossil fuels. However, the nuclear industry has failed to expand as expected (see Case study 5.1). Today, the nuclear option is back on the agenda as it is

Plate 5.5 The Chernobyl disaster spelled the end for nuclear power as a clean and safe alternative source of energy.

(Ria Novosti/Science Photo Library)

Thematic Case Study 5.1

Nuclear power on the re-bound?

During the 1950s the peaceful use of nuclear power was heralded as the great hope for the world's energy needs. Harnessing the power of the atom promised a cheap and clean source of electricity. The energy crisis of the 1970s increased interest in a nuclear solution. By developing a nuclear power industry, the developed world hoped that it could reduce its reliance upon OPEC-supplied oil. The 1970s and 1980s saw a rapid expansion in the generating capacity of the world's nuclear power plants. Between 1960 and 1990 total global electrical generating capacity had grown from zero to 328 gigawatts. In the last decade or so growth has stalled (total capacity in 2005 was 368 gigawatts) and there is now the real possibility of a decline in total capacity. What went wrong and have conditions changed such that a return to nuclear power is now back on the agenda?

In short, as the nuclear power industry developed, expanded and matured it became increasingly apparent that the supposed benefits were far outweighed by the environmental problems it posed. Reddish and Rand (1996: 79) have listed four criticisms levelled at nuclear power by environmental groups and these remain the major concerns today:

1. abnormal radiation levels from normal operations will cause cell damage, malignant cancers, genetic diseases, etc.;
2. reactor operations and transport of irradiated fuels cannot be guaranteed safe against catastrophic accidents;
3. radioactive waste has to be disposed of safely and reactors have to be decommissioned, all of which is extremely costly;
4. the radioactive materials needed to fuel the reactors need to be kept safe and secure from theft or misappropriation, to make nuclear weapons (the so-called proliferation problem).

Over the years public trust in nuclear power has been significantly eroded by a number of serious accidents such as those at Windscale, Cumbria (United Kingdom) in 1957, Three Mile Island (USA) in 1979 and Chernobyl in the Ukraine in 1986. Furthermore, as the true cost of building, maintaining and decommissioning nuclear power plants has become apparent many have questioned the economics of nuclear power.

Despite these problems, in 2005 nuclear power accounted for 15 per cent of global electricity generation; but until recently it has been very much an option for the developed world, partly because of cost and technology, but also because of concerns over the non-proliferation of nuclear weapons. In 1996, OECD member states accounted for 86.6 per cent of total nuclear power production, but by 2005 that share had fallen slightly to 83.8 per cent (IEA 2006: 437). In 2005 the top three countries are the United States 26.7 per cent, France 17.2 per cent and Japan 13.0 per cent. In the same year, nuclear power accounted for 78.5 per cent of total domestic power generation in France and 45.4 per cent in Sweden. But Sweden has decided to phase out its nuclear power stations and to switch to conventional thermal power stations. Until recently, the high cost of constructing nuclear power stations, rising public opinion against them and the increasing availability of cheaper alternatives, such as LNG, meant that substantial expansion seemed unlikely. However, the current concerns about climate change and the associated desire to reduce CO_2 emissions, plus the tight supply of oil and gas and associated high-energy prices have changed government attitudes towards nuclear power.

For the emerging markets of China, India and Russia the expansion of nuclear power is seen as a key element of policies to meet the energy demand associated with rapid economic growth. Russia, for example, has a programme to increase the share of nuclear power in electricity generation from 15 per cent now to 25 per cent by 2020. China has set a target to build 40 gigawatts of nuclear generating capacity by 2020 and India has a target of 40 gigawatts by 2030. Many OECD countries are also reconsidering the nuclear option because, first, it addresses the need to reduce greenhouse gas emissions and second, increased reliance on imported oil and gas is raising concerns about energy security. In particular, the European Union is concerned about its increasing reliance on supplies of natural gas from Russia. In May 2007 the British Government published an Energy White paper and launched a consultation document on the future of nuclear power (visit http://www.dti.gov.uk/energy/whitepaper/). The White Paper notes that 40 per cent of global carbon dioxide emissions are created by the generation of electricity. In the United Kingdom, nuclear power's contribution to total electricity generation peaked at 30 per cent in the 1990s; but, by 2023 all but one of the United Kingdom's current nuclear power plants is due to be closed. The question is: how can that capacity be replaced without generating higher CO_2?

→

The UK Government 'believes that it [nuclear power] has a potential contribution to make alongside other low-carbon generating technologies' (DTI 2007: 186). Given the high costs and long lead times to build nuclear power plants, a decision needs to be made now to avoid energy shortages in the future. The environmental movement remains vehemently against nuclear power as it argues that none of the long-standing problems have been adequately addressed. Thus, the nuclear debate is being revisited in the United Kingdom and elsewhere in the OECD, as the disadvantages of the nuclear option are set against the possible benefits it might bring in terms of addressing climate change and promoting energy security (see Case study 5.3). Regardless of the outcome of the debate in the United Kingdom, the United States' Energy Information Administration (EIA 2007: 8) predicts that total installed nuclear capacity will rise from 368 gigawatts in 2004 to 481 gigawatts in 2030 and that non-OECD Asia will account for 68 per cent of the total projected increase in capacity (China 17 per cent, India 17 per cent and Russia 20 per cent). Therefore, nuclear power is on the rebound when it comes to meeting growing global energy demand.

one source of energy that does not generate CO_2; however, many of the concerns about high costs, safety, the disposal of radioactive waste and the decommissioning of power stations remain. In the next few years we can expect a heated debate about the ways in which nuclear power can contribute to mitigating climate change.

The other alternative to fossil fuels is the development of 'new' renewable sources of energy. Many of these renewable sources involve applying new technologies to historic sources of energy supply (see Elliott 2003 for further discussion of renewable energy sources and their prospects):

- *Biomass energy* involves the burning of plant and animal residue to produce heat. In much of the world this is still the traditional source of energy. According to the IEA's Biomass Project, the share of biomass energy in Africa, Asia and Latin America averages 30 per cent. However, in some of the poorest countries of Asia and Africa it can be as high as 80–90 per cent. It is estimated that for some 2 billion people biomass is the only available and affordable source of energy. More recently, the use of biomass to produce alternative fuels, so-called biofuels, has gained popularity in the developed world; but it is increasingly apparent that biofuels production creates as many problems as it solves. The land being taken up to produce crops for biofuel production is either a natural habitat or agricultural land that could be used for food production. At the same time, increased demand for crops such as corn is increasing the cost of animal fodder and basic foodstuffs. Further, much of this production requires the use of fertilizers and pesticides that are hydrocarbon-based and consume considerable amounts of energy when they are produced. The net result is that many biofuels may actually consume more energy to produce than they generate when used to run a vehicle.
- *Solar energy* usually involves capturing the sun's rays in order to warm buildings and heat hot water. As the efficiency of solar panels increases and their cost declines, solar energy is becoming a viable small-scale local solution. However, it is subject to daily or seasonal variations and is weather-dependent, but it can provide an efficient supplemental source of energy for heating domestic hot water, and so on (visit the website of the International Solar Energy Society at www.ises.org).
- *Wind power* is of historical significance and has now made a return in the form of 'wind farms'. Improved design has increased the efficiency of modern windmills and they are becoming an increasingly common sight in exposed coastal and upland locations in the developed world. Wind farms can generate electricity for local use or for the power grid. However, some would argue that modern wind farms lack the aesthetic qualities of traditional windmills and therefore many see them as a source of both visual and noise pollution (for more information on wind power see www.windpower.org). In 2003, the UK government announced plans for huge offshore wind farms to be built off the UK coast in the Thames Estuary, the Greater Wash and the North-west. The government is aiming to source 10 per cent of UK electricity from renewable power by 2010 and the energy from these new wind farms is expected to power 15 per cent of UK households.
- *Hydropower generation* also pre-dates the industrial revolution, but the watermills that powered the early machines have now been replaced by dams and hydroelectric schemes. In many regions of the world

hydroelectricity has become an important source of energy, but, in all but a few cases, it cannot offer a large-scale solution to a country's energy needs. The building of dams to store water and build a 'head' to drive the turbines that generate the electricity floods large areas of land; it also starves river systems of flow and damages deltaic and riverine environments. The construction of Three Gorges Dam on the Yangtze river in China involves planning the displacement of 1.2 million people and the creation of a reservoir occupying 111,000 hectares. Other forms of water-power, such as wave and tidal power, are also being developed; again, they cannot provide a viable alternative to fossil fuel power generation. At present, there is only one large-scale commercial tidal power station in the world, the La Rance Tidal Power Plant in Brittany, France. However, plans have been announced to build a 10-mile long barrage across the mouth of the River Severn between England and Wales. The barrage would exploit the Bristol Channel's 14-metre tidal range, the second highest in the world, and could generate 5–6 per cent of the electricity needed by England and Wales. However, environmental groups are concerned that the scheme will damage the region's ecology and the cost of the scheme is put at £15 billion (US$30 billion).

- The final form of renewable energy is *geothermal energy*. As the name suggests, this form of energy uses naturally occurring heat in the earth's crust and is only really viable in areas of volcanic activity. There are two forms of geothermal energy. One is wet-rock geothermal energy where steam or hot water is trapped from boreholes or surface vents and used to heat buildings or generate electricity. The other is hot–dry geothermal energy, which is accessed by boreholes drilled into hot dry rocks. Water is then forced down and the steam used to generate electricity. Geothermal energy, by its very nature, is a highly localized energy source and requires careful management. For example, there is now evidence from New Zealand that excessive extraction of geothermal energy has reduced the level of geyser activity, creating a conflict between the desire to generate power and heat houses and the need to maintain the revenues from tourists attracted by the volcanic springs. In 1998 geothermal energy accounted for 15 per cent of New Zealand's energy production.

In nearly all of the above cases, renewable energy seems capable of providing localized, small-scale solutions to energy needs. At present, renewable energy sources do not provide a viable alternative to fossil fuels. Furthermore, most of these alternatives require considerable capital investment and access to advanced technology; therefore, they do not offer a solution to the fast-growing energy needs of the 'developing world'. That said, the 'developed world' can afford to develop renewable energy technologies and considerable progress has been made. The European Union aims to increase the contribution made by renewable energy from 7 per cent today to 20 per cent by 2020 (Commission of the European Communities 2007: 13). If fossil fuel prices remain high, it will stimulate investments in renewable energy and it will become a viable alternative for those that can afford it. But, to provide a more global solution, those technologies will need to be made available to the regions experiencing the most rapid increases in energy demand, the global south.

Plate 5.6 Wind farms: a sustainable energy source for some, an eyesore for others.
(Steven gillis hd9 imaging/Alamy)

5.2.3 The changing energy mix

It follows from the discussion above that the relative importance of the various sources of energy has varied across time and space. The balance between the various sources of energy is known as the energy mix. During the last century, there was a substantial change in the global energy mix, but as Table 5.2 shows, there still remains substantial regional variation. The evolution of the energy mix tends to be linked to the development of the industrialized world, simply because it accounts for most of the world's energy consumption (Table 5.3). At the beginning of the twentieth century, coal was by far the most important energy resource in the industrialized world. The oil industry was still in its early stages; the first modern oil well was drilled in Pennsylvania in the United States in 1859. Natural gas was not recognized as a resource at all. It was not until after the Second World War that the dominance of coal was threatened. Advances in transportation technology, such as the invention of the jet engine, and the emergence of a petrochemicals industry led to a rapid increase in the demand for oil in the post-war boom years. In 1950 coal accounted for 61 per cent of the world's commercial energy consumption and oil

Table 5.2 Regional variations in energy balance, 2006 (per cent of total consumption)*

	Oil	Natural gas	Coal	Nuclear	HEP	Total (mtoe)
OECD	40.7	23.2	21.1	9.7	5.4	5,553.1
EU 25	41.0	24.4	17.7	12.7	4.1	1,722.8
North America	40.1	25.1	21.8	7.6	5.4	2,803.0
Europe & Eurasia	32.0	34.1	18.3	9.5	6.1	3027.2
S. & C. America	44.7	22.2	4.1	0.9	28.0	528.6
Africa	40.3	21.0	31.7	0.7	6.2	324.1
Middle East	50.5	47.0	1.6	0.0	0.9	554.2
Asia Pacific	31.5	10.8	49.2	3.5	4.9	3641.5
World	35.8	23.7	28.4	5.8	6.3	10,878.5

*Includes commercially traded fuels only. Therefore excludes wood, peat and animal waste as fuels.

Source: BP *Statistical Review of World Energy* (2007: 41). (Also available at: www.bp.com)

Table 5.3 Global distribution of primary energy consumption, 2006 (per cent of total world consumption)*

	Oil	Natural gas	Coal	Nuclear	HEP	Total
OECD	58.1	50.0	37.9	84.5	43.5	5,553.7
EU 15	18.2	16.3	9.9	34.5	10.4	1,722.8
North America	38.9	27.3	19.9	33.4	22.1	2,803.0
Europe & Eurasia	24.9	40.1	17.9	45.3	26.8	3027.2
S. & C. America	6.1	4.6	0.7	0.8	20.7	528.6
Africa	3.4	2.6	3.3	0.4	2.9	324.1
Middle East	7.2	10.1	0.3	0.0	0.7	554.2
Asia Pacific	29.5	15.3	58.0	20.2	25.9	3641.5

*Includes commercially traded fuels only. Therefore excludes wood, peat and animal waste as fuels.

Source: BP *Statistical Review of World Energy* (2007: various pages). (Also available at: www.bp.com)

27 per cent. By 1970 coal had fallen to 30 per cent, oil had increased to 44 per cent and natural gas to 20 per cent. The absolute level of coal production did not decline (in fact at a global scale coal production continues to increase). The rapid economic growth experienced by the developed world was instead increasingly underpinned by oil and, more recently, by natural gas. The recent 'dash-for-gas' in the developed world is being driven by a desire to reduce dependence on oil imported from the Middle East and by the fact that using natural gas to produce electricity produces less greenhouse gases than coal or oil. Thus, in the past hundred years, the energy mix in the industrialized world has gone through two major transitions, from coal to oil and from oil to gas.

These energy transitions have not been a universal experience. As Table 5.2 illustrates, the energy mix in the developing world is somewhat different. For the most part the developing world remains more reliant upon oil for its commercial energy and in some regions coal remains the dominant resource. For example, in China in 2006 coal accounted for 70 per cent of commercial energy production and China accounts for 38.5 per cent of global coal consumption. In fact, over the past few years China has met its growing energy needs by burning more coal, which has raised obvious concerns about climate change. Elsewhere dependence upon oil has made the developing world particularly vulnerable to the sudden increases in oil price that have been experienced at times over the past three decades. But, unlike the developed world, these economies lack the capital and technology to develop their own energy potential or to diversify their energy mix. However, as the developed world has depleted its own sources of oil and gas it has become increasingly dependent upon supplies from the developing world. Thus there has been a marked change in the geography of energy production and, to a lesser extent, consumption. The transnational oil companies have been key actors in developing the energy resources of the developing world, but increasingly the governments of the oil-producing states in the developing world have seized control over their oil and gas industries. Now the leading oil majors, such as BP, ExxonMobil, ChevronTexaco and Shell, find themselves working in partnership with regimes whose actions in relation to human rights and environmental protection are condemned by governments and NGOs in the developed world. Thus, for the international oil companies the age of 'easy oil' is over and they must now make do with access to the most technologically and environmentally challenging fields.

5.2.4 The changing geography of energy production and consumption

Whereas in 1950 North America, Europe and the Soviet Union accounted for nearly 90 per cent of world energy demand, by 1990 their share had fallen to two-thirds (Jones and Hollier 1997: 181). In 1997 the member states of the Organization for Economic Cooperation and Development (OECD) accounted for 58.4 per cent of global energy consumption but by 2006 this had fallen to 50.1 per cent (Table 5.3). The redistribution of the global pattern of energy consumption has not been due to decline in the developed world, although during the 1990s this has been the case in the so-called 'transition economies', but to increased demand in the developing world. Table 5.4 provides information on energy consumption and efficiency by major world region and income group. The table highlights the substantial variations that exist in terms of level of per capita energy use and energy intensity. The high-income countries have by far the highest level of energy use. The high levels of usage in the middle-income countries and in Europe and Eurasia reflect the energy legacies of the post-socialist states in Central Europe and the former Soviet Union. The World Bank (2007: 6) observes that 'in 2003 nearly half of the world's carbon dioxide emissions from cement manufacturing and the combustion of fossil fuels came from the high-income countries. The largest emitter is the United States. In the developing world China and India are the major emitters.' Over the past decade or so, there has been a global shift in the energy system. The growing populations and developing economies of the countries like Brazil, India and China is driving the growth of energy demand, while the developed economies are seeking to reduce their levels of energy consumption. According to BP, in 2005–6 Chinese energy consumption grew by more than 8 per cent, taking the country's share of global energy production to more than 15 per cent. This differential growth rate, together with the decline in indigenous production in the developed world, has major implications for the geography of energy production and consumption and for global geopolitics.

There are at least three elements to the geography of energy production: the global distribution of reserves, the distribution of energy production and consumption and the resultant pattern of trade between energy-surplus and energy-deficit regions. As we noted earlier, the potential to produce coal, oil and natural gas is geologically determined, but whether or not a given

Table 5.4 Energy consumption and efficiency

	GDP per unit of energy use (2000 ppp $/kg oil equivalent)	Energy use per capita (kg oil equivalent)	CO_2 emissions per unit of GDP (2000 ppp $/kg oil equivalent)
World	4.7	1,793	0.5
Low income	4.4	513	0.4
Middle income	4.2	1,451	0.6
Lower middle income	4.5	1,175	0.6
Upper middle income	3.7	2,583	0.7
Low & middle income	4.3	1,068	0.6
High income	5.2	5,511	0.5
East Asia & Pacific	4.4	1,124	0.6
Europe & Central Asia	2.8	2,847	0.9
Latin America & Caribbean	6.2	1,186	0.3
Middle East & N. Africa	4.2	1,189	0.7
South Asia	5.5	486	0.4
Sub-Saharan Africa	2.8	703	0.4

Source: World Bank (2007: 12–26). *The Little Green Data Book 2007*. (Also available at: www.worldbank.org)

deposit is exploited depends on a whole host of factors. The regional distributions of global reserves of coal, oil and natural gas are presented in Tables 5.5, 5.6 and 5.7. Each table also includes the reserve to production ratio at the end of 2006. This is a measure of how many years the current proven reserves would last if current levels of consumption were maintained. From a comparison of the reserve to production ratios for the three fossil fuels, it is quite clear that coal is the most abundant fossil fuel, followed by natural gas and then oil. At the end of 2006 the world had sufficient reserves of coal to last 147 years at current rates of consumption, while current oil reserves would last 40.5 years. These tables also reveal some interesting regional variations in the level of reserves. While the OECD countries have more than sufficient coal to meet their needs, it is these

Table 5.5 Coal reserves at end of 2006 (million tonnes)

	Anthracite and bituminous	Sub-bituminous and lignite	Total	Share of total %	R/P ratio[1]
North America	115,669	138,763	254,432	28.0	226
S. & C. America	7,701	12,192	19,893	2.2	246
Europe & Eurasia	112,256	174,839	287,095	31.6	237
Africa & Middle East	50,581	174	50,755	5.6	194
Asia Pacific	192,564	104,325	296,889	32.7	85
Total World	478,771	430,293	909,064	100.0	147
Of which OECD	172,363	200,857	373,220	41.1	177

[1]Reserve to production ratio

Source: BP *Statistical Review of World Energy* (2007: 32). (Also available at: www.bp.com)

Table 5.6 Proved oil reserves at end of 2006

	Thousand million tonnes	Share of total %	R/P ratio[1]
North America	7.8	5.0	12.0
S. & C. America	14.8	8.6	41.2
Europe & Eurasia	19.7	12.0	22.5
Middle East	101.2	61.5	79.5
Africa	15.5	9.7	32.1
Asia Pacific	5.4	3.4	14.0
Total World	164.5	100.0	40.5
Of which OECD	10.4	6.6	11.3
OPEC	123.6	74.9	72.5
Non-OPEC[2]	23.2	14.4	13.6
Former Soviet Union	17.7	10.6	28.6

[1]Reserve to production ratio
[2]Excludes Former Soviet Union
Source: BP *Statistical Review of World Energy* (2007: 6). (Also available at: www.bp.com)

Table 5.7 Proved natural gas reserves at end of 2006

	Trillion cubic metres	Share of total %	R/P ratio[1]
North America	7.98	4.4	10.6
S. & C. America	6.88	3.8	47.6
Europe & Eurasia	64.13	35.3	59.8
Middle East	73.47	40.5	Over 100 years
Africa	14.18	7.8	78.6
Asia Pacific	14.82	8.2	39.3
Total World	181.46	100.0	63.3
Of which OECD	15.90	8.8	14.7
European Union 25	2.43	1.3	12.8
Former Soviet Union	58.11	32.0	74.6

[1]Reserve to production ratio
Source: BP *Statistical Review of World Energy* (2007: 22). (Also available at: www.bp.com)

economies that have reduced their reliance upon coal the most. Thus, the developed world is dependent upon the very fuels for which it has the lowest reserves to production ratios. What this means is that the developed world is likely to become even more dependent upon imports of energy from the developing world. In the case of oil reserves, the Middle East still accounts for the vast majority of the world's oil. Natural gas reserves are provided by two dominant regions, the Middle East and the former Soviet Union (predominantly Russia). Thus, the majority of the world's oil and gas reserves are in regions that are presently politically and/or economically unstable.

The relationship between production and consumption is illustrated in Figures 5.5 and 5.6. These diagrams compare the share of world production

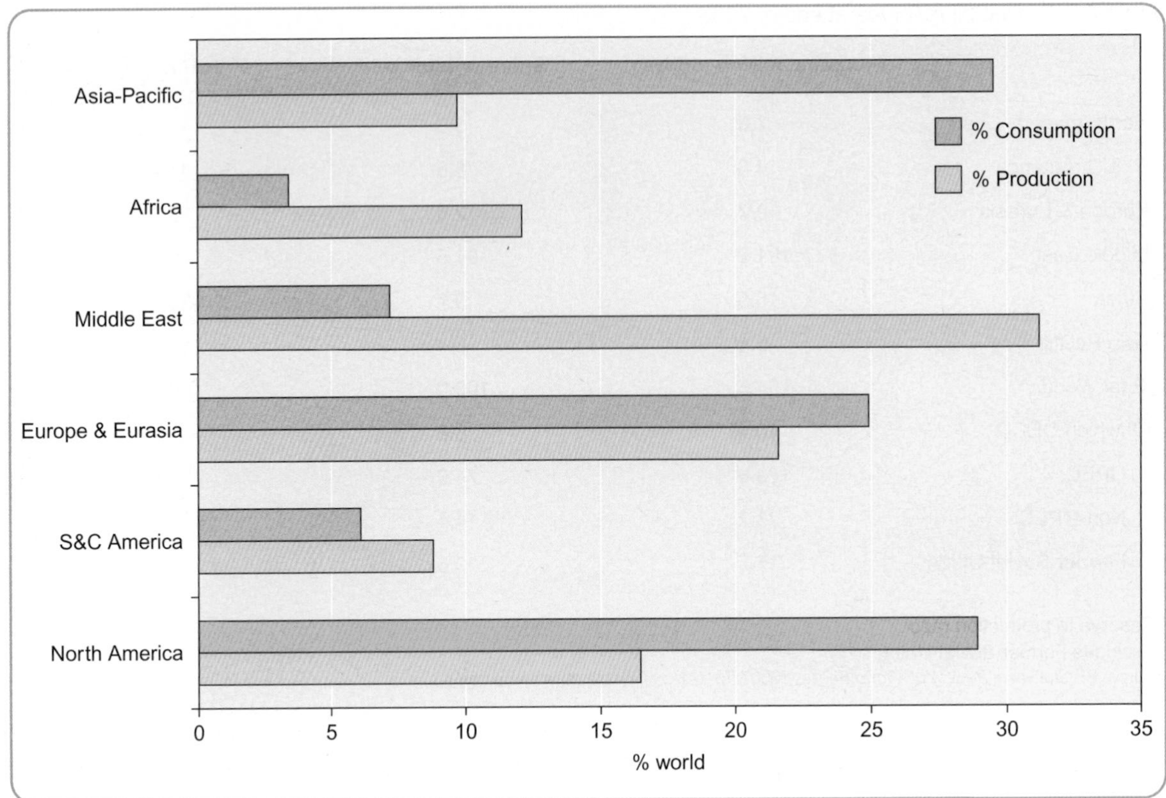

Figure 5.5 Distribution of oil production and consumption in 2006.
Source: BP *Statistical Review of World Energy* (2007: 8, 11). (Also available at: www.bp.com)

and consumption by major region. The comparative distributions of oil production and consumption show that Asia-Pacific, Europe and North America are the major deficit regions with Africa, the Middle East, the former Soviet Union and South and Central America as the major surplus regions. The individual countries involved in the resultant trade are identified in Table 5.8 and the movements of oil are shown in Figure 5.7. Despite being the second largest producer in 1997, the United States still imports almost as much oil as it produces. Likewise, North Sea production falls well short of meeting Europe's demand for oil and is now past its peak production. The result is a movement of oil production dominated by Middle Eastern supply to Europe and North America. Figure 5.7 also reveals substantial movements to Singapore and Japan.

The major exporters and importers of natural gas are listed in Table 5.9, while Figure 5.8 shows the movements of natural gas. A comparison of the two figures reveals the less 'transportable' nature of natural gas; for the most part there is a balance within each region. The notable exceptions are Africa and the former Soviet Union (Russia and Turkmenistan), which have a surplus of natural gas, and Europe, which has a deficit. Natural gas supply into Europe is via pipeline from West Siberia (Russia) and North Africa or as LNG from North Africa. There is also a growing movement of natural gas in the form of LNG to Japan and South Korea. In 1995 LNG accounted for 14.3 per cent of Japan's mineral fuel imports (Japan Institute for Social and Economic Affairs 1996: 74–7). At present most of this LNG is supplied from within the Asia-Pacific region, but there is likely to be an increase in supply from the Middle East to Asia and the west coast of Africa to Europe and the United States. In 2008 LNG from Sakhalin Island in the Russian Far East will be delivered to the west coast of Mexico. There is also the possibility of pipeline supply of natural gas from Russia to China, Japan and South Korea.

The geographies of production and consumption described above are the result of a complex interaction of economic and political factors. Not surprisingly, there are substantial regional differences in the cost of oil production. The record-low oil prices of the late 1990s focused attention upon the actual cost of producing oil. At the time, a report by Cambridge Energy Research Associates put the cost of production in the Middle East as low as US$2/barrel (*Economist*, 4 March 1999: 29).

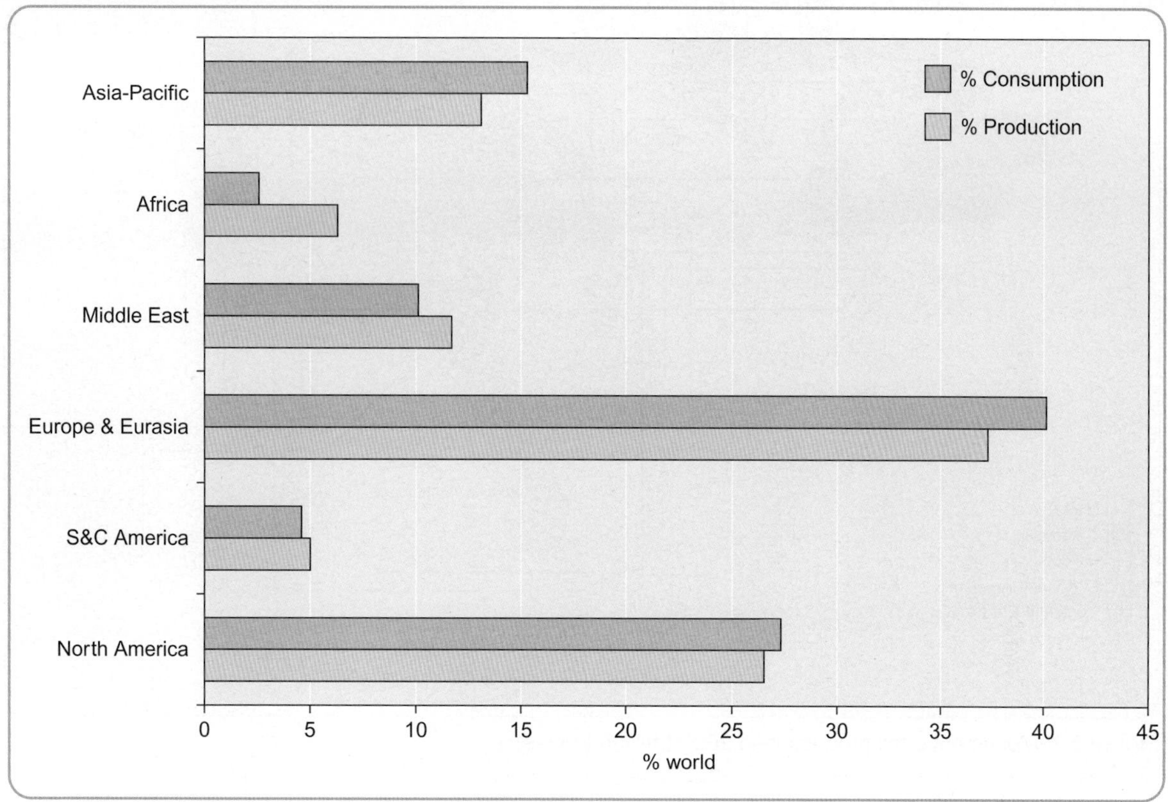

Figure 5.6 Distribution of natural gas production and consumption in 2006.

Source: BP *Statistical Review of World Energy* (2007: 24, 27, 25). (Also available at: www.bp.com)

Table 5.8 Producers, exporters and importers of crude oil

Producers in 2005	Mt	% of world	Exporters in 2004	Mt	Importers in 2004	Mt
Saudi Arabia	519	13.2	Saudi Arabia	346	United States	577
Russia	470	12.0	Russia	258	Japan	206
United States	307	7.8	Norway	132	China	123
Iran	205	5.2	Nigeria	123	Korea	114
Mexico	188	4.8	Iran	122	Germany	110
China	169	4.8	Mexico	105	India	96
Venezuela	162	4.1	UAE	93	Italy	93
Canada	143	3.6	Venezuela	94	France	85
Norway	139	3.5	Canada	79	United Kingdom	63
Nigeria	133	3.4	Iraq	75	Netherlands	60
Rest of World	1,474	37.7	Rest of World	716	Rest of world	708
World	3,923	100	World	2,513	World	2,235

Source: OECD/IEA (2006) *Key World Energy Statistics* (Also available at: www.iea.org)

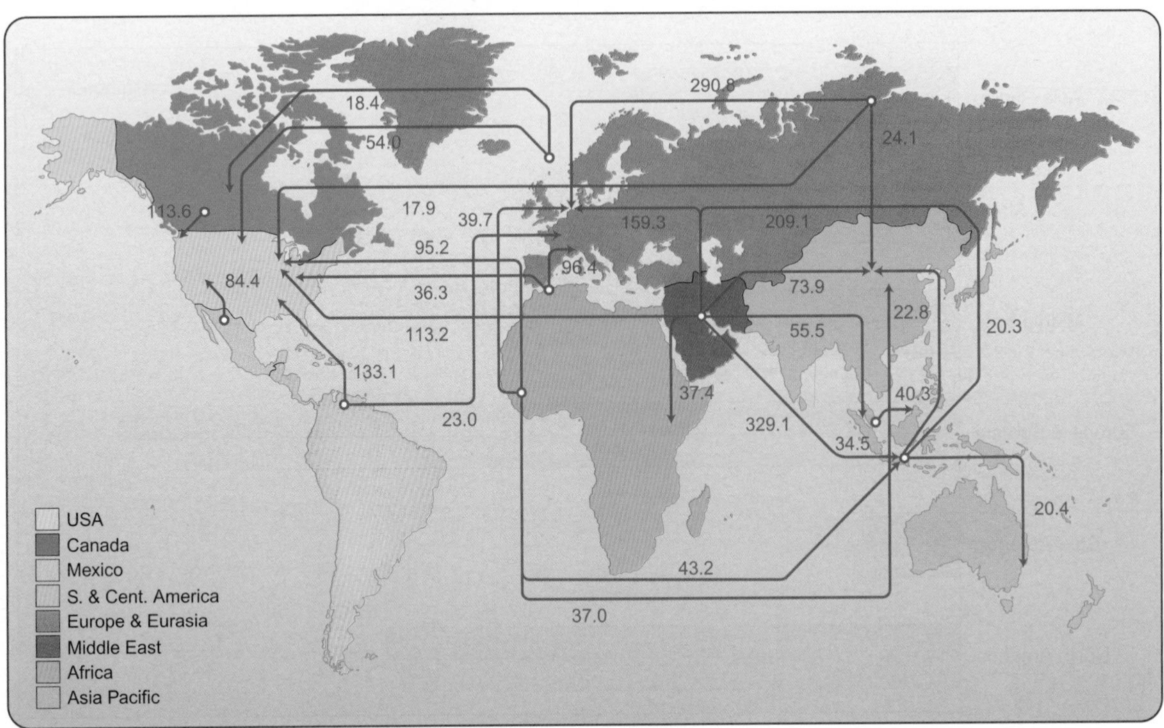

Figure 5.7 Movement of oil production in 2006 (million tonnes).

Source: BP *Statistical Review of World Energy* (2007: 13). (Also available at: www.bp.com)

Table 5.9 Producers, exporters and importers[1] of natural gas

Producers in 2005	Mm³	% of world	Exporters in 2005	Mm³	Importers in 2005	Mm³
Russia	627,466	21.8	Russia	203,727	United States	121,348
United States	516,614	18.0	Canada	106,353	Germany	90,700
Canada	187,164	6.5	Norway	82,801	Japan	80,915
Algeria	92,797	3.2	Algeria	68,638	Italy	73,460
United Kingdom	92,045	3.2	Netherlands	52,355	Ukraine	62,132
Norway	89,559	3.1	Turkmenistan	49,423	France	46,975
Iran	83,535	2.9	Indonesia	36,146	Spain	33,118
Netherlands	78,804	2.7	Malaysia	32,614	Korea	29,494
Indonesia	77,305	2.7	Qatar	27,992	Turkey	26,572
Saudi Arabia	69,500	2.4	United States	22,288	Netherlands	23,025
Rest of World	957,004	33.3	Rest of World	165,646	Rest of World	188,338
World	2,871,773	100.0	World[2]	847,983	World	703,766

[1]Exports and imports include pipeline gas and LNG
[2]World trade includes intra-trade of former USSR

Source: OECD/IEA (2006) *Key World Energy Statistics*. (Also available at: www.iea.org)

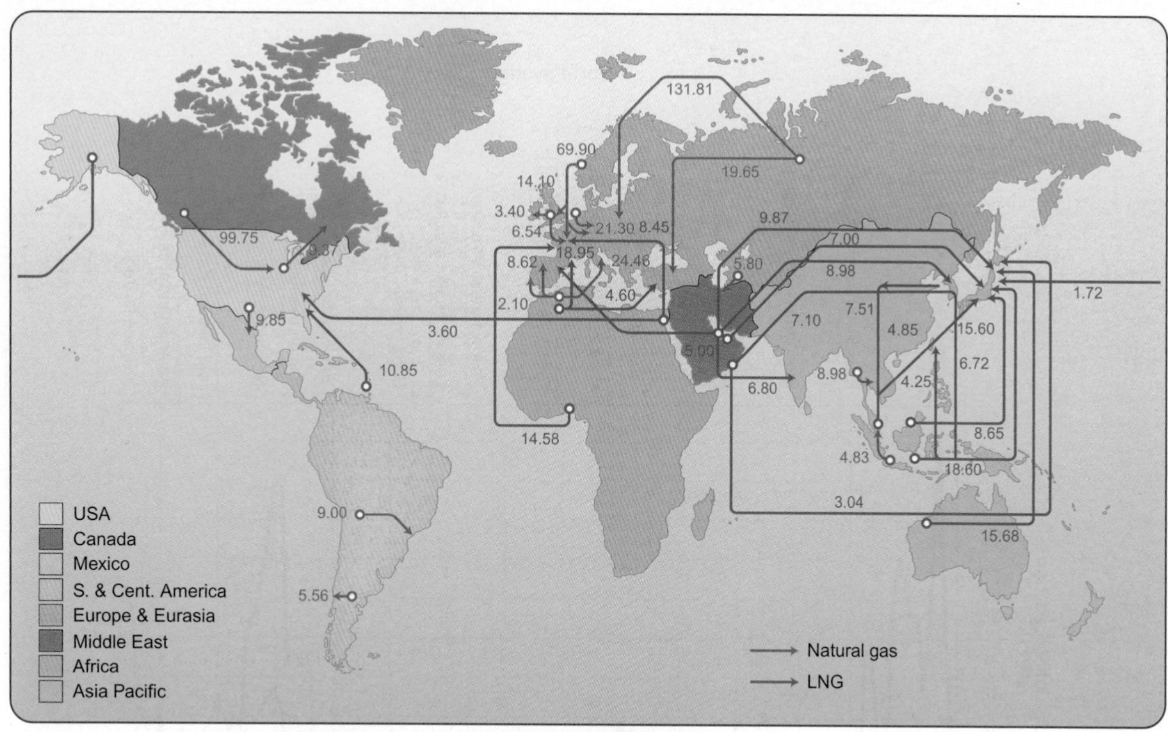

Figure 5.8 Movement of natural gas in 2006 (billion cubic metres).

Source: BP *Statistical Review of World Energy* (2007: 31). (Also available at: www.bp.com)

That same report estimated the cost of production in Indonesia at US$6, Venezuela and Nigeria at US$7, Mexico at US$10, the United States and the North Sea at US$11 and Russia at US$14. The current period of high oil price is promoting exploration activity in high-cost locations and it also promotes the exploitation of high-cost tar sands, such as found in western Canada. These production costs pose two questions: why develop high-cost fields when the Middle East has massive reserves and very low production costs, and what will happen to the high-cost producers if oil prices fall?

The answers lie in the realms of geopolitics as much as economics. Prior to the first oil shock of 1973–4 the industrialized world had become increasingly dependent upon supplies of cheap oil from the Middle East (Odell 1989). When the Middle Eastern oil producers formed OPEC (the Organization of Petroleum Exporting Countries, see www.OPEC.org) and used oil as a political weapon to punish the industrialized West for supporting Israel in the Arab–Israeli war this brought about a rapid increase in the price of oil. The first oil shock was followed by a further shock in 1979–80 following the Iranian Revolution and a mini-shock in 1990–1 because of the Gulf War (see Figure 5.9). Following the second shock prices started to decline. There were at least three reasons: first, high energy costs promoted conservation; secondly, the high price of oil and the actions of OPEC promoted production by high-cost non-OPEC producers; and thirdly, recession and economic restructuring reduced the growth of demand in the developed world. The economic crisis in Asia further dampened demand for energy. As a response to very low prices, OPEC members agreed to cut back production and this action, combined with global economic recovery, saw prices recover during 1999. However, subsequent events dramatically illustrate the volatile nature of oil and gas markets.

In the aftermath of 11 September 2001, a global economic downturn kept oil prices low. Then the global geopolitical situation changed dramatically with the war in Afghanistan and then the second Iraq War. As this book goes to press, the price of oil has topped $105 a barrel and there is every indication that prices will remain high. A surge in demand from emerging markets, principally China and India, plus tight supply and continued instability in the Middle East have all contributed to a short-term 'supply gap'. In the longer term, increased production could ease supply concerns, or global recession could reduce demand.

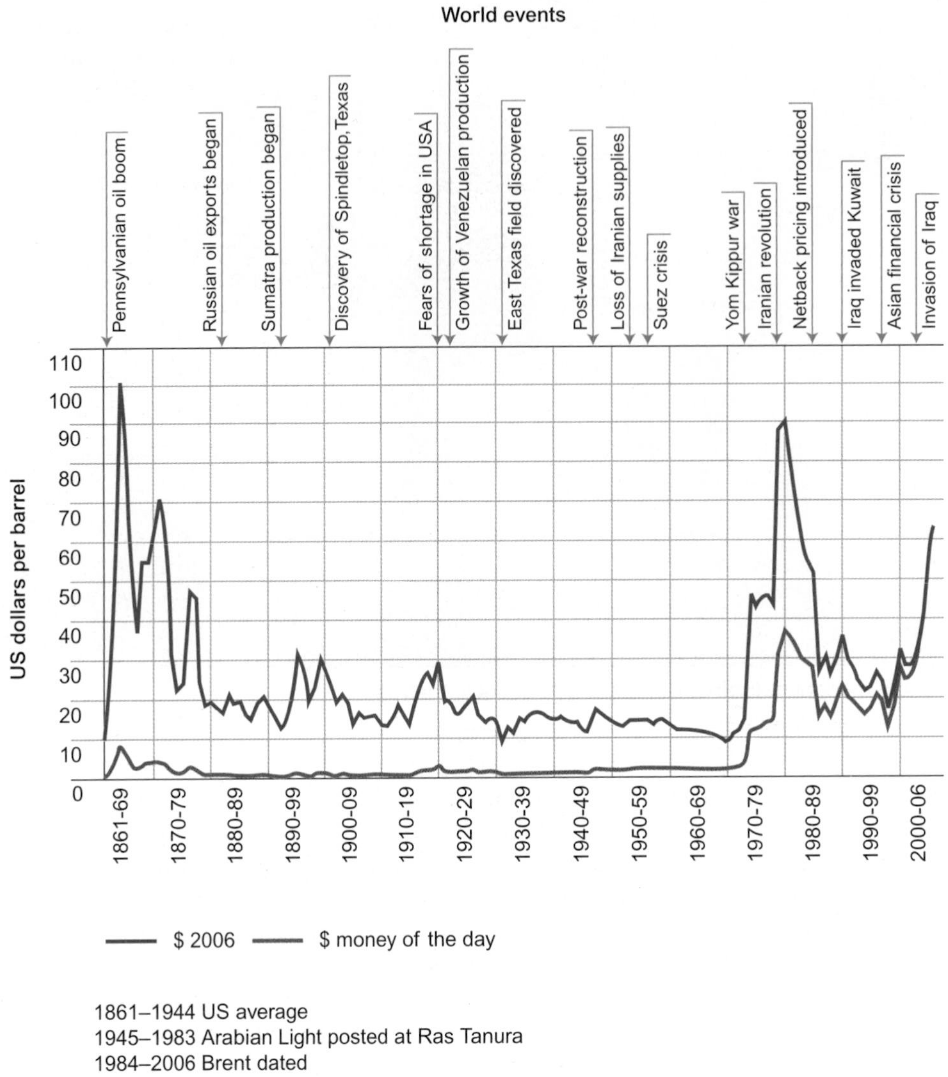

World events

Pennsylvanian oil boom
Russian oil exports began
Sumatra production began
Discovery of Spindletop, Texas
Fears of shortage in USA
Growth of Venezuelan production
East Texas field discovered
Post-war reconstruction
Loss of Iranian supplies
Suez crisis
Yom Kippur war
Iranian revolution
Netback pricing introduced
Iraq invaded Kuwait
Asian financial crisis
Invasion of Iraq

US dollars per barrel
0 10 20 30 40 50 60 70 80 90 100 110
1861-69 1870-79 1880-89 1890-99 1900-09 1910-19 1920-29 1930-39 1940-49 1950-59 1960-69 1970-79 1980-89 1990-99 2000-06

$ 2006 $ money of the day

1861–1944 US average
1945–1983 Arabian Light posted at Ras Tanura
1984–2006 Brent dated

Figure 5.9 Historical trends in the world crude oil prices, 1861–2006.

Source: BP *Statistical Review of World Energy* (2007: 16). (Also available at: www.bp.com)

Thematic Case Study 5.2

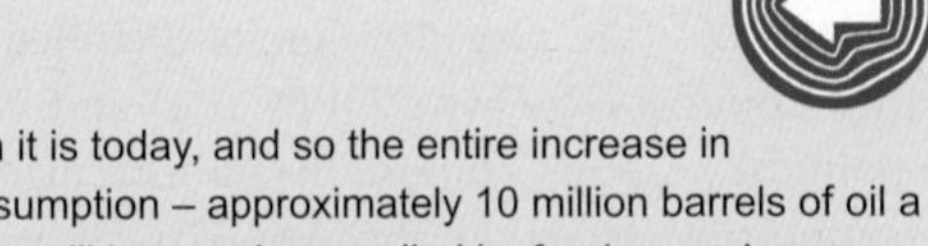

Energy security

At present, the United States – with something less than 5 percent of the world's total population – consumes about 25 percent of the world's total supply of oil. In 2025, if current trends persist, we will be consuming half as much petroleum again as we do today; however, domestic production will be no greater than it is today, and so the entire increase in consumption – approximately 10 million barrels of oil a day – will have to be supplied by foreign producers. And because we can't really control what goes on in those countries, we become hostage to their capacity to ensure an uninterrupted flow of petroleum.

Klare (2004: 11)

A generally accepted definition of 'energy security' is the reliable supply of energy at reasonable prices. As we know from discussions earlier in this chapter, resource scarcity is seldom about the physical shortage of a particular resource: this is particularly true of oil and gas. When the first edition of this book was in preparation we were in a period of very low oil prices; today we face sustained high oil prices. Furthermore, there seems to be a growing consensus that in the longer term we have to reduce our reliance on the consumption of hydrocarbons to avoid the more catastrophic consequences of climate change.

At present, we seem to be experiencing a period of pressure on global oil supplies because of increased demand from growing economies, such as China and India, combined with supply interruptions caused by hurricanes in the Gulf of Mexico, labour unrest in Nigeria, declining production in long-established fields and delays in bringing on new fields. Thus, there is increased competition for access to supplies of oil and gas. Furthermore, an increasing share of the oil and gas consumed by the developed world (OECD) is imported from non-OECD and OPEC suppliers, particularly in the Middle East. In such a context, it seems that it is the business of national governments to devise policies to promote energy security. These may take the form of seeking to promote energy conservation, promoting other sources of energy supply, promoting cooperation with, and investing in, oil and gas producing states (often via support for the activities of state and private oil companies) and *in extremis* taking military action to protect the supply of oil. Many would argue that since the so-called 'Carter Doctrine' of 1979, when President Jimmy Carter, in the wake of the Soviet invasion of Afghanistan, 'declared that any move by a hostile power to acquire control over the Gulf region would be regarded "as an assault on the vital interests of the United States of America", which would be opposed "by any means including military force"' (quoted from Rutledge 2006: 48), US foreign policy in the Middle East has been all about oil (Harvey 2003; Klare 2004). Both President Bush and Prime Minister Blair denied that the invasion of Iraq was motivated by a desire to secure Iraqi oil production; even if it was, it failed as Iraqi production is now lower than before the war and this is helping to keep the oil price up. Whatever the motivations for military intervention in the Middle East, it is a fact that the United States, Europe and Asia are dependent on supplies of oil from that region. It is also true that Europe has sought to reduce this reliance by importing oil and gas from Russia and that this is now also proving a source of geopolitical concern as Russia is using its perceived status as an 'energy superpower' to pursue its own political agenda.

The combination of concerns about the geopolitical dimension of energy security, becoming increasing reliant on unreliable and potentially hostile sources of supply, and about climate change, the need to reduce levels of hydrocarbon consumption, have created a new energy paradigm (Helm 2007) and have also prompted a need to rethink the concept of energy security (Yergin 2006). As Daniel Yergin (2006: 70) puts it in a highly influential article in the journal *Foreign Affairs*: 'a wider approach [to energy security] is now required that takes into account the rapid evolution of global energy trade, supply-chain vulnerabilities, terrorism, and the integration of major new economies in the world market.' According to Yergin, this new approach or framework must abide by the following principles:

- Diversification of supply to reduce the impact of a disruption in supply from one source.
- A 'security margin' in the energy supply system that provides a buffer against shocks and facilitates recovery after disruptions.
- Recognition of the 'reality of integration', that there is only one global oil complex and a worldwide system that moves 86 million barrels of oil every day. For all consumers stability resides in the stability of this market.
- The importance of information that underpins well-functioning markets.

In addition to these principles, Yergin (2006: 70) maintains that there is a need to recognize the globalization of the energy security system, which means that the developed world must engage with the likes of China and India, and a need to expand the concept of energy security to include the entire energy supply chain and infrastructure. In the longer term, it may be the case that the solution lies in developing an alternative to today's hydrocarbon economy; however, for the next 30 years or more we will depend on oil and gas for the majority of our energy needs and the competition for increasing scarce resources means that concerns about energy security can only grow. Next time you switch on the light think about the complex set of issues that now lies behind keeping the lights on!

High oil prices have now resulted in renewed interest in non-renewable sources of energy and even a growing enthusiasm for nuclear power. Furthermore, continued instability in the Middle East is encouraging Western governments and the multinational oil companies to look for oil and gas elsewhere, such as off the west coast of Africa. All of these developments illustrate the complexity of the notion of resource scarcity. But what does the current situation mean for the developing world? Much depends on whether or not the developing world follows the same pattern of economic development as the developed world; this is the subject of the final part of this chapter.

5.3 Energy and development

What is the relationship between energy consumption and economic development? The experience of the developed world suggests that in the initial phases of industrialization there is a direct link between increased energy consumption and economic development. That is, as industrial activity grows it consumes more and more energy. However, in the past thirty years it has become increasingly apparent that in the developed world the relationship between economic development and energy consumption, has changed. This was a direct consequence of the OPEC-inspired energy crisis of the 1970s and is largely due to the processes of de-industrialization and economic restructuring that have seen the less energy-intensive service sector replace heavy industry and, to a lesser degree, manufacturing as the major generators of wealth. At the same time, conservation and technological change has made industry more energy-efficient. Similarly, a change in transportation technology has increased energy efficiency. Thus, today's post-industrial societies have decoupled the link between economic growth and increased energy consumption. Nevertheless, it is still the case that the developed world consumes the lion's share of the world's energy. Indeed, just one country, the United States, which accounts for only 5 per cent of the world's population, consumes almost 25 per cent of total world energy consumption (Mather and Chapman 1995: 140).

5.3.1 Energy consumption, economic development and climate change

The relationship between energy consumption and economic development is usually depicted as a scatter plot between the level of per capita energy consumption on the one hand, and GNP per capita on the other hand (Figure 5.10). The data do indeed suggest a clear relationship between energy consumption and GNP per capita; the higher the level of energy consumption the higher the level of GNP per capita. There are obvious outliers, but these are easily explained away. Some countries traditionally have higher levels of energy consumption than GNP. This was the case in the

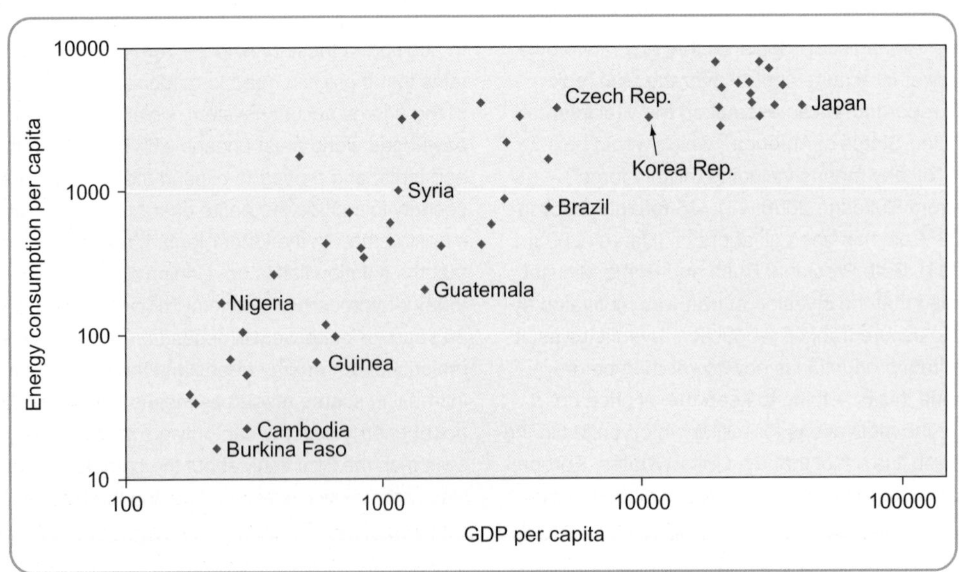

Figure 5.10 Relationship between energy consumption and economic growth.

Plate 5.7 When 'muck was money'.
(FOX PHOTOS/Getty Images)

Plate 5.8 Not a smokestack in sight on this new industrial estate.
(Realimage/Alamy)

Soviet-type economies that had a bias towards heavy industry and were notoriously wasteful in their use of energy. Others, such as Japan, have low levels of energy consumption, given their high GNP. This is because they have introduced energy-saving technologies and have exported energy-intensive industry offshore. Finally, there are some countries, such as Canada, where climatic extremes (both hot and cold) require a large amount of energy to be used for heating or for air conditioning. There are also other factors that affect the level of energy consumption: for example, the level of urbanization. But it has also been suggested that there is a similar relationship between economic development and urbanization as there is between the former and energy consumption. Finally, the size of a country may also affect the level of energy consumption, as larger amounts of energy are required to move between places. Such an argument could be marshalled to explain the very high levels of energy consumption in the United States, as could climatic factors.

On the face of it the relationship between energy consumption and economic development seems unproblematic. However, as Chapter 8 reveals, the whole question of what stands for 'development' is problematic and contested. The measurements used in Figure 5.10 systematically understate the relative position of the developing world. First, because the use of commercial energy as a measure of energy consumption ignores the role of non-commercial biomass energy sources, as a consequence it only measures the 'modern' sectors of the economy. Secondly, the monetary measure of GNP per capita is an inadequate measure of the 'human condition' in many of the world's poorest countries. That said, a re-evaluation of the relative position of the poorest countries would only move them slightly 'up the curve'. It would not alter the fact that there is a clear relationship between industrialization and energy consumption. The Stern Review (2006: xi) into the economics of climate change notes: 'CO_2 emissions per head have been strongly correlated with GDP per head. As a result, since 1950, North America and Europe have produced around 70 per cent of all the CO_2 emissions due to energy production, while developing countries have accounted for less than one-quarter. Most future emissions growth will come from today's developing countries, because of their more rapid population and GDP growth and their increasing share of energy-intensive industries.' The Stern report suggests that we need to 'decarbonise' development so that the developing world can enjoy the benefits of economic growth without substantially increasing global CO_2 emissions.

The historical relationship between industrialization and energy consumption may seem to suggest that all the countries of the developing world will eventually follow the same energy and development trajectory as that experienced by the developed world. Thus, the patterns of energy and development can be equated to a stage model of the kind proposed by W.W. Rostow. Such a model is presented in Figure 5.11. This model examines the change in energy ratios over time as an economy develops. The energy ratio is the relationship between the rate of change in energy consumption and the rate of change in economic growth (Mather and Chapman 1995: 154). In the pre-industrial phase, the energy ratio is less than one as these are 'low-energy' societies, mainly dependent on subsistence agriculture. As discussed earlier (also see Chapter 2), the industrial revolution resulted in a change in the relationship between society and resource consumption, and the harnessing of new sources of energy was at the heart of this process. As economies industrialize, the energy ratio begins to exceed one. In the early phases of industrialization energy efficiency is low and economies are dominated by energy-intensive industries. As the industrial economy matures so the energy ratio declines. This is due to increased efficiency, increased energy costs or the actual decline in the level of economic activity. This stage is typified by the developed world during the 1970s. Eventually, the economy moves into a post-industrial phase and the ratio falls below one. If the model seems very familiar, it is much

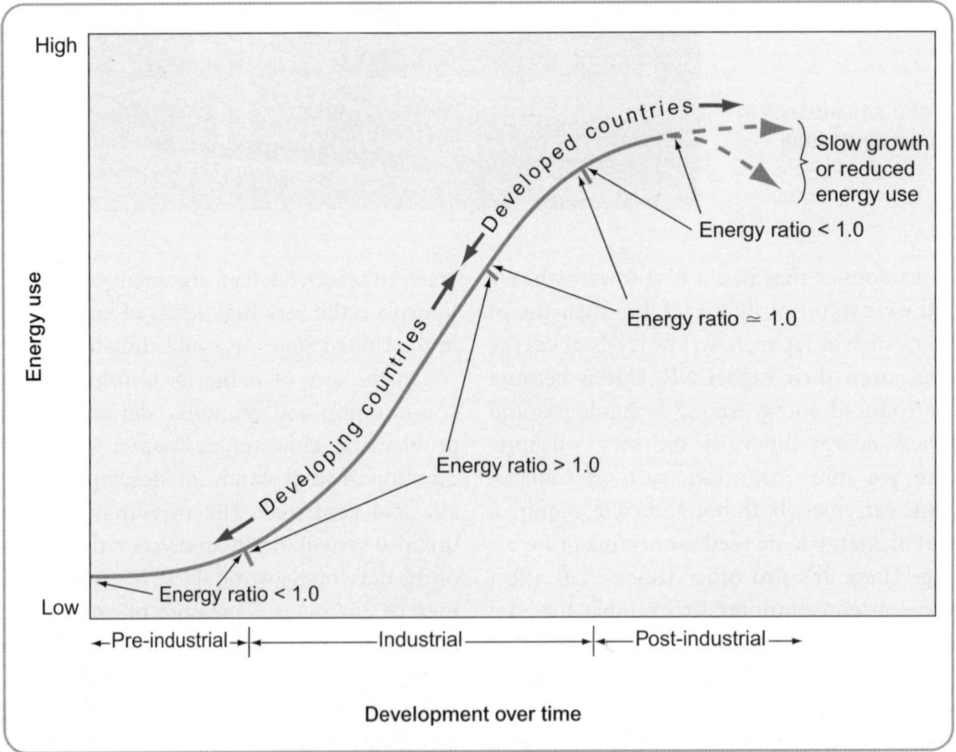

Figure 5.11 Energy ratios and economic development.

Source: *Environmental Resources*, Mather, A.S. and Chapman, K., Pearson Education Limited © Prentice Hall (1995)

like the model of the demographic transition (see Chapter 4) because it describes the evolution of the energy economy as it actually occurred in Western Europe and the United States.

What evidence is there that the rest of the world will inevitably have to go through the same stages? It could be argued, with some justification, that the Newly Industrialized Countries (NICs) of Asia have followed this pattern of industrialization; however, the time taken to move through the phases here seems to have been compressed as the NICs have tried to diversify their economies and promote service sector growth. They have also tried to industrialize without the advantage of substantial indigenous energy resources, a fact that has made them vulnerable to the effects of price increases. Somewhat ironically, the financial crisis in Asia was largely responsible for the fall in energy prices in the late 1990s. By contrast, the transition economies of the post-socialist world (East-central Europe and the former Soviet Union) pose a different challenge to the model. Prior to 1989–90 these economies were undoubtedly on the steepest part of the curve, consuming large amounts of energy and producing a relatively low level of economic growth. Isolated from the oil price shock by cheap supplies of Soviet oil, these economies had failed to increase their energy efficiency during the 1970s and 1980s. This failure was a major reason for the collapse of the Soviet-type economic system. Transitional recession has followed the collapse of the socialist economies and this has depressed energy consumption. The question now remains: what will the relationship be between energy consumption and economic recovery in the transition economies? Will they follow the path of post-industrialization (see Chapter 4) or will they re-industrialize? The likely answer is that individual economies will follow their own paths (OECD 1999). As the Stern Review notes, a much bigger question mark hangs over the developing world. Is it inevitable that it will have to industrialize to improve living standards? If it is, then the availability of energy resources and/or energy-saving and energy-efficient technologies will be very much centre stage for world economic development in the twenty-first century. Furthermore, as we have seen in this chapter, many of those same countries are also likely to be the major suppliers of oil and gas and other minerals and there is plenty of evidence to suggest that such resource-based development often fails to provide a sustainable basis for improving living standards (see Case study 5.3) and can generate conflict and instability. The first decade of the twenty-first century suggests that the global shift in economic growth and energy demand will generate a whole series of challenges to the existing world order.

Thematic Case Study 5.3

Resource abundance: blessing or curse?

From all that has been said in this chapter, one would assume that having a relative abundance of natural resources would convey an advantage in terms of prospects for economic development. However, the reality is somewhat different. There is now substantial evidence to suggest that resource-abundant economies have actually performed worse in terms of rates of economic development than resource-poor economies. The geographer Richard Auty (1993, 2001) was the first to term this phenomenon the 'resource curse thesis'. Sachs and Warner (2001) have weighed up the evidence and conclude that there is sufficient evidence to suggest that 'high resource intensity tends to correlate with slow growth'. In other words, economies that have a high degree of dependence on the resource sector tend to grow more slowly than resource-poor economies. While the 'resource curse thesis' is now generally accepted, there are exceptions to the rule, those usually cited being Botswana, Chile and Malaysia, and there is no single explanation for this under-performance. Sachs and Warner suggest that most forms of explanation follow some form of 'crowding out' logic, whereby the dominant resource sector inhibits the development of the non-resource sector. Resource economies are particularly susceptible to fluctuations in income because of the volatility of resource prices, added to which, as we know from earlier discussions, the resource base itself is often soon depleted. Once the 'resource boom' is over, the economy is not sufficiently developed or diversified to sustain living standards and a period of 'bust' often follows. Such a 'boom and bust' cycle is by no means inevitable: effective government policy can

→

use resource income to promote a more diversified economy that can sustain living standards once the boom has passed. A review of the literature (Stevens 2003) suggests that there are number of dimensions to the 'resource curse' (though Stevens favours the term 'resource impact'):

- In many resource-rich economies there has been a failure to save income during boom periods to cover periods of bust (some states have created so-called 'stabilization funds' to save for a rainy day), plus a tendency to spend income on consumption (usually through increased imports) and on prestige projects.
- There is also a failure to redeploy income from the resource sector to promote a more sustainable pattern of economic development. In some instances there is also a tendency to use resource income to subsidize and protect the activity of inefficient producers in the non-resource sector. Later, when the resource income dwindles and the economy is opened up to international competition, these inefficient producers then fail.
- In many instances, often as a consequence of a colonial heritage, the resource economy remains relatively isolated from the rest of the economy. This minimizes the multiplier impact of large-scale resource-based investment projects, beyond their payment of taxes. One solution is to impose a 'local content' requirement on the resources companies, forcing them to use local suppliers of goods and services. However, given the relative lack of 'economic development' in many resource economies it is often difficult to source goods and services locally.
- The increase in export income associated with a resource boom tends to result in a strengthening of the value of the domestic currency of the resource economy. This can have the effect of making the cost of domestic production in the agricultural and manufacturing sectors higher than the cost of imports. This is known as 'Dutch Disease' following the experience of the Netherlands; the net result is a decline in the competitiveness of the non-resource sector, which aggravates the problems discussed above.
- Finally, there is increasing evidence that suggests that a sudden influx of resource income tends to promote crime and corruption, armed conflict and an abuse of human rights. Such problems not only have a direct impact on the welfare of individuals, but also promote increased social inequality and undermine the effectiveness of the state (see Ross 2001; Renner 2002; Le Billon 2007).

The continuing underperformance of resource-rich economies has led the World Bank and the major resource companies to reassess the impacts of resource development. The World Bank through its Extractive Industries Review (www.eireview.org) is reconsidering whether it should be promoting resource-based development as a means of improving living standards and promoting sustainable development. The EIR's final report, called *Striking a Better Balance*, concludes: 'the Extractive Industries Review believes that there is still a role for the World Bank Group in the oil, gas and mining sectors – but only if its intervention allows extractive industries to contribute to poverty alleviation through sustainable development and that can only happen if the right conditions are in place'. The EIR was prompted by the World Bank's decision to finance the construction of the Baku-Tbilisi-Ceyhan (BTC) pipeline from landlocked Azerbaijan to the Mediterranean coast of Turkey. The collapse of the Soviet Union has prompted a dash to gain access to new energy resources in the Caspian and Central Asia, and more recently in Siberia and the Russian Far East (Kleveman 2003). The BTC pipeline also prompted the UK Government to champion the Extractive Industries Transparency Initiative (http://www.eitransparency.org/), which is 'a coalition of governments, companies, civil society groups, investors and international organizations. The EITI supports improved governance in resource rich countries through the full publication and verification of company payments and government revenues from oil, gas and mining.' High oil and gas prices and concerns about energy security have also promoted increased oil and gas activity off the shores of West Africa and elsewhere (Gary and Karl 2003; Rowell et al. 2005; Ghazvinian 2007) as the developed and emerging economies (China in particular) seek to gain control of sources of energy. Sustained high energy prices is resulting in a new influx of resource revenues and the expansion of oil and gas production into new areas will create a new set of resource-abundant economies and regions; NGOs such as Revenue Watch (www.revenuewatch.org) are now closely monitoring the flow of revenues arising from these new resource projects (see Caspian Revenue Watch 2003). The dangers of the resource curse are now well recognized, but only time will tell if the international community can assist the newly resource-rich economies to avoid the pitfalls of the past.

5.4 Conclusions: global energy dilemmas

This chapter has considered the nature of resources through a detailed case study of the relationship between energy and economic development. The chapter concludes by considering the energy dilemmas that now confront the world. It is no longer the relatively simple matter of whether or not the world has sufficient resources to meet demand. In the short term, there is little danger of physical scarcity at a global scale. Rather, the essential dilemma is whether the global ecosystem can absorb the consequences of continued increases in resource production and consumption. Furthermore, the exact nature of the resource dilemma varies across the globe. Therefore, there is no single resource problem, nor a universal solution.

In the advanced post-industrial societies of the developed world, the dilemmas relate to the economic and political costs of geopolitical scarcity (see Table 5.1). Increasingly, the economies of the developed world are finding themselves dependent upon what they perceive to be 'hostile' and 'unstable' sources of supply in the developing world. The response has been to develop high-cost resources close to home, to promote increased efficiency and conservation and to seek alternatives, such as renewable energy. This strategy imposes an economic cost as it forces the developed world to pay more for its energy than it needs to. However, decoupling economic growth from energy consumption, increasing energy efficiency and reducing carbon dioxide emissions compensate for the increased cost of energy. The developed world can afford to seek technological solutions to its resources problems, particularly in the area of energy supply.

In the transition economies of Central Europe and the former Soviet Union resource dilemmas are somewhat different. The Soviet economic development strategy imposed a resource-intensive form of industrialization upon these countries that made them dependent upon cheap supplies from the Soviet Union (predominantly Russia); it also caused widespread environmental degradation. The economic transformation now taking place in the post-socialist world has closed many of the smokestack industries; however, the newly independent states in this region now face the problem of having to secure access to resources. The resource-rich states, such as Russia, hope that economic recovery will bring a resource bonanza as regional demand rebounds and prices increase. It remains to be seen whether economic recovery in Central Europe and the former Soviet Union will create increased demand or whether these economies will follow the path of the developed world and achieve economic growth without an equivalent increase in resource demand. There is already evidence that the fast-reforming economies of Central Europe and the Baltic States are likely to follow the model of Western Europe; while the member states of the Commonwealth of Independent States will follow a different, and as yet undetermined, path. Russia's reluctance to sign the Kyoto Protocols (see Chapter 5) was partly due to uncertainty over the future structure of the domestic economy and thus its environmental footprint.

In the developing world there is considerable diversity in terms of resource dilemmas. The NICs of Southeast Asia have succeeded in industrializing, but in doing so they have become increasingly dependent upon resource imports. Elsewhere in the developing world a dualistic pattern of resource consumption seems to have developed. On the one hand, there remains a 'low-energy' society dependent upon biomass energy and subsistence agriculture, in which the majority of people exist. Here population pressure is leading to environmental problems as forests are stripped for fuel wood and land ploughed up or over-grazed. On the other hand, there exists a growing 'high-energy' society linked to increasing commercial, agricultural (see the next chapter for more on this) and industrial activity and the spread of urbanization. Much of this activity is aimed at supplying resources to the developed world. The two sectors of society combine to create an increasing demand for resources, and pressure on the environment. The balance between the sectors also varies greatly among the countries of the developing world.

Given that the majority of the world's population lives in the developing world, this growth in demand will be the major factor contributing to increases in global resource demand. In its *Global Environment Outlook 2000*, the United Nations Environment Programme (UNEP 1999: 2) concluded that: 'A tenfold reduction in resource consumption in the industrialized countries is a necessary long-term target if adequate resources are to be released for the needs of developing countries.' It may be that in the future it is increasing demand in the developing world that raises the spectre of resource scarcity in any number of forms. However, the developed world already recognizes this threat and is using the issue of global warming to try to curb increased resource consumption in the developing world. For this to happen without curtailing the life chances of the people of the developing world, the developed world must help to shape an entirely new relationship between resources and development.

Learning outcomes

Having read this chapter, you should understand that:

- The notion of what is a resource is socially constructed and varies through time and across space.
- There are a variety of different ways of classifying resources.
- There are a complex set of factors that influence the availability of resources.
- There are substantial regional variations in energy production and consumption and a growing mismatch between the countries that consume most of the world's energy and those that consume it.
- There is a complex and increasingly challenging relationship between energy consumption, economic development and climate.
- Different types of resource dilemmas confront the different regions of the world.

Further reading

Elliott, D. (2003) ***Energy, Environment and Society***, 2nd edition, Routledge, London. Provides a good overview of the key issues and energy alternatives.

Jones, G. and Hollier, G. (1997) ***Resources, Society and Environmental Management***, Paul Chapman, London. Provides more detailed discussion of the issues raised by this chapter and deals with some essential issues, such as water resources, which have not been discussed in any detail.

Mather, A.S. and Chapman, K. (1995) ***Environmental Resources***, Longman, London. Provides a detailed, technical analysis of the resource issues raised by this chapter with numerous supporting examples.

Middleton, M. (2003) ***The Global Casino: An Introduction to Environmental Issues***, 3rd edition, Arnold, London. Provides concise analyses of numerous environmental issues; each chapter contains a list of suggested readings and recommended websites.

O'Riordan, T. (ed.) (2000) ***Environmental Science for Environmental Management***, 3rd edition, Prentice Hall, Harlow. An essential guide to the science behind the issues, plus up-to-date discussion of key challenges such as sustainable development and climate change.

Useful websites

There is a huge amount of material available on the Web that relates to the issues discussed in this chapter. Some of the key reference sources are listed below; additional websites have been referenced in the text.

www.bp.com/centres/energy Home of the BP *Statistical Review of World Energy*. This website, which is updated on an annual basis, has a wealth of statistical information on the energy sector, much of which is downloadable in Excel format or as PowerPoint slides.

www.iea.org The official site of the International Energy Agency. The site contains information on the Agency's operations and publications, as well as statistics on energy production and consumption. Also has links to other international organizations.

www.eia.doe.org The official site of the US Government's Energy Information Administration. In addition to information on US energy matters, it contains reports on individual countries and a massive list of links to other US government agencies, international agencies, foreign governments and commercial company sites.

www.worldbank.org The official site of the World Bank. The site contains information on World Bank operations and publications, as well as downloadable statistics and briefing documents.

www.wri.org The World Resources Institute is an independent centre for policy research and technical assistance on global environmental and development issues. Its website provides information on its own activities and publications, some of which are downloadable, as well as links to other organizations in the same area. The Institute, together with the UNEP, UNDP and World Bank, produces the biennial *Resource Report* that is an indispensable reference guide to the state of the earth's resources.

For annotated, clickable weblinks and useful tutorials full of practical advice on how to improve your study skills, visit this book's website at **www.pearsoned.co.uk/daniels**

VALUING THE ENVIRONMENT

Chapter 6

Jenny Pickerill

Topics covered

- How we value and understand the environment
- The evolution of sustainable development
- The complexity of scale of environmental issues
- Different strategies for action
- Challenges to how we conceptualize the environment
- Positive steps towards an ecological future

No one is untouched by the emerging environmental agenda; everyone has an interest in shaping the development of environmental politics (for good and bad).
(Connelly and Smith 2003: 358)

When I was once asked to discuss at length how individuals could best help resolve the environmental issues, I responded 'stay put'. Only by rediscovering a sense of place, a commitment to a particular piece of ground will we be able to redefine our relationship to the planet. (Anon. 2001)

There is no survival value in pessimism. A desperate optimism is the only attitude that a practical environmental philosopher can assume.
(Callicott, quoted in Callicott and Rocha 1996: 157)

6.1 What kind of world do you want?

What kind of world do you want? If we don't act now, we won't have a choice. Do you want a world where climate change continues to devastate our environment?

(Friends of the Earth flyer, June 2006)

There is a need for utopianism when thinking about the environment. Environmentalists are often accused of being unrealistic and demanding the impossible. Yet we seldom ask ourselves the question 'What kind of world would we like?' If we considered it a little more we might be taking bigger steps towards curbing our environmental impact and seeking environmental justice. Environmental issues affect all places and societies – the urban and the rural, the rich and the poor – nobody is unaffected. The environment matters and our impact upon it is obvious, from the air that we breathe to the energy that we use.

The rise of environmental activism has highlighted the importance of environmental protection with some notable successes. Britain's road-building programme of the 1990s was curtailed, and the logging of old-growth forests has been all but stopped in Western Australia. Britain has the 'oldest, strongest, best-organized and most widely supported environmental lobby in the world' (McCormick 1991: 34), and there are examples worldwide of people taking a stand and making changes to protect the environment. However, 'there is no doubt that environmental issues have had a big impact on contemporary politics, and yet the frequency with which governments adopt a business-as-usual response to environmental problems raises the cynical thought that perhaps nothing much has really changed' (Carter 2001: 2). Thus, despite high levels of environmental concern and an active environmental movement, we are faced with a wide array of environmental problems – from global **climate change** to river pollution, from water supply issues to genetic crop modification. To understand why this is, we need to explore: how we value the environment and the different philosophical approaches to environmental issues; the complexity of scale and responsibility; different strategies for action; what a focus on 'the environment' ignores, and finally what positive steps we should take.

6.2 Valuing the environment

We each value the environment in different ways and for different ends. Historically we have always debated what parts of our human and non-human world have what values. These values are used to answer ethical questions about the environment, for example whether we should eat meat, or whether it is justifiable to cut down a tree in order to build a house. Understanding values helps us understand why some people care greatly about the environment and others less so.

We can use three categories of value when talking about the environment. As summarized in Table 6.1 these are intrinsic, inherent and instrumental values. An object, or living being, has an **instrumental value** when we view it as a resource, or as of use to us for a specific end. For example, a forest has an instrumental value if we view it just as a source of wood for fuel. At the other end of the continuum an element has **intrinsic value** if we view it as important in and of itself, with no reference to how we might use it or to how it might make us feel. For example, we view human life as intrinsically valuable.

In between these, **inherent value** (sometimes referred to as 'weak' instrumental value) refers to how we value an element beyond its use as a resource but still relate it to how it makes us feel. For example, we might value a woodland because we enjoying walking through it and for its role in atmospheric stability. In all of these discussions of value we must remember that 'humans will always be the distributors of value, but they are not necessarily the only bearers of value' (Carter 2001: 36).

We can understand this better if we consider the rule of no-substitution. If something is instrumentally valuable, its value will disappear if there is a

Table 6.1 Defining value

Instrumental value	Value which something has for someone as a means to an end
Inherent value	Value something has for someone, but not as a means to a further end
Intrinsic value	Simply the value something has. No appeal need be made to those for whom it has value

Source: Carter (2001: 15)

(preferable) substitute. For example, if we could solve eyesight problems with a pill, we would no longer need opticians. But if something is inherently valuable for itself, and not as an end, it is not substitutable in this way. Hinchliffe and Belshaw (2003) use the example of the wild tiger to illustrate the consequences of these values. The wild tiger population is threatened by the international trade in tiger parts, often for use in Chinese medicine. One solution is to establish tiger farms. However, if we find tiger farms acceptable then there is room for instrumental valuations. If tigers' inherent value is paramount, then the notion of farming becomes objectionable. The implications of such values are also explored in Case study 6.1.

We can also consider the values we give the environment by understanding the terms 'anthropocentrism' and 'ecocentrism'. The **anthropocentric** approach argues that only humans have intrinsic value and the environment is only of instrumental value. In other words the environment is only useful as a resource for humans. In this approach we do not have any responsibility to ensure the environmental sustainability of our actions. **Ecocentrism** critiques anthropocentrism and suggests that non-human entities have intrinsic value. However, for many the situation is not this black and white. There is a large intermediate area within which some non-human entities have value, others do not, but where humans are valued as *more* important than

Plate 6.1 Logging in Styx Valley, Tasmania, Australia
(Jenny Pickerill)

Thematic Case Study 6.1

Old-growth logging in Tasmania, Australia

The logging of old-growth forests is a particularly controversial issue in Australia, with views sharply polarized (Hutton and Connors 1999). Old-growth forests have had only minimal human interference, take hundreds of years to grow and contain the tallest plants on earth.

The state government of Tasmania has shown little inclination to curb the logging of old-growth forests. This is despite a consultation process (Tasmania Together) where the population voted to end **deforestation**. In May 2003 the government lifted a ban on logging in the Tarkine – the single greatest stretch of temperate rainforest in Australia. The forests of the south-east and north-east are also threatened by expanding logging operations. Every year approximately 15,000 hectares of old-growth forest are logged in Tasmania.

In a country which draws much of its wealth from its natural resources (forests, minerals, uranium, fishing and agriculture), logging is often justified by the need to provide jobs in rural areas and to provide much-needed export revenue. Loggers argue that Australia is limited in its resources and consequently needs to make the most profitable use of them; that they have historically 'managed' forests through selective logging, consequently arguing that all forests in Tasmania have been influenced by humans for generations.

Environmentalists have argued that logging is simply a waste of resources. Logging in Tasmania is heavily subsidized and uneconomical (not only in price

obtained for products but in its negative impact upon tourism). Most of the wood is sold as woodchips to be made into paper. Moreover it is unsustainable and irreversible and many of the trees destroyed would take over 300 years to regrow, e.g. the Houn Pine in the south-west of the island takes 3,000 years to grow to full size. Old-growth forests cannot simply be replaced; their destruction results in the loss of **biodiversity** of a barely understood and fragile ecosystem. Deforestation does not only have localized effects but also plays a key role in CO_2 absorption and can upset hydrological systems beyond the area of logging. Environmentalists also believe that the largest trees are important in and of themselves. The Wilderness Society, a key environmental group in the Australian logging debate says, 'our forests are places that inspire and rejuvenate us all – nurturing us as well as the delicate ferns and mosses on the forest floor, or connecting us to timeless grandeur as we touch the tallest flowering plants on Earth' (2006).

The campaigns to save the old-growth forests have ranged from small-scale local protests to garnering international support. The Styx Valley in the south-east is an example of an iconic campaign. It contains the world's tallest hardwood trees – *Eucalyptus regnans*. Many are taller than a 25-storey building. The forest is also home to many native species of wildlife, including the majestic Wedge-tailed Eagle and the Yellow-tailed Black Cockatoo. While a few of the largest trees in the valley have been protected, between 300 and 600 hectares are being logged each year. Furthermore the main form of logging is destructive clear felling and burning (not selective logging).

Action on a local level has involved occupying threatened trees and targeting the woodchip mills. Regionally, environmentalists have been using their research to argue against clear fell logging, the use of 1080 poison (which is used to kill possums and wallabies that may graze on re-sprouting saplings thus preventing regrowth of the forest), and to prove the environmental implications of logging. Environmentalists have appealed to the federal government and the Australian populace for support through media campaigns and advertising. These tactics have only proved moderately successful in Tasmania thus far. In 2004 the state government promised to protect 17,000 hectares of forest but did not clearly identify which areas this would include, or commit to stopping the destructive practice of clear felling.

Analysis of the conflicts about protection of forests enables us to understand how the environment is valued differently within a community, in other words why some argue for its protection and others choose its destruction. The state government and logging companies clearly view the environment as a resource for use by humans, thus they believe the environment has an instrumental value. They will only be interested in managing its sustainability if that will add value to its use and exploitation in years to come. For environmentalists it is more complex. On one level, many acknowledge the instrumental value of forests (for making furniture, for example) but think the current methods of forestry are wasteful. But in addition environmentalists believe the environment has an inherent and an intrinsic value which is not related to human need. In this way they believe it should be protected from all but recreation, and left untouched for future generations to enjoy.

Q How do you value forests like those in Tasmania and how does that affect your perspective as a geographer?

the environment. Here the environment has some value beyond human use. Thus there is a continuum between these approaches and in 1973 Arne Naess distinguished between 'deep' and 'shallow' ecology, identifying a continuum between the two (Naess, 1973). Shallow ecology is an anthropocentric position. **Deep ecology**, as defined by Naess, is a form of ecocentrism but one where everything has intrinsic value (see Table 6.2 for the eight basic principles of deep ecology). The approach asks us to consider humans not only as part of nature but also of *equal* value to other non-human entities. This is a holistic vision whereby all human and non-human entities are interconnected and interdependent. A holistic perspective enables us to understand that if we upset one element it will have an impact on all other elements.

What are the consequences of valuing non-human entities in this way? If we were to adopt such an approach it would require us to make major and radical changes to the way we live. No longer could we see the non-human world as merely providing resources for our existence: rather, deep ecology asks us to radically reduce our population, restrict economic growth, and to live a simpler life by giving up excesses and 'to touch the Earth lightly'. There is a strong element of self-sacrifice to this in order to ensure survival of the

Table 6.2 The eight basic principles of deep ecology

1	The well-being and flourishing of human and non-human life on Earth have value in themselves (synonyms: intrinsic value, inherent value). These values are independent of the usefulness of the non-human world for human purposes.
2	Richness and diversity of life forms contribute to the realisation of these values and are also values in themselves.
3	Humans have no right to reduce this richness and diversity except to satisfy vital needs.
4	The flourishing of human life and cultures is compatible with a substantial decrease of the human population. The flourishing of non-human life requires such a decrease.
5	Present human interference with the non-human world is excessive and the situation is rapidly worsening.
6	Policies must therefore be changed. These policies affect basic economic, technological and ideological structures. The resulting state of affairs will be deeply different from the present.
7	The ideological change is mainly that of appreciating life quality (dwelling in situations on inherent value) rather than adhering to an increasingly higher standard of living. There will be a profound awareness of the difference between big and great.
8	Those who subscribe to the foregoing points have an obligation directly or indirectly to try and implement the necessary changes.

Source: Naess (1973: 95–100)

non-human world. Naess has been criticized for asking humans to make such sacrifices when we would not ask animals to do the same. However, Naess 'accepts that human beings . . . are sometimes justified in lending greater moral weight to human well-being than to that of non-human life' (Cooper 2001: 214). What deep ecology really challenges us to do is to value non-human life as we would humans and thus truly consider how our daily actions could be different. Although some of the sacrifices might appear difficult and controversial (such as radically reducing populations), Naess clearly illustrates the extent of actions required to put into practice a deep concern for the environment.

An examination of values has illustrated some of the potential conflicts in how we view the environment. Environmentalists often argue that we need to rethink humans' relationship with the non-human world and they use values to argue that non-human elements hold greater value than often assigned them.

6.3 Limits to growth and the challenge of capitalism

This rejection by environmentalists of the anthropocentric approach whereby the environment is viewed instrumentally, and that instead it has an intrinsic value, is a cornerstone of modern environmental thought. A second key insight is that there are **limits to growth**. The earth is a finite system and thus eventually growth will have significant impacts upon the ability of the environment to sustain future generations. This notion was introduced by Meadows et al. in 1972 with their publication of *Limits to Growth*. Using computer models they sought to project trends in resource use and population growth into the future. They predicted natural limits on future growth and a doomsday scenario of environmental catastrophe. Although we now know these predictions to be overly pessimistic the idea of limits and finitude (scarcity rather than abundance) struck a chord.

However, it also became apparent, for people in the developed world at least, that it was possible to maintain economic growth and reduce resource consumption per unit of wealth created. This argument saw the emergence of an alternative view to **ecocatastrophism**: a view that would grow into the concept of sustainable development.

Connelly and Smith (2003) identify two broad interpretations of what an environmentally sustainable society would look like: a deep green approach, and the more reformist **ecological modernization**. The deep green interpretation takes many of Naess' principles and seeks to put them into practice. It is a radical approach, a strong interpretation of sustainable development that calls for change and restructuring at all levels of society in political, economic and social arenas.

Spotlight Box 6.1

Sustainable development

The 1987 **Brundtland Report** adopted the position that it was possible to pursue economic growth without compromising the environment, and it is generally accepted as introducing the concept of sustainable development (the seeds of which had been sown some 15 years earlier). It also provided the first, and still widely used, definition of sustainable development: 'development which meets the needs of the present without compromising the ability of future generations to meet their own needs'.

However, its meaning has now become highly contested and vague, leading to weak interpretations. For example, how can present and future needs be determined, and whose needs does it refer to? Many attempts have been made to refine the concept, which have led to numerous, often contradictory, definitions. The result is that 'sustainable development' is often little more than a soundbite, signalling recognition that the environment matters but failing to provide substance.

The real success of the Brundtland Report was the emphasis it placed on the relationships between economic growth, social conditions and environmental degradation, an emphasis that placed sustainable development firmly on the global political agenda, and that over the next five years would see a concerted effort by UN agencies and many governments towards a global summit on environment and development. In 1992 – after five years of negotiations and preparatory meetings – the UN Conference on Environment and Development (also known as the **Earth Summit**) was held in Rio de Janeiro.

Q Is it morally justifiable and politically plausible for the developed world to try to dictate the terms of sustainable development and good environmental management to the rest of the world?

In essence it is a critique of everything about the way we currently live today, a critique of contemporary industrial society. Accepting that there are limits to growth means changing our notions of what makes a good life and learning to live more simply (thus slowing growth): for example, consuming less and more locally produced goods, producing less waste, travelling less and being conscious of our impact on the environment in our daily lives. The result is an emphasis on 'localization', living small-scale, perhaps more communally to share the load of everyday tasks, in energy-efficient buildings, and often with an emphasis on producing for your own needs, for example, through growing your own vegetables. See Case study 6.2 for a British example of low impact living.

The ecological modernization interpretation is far less radical and has been adopted by many more in mainstream politics. Supporters of ecological modernization criticize the deep green vision as naive and utopian with an unnecessary and ineffective focus on

Plate 6.2 Low impact house: Tony's Roundhouse, Brithdir Mawr, Wales.
(Jenny Pickerill)

Thematic Case Study 6.2

Low impact development in Britain

Low impact development (LID) is a form of low impact living, a deep green vision where we minimize our environmental impact in all aspects of our daily lives. LID is a radical form of housing and livelihood that works in harmony with the landscape and natural world around it. Simon Fairlie, a pioneer of LID in Britain argues that 'a low impact development is one that, through its low negative environmental impact, either enhances or does not significantly diminish environmental quality' (1996). It is also an example of a mixture of mitigation and adaptation to climate change – building homes that will reduce carbon usage and cope with a changed climate.

An LID has a low visual impact by blending with its surroundings, is built from local, recycled or natural materials, is small-scale and environmentally efficient. They are autonomous in the sense that they generate their own energy (through wind, solar or water power) and deal with their waste (through recycling and composting), thus they are not connected to mains supplies. In addition LID is about more than just the construction of environmentally friendly dwellings: it is about the *way* we live. Implicit in the LID approach is an emphasis on minimizing vehicle use, creating livelihoods largely from the land on which we live, reducing consumption and purchasing other items locally. Moreover, LID can positively enhance biodiversity and is reversible.

Why do we need such a radical approach? In Britain our '21 million homes are responsible for 27% of [our] CO_2 emissions' (BBC 2006b), and the government has finally acknowledged the need to make changes in the way we construct houses. In a 2006 consultation document *Building a Greener Future* it has called for significant energy efficiency improvements and a move towards zero carbon development by 2016 – zero net emissions of carbon dioxide from all energy use in the home. While this is a welcome move, for many environmentalists it is too little too late. It fails to take account of existing housing stock efficiency, energy used in construction materials and the build process, and of the ways people live and consume. Moreover, to improve energy efficiency of a standard house can be costly and thus off-putting, whereas a LID is often low-cost *by design* because of its focus on appropriate scale and materials.

In Britain there is a growing number of LIDs. Hockerton Housing Project is a LID in Nottinghamshire which is an earth-sheltered, self-sufficient terrace of five single-storey houses. Built into the side of a hill with large windows to the south, they rely on passive solar heating (and no other form of heating) and generate all their electricity through wind turbines and photovoltaic systems. The houses are highly energy-efficient and waste is disposed off through a reed-bed system into a large lake. They also grow their own vegetables and generate income through education. Another example is Tony Wrench's roundhouse at Brithdir Mawr, south-west Wales. This is built from home-grown timber, straw and earth. It is built into the earth with a turf roof so that, from behind, you can hardly see it. It is highly insulated, uses recycled windows and uses passive solar heating and a wood burner for heating.

Until recently it has been very difficult to get planning permission in Britain for LIDs. Many communities have secretly constructed their dwellings or fought long retrospective planning battles to keep their homes. However, on 13 July 2006, eight years after the discovery of roundhouses at Brithdir Mawr and the ensuing planning controversy, Pembrokeshire County Council and Pembrokeshire National Park adopted an innovative policy on low impact development. This policy explicitly allows rural greenfield development under a number of conditions aimed to ensure developments remain low impact. Under the guidelines, development must be highly sustainable, using local, renewable, recycled, and/or natural materials and built to high standards of eco-design, with the emphasis on 'low impact', including visual effects. Moreover, the proposal must offer positive environmental, social and/or economic contributions with public benefit. These could include services to the community, economic diversification for the area, opening paths for walkers and improved biodiversity. Furthermore, residents must prove a need to be on the land (rather than clustered within existing rural settlements). Thus residents' livelihoods will be 'tied directly to the land on which it is located', through, for example, agriculture, forestry or horticulture and these land-based activities will substantially meet household needs.

Q What measures should we put in place to ensure our buildings, and the way we live in them, have minimal environmental impact?

localization. In essence this approach argues that economic growth does not need to be slowed to ensure environmental protection. It does not undermine the limits to growth thesis entirely, but suggests that we can 'reorient' economic growth and use technological solutions for environmental problems, thus overcoming the environmental impact of growth. It is a very weak interpretation of sustainable development that allows for a reformist response where the current dominance of free-market capitalism is not challenged; 'the future of environmental politics on this reading is a technical question of how to make capitalism more environmentally sensitive' (Connelly and Smith 2003: 359).

Table 6.3 Evolution of environmental issues

First generation: Preservation and conservation (pre-1960s)	• Protection of wildlife and habitats • Soil erosion • Local pollution
Second generation: 'Modern environmentalism' (from 1960s)	• Population growth • Technology • Desertification • Pesticides • Resource depletion • Pollution abatement
Third generation: Global issues (late 1970s onwards)	• Acid rain • Ozone depletion • Rainforest destruction • Climate change • Loss of biodiversity • Genetically modified organisms

Source: Carter (2001: 4)

6.4 The complexity of scale and responsibility

The scale of response to environmental issues is a key point of debate. As we have seen, deep greens argue that localization is one of the most important elements to enacting environmental sustainability. If we adopt a more localized lifestyle then we can minimize our environmental impact. However, our environmental problems are far more complex, and global, than the notion of 'localization' suggests. To understand this we need to explore the evolution of environmental concern from one of local problems to global environmental issues.

There is no clear beginning to environmental concern. Plato, Lucretius and Caesar all noted the problems of soil erosion. We now know that many of today's most pressing environmental issues were first identified in the nineteenth and early twentieth centuries. For example, in his 1869 book *Man and Nature*, George Perkins Marsh presented evidence for consequences of tropical deforestation. Carter (2001) has categorized the evolution of environmental issues (Table 6.3) into three key periods. Pre-1960s a great deal of environmental concern was focused upon small-scale localized problems such as local pollution and wildlife protection. In the 1960s concern grew about the interrelation between human actions and more complex environmental implications. For example, *Silent Spring*, written by the biologist Rachel Carson in 1962, distilled the essence of the broad concerns of the era, and is generally regarded as a milestone in modern **environmentalism**. Carson pointed out the long-term ecological consequences of intensive agriculture, in particular the ways in which synthetic pesticides (notably **DDT**) persisted in the food chain and poisoned birds and mammals (including humans). At the same time there were growing concerns about so-called '**overpopulation**'. Paul Ehrlich, Garrett Hardin and Barry Commoner were considered **neo-Malthusian** because they saw environmental problems as a consequence of population growth, following the arguments set out by Thomas Malthus in the late eighteenth century. The misguided simplicity of neo-Malthusian explanations of environmental problems is shown in studies that have tried to explain land degradation in terms of population dynamics (Boserüp 1990). Not only is the evidence contradictory, for example population growth in northern China has been linked to increased land degradation (Takeuchi et al. 1995), whereas in northern Yemen (Carapico 1985) and Kenya's Machakos Hills (Tiffin et al. 1994) it has been related to population decline, but also it is clear from these studies that linking land degradation to one explanatory parameter – population size – masks the complexity behind population dynamics and labour–land relationships (Millington et al. 1989).

By the early 1970s there was recognition not just of the interconnectedness of environmental issues, and of the problems of cumulative damage, but also that these were global problems that needed internationally co-ordinated responses. Environmental scientists identify two types of global environmental change. On the one hand, systemic change occurs when there is a direct impact on a physically interconnected, global system, i.e. the atmosphere or the oceans: for example, the effects that increased emissions of greenhouse gases have on

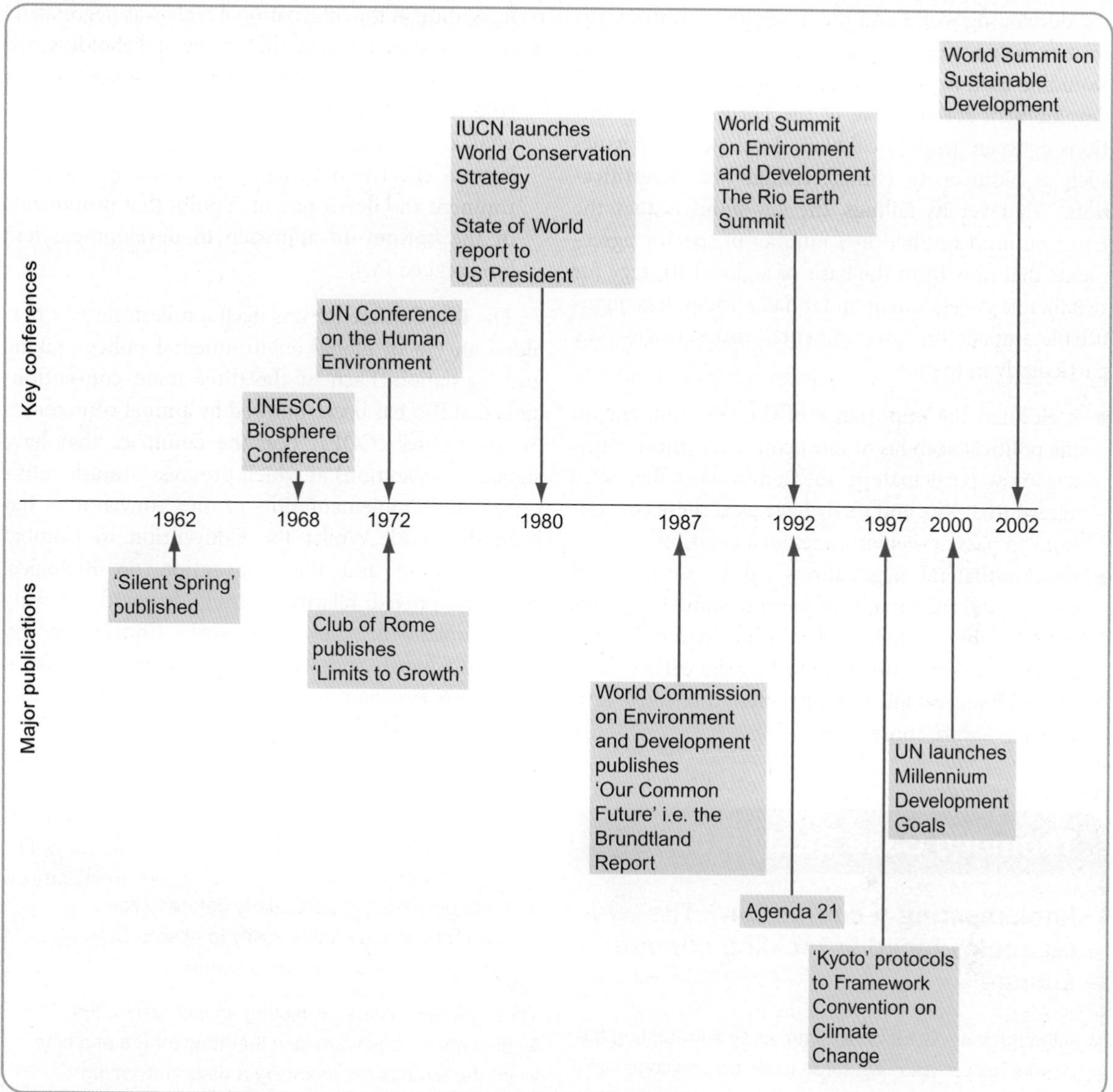

Figure 6.1 Timeline of key environmental conferences and publications.

global climate. Cumulative change, on the other hand, occurs when many discrete events become significant because their distribution is global or because, added together, their impact is felt across a large proportion of the globe. For example, across the Americas, Africa and Asia-Pacific, the relatively small amounts of tropical forest lost in each area being converted from forest to farmland, or being lost to a mining operation, add up to a global-scale problem affecting humid tropical forests, one of the most biodiverse biomes found on earth: forests of vital significance to the global water and carbon balances. Recognition of this duality is important in formulating policies and strategies. Those actors addressing systemic issues need to be global, with all countries agreeing and adhering to a particular policy or strategy (see Figure 6.1). Those focusing on cumulative issues require both global approaches (e.g. in the fields of data exchange and comparative research) and local approaches (to reduce the impacts of a problem because slightly different causes exist in different geographical locations) (Middleton 2003).

In 1992, after five years of negotiations and preparatory meetings, the UN Conference on Environment and Development (also known as the Earth Summit) was held in Rio de Janeiro. It brought together 178 government delegations, over 120 heads of government, over 50,000 **NGO** representatives and a press corps of around 5,000. A rift opened up between the representatives of the developing world and the developed countries, the former being worried that environmental concerns would be used by the developed world to 'limit [their] development for the sake of the Earth'. Shifting tensions between geopolitical blocs developed over issues such as biodiversity, climate change and desertification. Whatever its failings, the key point is that the Earth Summit resulted in a number of specific agreements that now form the basis of a global strategy for sustainable development and it had a wider, less quantifiable impact on environmental issues worldwide, particularly in that:

- it elevated the importance of the environment on the political agendas of most countries (most countries now participate in the conventions that were agreed to at Rio, and many have used these conventions to pass new environmental legislation)
- the multilateral negotiations led to some issues gaining internationally important status (perhaps most notably biodiversity loss, which had only been recognized by scientists some 12 years earlier), and which changed long-standing, often ineffective attitudes to nature conservation
- legislation at the international scale was negotiated
- there was recognition that many stakeholders had legitimate voices in environmental policy-making (the recognition of the stakeholder principle was vital in allowing actors outside government to become effective in bringing about changes in environment and development, a point that proponents of the bottom-up approach to development had long argued for).

The Rio Conference was itself a milestone as it heralded an era of global environmental policy-making and legislation. Each of the three main conventions agreed at Rio has been followed by annual conferences of the parties (COPs) (i.e. the countries that have signed a convention) at which progress towards ratification and implementation of the conventions has been discussed. Whilst the Convention to Combat Desertification and the Convention on Biological Diversity proved relatively straightforward to ratify and implement, the Framework Convention on Climate Change and its most important protocol – the **Kyoto Protocol** – have proved difficult to ratify (see Case study 6.3).

Thematic Case Study 6.3

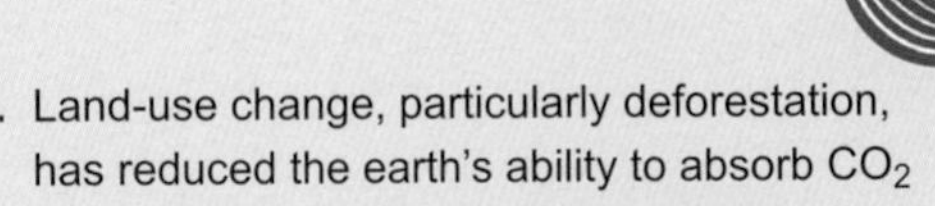

Implementing a convention? The obstacles faced in tackling climate change

Although there is residual controversy surrounding the issue of global warming, three basic propositions can be accepted:

1. There is a natural greenhouse effect that warms the earth's surface by some 33°C. Without this human habitation would be impossible.
2. Since the onset of industrialization, some 150 years ago, the amount of CO_2 released into the atmosphere has increased markedly. Carbon dioxide is a greenhouse gas – that is, it traps the long-wave radiation emitted by the warming of the earth's surface by the sun and, in doing so, heats up the atmosphere. Industrialization has also increased the amounts of other greenhouse gases, e.g. methane and nitrous oxide, and has introduced new compounds such as chlorofluorocarbons (CFCs), which promote global warming and, in some instances, deplete the ozone layer.
3. Land-use change, particularly deforestation, has reduced the earth's ability to absorb CO_2 through the process of photosynthesis.

These processes are promoting an increase in the level of greenhouse gases in the atmosphere at a time when the earth is experiencing a natural warming trend. While natural processes account for some of the warming trend, it is now accepted that elevated levels of CO_2 in the atmosphere are a major cause of global warming. The single largest anthropogenic source of CO_2 is the burning of fossil fuels (for details on global warming see O'Riordan 2000, Chapter 7).

Solutions to the problem of global warming lie in reducing fossil fuel combustion in transportation, industry and power generation. The question is how?

The United Nations Framework Convention on Climate Change (UNFCC) was signed at Rio in 1992 (visit the UNFCC website at www.unfcc.int for further details). The ultimate objective of the UNFCC was to 'stabilize greenhouse gas concentrations in the atmosphere at a level that would prevent dangerous anthropogenic interference with the climate system. Such a level should be achieved within a time-frame sufficient to allow ecosystems to adapt naturally to climate change, to ensure that food production is not threatened and to enable economic development to proceed in a sustainable manner'. Conferences of the parties (COP) are attended by all the states that have ratified or acceded to the Convention (over 194 by February 2004) every two years. The third COP held in Kyoto, Japan, in 1997 saw agreement on the 'Kyoto Protocols', the first actions to address greenhouse gas emissions. The breakthrough at Kyoto was due to a flexible approach that allowed individual countries to set their own targets, while aiming at a 5 per cent global reduction in CO_2 over 1990 levels. The target date for individual countries was set as some time between 2008 and 2012. In addition, the developed countries agreed to reduce emissions in 2000 to their 1990 level, a target few have met. The European Union has an emission reductions target of 28 per cent, the USA 27 per cent and Japan 26 per cent. The transition economies of Central Europe and the former Soviet Union were granted concessions; for example, Russia has a zero-change target. Countries seeking admission into the European Union – such as Poland – accepted reductions. Not surprisingly, a complex set of rules and regulations surrounds the targets, making it possible to trade 'emissions reductions' between states. Thus, for example, if Japan helped Russia to reduce its emissions below the 1990 level, that reduction would be set against Japan's target.

For the protocol to come into force, the countries responsible for 55 per cent of the greenhouse gas emissions in 1990 have to ratify it. A total of 164 countries had ratified the Kyoto Protocols by July 2006, their total emissions amounting to 61.6 per cent of 1990 emissions (http://unfccc.int/files/essential_background/kyoto_protocol/application/pdf/kpstats.pdf). However, the USA and Australia have refused to ratify, and in 2005 the USA was the largest single emitter of CO_2.

In Britain there has been mixed leadership. On the one hand we are likely to meet our Kyoto pledge and on 15 November 2006 the Government announced plans to introduce a Climate Change Bill which potentially will enforce a yearly CO_2 reduction of 3 per cent. The Government has pledged to reduce carbon dioxide emissions from 1990 levels by 20 per cent by 2010 and 60 per cent by 2050. However, at the same time the Government continues to subsidize air travel, allow airport expansions, and fails to promote change at an individual level. This was exemplified by Tony Blair, Prime Minister at the time, in January 2007 when he said 'I'm not going to be in the situation of saying I'm not going to take holidays abroad or use air-travel. It's just not practical . . . it's like telling people you shouldn't drive anywhere'. Blair's lack of commitment to setting an example was seen as further evidence of a lack of commitment to tackling climate change. Air travel is the most environmentally damaging form of travel (Whitelegg 1993). The Netherlands Centre for Energy Conservation and Environmental Technology has calculated that 'One long-haul return flight can produce more carbon dioxide per passenger than the average UK motorist in one year'. Yet a Government White Paper *The Future of Air Transport* (December 2003) predicted that by 2030 UK airport passengers would triple and thus there was a need for national and regional level airport expansion. Planned expansion includes a third runway at Heathrow by 2015/20, a second runway at Stansted to open by 2011/12, new runways at Birmingham, Edinburgh, Glasgow, and considerations of runway extensions at Bristol, Leeds/Bradford, Liverpool and Newcastle. This expansion is inconsistent with government's own policies on CO_2 reductions.

At the same time as the developed world committed itself to emissions reductions, the developing world was persuaded to accept the concept of the **clean development mechanism (CDM)**. The developing countries feel that emissions reductions in the developed world should be on such a scale that would compensate for increased future emissions because of economic growth in the developing world. CDM is aimed at enabling economic growth without a commensurate increase in greenhouse gas emissions. However, this will require transfer of considerable amounts of capital and technology. In other words, the developed world will have to help the developing world implement the CDM. It is unclear how this will happen. It is proposed that developed countries can earn credits towards their own reduction targets by assisting in the implementation of CDM in the developing world. O'Riordan (2000: 202) notes that 'the CDM mechanism denies the right of Third World nations to select their own CO_2 future. This is "ecological colonialism" by another name. While the developed world has access to the capital and technology to achieve the Kyoto targets, it still requires the political will. In the developing world, political will amounts to little without the material requirements to balance the economic growth with environmental needs.'

→

There is an alternative on the table: this is known as contraction and convergence (C&C). At COP9 in Milan many representatives admitted privately that 'C&C [was] what [they] had been waiting for' (Pearce 2003: 6). Contraction means that greenhouse gas emissions would be reduced globally, resulting in dramatic cuts during the next half century. Convergence would see each country's emissions reduced, until by 2050, according to authorities in the United Kingdom and Germany, everybody in the world would have an equal right to pollute – the amount being 0.3 tonnes C per person. Carbon trading permits would help the heaviest polluters to reduce their emissions rapidly. C&C also overcomes the USA's objection to Kyoto, which is that their Asian economic competitors such as China have no emissions reductions targets. Some environmentalists and politicians are now beginning to regard Kyoto as an obstacle, not a solution; it remains to be seen whether they say the same about C&C if the world adopts that route.

Q What will it take for all developed world countries and elites everywhere to cast aside their domestic concerns and privileges in favour of the global and collective good?

Although a few governments have found it difficult to ratify conventions and make efforts to achieve targets, the implementation of sustainable development actions by local administrations – **Local Agenda 21**s – has been far more successful. The International Council for Local Environment Initiatives (2002) reported that 6,416 LA21s were underway or committed to in 113 countries, that national campaigns were underway in 18 countries, and that formal stakeholder groups had been established in 73 per cent of administrative units with LA21s.

The problem of scale can make it hard to determine how to take the best action. Not only do we need to unpack the causes of environmental change and identify likely solutions, but also 'those making decisions concerning global environmental problems need to consider the ramifications for national economies and local populations' (Harris 2004: 13). In other words, we (individuals, business, nations etc.) need to take responsibility for our environmental impacts *and* understand the consequences of mitigating these problems for other people (local populations, the economy etc.). For example, if we want to reduce the negative consequences of driving cars we reduce individual use, encourage companies to design more fuel-efficient vehicles, and encourage governments to increase the price of petrol through tax rises, subsidize eco-alternatives and introduce road pricing. However, in practice, governments of countries like the USA have argued in the past that it would damage their economy, making them less globally competitive (leading to job losses). It was only in 2007 that US President George Bush began to make pledges to tackle climate change. In his State of the Union Address, January 2007, he argued that the USA must 'reduce gasoline usage in the United States by 20 percent in the next 10 years . . . increase the supply of alternative fuels . . . reform and modernize fuel economy standards for cars' (Bush 2007). It is obviously too early to judge the success of these promises and China and India argue that they should be allowed to benefit from economic development through the use of fossil fuels, just as the USA already has (see Chapter 5).

All these complexities are shaped by how we value the environment, at what scale we view these problems, and thus whose responsibility we believe it is to make changes. It is clear, then, that we all need to take responsibility for our actions but that a solely localized approach by some of us will allow the harmful environmental consequences of others to go unchecked. Multi-scalar action is required, from local to international, involving all of us in the processes of change. But how have people sought to make such change happen and how does it work in practice?

6.5 Strategies for change

There is a wide variety of ways in which action is taken, through international agreements, the legislature (i.e. formal party politics), lifestyle changes, community living, and through direct action (Dobson 2001). We can broadly conceive of these as a continuum between reformist and radical approaches (see Table 6.4). A reformist approach advocates working within the current political and economic system. It reflects the shallow green perspective that it is possible to modify our actions to reduce environmental impacts. The radical approach advocates major changes to how we act,

Table 6.4 Reformist and radical approaches to environmentalism

Reformist	Radical
1. Modified sustainable economic growth/ Ecological Modernization.	1. Limits to, and undesirability of, economic growth.
2. Large role for technological development as a provider of solutions for environmental problems.	2. A distrust of scientific and technological fixes.
3. Environmental solutions can co-exist with existing social and political structures.	3. Radical social and political change necessary: either authoritarian or decentralized and democratic political organization.
4. Anthropocentrism and a commitment to intragenerational and intergenerational equity.	4. Intrinsic value of nature or, at least, a weaker version of anthropocentrism; a commitment to social justice within human society and between humans and non-human nature.

Source: Garner (2000: 11)

which cannot be accommodated within the current system: a deeper green world-view; 'the most radical [green aim] seeks nothing less than a non-violent revolution to overthrow our whole polluting, plundering and materialistic industrial society and, in its place, to create a new economic and social order which will allow human beings to live in harmony with the planet. In those terms, the Green movement lays claim to being the most radical and important political and cultural force since the birth of socialism' (Porritt and Winner 1988: 9). Both approaches are part of the broad and diverse environmental movements. We can use some examples to explore these strategies in practice: in national party politics (the German Green Party), political lobbying (Friends of the Earth UK), direct action (Earth First!), individual actions and finally business responses.

The West German Green Party (now the German Green Party) – Die Grunen – made its entrance into formal politics in 1983 with 5.6 per cent of the national vote and 27 seats in Federal Parliament (Doyle 2005). The party emerged from a vibrant environmental movement that was particularly active on anti-nuclear issues. When it was first launched the party was closely aligned to those seeking radical change. Petra Kelly, an early leader of the party, described Die Grunen as 'half party and half local action group – we shall go on being an anti-party party' (Kelly 1984: 21). Thus the party sought to be part of formal politics but simultaneously subvert the process by seeking to practise more **participatory democracy** and pushing for fundamental change. They did this through the legislation they sought, funding for environmental groups and organizations, and structurally by rotating their representatives sitting in parliament by limiting the length of time to two years, limiting the length of office of their leaders and practising participatory decision-making processes.

However, over time 'the party has been colonized by the demands and temptations of parliamentary activity' (Dobson 2001: 127). The party has dropped many of its earlier commitments (environmental sustainability, disarmament, social justice and democracy) and abandoned 'their experimental attempt at institutionalising direct democratic structures within the framework of **representative democracy**' (Poguntke 1993: 395). The radical faction lost out to those in the party who believed in reformist change. This approach led to the party forging coalitions with other parties, eventually forming an alliance with the Social Democrats to form the national government in 1998, consequently weakening their stance on war and nuclear power, and taking a stance *against* those undertaking environmental direct action (Doyle 2005). The German experience is a cautionary tale about the potential for environmental action to take place through formal political arenas. However, just as a party might become institutionalized, this acts as a stimulus for other non-political party action; 'the very fact that social movements tend to be absorbed by the system provides a motive for the radicalisation of other groups' (Rucht and Roose 1999: 91). So although forming a political party might not be the most effective strategy for action (and in countries like Britain and Australia without proportional representation, it is almost impossible for Green Parties to gain any form of power), in practice few environmentalists follow such a singular path: rather they will be involved in and support other forms of action simultaneously.

Environmental NGOs and related groups serve two key roles in society: they often identify environmental issues and, importantly, they seek to translate environmental concern into practical strategies for change. In 1971 Friends of the Earth (FOE) was launched in London; it has become one of the biggest and most influential of the British environmental NGOs, and is part of an international FOE network. Its formation reflected a frustration with the staid conservation movement's lack of action against those perpetrators of environmental degradation (Rawcliffe 1998). They have a five-pronged strategy: to lobby those in political power and industry, often using legislative activity to press for change; to generate scientific research and publish it in accessible formats; to employ the media to attract attention to particular issues; to mobilize the public through local groups; and to coordinate and cooperate with other groups to run large-scale campaigns.

By the early 1990s however, FOE had evolved from a radical group to one that seemed less keen to press for far-reaching change. In effect it had become reformist – pragmatically seeking inclusion of environmental concern within the current political and social system. Many saw this reformist approach as being ineffectual, slow and hierarchical. Drawing inspiration from other movements and other countries, smaller, more radical environmental groups (or informal networks) began to proliferate. The visions they espouse are often idealistic and the steps to realizing their dreams often challenging, such as dramatically reducing consumption, abandoning the use of cars, or changing our eating habits. They pose a systematic challenge to existing societal practices, often rejecting 'representative' democracy and formal politics and promoting grassroots participation in environmental decision-making and a 'do-it-yourself' (DIY) approach. The tactics employed reflect their belief in the need for radical change and often involve the use of non-violent direct action. Direct action is 'intended to have an immediate effect on a situation, as distinct from political activity which might have a roundabout effect through representatives, or demonstrative activity whose effect was to get publicity' (Rooum 1995: 27). Examples include destroying **GM** crops, road blockades and office occupations. Earth First! (EF!) formed in the United States in 1979 and spread to the United Kingdom in 1991. Based upon anarchist ideology, it is a network of autonomous groups that eschew formal membership and hierarchy (and thus leadership) and espouse consensus decision-making structures. It is based on the belief that actions speak louder than words, and has become synonymous with (often illegal) non-violent direct action and the DIY approach: one of their slogans is 'If not you, then who?' (Wall 1999; Doherty 2002). EF! has been influential in British anti-roads protests, e.g. Twyford Down (1992), Newbury (1995–6) and A30 Devon at Allercombe and Fairmile (1994–7). They used protest camps with tree-sits, tunnels and lock-ons with the aims of (i) physically preventing road construction; (ii) generating media publicity; (iii) educating the public; and (iv) acting as a catalyst for mass mobilization. EF! activists have expanded actions against issues such as GM crops and the arms trade. The defining characteristics of such activism are the continuously evolving creativity in tactics coupled with a broad concern for a variety of issues, for example 'they see exploitation of the Third World, the global poor, women, animals and the environment as a product of hierarchy, patriarchy, anthropocentrism, racism and, most prominently, capitalist economic relations' (Seel and Plows 2000: 114).

Plate 6.3 An anti-car stunt calling for radical environmental activism.

(Jenny Pickerill)

Such actions have been influenced by many others and replicated worldwide. In Australia, environmentalists have employed direct action since the late 1970s. Actions at Terania Creek (New South Wales), Daintree (Queensland) and surrounding the Franklin Dam (Tasmania) all focused on preventing the logging of old-growth trees. Activists set up protest camps and used tree-sits and roadblocks to deter environmental destruction (Cohen 1997; Doyle 2000). Such tactics are still employed today at anti-logging protests at East Gippsland (Victoria) and Styx Valley (Tasmania). India has an even more extensive history of *satyagraha* (non-violent action) inspired by Mahatma Gandhi. In the early 1970s, villagers (mostly women) from Uttar Pradesh physically began to try to prevent local logging – resulting in the Chipko Andalan movement (McCormick 1989). More recently, villagers threatened with losing their land, homes and livelihoods by the construction of the Narmada Dam have committed to drowning rather than forced resettlement (Shiva 2002). Such examples clearly illustrate the implicit link between environmental concern and social justice.

A key tenet of many environmental approaches is the need for individuals to take action. This is both because it is relatively easy for individuals to make changes to their lifestyles *and* because it is only with all individuals making changes that bigger changes can occur: 'the only possible building blocks of a Greener future are individuals moving towards a Greener way of life *themselves* and joining together with others who are doing the same' (Bunyard and Morgan-Grenville 1987: 336). Thus individuals are encouraged to 'reduce, reuse, recycle'. The three R's are deliberately in the order of magnitude of impact. Thus, for example, reducing our consumption (see Case study 6.4) will have a far greater environmental benefit than recycling, which is time, energy and financially inefficient. However, 'it seems unlikely that a massive number of individuals will experience the conversion that will lead to the necessary changes in their daily behaviour' (Dobson 2001: 131).

Thematic Case Study 6.4

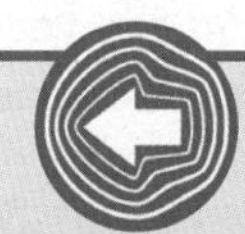

Consumption and the environment

Most of the luxuries, and many of the so-called comforts of life, are not only indispensable, but positive hindrances to the elevation of mankind. With respect to luxuries and comforts, the wisest have ever lived a more simple and meagre life than the poor . . . simplicity, simplicity, simplicity! (Henry David Thoreau, *Walden*, 1854)

As Thoreau illustrates, environmentalists have long argued for the need to consume less. There is an explicit link between economic growth, increased consumption, and negative environmental impact, and 'if the global population adopts the lifestyle currently seen in the developed countries, the environmental effects would be enormous' (Harris 2004: 271). Although environmentalists argue for a need to limit economic growth per se there is a particular focus on consumption, because if we were able to reduce demand for environmentally destructive goods it follows that production would also fall. Moreover, reducing consumption can be a personal act, available to all, and is part of the broader green vision of a life where we spend leisure time absorbed by non-material pleasures such as socializing, exercise, reading, etc., rather than shopping.

Like Thoreau, British environmentalist Jonathan Porritt, Chair of the Sustainable Development Commission (the government's advisory body on sustainable development) argues for a life of 'voluntary simplicity' (1984: 204). But what would a simple life look like, and how can we define what are justifiable needs and excessive luxuries? Although we can agree on what constitutes basic needs (shelter, food, clothing), there is disagreement as to what other items would be part of a greener vision. For example, are cars allowable? If so, what type of car? This ambiguity has led to the rise of green consumerism – purchasing goods with less environmental impact. In recent years there has been rapid growth in this sector, for example, eco-friendly washing liquids, organic food, or 'natural' shampoos. However, as Dobson notes, 'there is nothing inherently green . . . in green consumerism . . . none of it – absent other strategies – can bring about the radical changes envisaged by ecologism' (2001: 131–2). The trouble is that green consumerism is not about buying *less*, just purchasing different products; 'The underlying aim of this green consumerism is to reform rather than fundamentally restructure our patterns of consumption' (Porritt and Winner 1988: 199). A similar argument can be made for why recycling in

→

and of itself is limited. Recycling is not about *reducing* what waste we produce in the first place, just dealing with our waste more appropriately.

Green consumerism and the notion of 'sustainable consumption' are becoming mainstream practices partly because they do not require us to make major changes in our lifestyles (Burgess et al. 2003). In contrast, smaller-scale models of production and consumption, such as local organic food networks, or LETS and Time Bank schemes (which use alternative currencies to encourage non-monetary transactions) are being advocated as more environmentally sensitive and sustainable (Seyfang 2002 and 2006). Moreover, there is now an annual international 'Buy Nothing Day' which encourages people to 'host swap shops – free concerts and shows – credit card cut-up tables – create a shopping free zone with carpet and chairs – conga against consumerism – protest against sweatshops . . . There is only one rule – BUY NOTHING!' (www.buynothingday.co.uk).

For more information on consumption see the report by the Sustainable Development Commission, 2006 'I will if you will – towards sustainable consumption' at www.sd-commission.org.uk

Q 'If something cannot be manufactured, built or grown without causing irreparable ecological damage, can't we strive to create something to take its place, or simply decide to do without it?' (Tokar 1994: 80). What could you do without in your consumption and what would make you willing to give it up?

Finally, businesses can play a key role in strategies for environmental change, both good and bad (Doyle and McEachern 2001). In the good, businesses have sought to develop technologies that reduce environmental impacts (wind farms are one example), establish ethical banking practices (such as the Co-operative Bank), and practise Corporate Social Responsibility throughout their operations. These are models that attempt to marry capitalist approaches with softer environmental costs. However, some companies have lobbied hard against any climate change mitigation strategies that they perceive would impact upon their profits. For the automotive industry in the USA this has included lobbying against tax increases on gasoline or for producing smaller more fuel-efficient vehicles. Chrysler's chief economist Van Jolissaint has even got as far as questioning the existence of climate change and 'quasi-hysterical' policies of the Europeans (Schifferes 2007). Those that market themselves as being 'green' also face accusations of greenwashing – appearing green while changing none of their fundamental approaches, for example see Spotlight box 6.2. Moreover, businesses' entry into green markets can lead to a dilution of the original principles and eventually subvert the environmental benefits. For example, a core element of organic food production is to reduce pollution. This principle is enhanced when people purchase organic food locally (for example, through a box scheme from their local farm). However, when supermarkets saw organics as a market opportunity, they

Spotlight Box 6.2

Carbon offsetting

Carbon offsetting is the practice of financially contributing to projects which lead to a reduction in carbon dioxide emissions. This contribution is then counted as a reduction to the purchaser's carbon footprint. In other words it is offset against their actual carbon emissions. It is an indirect form of reducing carbon usage. Companies increasingly seek to buy carbon credits through funding energy efficiency, reforestation or renewable power projects, often in majority world countries. In 2005 HSBC, one of the world's biggest banks, asserted that it had gone 'carbon neutral'. It achieved this by

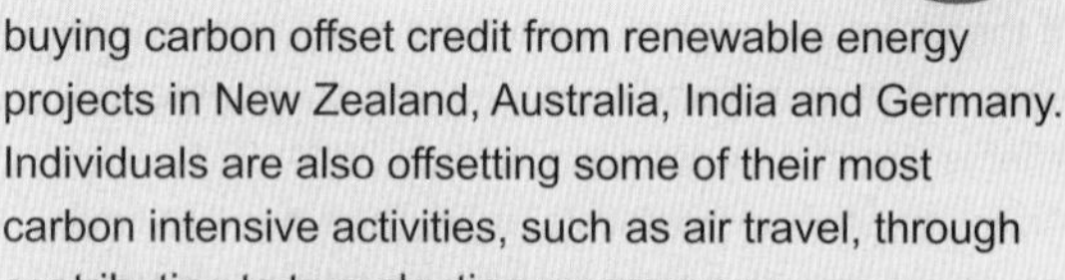

buying carbon offset credit from renewable energy projects in New Zealand, Australia, India and Germany. Individuals are also offsetting some of their most carbon intensive activities, such as air travel, through contributing to tree planting programmes.

Critics of carbon offsetting argue that it is a poor substitute for actually reducing our carbon emissions and is another way that companies and individuals can buy their way out of responsibility for the environment. There is significant debate as to its validity and effectiveness, with various different forms of measurement being used and the time-frame of, for example, forest growth still yet to be taken properly into account.

began to import organically produced food from Africa and Latin America, undermining many of its environmental benefits (Hughes 2005, 2007). Here supermarkets failed to take account of the importance of reducing the 'air miles' that food has travelled as part of the measure of the environmental impact of food consumption.

The dilemma remains how best to enact change, be it through NGOs that are able to lever access to the halls of power but in doing so can appear elitist, or through radical direct action which encourages grassroots participation and individual responsibility but which in turn can be small in scale and effect. For now, the answer appears to lie in the vibrancy of environmental movements composed of a variety of contesting, challenging and supportive groups, incorporating international NGOs alongside more radical small-scale groups.

What can one person, one household, one community or one administrative unit do when governments of some of the world's most economically powerful states cannot be brought onside? The environmentalist's answer is to be a good global citizen, to put pressure on administrations, governments, corporations and supranational organizations through political and non-political means, and to educate others. We have gone some way down this route in the past three decades, but significant problems remain.

6.6 What is missing from our focus on 'the environment'?

One of those problems is the way in which we conceptualize the environment. Explanations by scientists of the human causes of environmental degradation have, until recently, almost entirely focused on direct causes. Middleton (2003), for example, develops a fourfold typology of society–environment relationships (see Table 6.5). However, we are beginning to acknowledge the complex web of relationships between humans

Table 6.5 A typology of society–environment relationships

Type of relationship	Examples
Deliberate environmental manipulation	1. Diversion of watercourses for irrigation which leads to changes in river flows. Positive benefits include flood control and water availability for agricultural production over a long period, e.g. the Indus Irrigation Project in Pakistan. 2. Construction of large dams which leads to forced relocation of people, loss of productive land and wetland habitats, and changes in water resources and ecology downstream, e.g. Three Gorges Project in China.
Unanticipated impacts of activities that have been carried out to improve society	1. Irrigation in the former Soviet Union led to the desiccation of the Aral Sea because of diversion of too much water from the rivers that flowed into it. This led to the loss of the Aral Sea commercial fishery and health problems related to wind erosion of agro-chemicals trapped in former lake-bed sediments. 2. Expansion of cereal cultivation onto semi-arid grasslands has led to unexpected episodes of severe soil erosion (e.g. the US 'dustbowl' in the 1930s).
Complex problems in which cause and effect are not obvious. These are often characterized by human activities and natural forces combining synergistically	This applies to climate change, where there is still debate as to what proportion of elevated temperatures is due to natural climate change as against the effects of burning fossil fuels.
Linkages characterized by overlap and interaction	Contemporary deforestation is causing changes to people's livelihoods, losses of biodiversity, changes in the regional water balance and changes in global climate. All of these impacts overlap at scales ranging from local to global.

Source: After Middleton (2003)

Table 6.6 United Nations' Millennium Development Goals

Goal	Targets associated with goal
1. Eradicate extreme poverty and hunger	1. Halve, between 1990 and 2015, the proportion of people whose income is less than $1 a day. 2. Halve, between 1990 and 2015, the proportion of people who suffer from hunger.
2. Achieve universal primary education	3. Ensure that, by 2015, children everywhere, boys and girls alike, will be able to complete a full course of primary schooling.
3. Promote gender equality and empower women	4. Eliminate gender disparity in primary and secondary education, preferably by 2005 and in all levels of education no later than 2015.
4. Reduce child mortality	5. Reduce by two-thirds, between 1990 and 2015, the under-five mortality rate.
5. Improve maternal health	6. Reduce by three-quarters, between 1990 and 2015, the maternal mortality ratio.
6. Combat HIV/AIDS, malaria and other diseases	7. Have halted by 2015 and begun to reverse the spread of HIV/AIDS.
7. Ensure environmental sustainability	8. Integrate the principles of sustainable development into country policies and programmes and reverse the loss of environmental resources. 9. Halve, by 2015, the proportion of people without sustainable access to safe drinking water and basic sanitation. 10. Have achieved, by 2020, a significant improvement in the lives of at least 100 million slum dwellers.
8. Develop a global partnership for development	11. Develop further an open, rule-based, predictable, non-discriminatory trading and financial system (includes a commitment to good governance, development and poverty reduction – both nationally and internationally). 12. Address the special needs of the least developed countries (includes tariff- and quota-free access for exports, enhanced programme of debt relief for HIPC and cancellation of official bilateral debt, and more generous ODA for countries committed to poverty reduction). 13. Address the special needs of landlocked countries and small island developing states. 14. Deal comprehensively with the debt problems of developing countries through national and international measures in order to make debt sustainable in the long term. 15. In cooperation with developing countries, develop and implement strategies for decent and productive work for youth. 16. In cooperation with pharmaceutical companies, provide access to affordable, essential drugs in developing countries. 17. In cooperation with the private sector, make available the benefits of new technologies, especially information and communications.

Source: www.un.org/millenniumgoals. The United Nations is the author of the original material.

and the environment, and its relation to poverty, development, injustice and democracy.

Since Rio the emphasis has changed to one that seeks to integrate environment and development more closely and for the environmental agenda to be led by concerns about poverty. These new lines of thinking are underpinned by the World Bank's Poverty Alleviation Strategies, the Millennium Development Goals which were agreed by the UN General Assembly in September 2000 (www.un.org/millenniumgoals) (see Table 6.6), and as a general response to lack of progress on poverty and debt in the Third World. For many in the majority world the links between poverty, development and the environment are explicit. It is not a simple case of poverty leading to environmental damage: there are countless examples of poor people being key actors in environmental campaigns in India, Indonesia, the Philippines and across South America. Certain forms of development have significant environmental consequences – such as dam building, fossil fuel power generation and natural resource extraction. But we should understand these developments within the context of a history of colonial exploitation and contemporary processes of competitive economic globalization.

Plate 6.4 Teaching indigenous knowledge about the land, Australia.

(Jenny Pickerill)

Moreover, **ecological modernization** has failed to consider international justice (Connelly and Smith 2003). Where industrialized nations have sought to reduce their environmental damage they have in many cases simply exported their environmental problems to less industrialized countries: for example, the export of harmful computer wastes to China (Shabi 2002). This process has also occurred internally, see Spotlight box 6.3. Furthermore, environmental protection has resulted in the dispossession of indigenous peoples from their land to make way for national parks (see Case study 6.5). This practice results from two mistaken beliefs: first, that humans will only ever have a detrimental impact on their surroundings and second, that preservation of an environment is more important than human ties to a particular place, and in some cases, human survival.

Spotlight Box 6.3

Environmental justice

The environmental justice movement called for acknowledgement that it is often the already marginalized who suffer most from environmental degradation. Robert Bullard in his book *Dumping in Dixie* (1990) argued that toxic dump sites were more likely to be near neighbourhoods of black Americans than whites – race being a major factor in the quality of a person's environment. This is a form of environmental racism and environmental injustice.

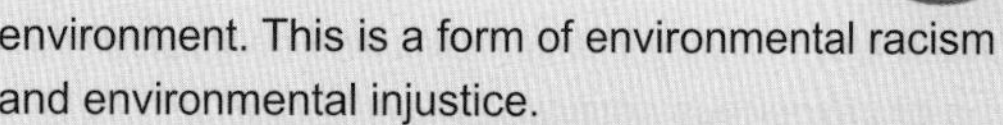

It is also a critique of environmentalists' focus on wilderness preservation rather than the plight of the underprivileged, and 'has dragged other parts of the US environmental movement into a political place where the social ramifications of being green have to be confronted' (Doyle and McEachern 2001: 69). It remains a largely US approach.

Thematic Case Study 6.5

Wilderness and conservation debates in Australia

Environmental campaigns worldwide are often framed as conserving 'wilderness' and preventing exploitation of natural resources. Organizations lobbying for protection of vast tracts of Alaska, Canada and Australia employ an emotive language using words such as 'pristine', 'untouched', 'undisturbed', 'intact expanse' and 'wild frontiers' to rally support. Photographs of lands empty of people or any human structures often accompany these words. Whether directly, through hunting for example, or indirectly through their development activities, people are usually held responsible for losses of wildlife. Until the 1980s, conservationists' solutions relied on controlling people's actions through strategies like hunting bans and establishing protected areas. These strategies – collectively known as protectionism – are overtly neo-Malthusian and have dominated nature conservation. This approach has proved successful in arguing for the need to place legal boundaries around tracts of land through World Heritage status or as National Parks.

Despite this, there is, increasingly, recognition not only that indigenous people have historic rights of possession to some of that land, but also that indigenous environmental knowledge and land management practices can be beneficial for conservation outcomes. This recognition and the legal changes that have accompanied it (such as the development of Native Title in Australia) have forced environmental and conservation groups to reconsider their approach. Simultaneously a global movement has emerged, focused on environmental justice and environmental racism. Yet this operates in tandem and separate from those groups concerned with biodiversity and wildlife preservation. Many environmental groups have been slow to change their language or practices when it comes to viewing areas as 'wilderness' devoid of people, cautious of meddling with their model of success and entering the complex and contested arena of indigenous politics.

There are many reasons to urge such groups to change: such practices are a continuation of an unjust colonial system complete with racism and stereotypes; they not only fail indigenous populations but lead to the loss of valuable environmental knowledge and academics have long called for the need for a more nuanced understanding of nature – not viewing it as a separate entity from social and cultural influences (Castree 2003). Moreover, while progress at the global level through UN resolutions to ensure that indigenous people are not forcibly removed from their lands has been slow, small changes can occur at a local scale with positive effects for indigenous communities. In other words, environmental groups have the power and capacity to markedly improve the situation for the betterment of conservation outcomes *and* indigenous people.

Language is a key articulation of power in understandings of the 'environment'. The term 'wilderness' and its relation 'wild' is highly problematic when talking of any landscape, but especially so in Australia, a land inhabited by the indigenous population for tens of thousands of years. Yet these terms are still employed. Such a romanticization of the environment draws upon the writings of Henry David Thoreau, John Muir and Miles Dunphy, and is used to sell Australia and garner support for environmental protection.

Indigenous people have been highly critical of the use of such a word. It has obvious colonial and racist connotations. This narrow vision, propagated by environmental groups, of what 'wilderness' entails is dualistic. Indigenous Australians argue that there can be no such division between the environment and culture and 'land is a much more energetic configuration of earth and air, water and minerals, animals and plants, as well as people, than a surface area contained by lines on a map' (Whatmore 2002: 71). It is precisely because of these interrelations between environment and culture that indigenous people need to live on their land in order to care for their country; 'the land needs the people and the people need the land . . . these are important cultural environments that require people to manage them according to tradition and culture, to maintain . . . and to encourage species' (Damian Britnell, Chief Executive Officer, Bamanga Bubu Ngadimunku Inc., Mossman Gorge Aboriginal Community).

The indigenous critique of 'wilderness' and approach to 'country', especially calls to live on land, is controversial. However, the majority of environmental groups in Australia now have (albeit only very recently) a vague policy of supporting indigenous rights. Some groups have attempted to go further and alter their language and practices to implement this policy. The Wilderness Society has shifted away from large-scale agreements and broad alliances with indigenous groups to grounded community-level cooperation. This approach appears more successful. Although they have not discarded the contentious goal of creating National Parks, they are working more in collaboration with traditional owners, supporting their attempts to gain Aboriginal Freehold land and establish Indigenous Protected Areas (Schneiders 2006: 27). They are developing 'tenure-blind' conservation measures: less concerned with who owns the land and more focused on good land-management practices.

The Australian Conservation Foundation has taken a different path. They have instigated a northern Australia Programme which aims to be a long-term proactive approach towards creating a bicultural organization. It is focused on coming up with joint goals rather than finding campaign partners for issues they think are important. To this end ACF has even opposed the creation of new National Parks when traditional owners have objected. In terms of language, ACF avoids the words 'wilderness' and 'biodiversity' altogether and instead talks about 'nature' and 'culture'.

Both these organizations have made progress in modifying their language and practices in response to the charge that they were party to 'a form of ecological imperialism' (Langton 1998: 18). However, two key problems remain. First, the majority of indigenous engagement by these two groups has been confined to northern Australia. A view persists that there is no need to engage with indigenous politics further south, perpetuating the myth that only those indigenous people who have a more apparent and historic (according to non-indigenous adjudicators) connection to their homeland need consultation. Secondly, although the language has changed, the underlying premise of why a landscape is of enough value to protect has not altered. Both still employ a biophysical-based and scientific method to determine value. We are a long way from bridging the gap between 'environment' and 'culture' and widening the ways in which we value landscape.

(Adapted from Pickerill (2008))

Q How would an Indigenous Australian approach change how we looked after the environment?

Thus the industrialized world's approach to 'the environment' often fails to understand majority world experiences, the need to incorporate humans *into* our understandings of the environment, and to acknowledge that some people have limited environmental choices. Recent work has sought to change how we view the environment, suggesting that the division between human and non-human is an artificial and unhelpful one (see Chapter 12). Environmental concerns have been recast to be ones of social survival. For example, the impact of sea-level rises on Pacific Islanders will be so catastrophic that it will potentially wipe out entire nations. Rather than being seen as a hindrance to wildlife preservation, indigenous people's environmental knowledge and management skills are slowly beginning to be valued. Moreover, their often alternative ways of viewing the environment, in the case of indigenous Australians for example, as being a melding of culture, land and sky, challenge our conceptualizations. The need to understand how we can trigger behavioural changes and the psychology behind environmental concern are becoming priority areas for environmental research (Hulme 2004).

6.7 Environmental futures

Society is taking action: policies are being designed and implemented at global, regional and local scales. But there are issues that still dog progress:

- Remediation and restoration of many types of environmental damage takes years.
- The costs of environmental remediation and protecting environments are often enormous.
- Our incomplete knowledge about many aspects of the environment, and how environment–society linkages work, can act as a barrier to finding effective solutions.
- The speed of economic and cultural change is rapid, the desire and drive for economic development strong; in contrast, the negotiation and implementation of global and regional environmental policies are slow.

Despite these obstacles it is up to us to decide what kind of world we want and to work towards it. We have explored a range of ways in which we understand

and value the environment, and different ways in which we could 'reconceive the environment', and strategies, for change. In summary, there are four key steps we can take towards reducing our environmental damage.

First, we need to understand our environmental impacts better: globally, regionally, locally and individually. At an individual level we can use the concept of an ecological footprint to evaluate our resource use (see Case study 6.6). Secondly, we can *participate* in

Thematic Case Study 6.6

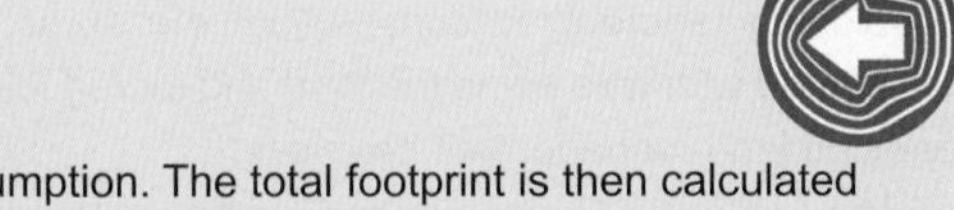

What is your ecological footprint?

Ecological footprinting is an accounting system to measure and compare our resource use. Wackernagel and Rees from the University of British Columbia developed the concept in their 1998 book *Our Ecological Footprint*. Looking at the world we can see that the total land area is approximately 15 billion hectares. Of this the total 'productive land' is approximately 10.3 billion hectares, the total 'productive sea' is approximately 2.9 billion hectares, giving us a total of 13.1 billion hectares to provide all the resources for all the people on the planet AND all the other living creatures that we share it with.

So 13.1 billion hectares between 6 billion people equals 2.18 ha each; 2.18 ha minus 12 per cent for other life and we have *1.9 ha each* now.

So with an idea of what our fair share might be, to achieve sustainability we need to live by consuming in one year only the products and services that can be produced by that land's biological growth in one year AND producing only wastes that can be absorbed by the natural recycling processes that take place on that land in one year. We could call our land and its resources our 'Biological Capital' and the sustainable produce of that land over a year our 'Biological Interest'.

We all use resources from many different parts of the world and it is not always possible to get what we need locally. The important thing is that ecological footprinting helps us to determine how much of our fair share we are using even if we consume something from the other side of the world.

Calculating how much land we are actually consuming from

Component-based calculation produces a breakdown of footprint by activity. It is applied to individual lifestyles, families, schools, communities, businesses and projects. The ecological footprint values for components are pre-calculated using regionally factored, direct and indirect lifecycle impact data. Consumption is then monitored for 24 major components which have been shown to account for the majority of consumption. The total footprint is then calculated as the sum of the components' footprints.

The main disadvantage of this approach currently is linked to the variability and reliability of lifecycle analysis data for components, which can make national and international comparisons problematic. However, for different consumption choices, whether for a family or a project, provided the same assumptions are used for all choices, the results will help compare the different options in relative terms.

How do we calculate a component's ecological footprint?

In essence you examine your component. Make decisions as to which land categories are utilized to provide its raw materials and to absorb the CO_2 emissions from the fuels/energy necessary to process and transport the materials/component throughout its lifecycle, including what happens to it at the end of its life/use. The land categories used are: energy land, built or degraded land, bioproductive land, bioproductive sea and biodiversity land.

For example: FOOTPRINT OF A LOAF OF BREAD

How much grain is required for 1 kg of bread? Based on recorded figures for cereal yields, this will relate to a certain area of land (bioproductive land) required to grow the grain.

We then have to examine the energy used to:

(a) Grow the grain on the farm and harvest it.
(b) Transport and mill the grain to flour.
(c) Transport the flour to the bakery and bake the bread.
(d) Transport the bread to the retailer and sell it.

Depending upon the various fuels used this will relate to a certain area of forest growth (energy land) required to sequester the CO_2 produced. The sum of the two areas is then the EF for your 1 kg of bread.

(Calculate your basic ecological footprint at www.ecologicalfootprint.com)

(Adapted from Radical Routes (2006) *How to Work Out Your Ecological Footprint*)

Plate 6.5 A Reclaim the Streets demonstration in Hull, May 1999, which illegally occupied the streets for a day in protest at a car-obsessed culture.
(Jenny Pickerill)

environmental decision-making. Connelly and Smith (2003: 361), through their concept of **ecological democratization**, argue that the only way to achieve environmentally sustainable practices is through 'a commitment to both justice and participation'. They advocate extensive citizen participation and development of democratic institutions to tackle environmental issues at all scales. In many ways this is an extension of many pre-existing local initiatives. It is also a step beyond advocating education. As Pepper notes, 'people will not change their values just through being "taught" different ones. What, then, is the real way forward, if it is not to be solely or even largely through education? It must be through seeking *reform at the material base of society, concurrent with educational change*' (1984: 224, emphasis in original). Thirdly, we need to recognize and make use of innovative grassroots solutions. Seyfang and Smith (2007) argue that it is the grassroots-community level where significant innovation occurs that can offer us inspirational visions of alternative futures. We have already seen the challenge posed by **low impact development**. Some of these innovations are small-scale in their approach, others, such as **permaculture**, are now being adopted globally. Fourthly, and least radically, we need to improve our management of the current situation. We have existing environmental issues that we need to mitigate immediately, even while we begin to explore alternative futures. For example, in Britain we have growing numbers of car users. A 2006 government report, The Eddington Transport Study, supported road pricing, rather than more road building, in order to overcome existing congestion problems, and that 'all transport users should meet all their external economic, social and environmental costs' (see http://www.hm-treasury.gov.uk/independent_reviews/eddington_transport_study/eddington_index.cfm). However, there was little support in the report for building new high-speed rail links or limiting airport growth. We have to decide if this is an appropriate management strategy, or whether advocating having a travel allowance, akin to rationing, would be more efficient. Given these choices, which pathways forward are you going to take?

Learning outcomes

Having read this chapter, you should understand:

- How we value human and non-human life/the environment.
- The evolution of environmental concern.
- The complexity of scale when looking for solutions to environmental problems.
- The possibilities and limitations of different strategies for action.
- How we can begin to make progressive steps towards reducing our environmental impacts.

Further reading

Bingham, N., Blowers, A. and Belshaw, C. (2003) ***Contested Environments***, Open University Press, Milton Keynes. This book explores why environmental issues are so often controversial, and uses concepts of value, power, and action to unpack environmental concerns.

Doyle, T. (2005) ***Environmental Movements in Majority and Minority Worlds: A Global Perspective***, Rutgers University Press, London. A good introduction to a variety of environmental movements in the USA, Australia, Britain, Germany, Philippines and India, incorporating protection of forests, rivers and wilderness areas, and movements against mining, road building and nuclear power.

Dryzek, J. S. and Schlosberg, D. (eds) (2004) ***Debating the Earth: The Environmental Politics Reader***, Oxford University Press, Oxford. A useful reader incorporating a wide variety of contemporary and historic authors on key themes of environmental politics such as democracy, justice, wilderness and the place of people.

Harris, F. (ed.) (2004) ***Global Environmental Issues***, John Wiley and Sons, Chichester. Taking a global perspective this book examines environmental problems as complex issues with a network of human and biophysical causes.

Middleton, N. J. (2003) ***The Global Casino: An Introduction to Environmental Issues***, 3rd edition, Arnold, London. An up-to-date introduction to a wide range of environmental issues in 22 chapters – from tropical deforestation to climate change, from urban environments to food production, as well as chapters on sustainable development and war. It provides an excellent starting point for delving deeper into individual issues.

Palmer, J. A. (ed.) (2001) ***Fifty Key Thinkers on the Environment***, Routledge, London. A useful summary of key environmental thinkers.

Peterson Del Mar, D. (2006) ***Environmentalism***, Longman, New York. A short thought-provoking book which examines the complex relationship between prosperity and environmentalism.

Useful websites

http://www.climatecrisis.net provides additional information, science and 'take action' suggestions for Al Gore's film *An Inconvenient Truth* about climate change.

http://www.lowimpact.org Low Impact Living Initiative. A non-profit organization whose mission is to help people reduce their impact on the environment, improve their quality of life, gain new skills, live in a healthier and more satisfying way, have fun and save money.

http://www.wilderness.org.au The Wilderness Society Australia. Contains information on old-growth logging in Australia and the wide variety of other environmental campaigns in which this Australian environmental advocacy organization is involved.

http://www.treesponsibility.com/index.html Treesponsibility – a small group in Calderdale, Yorkshire who plant trees to act as a carbon sink.

http://www.unep.org/GEO/geo3 *GEO-3: Global Environment Outlook 2003*, Earthscan, London. The third book in a series of three-yearly syntheses of the global environment produced by the United Nations Environment Programme.

http://www.zerofootprintoffsets.com/calculator.aspx Zeroprint Carbon Calculator: enables you to estimate your carbon footprint.

For annotated, clickable weblinks and useful tutorials full of practical advice on how to improve your study skills, visit this book's website at **www.pearsoned.co.uk/daniels**

CHANGING GEOGRAPHIES OF FOOD CONSUMPTION AND PRODUCTION

Chapter 7

Brian Ilbery and Damian Maye

Topics covered

- Thinking about food
- Contrasting cultures of food consumption
- Farming, food chains and globalization
- Alternative geographies of food
- Towards critical food geographies

7.1 Thinking about food

This chapter introduces some recent debates about the changing nature of food consumption and provision and presents ideas and case studies to prompt critical reflection. Food has become a topic of great interest to human geographers in recent years (Atkins and Bowler 2001; Winter 2003a; Hughes and Reimer 2004; Cook 2006; McCarthy 2006; Maye et al. 2007) and the chapter begins by identifying its value as a tool for geographical analysis. Thinking about food generally, it is useful to consider the following quote from the food writer, Nigel Slater, who, in one of the first editions of the *Observer Food Monthly* magazine, argues that:

> Never has there been more interest in what we eat and drink. But never have we needed to know more either. The bottom line is that the food we eat becomes part of us. Each blood-stained bite of meat, every cool sip of water, each spoonful of organic cabbage and every additive-encrusted potato chip; the salads, the Mars bars, the wine and the beer, all play their small part in what we are. Our bodies take both the virtues and horrors of every forkful – for better or for worse – and turn it into Us.
>
> (April 2001)

In this quote, food extends well beyond its obvious biological significance as the protein for sustaining life. It is something that one relates to on a very human scale – it reflects who we are. As Atkinson (1991) coins it, food is a 'liminal substance' that links humans and nature. Food also reveals important social, cultural and economic geographies. On the social side, there are inequalities in terms of food access, good nutrition and the reported problems with obesity and diet, often tied to socio-spatial inequalities in terms of wealth and education. Places where access to affordable, healthy food is poor are described as **food deserts**, which are particularly prominent in urban areas (Wrigley 2002). Food also has cultural significance. As Visser (1986: 12) puts it, 'food shapes us and expresses us even more definitively than our furniture or houses or utensils do'. Think, for example, about the different places associated with eating food (e.g. roadside café, Burger King, the Savoy Grill) and the different behaviours associated with them. As Bell and Valentine (1997) note, if 'we are what we eat' we can also say that 'we are *where* we eat'. There are also cultural associations between different cultural groups and ethnic or national cuisines (e.g. the favourite dish in the United Kingdom is Tikka Masala), as well as a range of cultural images used to market and sell food. Finally, the economic significance of food is enormous: for example, one only has to think about the investment and number of people involved in the production, processing and retailing of food.

Thus it is quite easy to make a case for why food matters as an object of study in human geography. However, tracing material and metaphorical associations with food is less obvious because they are connected together in complex ways and often 'hidden' as incidental parts of everyday life. For example, drinking a cup of coffee has numerous social and geographic connotations (Giddens 2001): it has symbolic value (as part of day-to-day life); it is a drug (caffeine provides 'extra lift', but coffee drinkers would not be seen as 'drug users'); it represents past social and economic relationships (e.g. in terms of colonization, mass consumption); and it is a symbol of globalization and world trade links (e.g. global brands like Nescafé and Starbucks). Another example of this global–local 'inter-connectedness' would be a visit to a local supermarket where one can identify 'the world on a shelf', with products such as apples from South Africa, spices from India and coffee from Columbia. For shoppers in Western society, it is easy to forget the complex array of economic and social ties that connect these local retail spaces to other parts of the world. Geographers argue that one cannot separate these localized, mundane acts from larger social settings that extend around the world. This realization about the need to trace *'connections'* between food production and consumption, sparked in part by ongoing public debates about obesity, food risks and so on (e.g. Jamie Oliver's School Dinners), has become central to agri-food studies in recent years.

Geographical research in this area is increasingly focused on different types of food supply, which involves thinking about where and how food is produced, how it is retailed and how and where it is consumed. The routes traced by particular foodstuffs from 'farm to fork' are often referred to as a **food chain** (or 'network') and geographers have attempted to 'map' the system of connections for different products which may vary both in complexity and geographical coverage. Hartwick (1998: 425) defines food chains as 'significant production, distribution and consumption nodes, and the connecting links between them, together with social, cultural and natural conditions involved in commodity movements'. Geographers have thus adopted the supply chain metaphor to literally trace and follow the nature of 'connections' for particular commodities (Cook 2006). Different food supply chains will be examined later in the chapter. At this stage, Spotlight box 7.1 shows how the food chain concept is also relevant in a wider policy context (Jackson et al. 2006). This example highlights two key things: first, it notes that the food chain is not a new concept, recast as a 'food system', 'food circuit' or 'food network' (see also Maye and Ilbery 2006 who in addition draw reference to a 'food convention'); and secondly, it demonstrates how the notion of a food chain is defined and 'mobilized' in different ways by contrasting institutions to restructure food provision, on the one hand, in econometric terms (the Food Chain Group) and, on the other, in environmental terms (Sustain). Crucially, Jackson et al. (2006) argue that this seemingly chaotic application does not render the term useless, but instead shows its continued symbolic and applied values, whilst recognizing the importance of political context. This new growth in food chain research is a key theme of this chapter.

Food, therefore, is a geographical topic, including its role in society and economy, and the food chain can be used as an overall organizing framework to explore changing geographies of food consumption and production. The rest of the chapter is divided into four parts: first, the focus is on the eating side of the food

Spotlight Box 7.1

Mobilizing the food chain

The food chain concept has become central to agri-food studies in recent years, especially in relation to specialist/local food production. The interest in food provenance, for example, is often conceptualized as offering small-scale producers the potential to develop **short food supply chains** (SFSCs) to better capture value added and 'connect' with end consumers. Analysing institutional responses to the 2001 outbreak of foot-and-mouth in the United Kingdom, especially responses to recommendations made by the Policy Commission on the Future of Farming and Food (2002), Jackson et al. (2006) argue that the food chain is also important in policy contexts. However, they add that the way a food commodity chain is defined and used (or mobilized) may differ between institutions, as shown in the following two examples from the Food Chain Group (set up jointly by Defra and the Institute of Grocery Distribution) and Sustain (an environmental campaigning group). The authors argue that this seemingly chaotic application does not render the term useless, but instead shows its continued symbolic and applied values, whilst recognizing the importance of political context.

The Food Chain Group

The Food Chain Group was established in 1999 and included senior representatives from the food and farming industries. From the industry's perspective, a food chain approach was useful to help improve understanding of the cost structures at each stage of the chain, as well as to enable producers to improve their production and marketing systems. The work of the Food Chain Group was given greater impetus when the

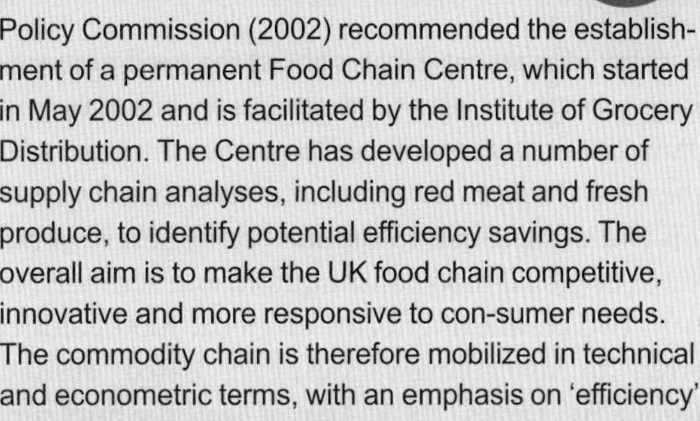

Policy Commission (2002) recommended the establishment of a permanent Food Chain Centre, which started in May 2002 and is facilitated by the Institute of Grocery Distribution. The Centre has developed a number of supply chain analyses, including red meat and fresh produce, to identify potential efficiency savings. The overall aim is to make the UK food chain competitive, innovative and more responsive to con-sumer needs. The commodity chain is therefore mobilized in technical and econometric terms, with an emphasis on 'efficiency' and 'integration' along the chain.

Sustain

Sustain (the alliance for better food and farming) was formed in 1999 and represents over a hundred national public interest organizations working at international, national, regional and local levels. As a campaigning organization, it advocates 'food and farming policies and practices that enhance the health and welfare of people and animals, improve the working and living environment, promote equity, and enrich society and culture' (www.sustainweb.org). Their overall goal is to support the small family farm against larger corporate interests, encouraging shorter, more direct supply chains as a means of reducing 'food miles'. The food chain and associated concept of food miles are used to place an emphasis on sustainability at all geographical scales. The 'Sustainable Food Chains' project is one of a number of projects designed to promote sustainable food provision and tackles three policy themes: (i) local food economies; (ii) food miles, and (iii) public procurement. Sustain thus advocates a much more circular and expansive view of commodity chains, including social and environmental factors, in addition to economic costs.

chain and on identifying two contrasting cultures of food consumption; secondly, the current modes of global food supply are examined, including the nature of global food chains and the resulting 'disconnection' between producers and consumers; thirdly, some of the ways in which producers and consumers are beginning to establish alternative systems of food provision are outlined; and finally, the chapter summarizes some of these recent food chain trends and recognizes the need for geographers to continue to interrogate critically the conceptual foundations of food supply, including terms like 'local' and 'alternative'.

7.2 Culinary journeys: contrasting cultures of food consumption

In keeping with recent debates about the need to recognize consumers in agri-food studies (Goodman and DuPuis 2002), this section begins at the consumer end of the food chain. In the space available, the focus will be on two things: first, to outline briefly some overall patterns of global food consumption; and secondly, to present two contrasting forms of global food culture, referred to as 'culinary journeys'. Based around Millstone and Lang's (2003) *Atlas of Food*, the brief survey of global food-consumption patterns examines five trends: changing diets, the role of retailers, branding and advertising, the growth of organic food sales, and eating out. It goes without saying that diets vary around the world. That said, some would argue that world diets are becoming increasingly homogeneous. Diets in developing countries have traditionally been dominated by a single staple food. This is no longer the case in industrialized societies, where the variety of foods on offer has never been greater. Here, diets usually contain more food of animal origin and less food of plant origin. In the USA, for example, only about 20 per cent of the total dietary energy supply comes from cereals, with almost as much being supplied by sugar and sweeteners and by fat. Millstone and Lang (2003) note that a 'nutrition transition' is affecting urban populations in many developing countries. Although a shortage of even staple foods continues to be a problem for sectors of the population in these countries, an expansion in food trade, improvements in global communications and the penetration of new markets by food corporations are all having an effect on diet. These changes can have significant health impacts, leading to an increase in diet-related diseases, such as some forms of cancer and cardiovascular diseases.

The power of food retailers is well documented. This includes influencing the types of food we eat. In the United Kingdom, it is estimated that half of the food consumed by around 60 million people is bought from just 1,000 stores. The 1980s saw the beginning of a new era of 'hypermarket economics', led by a small number of large companies (see Wrigley and Lowe 2002). The US-based Wal-Mart, for example, has 3,500 stores in the USA and a yearly turnover of $200 billion. It started its overseas expansion in 1991 and in 2002 invested $9.2 billion on increasing its sales worldwide. As a result of this dominance, most people, at least in developed market economies, now buy their food from supermarkets. Food choices are also significantly influenced by branding and advertising. The global food industry spends $40 billion each year on advertising. More often than not it is branded and processed products that are heavily advertised. Coca Cola is the most heavily advertised brand – with an annual advertising spend of $1.5 billion worldwide. Millstone and Lang (2003) regard brands as moulders and reflectors of mass consciousness. Food processors, who may own several brands, invest heavily in advertising so that they are instantly recognizable and desirable.

These trends are not all that surprising. Perhaps less predictable is the growing demand for organic produce in industrialized countries. The total world retail market for organic food and drink is estimated to have grown from $10 billion in 1997 to $17.5 billion in 2000. In response, retailers are promoting organic food more vigorously than ever before. The reasons why consumers demand organic foods are significant: partly a reaction to growing awareness about the environmental and health issues associated with certain types of farming and also because higher disposable incomes enable people to make 'lifestyle choices' about what they eat. Recent years have also seen a significant rise in the number of people eating out, reflecting lifestyle changes, especially the increase in the number of women in paid work. Americans spend almost $1,400 per person a year on eating out. In most developed countries, annual expenditure on eating out exceeds $400 per person. In particular, **fast food** is becoming available worldwide and offers a quick and accessible way to eat food. The rest of this section uses these facts about food consumption as a context to introducing two contrasting cultures of food consumption.

7.2.1 Culinary journey 1: fast food culture

Fast food is part of all consumer cultures: a meat burger in a bun, a frankfurter sausage, fish and chips, a pizza slice, samosas, etc. It is usually eaten out of the home and out of our hands. Millstone and Lang (2003) argue that the global spread of the three big US fast-food corporations – McDonald's, Burger King and Tricon (Kentucky Fried Chicken and Pizza Hut) – has come to represent fast food and has most influenced eating habits worldwide. In particular, fast food is central to the American way of life. On average, one in five Americans eats in a fast-food restaurant each day. Elsewhere, the figures are lower, but everywhere the trend is upwards, with Latin America and Asia experiencing the biggest increase in McDonald's restaurants in the late 1990s (Millstone and Lang 2003).

The classic book on fast-food culture is *The McDonaldisation of Society* by the American sociologist George Ritzer (2000). For Ritzer, McDonald's has succeeded because it offers consumers, workers and managers four key things. The first element of McDonald's success, he argues, is *efficiency*. As Ritzer (2000: 12) puts it:

> For consumers, McDonald's offers the best available way to get from being hungry to being full. In a society where both parents are likely to work or where a single parent is struggling to keep up, efficiently satisfying hunger is very attractive. In a society where people rush, from one spot to another, usually by car, the efficiency of a fast-food meal, perhaps even a drive-through meal, often proves impossible to resist.

The second important element is *calculability*. This puts emphasis on the quantitative aspects of products sold (portion size, cost) and the time it takes to get the product. For example, people order the Quarter Pounder, the Big Mac, the large fries, with options to 'go large' for often as little as 50p. Customers can quantify these things; it feels like a good deal. Ritzer also observes that people tend to calculate how much time it will take to drive to McDonald's, be served the food, eat it and return home; they then compare that interval to the time required to prepare food at home and, rightly or wrongly, usually go for the fast-food option.

The third element is *predictability*: the assurance that products and services will be the same over time and in all locales. As Schlosser (2001: 5) notes in *Fast Food Nation*, 'customers are drawn to familiar brands by an instinct to avoid the unknown. A brand offers a feeling of reassurance when its products are always and everywhere the same'. For example, the Egg McMuffin will, for all intents and purposes, be identical if consumed in New York, London or Moscow. Also those eaten next week will be identical to those eaten today. The behaviour of workers, in terms of what they say and do, is also highly predictable. The fourth element is *control*, which is exerted over the people who enter McDonald's restaurants through limited menus, few options, queue lines and uncomfortable seats designed to control behaviour and encourage customers to eat quickly and leave. The people who work in McDonald's are also controlled, performing a limited number of jobs in precisely the way they are told to do them.

Although persuasive, not all academics accept Ritzer's view that cultures and traditions have become homogenized as a result of the extended reach of global, standardized products. Jackson (2004), for example, argues that local consumption cultures are quite resilient to global forces. Using examples which include McDonald's menus in India (noting, for instance, the addition of garlic and chilli sauces to McDonald's burgers) and Cadbury's chocolate recipes in China (which were changed, with a reduced sugar content and increase in cocoa solids), Jackson argues that 'rather than simply rolling out their existing products across a geographically undifferentiated market, producers have had to adapt their global brands to a variety of local conditions' (p. 167). Places thus create specific consumption cultures and local cultures still influence the style of food on offer in each country and, in some cases, use food to defend that sense of local identity.

7.2.2 Culinary journey 2: slow food culture

Urry (2000: 43) observes that 'global flows engender multiple forms of opposition to their various effects'. The second culinary journey represents a long-standing form of opposition to McDonald's as a global network and echoes the earlier raised point about the importance of place. It concerns Slow Food, a consumer movement established in Italy during the mid-1980s in response to the opening of one of the first McDonald's restaurants there, in the famous Piazza di Spagna in Rome (Murdoch and Miele 2004). The opening of the restaurant raised the possibility that traditional eating habits might be threatened by Americanized fast food. In protest, the food writer Carlo Petrini gathered chefs, authors and journalists together to discuss the best means of countering the spread of fast food in Italy. This meeting, in 1986, gave birth to Slow Food which is devoted to the promotion of an 'anti-fast food' culture.

Following Murdoch and Miele (2004), the main elements of this movement can be summarized. Its initial aspiration was a celebration of *cultural connections* surrounding local cuisines and traditional products. The movement targets discerning consumers in order to heighten their awareness of 'forgotten' cuisines and the threats that they currently face. Its main means of reaching potential customers were via a publishing house (Slow Food Editore) which disseminated material on previously unknown or neglected foods. The first edition of the movement's magazine, Slow, explicitly sought to oppose the spread of McDonald's and other fast-food chains. Adopting the snail as the movement's logo, Petrini (1986) explained: 'it seemed . . . that a creature so unaffected by the temptations of the modern world had something new to reveal, like a sort of amulet against exasperation, against the malpractice of those who are too important to feel and taste, too greedy to remember what they had just devoured' (quoted from Murdoch and Miele 2004: 241).

The central objective of Slow Food is to *decelerate* the food consumption experience so that alternative forms of taste can be reacquired. The movement also has spatial significance: it wishes to *embed food in territory* and bring consumers closer to these foods, reasserting also the natural bases of food production (e.g. seasonality) and the role of cultural context (e.g. culinary skills, tacit knowledge). Its prime concern is for 'typical' or 'traditional' foods, although increasingly recognizing that some regional foods are disappearing because they are too embedded in local food cultures. While the roots of the movement lie in Italian food cultures, Slow Food also promotes 'typicality' further afield. In 1989, it launched itself as an international movement and has now spread to about 40 countries, with 70,000 members worldwide and organized at the regional level into a convivium (or consumer club). These clubs identify restaurants that promote 'slowness', organize tastings/talks, and promote local food quality in both the area and local schools (for details see: www.slowfood.com).

This necessarily brief account of Slow Food provides a useful comparison to the fast-food journey. It is possible to contrast the core elements of the two culinary journeys as follows:

Fast food	Slow food
• Efficiency	• Slowness
• Calculability	• Typicality
• Predictability	• Differentiation
• Control	• Openness

So, while McDonald's imposes a standardized format upon each locality, Slow Food encourages and supports differentiation. Similarly, whereas McDonald's is based on the dissemination of a simple formula (the Speedee Service System), Slow Food is built on an appreciation of diverse/open food cultures. Thus these culinary journeys represent contrasting styles of food consumption and the value of place in terms of understanding their construction. They also represent different ways of producing and processing food and it is to these aspects of food provision that the chapter now turns.

7.3 Agrarian questions: farming, food chains and globalization

This section examines farming, food production and global food supply. Two processes are significant: the *industrialization* of farming and the *globalization* of food supply. After introducing some concepts, case studies of two global food commodities are explored before identifying a series of questions that emerge in response to these particular food chain practices. Globalization is often defined in terms of the integration of systems among geographically dispersed places (see Dicken 2007 for details). Crucially, this process of global integration has been guided by powerful transnational corporations (TNCs), institutions and actors, which in this context led to a new political economy of agriculture, epitomized by the mass production of manufactured food. Despite inequalities, developed and developing countries are reportedly linked together in 'highly industrialised and increasingly globalised networks of institutions and products, constituting an agri-food system' (Whatmore 1995: 37). The case study of McDonald's is a clear expression of this.

However, it would be a mistake to see the global food economy as new. For example, Friedmann and McMichael (1989) argued that relations between agriculture and industry have historically been more global than generally thought. Using the concept of 'food regimes', they linked international patterns of food production and consumption to the development of the capitalist system since the 1870s. Their paper identified three food regimes, each one representing the modern food system of its time (Table 7.1). Thus there has been a global dimension to the geography of food supply for some time (for a useful review of food regimes see Robinson 2004).

The main criticism of this conceptualization is its view of the globalization of agriculture as a logical progression (similar in this respect to modernization

Table 7.1 Global food regimes since the 1870s

First regime: pre-industrial (1870s–1920s)
This involved settler colonies supplying unprocessed and semi-processed foods and materials to the metropolitan core of North America and Western Europe. Characterized by *extensive* forms of capital accumulation, the main products were grains and meat. The regime slowly disintegrated when agricultural production in developed countries competed with cheap imports and trade barriers were erected.
Second regime: industrial (1920s–1970s)
This regime relates to the productivist phase of agricultural change, focused on North America and the development of agri-industrial complexes based around grain-fed livestock production. Characterized by *intensive* forms of capital accumulation, the second regime incorporated developed and developing nations into commodity production systems. Agricultural surpluses and environmental disbenefits undermined this phase of production in the 1970s.
Third regime: post-industrial (1980s onwards)
This regime refers to the crisis surrounding industrialized farming systems and involves the production of fresh fruit and vegetables for the global market, the continued reconstitution of food, and the supply of inputs for 'elite' consumption in developed countries. Characterized by a *flexible* form of capital accumulation, this regime is dominated by the restructuring activities of agribusiness TNCs and corporate retailers.

Source: based on Robinson (2004), as derived from Friedmann and McMichael (1989)

theory). In reality, it is much more unstable. Returning again to McDonald's, its global presence is in fact atypical of the complex and highly uneven process of globalization that has reshaped food production since the post-war period. Much less contentious in these debates is identifying the key agents involved. FitzSimmons (1997), among others, identifies TNCs as the primary agents of globalization in the agri-food sector; to her mind, they sit at the centre of webs of relations that link farming, processing and marketing. As already shown, food retailing in Europe and America is concentrated in the hands of a few supermarkets and these agents also play an important role in developing regulatory systems that ensure their dominance over the supply of key food products. It is also agreed that TNCs and corporate retailers played significant roles in the 'industrialization' of farming and food production activities, which, initially at least, were mostly concentrated in the developed world.

7.3.1 Industrialized agriculture in the developed world

Millstone and Lang (2003: 38) note that, 'encouraged by governments as a response to food shortages and the need to raise productivity after World War II, industrial farming has dominated . . . agriculture in the EU and North America since the 1950s'. The industrial model of farming, key features of which include a specialization of labour, product specialization and intensification, and assembly-line type production, led to three important trends in the developed world in terms of food production: first, the concentration of agricultural production on a limited number of large-scale farms; second, an increase in capital expenditure on major agricultural inputs like chemicals; and third, a growth in the processing and manufacturing of food. These developments explain why it is now vital to view farming in the wider context of an *agri-food system*, where the production sector itself is inextricably linked to various 'upstream' (input supplies) and 'downstream' (processing, distribution and marketing) industries. In many cases, the food supply system is also dominated by large **agribusinesses** which, according to Davis and Goldberg (1957), are the sum of all operations involved in the manufacture and distribution of farm supplies, the production operation of the farm, storage, processing and distribution of farm commodities and items made from them. These agribusinesses often develop commodity chains beyond national boundaries (Wallace 1985).

A number of papers have charted the industrialization of agriculture, especially in the United States. Two short examples are presented here. The first is FitzSimmons's (1986) seminal paper on the Salinas

Valley in California. This shows how lettuce production was transformed between 1950 and 1980, identifying two key things: first, *intensification*, with a shift to more investment in intensive crops, increased labour productivity and intensified planting; and secondly, *restructuring*, whereby large farms dominated sales. By 1978, the ten largest grower–shippers in the region sold 65 per cent of all lettuce. Over time, therefore, power and control shifted towards the largest firms. The paper also identified a process of vertical disintegration, whereby segments of the production process were subcontracted out to smaller-scale growers. The second example is more recent and taken from Hendrickson and Heffernan (2002) who examined some well-known US agribusinesses, including ConAgra, one of the three largest flour millers in North America. The supply chain for ConAgra is reproduced in Figure 7.1 and shows that the company produces its own livestock feed, slaughters cattle, is involved in pork and broiler production and processing, crop protection and fertilizers (upstream), and the retail of foods (downstream). Foods produced through this vertically integrated chain are sold under a number of well-known US food labels, including Healthy Choice and Peter Pan Peanut Butter.

In these and other examples, it is often the non-farm sectors of the agri-food system that have become most industrialized and dominated by TNCs. This has occurred through two processes (Goodman et al. 1987): **appropriationism**, where certain agricultural inputs are replaced by 'industrial' alternatives (e.g. synthetic chemicals replacing manure); and **substitutionism**, which focuses on outputs rather than inputs and is concerned with the increased utilization of non-agricultural raw materials and the creation of industrial substitutes for food and fibre (e.g. sweeteners for sugar). Agri-food systems are different from other production systems because agriculture is bound by biological processes and cycles, and the essential aim of appropriationism and substitutionism is to 'replace' nature. Agribusiness TNCs also attempt to increase their influence over farming indirectly through a process of *formal subsumption*, where arrangements or contracts are made with farmers to provide 'raw materials' for their value-adding food-manufacturing activities.

The overall key pattern of agricultural industrialization is *concentration* in particular sectors and regions characterized by large farm businesses that have adopted intensive farming methods and become integrated into global food networks. In the European Union, 80 per cent of agricultural production is concentrated on less than 20 per cent of farms in particular 'hot spots', including East Anglia, the Paris Basin, southern Netherlands, and Emilia Romagna in north-east Italy. A similar pattern of concentration is evident in the United States. Agricultural industrialization is also increasingly global in scale, where food commodity chains lengthen and producers become 'distanced' from consumers. In response to various issues in the developed-world market, another key feature of TNCs and major retailers has been their attempt to relocate the production function of the agri-food system to new agricultural spaces, often in Less Developed Countries (LDCs) and usually in association with the intensive production of high-quality, high-value food commodities.

7.3.2 Developing-world agriculture

Goodman and Watts (1997) argue that the classic export commodities (e.g. coffee, tea, sugar, cocoa) associated with the LDCs have recently been complemented by high-value foods, including fruits, vegetables and shellfish. By the 1990s, 24 low-middle income countries annually exported over US$500 million of high-value foods; just five countries (Brazil, Mexico, China, Argentina and Kenya) accounted for 40 per cent of such exports from the LDCs. The growth in high-value food exports reflects various things, including: technical changes in the food industry, the liberalization of world trade, and dietary changes in the developed world, with high-value foods produced to satisfy consumer tastes. Agribusiness TNCs are also attracted to high-value food production in newly agriculturalizing countries because of low-cost labour, government support (via Structural Adjustment Programmes), good global communication links and the ability to produce high quality/value products for the developed world market. Detailed accounts of the export trade in high-value foods and cut flowers are now available, especially in an African context (see, e.g., Barrett et al. 1999; Hughes 2000; Dolan and Humphrey 2002; Freidberg 2004). However, in the space available, attention is turned to South America and, more specifically, Chile.

7.3.3 Salmon farming in Chile

Salmon farming in Chile is mostly located in Chiloe, a large island that runs along the southern third of the Chilean coast and is characterized by a mild, rainy climate and protected inlets. Since the 1990s, the salmon farming industry has been growing rapidly in Chile. The following analysis is based on a paper by Phyne and

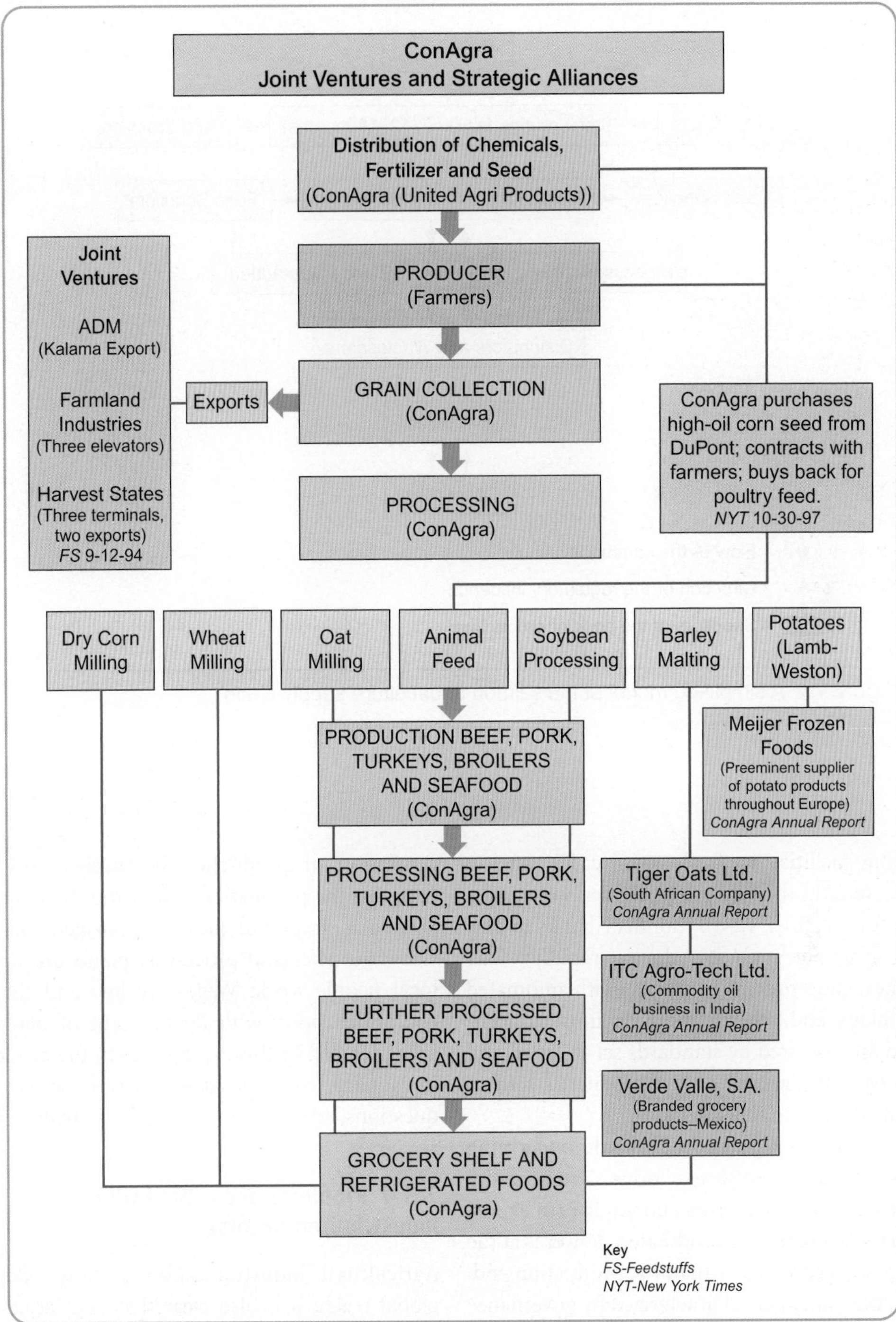

Figure 7.1 The ConAgra food chain cluster (situation in 1999).

Source: Hendrickson and Heffernan (2002)

Mansilla (2003). In 1987, Chile had 2,000 tonnes of Pacific salmon, but no production of Atlantic salmon. In 1992, Chile produced less than 10 per cent of Atlantic salmon, but by 2000 this was 17 per cent (second to Norway). Chile also has dominance of Pacific salmon (1992: 37 per cent and 2000: 76 per cent). Figure 7.2 shows a simplified model for the salmon aquaculture supply chain. The process proceeds as follows: eggs are hatched and the fingerlings raised near the hatchery; once fingerlings become smolts, they are raised in

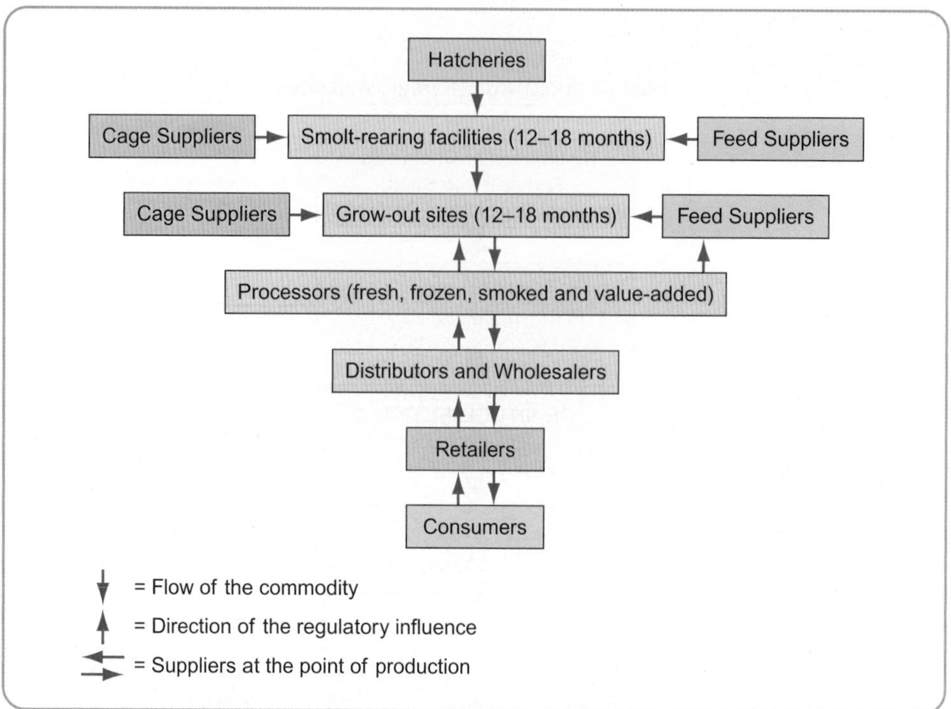

Figure 7.2 A simplified model of the salmon aquaculture supply chain.
Source: Phyne and Mansilla (2003)

smolt-rearing facilities in fresh-water cages; after a period of 12 months they are shipped to sea-water cages and raised for a further 12–18 months (thanks to new feeds, this grow-out stage is reducing rapidly); harvested salmon enter processing plants (with automated feeding systems and where labour is most concentrated) and are prepared by standards set in Japan and America. Over 80 per cent of the product is sold to Japanese and American retail markets.

This section shows how the Chilean salmon farming industry is inserted into a **'buyer driven' commodity chain**, where major distributors and retailers in export markets shape the nature of production. Power is at the retail end of the chain rather than the production end. Essentially, the industry is characterized by governance from lead firms and external authorities. Similar conclusions were noted in Barrett et al.'s (1999) analysis of high-value foods in Kenya, Dolan and Humphrey's (2002) analysis of fresh fruit and vegetables in Kenya and Zimbabwe, and Cook's (2004) analysis of papaya in Jamaica (see Spotlight box 7.2 for details). In the middle part of the salmon supply chain, there is also clear evidence of concentration and squeezing, with forward integration by feed giants and increasing concentration of production by foreign and domestic firms. At the production end of the chain, benefits are skewed in favour of those in management positions. Grow-out sites and processing plants are where most local people work. Wages are low and the working conditions poor, with 80 per cent of plant workers being female. In this example, as in the case of papaya in Jamaica, global food production raises significant questions, especially in terms of who truly benefits.

7.3.4 Problems associated with industrialized farming

Agricultural industrialization is thus extending its global reach; it is also clear that large-scale agribusinesses and corporate retailers have successfully linked regional economies and food sectors of some LDCs to a global system of food production and consumption. However, these systems of provision also raise questions about their suitability to LDCs, especially their contribution to sharp inequalities in income, productivity and technology compared to the sector producing domestic staples, as well as the unsustainable nature of agricultural practices often favoured by agribusiness

Spotlight Box 7.2

Follow the thing: papaya

In a recent review paper, Cook (2006: 655–6) picks up on a recent trend in food geography to quite literally 'follow' things as a method to unpack those hidden food chain stories which foods might tell if they could talk. One of the best examples of this is an earlier paper by Cook (2004) which followed the supply chain for papaya (a large green fruit from the West Indies). It provides detailed accounts from multiple sites along this particular global fruit chain, including the papaya buyer (based just outside London), the papaya importer, the papaya farmer, the farm foreman, the papaya packer (all based in Jamaica) and the papaya consumer (in North London). The fruit itself is no passive agent in this story, secreting an enzyme that eats into the skin of pickers and packers; changing the sex of its flowers according to the weather – with only hermaphrodites producing exportable fruits; requiring specialist, intensive labour if fruits of the right shape, size and quality are to be produced for market; and dying once harvested, thus increasing pressures to get the fruit packed and on a plane to supermarket shelves in the United Kingdom and United States.

The research sets out to 'de-fetishize' commodities, but is unusual in that it chooses not to impose any specific theoretical framework, offering instead an apparently descriptive ethnographic account of individuals at different stages in the supply chain, interweaved with historical and contemporary contexts about the place, plant and its trade routes. The challenge Cook sets is for us to read and draw our own conclusions. Various important things emerge for consideration. First, the paper shows the complex and entangled nature of global food-supply chains, linking in this example farmers and labourers in the Caribbean, fruit buyers, supermarkets and technicians in the United Kingdom, and First World consumers. Secondly, the role and evidence of power in these fruit chains are manifested in different ways and at different stages along the chain. Most notable are the pressures on growers and buyers to supply products that meet specific retail standards and are delivered on time. This confirms the buyer-driven rhetoric of most global food chains, but is equally disturbing in the way power games trickle down to the farm level, with the farm owner and foreman using surveillance to monitor workers' performance. Thirdly, disconnections are highlighted between the peoples and places involved with papaya, including the farm workers who only know the 'consumer' as one vague entity that dictates market pressures, as well as the product itself which changes from plant to agricultural commodity to exotic fruit to beer by-product. Above all, it makes a clear case for alternative food systems that can better the livelihoods of workers and pickers at the bottom of the chain.

TNCs. In developed world economies, questions are also being asked about the logic and impacts of industrial systems. Two areas of concern are notable. The first relates to food production. Questions about intensive agricultural production hit public imaginations in large part because of mad cow disease, which emerged in the United Kingdom in the 1980s. This disease is caused by a protein called prion, which entered the cattle-feed chain when animal carcasses were rendered down into meat and bonemeal and incorporated into cattle feed without being adequately decontaminated. The disease is a direct consequence of the large-scale industrialization of food production and also represents a direct threat to human life in a novel form of the disease called variant Creutzfeld–Jakob Disease or vCJD. Similar questions were being asked about the industrialization of agriculture after the outbreak of foot-and-mouth disease in the United Kingdom in 2001 and again after the current spate of bird flu incidents reported in south-east Asia, Eastern Europe and, most recently, East Anglia in the United Kingdom.

The second area of concern relates to fair trade and environmental responsibility. International trade in food has expanded significantly in recent decades. Between 1961 and 1999, there was a fourfold increase in the amount of food exported. In 2000, eight of the wealthiest countries accounted for almost half of all agricultural exports (Millstone and Lang 2003). One of the drivers of this process has been the World Trade Organization (WTO). The WTO's Agreement on Agriculture promoted trade liberalization through reductions in agricultural subsidies, tariffs and import quotas. However, concerns have been voiced about the unfairness of global trade because of the continued use of subsidies and tariffs. As well as unfair trade, the environmental impact of a major increase in world food trade is also being felt around the world. For example, trade-related transportation is one of the fastest growing sources of greenhouse gas emissions. The distance food travels is expressed in terms of **food miles**. As well as distance, the mode of transport (e.g. air freight vs. sea freight) is also an equally important, but sometimes forgotten, variable. Milk provides a useful example of unnecessary food trade. Until the 1960s, most people consumed milk produced locally, but between 1961 and 1999 there was a fivefold increase in milk exports, with many countries importing and exporting large quantities, resulting in millions of extra food miles. In relation to food miles, Pretty et al. (2005) conclude that an important policy question centres on what can be done to alleviate these costs through the adoption of more sustainable methods of food production.

7.4 Alternative geographies of food: concepts and case studies

One important way to respond to some of the above questions is to establish **alternative food networks** (AFNs). There has been a huge resurgence of interest within many developed market economies in foods of local and regional provenance (Ilbery and Kneafsey 1998; Parrott et al. 2002; Allen et al. 2003; Hinrichs 2003). These include consumer initiatives like the Slow Food movement and fair trade, as well as a growth in food purchases from 'alternative' supply chains rather than supermarket outlets, including farmers' markets (FMs), box schemes, community supported agriculture and home deliveries (Cone and Myhre 2000; Renting et al. 2003; Goodman 2004; Kirwan 2004; Ilbery and Maye 2005; Renard 2005). This interest is often seen as a response to the environmental and socio-economic disbenefits associated with agricultural industrialization and global food supply. In an examination of consumers' confidence in Austria, Italy and the United Kingdom, Sassatelli and Scott (2001), for instance, interpreted the growth in novel food markets as confidence-building strategies to address deficits in disembedded trust, widening chains of interdependence, food scares and the introduction of GM technologies.

Many authors have referred to the distinctions drawn between 'conventional' and 'alternative' agri-food systems. Some of these contrasts are listed in Table 7.2, where a number of binary opposites are depicted. Thus words such as 'quality', 'embedded',

Table 7.2 Distinctions between 'conventional' and 'alternative' food supply systems

Conventional	Alternative
Modern	Post-modern
Manufactured/processed	Natural/fresh
Mass (large-scale) production	Craft/artisanal (small-scale) production
Long food supply chains	Short food supply chains
Costs externalized	Costs internalized
Rationalized	Traditional
Standardized	Difference/diversity
Intensification	Extensification
Monoculture	Biodiversity
Homogenization of foods	Regional palates
Hypermarkets	Local markets
Agrochemicals	Organic/sustainable farming
Non-renewable energy	Reusable energy
Fast food	Slow food
Quantity	Quality
Disembedded	Embedded

Source: based on Ilbery and Maye (2005)

'sustainable', 'traditional' and 'natural' characterize alternative food production systems (Ilbery and Maye 2005). In reality, these binary opposites are not as simple and clear-cut as this. For example, while organic food may be regarded as 'alternative', most organic sales still occur through 'conventional' supermarkets. Interest in the alternative food economy has also led some geographers to proclaim the emergence of 'alternative geographies of food', which revolve around changing production and consumption relations that give rise to new regional and local food 'complexes'. Parrott et al. (2002: 243) thus argue that alternative geographies of food may be associated with agriculturally peripheral regions because such regions have 'for a variety of reasons, failed to fully engage with the productivist conventions that have predominated in the agri-food system in the second half of the twentieth century'. One can equally talk about alternative geographies of food in a global sense, notably through the international Fair Trade and organic food movements which challenge exploitative relations in agri-food systems (see also Spotlight box 7.3).

Goodman (2003) draws a useful comparison between American and European alternative food practices, framing US alternatives more as social and oppositional movements, in contrast to the European Union's more endogenous reclaims on historical and cultural traditions of product and place. These distinctions will now be examined a little further, via some empirical case studies; hopefully, they will also help you to think

Spotlight Box 7.3

Towards conventionalization? Fair trade and organic foods

Conversations about alternative food networks often draw reference to fair trade and organic foods because they offer ethical and ecological possibilities that counter some of the negative externalities associated with conventional food supply. International trade often appears as a remote concern, but when commodity prices fall it has devastating impacts on the livelihoods of millions of small producers. The prices paid for coffee, for example, have not increased in real terms in the past forty years, whereas the costs of inputs like fertilizers and machinery have. Low coffee prices in the early 1990s had catastrophic impacts on the lives of millions of small farmers, mostly in LDCs, who were producing coffee at a loss. However, despite the positive benefits of fair trade and organic foods, both sectors have been subjected to recent debates about 'mainstreaming' and 'conventionalization'.

Fair trade offers an alternative to overcome the injustices of free trade, guaranteeing producers a fair price (http://www.fairtrade.org.uk/about_history.htm). It was started some fifty years ago by development agencies and charities like Oxfam and Traidcraft who realized the important role consumers could play in improving the lives of impoverished producers. Charities bought direct from the farmers at better prices and so were able to offer consumers the chance to buy products that were bought on fairly traded terms. However, to generate greater sales it became necessary to involve commercial manufacturers and to get these products into supermarkets where most people shop. Some twenty countries now have fair-trade labelling initiatives, mostly in Europe and North America, and products include coffee, drinking chocolate, chocolate bars, orange juice, honey, tea, sugar and bananas.

Fair trade is now available in most European supermarket chains, with some products achieving 15 per cent of the national market share. It is, quite rightly, a global success story and fair trade consumerism is celebrated as enabling better everyday ethical practice. The success of these networks depends upon the ability and willingness of Northern consumers to pay redistributive premiums for such commodities. However, Renard (2005) suggests that the rapid growth of alternative commodities may be levelling off, with relatively few consumers willing to pay premiums and then only for products of seemingly better quality and taste. Some fair trade organizations have responded by competing on price, lowering premiums paid to producers and aggressively pursuing 'mainstreaming' strategies. This parallels developments in organic farming where some have argued that organic supply chains have become 'conventionalized' (especially in terms of rent structures, the size of businesses controlling production, conventional patterns of marketing and distribution), suggesting that it is often very difficult for alternative economies to maintain their differences from the global capitalist economy (Guthman 2004). As Mike Goodman (2004) suggests, such networks unveil the commodity fetish and simultaneously construct fetishes of their own.

critically about the blurring between 'conventional' and 'alternative' systems of food provision. Before doing so, a summary of four concepts central to an understanding of the development of so-called 'alternative geographies of food' is provided:

1. **Short food supply chains** *(SFSCs):* The key characteristic of SFSCs is that foods reach the final consumer having been transmitted through a supply chain 'embedded' with value-laden information concerning the mode of production, provenance and distinctive quality assets of the product. In many cases, the number of nodes between the primary producer and the final consumer will also be minimized (Renting et al. 2003). While this 'reconnection process' is best demonstrated through forms of direct marketing and thus face-to-face contact between producer and consumer, Marsden et al. (2000) identified two further types of SFSC: spatially proximate and spatially extended. The former is where products are sold through local outlets in the region, locality or place of production, so that the consumer is immediately aware of the locally embedded nature of the product at the point of retail. In contrast, the latter occurs when products are sold to consumers (e.g. via the Internet) who are located outside the region of production and/or have no personal knowledge of the area.
2. **Social embeddedness**: This propagates the idea that economic behaviour is embedded in, and mediated by, a complex and extensive web of local social relations. In the case of local ('alternative') foods, both economic (e.g. price, markets) and social (e.g. local ties, trust) relations are vital for success (for details see Hinrichs 2000). In stressing the role of social relations in generating the trust necessary for economic transactions to take place, it is easy to make the false assumption that social embeddedness relates just to alternative food systems. In reality, all economic relations are socially embedded in a range of contrasting ways and so there are different degrees of embeddedness in all food supply systems (Winter 2003b). Nevertheless, social interaction between producer and consumer can make the difference between success and failure for local food businesses. This can take the form of acknowledgement, attention, respect, friendship and sociability, often subsumed within the concept of 'regard' (Sage 2003; Kirwan 2006).
3. *The 'turn' to quality:* Closely linked to the ideas of SFSCs, embeddedness and regard is the notion that local (alternative) foods are of higher quality than products produced under more conventional farming systems. In other words, quality and locality are inextricably linked. Yet, 'quality' is a social construction and thus means different things to different people (Ilbery and Kneafsey 2000a). Product labelling is an important dimension of marketing quality and, in the case of local foods, labels tend to focus on different combinations of the three 'Ps': product (e.g. local cheese), process (e.g. organic) and place (e.g. Stilton) (Ilbery et al. 2005). The turn to quality has led to a growth in quality assurance schemes (QAS). However, the recent proliferation of QAS has been dominated by supermarkets, thus further blurring the distinction between 'conventional' and 'alternative' food supply systems.
4. **Defensive localism**: It is beyond the scope of this chapter to enter into debates surrounding the definition of terms such as 'local', 'locality' and 'regional'; nevertheless, it is clear that the literature is divided about whether 'quality' or 'local' is more important in local food production (cf. Winter 2003b; Weatherell et al. 2003). In fact, Winter (2003b) used the concept of defensive localism to suggest that the turn to local is more important than a turn to quality based on, for example, organic or ecological principles. In his survey of consumers in different rural districts of England and Wales, he noted a clear preference for the purchase of local over organic food. This reflected support for local farmers and the local economy, irrespective of whether the food was produced using conventional or alternative methods. In other words, the turn to local is not just about alternative food systems; instead, it can cover different forms of agriculture (including conventional) and a range of consumer motivations.

It should be clear that the processes affecting 'alternative' food chains are complex and that there is considerable blurring between these and 'conventional' food production systems. The literature also suggests that a simple 'alternative geography of food' does not exist. Spotlight box 7.3 introduces one way of critically examining these processes, referring to the 'conventionalization' of two well-known 'alternative' food examples: fair trade and organic foods. The chapter now explores some of these issues in more detail by presenting four case studies, all of which somehow attempt to 'reconnect' food production and consumption and make a statement about the alternative food economy, especially in terms of the nature of the food chain.

7.4.1 Quality food

One of the dominant features of AFNs, particularly in Europe, has been the attempt to link 'product and place' in order to add value to agricultural products (Ilbery and Kneafsey 1998; Watts et al. 2005). This is often defined as a process of relocalization, in which locally distinctive quality food products (for an example, see Plate 7.1) are transferred to regional and national markets as a mechanism to provide a valuable economic stimulant and reduce the deleterious impact of national and EU subsidy reforms and increasing trade liberalization. In particular, regional speciality food products have been linked to particular places in this way, especially via a formal system of quality food labels and the establishment of regional speciality food groups. The most notable example of this is the Protected Designation of Origin (PDO) and the Protected Geographical Indication (PGI) quality labels introduced by the European Union in the early 1990s to 'protect' and 'promote' food and drink products with a recognizable geographical origin (Ilbery and Kneafsey 2000b; Parrott et al. 2002). Groups of producers in Europe can thus apply for either a PDO label (product originates from a specific place and is linked to its natural environment) or PGI label (linked to place, but not necessarily in terms of raw materials). These labels

Plate 7.1 Speciality potato crisps in the UK.
(Image courtesy of The Advertising Archives)

protect producers from attempted copies and act as a marketing device. These changes have been interpreted as a symbol of a 'quality turn' (Goodman 2003), with particular emphasis on specialist food production, especially in 'marginal economies'.

In terms of geography, the United Kingdom already has over 30 PDO/PGI labels. Specialist cheeses dominate the PDOs, indicating that these are made exclusively from local raw materials (e.g. Stilton cheeses). In contrast, drinks dominate the PGIs. For example, Newcastle Brown (a beer) was awarded PGI status because, although associated with the city of Newcastle, not all of its ingredients come from the region. In August 2005, the brewery (Scottish and Newcastle) closed the Tyne Brewery and moved production across the river to the Federation Brewery in Dunston, Gateshead. As the beer is no longer brewed in Newcastle, the future of this particular PGI status is uncertain. There are now more than 350 PDO and 200 PGI designations in the European Union (Parrott et al. 2002), with a southern concentration in places like France and Italy (over 100 each) and Greece and Portugal (over 80 each), compared with limited numbers in northern member countries (e.g. Finland, Denmark and Ireland). This implies that regional/speciality food has more cultural significance in some European societies than in others (Parrott et al. 2002).

7.4.2 Virtual food

The second case study emphasizes online retailing, a virtual space that also gives access to a material food product (i.e. 'virtual food'). It represents a second example of an extended SFSC. Thanks to the Internet, more and more people now sell and buy food in this way. Based on work by Holloway (2002), two such schemes are introduced, both of which give clients a say in the growing of food that is then delivered to their door. In the process, distinctions between the virtual and the real, and 'production' and 'consumption', become blurred.

The first, 'My Veggie Patch', was an online service offering urban customers in West London the opportunity to grow (or have grown for them) vegetables on a farm in rural Suffolk. The scheme opened in 2000 and closed in 2002. It nicely illustrates the complex nature between production–consumption, as demonstrated in the opening text on the website:

> What is My Veggie Patch? – it is about *you* having *your* vegetables delivered to *your* door – *you* decide what you want to grow and how they are to be grown – we do all the hard work and deliver the produce *direct to you. My Veggie Patch – controlled by you, grown for you!* (from Holloway 2002: 73; emphasis in original).

Here consumers could engage with the materiality of food production through the virtual medium of the Internet, without doing anything to grow it.

The second scheme, currently running and based in Italy, is 'Adopt a Sheep'. This scheme has an international customer base, but again demonstrates that 'closeness' and 'connectedness' can be achieved through the Internet. Clients 'adopt' a sheep, have some input into its management and receive, along with adoption documents, produce including sheep's cheese, woollen socks and salami from the yearly lamb. The website text and photographs attempt to instil a sense of connectedness with customers, evoking the smells, sounds, sights and textures of the region, and suggesting that adoption of a sheep contributes to the conservation of 'nature', local sheep breeds and 'traditional' ways of life. This idea about 'reconnection' is nicely emphasized in a quote from one of the scheme's owners:

> The idea is to give people in the cities a sense of faith in what we are doing . . . Clients will once again have direct contact with the origins of what they eat . . . this project recreates a direct contact between the producer and the buyer, and restores the client's confidence in the quality of produce – something that was lost with mass production and distribution (Carroll 2000; quoted in Holloway 2002).

Growing virtual vegetables and/or rearing an adopted sheep thus offers consumers 'closeness-at-a-distance', a technologically mediated and clean connectedness to the food and the place where it was produced.

7.4.3 Direct food

Studies of the local food sector in the United States, United Kingdom and other parts of Europe typically include some of the archetypal examples noted earlier, especially farmers' markets, box schemes, farm shops and on-farm butchers (see Watts et al. 2005 for a review). Significant in all these examples is the direct nature of the supply chain and the important social and economic benefits which accrue from these types of food transaction. This is usefully illustrated in Plate 7.2, which shows the *direct interaction* between producers and consumers at a small, rural farmers' market in Hawick in The Borders. The third case study focuses on an organic meat producer business examined as part of

Plate 7.2 Farmers' market in Hawick, The Borders.
(Authors' survey)

a wider research project on the nature of direct marketing channels (Ilbery and Maye 2005).

The organic farm and on-farm butchery is located in the Northumberland uplands. Established in 1998, the business produces and retails a range of organically reared meat products, using local branding and traceability. The business supply chain is reproduced in Figure 7.3 and, crucially, records both the upstream inputs coming into the business and the downstream links in terms of how and where final products are sold. This *'whole chain'* approach extends previous conceptualizations of SFSCs which tended to focus only on downstream elements of the business supply chain (cf. Marsden et al. 2000).

Most of the inputs coming into the farm are from local suppliers. Constructed to add and retain value from primary production, the business sources some cattle, sheep, pigs and poultry from other organic farms in the region. All livestock are slaughtered at the (organically accredited) abattoir in Whitley Bay and then delivered to the on-farm butchery to be processed and packaged for retail. Businesses like this have to 'dip in' and 'dip out' of 'conventional' supply chains because of the ways in which the dominant agri-food system is currently structured. The meat products are sold through various SFSCs, for example direct sales, local/regional specialist food shops and caterers, and mail order. Meat sales over the Internet are minimal. Over time, the business reduced sales output at farmers' markets and stopped supplying (specialist) butchers in the region; the preference is for more 'stable' alternatives (e.g., independent retail and catering). Thus the producer adjusts the business supply chain in a bid to establish as much control of the chain as possible.

7.4.4 Community food

Attempts to relocalize food production and establish 'alternative' supply chains also extend beyond farm-based production. This is an important point. The final case study is therefore an example of a community food scheme, called the Kansas City Food Circle (KCFC). It is taken from Hendrickson and Heffernan (2002), who also examined the ConAgra food chain. This is significant because the KCFC is presented as a (local) site of resistance within an industrialized food system dominated by global corporations like Cargill, Monsanto and ConAgra and whose food chain arrangements are typified by joint ventures and strategic alliances.

The KCFC organization began in November 1994, seeking to educate the public about the consequences of the industrial agricultural system and to persuade more people to participate in local, sustainable alternatives. It argued that industrial agriculture had eroded soils, made water unfit to drink, and increased pesticide resistance in insects; the food system had become so centralized that citizens lacked control over their food choices. For KCFC, the industrialized food system was unhealthy, unjust, unethical and economically unviable.

The explicit political alternative that emerged from this critique was an attempt to create a local,

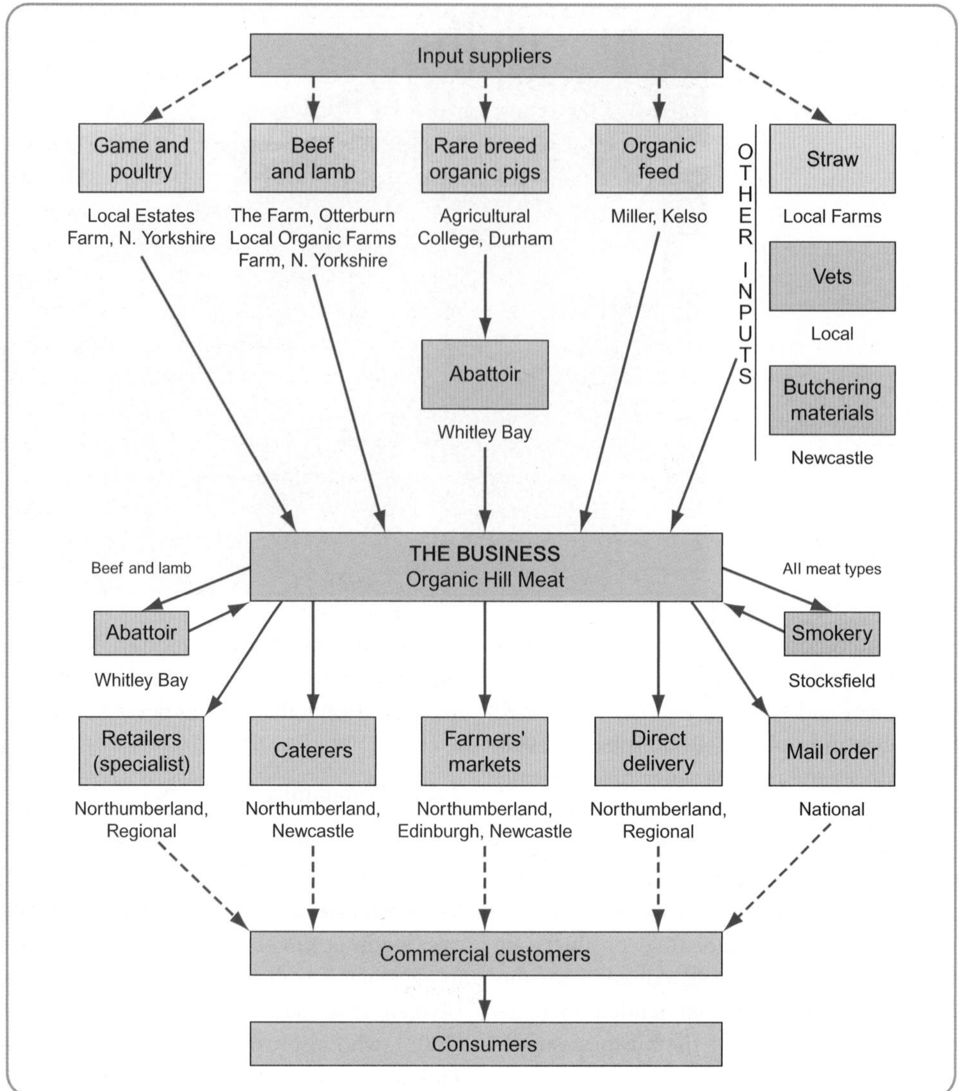

Figure 7.3 Business supply chain diagram for organic hill meat producer.
Source: Ilbery and Maye (2005)

organic food system where consumers can get seasonal, fresh food at a price that supports farmers who use sustainable practices. The KCFC thus connects all actors in the food system in a way that sustains and returns control to local communities. Effectively, the KCFC is about creating a new kind of community that recognizes the interconnectedness of people through the production and consumption of food. The Circle continued to grow, with 30 organic growers in the region surrounding Kansas City signing up to supply food, and roughly 600 consumers. The trust-based relationships that the KCFC is trying to cultivate between farmers and consumers mirror those of the Slow Food movement.

7.5 Towards critical food geographies

This chapter has examined the changing nature of food consumption and production, especially in developed market economy contexts. It has highlighted the important contributions that geographers have made to debates concerning the long-term sustainability of increasingly global agri-food systems. Some commentators argue that a third food regime exists – one which is flexible in nature and focused essentially on the supply of high-value food for 'elite' consumption in the developed world. Unfortunately, this only serves to exaggerate the huge inequalities that exist in global patterns of food supply and consumption. These kinds of inequality reflect the prevailing capitalist mode of production, where large agri-food companies and corporate retailers have developed considerable power and effectively exert control over the entire food supply system. Such businesses also concentrate their activities in places, in both the developed and developing world, where industrialized methods of production are most suited and where farming represents one 'cog' in a larger agri-commodity chain. The majority of other farms and regions are left out of these agri-food chains and become marginalized. With increased pressure to both liberalize world trade in agricultural products and establish market-oriented agricultural policies, including proposals to abolish subsidy payments in Europe and America, it seems that international competition in food production will further intensify and the power of agribusiness TNCs and corporate retailers will strengthen.

However, responses to these developments are not going unchallenged. Thus consumers are beginning to 'bite back' by demanding, and in some cases establishing, alternative methods of food supply (noted, for example in Spotlight box 7.3 – fair trade). Farmers and other producers are also embarking upon more sustainable and 'alternative' forms of agriculture, based on territorial association and direct consumer contact. As shown in the first 'alternative' case study, a focus on high-quality, authentic and traceable regional/local food products for niche markets is one possible pathway. In some cases, these systems of provision also directly challenge 'mainstream' forms of food provision, a point emphasized in the example of the Kansas City Food Circle (section 7.4.4). Such examples are thus offering their own critical geographies.

Given the range and diversity of work on 'alternative geographies of food', these mechanisms of food supply can also be critiqued, especially in terms of the privilege assigned to terms like 'local' and 'alternative'. Geographers and other agri-food researchers are thus beginning to ask what is alternative about the alternative food economy (Whatmore et al. 2003; see also various chapters in Maye et al. 2007). This is most evident in the critical tone of the 'conventionalization thesis' introduced in Spotlight box 7.3 which raises important questions about the alternative status of some fair trade and organic food supply chains. Indeed, recent work by Venn et al. (2006) suggests that the actors involved in what are described as 'alternative' projects very rarely describe themselves as 'alternative'. This disjuncture between academic and lay discourses suggests that such systems need to be analysed with a considerable degree of caution. In this vein, Watts et al. (2005) distinguish between 'weaker' and 'stronger' alternative systems of food provision. The former place emphasis on quality and the labelling features of locality *food* networks (i.e. the product is key), whereas the latter focus on the revalorized and embedded characteristics of local food *networks* (i.e. the supply chain/network and nature of relations are key). In the four examples of AFNs, the first case study falls within this weaker classification, whereas the other three are stronger because of their emphasis on establishing alternative food chains. The status of fair trade or organic products in this schema is less clear cut and is best judged on the basis of individual schemes/projects, as in section 7.4.3 – an organic farmer who sells his meat direct to the public, rather than fair trade or organic product sectors per se. As Watts et al. (2005: 34) conclude, 'alternative food networks can be classified as weaker or stronger on the basis of their engagement with, and potential for subordination by, conventional food supply chains operating in a global, neoliberal polity'.

However, one could argue that a singular 'alternative food system' does not exist. While this may appear counter to some of the points raised in this chapter, it does not deny the values associated with 'alternativeness', at least in an ideological sense. Instead, it suggests that categorizing spaces of economic activity into either 'alternative' or 'conventional' systems of supply is too simplistic. Indeed, food consumption and production are best understood in terms of the construction of complex, changing and multiple sets of relationships. Finally, as with more industrial agri-food systems, one must also ask who these 'alternative' systems of provision truly benefit. It is in this context that DuPuis and Goodman (2005) argue that local food systems must not be interpreted uncritically. As research under the banner of 'alternative' proliferates to include healthy living, community development, sustainability, organic, ethical and green consumption, these critical questions about benefits and terminology become all the more important.

Learning outcomes

After reading this chapter, you should have acquired:

- An understanding of the importance of food to modern economies and societies and what is meant by a food chain and its application in different contexts.
- An understanding of recent trends in food consumption and be able to recognize contrasting food consumption practices.
- An awareness of how and why the agri-food system has become globalized – especially the dominance and spatially uneven penetration by transnational agribusinesses.
- An awareness of the emergence of quality-based commodity food production in some parts of the developing world and the reasons why this has happened.
- An understanding of so-called 'alternative food networks', including some of the main reasons why they have arisen, how they are conceptualized and case study examples to explain their diverse character.
- An ability to begin to think critically about changing geographies of food consumption and production, including what actually constitutes an 'alternative food economy'.

Further reading

Atkins, P. and Bowler, I. (2001) ***Food in Society: Economy, Culture and Geography***, Arnold, London. A collection of fairly short chapters on many of the global food production issues raised in this chapter, as well some useful introductory chapters on food consumption.

Friedberg, S. (2004) ***French Beans and Food Scares: Culture and Commerce in an Anxious Age***, Oxford University Press, Oxford. A first-rate comparative case study of global food commodity networks, tracing the supply of green beans from production through to consumption for two culturally specific trade links – France/Burkina Faso and England/Zambia.

Hughes, A. and Reimer, S. (eds) (2004) ***Geographies of Commodity Chains***, Routledge, London. An excellent set of theoretically informed chapters on the complex socio-spatial nature of global commodity chains – including different food commodity examples. It also contains a very useful introduction to the commodity-chains literature.

Maye, D., Holloway, L. and Kneafsey, M. (eds) (2007) ***Alternative Food Geographies: Representation and Practice***, Elsevier, Oxford. This book reflects on the diversity of debates and practices surrounding efforts to establish 'alternative' systems of food provision in different parts of the world. It includes empirically rich case studies from Europe, North and South America, Australia and Africa.

Millstone, E. and Lang, T. (2003) ***The Atlas of Food***, Earthscan, London. A great introductory text that reveals the often bizarre facts about the way we make, process, ship, trade and eat foods. If you want to know the basic facts about food, this is the place to start. The graphics are particularly well done and easy to digest.

Robinson, G. (2004) ***Geographies of Agriculture***, Pearson, London. The most recent and up-to-date textbook on geographical aspects of agriculture and food supply systems. It is particularly good on the production side of the food chain equation.

Also see suggestions in Chapters 11, 15 and 18.

Useful websites

www.europa.eu.int The official site of the European Union. Within this very large site, one Directorate General (DG) has responsibility for Agriculture. This DG site (labelled under activities) contains information on the European Union's Common Agricultural Policy, key speeches and policy papers, statistics, links to member-state sites and an Information Resource Centre. You can also follow up some useful links, including the Commissioner for Agriculture and Rural Development's (Mariann Fischer Boel) recently established web-log (or see: **http://blogs.ec.europa.eu/fischer-boel**).

www.defra.gov.uk The site of the UK government's Department of Environment, Food and Rural Affairs. The site contains information on UK agricultural policy (including the most recent round of mid-term reforms), statistics and links to other government departments, non-ministerial departments and NGOs. Similar sites exist for other national governments (see, for example, **www.usda.gov,** home of the US Department of Agriculture, including useful details about the 2007 Farm Bill).

www.sustainweb.org The site for Sustain – the alliance for better food and farming, an NGO umbrella organization which campaigns for more sustainable food-chain practices. Contains information about various ongoing projects, including, for example, the 'Good Food on the Public Plate' project, a public procurement initiative to improve the quality of foods supplied to hospitals, as well as a raft of challenging position papers and responses to various food-related topics.

www.soilassociation.org This site is home to the UK organization dedicated to promoting organic food and farming. The site contains an online library with useful papers on various topics, including local/regional food schemes, food and farming policy, GMs and animals, as well as an excellent 'links' page which lists other useful websites related to food and farming. You can also visit their consumer website, **www.why.organic.org,** which contains useful information about the nutritional benefits of organic food.

www.fairtrade.org.uk Site for the Fairtrade Foundation. Contains information about the fair trade movement, products sold, suppliers, etc; it also has a useful resources section with links to downloadable position papers and case studies. For those interested in ethical consumerism more broadly, see also **www.ethicalconsumer.org,** which examines the ethical credentials of individual products/organizations.

For annotated, clickable weblinks and useful tutorials full of practical advice on how to improve your study skills, visit this book's website at **www.pearsoned.co.uk/daniels**

WORLDS APART: GLOBAL DIFFERENCE AND INEQUALITY

Chapter 8

Marcus Power

Topics covered

- The importance of the 'three worlds' schema
- Development as knowledge and power
- The view from 'the South' and the view from 'below'
- Alternative geographies of global development and inequality

What is the geography of the Third World? Certain common features come to mind: poverty, famine, environmental disaster and degradation, political instability, regional inequalities and so on. A powerful and negative image is created that has coherence, resolution and definition. But behind this tragic stereotype there is an alternative geography, one which demonstrates that the introduction of development into the countries of the Third World has been a protracted, painstaking and fiercely contested process.
(Bell 1994: 175)

Perhaps one of the most interesting features of the word 'development' is that it has produced a bewildering array of labels for people and places that are not considered 'developed'. As Morag Bell suggests in the above quote, the space of the **Third World** has lots of tragic stereotypes of famine, poverty, drought, etc., to animate it and make it seem coherent and defined. Learning to understand and appreciate social or economic differences relative to the area of the world in which we live is a difficult and value-laden process. The term 'development' is notoriously hard to define and often refers simply to 'good change', a positive word that in everyday parlance is practically synonymous with 'progress' and is typically viewed in terms of increased living standards, better health and well-being and other forms of common good that are seen to benefit society at large (Case study 8.1). Conditions of poverty clearly vary between different areas, as does the way poverty is experienced, so simplistic stereotypes of 'Third World poverty' are of little use.

In recent years there has been an upsurge in the categorization and labelling of non-Western peoples and spaces as demands for aid effectiveness and concerns with inequality grow (see for example the World Bank 2006 *World Development Report*). Thus countries are variously seen as belonging to the 'Third World' or the 'Global South', to be somehow 'lesser developed', 'underdeveloped' or even 'backward' in some way. In this way labels are about power and politics, they can impose boundaries and can be incredibly bureaucratic. Constructing categories of peoples and places is often seen as an efficient way to manage resource allocation and for tracking whether disadvantaged people are 'benefiting' from 'development' (Eyben et al. 2006). In some ways the lack of an agreed set of international development indicators and measures or of common systems of data collection tells its own story of the failure of international development since 1945. A major problem with the geographies of development produced from these statistics is that they have allowed some observers to label whole areas as 'Third World' or 'lesser developed' as if the same could be said of all its constituents (Wood 1985). As Morag Bell puts it:

> In emphasising what people are deprived of (as is implied by poverty), they [statistics] impose a negative uniformity upon non-western societies. As objects of study poor people become categories and are labelled as a homogeneous group.
>
> (Bell 1994: 184)

Thus people's life stories and multiple identities are reduced by imposing labels such as 'poor' or 'refugee', they can be 'top down' and can lead to particular interpretations of the underlying problem. Further, labelling whole regions and spaces as 'developing' or 'developed' reduces and overlooks the political, economic, social and cultural diversity of the places and communities included within these gross generalizations, simplifications and aggregations (Wood 1985). Labels can stigmatize people and places but they can also be resisted and can serve to mobilize people (e.g. through labels like 'citizen' or 'disabled'). This chapter seeks to explore the notion of 'Third World' development and outlines some of the ways in which this has been theorized and practised. In order to do this, the chapter examines the historical and geopolitical dimensions of development and focuses on the need to contextualize our studies of development. Rather than seeing poor people as 'objects of study' we need to attend to the ways in which poor people themselves understand and make sense of their experiences of poverty.

Moving beyond the labelling of 'Third World' peoples and places as a homogeneous group, it is important to grasp how places and peoples are spatially and socially *differentiated* through development and inequality, experiencing progress and 'good change' in a variety of ways. We need to encourage a diversity of ways of enframing the problems of 'development' and resist the temptation for neat categories and easy quantification. The key question here is how to listen to and understand the differences that define and distinguish how development is lived in different sorts of places and amongst different groups of people. Do we always need to rely on quantification and statistics to understand these issues or might we look to learn from other kinds of stories and narratives? Thus what constitutes 'knowledge' in development policy and practice is often defined in relation to quantitative measures, statistical data, formal academic research and Western 'science' but we can also learn a great deal from fictional writing and literary perspectives in order to reveal a different side to experiences of development (Lewis et al. 2005). This chapter is divided into four further sections. The first discusses two of the most important conceptual perspectives that have been formulated on the relations between development and inequality: the *modernization* and *dependency* schools. Although there have been many other different strands of development thinking, both these perspectives have been widely influential and remain relevant to an understanding of theory and

Regional Case Study 8.1

The United Kingdom and Uganda

A comparison of the United Kingdom and Uganda is useful here in allowing us to evaluate the relative value of global indicators of inequality (Figure 8.1). Both countries are roughly the same size but their populations (around 28 million in Uganda and 61 million in the United Kingdom) clearly do not have the same levels of access to water, health services or education. Uganda spent US$198 million on defence in 2005 (2 per cent of its GDP) (House of Commons, 2005) compared to the UK defence budget in 2005/6 of some £30.1 billion (MOD 2006). The proportion of the UK population living in poverty (defined as having a household income that is less than 60 per cent of the national average) is one in five (British Broadcasting Corporation 2005a) whereas in Uganda this is nearer to one in three. Uganda has a per capita income of US$338 (DFID 2005) whilst in the United Kingdom per capita income in 2003 was US$ 27,770. In 2006 the UK per capita GDP was US$31,400 whereas in Uganda this was US$1,800 in the same year. The UK's infant mortality rate is 5.08 deaths per 1,000 live births whereas in Uganda this is 66.15 deaths per 1,000 live births (CIA 2005). These statistics clearly suggest massive inequalities between the two countries such that UK citizens have a life expectancy of 79 years compared to 53 years for their Ugandan counterparts. In understanding how this is possible and how to respond to this unevenness there must be agreement on the significance of these inequalities and their causes or origins. There must also be an appreciation of history (Uganda was a British colony) and of the fact that not every Ugandan citizen has a life expectancy of 53 years or earns US$338. Similarly poverty is still very much a problem in the United Kingdom, despite its supposed status as a 'developed country'. Ultimately simple measures like these tell us very little about wider social and economic differences *within* societies or about how these national and international inequalities are produced or experienced.

Figure 8.1 A comparison of the UK and Uganda in the late 1990s: some key statistics.

Source: From International Broadcasting Trust (1998)

practice today. Section 8.3 outlines the need to view development historically and formulate a sense of how it has been redefined through time. How have historical forces shaped our understanding of the geography of development and in what ways are the legacies of the past important to understanding contemporary global economic difference and inequality? The next section focuses attention on people and places and explores how the subjects of development experience developmental interventions. Three major 'brokers of development' (non-governmental organizations (NGOs), states and international institutions) dominate the literature on development, but how do these agents of the development process understand the needs and aspirations of intended beneficiaries? The final concluding section of the chapter poses questions about the political, cultural and economic implications of globalization for the future of development theory and practice and examines possible alternatives to development based around local cultures and knowledges.

8.1 Geography and 'Third World' development

By whom is development being done? To whom? These are potent questions. They should particularly be asked when 'solutions' are put forward that begin 'We should . . .', without making clear who 'We' are and what interests 'We' represent.

(Allen and Thomas 2000: 4)

The principal problem with labelling vast regions as 'lesser developed', 'developing' or 'Third World' is that a set of negative assumptions are often made and a 'negative uniformity' is imposed. Thus the picture of unevenness and injustice in the contemporary world that comes to us through these labels is not always a sharp, coherent and precise one and often this unevenness is not effectively conveyed in the statistical measures that are taken as indices of what constitutes 'development' (see Case study 8.2). Crucial then to the construction of a 'Third World' is a process of setting worlds apart and a politics of labelling that it is necessary to understand and be critically aware of. All too often the 'developing world' has been defined as a 'problem' for Western governments that can only be resolved with the intervention of Western 'experts', donors, technology, expertise or ideology (Case study 8.2).

Definitions of the 'Third World' have been contested, as have the origins of the phrase (Mountjoy 1976; O'Connor 1976; Auty 1979; Pletsch 1981; Wolfe-Phillips 1987), yet the concept of 'three worlds' can hardly be said to convey a precise meaning. The geographical boundaries of the 'Third World' are thus incredibly vague and open-ended (Chaliand and Rageau 1985; Griffiths 1994; *New Internationalist* 1999) and the idea has different meanings in different cultures. The 'three worlds' schema is very much a Cold War conceptualization of space and is strongly associated with the global social and political conflict between capitalism and communism, between the USA and USSR, in the second half of the twentieth century. The term posited a first world of advanced capitalism in Europe, the USA, Australia and Japan, a second world of the socialist bloc (China's position within this has been much debated) and a third world made up of the countries that remained when the supposedly significant spaces of the world had been accounted for. This extremely rudimentary schema of classification has provided many observers with a convenient shorthand expression and is undoubtedly linked to the formation of 'mental maps' and 'imagined geographies' of inequality (Bell 1994). These 'mental maps' and 'imagined geographies' are often created where people have no direct experience of encountering the regions or peoples they are describing or referring to. In an interesting study of student travellers to the 'Third World', Desforges (1998) illustrates the links between travel and identity for many students who sought to 'collect places' that offered authentic individual knowledges and personal experiences that could be gained through travel. This involved a framing of 'Third World' others and places as different and the assumption that it was possible to 'collect' experiences of Third World places. In this way, travel can be understood as one important means by which 'youth identities "stretch out" beyond the local to draw in places from around the globe' (Desforges 1998: 176). Thinking critically about the stretched-out connections and identities that link 'North' and 'South', 'first world' and 'third world' is crucial therefore. For example, have you ever stopped to wonder where the food you consume actually comes from? Much of our food is sourced locally but an increasing amount of food is drawn from global sources (e.g. rice from Thailand, bananas from Costa Rica, cashew nuts from Mozambique), which links consumers in rich Western countries directly to the producers of these foodstuffs in the 'South'.

Thematic Case Study 8.2

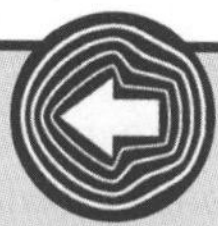

The 'strategic interests' of foreign aid

In many ways, examining the case for aid can tell us a great deal about the history of development theory and practice and provides a useful opening on to wider discussions of North–South relations. In 2002 total aid amounted to about US$56 billion from 21 major OECD countries and some others (Maxwell 2002). Two-thirds of this is government-to-government or *bilateral aid* and the remainder is *multilateral aid* disbursed by agencies like the World Bank group, the UN bodies and the European Union. Critics of foreign aid have argued that it has been less effective than private investments and commercial loans in stimulating long-term economic growth. Other critics point to the dubious Cold War record of foreign aid and its subsidizing of autocratic regimes and inflammation of regional conflicts. Thus, many critics of foreign aid have sought to highlight the 'strategic interests' at work in its distribution as well as the inequality and unpredictability of aid provision. During the current 'War on Terror', for example, there is evidence that the USA has used foreign aid donation to secure support from a number of states in the South such as Pakistan and Angola. Strangely, except for a brief period during the mid-1970s, anti-poverty measures have not been an important focal point of foreign aid, whilst aid has led to many reversals as well as to advances. Seen as a (simultaneous) remedy to problems of growth, governance, poverty and inequality, it has become (not unlike the idea of development) overburdened with expectations (Sogge 2002) and an overambitious enterprise (Rist 1997). Successive UN 'Development Decades' have recognized that economically advanced states like the USA have a responsibility to contribute to the financing of development and have specified that each 'developed' country should provide *at least* 0.7 per cent of its annual GNP as official development assistance. While still the world's biggest aid donor, the USA has slashed overseas development assistance since 1990 by some 34 per cent and foreign aid accounts today for just 0.1 per cent of gross domestic product – the lowest level of any industrialized nation (Engardio 2001). There are also plans to further limit foreign aid in order to minimize the US budget deficit (Zouza 2002). According to Sogge (2002) aid involves (1) a financial services industry promoting loans on easy terms, (2) a technical services industry improving infrastructure and know-how, (3) a 'feel-good' and image industry that can relieve guilt, (4) a political tool-shed stocked with carrots and sticks to train and discipline recipients, (5) a knowledge and ideology industry and (6) an industry that sets policy agendas and shapes norms and aspirations. The expensive 2002 Earth Summit (RIO+10) held in Johannesburg (South Africa) in August and September 2002 has been criticized by many for being no more effective at pioneering new and alternative perspectives on development aid than its predecessors (except through the social/protest movements that mobilized around it).

Given that definitions of development have often been contested, these 'imagined geographies' of the 'Third World' often overstate a set of (tragic) stereotypical and supposedly 'common' features such as overpopulation, environmental degradation or political chaos. In discussing unevenness and inequality, therefore, a representation or an imagination of the people and places subject to the development process is constructed which is only partial and provisional. An 'alternative geography of the Third World' is proposed here, based on a recognition that development has been a 'protracted, painstaking and fiercely contested process' (Bell 1994: 175). This chapter tries to offer such a critical 'alternative geography' that aims to challenge some of the 'representations of peoples and places upon which grand [development] theories are based' (Bell 1994: 193–4).

It is extremely difficult to generalize about the people and places of the 'Third World' since relations between different regions of the world are imagined and simplified by geographical expressions like 'North' and 'South', 'rich' and 'poor', 'developed' and 'underdeveloped', 'First World' and 'Third World'. These terms have to be approached with some caution. Becoming particularly popular after the publication of the Brandt Report in 1980 (Brandt 1980), the idea of 'North' and 'South' is widely seen as preferable to the 'three worlds' scheme. However, no definitions were mentioned in the report and some of the boundaries it delineated between 'North' and 'South' were considered erroneous. Subscribers to this, and the three worlds scheme, have been criticized for the simplicity of these divisions and their failure to recognize diversity and difference within these spaces; the world does not consist of a series of individual national economies in the way often suggested in United Nations and World Bank reports. Many commentators have argued instead that there is a single global economic system with ever-closer degrees of interdependency (Allen 1995). Our task is to think critically about the uneven and unequal geography of the world economy and the ways in which the global capitalist economy *actively produces* inequality and uneven development. For some observers, a big part of the economic development and wealth of the rich countries is wealth that has been directly imported from the poor countries. The key question here then is: to what extent does the world economic system *generate* inequality and run on inequality? Where does this leave the primary liberal idea of 'development for all'? Is it potentially fraudulent and misleading to hold up an image of wealthy Western living conditions as something that is available to all and a realistic or attainable objective? If the global economy is a pyramid and everyone is encouraged to stand on top, how is this to be arranged?

Development equality – 'catching up' with the wealthy through economic activity – increasingly seems less likely at the start of the twenty-first century and as evidence grows of widening and deepening global inequality. 'Development' has served then in part as a kind of 'lighthouse' (Sachs 1992) or as a 'lodestar' (Wallerstein 1991) into which several different movements, governments and institutions have invested faith and meaning. Like the term 'Third World', development partly represents a geographical imagining, a representation of a better world, and a belief in the idea of correctable inequalities/injustices between nations, states and regions and within the existing global economic orders. Escobar (1995) argues that the idea of the 'Third World' also gives enormous power to Western development institutions to shape popular perceptions of Africa, Asia or Latin America. The Third World thus is defined by and becomes intelligible through the languages and representations of the agencies and institutions of global development.

In a famous speech in 1949 US President Harry Truman spoke of the emergence of an 'underdeveloped' world that presented a 'handicap and threat both to them and the more prosperous areas' (Truman 1949). Born of the ashes of colonialism and of the fires of anti-communism in the United States, 'development' occupied centre-stage as an ideal in the global arena after the Second World War and in particular during the Cold War. Truman went on to explain the need for 'modern, scientific and technical knowledge' as a pathway to overcoming this handicap of underdevelopment and announced the beginning of a 'bold new program' within the 'developed world' to resolve inequality and remedy impoverishment in 'backward' areas (Sachs 1992). Since then 'development' has become

> an amoeba-like concept, shapeless but ineradicable [which] spreads everywhere because it connotes the best of intentions [creating] a common ground in which right and left, elites and grass-roots fight their battles.
>
> (Sachs 1992: 4)

Ideological interpretations of development have varied but they have usually always claimed 'the best of intentions'. The idea of development is often discussed in relation to 'developing countries' (itself a value-laden expression) but given its 'amoeba like' nature, development has taken numerous forms in numerous locations and is still very much relevant in societies which proclaim themselves to be 'developed' (see Case study 8.3). A sense of development being simultaneously important in both 'developing' and 'developed' societies was recently formulated in a Human Development Index (HDI) (UNDP 1998). Again assuming that 'development' can be depicted in a single measure, the UNDP Index combines data (by country) on life expectancy, literacy, income, environmental quality and political freedom. Development is defined here as a series of numerical stages with a linear progression indicated (statistically) as degrees of departure from Western norms and ideals. Human development is represented as a series of league tables that measure progress with the unquestioned underlying assumption that every country can reach the highest levels of performance set in 'the West' or have access to the 'purchasing power' of the 'upper income' world. As in the case of GNP per capita, the HDI does in many ways (albeit at a very crude level) point to growing gaps between different areas of the world. Its significance is undermined by the failure of its 'generalized' indicators. However effective statistical measurement of development can become, it is necessary to be critical of the way in which statistics are collated and definitions formulated and to be careful in the assumptions and prescriptions made as a result. It is through the amassing of statistics and surveillance by global development agencies that the image of an underdeveloped, primordial, traditional and war-ravaged 'third world' is (re)produced (Ahluwalia 2001).

Thematic Case Study 8.3

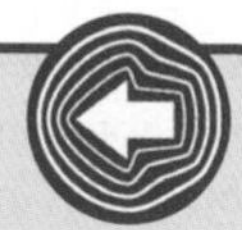

Hurricane Katrina and the USA as 'Third World'

One of the costliest and deadliest hurricanes in the history of the USA, Hurricane Katrina, devastated much of the north central coast of the United States in 2005, killing 1,836 people and causing US$81.2 billion worth of damage. Over 60 countries pledged monetary donations or other assistance to the USA including Kuwait (US$500 million), Qatar (US$100 million), South Korea (US$30 million), China and Pakistan (both US$5 million) and Bangladesh (US$1 million). Cuba and Venezuela were the first countries to offer aid, in the form of US$1 million, 1100 doctors and 26.4 metric tonnes of medicine but both offers were refused by the US government (British Broadcasting Corporation 2005b). Media reporters embedded in the mayhem caused by the Hurricane reported that a 'third world country' had suddenly appeared on the Gulf Coast or noted that the scenes in cities like New Orleans resembled 'third world refugee camps'. The *U.S. News & World Report* referred to 'the Third World images of death and devastation reeled across the nation's TV screens', whilst on *Fox News*, anchor Shepard Smith lamented that things would not be the same anymore, noting that Americans would 'forever be scarred by Third World horrors unthinkable in this nation until now'. As Brooks (2005) has argued however, the 'Third World' did not somehow 'sneak into' the USA with this hurricane, it had been there for a long time already, given that (at a conservative estimate) around 13 per cent of Americans live in poverty (around 18 per cent of children). According to Brooks, 'when you break down key economic development indicators by income group and race you find that conditions for poor Americans rival those in developing countries'. Just as it has often taken a mass famine or tsunami to generate media attention for the 'third world' in Western countries like the USA, so it has taken the destruction of a US city for the media to notice the 'third world' within the United States. It was almost as if by displacing disasters and human suffering to the 'Third World', this natural disaster was seen as not really happening in the United States (Wa Ngugi 2005). The disaster illustrates some of the ways in which the media shape and construct an imaginary 'third world' but also how the concept of 'three worlds' can produce a kind of distancing where problems of poverty, hunger and forced migration are seen to be worlds away, almost to be happening somewhere else. Hurricane Katrina also cruelly exposed the myth of the United States as a fully 'developed country' and highlighted the inequality and poverty that exists within US cities.

8.2 Conceptualizing development

The idea of progress has become an article of faith in definitions of 'development', which have themselves taken on a 'quasi-mystical connotation' in the sense that seeking to become 'developed' has been constructed as an objective that is unquestionably positive and beyond reproach. 'Development thinking' (Hettne 1995), or the sum total of ideas about development theory, ideology and strategy, has however often been caught in a 'Western' perception of reality. Conceptualizing development is thus partly about the negotiation of what constitutes progress and improvement and the definition of what constitutes 'appropriate' intervention in the affairs of 'poor' or 'lesser developed' countries. Since all-encompassing definitions have been contested and controversial, little consensus exists today but some core conceptions have emerged, many of which have continued relevance in the contemporary world.

Although there are many different strands of development thinking to explore, the modernization and dependency approaches have been two of the most influential in the twentieth century. In discussing these different conceptions it is important to think about where and when they have emerged (Slater 1993). Most reflect some of the priorities of development thinking characteristic of their era. The formation of development theories therefore depends on different perceptions of 'development challenges' at different times (see Case study 8.4).

8.2.1 The modernization school: an anti-communist manifesto

With the end of the Second World War and after the United Nations was established, conceptualizations of development received a decisive stimulus. Between 1945 and 1981, UN membership rose from 51 to 156 nation-states (Berger 2001). With many new states

Regional Case Study 8.4

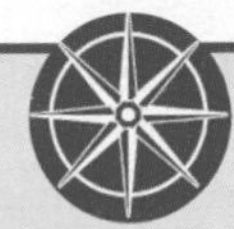

Bandung, non-alignment and the 'Third World'

The Bandung conference was a meeting of representatives from 29 African and Asian nations, held in Bandung (Indonesia) in 1955, which aimed to promote economic and political cooperation within the 'Third World' and to oppose colonialism. The conference was sponsored by Burma, India, Indonesia, Ceylon (Sri Lanka) and Pakistan and tried to cut through the layers of social, political and economic difference that separated nations of the 'Third World' in order to think about the possibility of *common agendas and actions*. The aims of the 29 nations that attended included a desire to promote goodwill and cooperation among Third World nations, to explore and advance their mutual as well as common interests. Bandung was in many ways the 'launching pad for Third World demands' where countries distanced themselves from the 'big powers seeking to lay down the law' (Rist 1997: 86). It was not hard for countries with shared histories of colonial exploitation to find something in common, since the 'agenda and subject matter had been written for centuries in the blood and bones of participants' (Wright 1995: 14). In his opening speech to the conference on 18 April 1955 President Sukarno of Indonesia urged participants to remember that they were all united by a common 'detestation' of colonialism and racism (Sukarno 1955: 1) and pointed out that colonialism was not dead or in the past but also had its 'modern' (neo-colonial) forms. The conference was especially successful in hastening the arrival of new international institutions explicitly dealing with 'development' (Rist 1997). Samir Amin coined the term 'Bandung Era' in a discussion of the national approaches to development that emerged from this period (Amin 1994; Berger 1994, 2001). The Non-Aligned Movement (NAM) was then founded (among 29 states most of which were former colonies) in 1961 on a principle of non-alignment with either the USA or the USSR in the Cold War. A year after the NAM was set up, the 'Group of 77' was then established (mostly made up of the attendees of previous conferences) to press for changes from UN agencies in the ways they viewed 'developing countries'. The NAM has continued to function as a space through which nations of the so-called 'Third World' meet and discuss common issues and objectives and now has some 115 member nations. The organization has been restructured since the end of the Cold War and continues its commitment to founding principles whilst also now incorporating a focus on reforming the international economy. South Africa took over as Chair in 1992 when the organization met in Jakarta to decide on future policy directions, whilst NAM today cooperates with NGOs, the United Nations, the European Union and the G8. At its 40th anniversary in New York in November 2001, NAM restated its commitment to work with the major global development agencies to help bring about a reform of the HIPC initiative concerning debt.

being formed after the end of colonialism and in the context of the Cold War between the USA and the USSR, theorizing development became a more complicated and contested enterprise. In some ways the task of development was seen by some as seeking to provide 'an ethos and system of values which can compete successfully with the attraction exercised by Communism' (Watnick 1952–3: 37). In this regard many observers called for the modernization of 'underdeveloped areas' and painted a picture (following President Truman) of 'underdeveloped peoples' confined to 'backwardness' but torn between the appeal of communism and the prospect of Western modernization. This was an essential characteristic of the *modernization school*, which was often dualistic, opposing 'traditional' to 'modern' life-styles, 'indigenous' to 'westernized', as if no country or citizen could belong to both categories.

Hirschmann (1958) was a key proponent of modernization ideas, voicing the optimistic view that the forces of concentration ('polarization') will 'trickle down' from the core to the periphery at national, regional and global scales. Hirschmann explained how development economics might 'slay the dragon of backwardness' (quoted in Rist 1997: 219). Friedmann's (1966) core–periphery model adopts many of the same assumptions about the polarization of development in 'transitional societies' and the 'trickle-down' effect of development. This model suggests that a number of stages exist in the national development of countries or cities leading to a final stage that represents the culmination of the development process. Geographers at the time sought to contribute to the mapping of modernization geography, seeking to look at how progress trickled down along urban hierarchies, through transport systems or with the introduction of modern technology. The message from these modernization geographies was that underdeveloped countries could move briskly into the modern tempo of life within a few years, whilst the state (concerned with macro issues and the national economy) would be the key monitor and broker of development.

One of the modernization theorists who identified stages of growth in the development process was the anti-communist writer W.W. Rostow (1960). Again the focus was on a top-down 'trickling' of capitalist development from urban-industrial areas to other regions (Stöhr and Taylor 1981). Rostow (1960) predicted that nations would 'take off' into development, having gone through five stages, which he likened to the stages an aeroplane goes through before take-off, from taxiing on the runway to mid-flight. Ranging from stage one, 'traditional society', to stage five, the 'age of high mass consumption', the theory takes our faith in the capitalist system for granted since Rostow assumes that all countries will be in a position to 'take off' into development. Further, in common with other modernization approaches, Rostow's model devalues and misinterprets 'traditional societies' which represent the 'lowest' form or stage of development. The advanced state of modernization is always represented as 'Western modernization'; traditional societies seem like distant, poor relations.

Geographies of inequality and development cannot be neatly summarized as a set of prescriptive stages. Urban areas, for example, are themselves subject to uneven development and inequality. Modernization approaches have also largely failed to address the importance of gender, assuming that men and women occupied equal positions in terms of power relations and decision-making. The 'trickle-down' effect has often failed to materialize among those who have been the subjects of modernization projects. This approach seemed to suggest that development could be mimicked, copied and replicated and that underdeveloped countries should try to reproduce the development paths of richer developed nations like Britain or the USA. As Gunder Frank has argued, this approach can also be understood as reflections of the 'Sinatra Doctrine':

> Do it my way, what is good for General Motors is good for the country, and what is good for the United States is good for the world, and especially for those who wish to 'develop like we did'.
>
> (Frank 1997: 13)

In the modernization schema, it is implied that there is nothing before the beginning of development in a 'developing country' that is worth retaining or recalling, only a series of deficiencies, absences, weaknesses and incapacities (Abrahamsen 2000; Andreasson 2005). The approach was also in a sense very much based around a 'top–down' rather than a 'bottom–up' approach, implying that the process could be brokered by states or development institutions rather than emerge from the 'grass-roots' struggles of 'Third World peoples' as had been called for in some more radical approaches. In terms of common criticisms, the division of the world into modern and traditional has often been seen as problematic (Pletsch 1981). Modern societies were much more fractured and were divided by ethnicity, class and politics and were not as united and responsive to the blueprints of planners as was often assumed. The scale of modernization programmes was also often a problem in that they assumed that 'big is beautiful' (involving large dam-building and irrigation projects, for example). Furthermore, the school and its practitioners often depoliticized development, making few if any references to history and culture. Rostow was able to say little about the 'final stage' of his organic model of development since his underlying principle was that growth had no limits (Rist 1997) and instead he simply discussed a number of very generalized scenarios. Like so many theorizations of development that followed, it ended with a creed, a set of principles about what was to be done, and heavily invested faith in the goals of mass consumption and westernization.

8.2.2 The dependency school: beyond 'core' and 'periphery'?

One of the major weaknesses of the modernization approach was that the notion of a 'trickle-down' diffusion of development implied that there were precise demarcations available of where the 'core' ends and the 'periphery' begins when this has never really been the case. Radical **dependency approaches** that emerged in the 1960s and 1970s challenged this notion of positive core–periphery relations, identifying instead exploitation between 'satellites' and 'metropoles'. Whilst the modernization approach was taken up by many international institutions and bilateral donors, dependency approaches comprised all those opposed to US post-war imperialism and allied in some way to the movement of 'third worldism' (see Case study 8.4). The dependency school is most commonly associated with Latin America but also emerged in Africa, the Caribbean and the Middle East. Drawing upon Marx's writings about the unevenness of capitalist development, dependency scholars such as Celso Furtado (1964) and Milton Santos (1974) drew attention to the mode of incorporation of each country into the world capitalist system, identifying this as a key cause of exploitation. The dependency school is most commonly associated with the work of André Gunder Frank who published a number of seminal pieces on 'The development of underdevelopment' in the late 1960s (Frank 1966). This thesis is relatively uncomplicated in that it views development and underdevelopment as opposite sides of the same coin; the development of one area often necessitates the underdevelopment of another. Dependency on a metropolitan 'core' (e.g. Europe, North America) increases the 'underdevelopment' of satellites in the 'periphery' (e.g. Latin America, Africa). Unlike many modernization approaches, dependency theorists (*dependistas*) sought to view development in historical contexts, and argued, for example, that colonialism helped to put in place a set of dependent relations between core and periphery.

These peripheral satellites, it was argued, were encouraged to produce what they did not consume (e.g. primary products) and consume what they did not produce (e.g. manufactured or industrial goods). Thus, rather than likening the process of economic growth to an aeroplane setting off into a blue sky of urban-based, Western life-styles and consumption, the dependency school was arguing that many 'underdeveloped' areas had been stalled on the runway by unequal relations and a history of colonialism, denying them a chance of ever being airborne, 'modern' or 'industrialized'.

In many ways, dependency approaches were so directly opposed to modernization approaches (almost point by point) to such an extent that eventually both 'seemed to checkmate each other' (Schuurman 2001: 6). Dependency even seemed to preserve the dualistic and binary classification of the world into developed/underdeveloped, First World and Third World, core and periphery, and also lacked a clear statement of what 'development' actually is. Key criticisms directed at the *dependistas* were that the theory represents a form of 'economic determinism' and also overlooks social and cultural variation within developed and underdeveloped regions. The dependency framework seemed to leave the simplistic impression of an 'evil genie who organizes the system, loading the dice and making sure the same people win all the time' (Rist 1997: 122). Another point of contention was that the dependency theorists seemed to be calling for a delinking from the world capitalist economy at a time when it is undergoing further globalization and economic integration. Elements of the dependency writings were, however, quite thought-provoking and remain relevant, particularly their contention that the obstacles to development equality were *structural*, arising not from a lack of will or poor weather conditions but from entrenched patterns of global inequality and 'dependent' relationships. A brief comparison of the two schools of development thinking is useful here, particularly given that dependency was primarily a critique of the classical traditions of modernization. According to Bell (1994) both perspectives deal in dualistic either/or scenarios:

> these contrasting schools of thought share a common starting point – poverty is viewed in deprivationist terms while their different views of development have a positive connotation.
>
> (Bell 1994: 184)

Concepts of development like the modernization and dependency perspectives are far from being static, uniform or unified, however. Neither represents a singular commonly agreed approach and both are complex and varied. In some ways it is the ability of each school of thought to capture and accommodate difference and diversity between states and within regions that is most important here. As we shall see in the following section, in order to do this it is important to further consider theories and practices of development in their specific historical and geographical contexts.

8.3 Development practice (1): the historical geography of development

Whilst both modernization and dependency approaches alluded to the importance of 'tradition', many early writings about development lacked a sense of historical perspective (Rist 1997). As Crush (1995) points out, development is primarily 'forward looking', imagining a better world, and does not always examine issues of historical and geographical context. It is particularly important to examine the significance of Empire in the making of international development. Between 1800 and 1878, European rule, including former colonies in North and South America, increased from 35 per cent to 67 per cent of the earth's land surface, adding another 18 per cent between 1875 and

1914, the period of 'formal colonialism' (Hoogevelt 1997: 18). In the last three decades of the nineteenth century, European states thus added 10 million square miles of territory and 150 million people to their areas of control or 'one fifth of the earth's land surface and one tenth of its people' (Peet with Hartwick 1999: 105). Colonialism has been variously interpreted as an economic process of unequal exchange, as a political process aimed at administration and subordination of indigenous peoples, and as a cultural process of imposing European superiority. According to the dependency theorists it was in this period that the periphery was inserted and brought into an expanding network of economic exchanges with the core of the world system. A new sense of responsibility for distant human suffering also first emerged during this time as the societies of Europe and North America became entwined within global networks of exchange and exploitation in the late eighteenth and early nineteenth centuries (Haskell 1985a, 1985b). Thus the origins of a humanitarian concern to come to the aid of 'distant others' lay partly in response to the practices of slavery in the trans-atlantic world and to the expansion of colonial settlement in the 'age of empire':

> Not only did colonisation carry a metropolitan sense of responsibility into new Asian, North American, African and Australasian terrains, it also prompted humanitarians to formulate new antidotes, new 'cures' for the ills of the world.
>
> (Lester 2002: 278)

These new antidotes, cures and remedies were to have enduring significance for the shaping of twentieth-century global development theory and practice, which also often carried an implicit 'metropolitan sense of responsibility'. Colonial development was also associated with an unconditional belief in the concept of progress and the 'makeability' of society, being heavily conditioned by the dominance of the evolutionary thinking that was popular in Europe at the time. Imperialism was viewed as a cultural and economic necessity where colonies were regarded as the national 'property' of the metropolitan countries and thus needed to be 'developed' using the latest methods and ideas. With this came a missionary zeal to 'civilize' and modernize the colonized and their ways of life. An important contention here then is that colonialism 'conditioned' the meanings and practices of development in a number of important ways.

After 1945 and under US President Truman, 'underdevelopment' became the incomplete and 'embryonic' form of development and the gap was seen as bridgeable only through an acceleration of growth (Rist 1997). As we shall see, there was an important sense in which, after 1945, 'development' was seen as a process that was entrusted to particular kinds of agents who performed the acts and power of development upon distant others. Globally, development would have its 'trustees', guiding 'civilized' nations that had the 'capacity' and the knowledge or expertise to organize land, labour and capital in the South on behalf of others. This is an important idea then that quite a paternal and parental style of relationship was established through the imperial encounter between colonizer and colonized in ways which continue to have a bearing on the definition of North–South partnerships in the 'post-colonial' world. Additionally, what is also relevant here is that many 'post-colonial' states continue to maintain important political, cultural and economic ties with their former colonial rulers (see Figure 8.2).

Decolonization is thus simultaneously an ideological, material and spatial process, just as complicated as colonization (Pieterse and Parekh 1995). Colonialism put in place important political and economic relations but the cultural legacies of colonialism bequeathed deep social and cultural divisions in many societies. In the process of decolonization, 'development' became an overarching objective for many nationalist movements and the independent states they tried to form. Although experiments with development were tried in many colonies, the idea of development was invested with the hopes and dreams of many newly emerging states who wanted to address these inequalities and divisions in their societies (Rahnema 1997). An important issue here concerns the extent to which colonial state machineries were reworked and transformed after independence (Power 2003). The colonial state had rested on force for its legitimacy, a legitimacy that was thus highly superficial. Colonial states also had a role in creating political and economic communities, defining the rules of the game and the boundaries of community whilst creating power structures to dominate them. The colonial state was also the dominant economic actor, creating a currency, levying taxes, introducing crops, developing markets, controlling labour and production. Above all, colonial state administrations sought the integration of the colonial economy into the wider economies of empire, to make linkages with the metropole and to establish flows of peoples and resources. After the formal end of colonialism, new states have had to formulate alternative methods of garnering legitimacy for their authority (i.e. other than the use of force preferred by the colonists).

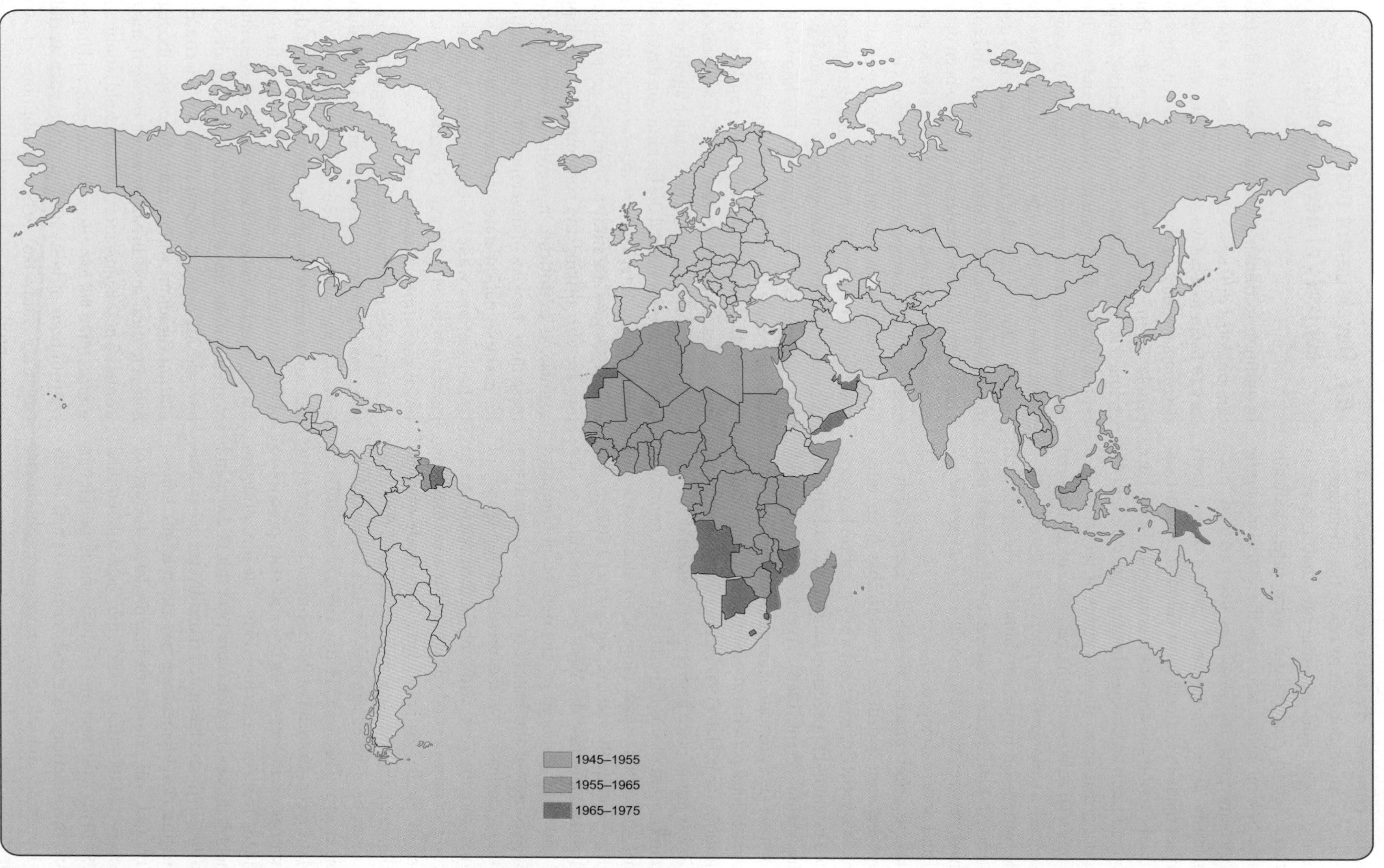

Figure 8.2 Decolonization and the proliferation of independent states, 1945–1975.

As the European capitalist system expanded and became ever more global in its reach, the structures of economic, social and political life that existed in colonies (before colonialism) were often radically remade. Several recent shifts in the global economy and the emergence of what has been termed 'neo-colonialism' have undermined attempts to meet promises made in the early post-colonial era. Particularly in the past three decades, a number of issues have served to sustain the linkages between the 'core' and the 'periphery', usually to the detriment of the latter. In a way, imperial rule created the very idea of an imperial centre and a colonized periphery and established a whole variety of important binaries, divisions and constructed boundaries between civilized and non-civilized, between the West and the rest, between developed and underdeveloped.

The historical process by which 'gaps' began to emerge between 'North' and 'South' has been interpreted in a variety of ways, but a key question has been: to what extent did European expansion and colonialism 'underdevelop' (Frank 1966) large areas of the world? The impact of European expansion was not uniform; the geographical patterns of expansion varied, as did the motivations for it. Hall (1992) argues that an important divide was put in place between 'the West' and 'the rest' as a direct result of this imperial expansion, reminding us that 'the West' is much more of an idea than a geographical reality. By accelerating contact between cultures and economies 'the West' was presented as 'the best' and most advanced or 'civilized' of all humanity. Many accounts of the history of European expansion are thus dominated by the presumed supremacy of 'Europe' and 'the West' with only limited references to the complex histories and cultures of the areas that were colonized. Similar criticisms have been made of some development theories (e.g. modernization) that have also often been based around 'Western' philosophies, experiences and histories (Hettne 1995; Power 2003).

Many recent histories of development have dated its beginnings as an area of theory and state practice to President Truman's speech of 1949. The idea of development is, however, much older than this (Cowen and Shenton 1996) and has much more diverse geographical origins. Development was not a simple 'gift' following contact with Europeans but predates the 'age of discovery' (1400–1550) and the 'age of empire' (1875–1914) (see Section 1). Indigenous peoples in Africa, the Americas, Asia and Australasia had highly developed and sophisticated cultures and technologies prior to colonization (Dickenson et al. 1996).

8.4 Development practice (2): subjects of development

In addition to thinking historically about the production of global difference and inequality, it is also necessary to listen to what we might call the *other side of the development story*, or the views of women, the views of people in the South, the views of people that are excluded socially, and the view from 'below'. People are central to the development process and an integral element in all development strategies. However, their needs have often been ignored and there has been a general failure to consider the implications of development policies for individuals, households and communities. Development strategies have also been driven by economic objectives, relegating provision for people's basic needs, their empowerment or efforts to raise their living standards to second place (Potter et al. 1999). Development strategies often assume that 'people' and 'communities' are homogeneous and passive rather than differentiated and dynamic. The knowledge and skills of people in particular communities have often been overlooked in favour of broader and less sensitive approaches to 'modernization' that bypass local cultural traditions and histories. Three major agents of development have dominated the literature on development: nation-states, international development institutions and NGOs (as Case study 8.5 explains, geopolitical factors are also driving the contemporary development agenda). It is difficult, however, to view any one of these three 'brokers of development' in isolation since, for example, nation-states are the primary focus of the United Nations while NGOs often try to work at the interface between national governments and international governmental institutions.

Many international development organizations have often made erroneous assumptions about the distribution of power and decision-making within the nations and communities where they operate. Consider the level of the household, for example. It is taken for granted that an equal division of labour exists between men and women within households and communities. As we have already noted, national statistics often conceal significant variations within and between specific regions, between rural and urban areas or between men and women. These aggregations of household labour and resources often fail to capture the changing and important contributions of household members such as children (see Case study 8.6). The contributions of women to the productive and reproductive activities of the

Plate 8.1 The environmental pollution left by Shell's oil refineries in Nigeria has had particular implications for the people of Ogoniland who have seen very little of the profits generated by oil reinvested in their state. This farmer has lost all of his productive farmland to an oil spill. Some observers have argued that Shell's relationship with the Ogoni people and the Nigerian state is one of 'neocolonialism'.
(Martin Adler/Panos Pictures)

household are still widely overlooked in national 'official' development statistics (Momsen 1991; Østergaard 1992). Nonetheless, the household has been and is likely to remain the basic living unit in many parts of the global South, the key to controlling and organizing production, consumption and decision-making. Household sizes and structures differ considerably from those of the global North, often including an extended family of grandparents, aunts, uncles and other more distant relatives. Unequal access to resources means that all

Thematic Case Study 8.5

States and the geopolitics of development

At the time of the terrorist attacks in Washington, DC and New York City in 2001, world leaders made numerous comments about the links between international poverty and terrorism, suggesting that this was to become *the* key issue of the twenty-first century. The ensuing *Global War on Terrorism* led by states defining themselves as the 'free' or 'civilized' world began in Afghanistan and has shaped contemporary development thinking and practice in many important ways. These recent events highlight the need to understand more fully the geopolitics of international development and in particular the way in which many states in the South are seen to lack necessary degrees of 'order', hierarchy and societal organization. In thinking about geopolitics it is important to explore the rights and abilities of each state to control the territory encompassed by its boundaries or to gain international recognition for its territorial authority. This also means thinking about the way in which foreign aid is used by Western donors to solicit support for the war on terrorism from states in poorer countries. The state in Angola, for example, was offered a wide range of incentives in 2003 in order to gain Angolan support for the USA when the African country had a temporary seat on the UN Security Council. Foreign aid has sometimes been seen as a kind of political narcotic, fostering addictive behaviour among states that receive it and thus come to depend on it. States are thought to exhibit the symptoms of dependence – a short-run 'fix' or benefit from aid, but external support sometimes does lasting damage to the country. During the Cold War, the USSR in particular played on its ambiguous position as both 'inside' the international system and 'outside' it, presenting itself as the natural 'midwife' for completing the independence of newborn states (Laïdi 1988). With the accelerating pace of decolonization and the creation of independent states in the South, geopolitical questions were addressed from a set of new or 'Third World' perspectives, new forms of North–South geopolitics which had emerged from the legacies of colonialism and the perception that 'underdeveloped' countries had distinct geopolitical considerations from those of Western societies (see also Chapter 21).

Thematic Case Study 8.6

Children and development

One of the clearest examples of global inequalities comes from a comparison of the relative position of children in societies of the global 'North' and 'South'. In wealthier societies children are often well fed, have clean clothes and access to an education until the age of 16 or older and have good protection against ill health. Their counterparts in some of the world's poorest countries have few of these assurances. Child poverty is also common in 'developed' countries as well, however: UNICEF estimates that one in six children in rich nations is poor (UNICEF 2000). A report from ActionAid (1995) clearly shows that children and children's work have been neglected research areas despite the fact that, as Robson (cited in Earthscan 1998) has argued, '[b]y the age of 10–12 years, some children may contribute as much to household sustenance as adults'. Again, gender is crucial in determining the roles and responsibilities of children, with girls often expected to provide assistance with household duties and boys involved in more public forms of labour. Access to education for boys and girls is also often unequal, with the continuation of girls' education being seen as much less of a priority for many poorer households (Momsen 1991). Children have also often been caught up in the political conflicts that divide many societies and make development a more complicated proposition. Graça Machel's (1996) study on the impacts of armed conflict on children makes clear that 'in the course of displacement, millions of children have been separated from their families, physically abused, exploited and abducted into military groups, or they have perished from hunger and disease'. The 2006 *State of the World's Children Report* (UNICEF 2006) focuses on the world's most vulnerable children, arguing that many children grow up beyond the reach of development campaigns and are often invisible in public debates, legislation, news stories and statistics. Greater investments (particularly by states) to protect the human rights and well-being of children and young people and to allow them to participate more fully in development are key to eradicating world poverty.

these households do not have the same level of access to income, assets (e.g. land) or local and national power structures. It is important for international development organizations to understand the different roles and responsibilities of men and women within these diverse households. The contribution of women in particular to household income and welfare needs to be much more carefully understood (Moser 1993; Kabeer 1994; Sweetman 2000; Perry and Schneck 2001).

In development policy and practice it has often been assumed that the household in the 'global South' corresponds to the Western model of the nuclear household, yet these households are far from homogeneous in their structure and in the amount of control over decision-making and resources that women enjoy. The contribution of women to the development process, of their knowledge, time, labour and capital, has often been underestimated by development 'experts'. Mohanty (1991) rejects the idea of the stereotypical 'Third World woman' as the passive recipient of social services that has often been constructed by development planners and theorists alike. Similarly, the visibility of women's productive, reproductive and community-management roles in the recording of national development statistics has been widely criticized by gender and development (GAD) researchers (Rathgeber 1990; Moser 1993; Visvanathan et al. 1997). Women were often not defined as 'productive' contributors to their societies, as it was abstractly assumed that women were largely confined to the private, reproductive or domestic sphere (Marchand and Parpart 1995). In agriculture, for example, access to credit, rural extension services, land ownership and farm technology has often been denied to women, despite the fact that women produce more than half the world's food (Koczberski 1998). In this regard it is important to note that many countries in the South have attempted to create indigenous versions of European welfare states, which has proven economically very difficult for many states, leading to the uneven development of citizenship benefits amongst different socio-economic groups and between men and women (see Chapter 20). Again, however, we see the problems created by trying to draw on a universal set of standards based on the experiences of Western liberal democracies and the problems created by ignoring or overlooking the particularity and diversity of peoples and places. Development planners have often neglected the importance of women's 'double day' (Moser 1993) or the limitations placed on their time by the need to combine reproductive (e.g. childcare) with productive (e.g. farming) activities. In addition to these two important roles, women often participate in the management of their communities. They organize political, social and economic structures and play a crucial role in the management of the environment and natural resources (Dankelman and Davidson 1988; Sondheimer 1991; Braidotti et al. 1994; Shiva and Moser 1995; Buckingham-Hatfield 2000).

Although the importance of women's participation in development planning has increasingly been recognized by international institutions, the capacity of national governments to respond effectively to the needs of their societies at the 'grass roots' has been severely curtailed by indebtedness, reduced commodity prices, dependence on international donors, recession and protectionism in the global economy. The state has typically been seen as the primary agent of development, particularly since the 1960s when many countries attempted ambitious large-scale modernization projects with the backing of international development institutions and commercial lenders. The formation of new states, particularly in sub-Saharan Africa, followed at a time when development ideals were becoming increasingly popular worldwide. The 'age of great optimism' imbued these emerging states with a sense of the importance of development in the post-colonial era. Again, as with so many other aspects of 'development', there are no universal definitions of the state and states do not behave in any predetermined way. States in the South have become subject to enormous pressures from the societies within their borders, becoming sites of protest and contention when their activities impinge on the lives of their subjects and citizens.

Ideology also has an important bearing on the way the role of the state in development is viewed. For example, reflecting the World Bank's 'neoliberal' framework, the annual World Development Reports (WDRs) often view an effective state almost exclusively 'through the lens of economic efficiency', with states seen as providing either 'barriers' or 'lubricants' to free-market economic reforms (see Case study 8.7). Neoliberalism promotes and normalizes an economic growth-first strategy where the social and welfare concerns of citizens are seen as less of a priority. Neoliberal development models also naturalize an image of international markets as fair and efficient, privileging 'lean government', deregulation and the removal of state subsidies. More importantly, in many ways neoliberalism forecloses alternative paths of development, narrowing the ideological space in which it is possible to think outside the 'development box' (Power 2003). Behind neoliberalism is the core assumption that the economy should dictate its rules to society rather than the other way around.

The Bank's neoliberal vision of development has emerged clearly in debates about globalization and the world economy. Globalization is a key means through

Thematic Case Study 8.7

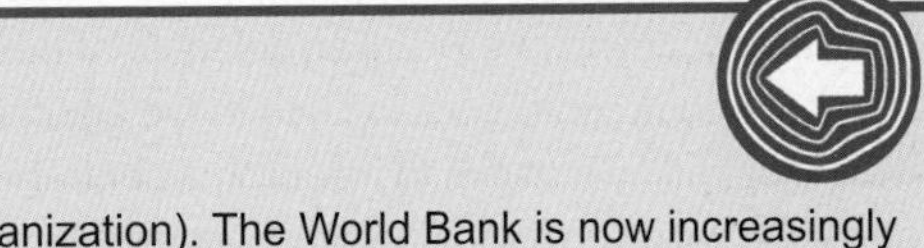

The World Bank and the IMF

Both the International Monetary Fund (IMF) and the World Bank are based in Washington, DC. The Fund was set up in the 1940s as a result of the Bretton Woods conference to maintain currency stability and develop world trade by establishing a multilateral system of payments between countries based on fixed exchange rates and complete convertibility from one currency to another. The IMF is not controlled by the people or the governments of low-income countries and is not like other UN agencies where every government has a vote (Wood and Welch 1998). Instead the fund is run by a select band of high-income countries that fund its operations (the USA has an 18 per cent share of votes and has de facto control of the organization). The World Bank is now increasingly involved in macroeconomic issues and is always chaired by an American, while the President of the IMF is usually European. The latter is dominated by OECD countries and in particular by the USA and is organized in quite a similar way to the Bank.

The International Bank for Reconstruction and Development (better known as the World Bank) was founded as a means of reviving war-damaged European economies, a mandate that was later extended to developing nations. The Bank is funded by dues from members and by money borrowed in international markets. It makes loans to member nations at rates below those of commercial banks to finance development 'infrastructure' projects (e.g. power plants, roads, hydro-dams) and to help countries 'adjust' their

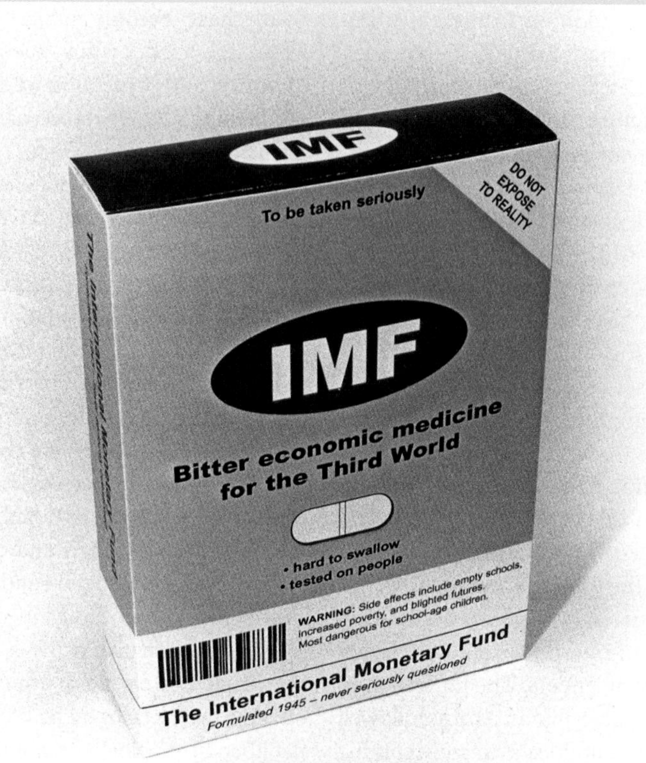

Plate 8.2 The IMF and its hard to swallow 'bitter economic medicine' has had major economic implications for 'Third World' countries depicted in this Oxfam campaign. Its side effects include empty schools, increased poverty and 'blighted futures'.
(Oxfam GB)

economies to globalization. Like the IMF, the Bank was an early and enthusiastic supporter of the neoliberal agenda. According to a Brazilian delegate at the Bretton Woods conference in 1944 these agencies were inspired by a desire to establish a central lending source to which governments would contribute such that 'happiness be distributed throughout the world' (IBT 1998: 3). In 1990 the Bank's then President Barber Conable remained confident of alleviating world poverty: 'we know a great deal about who the poor are, where they are and how they live. We understand what keeps them poor and what must be done to improve their lives'. The 2000/2001 World Development Report prepared by the Bank was entitled *Attacking World Poverty* (World Bank 2001a) and promised 'the most detailed-ever investigation of global poverty' (World Bank 2001b). The result of more than two years of research, the WDR draws on a 'background study' that sought the 'personal accounts' of more than 60,000 men and women living in poverty in 60 countries around the world. This approach allows the Bank to claim that their vision of development is based upon 'the testimony of poor people themselves'. The 2002 WDR *Building Institutions for Markets* (World Bank 2002) focuses on weak domestic institutions 'which hurt poor people and hinder development'. Instead, the Bank argues, institutions suitable for (supporting) markets should be promoted so that all poor people can feel the benefits of the free market and free trade. Out of an annual budget of US$15 billion the Bank reserves about US$12 billion for project grants, with adjustment lending coming to an estimated $3 billion. Additionally, the Bank annually concludes about 40,000 contracts in projects it finances in 95 countries worldwide (*Geobusiness Magazine* 2001: 61). The Bank has contacts and business dealings with some 300 businesses across Europe, representing 47 per cent of its contracts. Consultants in borrowing countries cream off another US$1.2 billion a year (*Geobusiness Magazine* 2001). While the agency listens to the 'voice of the poor', consultants are paid handsomely for their work and perspectives on development projects.

which many people seek to understand transitions amongst human societies at the start of the third millennium (see Section 4). The Bank attaches a great deal of importance to the impacts of globalization and the increased interconnectedness of the world economy and the ease of travel and communication that characterize the contemporary world. In some ways, however, globalization has been 'fetishized' by the Bank and other international development agencies to the extent that as an idea it now has 'an existence independent of the will of human beings, inevitable and irresistible' (Marcuse 2000: 1). There has also been a tendency to speak of the 'end of geography', envisioning a time when all places will have similar social and cultural characteristics as global corporations and as media agencies disseminate similar products and images across the globe (Holloway and Hubbard 2001). It is necessary, however, to avoid the view that globalization is a relatively new and recent phenomenon, led by the 'unstoppable' juggernaut of global changes in information, knowledge and technology. What is often not clear in these debates is what role there is for the people of the South to shape their own destinies and the specific impacts on places, territories and communities.

Not surprisingly (given its economistic focus) the World Bank often 'locks society out' (Hildyard 1997: 43) in its definition of the state (Figure 8.3). Although

Figure 8.3 This illustration depicts the idea that World Bank 'experts' view the role of the state in development as comprising a simple formula or recipe, with several distinct 'ingredients'.

Source: *The World Bank and the State: a Recipe for Change?* by Nicolas Hildyard, Bretton Woods Project, www.brettonwoodsproject.org

the WDR calls for greater participation in decision-making in order to bring 'government closer to the people', a number of key policy areas are seen as 'off-limits'. Despite its dismal environmental record (where lending policies have often financed ecological destruction), the World Bank has sought to appease its critics by including environmental NGOs (non-governmental organizations) in its operations in recent years. Thus the Bank has co-opted its environmental critics in order to 'green' itself and illustrate its commitment to environmental conservation (Sheehan 2000). According to Chossudovsky (1997) the World Bank has claimed to have shifted policy focus towards 'sustainable development' and 'poverty alleviation' in recent years, but neoliberal policy prescriptions continue their dominance within the international development industry whilst the Bank has been able to maintain many aspects of its core policy focus by co-opting critics and cloaking itself in the guise of green environmentalism.

The term 'empowerment' is often used by the Bank and other sections of the 'development industry' (Ferguson 1990) to describe their commitment to popular participation in decision-making for development. Facilitating 'people's participation' is now on the agenda of most international development institutions. Again, however, some development agencies have manipulated this agenda in order to present themselves in more positive ways, as 'listening' to the voice of the people and as facilitating popular and local 'ownership' of development plans and policies (Cooke and Kothari 2001). For Rahnema (1997), the increasingly widespread acceptance of the idea of participation suggests that it has been severely diluted and has lost some of its more radical potential:

> governments and development institutions are no longer scared by the outcome of people's participation . . . there is little evidence to indicate that the participatory approach, as it evolved, did, as a rule, succeed in bringing about new forms of people's power.
>
> (Rahnema 1997: 118–23)

Escobar (1995) also seems somewhat disillusioned with the development concepts of 'participation' and 'empowerment', pointing out that development planners and politicians have often tried to manipulate experiences of participation to suit their own ends. Munslow and Ekoko (1995) similarly argue that empowerment has been stronger on rhetoric than in reality and that advances in widening political participation have been slow, particularly in Africa, where the 'democracy and development' agenda has been manipulated by political movements across the continent (Baker 1998). Empowerment speaks of making people the agents of their own development while doing little about the causes of inequality, and without transforming the nature of ties with international development institutions and their ideologies.

The ideals of empowerment and popular participation and their increasing acceptance in development circles is partly a consequence of the rising influence of NGOs which have often been highly localized and focused on particular issues. NGOs have been defined as a 'residual' category that includes a wide variety of formal associations that are not government-led, even though definitions vary enormously and many NGOs do receive finance and support from their home governments. They can operate at a local, national or international level, but there is often little overall coordination between different NGOs even though they usually work on similar issues, sometimes in the same country or city. The role of NGOs as agents of development has grown vastly in the past thirty years. NGOs have a particular geography of power, with command and coordination functions usually located in the North/West (decision-making and fundraising) and network power in the South (the necessity to be represented in capital cities and 'on the ground' in provinces). A key issue that has been raised in critiques of NGO activity is the closeness of their contact with local cultures and knowledges. Although many have argued that the work of NGOs is often poorly monitored and sometimes imposes foreign solutions in delicate situations, many observers have argued that they can provide an effective link between state and society, often working in 'very practical ways' (Edwards and Hulme 1992: 14). The relationship between the state and NGOs is crucial to the prospects of inclusive democratic development. Particularly in situations of political instability or conflict, NGOs can provide critical linkages in the absence of effective or disputed state administration.

8.5 Conclusions: geography, unevenness and inequality

Radical development geography must adopt a framework that is liberated from the tyranny of dualism and allows for changes in the world economy and variations between states – one that accommodates alternative historical and contemporary geographies, cutting across the North–South divide. (Bell 1994: 188)

Development has nearly always been seen as something that is possible, if only people or countries follow through a series of stages or carry out a series of prescribed remedies and recipes. It is also often assumed that an organic and inherently positive and 'natural' process of evolution is somehow at work, one that is both progressive and forward-moving: 'from little acorns do giant oaks grow'. Development also often means simply 'more': whatever we might have some of today we might or should have more of tomorrow (Wallerstein 1994). After nearly six decades of global development the promise that tomorrow will bring something 'more' is wearing very thin for many poor people.

As Radcliffe (1999: 84) has argued, '[r]ethinking development means reconsidering the categories we use in development geography, and unpacking the power relations that shape them'. This question of the power of development and the unpacking of the power relations that are inscribed in development practices is a critical one. A rethinking of development is also necessary because the lines that have so far divided North and South are now present within *every nation-state* and are making ever less appropriate the conventional language used to interpret the geography of development in the world economy (rich/poor, North/South, First World/Third World, developed/developing) (Power 2003). It is also necessary to (re)formulate a view of development that focuses on the relationships to households and communities and not just (as often happens) on 'formal' institutions like the state, the transnational corporation, the international development agencies or NGOs. This is an incredibly difficult balance to maintain consistently since development operates across so many spatial scales.

At the very core of development studies and development geography there has been a normative preoccupation with the poor (saying what 'ought' to be *done*) and with the 'marginalized and exploited people of the South' (Schuurman 2001: 9). Perhaps we should also ask ourselves what ought to be done about wealth creation and the wealthy? As Hart (2001) suggests, a distinction can usefully be made between 'big D' Development and 'little d' development. The former refers to a post-Second World War project of intervention in the 'Third World' that emerged during the time of decolonization and the Cold War. The phrase 'little d' development, however, points to the development of capitalism as a 'geographically uneven, profoundly contradictory set of historical processes' (Hart 2001:

Plate 8.3 Bob Geldof salutes the crowd at the Live 8 concert in London in 2005.
(PA Photos/PA WIRE).

650). D/development can thus be viewed simultaneously as both a project and a process, with overt and covert forms, shot through with contradiction and unevenness.

In recent decades one of the major trends has been that the world in which we live has become ever more global in character and orientation (globalization is the subject of Section 4). Global corporations and global marketing activities have resulted in the availability of standardized projects and products, and communications have significantly improved, creating new regimes of 'interconnectedness' (Allen 1995) (Case study 8.8). In studying globalization and its impacts it is necessary to consider its economic, political and cultural strands (Allen 1995).

Globalization provides important historical contexts and frameworks through which development has been understood and experienced. Similarly, globalization has to be viewed in relation to the conditions of 'unequal development' that characterize the contemporary world. The notion of a 'shrinking world' has been important in discussions of the impacts of globalization, but patterns of global inequality have not shrunk after 50 years of development theory and practice. Ake (1996) reminds us that for Africa and other 'peripheral world regions', globalization is a highly uneven process that has not affected all areas of the so-called 'Third World' in the same way. Development in Africa, for example, has never failed; it just never really got started (Ake 1996). In addition, although development thinking has itself been far from monolithic, many conceptions had the habit of 'robbing people of different cultures of the opportunity to define the terms of their social life' (Esteva 1992: 9). It is crucial then to avoid painting a picture of development (and its subjects) as monolithic and to avoid repeating this habit of robbing people of their cultural and historical difference and diversity. The *Dictionary of Human Geography* (Johnston et al. 2000) defines development as a process of 'becoming' and as a 'potential state of being', but also mentions how:

> The achievement of a state of development would enable individuals to make their own histories and geographies *under conditions of their own choosing.*
>
> (Johnston et al. 2000: 103; emphasis added)

International development institutions have often overlooked local perspectives and knowledges of development and state approaches have often been heavily centralized and 'top–down'. Escobar (1995), for example, has argued that development has 'created abnormalities' such as poverty, underdevelopment, backwardness and landlessness and then proceeded to address them in a way that denies value or initiative to local cultures. Whilst many sections of the 'development industry' have in recent years come to acknowledge the importance of people power and popular participation in development (Friedmann 1992), 'empowerment' has been co-opted by the mainstream of development theory and practice and has been stripped of its radical potential. UN agencies in particular have been criticized for their neglect of participatory approaches and their excessive

Thematic Case Study 8.8

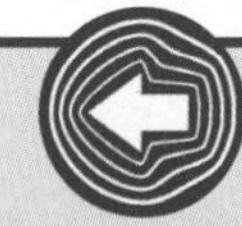

Globalization, agriculture and development

The key political and economic trends that together summarize what we usually take the term 'globalization' to mean include: (1) the increasing importance of the financial structure and the global creation of credit, leading to the dominance of finance over production; (2) the growing importance of the 'knowledge structure', with knowledge defined as a significant factor of production; (3) the transnationalization of technology and the increasing rapidity with which technologies become redundant, with increasing emphasis on 'knowledge industries'; (4) the rise of global oligopolies in the form of TNCs: corporations must 'go global', acting simultaneously in a number of different contexts; (5) the globalization of production, knowledge and finance leading to a decline in the regulative power of nation-states; and (6) the new 'freedom' of capital from national regulative control and democratic accountability, which is said to have led to increasing poverty, environmental destruction and social fragmentation. In thinking about how globalization has impacted upon countries of the South we can take the example of the world food economy, which is very much a global reality that shapes our lives in profound cultural, ideological and economic ways (Goodman and Watts 1997: 3).

Since the United Nations staged the first World Food Conference in Rome in 1975, amidst growing anxiety about food availability and food prices, the world food economy has changed in many different ways. At the level of *production and distribution*, there has been a radical restructuring. These areas have been internationalized or have become more global in scope and character, particularly since the emergence of transnational corporations (TNCs). By the second World Food Conference in Rome in 1996, UN agencies were becoming increasingly aware of the ways in which TNCs are (re)shaping the world food economy in ever more global ways, fundamentally reorganizing the division of labour in the global agri-food system over the last 25 years (see Chapter 7).

Transnational biotechnology corporations like Monsanto/Pharmacia, Hoechst, Pannar and Calgene have claimed that the recent emergence of GM or genetically modified crops has the potential to eradicate global food shortages and malnourishment, offering sustainable food security for the poorest parts of the world. These claims have, however, been widely contested in the 'global South'. In March 1999, leading scientists and activists from 12 countries (including Malaysia, India and Bangladesh) met at a conference in Delhi to discuss the prospect of 'biodevastation' associated with the introduction of GM crops and vowed to bring down Monsanto and other biotech TNCs and to build a global mass movement for sustainable and organic agriculture (British Broadcasting Corporation 1999).

In April 1999 an alternative year-long seed *Satyagraha* was launched in India in direct opposition to Monsanto's GM seeds. Vandana Shiva, Director of the Research Foundation for Science, Technology and Ecology that helped to launch *Satyagraha*, described Monsanto's seeds as 'terminator technology' and has promised that '[w]e will drive Monsanto out of this country. It will be destroyed. Free trade in food is a recipe for famine and farmer's suicide' (Shiva 1999: 9). Almost 1,500 organizations, many of them established by women, have come together across India to reject the new GM seeds – like Monsanto's Bollgard Cotton, which can destroy beneficial species and can create superior pests. Many Indian environmentalists described the recent patent crisis as a 'neo-colonial' ploy by the West and the World Trade Organization to (re)colonize the 'developing world' (Shiva 1999: 5).

bureaucracy, but prospects for reform look bleak. In some ways, the policy prescriptions of macroeconomic global institutions like the World Bank and the IMF claim to be leading the campaign to eradicate world poverty, but a wide variety of 'anti-capitalist' and 'anti-globalization' protests in recent years have shown that these prescriptions are inflexible and ignorant of the particularities of place and country-specific contexts. There have also been transnational movements like the *Jubilee 2000* and *Make Poverty History* campaigns or the series of 'Live 8' music concerts in 2005, which have all sought to put pressure on Western political leaders.

Many recent critiques of development appear disillusioned with the future of the development industry and its capacity to understand and alleviate world poverty. Further, many 'Western' politicians are only gradually beginning to wake up to the realities of these contemporary global inequalities. A number of them see these concerns as those of distant geographies, a world of problems pushed and 'worlded' beyond the universe of immediate moral concern:

> Elite western policy-makers seem to regard the growing income equality gap as they do global warming. Its effects are diffuse and long term and fears of political instability, unchecked migration flows and social disruption are regarded as alarmist.
>
> (Wade 2001: 80)

Thus poverty and inequality are seen as too diffuse and so long-term as to be of no immediate concern. Additionally, rather than thinking of a single, interconnected world economic system, this impoverishment and inequity is constructed as somehow unique or exclusive to the peoples of the 'Third World'. If our concern then is to build a more radical development geography it needs to be understood that poverty also occurs in 'developed' countries and that the aid and 'development' policies of such countries, far from being a part of the solution, may actually be considered a part of the problem. Marginality and deprivation (or for that matter, excessive consumption amongst the affluent) in Europe, North America or Russia and other post-communist 'transition economies' should also be seen foremost as issues of 'development' (Woods 1998; Jones 2000). Radical development geography must be 'liberated from the tyranny of dualism' and the constant stratification of the globe into three distinct worlds. This also means moving away from the developed/developing or lesser/more developed dualisms, which see a simplistic bipolar world of 'haves' and 'have nots'. It is important therefore to allow for variations between states (historically and geographically) and to move beyond the tragic stereotypes of a single condition of 'third world poverty' and a single 'geography of the third world'. This could make a world of difference.

Learning outcomes

Having read this chapter, you should understand:

- The importance of critical and alternative approaches to the study of global development.
- The historical forces and pressures that shape development theories and practices.
- The power and politics of 'labelling' and categorizing poor peoples, places and the spaces of development.
- The role of some of the key brokers of development such as states, United Nations organizations, communities and NGOs.
- The differentiation and diversity of peoples, places and spaces of development at the beginning of the twenty-first century.
- The importance of a people-centred view of development and the need to understand the *other side of the development story*, or the views of women, the views of people in the South, the views of people who are excluded socially, and the view from 'below'.

Further reading

Desai, V. and Potter, R. (eds) (2008) ***The Arnold Companion to Development Studies***, 2nd edition, Arnold, London. Contains short and accessible chapters on a wide variety of development themes. A useful companion text in studying development, with suggestions for further reading on each topic.

Kothari, U. and Minogue, M. (eds) (2002) ***Development Theory and Practice: Critical Perspectives***, Macmillan, London. Explores a wide range of perspectives on critical approaches to the study of development, looking at globalization, governance, social development, participation, feminism and post-colonialism.

Potter, R., Binns, T., Elliott, J.A. and Smith, D. (2003) ***Geographies of Development***, 2nd edition, Longman, London. Updated and revised in a second edition, this book offers a wide-ranging discussion of theories of development, urban/rural spaces and the important institutions of global development.

Power, M. (2003) ***Rethinking Development Geographies***, Routledge, London. Draws out the spatial dimensions of development and outlines how the discipline of geography has been implicated and involved in the theory and practice of 'development'. The book offers a critical and stimulating introduction to the imperial and geopolitical dimensions of development, looking at cold war and colonial constructions of 'The Tropics' and the 'Third World'.

Radcliffe, S.A. (2005) ***Rethinking development***, in Cloke, P., Crang, P. and Goodwin, M. (eds) ***Introducing Human Geographies***, Arnold, London: 84–92, 2nd edition. Presents a critical discussion of the role of state power in development and of the importance of gender relations. Forms part of a larger three-chapter section, 'On Development Geographies' with contributions from P. Routledge and S. Corbridge.

Willis, K. (2005) ***Theories and Practices of Development***, Routledge, London. Explores the development theories behind contemporary debates such as globalization and transnationalism and traces the main definitions of 'development' and 'development theory' over time. It contains student-friendly features, including boxed case studies, with examples, definitions, summary sections, suggestions for further reading, discussion questions and website information.

Useful websites

www.id21.org or www.ids.ac.uk/eldis The first site is a useful development research gateway with regular updates and short summaries of recent research topics; the second site is prepared by the Institute of Development Studies (IDS) and offers an indispensable gateway to a range of relevant development resources. The site allows detailed searches (by country or theme).

http://wsf2007.org The website for the World Social Forum meeting in Nairobi, 2007.

http://europa.eu/pol/dev/index_en.htm This site provides an overview of EU activities in the area of 'development'.

www.eldis.org The ELDIS Development information research gateway site with a useful country search facility and good web links.

www.Jubileesouth.net or **www.studentsforglobaljustice.org** The first site concerns the global debt crisis and the second links student resistance to a variety of anti-war, anti-capitalist and global economic justice campaigns.

www.twnside.org.sg or **www.anotherworldispossible.com** The first site is the home of the Third World Network (TWN) and the second relates to the 'Another World is Possible' campaign.

www.globalissues.org or **www.focusontheglobalsouth.org** The first site contains useful material on global issues and has introductory material on geopolitics; the second is the home of 'Focus on the Global South'.

www.nam.gov.za The home of the Non-Alignment Movement (NAM).

For annotated, clickable weblinks and useful tutorials full of practical advice on how to improve your study skills, visit this book's website at **www.pearsoned.co.uk/daniels**

SOCIETY, SETTLEMENT AND CULTURE

Section 3

Edited by Denis Shaw

Geographers are interested in the ways places differ from one another and also in the similarities and interconnections between places. But the complexities of the changes that human societies have experienced since the late twentieth century render the geographer's task of making sense of the world increasingly difficult. Old dichotomies, like that between 'urban' and 'rural', and the lines which geographers have customarily drawn on their maps to divide country from country, region from region and culture from culture, seem to make less and less sense in a globalizing environment. Even so the results of such processes seem not to be a world that is more uniform. While distant places interconnect and interact with increasing intimacy, new spaces appear where people are excluded from such developments, and almost everywhere some individuals who share spaces with others are denied the rights and advantages that the latter enjoy.

It is such complexities, making themselves felt in urban and rural spaces and in the sphere of culture, that form the subject of this section. The emphasis is on the multifaceted effects of globalization on human societies and the way people think. Thus the world has become more urban – most of the world's population now lives in cities and only three major regions of the world (sub-Saharan Africa, south and much of South-east Asia, and China) remain predominantly rural (though this still represents a substantial proportion of the world's population). But the experience of urbanization differs radically between cities, spaces within cities, and individuals. The corollary is that the world has become less rural – rural spaces may be urbanized, linked into global networks, or lose population to the cities – but again those spaces, and the individuals within them, are affected in very different ways by such processes. And the world has become more 'cultural' in the sense that present-day changes are as significant for their cultural effects as for any others. Issues like identity and the meanings and ways of life that are shared by, or that divide, individuals and groups become matters of concern, contestation and even conflict.

→

→ Chapter 9 on cities considers how the term 'urban' is to be defined and the implications of urbanism for the ways in which people live and interact. The significant issue of the role cities now play in global and regional economic and power networks is addressed before the chapter turns to examine the issue of urban governance and how urbanization processes are and can be regulated. Chapter 10 on social inequalities argues that all modern societies are riddled with inequalities, some of which are more fundamental than others. The chapter demonstrates that one of the clearest ways in which social disadvantage is reflected is through its geographical expression, particularly in the existence of spaces described as being 'on the margins'. Chapter 11, entitled 'Rural worlds', is similarly concerned with definititions of the 'rural', and analyses the alternative consequences for both rural spaces and individuals of the processes of urbanization and globalization. Chapter 12 on social constructions of nature argues that nature is not some pre-given physical reality, existing totally outside ourselves, but is in fact an idea which is constructed by society. It is never easy to draw a line between things that are deemed 'natural' and things that are 'social', and different ways of conceiving of what is 'natural' can be used to advance particular political agendas. Even science is not immune from this process – as the chapter claims, 'science constructs nature as a mirror to reflect the things that society wants to see in itself' – and therefore we should always be ready to challenge the assumptions lying behind theories about 'nature' and the political stances and policies to which they give rise. Chapter 13 on culture defines the latter as a process and as a system of shared meanings. The argument is that globalization is not producing a single global culture although the idea of a global culture is becoming important. Global cultural processes interact with processes occurring at national and regional levels to produce a complex world in which it is necessary to foster approaches that will welcome and celebrate cultural differences rather than shun and fear them.

This section makes clear the fact that the geography of a globalizing world is as much a geography of exclusion as it is one of interconnection and interlinkage. It shows that it is a dynamic world where social and cultural change is very much the order of the day, and yet it is a world punctuated by continuities. Finally it suggests that it is a world in which people can make a difference, where they can be more than mere ciphers dominated by impersonal forces like global financial and power networks, and where ideas, attitudes and relationships matter.

CITIES: URBAN WORLDS

Chapter 9

Allan Cochrane

Topics covered

- The present-day importance of cities
- Defining urbanism
- Wirth's definition of urbanism
- Decentralization and suburbanization
- The postmodern city
- European cities
- Megacities and ordinary cities
- Global urban systems
- Urban heterogeneity and its implications
- Social polarization and social diversity
- Cities of connection and disconnection

Most of the world's population now lives in cities and, according to UN-HABITAT (2006: 4), by 2030, five billion people will be living in urban areas.

It is no longer possible – if it ever was – to divide the world into a developed world, with big cities, and an 'underdeveloped' or 'Third' world, without them. Certainly some areas of the world (North America, Europe, Japan) are more heavily urbanized than others, but there is no neat line that can be drawn on the globe to divide a world of urbanites from a world of peasants. Indeed, although the most heavily urbanized countries are still those of the global North, urbanization is taking place fastest and most dramatically in countries that might previously have been described as 'under-developed', and some of the biggest cities in the world are to be found in them. At the same time some older industrial cities (particularly, but not only, in Central and Eastern Europe) are actually losing population, attracting the epithets 'shrinking city' (Oswalt 2004) or 'weakening mature city' (Hall and Pfeiffer 2000). São Paulo, Mexico City, Shanghai, Mumbai (Bombay), Jakarta and Beijing are all among the ten cities with the biggest populations in the world, and Manila, Cairo and Kolkata (Calcutta) are not far behind (see Table 9.1), even if the fastest growing cities are now the smaller and middle sized cities in the same countries.

The dramatic global growth of cities makes it very important for us to understand their geographies and the ways in which they are constituted. It represents a significant shift in the spatial organization of society – in the geographies of everyday life. The changing spatial organization of our lives represented by urbanization also fundamentally shapes the ways in which we live our lives and the social relations of which we are a part.

This chapter starts out by looking for ways of defining cities. It explores the features of urban life that distinguish them from other forms of human settlement, and other forms of socio-spatial organization. It considers the implications of some of these features for the ways in which people live their lives in cities, before going on to discuss their key role in wider global and regional networks of social and economic power.

Table 9.1 Cities with the largest populations in the world (millions of people), projections for 2015.

1	Tokyo/Yokohama	28.7
2	Mumbai (Bombay)	27.4
3	Lagos	24.1
4	Shanghai	23.2
5	Jakarta	21.5
6	São Paulo	21.0
7	Karachi	20.6
8	Beijing	19.6
9	Dhaka	19.2
10	Mexico City	19.1

Source: *Philip's Great World Atlas* (2003: 27). These figures are for urban agglomerations.

9.1 Trying to define cities

Recognizing the importance of urban development in general terms is only a starting point. Most of us probably feel that we have a reasonable grasp of what makes some places cities, while others are towns, villages or rural areas. But it is important to explore rather more systematically what it means to make these distinctions. This should help us to understand what is shared between cities (and the social relationships associated with them) as well as being able to assess the significance of the ways in which they differ. A large population, for example, may make a place a city (even a 'megacity'), but only some cities (such as Tokyo, New York and London) are key nodes of financial and economic power within the global economic system, to the extent that they are often referred to as global cities (or world cities) (see, for example, Sassen 2001). Since one might also expect a higher concentration of wealth and of members of global elites in those cities, their social structures are also likely to be quite different from those of other large cities, such as Mexico City, or Mumbai.

So, are there any key – defining – features of cities?

In the 1930s, Lewis Wirth, a sociologist at the University of Chicago (see Plates 9.1–9.4), set out to specify them, in ways that incorporate what are probably still among the dominant commonsense understandings of urban life. For him the city was 'a relatively large, dense and permanent settlement of socially heterogeneous individuals' (Wirth 1938: 50). In other words, cities could be defined through four main characteristics. Wirth believed that a series of more or less logical consequences flowed inexorably from these characteristics.

Plates 9.1–9.4 Images of 1930s Chicago: the slums; the stockyards; gangsters; the Wrigley Building
(© Bettman/CORBIS; © Hulton-Deutsch Collection/CORBIS; © Underwood & Underwood/CORBIS; © Hulton-Deutsch Collection/CORBIS)

- *Size.* The large population size of cities was held to mean that contact between urban residents would generally be impersonal, superficial and transitory, since no one person could know all the others in the city. Wirth's arguments build on a longer sociological tradition, which emphasized the difference between traditional society (characterized by what was called ***Gemeinschaft*** or community) and industrial – and urban – society (characterized by ***Gesellschaft*** or association) (Toennies 1887). The former was associated with closeness and stability, in which there was a high degree of informal personal contact; by contrast, the latter was characterized by transitory and superficial relationships, in which individuals only came together in groups to achieve particular ends (or defend particular interests). Wirth similarly argued that in cities human relationships were fragmented along the lines of the services that each individual might perform for another.
- *Density.* The density of population was also said to produce increased differentiation between individuals, an increasingly complex social structure, social segregation and a dissociation of workplace from

residence. It fostered a spirit of competition, aggrandizement and mutual exploitation, requiring complex rules to manage the resulting friction.

- *Permanence.* It was the permanence of cities that created the spaces within which the social processes identified by Wirth worked themselves out. In a sense it is also that permanence which confirms the importance of 'urbanism as a way of life' (the title of Wirth's seminal article). Urbanism gains its status because it represents a particular stage in the development of human society, as the dominant form of social organization in the industrial (or modern) era.
- *Heterogeneity.* In some ways, for Wirth, this is the most important (and distinctive) feature of cities. He argues that: 'The city has . . . historically been the melting pot of races, peoples and cultures. It has not only tolerated but rewarded individual differences. It has brought together people from the ends of the earth because they are different and thus useful to one another, rather than because they are homogeneous and like-minded' (Wirth 1938: 50). Heterogeneity, said Wirth, resulted in the creation of a complicated caste and class structure reinforcing the effects of high population density. No single group had the full loyalty of individuals.

Wirth's arguments have been criticized for failing to capture some key features of urban life and exaggerating others. Some have pointed out that the contrast between urban and non-urban is hard to sustain, particularly if it carries with it the implication that the urban is somehow modern and the non-urban traditional or even primitive – instead it is important to recognize how the two not only exist alongside each other, but may also co-constitute each other in cities. This is highlighted by the contemporary experience of cities in poorer countries of the world, since the spread of urban culture into non-urban spaces and the integration of non-urban culture in urban spaces is particularly apparent there. Urban residents, suggests Robinson (2005a), are connected into multiple networks, only some of which directly link them into the economic spaces of global corporations, while others do so more ambiguously or not at all. Sustaining a fundamental distinction between these different aspects of urban life is unhelpful. So for example, the **informal economy** – in which goods and services are provided without being captured in the national statistics of income and growth – may be just as much an expression of urbanity (and hence of 'modernity') as the formal economy.

The growth of the **suburbs** also raises questions for Wirth's approach (see Plate 9.5), since they meet many of his broader criteria for inclusion as urban without exhibiting quite the same symptoms in their social relations. So, for example, Gans (1968: 40) points out that (at least in the case of the US suburbs of the 1960s) 'there is little anonymity, impersonality or privacy' of the sort identified by Wirth, nor are they noted for having a heterogeneous population. This argument has been taken further by those who point to the emergence of **edge cities** (again particularly in the USA, but said to be a global phenomenon) – so-called because 'they contain all the functions a city ever has, albeit in a spread out form that few have come to recognize for what it is. Edge, because they are a vigorous world of pioneers and immigrants, rising far from the old downtowns, where little save villages or farmland lay only thirty years before' (Garreau 1991: 4).

In the last decades of the twentieth century, it has been suggested that the suburbs or periphery have come to dominate. Drawing on the experience of Los Angeles and Southern California, Dear (2000, 2002) argues that a postmodern urbanism has effectively replaced Wirth's modernist urbanism, promising quite different 'ways of life'. From this perspective, rather than viewing urban development as an 'organic' (almost biological) process growing out from a core, in the contemporary – or postmodern – city, it is the urban periphery which organizes and comes to define the centre: 'the imperative towards decentralization . . . has become the principal dynamic in contemporary cities' (Dear 2002: 3).

Understanding this 'postmetropolis' requires us to accept the coexistence of a series of overlapping and apparently contradictory sets of understandings, each of which may nevertheless be helpful in understanding key aspects of the new urban world: Soja (2000) identifies six of these, without implying that his listing is exhaustive. The first focuses on economic restructuring ('the postfordist industrial metropolis'); the second on cities within globalization; the third on 'exopolis' (that is, the decentralized urbanism of edge cities); the fourth on the complex mosaic of social inequality which has replaced the more ordered maps of social inequality associated with the modern city ('the fractal city'); the fifth is concerned with the disciplinary, policed and gated spaces of the city; while the sixth reflects on ways in which cities are imagined as virtual communities.

Graham and Marvin (1996) similarly point to the contradictory ways in which telecommunications and

Plate 9.5 The spreading suburbs: Leavitown, USA.
(© Bettmann/CORBIS)

the construction of virtual spaces of interaction may find an expression in cities – with decentralization of some functions at the edge blurring into wider metropolitan regions, alongside a continued centralization of high-level managerial and other functions. While the city of modernity outlined by Wirth may not survive in this new context, they point to the emergence of 'splintered cities' which produce protected spaces for elite and middle-class groups in the city that link work, home and leisure (Graham and Marvin 2001).

One way of responding to such developments might simply be to deny that the phenomena being described can any longer be understood as cities – or expressions of 'urbanism'. From such a perspective, following Wirth, they might even represent a move to a different form of social organization, one in which the ambition is to eliminate or marginalize the tensions of urban life, to separate out different geographies and histories, and to de-intensify urban social interactions (Pile 1999: 30). But that does not really help, since we still want to know what is happening here, whatever the places are now called, in these locations that most of us still understand to be cities.

Another response might be to deny that the US model – and particularly the Californian model – is helpful as a way of interpreting the experience of other cities. It has, for example, been argued that European cities stubbornly continue to sustain rather different forms of urbanism (unlike either Chicago or Los Angeles), both because they are not stretched out over space in the same way, and because they have longer histories and continuing coherence (see, e.g., Le Galès 2002; Savitch and Kantor 2002). It is in this context that arguments for the 'compact' city have been mobilized, to suggest that denser forms of urbanism are more 'sustainable', both environmentally and socially (see, e.g., Rogers 1999; Rogers and Power 2000). Le Galès focuses his attention on Europe's middle-sized cities, emphasizing both their longevity and their role as 'collective actors' and local societies. 'Although the city is an unceasing movement', he says, 'it is also the locus of collective action, of innovation, of interest-aggregation mechanisms, of negotiation, and of conflict' (Le Galès 2002: 25).

There is certainly some validity to such arguments, although even in Western Europe there is evidence that traditional urban forms are being eroded. Hall and Pain (2006) identify a series of eight polycentric mega-city regions in Europe, including the south-east of England, the Rhine Ruhr, the Randstadt, northern Switzerland and Paris, which seem to share some of the key features identified by the theorists of the post-modern city. De Ru (2004) goes further to describe an extended urban region that she calls Hollocore. She delineates a region that stretches between, and incorporates, Brussels, Amsterdam and the Ruhr Valley, with a population of 32 million, yet without a single city of over 1 million (de Ru 2004: 336). Her summary of the position is a powerful one:

> established cities lose residents while thinly populated areas gain. In the name of identity, city centres are stripped down to their historic pedestrian shopping streets, and appear more village-like than ever – frozen in a time that never was. Meanwhile the periphery fills up with a mix of business-, commerce-, leisure-, industry-, logistic-, villa-, office-, or brainparks, generic urban matter embedded in massive inversions of green. In the Hollocore, the city has become the void left in the wake of its own expansion.
>
> (de Ru 2004: 338–9)

Plates 9.6–9.7 Making a megacity: Hong Kong and the Pearl River Delta.

(Amanda Hall/Robert Harding; Thomas Schlegel/transit/STILL Pictures)

Regional Case Study 9.1

The Pearl River Delta, China

In 1995, this spatial system, still without a name, extended itself over 50000 km^2, with a total population of between 40 and 50 million, depending on where boundaries are defined. Its units, scattered in a predominantly rural landscape, were functionally connected on a daily basis, and communicated through a multimodal transportation system that included railways, freeways, country roads, hovercrafts, boats and planes . . . [it] is not made up of the physical conurbation of successive urban/suburban units with relative functional autonomy in each one of them. It is rapidly becoming an interdependent unit, economically, functionally and socially. . . . But there is considerable spatial continuity within the area, with rural settlements, agricultural land, and undeveloped areas separating urban centers, and industrial factories being scattered all over the region (Castells 1996: 407–9).

The architect Rem Koolhas says that 'In the Pearl River Delta, we are confronted with a new urban system. It will never become a city in the recognizable sense of the word. Each part is both competitive with and has a relationship to each other part. Now these parts are being stitched together by infrastructures so that every part is connected, but not into a whole . . . A city that does not imply the stability of a definitive configuration because each part is fixed, unstable and in a state of perpetual mutual adjustment defining themselves in relation to other parts' (quoted in Hanrou and Obrist 1999: 10–11).

In some respects, Wirth's criteria appear to fit particularly well with the massive new cities of the 'Third World' – their populations are even larger than those in the richer countries of the world; their populations are dense and heterogeneous, too. However, the notion of permanence is, perhaps, rather more ambiguous, because of the extensive temporary and informal housing and fast-changing population, both in the core of cities and in extensive shanty towns or *favelas* on the outskirts. Castells's description of an emergent megacity around the Pearl River Delta in China – what he calls the Hong Kong–Shenzen–Canton–Pearl River Delta–Macau–Zhuhai metropolitan regional system – highlights some of the difficulties in uncritically adopting Wirth's approach (Case study 9.1).

But the drivers of this new urbanization also seem to be rather different from those that characterize development in Southern California. There continues to be a close relationship between industrialization and urbanization, even if on a much bigger scale and with growth at a rate which far exceeds that of traditional urbanization (see Plates 9.6–9.7). As Robinson (2002, 2005a) notes, theorizing about the city tends to bypass and marginalize the experience of urbanization in poorer countries. Even the obsession with megacities shifts attention from the still faster urbanization taking place in smaller cities. The danger is that the cities of the global South, for all the fundamental differences between them, are explored through the lens of Western cities, or debates about those cities. Here too, just as much as in Europe, she argues that there is a need to track 'a diversity of longer term historical influences on city life' (Robinson 2002: 543).

Wirth's approach to understanding the city has some real strengths, because of the way in which it systematically explores the implications of the features that most of us would probably take for granted as elements in urban life. However, his identification of certain more or less automatic linkages between urban form and the lived experience of cities is difficult to sustain. Apparently similar forms of (urban) geographical organization may be associated with rather different experiences of urban life, while apparently similar (urban) experiences may be associated with significantly different urban forms. There is a danger of losing the variety and excitement (as well as sometimes the tedium) of urban life in the search for a conclusive definition.

9.2 Dynamism and global networks

By starting with Wirth, we have also effectively started from an understanding of the city as a more or less free-standing entity which develops according to some internal (almost evolutionary) logic. The discussion is focused on internal processes – the divisions and connections within the city. And this is helpful in highlighting some fundamental aspects of the urban experience. But cities are also the products of flows and interconnections that link them into wider sets of economic and social relationships. They are, as Massey (1999) argues, 'the intersections of multiple narratives' or multiple space–times. The relationships that define cities stretch across space since each city is made up as much through its connections with other cities and what Taylor (2004: 101–2) calls its 'hinterworld' (that is the stretched-out linkages to suppliers, sources of information and ideas etc.) as with any more territorially contiguous 'hinterland'.

In other words, approaches which start from the existence of individual cities and only then construct their linkages to other cities (or places) fail to grasp the extent to which cities are defined by their position within wider – global – networks. So, for example, Allen (1999: 204) notes that: '**global cities** – or powerful cities – would not exist but for the connections which constitute them'. He explores the ways in which cities of power are defined through their positioning in global networks, arguing that the importance of cities such as New York and Tokyo stems 'from the fact that the world of banking and finance seems to have little choice other than to go through the financial districts of those cities. They stand at the intersections . . . of all that matters in global economic terms. In that sense, both cities are locations for something far greater, far more significant in scale, than their actual size would have led us to believe.' They are 'the points at which the lines of authority and influence meet' (Allen 1999: 187).

The broad recognition that cities can only be understood through their connections to others elsewhere in the world is, therefore, an important starting point, but it has been interpreted in different ways. Some identify explicit or implicit hierarchies, whether literally capable of being mapped and charted on the globe (with some cities identified as primary and others as secondary, within the core or semi-periphery) (Friedmann 1986) or powerfully summed up through the identification of a limited number of global or world cities (Sassen 1994, extended in Sassen 2001 to include a second tier of global cities, the 'metaregional' capitals of Paris, Frankfurt, Sydney, Hong Kong, São Paulo and Mexico City). As we have already seen, Castells has identified the growth of 'megacities' (Plates 9.6 and 9.7), which concentrate 'the directional, productive, and managerial upper functions all over the planet' (Castells 1996: 403), and Hall and Pain (2006) focus their attention on a series of European 'megacities', which they also define through networks of interconnection as polycentric functional urban regions.

Taylor (2004) goes further, starting not from particular cities or some broadly defined 'space of flows', but instead from a focus on what he defines as a global urban system (which he sees as a phenomenon stretching back to the thirteenth century). From this perspective, all cities are understood as 'global' cities (Taylor 2004: 25), or even part of a 'collective world city', since contemporary urbanism is understood to be a 'consequence of how local and inter-local flows of material and information . . . intersect in a rapidly converging globally integrated economy' (Dear and Flusty 2002: 79). In principle, Taylor is not concerned to identify a hierarchy, but, instead, to map sets of relations between cities. But in practice (drawing on Jacobs 1984) since he sees innovation and dynamism as the keys to understanding how cities work and should be defined, he ranks cities in terms of their connectivity in advanced producer services (financial and business services are identified as the leading economic sectors). The cities at the top of the hierarchy are the familiar ones (New York, London and Tokyo) even if it is not presented as a command and control 'hierarchy'. As a result Taylor, too, has been criticized for effectively removing some cities from the map or, at any rate, giving them a different status from those which are home to what he identifies as the leading economic sectors.

It is important to focus actively on the making of cities within a global context in a way that recognizes the importance of place or territory as well as understanding the systems within which they are embedded – the 'territorialization of the global political economy' (Robinson 2005b: 761). Storper (1997) stresses the importance of viewing cities as complex and differentiated 'ensembles' made up of (informal as well as formal) relationships and conventions that exist between the agencies (people, organizations, interests) which underpin their economic operation. He develops the metaphor of a 'crucible in which the ingredients, once put in the pot together and cooked, often turn out very differently from what we can deduce from their discrete flavours' (Storper 1997: 255). Scott develops similar arguments for city-regions arguing for the importance of 'clusters of interdependent activities whose mutual proximity to one another creates complex, dynamic flows of agglomeration economies' (Scott 2000: 25). This is world of diverse cities within which it is possible to experience many different 'urbanisms' (Robinson 2005a: 34). Robinson emphasizes the importance of seeing the 'city as both a place (a site or territory) and as a series of unbounded, relatively disconnected and dispersed, perhaps sprawling and differentiated activities, made in and through many different kinds of networks stretching far beyond the physical extent of the city' (Robinson 2005b: 763).

Cities are made up, as Amin and Graham (1997: 418) remind us, of the 'complex interlinkage between place-based relational webs and distanciated ones'. In other words, 'the "city" now needs to be considered as a set of spaces where diverse ranges of relational webs coalesce, interconnect and fragment' (Amin and Graham 1997: 418). Approaches which seek to define cities solely through their internal relationships will fail to grasp the significance of those which stretch between cities, but those which define them solely in terms of their position within global networks will struggle to understand those place-based relationships. Even if Amin and Thrift (2002: 2) are right to note that spatial propinquity may no longer be 'a central feature of the sprawling and globally connected city', surely they are also right to stress that 'The "cityness" of cities seems to matter'. The apparent tension between these two ways of thinking about cities is misleading, and the task is, rather, to see how they coexist in shaping the urban experience.

9.3 Cities of connection and disconnection

One of the tensions associated with urban life arises because cities are the sites where different networks, different webs of relationship, come together – existing alongside each other, but also potentially colliding or intermingling with each other. Social heterogeneity and the juxtaposition (or physical proximity) of different groups within cities ensures that cities become arenas within which different groups and classes find ways of learning to live with each other, engage in struggle for dominance, and search for forms of security in the face of challenge from others. Cities both bring different sets of people together, requiring them to live alongside each other, and at the same time generate new ways of reinforcing and developing segregation and division. From one perspective, urban spaces can be seen as quite remarkable because of the extent to which apparently antagonistic or, at an rate, disconnected people, groups and interests come together and negotiate over the shared spaces within which they find themselves, but they can equally persuasively be seen as sites across which conflicts are pursued and where the accommodations that are negotiated are crystallized in defended and policed places that reflect fundamental divisions of power and wealth.

The way in which cities bring together people from different classes and different cultural backgrounds as well as combining a wide range of activity spaces makes it possible to imagine them as places of excitement, as sites of innovation and dynamism. Nobody has put this more clearly than Jane Jacobs. In her classic book, *The Death and Life of Great American Cities* (published in 1961), she argues that:

> Although it is hard to believe, while looking at dull grey areas or at housing projects or at civic centres, the fact is that big cities are natural generators of diversity and prolific incubators of new enterprises and ideas of all kinds. . . .
>
> The diversity, of whatever kind, that is generated by cities rests on the fact that in cities so many people are so close together, and among them contain so many different tastes, skills, needs, supplies, and bees in their bonnets. (Jacobs 1961: 156–9)

Jacobs emphasizes the extent to which complexity and heterogeneity may create order out of apparent chaos and, above all, generate dynamism and innovation.

By contrast, a rather more dystopian vision of the urban emphasizes the way in which cities are defined through what Davis (1998: Ch. 7) has called an 'ecology of fear', in which a range of protected enclaves is constructed by the middle classes to exclude those who may challenge their security (see Plate 9.8 and Case study 9.2). It is in this context, too, that other areas of cities are identified as potential threats to the security of such groups in what Soja (2000) has identified as the carceral city – that is one which is policed in ways that effectively imprison those populations which are deemed to be a threat. Although this has sometimes been identified as a predominantly US phenomenon, similar arguments have been mobilized to explain divisions within European cities, for example with respect to the management of peripheral estates in French cities (Wacquant 1993; Dikeç 2006a).

Some of the consequences of the apparently insatiable search for 'defensible space' (Newman 1972) are rather paradoxical, however. Instead of creating greater feelings of security in suburban areas, or in areas with neighbourhood-watch schemes, or even in shopping malls overseen by closed-circuit television, the creation of more protection seems to have been accompanied by increased concerns about crime and increased insecurity, potentially reinforcing 'a decline in communal trust and mutual support' (McLaughlin and Muncie 1999: 135). The global search for security has also had the paradoxical effect of ensuring that cities across the world have increasingly become the sites of terrorist action and the targets of state violence through war and its surrogates (see, e.g., Graham 2004).

Plate 9.8 Making protected space in South London.
(© Dave Ellison/Alamy)

Thematic Case Study 9.2

The creation of defended enclaves

Davis (1990) vividly describes the creation of protected enclaves in Los Angeles: 'The security-driven logic of urban enclavization finds its most popular expression in the frenetic efforts of Los Angeles' affluent neighbourhoods to insulate home values and lifestyles . . . new luxury developments outside the city limits have often become fortress cities, complete with encompassing walls, restricted entry points with guard posts, overlapping private and public police services, and even privatised roadways. It is simply impossible for ordinary citizens to "invade" the cities of Hidden Hills, Bradbury, Rancho Mirage or Rolling Hills without an invitation from a resident. . . . In the once wide-open tractlands of the San Fernando Valley, where there were virtually no walled-off communities a decade ago, the "trend" has assumed the frenzied dimensions of a residential arms race as ordinary suburbanites demand the kind of social insulation once enjoyed only by the rich' (Davis 1990: 244–6).

Even in its most extreme form – that is the literal creation of gated areas, accompanied by private policing – Los Angeles is not unique. Similar arrangements are emerging in most of the world's major cities, in luxury developments in London, as well as Detroit and Johannesburg (Robinson 1999). In some cities in poorer countries the contrast is particularly sharp, since the physical distance created by suburbanization is often absent, so that the enclaves may be right up against the shacks or slums of the dispossessed. In São Paulo, for example, Caldeira (1996) describes one development (Morumbi) that was constructed in the 1980s. The new housing offers 'the amenities of a club, [they] are always walled, have as one of their basic features the use of the most sophisticated security technology, with the continual presence of private guards. . . . [The] luxury contrasts with the views from the apartments' windows: the thousands of shacks of the *favelas* on the other side of the high walls which supply the domestic servants for the condominiums nearby . . . for residents of the new enclosures, the inconvenience seems to be more than outweighed by the feeling of security they gain behind the walls being exclusively among their equals and far from what they consider to be the city's dangers' (quoted in Amin and Graham 1999: 39).

The social and economic inequalities that characterize urban life are discussed more fully by Phil Hubbard in Chapter 10, but here it is worth noting that social relations within cities also reflect the ways in which cities are defined by their positioning within wider networks. So, for example, within global cities such as New York, London and Tokyo, which operate as key nodes in a global economic and financial system, it has been argued that there is an expanding gap between rich and poor, powerful and marginalized, as there is a growth in employment in financial and other specialist services alongside a decline in manufacturing and traditional industry (Sassen 2001). At the same time as there is a growth in highly paid employment, there is a parallel growth in low-paid, low-skilled and casualized employment at the bottom end, servicing the high-end service sector and those who work in it (Green and Owen 2006).

The highly complex nature of the globalized social relations is reflected in the local experience of division. So, for example, Zukin (1995) reminds us of the ways in which New York's restaurants

> bring together global and local markets of both employees and clientele. . . . While it seems to be among the most "local" of social institutions, a restaurant is also a remarkable focus of **transnational** economic and cultural flows. As an employer, a restaurant owner negotiates new functional interdependencies that span local, regional, and global scales. (Zukin 1995: 158–9).

The heavy dependence on low-wage immigrant labour in restaurants is a reflection of the role cities play within global flows of migration, while the market for luxury restaurants is itself a reflection of cities' importance for global elites within worldwide networks of power. In these restaurants, the global elites find themselves being served, in 'fleeting ties of local interdependence', by those who are in rather a different position in the global pecking order – often migrant labour in casualized employment whose relationship to global networks is rather different but no less significant (Amin and Graham 1999: 14).

It has been strongly argued that the networked, interconnected world is one with significant 'holes' in it, what Castells has described as 'multiple black holes of social exclusion throughout the planet' which he says are 'present in literally every country, and every city, in this new geography of social exclusion. It is formed of American inner-city ghettoes, Spanish enclaves of mass youth unemployment, French banlieues housing North Africans, Japanese Yoseba quarters and Asian mega-cities' shanty towns' (Castells 1998: 164–5). If the megacities and some of their populations are connected into global networks, then, according to Castells, a further defining characteristic is that large sections of their population, and significant spaces within them, are disconnected and explicitly excluded.

Although the emphasis on 'holes' is helpful in undermining approaches which suggest that the world is becoming 'flatter', an undivided space of globalization, it may understate the extent to which the apparently isolated and disconnected are connected into other webs of relationships that stretch across space. It is all too easy to forget the importance of the global networks that link the poor as well as the rich, the migrants as well as the global elites – perhaps most obviously apparent in the remittances sent by migrants to families in the countries from which they come, but also in the connections associated with global religions such as Islam or Roman Catholicism. Research on global care chains – that is, the movement of care workers as migrants from poor countries to provide health and child care to the better-off in richer countries – suggests that the process relies on continuing linkages across national boundaries. Those providing care in one country also provide financial support to those caring for their own dependents (including children) in the places from which they have travelled (see, e.g., Yeates 2004). This is, of course, not an equal process, but it is one defined by forms of movement and interaction and not simply disconnection and rigidity.

Gilroy's discussion of the 'black Atlantic' is also a useful corrective to any overemphasis on disconnection (Gilroy 1993). He charts the emergence of a 'black Atlantic' political culture through music (soul, funk, reggae, hip hop and rap) which actively links black people in the cities of Kingston, New York and London. This is a global movement with its own networks and its own nodes. So, for example, Gilroy argues that hip hop emerged from the interaction of a DJ style drawn from Kingston in Jamaica, with the emergent dance music styles of the Bronx in New York (alongside some borrowings from Hispanic styles, too) (Gilroy 1993: 103). In other words, one of the most successful forms of global contemporary popular music came together from the active interaction of people and music through urban networks created by migration (and diaspora), even if it ended up as part of more familiar networks of urban power in major media corporations.

Plate 9.9 A gleaming new office building stands amidst the slum apartments in downtown Kowloon, Hong Kong, SAR.
(© Bohemian Nomad Picturemakers/CORBIS)

9.4 Conclusion

With the help of Wirth, this chapter set out by trying to identify some key defining characteristics of cities. It has stressed the importance of recognizing the complex, contested and overlapping, internal and external networks of relationships that help to define cities. It is, finally, important to remember that cities cannot simply be explained in terms of those networks (or in terms of the flows that characterize them). They are more than simply nodes in a system. On the contrary, it seems increasingly clear that the density of activities that is concentrated in cities remains of vital importance, not only in sustaining the networks but also for the people who live in them. They continue to be defined by the intensity of the social relationships within them, not only the excitement generated by the juxtaposition of people and groups with different cultures and understandings, but also the possibility of interaction between relatively large concentrations of people who share particular (sometimes specialist) skills, cultures and understandings. The lives of the world's major cities are simultaneously characterized by the experience of closeness alongside extreme processes of social differentiation.

Cities have been seen both as places of hope and places of despair. Amin and Thrift (2002: 157) neatly summarize what they see as the possibilities inherent in the modern city, which is 'so full of unexpected interactions and so continuously in movement that all kinds of small and large spatialities continue to provide resources for political invention as they generate new improvisations and force new forms of ingenuity'. But there is a continuing tension at the core of 'cityness' – between the search for security, the struggle for order and the celebration of difference, the emphasis on opportunity. Cities are, as McLeod and Ward (2002) note, 'spaces of utopia and dystopia'. This tension remains and cannot be wished away: cities are places of division, in which power relations are often expressed in built form and in social and political practices, as well as places within which it is possible to negotiate for alternative futures and to challenge what are sometimes presented as inevitable inequalities and hierarchies.

Learning outcomes

Having read this chapter you should be able to:

- Think geographicaly about cities as socio-spatial entities. Cities are the products of living social relationships, which have clear expressions both internally (as people come together within cities) and in terms of their changing position within wider economic and social networks of connection.
- Understand that the world is becoming increasingly urbanized – most people now live in cities. But, recognize that it is important to understand not only what links them together (what makes it possible to talk of 'cities' as recognizable phenomena) but also what helps to differentiate them.
- Systematically identify and critically explore key defining characteristics of cities, building on features drawn from Wirth (for whom cities are defined as being large, densely populated, permanent and socially heterogeneous).
- Recognize the implications of social heterogeneity and diversity for the ways in which people live their lives in cities, including the consequences of the existence of increasingly diverse and overlapping social worlds.

Further reading

Amin, A. and Thrift, N. (2002) ***Cities: Reimagining the Urban***, Polity Press, Cambridge. This is an attempt to open up new ways of thinking about the city, reimagining their potential and reflecting the complexities of urban life.

Bridge, G. and Watson, S. (eds) (2000) ***A Companion to the City***, Basil Blackwell, Oxford. A multidisciplinary collection of contemporary writings, which draws its examples from across the globe, and introduces a wide range of ideas about the city without seeking to impose any single approach on the reader.

Dear, M. (2000) ***The Postmodern Urban Condition***, Blackwell, Malden, MA. A clear exposition of the LA school as a way of thinking about contemporary cities, which might usefully be read alongside **Le Galés, P.** (2002) ***European Cities: Social Conflicts and Governance***. Oxford University Press, Oxford, which actively celebrates an alternative vision of European cities, with an emphasis on the importance of continuity and the role of embedded local elites.

Massey, D., Allen, J. and Pile, S. (eds) (1999) ***City Worlds***, Routledge, London. This provides a relatively brief (185-page) introduction to geographical ways of thinking about cities. It highlights some of the tensions and ambiguities of urban life, exploring both the problems and potential of cities.

Robinson, J. (2005) ***Ordinary Cities: Between Modernity and Development***, Routledge, London. This works well as a companion to the previous book. While sympathetic to approaches that seek to locate cities within a global system, it is also a powerful call to shift the focus of urban studies so that it is possible to incorporate the experience of a wider set of cities and to explore the ways in which networks of social relations overlap in cities.

Taylor, P. (2004) ***World City Network: A Global Urban Analysis***. Routledge, London. A coherent and comprehensive approach to the study of cities which locates them as part of a world system.

Useful website

www.lboro.ac.uk/gawc/ A wealth of data and reflections on the external relations of world global cities.

For annotated, clickable weblinks and useful tutorials full of practical advice on how to improve your study skills, visit this book's website at **www.pearsoned.co.uk/daniels**

SOCIAL INEQUALITIES AND SPATIAL EXCLUSIONS

Chapter 10

Phil Hubbard

Topics covered

- Studying society and space
- Social exclusion, poverty and marginalization
- Stereotyping, representation and the exclusionary urge
- Racial exclusions: real and imagined geographies of the 'ghetto'
- Gentrification as purification strategy

All societies are riddled by inequalities, some of which are more fundamental than others (for example, a person's skin colour, age, religion or income typically plays a critical role in shaping their status in contemporary society, whereas their hair colour or shoe size tends to be less significant). Quite why some social differences matter more than others is a key debate in the social sciences – one to which geographers are contributing in a variety of ways. One specific contribution here involves showing that social difference is observable in the spaces around us, with particular social identities becoming dominant in particular places, often to the detriment of others. In short, geographers have demonstrated there is a definite, if complex, relationship between society and space, with social order bequeathing spatial order, and *vice versa.* This relationship constitutes the primary focus of *social geography,* a sub-branch of the discipline concerned with elucidating the spatial imprints of social relations and, conversely, demonstrating the centrality of space in social life.

This chapter explores some of the key concepts in contemporary social geography. It does this by focusing on the geographies of the least powerful and advantaged in society, and their relationship with those who wield more influence and power. As this chapter will demonstrate, one of the clearest ways in which social disadvantage is expressed is through its geographical manifestations, particularly in the existence of spaces described as being 'on the margins' – such as 'slums', 'problem estates' or 'ghettos'. Such areas are a focus of simultaneous fascination and fear for most members of society (and many social geographers) because they represent spatial concentrations of those who fail to match mainstream ideas of how people should live and work. Even those who have never visited such spaces tend to be acutely aware of them, often regarding them as 'no-go' areas. As this chapter will demonstrate, the **stigma** surrounding these places compounds their poverty and isolation, as once a place is labelled as degraded and deprived, its residents find it hard to be accepted by mainstream society. Underlining the fact that space plays an *active* role in shaping people's standard of living, this chapter's focus on geographies of exclusion thus allows us to demonstrate why a sensitivity to place matters, and to illustrate some of the ways social geographers contribute to a continuing tradition of socially relevant and vital research.

10.1 Poverty and marginalization

While social geography has always had a strong interest in mapping social groups and their location in space, for many social geographers, this is a means to an end: namely, highlighting the importance of spatial processes in perpetuating social inequality. Social geography hence circles around questions of the **quality of life** that people experience in different spaces. In simple terms, quality of life can be defined as the extent to which our needs and desires (be they social, psychological or physiological) are fulfilled. Though this can only truly be measured through subjective assessments of the extent to which individuals are satisfied with their life, a plethora of studies by geographers have reached consensus that it is easier to attain an acceptable or good quality of life in areas that offer ample economic opportunities, good social facilities, and a quality environment. Given this, we can see that our life chances are closely related to our place of residence. One widely noted aspect of this is the so-called 'postcode lottery' around health services in the United Kingdom: depending on where you live, you will enjoy very different levels of medical provision through the National Health Service. Where you are born, and where you live, has a major bearing on where (and how) you die.

The same is true of prosperity. Geographical studies of wealth and income suggest that we should not just think about places with concentrations of rich people, but should recognize the existence of *rich places.* This distinction is subtle but significant, for it suggests that wealth is amplified in some places. To elaborate: people who are born in places characterized by high concentrations of affluence are more likely to be wealthy themselves because the concentration of money in the area means that the standards of education, health care and employment are relatively high. While house prices are high, housing standards are good, and may be an important source of inherited wealth. As Dorling (2004) notes, there is a circularity here whereby affluence breeds affluence.

Surprisingly, geographers have said little about geographies of affluence, despite the obvious relationship that exists between income and quality of life (Beaverstock et al. 2004). In contrast, much has been said about geographies of poverty, with social geographers often drawn towards this topic because they feel their relatively privileged position might allow them to draw attention to – and perhaps improve – the poor life chances of those living in poor areas. Poverty can be defined as the condition where individuals or households are unable to afford what might be perceived to be the normal necessities of life. Inevitably, judgments of what we need to consume to survive vary across time and space, meaning geographers are generally more interested in *relative* rather than *absolute* poverty. On a worldwide scale, for example, it is evident that indicators such as average income per capita provide only a very superficial insight into experiences of poverty in different nation-states (see Chapter 8, pp. 181–184). Consequently, recent attempts to measure poverty have generally defined people as being in poverty if they lack the financial resources needed to obtain the living conditions that are customary in the society to which they belong. For example, one study suggested 26 per cent of the British population were living in poverty in 1999, on the basis that their income would not secure at least three of the 35 items considered by at least 50 per cent of the British population as necessary for maintaining an acceptable standard of living (see Table 10.1). While some of these items may not have been considered a necessity in previous decades (and others remain luxuries that many in the global South can only dream of), comparison of these figures with previous studies suggests the incidence of poverty has actually risen sharply. In 1983, 14 per cent of British households lacked three or more necessities because they could not afford them, a figure that increased to 21 per cent in 1990 and over 25 per cent in 1999 (Pantazis et al. 1999).

Roughly translated, such figures suggest that nearly 10 million people in Britain cannot afford adequate housing, about 8 million cannot afford one or more essential household goods, and 4 million are not properly fed by today's standards. While such deprivation is particularly pronounced among those who are out of work, many of those in the lower strata of the labour market are also living in poverty (the persistence of low-paid, part-time or precarious forms of employment being significant here, alongside the restructuring of welfare provision for those on low incomes). Coupled with the rapidly rising incomes characteristic of managerial and professional occupations (particularly in 'knowledge-rich' sectors such as law and finance), the consequence is an increasingly polarized society. For example, between 1979 and 1991, the average British income grew by 36 per cent, yet for the bottom tenth of the population it dropped by 14 per cent. While the rate of income polarization slowed in the 1990s – partly because of government policy – the gap between the 'haves' and 'have-nots' continued to grow (Dorling and Rees 2003). By 2001, the average income of the top fifth of households was around 18 times greater than for those in the bottom fifth (£62,900 per year compared with £3,500) (see Figure 10.1).

While some nations do not exhibit the levels of social polarization apparent in the United Kingdom, it is important to note that levels of social inequality evident in many other nations are far in excess of this (see Spicker 2006). Moreover, in each and every case this social inequality is spatially expressed, whether as a division between rich and poor regions, an urban–rural divide or differences in wealth between cities. Yet some of the sharpest contrasts between wealth and poverty are *within* cities, and one of the major contributions social geographers have made to debates surrounding poverty is to draw attention to the existence of areas of acute need in urban areas. According to the United Nations (2001), over one billion of the world's population live in 'slum' urban environments without access to safe water, acceptable sanitation and secure, tenured housing of an acceptable standard. Ninety-five per cent of these concentrated slum environments are found in the 'Third World', and it has accordingly become something of a geographical cliché to juxtapose the assuredly affluent 'Westernized' city centres now typical of many of the world's rapidly expanding cities with the 'slum' dwellings that are often just a stone's throw away. For instance, profiling Jakarta (population 9 million), Cybriwsky and Ford (2001) contrast the city's Golden Triangle of prestigious residential districts (e.g. Cikini, Kuningan and Menteng) with the *rumah liar* ('wild houses') characteristic of its sprawling, chaotic *kampungs*. They thus conclude that Jakarta 'has extraordinary contrasts between the worlds of prosperity and poverty, and significant challenges ahead for continued development as a global metropolis' (Cybriwsky and Ford 2001: 209).

Yet such 'extraordinary' contrasts between landscapes of wealth and poverty are not just restricted to megacities or those cities living with the legacy of colonial rule: for example, most British cities are marked by contrasts between the affluent landscape of the central city and the 'pockets' of deprivation that surround it. Notable here are those peripheral areas of 1960s local authority housing, which, in the 1980s and 1990s, came

Table 10.1 Perception of adult necessities and how many people lack them (all figures show % of UK adult population)

	Percentage of respondents considering items necessary	Percentage of population unable to afford item
Beds and bedding for everyone	95	1
Heating to warm living areas of the home	94	1
Damp-free home	93	6
Visiting friends or family in hospital	92	3
Two meals a day	91	1
Medicines prescribed by doctor	90	1
Refrigerator	89	0.1
Fresh fruit and vegetables daily	86	4
Warm, waterproof coat	85	4
Replace or repair broken electrical goods	85	12
Visits to friends or family	84	2
Celebrations on special occasions (e.g. Christmas)	83	2
Money to keep home in a decent state of decoration	82	14
Visits to school, e.g. sports day	81	2
Attending weddings, funerals	80	3
Meat, fish or vegetarian equivalent every other day	79	3
Insurance of contents of dwelling	79	8
Hobby or leisure activity	78	7
Washing machine	76	1
Collect children from school	75	2
Telephone	71	1
Appropriate clothes for job interviews	69	4
Deep freezer/fridge freezer	68	2
Carpets in living rooms and bedrooms	67	3
Regular savings (of £10 per month)	66	25
Two pairs of all-weather shoes	64	5
Friends or family round for a meal	64	6
A small amount of money to spend on self weekly	59	13
Television	56	1
Roast joint/vegetarian equivalent once a week	56	3
Presents for friends/family once a year	56	3
A holiday once a year (not with relatives)	55	18
Replace worn-out furniture	54	12
Dictionary	53	5
An outfit for social occasions	51	3

Source: Pantazis et al. (1999)

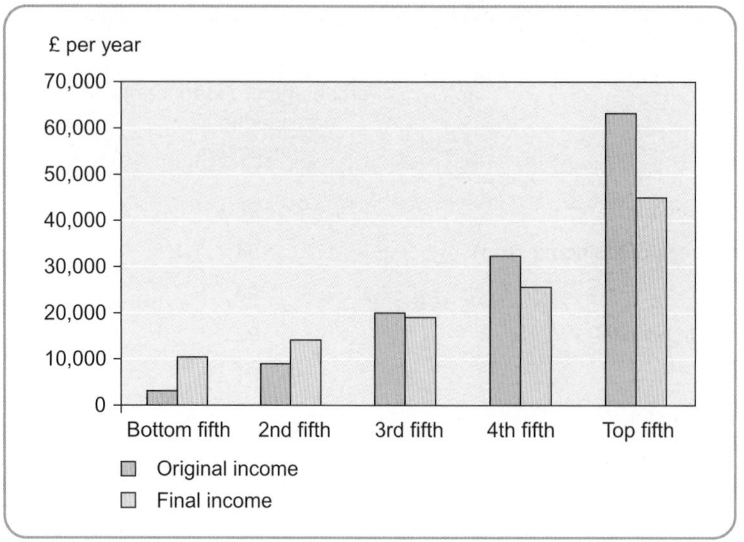

Figure 10.1 Average original and final income per household, UK, 2002–2003.

Source: National Statistics Online (2004)

to be characterized by high rates of economic inactivity. While these are not 'slums' according to the UN's definitions, evidence from 20 of these 'unpopular' housing estates suggests they possess unemployment rates, on average, two and a half times higher than the surrounding urban area (Power and Tunstall 1997), representing pernicious spatial concentrations of relative poverty (Table 10.2).

The particular combination of problems encountered in such deprived estates created national headlines in the 1990s as confrontations between police and local youths (especially joyriders accused of stealing cars) escalated into fully fledged riots. The spatial concentration of unemployment on these estates, especially among young men, has been hypothesized as the most important underlying cause of these disturbances. Figures for the 13 estates affected by serious disorder between 1991 and 1992 suggest that nearly 80 per cent of their occupants were either economically inactive or unemployed (Table 10.2). A simultaneous withdrawal of economic infrastructure – including businesses, shops and banks (Case study 10.1) – meant

Table 10.2 Social characteristics of 12 'riot' estates and 20 'unpopular' estates from 1991 UK Census of Population

	British average (%)	20 'unpopular' housing estates (%)	12 'riot' housing estates (%)
Unemployment	10	34	31
Economic inactivity	36	44	45
Population under-16	19	31	31
Population under-24	33	46	48
Lone parents	4	18	15
Ethnic minority population	6	26	11
Pupils aged 15+ gaining 5+ GCSEs (grade A–E)	43	20	20

Source: A. Power and R. Tunstall, *Dangerous Disorder: Riots and Violent Disturbances in Thirteen Areas of Britain, 1991–92*, Joseph Rowntree Foundation, 1997. Reproduced by permission of the Joseph Rowntree Foundation

Thematic Case Study 10.1

Geographies of institutional disinvestment

During the late 1990s, debates on poverty began to focus on the role institutional disinvestment played in exacerbating the problems encountered by those living in 'poor places'. One dimension of this was the withdrawal of financial services from poorer areas, with as many as 15 per cent of bank branches in the United Kingdom closed down in the early 1990s, the majority in low-income, inner-city areas (Leyshon and Thrift 1994). Another was the closure of once-vibrant neighbourhood shops as retail chains concentrated on the development of large superstores catering to mobile and affluent consumers (Williams and Hubbard 2001). Given that many of these were located in out-of-town and edge-of-town locations, this left many neighbourhoods described as food 'deserts', where a range of affordable and varied food was available only to those who had private transport:

> Food deserts . . . are those areas of cities where cheap, nutritious food is virtually unobtainable. Car-less residents, unable to reach out-of-town supermarkets, depend on the corner shop where prices are high, products are processed and fresh fruit and vegetables are poor or non-existent.
>
> (Laurence 1997)

This has raised serious concerns about the health of those in poor places, with the lack of fresh fruit and vegetables and a reliance on fatty, fast foods implicated in the making of significant health inequalities. Confirming the intensity of food-access problems in some of the large local authority housing estates in British cities, Wrigley et al. (2003) noted significant improvements in diet following the construction of a new superstore in the Seacroft district of Leeds, an area that had previously suffered considerable retail disinvestment. This suggests that large-scale corporate intervention can significantly improve retail access and diet for people living in stigmatized neighbourhoods, albeit that small-scale, community-based initiatives might also provide a solution in some cases (Cummins and MacIntyre 1999).

that there was little money circulating locally. The mythical black market (an informal economy based on petty theft, drug dealing or cash-in-hand trading) did not offer any sort of escape for the majority of residents, given that black economies rarely flourish when there are few people who can afford to buy stolen goods, let alone legitimate ones (Green 1997).

Even so, it has been suggested that the spate of vehicle thefts occurring on such estates was not primarily economically motivated, as Campbell's (1993) vivid account of the Blackbird Leys estate, Oxford, suggests:

> The lads had stopped making cars, but that did not stop them stealing them. The collapse in manufacturing work for men was succeeded in Oxford by the rise in car theft, a crime that was emblematic of the Eighties and Nineties. . . . Car crime on the estate was about much more than trade, however. It was about a relationship between young men, power, machinery, speed and transcendence. . . . It was the spectacle of the displays that vexed and humiliated the police, uniting thieves, drivers and audiences in alliance against the authorities.
>
> (Campbell 1993: 32–3)

The idea that car crime embodies a form of resistance suggests joyriders were not motivated simply by greed or want, but by a more complex desire to reassert their masculinity by confronting authority. Such actions can, however, only be understood in the context of the economic transitions that led to significant changes in the relationship between masculinity, work and domesticity on such estates. Here, Nayek (2003) suggests that the disappearance of a once-familiar **gender division of labour** organized around a male 'breadwinner' (and the expectation of women's domestic role) was the cause of much male anxiety in areas of high unemployment. With the shift to a service-based economy valuing 'feminized' attributes such as keyboard skills and communication proficiency over the robust 'masculine' qualities associated with the culture of manual labour, young men unwilling (or unable) to adapt to demands of service-based work found themselves increasingly cut off from the world of work. Without a regular source of income, or any real incentive to look for work, the traditional relationship between masculinity and spatiality broke down as these young men rarely left the estate for any particular reason. Yet, in the relatively secure space of a stolen car, Massey (1996) argues, masculine identities could be reasserted. Driving with speed – 'on the edge' – thus represented young men's attempt to reaffirm their masculinity, albeit in a manner that created conflict with other estate dwellers (as well as police).

While geographers have stressed that joyriding and other uncivil actions are intimately connected to (gendered) processes of economic restructuring and institutional disinvestment, the media largely ignored such arguments in favour of a rhetoric depicting Britain's peripheral estates as breeding grounds of immorality, characterized by high rates of teenage pregnancy, widespread drug abuse, alcoholism and antisocial behaviour:

> The sense of decline and neglect in many of these areas is palpable: the built environment in many of these areas has now taken on all the classic, ominous characteristics (boarded-up windows, barbed wire surrounds) of the enclaves of high crime and violence associated with Los Angeles and its ghettos in the months leading up to the 1992 riots. Public space is often colonised by young men in baseball caps and cheap khaki.
>
> (Taylor 1997: 6 © Guardian News & Media Ltd 1997)

While there are as many as 2,000 local authority estates in Britain, accounting for around two million people in total (Goodwin 1995), the term 'no-go' estate became a convenient way for the media and public alike to label a wide variety of social environments (only a small number of which experienced unrest). Such labelling exacerbated neighbourhood decline: those who could, moved out, leaving behind only the most vulnerable (Hastings and Dean 2003).

From a parlous position in the 1990s, however, Britain's most unpopular estates improved in terms of employment, education and health, thanks principally to substantial government investment (Tunstall and Coulter 2006). In other instances, problem estates have been abandoned by both state and market, blighted by a debilitating cycle of labelling and marginalization. An example here is provided by the *banlieues* (suburbs) of social housing surrounding Paris and other major French cities. In October 2005, three young men in one of these *banlieues* – Clichy-sous-Bois (NE Paris) – sought to evade a police check (a regular occurrence in the *banlieues*) by hiding in an electricity substation. Two were electrocuted, prompting an anti-police protest which rapidly spread across the country. In over a week of riots, 10,000 cars were torched and 3,000 people arrested. Video footage of burning cars was screened nightly by the world's media, prompting commentators to speculate why such scenes were happening. Whilst some sections of the press suggested these were simply copycat protests in which young men wanted to appear on television, more trenchant critiques suggested that at their source were structural inequalities that needed to be understood in geographical and historical context. For, as Dikeç (2006b) argues, the *banlieues* are not all deprived (and some are reasonably affluent). Yet they have always had negative connotations, identified as '*les quartiers difficiles*' which contain the social problems that have, by and large, been cast out of the central city. At the turn of the millennium, these areas of social housing possessed unemployment rates twice the national average, with 40 per cent of young people unemployed. The fact that some of these areas of social housing were explicitly designed to house North African immigrants in the 1960s is also highly significant here, given that the *banlieue* has always been understood as a 'problem area' (i.e. designed to solve the problem of housing new migrant labour). Writing before the riots, Tissot and Poupeau (2005: 5) argued that when one talks of poverty in France, one can talk of the 'socially excluded', the 'immigrants', or even of 'the young', but it is easier for reports to dwell on the places where it is found: '*les banlieues*'. Post-millennium, and in a context of considerable Islamaphobia in the West, the problem of the *banlieue* was transformed from one where inhabitants of the *banlieue* were a problem for one another to one where they were imagined as threat to the city (and the nation) as a whole.

The connections made here between ethnicity and space are significant, and we will return to these subsequently. At this point, it suffices to say that the white-dominated media in France could not accept that the protest was one in which French citizens were justifiably declaring their anger at the deteriorating housing, lack of transport, and poor schooling that characterized the *banlieue*. Rather, this was depicted as a riot in which 'outsiders' from the margins sought to attack the centre (Dikeç 2006b). Such arguments support Shields's (1991: 5) view that marginal places 'carry the image and stigma of their marginality which becomes indistinguishable from any empirical identity they might have had'. In this sense, it is clear that media stereotypes play a crucial role in creating and perpetuating social and spatial inequalities.

Plate 10.1 *Banlieue* riots.
(Associated Press/PA Photos)

10.2 Social exclusion and cultural stereotypes

The fact that the media exacerbated the deprivation experienced in the *banlieue* and outer city estates by drawing broad-brushed stereotypes of place suggests that when we explore social inequality, we need to think carefully about the way representations of people and place entwine. This involves consideration of **cultural** issues (see Chapter 13), and shifts attention from the economic manifestations of inequality to questions of why some groups become identified as threats or problems for mainstream society. No matter how widespread the rhetoric that we are all created equal has been, historically, it is difficult to identify a society that has not made distinctions between the social mainstream (or norm) and those Others who threaten the coherence of society. The notion of a scapegoat is certainly an ancient one, and history can tell of many groups – e.g. Jews, prostitutes, homosexuals – whose lives have been made intolerable because they are seen to pollute the body politic. Such groups have often been cast out, physically, as well as metaphorically, socially excluded on the basis that they disturb social and spatial orders. In situations where they have been allowed to remain, their occupation of space has often been fiercely contested. In many situations, the socially excluded have been unable or unwilling to occupy the spaces associated with mainstream society, carving out their own geographies on the margins.

Writing in the context of the contemporary urban West, Winchester and White (1988) suggest that socially excluded groups include the unemployed, the impoverished elderly, lone-parent families, ethnic minorities, refugees and asylum seekers, the disabled, illegal immigrants, the homeless, sexual minorities, prostitutes, criminals, drug users and students. If you are reading this book, the chances are that you are a student. If so, the identification of students as an excluded group may be causing you some puzzlement. Even though some students are highly indebted, may come from poor backgrounds or have to take part-time jobs to fund their (supposedly) full-time studies, few live in the conditions of poverty described above. In fact, in the United Kingdom the average student is from the South-east (the most affluent part of the United Kingdom) and has parents who are part of the professional or managerial class (Dorling 2004). This means most students are able to afford most of the adult necessities identified in Table 10.1 (and even when they are not, may be assisted through hardship grants or student loans). So why is it appropriate to describe students as an excluded group?

The answer is that students are identified as a distinctive and different group because their lifestyles are often perceived to lie outside the norms of society, making them infrequent or unwelcome visitors in many of the spaces that are the loci of mainstream social life. Historically, this is related to popular understandings of students as politically radical, embracing alternative fashion and music in a way that marks them out as part of the 'counter-culture'. In the contemporary context, it is also associated with stereotypes of students which suggest they transgress the established social boundaries between work and play (and day and night) by holding late night parties (at times when other people are sleeping) and sleeping all day (when other people are working). Now, as you read this you will no doubt argue that the reality of the situation is a good deal more complex: not all students enjoy drinking, clubbing and partying to all hours – and those that do are often able to do so while devoting considerable time to their studies. Further, while some students remain deeply committed to issues such as environmentalism, animal rights or the restructuring of Third World debt, stereotypes of student radicalism, drug-taking and 'free love' which were consolidated in the late 1960s are now woefully inaccurate as descriptors of contemporary student cultures.

That said, it is possible that you may have experienced exclusion in your own life as a student. No doubt there are parts of the town or city where you study where you are told students are not welcome and where you may have felt 'out of place'. For example, some pubs or clubs let students know, in subtle or not so subtle ways, that they are not welcome. You may even know of instances where students have been verbally abused (or even attacked) just for being students. Thankfully, such instances are rare. In contrast, opposition to **studentification** – the increased student occupation of the local housing market – is becoming very common. This phenomenon is mainly limited to university towns in the United Kingdom where on-campus accommodation is not sufficient to house burgeoning numbers of students and where particular areas become associated with high levels of rental accommodation targeting students (Smith and Holt 2007). Though student occupation can raise house prices (with the buy-to-let market proving extremely lucrative), student housing is often opposed by local residents who deem it responsible for neighbourhood decline. In some areas, the consequences of studentification have been reported to be nothing short of catastrophic, with lobby groups arguing for stronger controls on the licensing of student housing to prevent neighbourhoods becoming 'over-run' by young, transient populations, displacing longer-term residents (UK Universities 2005). Highly dramatized media stories evoke the environmental and social transformation occurring in districts undergoing student expansion, illustrated with images of overgrown gardens, rubbish left out and sheets used as makeshift curtains. Additionally, campaign groups opposing student occupation allege students cause noise and nuisance, displace local long-term residents and fail to contribute to community life (Hubbard 2008).

We can see here that media stereotypes of students and student housing are mutually reinforcing in a variety of ways, and certainly many opponents of student housing allege that students do not look after their house because they are too busy socializing. A counter argument here is that the deleterious state of many student houses is down to the absentee landlords who fail to look after them, and, in any case, not all student households are noisy and disregarding of their

Plate 10.2 A 'typical' student house.

(Janine Wiedel/Photofusion)

neighbours. No matter, for all it takes is one or two instances of antisocial behaviour to reinforce popular stereotypes and instigate campaigns aiming to reduce student occupation in particular areas (such as Storer in Loughbrough, Headingley in Leeds, Elvet in Durham and Selly Oak in Birmingham). In some cases, long-term residents even adopt the metaphor of 'studenticide', arguing that studentification needs to be reversed lest it destroys the urban social fabric: surveys suggest home-owners in UK cities now fear student neighbours more than squatters.

This discussion of studentification is highly relevant to our discussion of social exclusion, given it demonstrates that some groups may experience exclusion and stigmatization irrespective of their income or class. Yet talking about students as a marginal group remains problematic, not least because students exhibit their own exclusive geographies (principally in the form of clusters of student-centred bars and pubs) (Chatterton 1999). Moreover, students occupy their ambivalent social position for a short time only, and most proceed to relatively well-paid employment in time. In other instances, socially excluded groups suffer social stigmatization that is more likely to be associated with long-term financial exclusion or deprivation. For example, Gleeson (1998) suggests that people with a physical impairment or mobility problem typically find themselves excluded from workplaces designed around an able-bodied ideal. Here, the portrayal of the disabled as representing what Shakespeare (1994) describes as the 'imperfect physicality' of human existence is a major factor, with the able-bodied remaining anxious about those who are visibly different. In contrast, students typically move from a marginalized and indebted position to one of relative affluence within a few years as they convert their education and **cultural capital** into financial rewards.

Nonetheless, our consideration of studentification (and the associated development of student 'ghettos') shows us that highly diverse groups can become known through narrow social stereotypes, with these stereotypes shaping the relationship between them and dominant social groups. This suggests that while social exclusion involves the physical exclusion of individuals from the social institutions, rituals and practices enjoyed by dominant groups, it also has a symbolic dimension, being created in the realms of *representation*. Here, the term representation is taken to encompass the wide range of media – such as films, TV, internet sites and newspapers – through which we come to understand the world and our place within it (Woodward 1997). Inevitably, such media provide a partial and simplified view of the lifestyles of heterogeneous social groups. Although most people do not necessarily accept these stereotypes uncritically, they inevitably find themselves drawing on them in their everyday life. What is especially significant about these stereotypes is that they are ideological in nature, in the sense they are generally created by (and in the interests of) dominant social groups – typically white, able-bodied, heterosexual, middle-class men. It is those who do not conform with their view of the world who are society's Others.

A vivid example of how such stereotyped images contribute to social and spatial exclusion is provided by the media reporting of HIV/AIDS from the mid-1980s onwards. In the ensuing **moral panic** that followed diagnosis of the virus and its mode of transmission, a sometimes hysterical media began to focus on the groups most readily identified as at risk – non-white ethnic minorities, intravenous-drug users and gay men (Watney 1987). The presence of HIV infection in these groups was generally perceived to be no accident, but portrayed as a condition affecting those whose 'inner essence' diverged from that of normal society (and whose lifestyles were judged to be incompatible with the maintenance of 'family values'). Hence, sensationalist stories of sexual immorality, irresponsibility and wilful hedonism among those groups most affected by the virus were used to create social barriers between the 'healthy' and those regarded as sexually promiscuous, socially irresponsible and unclean.

Consequently, this symbolic marking of HIV-infected groups as deviant informed their social exclusion, encouraging widespread discrimination, prejudice and neglect. As Wilton (1996) has shown, this had clear spatial effects, with those infected with HIV exhibiting 'diminishing geographies' as their access to the workplace, the home and the street was subject to increasing constraint. In his analysis, Wilton stresses that these constraints were not a product of the physical onset of AIDS but of its social stigma. Symptoms of this stigma have included employers refusing to take on those with HIV, dentists refusing to treat them and, as Wilton relates, community groups opposing the construction of AIDS hospices in their neighbourhood.

Such neighbourhood opposition to community facilities and welfare services is so widespread in Western society, particularly in suburban landscapes, that few stop to question why people might object to such developments in their neighbourhood. Certainly, many NIMBY (Not In My Back Yard) campaigns are fought with reference to the detrimental environmental impacts of such developments (such as noise or air pollution). Such impacts are, to an extent, quantifiable; what is less measurable is the concern that homeowners have about the arrival of stigmatized populations. However, for Takahashi and Dear (1997), community opposition to facilities for those living with AIDS/HIV is indicative of the more general antipathy displayed towards Other populations. Their survey of homeowners in US cities reveals a 'continuum' of acceptance, where facilities for populations depicted as 'different' (such as homes for the elderly) are regarded more favourably than those for populations stereotyped as 'dangerous' or 'deviant' (such as those living with HIV) (Table 10.3). This frequently results in the concentration of facilities in inner-city areas where home-ownership rates are low and community opposition is least vocal (Case study 10.2).

This 'ghettoization' of facilities for those living with HIV, the homeless, asylum seekers and populations dependent on welfare demonstrates that the geographies of marginal groups are, to a lesser or greater extent, the product of dominant **imaginary geographies** casting minorities as 'folk devils' who need to be located elsewhere. As Sibley (1995: 49) contends, this elsewhere might be nowhere, as when the genocide of gypsies and Jews was undertaken by the Nazis, or it might simply be a space 'out of sight' of mainstream populations (such as the red-light districts that are the focus of sex work in many British cities – see Hubbard 1999). In *Geographies of Exclusion*, Sibley (1995) offers a theoretically informed account of how these imaginary geographies fuel exclusionary practices. Drawing particularly on psychoanalytical ideas about the

Table 10.3 Relative acceptability of human service facilities, based on a US survey of 1,326 respondents

Facility type	Mean acceptability, 1 = low, 6 = high
School	4.75
Day care center	4.69
Nursing home for elderly	4.65
Medical clinic for allergies	4.40
Hospital	4.32
Group home for mentally retarded	3.98
Alcohol rehabilitation center	3.80
Homeless shelter	3.73
Drug treatment center	3.61
Group home for mentally disabled	3.51
Group home for people with depression	3.47
Mental health outpatient facility	3.45
Independent apartment for mentally disabled	3.30
Group home for people living with AIDS	3.20

Source: Takahashi and Dear (1997: 83)

Thematic Case Study 10.2

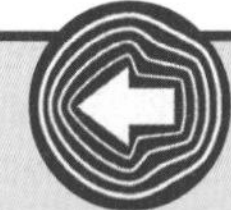

Community opposition to asylum seekers

In accordance with the United Nations' protocol on refugees, the United Kingdom is obliged to offer asylum to those in fear of persecution on the basis of their race, religion or culture. However, in the late 1990s, a mounting backlog of asylum applications, together with media stories of 'bogus' asylum seekers, combined to demonize this group in the public imagination. One strategy adopted by the government as it sought to diffuse concern about the 'wave' of asylum seekers 'flooding' Britain was the dispersal of asylum seekers away from the Channel ports where their visibility had been exploited by right-wing groups intent on exploiting racist fears of difference. As part of this process of planned dispersal, the Home Office proposed the construction of accommodation centres for asylum seekers in a number of rural locales. However, this policy often met with considerable local opposition. For example, in February 2002, residents of Newton in Nottinghamshire learnt that the Home Office was planning to convert nearby RAF Newton into an accommodation centre for asylum seekers. A series of protests

Plate 10.3 Protests against a planned asylum centre, Newton, Nottinghamshire 2002.
(Nottingham Evening Post)

organized by the Newton Action Group expressed vehement opposition to the proposal, with campaigners arguing that the development would cause increases in crime, vandalism and social disorder, and that property prices would fall. One resident claimed that parents would be 'terrified of letting their children play out', while another asked 'Can we be assured there are no child abusers, drug addicts or convicted rapists among them?' (see Plate 10.3). The fact that this is by no means an isolated example demonstrates how effective the media have been at exploiting public anxieties about the threat that asylum seekers pose to the sanctity and purity of the nation. As an extremely vulnerable group, asylum seekers have been less able than some to oppose these racist myths, and many have reported harrowing experiences of discrimination. For example, in the Sighthill area of Glasgow, the arrival of 1,200 Kurdish and Kosovan refugees in 2001 prompted a number of vicious assaults by local white youths who claimed the asylum seekers were being shown favouritism by the local council. Following a series of attacks, it took the murder of 22-year-old Kurdish asylum-seeker Firsat Dag to draw attention to the everyday racism fuelled by a media rhetoric that depicted asylum seekers as 'social security scroungers'.

importance of maintaining self-identity (literally, maintaining the boundaries of the Self), he argues that the urge to exclude threatening Others from one's proximity is connected to ideas about the importance of bodily cleanliness, many of which may be inculcated in early infancy. Developing psychoanalytical ideas, Sibley details how fears of the Self being defiled are consequently projected (or mapped) on to those individuals and groups depicted as polluting or dirty. In turn, Sibley argues that individuals adopt a series of exclusionary strategies designed to retain 'psychic distance' between themselves and **abject** groups, building symbolic, psychological and physical boundaries between Self and Other. This engagement with psychoanalytical theory offers geographers an important insight into processes of marginalization, suggesting that the exclusion of Other groups can only be understood with reference to the deep-seated urge people have to *purify* their surroundings.

10.3 Racial exclusions

The discussion above suggests that while society is highly diverse, some social differences are 'amplified' into significance through practices of representation. An example here is the division of society on racial lines. Like class, age, or religion, race is a **social construct**, the invention of so many discourses and images that suggest that our skin colour matters, and insist that cultural differentiation follows biological differentiation. Race, therefore, is intimately linked to racism. Unpacking the imaginary geographies that cast minorities to the margins, research in social geography has repeatedly exposed the way that media negativity identifies specific ethnic groups as threats to the social body, and hence as a group that needs to be excluded from mainstream space (see Smith 1989; Anderson 1991; Dunn 2001). In the urban West, such racist imaginaries have tended to result in the marginalization of those immigrant groups depicted as a threat to white privilege. Sibley (1999) accordingly argues that the equation of whiteness with purity, order and cleanliness in northern European cultures has been responsible for the creation of negative stereotypes of (for example) Africans, Afro-Caribbeans, people from the Indian subcontinent and Roma gypsies. He contends that each of these groups has consequently been allocated only a marginal role in national cultures because of their 'threatening' nature. In turn, such marginalization fuels spatial segregation, with the creation of ethnic-minority enclaves in the poorest neighbourhoods of many European cities being a product of personal prejudice in the job and housing markets that is fuelled by negative media stereotypes of minority groups (Wacquant 1993). In other contexts, this segregation has taken more institutional forms: for example, under apartheid in South Africa, the Group Areas Act prevented black people from straying into residential areas declared 'white', leaving a legacy of cities that are sharply divided on ethnic lines (see Case study 10.3).

While segregationist policies have long been illegal in the USA, whiteness remains profoundly encoded in the city–suburb distinction that exists there, with the suburbs seen to provide a refuge for the white middle classes from the dirt, disorder, and, above all, criminality of the coloured inner city (Valentine 2001). Indeed, the imagined association between areas of predominantly black residence and the welfare-dependent 'ghetto' has become so pernicious in the USA that it dominates white assumptions about Afro-American lifestyles, creating stereotypes that often bear little resemblance to black people's urban experiences. Events such as the Rodney King riots of 1992 (prompted by the acquittal of a white police officer for the filmed beating of a black Los Angeles resident) consolidated the reputation of the US ghetto as a crime-ridden environment;

Regional Case Study 10.3

Ethnic division in apartheid South Africa

Apartheid-era South Africa is often cited as representing an extreme example of socio-spatial segregation, with a white minority population passing a series of Acts in the mid-twentieth century designed to circumscribe the mobility of coloured, Indian and African people. Under apartheid laws, non-whites were denied access to major cities except if in possession of a pass that indicated they were gainfully employed in the city. As such, while the ruling white minority depicted non-whites as racially inferior, they recognized their importance as a source of unskilled and semi-skilled labour. Initially, non-white workers were housed in barracks or municipal hostels where they were subject to regimentation designed to curb the 'animal instincts' of natives and instill the importance of 'the proper habits of work and life' (Popke 2001: 740). The understanding that non-whites possessed useful and productive bodies, yet were prone to indolence and incivility, was clearly connected to colonial representations that associated non-whiteness with savagery, animality and nature, characteristics that were repressed in 'civilized' Western cultures (Anderson 2000).

However, the continued growth of non-white 'shanty towns' around the periphery of major cities fuelled white fears of '*swaart gevaar*' (black danger), and led to national legislation designed to construct buffer zones between the white city and non-white townships. Both physically and symbolically on the margins of South African cities, the townships thus became abject landscapes, depicted by the white media as spaces of vice, disease and lawlessness. Repeated attempts at creating racially sanitized, planned townships were thus part of an explicit attempt to modernize South African society while maintaining the distance between poorer non-white groups and the affluent white city.

Rather than suggesting that the apartheid city represents a racist distortion of the processes of segregation evident elsewhere in the world, more recent commentaries on the apartheid era suggest it illustrates the racist conceits that underpinned twentieth century modernism (Robinson 1997; Popke 2001). As such, it is possible to identify planning and 'improvement' schemes as underpinned by racist imaginaries throughout the world (and not just in those nations with a colonial history). Indeed, despite the abolition of apartheid Group Area Acts, racist fears of difference continue to fuel socio-spatial exclusions in South Africa, with the growth in gated communities being merely one manifestation of this (Jurgens and Gnad 2002).

although the areas of unrest were typically those with the poorest income levels, high population densities and high school-dropout rates, it was the 'racial character' of these areas that was highlighted in the subsequent reporting (Smith 1995).

This association between US minority populations and specific criminalized places has been cemented over the years in those media that have used 'ghetto' as a coded term for the imagined deviance of black people (especially black men). As McCarthy et al. (1997) argue, the American middle classes tend to learn more about black inner cities through 'long-distanced' media images than through personal everyday interactions. Stories of 'black-on-black' violence abound in the media, with black urban males depicted as the main criminal threat in the USA (Wilson 2005). The cultural responses – in the form of black media practices – have been ambivalent in their outcomes, with gangsta rap accused of banalizing the everyday violence and gun crime occurring in US cities. Even 'new wave' black cinema such as *Menace II Society*, *Boyz 'n the Hood*, *Jungle Fever* and *Straight Outta Brooklyn*, which has sought to show something of the reality of everyday life for Afro-Americans, has ultimately served to stereotype such places as landscapes of endemic violence and drug dependence (Benton 1995). In this context, 'urban' cinema, film and literature can be seen as a corroboration of dominant white myths that imagine drugs, guns and criminality to be part of everyday ghetto life. Such white fears and fantasies about the black inner city are acidly invoked in Wolfe's (1988) fictional account of the Wall Street trader, Sherman McCoy, who one night takes the wrong turning on his drive home only to be confronted with a side of New York that he remains cosseted from in his 'yuppie' housing development:

> At the next corner, he turned – west, he figured – and followed that street a few blocks. There were more low buildings. They might have been garages, they might have been sheds. There were fences with spirals of razor wire on top. But the streets were deserted, which was okay, he told himself. Yet he could feel his heart beating with a nervous twang. Then he turned again. A narrow street with seven or eight storey apartment buildings; no sign of people, not a light in a window. The next block, the same. He turned again, and as he rounded the corner – astonishing. Utterly empty, a vast open terrain . . . here and there were traces of rubble and slag. The earth looked like concrete, except it rolled down this way, and up that . . . the hills and dales of the Bronx . . . reduced to asphalt, concrete and cinders.
>
> (Wolfe 1988: 65)

This depiction of the Bronx as a wasteland inhabited by 'lots of dark faces' again reminds us of the ways in which fears of difference (in this example, based on skin colour) can provoke anxiety and feed the urge to exclude. From the perspective of the white narrator in Wolfe's book, the urban decay of the Bronx mirrors his perceived view of the city's black population as Other: they are street people rather than the 'air people' who inhabit his social world (adopting Raban's, 1990, memorable description of those who can afford to distance themselves from marginal places).

For many Afro-Americans, ghetto life is similarly typified by anxiety, but an anxiety originating from problems of poverty (in 1990 the average white household income was 1.72 times higher than that of black households, with one-third of black families living below the US poverty line). As the renowned feminist and cultural theorist bell hooks relates, the experience of living in the black ghetto is one that consistently reinforces feelings of subordination, marginality and inferiority:

> To be in the margin is to be part of the whole but outside the main body. As black Americans living in a small Kentucky town, the railroad tracks were a daily reminder of our marginality. Across those tracks were paved streets, stores we could not enter and restaurants we could not eat in, and people we could not look directly in the face. Across these tracks was a world we could work in as maids, as janitors, as prostitutes, as long as it was in a service capacity. We could enter that world, but we could not live there. We always had to return to the margin, to cross the tracks, to shacks and abandoned houses on the edge of town.
>
> (hooks 1984: 9)

Until the 1960s it was not uncommon to find specific ordinances and zoning laws controlling the development of black neighbourhoods. Although such laws have long since been overturned, the legacy of these remain in the racist practices of mortgage financiers, banks and estate agents, who may seek to maintain the blackness or whiteness of particular neighbourhoods in the interests of maintaining house prices (Short 1996).

It is thus unsurprising that the Otherness of the Afro-American population is mirrored in their continuing segregation, with black people remaining more spatially concentrated than any other US ethnic minority. However, Dissimilarity Indexes exploring how the Afro-American population is distributed relative to the white population suggest that their segregation has

Table 10.4 The five most segregated cities in the USA, measured by white/Afro-American Dissimilarity Index, 1980–2000

MSA/PMSA name	Year	Total population	Black or Afro-American population	Dissimilarity Index (D)
Detroit, MI	2000	4,441,551	1,045,652	0.846
	1990	4,266,490	942,794	0.874
	1980	4,387,613	892,347	0.874
Milwaukee-Waukesha, WI	2000	1,500,741	245,151	0.818
	1990	1,432,149	197,183	0.826
	1980	1,396,972	150,662	0.839
New York	2000	9,314,235	2,451,277	0.810
	1990	8,546,583	2,249,997	0.813
	1980	8,274,039	1,907,434	0.812
Newark, NJ	2000	2,032,989	473,829	0.801
	1990	1,915,600	424,037	0.825
	1980	1,963,131	409,244	0.827
Chicago, IL	2000	8,272,768	1,602,248	0.797
	1990	7,410,858	1,427,357	0.838
	1980	7,175,269	1,429,217	0.878

1 = complete segregation, 0 = complete integration

Source: US Census Bureau, www.census.gov/hhes

actually declined over the past twenty years (see Table 10.4). Yet the declines remain modest, and some metropolitan areas actually experienced an *increase* in residential segregation over the period 1980–2000 (the seven highest increases all in the southern USA, including Columbus, GA; Goldsboro, NC; Athens, GA; and Danville, VA). Wilson (2005) also notes a deepening of marginalization and poverty within the most ghettoized neighbourhoods, measured through the ratio of high- to low-income residents, with the ratio of rich to poor in Chicago's Englewood and Woodlawn neighbourhoods increasing by 40 per cent and 80 per cent in Cleveland Hough and Fairfax districts between 1990 and 2000. Workfare programmes fail to alleviate the situation, with the poorest forced into dead-end and demeaning employment for minimal wages.

While the disappearance of well-paid work from US inner cities (and the associated out-migration of the black middle class) has been postulated as the major cause of such concentrated black poverty in the USA, Mohan (2000) argues that such factors alone cannot explain the continuing segregation of Afro-Americans. Instead, he suggests it is necessary to consider the wholesale institutional abandonment that has accompanied the economic stigmatization of Afro-American communities. Coining the phrase 'desertification' to describe the planned shrinkage of essential health, welfare and emergency services in the Bronx from the 1970s onwards, Wallace and Wallace (1998, 2000) have suggested that such disinvestment has been part of a deliberate attempt to cut back on public spending by withdrawing services from the very districts most in need. As they detail them, the long-term consequences have often been catastrophic (including increases in infant mortality rates, low birth weights, cirrhosis, TB and AIDS-related deaths). At the same time, many areas stigmatized as pockets of black poverty have become insurance no-go zones where small businesses find it impossible to get insurance because premiums are so high, and large percentages of the population are denied car insurance because of excessive costs. It is for such reasons that some commentators feel justified in speaking of the 'apartheid' of American cities (Wallace and Wallace 1998).

10.4 Gentrification: reclaiming the margins?

As already noted, all societies possess marginal places, whether these are spaces of poverty, sites of illicit activities or areas occupied by stigmatized populations (and perhaps all three). Often, the boundaries of such places have remained consistent over time, even if the individuals and groups they contain have changed (an example here might be the persistence of the inner city as a disordered urban zone). This is clearly connected to the locus of power in society, with powerful social groups able to physically and symbolically claim the centres, marginalizing the less powerful:

> Centre/periphery distinctions tend frequently to be associated with endurance in time. Those who occupy centres establish themselves as having control over resources which allow them to maintain differentials between themselves and those in peripheral regions. The established may employ a variety of forms of social closure to sustain distance from others who are effectively treated as outsiders.
>
> (Giddens 1991: 131)

In this chapter we have already examined numerous processes which tend to spatially isolate and exclude less powerful groups – e.g. discrimination in the housing market, institutional racism, NIMBY politics and so on. Yet there are significant counter processes, with urban policy in particular often prefigured on the idea that marginal places need to be bought back into the mainstream in the interests of economic development as well as social cohesion. These policies are underpinned by an increasingly complex diversity of agencies and partnerships, blurring the distinction between public and private sectors as property developers and policy-actors work in tandem to regenerate deprived areas.

The impact of regeneration is mixed, to say the least. Suffice to say, many policies of regeneration attempt to 'take back' the margins from marginal people, rather than empowering marginal populations, effectively transforming landscapes of poverty and decay into prestigious (and *ordered*) space through selective investment and redevelopment. Hence, the net result of such improvement policy is often the onset of **gentrification**. A much-debated phenomenon in social geography, gentrification is the process by which poor neighbourhoods are transformed by an influx of capital in the form of affluent homebuyers and renters. In some instances, this involves individual gentrifiers buying existing properties and transforming them (often with governmental subsidy); in others, it involves the wholesale replacement of low-income housing with prestigious apartments by corporate developers. Either way, the net result is often one of displacement, with former residents frequently unable to afford to live in redeveloped gentrified areas. Marginalized populations may see their neighbourhoods enjoy a spectacular renaissance, but they often do not share in its subsequent prosperity.

For such reasons, central cities appear to be bifurcating throughout the urban West: either they are being abandoned to the urban poor or, conversely, are being gentrified. While the latter might instigate processes of improvement that have long-term consequences for marginal groups (and the jury is still out on 'trickle-down' urban theory), most geographers remain sanguine about its impacts. Gentrification, it seems, can be conceptualized as yet another form of profit-generation for the wealthy, representing a property wager rather than an investment in people. Indeed, one of the leading urban commentators, Neil Smith (1996), argues that gentrification ultimately only occurs if the gap between actual and potential land value is such that property development becomes lucrative. In such circumstances, the processes of disinvestment and abandonment associated with marginalization become the precursor to a subsequent wave of reinvestment. But for profits to be realized, the Other groups and marginalized land-users which occupy these devalued spaces must be driven out, so that potential developers can view them as safe investment opportunities. It is here that the role of the local state is crucial. Smith (2002: 442) points to this when he details how squatters, the homeless, squeegee merchants and 'street people' were ruthlessly dealt with in New York following the election of Mayor Rudolph Giuliani and appointment of Police Commissioner William Bratton. Espousing a rhetoric of Zero Tolerance for miscreants, these figures were pivotal in identifying the urban disadvantaged as a disorderly population. This urge to tame urban disorder was to trigger notorious police brutality against minorities, justified with reference to the need for improved quality of life, but actually intended, Smith argues, to make particular areas of the city safe for corporate gentrification (and the associated invasion of upper-income groups). One notable example of this process was the wholesale removal of the homeless and their supporters from Tompkins Square Park at a time when developers were seeking to sell the Lower East Side as a space for 'family' residence. The seizure and subsequent 'purification' of the park (and surrounding areas) serves to underline how exclusionary urges may turn a genuinely public space into a space reserved for those who accord with mainstream ideas of living and working (see also Mitchell 1996).

Regional Case Study 10.4

Gated communities in Uruguay

One of the most obvious symptoms of social division in our cities is the presence of gated communities. These communities currently enjoy a widespread popularity because their design promises security as well as a sense of belonging: residents typically sign up to a legally binding code of conduct that ensures all residents understand their rights and responsibilities as members of a community. Physically separated from the wider city, and often offering a privatized form of social life through the provision of enclosed gardens, gymnasia, shops and facilities, gated communities have been postulated as offering a new form of urbanism that may rekindle sociality and encourage diverse peoples to live together. In most cases, however, gated communities have been found to promote social dissociation and sharpen social segregation, being woven into processes of gentrification and urban revanchism by catering to middle-class tastes (and incomes).

In the case of Montevideo (Uruguay), there are still only a handful of gated communities (certainly if compared to its Latin American neighbours). Alvarez-Rivadulla (2007: 51) details that the majority of these are peripheral to the city centre in the Carrasco suburb, and offer a semi-rural setting in which the urban threats of downtown Montevideo are far removed: 'bird song replaces traffic noise, the grey of the city gives way to shades of green'. As is the case in most gated communities, however, the residents are remarkably homogeneous in terms of age, family status and class, as well as sharing certain cultural dispositions (such as their enthusiasm for golf). Their social life tended to be strongly family-oriented, and inward- rather than outward-looking. The occupants were manifestly not part of any global elite, but were significantly more affluent than the average for the city. Significantly, as Alvarez-Rivadulla (2007) notes, this was not a group that had ever mixed with less affluent groups (even if it shared the same urban public spaces and neighbourhoods with them). As such, gated communities reinforced already-existing social tendencies, feeding on fears of difference and myths of urban danger.

Alvarez-Rivadulla (2007) concludes that the reasons that the occupants of Montevideo's gated communities move towards them were not so different from the motivations listed by inner-city gentrifiers in studies in the United Kingdom and the United States: a desire to escape the unpredictability of the city, seek stability and put down roots. Located on the periphery of the city – an area traditionally associated by Montevideo's urban poor – these gated communities have effectively displaced working class and marginal populations in shanty dwellings, and raised land values in the vicinity, exacerbating contrasts between the traditionally affluent neighbourhoods and the neighbourhoods belonging to the urban poor. In a reversal of the trends we noted earlier in the chapter, an archipelago of affluent gated communities can be seen to be emerging within a sea of poverty (as opposed to cities in the United States and the United Kingdom, where we noted the existence of islands of deprivation increasingly dissociated from an otherwise mobile and affluent city).

The vicious **urban revanchism** that Smith documents is a symptom of the deep-rooted fear amongst white middle- and upper-class citizens of Other populations, be they the unemployed, sex workers, the homeless and immigrants. It is very common for mainstream middle-class groups to wax lyrical about urban living, and celebrate the diversity of the city, but all too few seem prepared to leave the suburbs unless it is for a similarly homogeneous gentrified inner-city district. Moreover, these tendencies are not merely isolated to the urban West, and it is easy to find instances of marginalized populations being displaced throughout the global South as shantytowns or informal settlements are bulldozed to make way for new middle-class developments, gated communities, shopping centres and highway developments (see Case study 10.4). Though often resisted, the net result is the corporate gentrification of city centres worldwide, to the extent that the CBD of Manila now looks very much the same as the CBD of Sydney (Winchester et al. 2003). While such processes of corporate gentrification may ultimately attract mobile consumer capital, and tie Third World cities into a network of world cities, the consequence is of course the production of new landscapes of exclusion, with displaced residents and workers being consigned to the marginal spaces that exist beyond the gentrified core.

The prognosis here for marginal groups is not good, as some of the main policies advocated in their name seem destined to serve interests other than their own.

But while such practices of regeneration tend to carve up space in favour of powerful middle-class elites, it is dangerous to suggest that poorer groups are unable to resist these processes. For example, some excluded groups may well resist forms of social closure by **transgressing** into the spaces of the powerful, challenging taken-for-granted expectations about where they should locate (Cresswell 1996). Such transgressions may trigger a moral panic (such as that which surrounded asylum seekers in the late 1990s or people living with HIV in the 1980s), encouraging new forms of social and spatial control. This control is often underpinned by the use of police power, as was the case with the French *banlieue* riots, where curfews were rigorously enforced in affected areas. In such cases, practices of spatial ordering and surveillance quickly reassert the social order, and remind people of their place. However, transgressions may also set in motion social changes by challenging assumptions about who or what belongs where. Parades, sit-ins, strikes, squats: all enact an opposition that, however fleeting, may bring about a change in social attitudes. For instance, gay pride marches in Western cities have often drawn attention to homophobia in society, and over time have encouraged the repeal of discriminatory legislation; perhaps less successfully, Muslim-identifying populations have pursued a variety of public actions designed to topple the Islamophobia that is rampant in the Western nations.

Developing these ideas on transgression and resistance, geographers have hence conceptualized places on the margin as sites from which the relatively powerless can organize themselves into self-supporting cultures of resistance and cooperation. For example, so-called 'gay ghetto' areas in the USA (e.g. Castro in San Francisco, or West Hollywood in Los Angeles) have been transformed from marginal spaces of persecution to relatively affluent centres of gay cultural life through political organization, creativity and activism, creating new gay identities in the process (Forest 1995). Likewise, alternative economies thrive in many racialized areas (particularly in the cultural industries – food, music, fashion, arts and media), allowing ethnic entrepreneurs to bring different values and ideas to the attention of wider audiences, making them more mainstream in the process. This gradual **demarginalization** process may, over time, bring excluded populations into the mainstream.

10.5 Conclusion

Social inequality is just one of many issues explored by social geographers. Nonetheless, it is one of the most important issues facing society in the twenty-first century. As we have seen, all societies possess their 'wild' and 'untamed' zones – ghettos, slums, *banlieues*, *favelas* – as well as less obvious sites of marginalization and exclusion. As such, research on residential segregation (e.g. Wacquant 2007), structural changes in urban retailing (e.g. Williams and Hubbard 2001), inequalities in health and well-being (e.g. Dorling and Shaw 2000), income polarization (e.g. Pinch 1994) and geographies of crime (e.g. Fyfe 2004) provides valuable insights into the particular combination of problems that beset our least well-off and most vulnerable populations. In many cases, such research also helps identify solutions to the problems faced by those living in the city's marginal and shunned spaces, noting the limitations of regeneration-fixated solutions that instigate gentrification. But work on geographies of social inequality does not just make a *practical* contribution to the policy debates surrounding exclusion and polarization; it also makes a key contribution to *theoretical* debates about the role of social processes in constructing categories of identity and difference (such as class, gender, sexuality and race). In the final analysis, concepts such as exclusion and resistance are not solely of relevance to the issues of social inequality discussed in this chapter: they are fundamental to making sense of the diversity that characterizes the contemporary world.

Learning outcomes

Having read this chapter, you should be able to:

- Understand that society and space are divided in ways that segregate stigmatized 'Other' groups from the mainstream.
- Appreciate the complexity of the processes encouraging this segregation, including people's desire to distance themselves psychically and physically from populations represented by the media as threatening or polluting.
- Identify the key characteristics of spaces of exclusion, which may lack economic, social or political infrastructure, and be typified by high crime rates, health problems and a poor quality of life.

- Recognize the importance of spatial stigmatization in creating a downward spiral of decline; once a place has obtained a reputation as a space of exclusion, it is unlikely to attract investment of a type that will benefit its inhabitants.
- Understand that efforts to reclaim spaces of exclusion in the name of improvement often instigate gentrification – an ambivalent process that often triggers new forms of spatial purification and exclusion.

Further reading

Caldeira, T. (2001) ***City of Walls: Crime, Segregation and Citizenship in São Paulo***, University of California Press, Berkeley. A fascinating account of the fear and anxiety that divides this Brazilian city, which makes frequent comparisons with the forms of racism and hate evident in North American cities.

Cresswell, T. (1996) ***In Place/Out of Place***, University of Minnesota Press, Minneapolis. The chapters of this lucid and lively book focus on the way that social orders are challenged by people acting 'out of place' (and how authority responds).

Dikeç, M. (2007) ***Badlands of the Republic: Space, Politics, and French Urban Policy*** Oxford, Blackwell, RGS/IBG Book Series. Develops ideas about the social and political imaginations that have marginalized French immigrant populations in the *banlieues* (the 'badlands' of his title), offering significant insight into the 2005 urban unrest in French cities.

Dorling, D. (2004) ***The Human Geography of the UK***, Sage, London. A student-centred overview of the widely diverging social and economic circumstances which characterize different regions of the United Kingdom, based on a rigorous census analysis.

Hamnett, C. (2003) ***Unequal City: London in the Global Arena***, Routledge, London. A detailed examination of the economic and social changes that have taken place in London over the past forty years, noting significant spatial shifts in occupational structure and income. Though generalizing from a world city is dangerous, Hamnett's case study makes a number of important points about the geographical imprints of polarization.

Holloway, L. and Hubbard, P. (2001) ***People and Place: The Extraordinary Geographies of Everyday Life***, Prentice Hall, London. Chapters 6, 7 and 8 of this textbook offer a fuller discussion of the issues discussed in this chapter.

Mitchell, D. (2003) ***The Right to the City: Social Justice and the Fight for Public Space*** New York, Guilford. An impassioned and provocative book which explores the impacts of urban revanchism and privatization on the disenfranchized populations of the USA, especially the street homeless.

Sibley, D. (1995) ***Geographies of Exclusion: Society and Difference in the West***, Routledge, London. A landmark theoretical statement on geographies of exclusion, bringing geographical ideas into dialogue with psychoanalytical literatures. The book is illustrated throughout with examples ranging from the marginalization of gypsies to the exclusion of non-white voices from the spaces of the academy.

Smith, N. (1996) ***The New Urban Frontier: Gentrification and the Revanchist City***, Routledge, London. Teases out many of the links between gentrification and the purification of space, with Smith's discussions of the 'regeneration' of New York having taken on added resonance in the post-September 11 era.

For annotated, clickable weblinks and useful tutorials full of practical advice on how to improve your study skills, visit this book's website at **www.pearsoned.co.uk/daniels**

RURAL WORLDS

Chapter 11

Warwick E. Murray

Topics covered

- Definitions and meanings of 'rural'
- Trajectories and gaps in rural geography as a sub-discipline
- The interlinking of rural change in the global North and South
- Demographic change in rural worlds: depopulation, counterurbanization and rural–urban drift
- Political change in rural worlds: countryside movements and rural resistance
- Cultural change in rural worlds: social constructions of the country, commodification of rural areas, social change
- Economic change in rural worlds: productivist vs. post-productivist agriculture, multifunctional rural areas

Ask somebody what he or she thinks of when you use the term 'rural' and the response you get will be influenced heavily by where that person 'comes from' in both a geographical and social sense. It is an assumption, but somebody in England might mention green fields, rolling hills, neat hedges, black and white houses, cider apples, cosy pubs and village fetes. Notions of what constitutes 'rural' are likely to be much less pleasant in Tonga, for example. There it might be suggested that poor infrastructure, absence of electricity, grinding poverty, chiefly and patriarchal governance systems, vulnerable housing and environmental deterioration characterize such areas (see Plates 11.1 and 11.2 for example). This is not to suggest that all is well with the countryside in the rich world (global North) and ill in the poor world (global South); marginalization and poverty are increasingly recognized to be part of rich-world rural areas, just as there are pockets of privilege in rural areas in poorer countries. Furthermore, within and between the global South and global North there are widely different rural geographical imaginations; in the USA the concept of 'rural' tends to be associated

Plate 11.1 'Shire Cottage' – a rural home in Tarrington, Herefordshire, England.
(Warwick E. Murray)

Plate 11.2 A rural home in Tongatapu Island, Tonga.
(Warwick E. Murray)

with wilderness, whereas in western Europe notions of the rural tend to be more idyllic. In the global South for example, Chilean notions of the rural – of unkempt green fields, horseback riders, unpaved roads, set against the snowy Andes – are very different to what might be imagined in sub-Saharan Africa, where drought and desert might be imagined. The point being made here is that imaginaries, together with the realities of the 'rural' are not homogenous across the globe, varying across time and space in ways that alter how the rural is defined, interacted with, inhabited, interpreted and socially reconstructed.

Much recent work in rural geography, in the Anglo/American/Australasian (**AAA**) tradition at least, has sought to show that there are in fact rural worlds within worlds and that peoples' identity and social position play a role in the way rural spaces are conceptualized and experienced. Furthermore, as already hinted at, the experience of the rural varies widely between the global North and global South, though much more geographical research on the latter is required. Although rural spaces differ vastly across the world, arguably they form interlinked parts of one global complex. As such, as globalization unfolds, the social, economic and environmental shifts that are occurring in rural worlds everywhere do not occur in isolation (Murray 2005). This chapter investigates a range of perspectives on rural worlds from different geographical viewpoints. In doing so, it draws on a wide range of examples from across the globe to paint a picture of the dynamic and increasingly interwoven spaces we call 'rural'.

From the vantage point of highly urbanized societies that tend to characterize the global North (and Latin America in the global South), it is easy to forget that approximately 50 per cent of the world's population lives in rural areas. In both absolute and relative senses the vast majority of these rural dwellers live in the poor world; in the richer countries there are just over 300 million rural inhabitants, representing approximately 26 per cent of total population there, whilst in the South there are over 3 billion rural inhabitants, accounting for just over 50 per cent of the total at present (UN 2006b). As such, it could be argued that rural geography, at the global scale, should be principally concerned with issues of development, poverty and inequality in poorer countries. This chapter seeks balance in this respect, building on examples from both the global North and South. The issues at stake in rural worlds across the world are serious and pressing, and concern billions of people. In the rural global South, poverty and deprivation are high and local environmental degradation is advancing rapidly. In the rural global North social justice concerns and local environmental issues, though arguably less pressing than those in the South, are considerable. The commodification of the countryside in the global North and the use of the rural idyll as a marketing tool to sell 'countryside' products mean that the rural is increasingly important all across society in richer countries. Far from being stagnant or in decline, as is sometimes assumed, the 'rural' is shifting rapidly – and these changes have implications for the whole of human society. This chapter looks first at how we might define the term 'rural'; it then turns to how geographers have grappled with the issues in such places. We then move on to consider changing rural geographies in terms of demographic, political, cultural and economic shifts.

11.1 Words and worlds: what is 'rural'?

What exactly is meant by the term 'rural'? As already discussed above, meanings and definitions vary across cultures and places. At the general level, there are two ways we can approach defining the rural. The first set of approaches might be termed *empirical*, and includes functional approaches including measuring land-use characteristics as well as demographic approaches that involve such things as population density measures. The second set of definitions can be termed *conceptual* and does not use directly quantifiable measures of rurality, drawing instead on **social constructs**, which have to do with how we *imagine* the countryside.

Governments have their own, generally empirical methodologies for delimiting what is meant by rural, which are very important for planning and development purposes. Using a functional definition assumes that there is a rural/urban dichotomy and that the rural is defined by what it is not – that is, not urban. But this rural/urban dichotomy will vary between places greatly, as both the rural and the urban are variously defined. In the Pacific Island nations many of the areas that are defined as urban in official statistics would likely be seen as rural areas to people from the West; such areas often have lush vegetation, urban gardens and relatively low-density populations in most cases. In this sense, using functional as well as demographic and other empirical measures is questionable when intended for use across societies. The point, again, is that defining the rural is not a precise science.

Human geographers realize that the rural and urban in fact overlap in many ways. It is common to refer to peri-urban, rururban and semi-urban areas. In this regard, some have argued that rural and the urban are best thought of as lying on a rural–urban continuum. Overall, it is now recognized that seeking to define the rural and the urban in opposition to each other (what are sometimes called binaries) is not the most useful approach. Both **political economy** and more recent cultural approaches to rural geography emphasize, in various ways, that processes that cut across them construct rural and urban spaces, and whilst the impacts of these processes on the ground will not be equal they are at least experienced in both. This leads to the second means of defining what is termed rural.

Conceptual definitions of rural are important, and not as obscure as they may appear at first. As we will see, economic opportunities, planning, infrastructure development and many other 'concrete' outcomes are predicated on how individuals, collectives and governments imagine the rural. Under such definitions rural is something in the mind – and thus rural 'social' space need not correspond with rural 'geographic' space. More will be said on this issue when we consider how rural geography has itself changed over recent years.

By way of summary, Cloke (2000: 718) defines the 'rural' as:

> Areas which are dominated by extensive land uses such as agriculture or forestry, or by large open spaces of underdeveloped land; which contain small, lower order settlements demonstrating a strong relationship between buildings and extensive landscape, and which *are perceived as rural by most residents.* (Emphasis added)

The emphasis is added above in order to highlight the fact that rural geographers have come to think of 'rurality' as much as a state of mind as one based on a specific configuration of functions. Cloke goes on to argue that while some rural areas are still defined functionally, in those closer to urban centres 'rural is more of a socially constructed and culturally constructed and therefore contested category' (Cloke 2000: 718). This latter point is most relevant to richer countries where the blurring between urban and rural is increasingly pronounced and where cultural commodification of the countryside and the construction of the rural idyll are widespread. As we will see, rurality in the global South generally has very different connotations.

11.2 Changing rural geographies

In order to understand the geographies of rural space, it is useful to review the evolution of rural geography as a sub-discipline. Rural geography is the study of the relationship between humans and the environment in rural areas, the nature of rural localities, economies, societies, cultures and environments and how this varies across space. Trajectories within the sub-discipline vary markedly across the world, but in the Anglo-American-Australasian (AAA) tradition there has been a relatively common path. Within the context of geography in general, rural geography has, until recently, been somewhat neglected. This may be because in richer countries the rural population is a minority. Also, it could be argued that much Western scholarship carries an urban bias, based on the assumption that it is in such areas that social progress is designed and takes place. The past two decades have seen a rise in interest in rural geography in the AAA tradition and rural areas in the global North are seen as spaces where *general* cultural and economic processes can be researched, understood and interpreted. Indeed, rural studies are increasingly interdisciplinary. Rural geographies of the poor world have been treated very differently and have tended to be conflated with development studies: thus there is very little cultural geography of the rural global South.

Early geography paid some attention to rural areas but this declined with the quantitative revolution and the rise of spatial science in the 1960s. Rural geography at this time was essentially agricultural geography, as 'rural' and 'agricultural' were – for good empirical reasons – more explicitly interchangeable at the time. Rural geography made something of a comeback in the 1970s and early 1980s, and during this period the sub-discipline could be described as functionalist, concerned with such issues as rural planning, land use change and urban encroachment. In hindsight, this approach can be seen as 'uncritical' and theory-free. The relevance of such rural geography was questioned in terms of its contribution to broader society in the late 1980s. Some also argued that by focusing on the distinctiveness of the countryside it ignored processes that cut across the increasingly blurred rural–urban divide.

By the early 1990s in the AAA tradition, there was resurgence in rural geography based on **political–economic** concepts. This focused in particular on a critique of the role of the 'restructuring' impacts of

globalization and – although not referred to as such then – **neoliberalism**. This approach saw the application of concepts from neo-**Marxist** and **world systems** perspectives and was focused on how circuits of capital across the world conditioned, and were conditioned by, the nature of rural space in different locations and how the state intervened in such flows. Some such concepts have been applied to agriculture in the South, particularly commodity chain analysis, although not always under the title of rural 'geography'. Themes considered during this period included diversification, environmental change, international food chains and deregulation, as well as attempts, in later forms at least, to bring together social and political issues with economic concerns (Cloke 2005a). The latter included work on commodification, **gentrification**, accessibility and **counterurbanization**. Much of the work in this area, particularly in **agri-food systems** has continued until the present. Indeed one might argue that in New Zealand and Australia, emphasis has remained on the political economy of agriculture and on the impacts of globalization, and has not shifted to cultural interpretations to the extent that it has in the United Kingdom. Furthermore, across the world traditional land use studies remain popular; Chilean rural geography is characterized by land use analysis, functional definitions, and quantification oriented towards planning and poverty alleviation, for example. In this sense it is more meaningful to talk of rural 'geographies' rather than geography (Roche 2002).

The most recent turn in rural geography has been profoundly influenced by the general cultural turn in human geography which itself was stimulated by the shift to **postmodern** ideas in the social sciences. This shift has by no means been homogenous and there are commentators who have argued that this has made our rural geographies less rather than more relevant (see Wilson and Rigg 2003). There can be little doubt that the new cultural geographies of rural areas are far less applicable to the global South, although the extent to which they have been explored there at all is woeful. The cultural approach, then, seeks to break down old binaries and structures, interrogating rurality as a **social construct**, which is participated in by persons of shared cultural, social and moral values. Crucially, the social space of the rural (which can be imagined and thus be located anywhere) need not necessarily overlap with the geographical space of the rural. Thus the new rural geography engages with ideas that have to do with the social construction of categories, and the way they are then represented and reproduced. Together with this there has been a move to explore the rural geographies of those who are 'marginalized', including geographies of rural women, the young, the old and those in poverty, for example. There have also been a number of more contemporary geographies exploring, among other things, rural sexuality, feminism, travellers and those who, because of dominant cultural constructions of the countryside (in the United Kingdom at least), have been severely 'othered' (Little and Leyshon 2003). Finally, there has been a shift in the AAA tradition towards the exploration and promotion of 'alternative' rural and agricultural production networks and sustainable rural livelihoods, including rural collectives and organic projects, for example.

Linking the new cultural rural geography and **political economy** rural studies is an important challenge (Philips 2002; Argent 2002). Furthermore, building in empirical and theoretical input from geographers in areas other than the United Kingdom and the United States is also very important. Most rural dwellers live outside the countries where the rural geography agenda is set, and in conditions that are very different to those found there. In this regard, more research on the rural geography of poorer countries is required (Wilson and Rigg 2003). Overall, the important thing to remember is that numerous traditions in rural geography coexist and that 'paradigms' never neatly succeed one another; talking across these world-views is a challenge for rural geography and geographers.

11.3 Shifting rural worlds

One of the core themes of this chapter is interaction, within rural worlds, between rural and urban spaces, and across continents as rural areas are bound together in global networks. Notwithstanding this, it is useful to subdivide the subsequent discussion into themes. In what follows we will consider geographical change in rural demographies, polities, cultures and economies. Relative weight is placed on the latter two categories, although this does not mean that they are more significant than other areas. Processes that operate in any one of these spheres overlap in various ways with processes in others, and reference to this will be made as we proceed.

Rather than considering the global North and South separately, examples from each and the links between them are emphasized within this framework. The forces that cut across these boundaries can be summarized as modernization (see **modernity**) (including increased urbanization and industrialization), and its latest incarnates, neoliberalism and globalization. These processes fall unevenly in different places, given local histories and respective positions in global networks. As global divisions of labour spread further across the planet and borders become increasingly permeable, rural and agricultural spaces in very different places form constituent parts of an evolving global network of rural spaces. As such, land use and identity and all other aspects of rural change in the North must be seen as intimately tied to change in the South.

11.3.1 Dynamic rural demographies

Rural demographies across the world are highly dynamic. This demographic change has been conditioned, to varying extents, by generally increasing levels of **urbanization**, the shifting economic base of the countryside away from agriculture combined with the rolling out of neoliberalism, and shifting cultures in rural areas. These shifts have been configured in very different ways across space but, in general terms, **depopulation** in the countryside has been an almost universal trend since the industrial revolution in the global North, and since the Second World War in the global South. Table 11.1 shows that whilst absolute world rural population rose from 1.7 billion to over 3.3 billion between 1950 and 2005, in relative terms it declined, falling from 71 per cent to 51.3 per cent of the total global population.

Table 11.1 also shows the breakdown of rural population by region, illustrating that in a proportional sense Africa and Asia are far more 'rural' – with levels of over 60 per cent – than the other global regions (see also Figure 11.1). Looking at regional data at this scale often hides important variations – for example, Oceania includes the highly urbanized countries of Australia and New Zealand, as well as relatively rural Pacific Island countries such as Vanuatu, Fiji and Tonga. Overall, however, it is clear that there is a significant North/South divide in terms of the demographic rural/urban dichotomy; in the richer countries there are just over 300 million rural inhabitants, representing approximately 26 per cent of total population, whilst in the South there are over 3 billion rural inhabitants, accounting for just over 50 per cent of the total population at the present time.

Since the industrial revolution the relative loss of population in the countryside in the West has proceeded as national economies became based first on

Table 11.1 Rural population by region and development grouping, 1950–2030.

	1950		1975		2000		2005		2030(*)	
	Population (million)	%	Population (million)	%	Population (million)	%	Population (million)	%	Population (million)	%
Africa	191	85.3	310	74.6	518	63.6	559	61.7	729	49.3
Asia	1162	83.2	1820	76	2313	62.9	2352	60.2	2236	45.9
Europe	271	49.5	232	34.4	206	28.3	203	27.8	152	21.7
Latin America	97	58	125	38.8	129	24.6	127	22.6	113	15.7
North America	62	36.1	64	26.2	66	20.9	64	19.3	53	13.3
Oceania	5	38	6	28.5	9	29.5	10	29.2	11	26.2
Global North	*390*	*47.9*	*350*	*33.1*	*320*	*26.8*	*310*	*25.9*	*240*	*19.2*
Global South	*1400*	*81.9*	*2210*	*73.1*	*2920*	*59.7*	*3000*	*57.1*	*3050*	*43.9*
World Rural	***1790***	***71***	***2560***	***62.8***	***3240***	***53.3***	***3310***	***51.3***	***3290***	***40.1***
World Total	**2520**		**4070**		**6090**		**6460**		**8200**	

* = projected

Source: based on data from UN (2006b)

Plate 11.3 The monthly market in Chile Chico, a small isolated rural settlement high in the Chilean Andes on the border of Chile and Argentina.
(Warwick E. Murray)

industry and then services, both of which are generally, though not exclusively, located in urban areas. This population loss has commonly been highly selective, often taking the young and sometimes the skilled, leaving depleted labour markets and service provision (see Plate 11.3). Indeed, as noted above, rural populations across the world are both in relative decline and ageing. Rates of urbanization are most rapid by far in the global South, while the level of urbanization remains low compared to the West. The most rapid rates of relative rural population decline, then, are found in poorer regions of the planet, which has significant and generally deleterious consequences for the rural society left behind and implications in terms of how they should be managed (Murray 2001).

There have been some exceptions to the rule of relative rural depopulation and one of the most studied demographic processes in the West with respect to

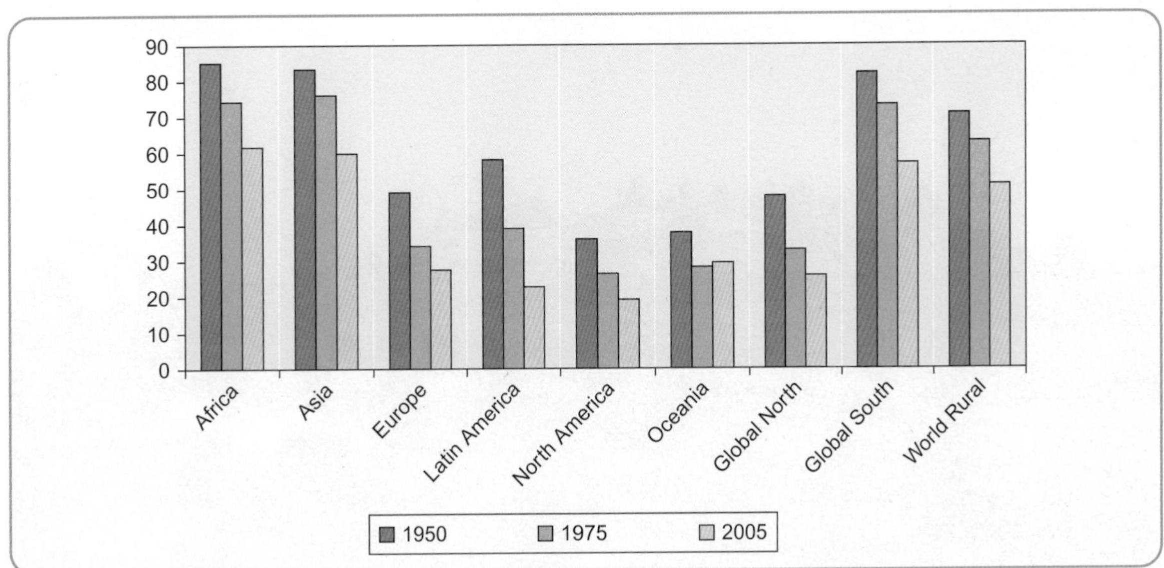

Figure 11.1 Rural population as a proportion of total population by region, 1950–2005.
Source: Based on data from UN (2006b)

rural transitions has been counterurbanization. In the 1970s in the USA for example, fuelled by rising net in-migration smaller towns grew at a more rapid pace than larger ones, reversing the historical trend in urbanization, towards large agglomerations. In the 1980s and early 1990s in the United Kingdom smaller settlements and some rural areas experienced growth whilst inner cities recorded net losses. Hereford, an agricultural centre in the west of England located on the Welsh border, became the fastest growing settlement in the country in the late 1980s, for example. The perceived diseconomies of scale of the city as well as changes in the nature of work and private mobility all combined to stimulate this outcome. Cultural shifts are also present in this demographic change. As people have tired of the diseconomies of scale inherent in urban living, some have sought out the idyll of rural life, closer community, and many other stereotypes that abound regarding the countryside in the global North. In New Zealand and Australia the rise of the 'lifestyle block', an acre or two of land in the countryside where the 'good life' can be practised, within commuting distance from a large city, is a growing phenomenon today (see Plate 11.4).

The universality of counterurbanization in the West has been called into question and some see it as very time- and place-specific: related to a period of manufacturing and service decentralization in Europe and USA that took place in the 1970s and early 1980s. Countries that have experienced periods of growth in the proportion of the total population living in rural areas include Germany (1970–85), United Kingdom (1950–70), Finland (marginally from 1980 onwards) and Australia (1975–90) (UN 2006b). There has been a clustering of relative rural population gain in Central and Eastern Europe since 1990 in places including Latvia, Estonia, Slovakia, Romania, Czech Republic and also Russia, for example. However, in many Western countries – including Spain, Holland, Belgium, Canada, Sweden and New Zealand, for example – counterurbanization has not occurred at all at a national scale. Outside of Europe there are virtually no examples of relative rural population gain over the past 60 years, except in very small countries or small island nations. However, the rural–urban drift has slowed in the global North; the rate of decline in the proportion of people living in rural areas has slowed from rates of 1.35 per cent per annum in the period 1950–55 to 0.68 per cent between the years 2000 and 2005 (UN 2006b). Rural depopulation is undoubtedly the dominant pattern in richer countries, albeit at lower rates than previously, particularly as inner cities are rejuvenated and people appear to be seeking settlement closer to their place of work. Recently, the rise of internet communications has led some to suggest that a new process of counterurbanization could well be imminent, as people can increasingly locate at a distance from their place of employment. Whilst possible in theory, such technology does not obviate the essential human need to meet face to face, and to date there is no evidence that cyber-technology is reversing rural depopulation anywhere on the planet.

Plate 11.4 A rural 'lifestyle' block near Lake Taupo, Taupo, New Zealand.
(Warwick E. Murray)

Although counterurbanization is not a worldwide phenomenon, and arguably represents a fascinating yet ultimately unimportant diversion from the principal trend of rural depopulation, there is no doubt that the social structure of the rural population in the global North is shifting because of demographic change. When one looks below the scale of the whole country, it has often been possible to detect smaller regional units of buoyant rural population growth based on migration patterns. This shift is influenced by urban-to-rural migration among certain groups and changes the nature of rural areas in a variety of ways. Some rural areas, particularly those close to larger urban settlements, have become effectively dormitory zones. When in-migrant populations in the rural United Kingdom – characterized inherently by high mobility – bypass local services such as small shops, local pubs and post-offices, the impact on the socio-cultural nature and economic base of rural areas can be dramatic. The influx of urban migrants in some areas has occasionally led to tensions between the 'indigenous' and in-migrant population, in terms of how rural settlements are governed and managed. Through the 1970s and 1980s Welsh nationalists, in response to the perceived invasion, burnt a number of holiday homes purchased by other British urban dwellers, for example. Interestingly, however, studies in the United Kingdom have shown that social collectives, such as playgroups and village football teams, have involved high participation by in-migrants, clearly keen to illustrate their commitment to their new home and live out the imagined community spirit that some might seek. Cloke (2005b) argues that differing densities of population and scale between countries leads to different impacts with respect to the processes outlined above. In the United Kingdom, for example, rural areas that are subject to immigration effectively become suburbs. This is contrasted to the situation in the USA, Australia and to an extent, New Zealand, where the migratory 'footprint' of any given city will be relatively dispersed. Again what should be clear from this discussion is that there are variable outcomes across space.

The demographic processes at work and outcomes in the rural areas of the global South are, generally speaking, very different to those outlined in the preceding discussion, although the theme of net population loss is common. As already noted this process is generally much more rapid in poorer countries. Between 2000 and 2005, the annual rate of the decline of the total population living in rural areas in the global South was 0.89 per cent, compared to a rate of 0.68 per cent in the global North, and 0.76 per cent in the world as a whole. Furthermore, this depopulation has become increasingly rapid over time, rising from 0.41 per cent per annum in the period 1950–55. Although rural–urban drift has ebbed and flowed in different places at different times, this shift has become one of the defining features of social change in the South over the past decades, leading to the abandonment of rural areas in extreme cases and the explosion of urban populations in receiving areas. Latin America has seen a particularly marked rural depopulation over the past 50 years in proportional terms, for example (see Figure 11.1), as has the Pacific Island region (see Case study 11.1).

A number of factors explain this rural–urban drift in the South, with the weight of each varying from place to place. Generally speaking, the move abroad of Northern transnational corporations beginning in the 1960s, the increasing mobility of capital from the 1980s onwards, and the concomitant rise in industrial employment, together with development theory and policy that see such modernization (i.e. industrial urban development) as beneficial and desirable, have combined to lead to this outcome (Potter et al. 2004). The process has been aggravated by the legacy of unequal landholding structures in rural areas, where peasant farmers often exist side by side with large landowners sometimes descended from colonial elites. The commercialization of agriculture and de-agrarianization of national economies at the same time have often led to fewer opportunities for small-scale producers, as larger players have access to the credit and collateral that is required to purchase the technology in order to compete. Thus smaller producers have either lost their land to become temporary labourers on large farms or have migrated to the city – leading to what is known as proletarianization in the case of the former, and de-peasantization in the case of the latter. The latter outcome has been the overriding consequence. There are instances when this rural population decline has been reversed, as is the case in the Monte Patria *comuna* (district) of northern Chile during the peak years of the grape export boom in the early 1990s, but this is not common.

In short, rural areas in poorer countries have become increasingly less attractive economically and socially and a large-scale drift to the cities has occurred. In some cases this has been coupled with severe environmental decline because of over-exploitation, rapid commercialization or climate change. Severe drought has increased over recent years in the sub-Saharan rural zones, for example, leading to a rise in rural–urban drift. Under such conditions, political circumstances,

Regional Case Study 11.1

Rural depopulation in the Pacific Islands: Niue

Niue provides a good example of rural depopulation in the global South. Niue is a very small island nation with a population under 2,000. Although Alofi, its capital, is defined as an urban area, functionally and demographically it is rural, being low-density and dominated by subsistence agricultural production. From a peak of 5,000, this rural island has been losing population since the 1960s, a process which accelerated when the former colonial power, New Zealand, offered dual citizenship for Niueans on independence in 1974. Owing to a mixture of natural disasters, environmental catastrophes, and misguided economic development policy, the islanders have moved consistently and in very high relative numbers to the cities of New Zealand, especially Auckland, and to a lesser extent to Australia. This loss has its positive side – remittances sent back from Niuean migrants, often second and third generation, form part of the backbone of the country's economy, with over 50 per cent of GDP accounted for by such flows.

Unlike remittance-based migration from some rural areas to urban areas in the South, this case is not circular in its nature: it is largely permanent. However, this loss of population means that Niue is the most rapidly depopulating rural zone and country in the world. The consequences of this demographic shift have been devastating: services in retail and transport have declined as they become uneconomic; whole settlements have been abandoned leaving empty shells where once were houses; and, the country's agriculture suffers from severe labour shortages. After the devastating cyclone Heta in late 2004, some have asked whether the country can survive and at what point it should be declared unviable. The disappearance of the rural society on the island of Niue would deny the world a unique and fascinating culture that contributes to the world's human diversity in an important way.

A similar story is true across much of the Pacific Islands as rural population is lost to the process of both intra- and transnational urbanization (Murray and Terry 2004).

including wars, and neoliberal economic policies such as export orientation, have often aggravated environmental tensions leading to full-scale food shortages and famine. Problems in rural areas in the South are made all the more significant when one considers that despite rapid relative population loss, the largest *absolute* rural populations are found in poor countries and this is set to be the case in the foreseeable future. Table 11.2 shows the largest 20 rural populations by country in 2005, projected into 2030, with the USA, Japan and Russia being the only countries in the global North featuring in the lists.

11.3.2 Dynamic rural polities

Power in the countryside in both the global North and South is shifting considerably in terms of how it is gained and exercised. As such, the governance of rural spaces is increasingly complex and contested. In Western countries rural areas have traditionally been over-represented in parliamentary politics, owing in part to the inherited importance of the landed elite in national affairs of state as well as the desire of some democratic governments to escape arguments of urban bias. Within the countryside itself this created space for various countryside interest groups, alliances, political parties and social movements that draw on particular imaginaries of the countryside to make their case. In New Zealand, for example, a small political party – the Outdoor Recreation Party NZ – won 1.3 per cent (approximately 25,000) votes in the 2002 general election on a platform of 'preserving' rural areas for hunting and fishing as well as broader rural-first policies. The Countryside Alliance in the United Kingdom is another case. With the motto 'love the countryside', the movement was formed in reaction to the single issue of the proposed ban on fox-hunting contained in the Labour Party 2001 election manifesto. The movement soon mushroomed to include a range of disgruntled rural groups who argued that central government was increasingly over-interventionist in rural affairs. The Alliance argues that over 100,000 individuals are members. The movement culminated in a massive march on UK Parliament in October 2002 involving approximately 400,000 people. The imagined conflict between city and country – traditional and modern,

Table 11.2 Twenty largest rural populations by country, 2005 and 2030 (millions).

Country	Rural population 2005	Country	Rural population 2030 (projected)
India	786	India	859
China	784	China	574
Indonesia	116	Pakistan	124
Bangladesh	106	Bangladesh	124
Pakistan	103	Ethiopia	94
Nigeria	68	Indonesia	84
Ethiopia	65	Nigeria	70
Vietnam	62	Vietnam	63
USA	57	Dem. Rep. of Congo	60
Japan	44	Uganda	57
Thailand	43	Egypt	49
Egypt	42	USA	47
Dem. Rep. of Congo	39	Kenya	41
Russia	39	Afghanistan	40
Myanmar	35	Thailand	40
Philippines	31	Tanzania	34
Brazil	29	Japan	32
Tanzania	29	Myanmar	31
Kenya	27	Russia	31
Mexico	26	Nepal	29

Source: UN (2006b)

and 'authentic' and 'fake' – was invoked in speeches during the march. Many groups that are truly marginalized in the countryside, such as travellers for example, were not represented to the same extent as the 'sport' of fox-hunting. This has led some to argue that the Alliance was a smokescreen erected by the elite, utilizing stereotypical interpretations of the countryside, to push a single issue pertinent to them, whilst drawing in support from the rural masses. Casting the particularities of the situation to one side, what is interesting is the way the Alliance used 'imaginaries' of the countryside in order to pursue a specific political point. It is also clear that these imaginaries – of fox-hunting and hounds and red suits on horses – are not necessarily shared, or even considered desirable, by the majority in the countryside in the United Kingdom.

But representations of the rural for political gain are not just made from within the countryside itself: arguments about rural identity are sometimes invoked in order to make broader political gains. Countries in the European Union, especially France, have used arguments concerning the importance of 'rurality' to make the case for the maintenance of subsidies for agriculture, claiming that such funds help sustain the lifestyles and landscapes that make a broader contribution to the character of the European countryside. More recently European governments have couched this argument in terms of the protection of the 'multifunctionality' of the countryside; that is to say sustaining agriculture has positive knock-on effects in other areas (such as environmental preservation, for example) that others in broader society also value. Some in the South see this as a form of protectionism, allowing the continuance of subsidized agriculture in core countries. At the same time global financial institutions insist on the adoption of free-market, non-subsidized, neoliberal policies in poorer countries, a situation which discriminates against millions of inhabitants of rural areas.

In the global South the practice of development (modernization) has often been described as 'urban biased'. The persistent neglect of rural development initiatives in favour of industrial urban policy has been written about widely and consistently, most notably in Africa and Latin America. In the latter, rural zones have been neglected by administrations that have been over-centralized, given the legacy of urban-based Hispanic society, combined with the opening of economies to foreign capital which exploits cheap labour pools in urban areas during the recent phase of neoliberalism. There are numerous examples of rural resistance to the neoliberalization of rural areas and concomitant urban bias (Kay 2001). One of the most notable examples is that of the Zapatistas in Chiapas, Southern Mexico (Murray 2005). In some countries where agro-exports have risen, population outflows have been stemmed, but the effective control of resources still remains in urban areas, resulting in a process of what could be called internal colonialism. Rural dwellers in the South have to endure the double impact of often over-centralized and urban-biased governments, and the economic control of large corporations based in capital cities or, often, outside of the borders of the country itself.

There are examples of resistance to 'modernist' rural production methods in the South by populations in the global North. Participation in alternative food networks, organic supply chains, and moral or fair trade linkages as a form of resistance has become a topic of some significance over the recent past. Some have conceptualized alternative projects as just part of neoliberalism, often controlled by retail power, where alternatives are mainstreamed and conventionalized quickly. Some fair trade NGOs maintain that consumers can play an important role in terms of caring at distance that can make a difference to rural spaces in poorer countries. McCarthy (2006) argues that to date alternative ethical networks have done little to alter existing North–South dynamics and power remains concentrated in the former.

11.3.3 Dynamic rural cultures

Traditional rural cultures are being altered, and in some cases threatened by the shifting relationship between the urban and the rural, although what we mean by 'traditional' varies across the world. As noted previously, the increasing intermingling of urban and rural people and processes raises questions as to what is meant by 'rural' at all. Indeed some have argued that the concept of rurality exists only in the mind and that the 'binary' of rural and urban is a false one. Thus rural *social* space and rural *territorial* space are not necessarily the same thing and 'rurality' is something that has to be deconstructed. Some have even argued that we have reached a 'post-rural' condition in that pure rurality no longer exists. Cultural geographers have been hard at work trying to tease out the way that meanings of rurality are perceived by different actors in the global North and how this affects behaviour, geographical or otherwise.

Postmodernists argue that today's world is a **simulacrum** where symbols or signifiers of an event replace direct experience. Reality is thus experienced through representations of it, creating a hyper-reality that is disembedded from territorial space. It is possible to argue that this has occurred in the case of the perception of the rural by urban consumers who wish to experience the 'rustic' and the relatively 'untouched', which they imagine is associated with the countryside. This might include: wearing country clothes, driving big four-wheel drive all-terrain vehicles around city streets, designing the interior of their homes like country cottages and so on. It could be argued that many people who live in the rural–urban fringe, extend this hyper-reality to their everyday existence. The idea that we are in a post-rural condition probably takes the argument too far; it certainly does so in the case of the global South where such notions are at best not applicable and, at worst, fanciful, irrelevant and elitist. There can be little doubt, however, that the cultural meanings associated with the rural have caused a shift in the relationship between society and geographical space, and between culture and nature.

The foregoing trends have influenced the cultural consumption of the countryside and the allied commodification of the rural experience. In the global North, rural imaginaries are used to sell things. When objects, traits and ideas of a particular culture are brought into capitalist circuits, or when new traits are invented in order to stimulate economic gain, 'cultural commodification' can be said to be occurring. Increasingly, cosmopolitan consumers seek out 'authentic' experiences that will help them differentiate themselves from the masses. There are further examples across the 'cultural' economy – world music, poverty tourism, 'ethnic' food aisles in supermarkets, cultural quarters: Chinatown, Little India and the like – being established in cities as diverse as Singapore, London and Auckland. Although the flow of cultural products from the periphery to the core is predated by imperial and, more recently, the globalized diffusion of westernization, there can be little doubt that there is a tendency towards the consumption of 'the Other' in richer markets (see Chapter 18).

In the case of rural areas, cultural commodification has been especially important. This is most obvious in the advertising world, which uses rural references in order to sell products that have little or nothing to do with the countryside. For example, the advertising campaign for Speights Beer in New Zealand involves images of typical rural South Island men who are tough in outlook and pioneering in spirit. As pioneers they are spurred on by the taste of Speights beer, which allegedly gives them energy, and which they prefer above all 'urban' temptations (such as a box at Eden Park rugby ground, home of the All Blacks, in one advert). The good-humoured rivalry between the North and South Islands of New Zealand, which is a proxy for urban and rural in some ways, is used to sell a product which 'real men' drink . In reality, although headquarters are in the southern city of Dunedin, most Speights beers are brewed in Auckland, a large urban conglomeration much like any other in the global North.

There is also a trend towards 'place-making' in rural areas across the global North, often for the purposes of creating niche tourism. This involves taking a trait, or inventing one, of a locality and using this for economic gain. Ludlow, in the English county of Shropshire, has reinvented itself as one of the principal gourmet destinations in the country based on the coincidental location there of a number of top restaurateurs in the 1990s. This has created a hub of restaurants, cafés, wine bars, and tea shops, which bears little relevance to the 'authentic' history of this small rural town. In New Zealand, Taihape has pronounced itself 'gumboot' (welly) capital of the world, has an annual world gumboot throwing competition, and has erected statues to make this obvious (Plate 11.5)! Towns that have been bypassed by the establishment of large highways, turn this around to their advantage advertising the 'best of rural life' to attract weary travellers, as in the case of Marton, New Zealand. These examples illustrate attempts to use rural imaginaries to create diversified livelihoods in areas where traditional agricultural production patterns are rapidly restructuring. In such efforts culture, invented or otherwise, plays an important economic role.

A phenomenon linked to cultural consumption is the creation of rural spectacle in order to market place (Cloke 2005b). The construction of 'fresh' and 'natural' experiences is important in this respect. Adventure tourism in New Zealand and Australia – including white water rafting, bungee jumping and zorbing, for example – is an interesting example of how rural areas have reconfigured their economic bases. At Kiwi 360 near to Tauranga, New Zealand, it is possible to visit orchards, take part in activities and tour a museum to kiwi fruit production. The multiple connotations of the word 'kiwi' – which refers to a native flightless bird, a fruit, and often used to describe a person of New Zealand citizenship – are clearly used in marketing

Plate 11.5 Taihape, New Zealand – gumboot capital!
(Warwick E. Murray)

Spotlight Box 11.1

The hidden others of the countryside

As mentioned previously, researching the 'hidden others' of the countryside in the West has now become more important in the AAA geographical tradition. This research has revealed a darker underside to the rural idyll where people are excluded because of various identity traits or socio-economic characteristics. Work has been undertaken on travellers, marginalized sexualities, the unemployed, the elderly and the poverty-stricken, for example. Poverty in the countryside in the West can be disturbingly high, and this has been particularly the case as the economic base has rapidly restructured over the past two decades under neoliberalism. In New Zealand, for example, the neoliberal restructuring of the 1980s and 1990s, led to the closure of many small and medium scale dairy reception plants, as the New Zealand Dairy Board (later Fonterra) rationalized its operation geographically. This, together with other rural economic shifts, led to very high levels of poverty in some rural regions such as Taranaki and the East Coast, rural decline and depopulation, and some unintended consequences such as the closure of rural rugby clubs, further undoing the social and cultural fabric of the countryside (Willis 2001). Despite limited state intervention it has proven impossible for many such areas to regain dynamism, and pockets of poverty, crime and work-poor cultures in small rural towns such as Patea and Waverley remain stubbornly present to this day.

One of the ironies is that many recent countryside movements and alliances, such as those discussed in a preceding section, in the global North represent anything but such disadvantaged groups, although the involvement of all classes in often elitist causes is frequently solicited in order to create the impression of a 'rural consensus'.

such rural spectacles. Related to the concept of rural spectacle is the evolution of cultural visits to rural areas. In New Zealand, in the area around Rotorua (central North Island), there are numerous 'authentic' Maori villages that have been developed for tourists to visit and experience indigenous culture. Today, most Maori live in urban areas and their lifestyle is far removed from the hyper-reality depicted in such locations, but such businesses provide an important source of income for those involved. Cultural commodification has the potential to bring economic gains but it can also straightjacket places into fossilized cultural representations, leading to the creation of damaging stereotypes. Notwithstanding this, some have argued that such ventures help maintain traits of Maori culture that might otherwise die out, thereby rejuvenating rural indigenous culture.

In the rural global South cultural issues are very different. Rural cultures are, of course, vibrant and enormously varied. In some cases 'national' culture is rooted in traditional and conservative imaginaries of the rural (see Plate 11.6), although this culture is

Plate 11.6 The annual bareback riding tournament high in the Limarí Valley, Norte Chico, Chile.

(Warwick E. Murray)

sometimes fossilized and invented. Notwithstanding this, very little is known of the rural cultural geography of the South, and geography in the West has tended to ignore it. There is a rich tradition of anthropology of rural areas in the Third World, but this tends to focus on the unusual rather the everyday cultures of the rural poor. In some post-colonial countries rural areas are often home to 'indigenous' cultures, or at least rural areas formed their traditional homes before the onset of rural–urban drift. This raises particular issues with respect to rural-development planning in poorer countries. In Chile, for example, many Mapuche – once the only unconquered indigenous group in Latin America – live in the south of the country in a marginalized economic and social environment. Those Mapuche that have moved to the capital city have often fared no better in the urban context. Far from the buzzing and primate metropolis of Santiago, the rural Mapuche have, until recently at least, been left behind by the economic and political progress made in the country over the past 15 years. In contrast to the global North, for such groups the experience of 'rurality' can be painful and poverty-stricken.

Across the Third World the **peasantry** is declining and this is leading to rapid change in the cultural life of rural dwellers. As neoliberalism unfolds, peasant farmers and other small-scale producers are forced from their land and, as discussed previously, leave for the cities. In Fiji, despite leaving their homes, urban migrants from the countryside still retain much of their rural culture – although it is metamorphosizing. Through the concept of *Vanua* indigenous Fijians are linked to the land; indeed they see themselves as indivisible from the land, and in particular the rural places they hail from. Second- and third-generation urban Fijians who live in the urban agglomeration based in and around Suva (population circa 350,000) might well still consider themselves as 'coming from' Taveuni, Ovalua, Tailevu or some other rural area. Notwithstanding some continuity, there are also significant changes; the influence of urban Fijians that return to the village, temporarily or otherwise, is contested and often bemoaned by those that stay behind, as they bring with them concepts, such as property rights, individualism and capitalism which hitherto were alien in such societies.

The marketization of the countryside in ways that increase tourism and so forth are rarely practised in the global South, although it is possible to find exceptions to this rule. Wine tourism in rural Argentina, for example, is increasingly popular. Similarly there are small rural towns across Africa and Latin America that specialize in handicrafts for tourists. In rural Tonga, women make tapa cloth, beautifully decorated bark cloth which is traditionally used for ornate and ceremonial purposes, for sale at tourist markets in the capital Nuku'alofa. In rural Sarawak (Malaysian Borneo) it is possible for foreign tourists to undertake longhouse visits. However, in general, agricultural production, combined with other diversified – yet often threatened – livelihood forms, are the mainstay of such localities, and culture as a commodity at least plays no role at all.

11.3.4 Dynamic rural economies

In the past the economy of rural localities was essentially agricultural. The terms 'rural' economy and 'agricultural' economy are no longer interchangeable, however. In this section we look at how the rural and the agricultural have decoupled and what has replaced agricultural livelihoods, at least in part. This shift has arisen as the world economy has 'de-agrarianized', involving a large-scale shift away from agricultural to industrial and service activities. Increasingly, and particularly in the North, there is a tendency towards pluriactivity (diversified livelihoods) in rural areas (see Plate 11.7). However, diversification away from agriculture, viewed at the global scale, has been uneven, leading to the creation of very different rural economic landscapes across the world. In order to understand this transition, below we consider 'productivist' agriculture, as well as newer concepts of 'post-productive' and 'multifunctional' rural spaces.

The contemporary agri-food system can be conceptualized as a complex that stretches across continents, linking producers, consumers, supermarket retailers, and government policy makers among other agents, representing nodes within commodity networks. Driving the agri-food system is agribusiness, which is associated with the industrialization and globalization of agriculture (Le Heron 1993). This has led to a shift in the nature of farming in many parts of the world involving a delocalization of activity and the creation of *long networks*, which see production oriented away from local and national markets and towards the global economy. The global agri-food network has not led to the homogenization of rural space. At the same time as we witness the industrialization of agriculture, especially in parts of the global South, we are also seeing the rise of niche, organic and alternative agriculture. Furthermore, in places such as Latin America and South-east Asia where large-scale commercial and

Plate 11.7 Pluriactivity in rural New Zealand – off this stretch of State Highway 1, north of Levin, you can buy petrol, book a river paddle tour and have a cup of tea in an aeroplane!
(Warwick E. Murray)

export-oriented agriculture is increasingly evident, we see the importance of subsistence farming.

Productivist rural economies

Productivist agriculture can be defined as the highly intensive production of a limited collection of primary commodities (McCarthy 2005). The agricultural sector has industrialized and commercialized, eclipsing non-capitalist agriculture. Agribusiness has expanded through vertical and horizontal integration, forming conglomerates that link the field to the supermarket. This has allowed companies to reduce the costs of transacting and internalize the risk of the inherently vulnerable business of farming. Consequently, agricultural sectors have increased in terms of ownership, and a new political economy of agriculture based largely on monopoly capitalism and mass production has emerged. Some people have referred to this as **Fordist** agriculture. Companies such as Monsanto, Nestlé, Fonterra and Bulmers are major players in this new economy.

Given the liberalization of the world economy from the 1970s onwards, agribusiness TNCs have increasingly invested abroad. Cheap labour and land as well as less strictly enforced environmental regulations provide incentives for agribusiness TNCs to diffuse the production component of their operations to poorer countries. This is especially the case in fresh fruit, horticultural and floricultural sectors, resulting in a proliferation of counter-seasonal exports to the Northern markets. As a network, for example, floriculture now generates over US$50 billion in sales per annum, and involves fresh-bloom supply from over eighty countries, most of which is channelled through the Netherlands to third countries in the rich world. Globalization has drawn Third World rural spaces into global capitalist circuits, yielding profound impacts. Export-oriented agriculture destined for the North has, in the case of Latin America, led to the evolution of monocultural localities which are extremely vulnerable economically, socially and environmentally (Murray 2006).

The evolution of agribusiness has profound implications for local rural socio-economies in the global North as well. It has, for example, led to the relative decline of the family farm (Whatmore 1995). Furthermore, agribusiness has altered the rural landscape, removing hedges and practising other policies intended to capture economies of scale. Rural choices have also been impacted by the consolidation of agribusiness (see Case study 11.2).

Post-productive agriculture and rural worlds

We are witnessing the return, in some places, of smaller-scale agriculture, sometimes as a direct resistance to the globalization of agriculture. Simultaneously we can also observe the various uses of rural spaces intended to diversify livelihoods and add 'off farm' components. These can be termed '**post-productive**' rural landscapes where the goals, aside from maximizing agricultural yields, include cultural preservation, optimizing ecological value, and stimulating niche-based agriculture.

Regional Case Study 11.2

Productivist agriculture in the Hereford cider industry

In the 1980s rapid growth in the UK cider industry led to a fundamental restructuring of the Herefordshire economy as the largest global producer of cider – HP Bulmers – began an aggressive campaign to dominate local supply networks and to secure the lucrative national market. The company purchased land from small-scale farmers and consolidated orchards in a way that allowed it to take full advantage of mechanized harvesting. At the same time it employed hundreds of medium-sized farmers on two-yearly contracts in order to shore up supply. Bulmers also took over a number of competing medium-sized cider producers including Symond's (producer of Scrumpy Jack). This granted Bulmers a regional bilateral **monopsony** position (single buyer and seller simultaneously) in the county (and for many miles beyond). Small-scale cider producers were out-competed, and the grubbing up of many old varieties of cider apples in place of the new mass-production bush-stock varieties accelerated the decline of small-scale farmhouse cider producers, thereby eroding diversity and choice.

The transition to post-productive agriculture is linked to the perceived costs and falling productivity of the globalized model and the regulatory crisis it is facing. Government intervention is being rolled back, and large-scale production operating at a loss can no longer be tolerated. There is some evidence to suggest that the main bastion of productivist protectionism, the European Union's Common Agricultural Policy, will be dismantled over the next decade. In this sense the transition to post-productive agriculture is consistent with neoliberalism and cannot be seen as resistance to it. Indeed, if we view the evolution of this part of the system as the consequence of the farming-out of the labour- and land-intensive parts of the agri-food production complex to the periphery, then the rise of post-productivism forms part of the broader evolution of global capitalism.

In contrast to the above, some commentators see the rise of post-productive agriculture as indicative of resistance to globalization, greater concern for the environment and the search for smaller scale, short network, sustainable and alternative food networks. Consumer demand has transformed Western consumption patterns, and food safety and quality issues also motivate elites in the South. This has led to a worldwide boom in organic farming, for example, that has impacted agricultural landscapes in the North and South. Ironically it is potentially easier to establish such sectors in the South. The concept of 'organic' farming in many parts of the Pacific and peripheral Latin America is what has been practised for centuries.

In reality, we face a situation in rural worlds where a mixture of productivist and post-productivist functions exist. At the global scale there is, of course, a division of labour that sees much productivist agriculture located in poorer countries because of favourable locational **comparative advantages**. Niche agriculture is undoubtedly becoming more common in Northern countries. However, this dichotomy blurs a complex reality. In the United Kingdom post-productive activities in the countryside are advanced, but in places such as New Zealand, Australia and the USA there is only limited evidence of a post-productive shift. New Zealand is considered one of the most cutting-edge of the agricultural economies but it is still heavily productivist, and it appears that because of a boom in demand from China large-scale dairy farming is in fact on the rise as more and more land is converted to dairy farming (see Willis 2005). In the global South the evidence for post-productivism is even scarcer. There are some examples, such as the rise of wine tourism in Chile, but in general we are seeing a shift *towards* productivism as neoliberalism is allowed to penetrate localities in the periphery. The rise of **GM** and its diffusion is likely to accentuate this trend. We thus have a significant North–South divide in agriculture, and rural geography might do well to revisit older ideas including **dependency** and **structuralism**, in order to interpret these differentiated outcomes within the one global system.

Wilson and Rigg (2003) have sought to 'test' the applicability of post-productive agriculture outside of the United Kingdom and the European Union. They argue that the concept of post-productivism (see **post-productivist transition**) is flawed because only when it was invented was the term 'productivitism' applied to the preceding period (Roche 2005). Taking an anti-modernization viewpoint they argue that the linearity of an assumed development towards post-productivism is unsound. In their view, 'multifunctionalism' is a much better way of characterizing what is occurring in the South, and arguably the North – and perhaps always has been, as we explore below.

Multifunctional rural worlds?

Post-productivism has been heavily criticized; use of the term is problematic because it makes assumptions about what is meant by 'productivist'. The nature of productivist agriculture has varied, and still does vary across the world; it persists in some places whilst in others it has never arrived, and probably never will. In places such as the Pacific Islands, subsistence agriculture plays a very important role in agricultural production and rural lives, and commercial networks bypass millions of rural dwellers. Although modernist discourses might portray this as backward, the socio-cultural and nutritional role of subsistence and semi-subsistence is critical. Yet agriculture of this nature is being squeezed by globalization, with one of the results being migration to large cities in the Third World. This illustrates the fact that rural transitions are not uniform, and human geography more than many other disciplines should be sensitive to this.

In recognizing the problems associated with the concept of post-productivism, the term 'multifunctional' has become widely employed in rural studies. Wilson argues that the productivist debate has been 'conceptualized from a UK-centric perspective that has largely failed to discuss whether the concept has applicability in Europe or beyond' (2001: 77). Instead he suggests that the use of 'multifunctional agricultural regimes' is a better way to characterize the shifts that are occurring in European rural space. Multifunctionalism refers to the notion that 'rural landscapes typically produce a range of commodity and non-commodity use values simultaneously and that policy ought to try and recognize and protect the entire range of values' (McCarthy 2005: 774). It is argued that this concept is more useful than post-productivism as it offers a positive characterization: it recognizes the continued importance of commodity production and is sensitive to geographical difference.

Multifunctionalism implies that the countryside across the world is used for both productive and post-productive purposes and that the combination of uses, including commodification and conservation, varies from place to place. Whilst this formulation reflects complex rural worlds much more accurately than previous constructions, it is also vulnerable to criticism. In some ways the concept says everything and nothing at the same time. Furthermore, commentators from the South have argued that its use in policy formation in the European Union, for example, is a disguised form of protectionism. Governments, particularly from France, have argued that subsidies to agriculture create positive social externalities by supporting other, often less tangible, functions of the countryside which helps maintain 'rurality'. Meanwhile, Southern countries are expected to open their markets and perform the productivist element of the global agri-food complex. Indeed, it could be argued that there is a move away from multifunctionalism in the South, as monocultural neoliberalized sectors come to dominate.

11.4 Conclusion

'Rural' is a relative term; it shifts across time and space. It is also an enormously diverse category and therefore difficult to make generalizations about. It is increasingly difficult to define rural spaces, be they social or geographical. In both the South and North rural spaces are increasingly politicized, often in ways that move beyond 'normal' electoral politics as new political movements abound. Rural identities have conditioned this political change and have themselves shifted in response to new social, economic and cultural configurations. Environments are threatened in all worlds because of neoliberalism and the rapid shift in the economic exploitation of rural areas. Certain constructions of rural culture are used to sell the countryside and to represent its interests while 'other' cultures remain marginalized or are disappearing. In short, rural spaces are increasingly dynamic demographically politically, culturally, environmentally and economically. Rural geography has had to move swiftly in an attempt to capture the complex changes.

This general shift, at the global level, has been influenced by the unfolding of neoliberalism and globalization. At the world scale, whilst it is true that rural spaces in the global North are becoming increasingly multifunctional and are shifting away from 'Fordist' agriculture, in the South we are seeing an erosion of multifunctionality and pluriactivity. This is certainly the case where neoliberal agriculture penetrates and the diversified livelihood-sustaining activities that have often been built up over decades and centuries are eroded. These are dynamic times for rural spaces and populations all across the world, as the insecurities, socio-economic and environmental, associated with the acceleration of globalization abound. This is especially the case in the South where neoliberal governance runs unrestrained. Although rapid advances have been made in rural geography in and of the West, it is imperative that geographers turn attention to rural spaces of the South, which have so often borne the brunt of the evolution of global capitalism. Furthermore, by shifting focus in this way geographers can contribute to a more balanced and democratic human geography that is relevant to rural society as a whole.

Learning outcomes

Having read this chapter you should be able to:

- Discuss the contested definitions of 'rural' from both a functional and conceptual point of view.
- Account for the evolution of dominant trends in rural geography in the AAA (Anglo-American-Australasian) tradition and how these have influenced interpretations of rural processes.
- Appreciate the impact of neoliberalism and globalization on rural spaces and how change in the global South and global North is linked.
- Consider the evidence for rural depopulation across the world and outline the arguments for counter-urbanization in richer countries.
- Understand shifting rural polities and the nature of the alliances and resistance movements that have evolved over the recent past across the world.
- Interpret the commodification of the 'rural' in richer countries and the contribution of culture to the economic dynamism of the countryside in the global North.
- Comprehend the concepts of 'productivist', 'post-productivist' and 'multifunctional' rural economies, and debate the shifts between the three in various parts of the world.
- Recognize that more research on issues in the rural global South is required if a more democratic rural geography is to be attained.

Further reading

Chambers, R. (1982) ***Rural Development: Putting the Last First***, Longman, London. This is a classic book on rural development in the Third World which argues that rural issues are often misperceived by outsiders and that participatory approaches need to be pursued in order to solve rural poverty.

Cloke, P. and Little, J. (eds) (1997) ***Contested Countryside Cultures***, Routledge, London. This is a useful collection of chapters on rural 'others'.

Cloke, P. J., Marsden, T. and Mooney, P. (eds) (2005) ***Handbook of Rural Studies***, Sage, London. This is a very broad-ranging and complete edited collection of viewpoints from some of the top writers in the field of rural studies that is particularly strong on cultural perspectives.

Robinson, G. (2003) ***Geographies of Agriculture: Globalization, Restructuring and Sustainability,*** Prentice Hall, Harlow. A useful survey.

Roche, M. (2002) Rural geography: searching rural geographies, ***Progress in Human Geography***, 26(6): 823–29. This review article provides an excellent stock-take and discussion of the main trends and major scholarly articles in rural geography through the 1980s and 1990s.

Wilson, G. and Rigg, J. (2003) 'Post-productivist' agricultural regimes and the South: discordant concepts?, ***Progress in Human Geography***, 27(6): 681–707. This article is an excellent critical review of the non-applicablility of post-productivist rural geography concepts to agriculture in developing countries.

Woods, M. (2005) ***Rural Geography,*** Sage, London. A useful way to follow up arguments here, though focuses mainly on the British case.

For annotated, clickable weblinks and useful tutorials full of practical advice on how to improve your study skills, visit this book's website at **www.pearsoned.co.uk/daniels**

SOCIAL CONSTRUCTIONS OF NATURE

Chapter 12

James Evans

Topics covered

- The concept that nature is not a pre-given physical reality, but an idea that is constructed by society
- The politics of representing things, behaviours and landscapes as natural
- Environmental myths
- Science and the construction of human nature
- Nature and environmental change in the media

12.1 Questioning nature

The one thing that is not natural is nature.

(Soper 1995: 7)

What is nature? There is an easy answer. It is the birds and the bees, the plants and the landscapes around us – from the familiar things in our lives, like pet cats and dogs and the park at the end of the road, to more distant things that we see on television, like Amazonian rainforests and Giant Pandas. Nature is the set of things that are *separate* from human society. We have the social world on the one side, with its politics, injustices and cultural achievements, and on the other we have the natural world, a pre-given set of biological entities – the domain of natural scientists – that have no politics and no culture . . . things that are simply *there.*

Albert Einstein once quipped that 'the environment is everything that isn't me'. But is it really this simple? Can we draw lines between ourselves and the surrounding world so easily? The microbes that occupy our gut to the food like fruit and vegetables that we eat every day to sustain and reproduce our bodies suggest that it is hard to draw clear lines between our own bodies and the surrounding environment. Food is itself far from natural. Humans use fertilizers, pesticides and mechanized agriculture to give nature a helping hand. Biologically, food represents the culmination of thousands of years of selective breeding to form more productive breeds of plant and animal. In more recent times humans have created new organisms using genetic modification in biotech laboratories.

Similarly, it is hard to draw a line between human society and nature at larger scales. Since our ancestor *Homo erectus* began using primitive tools some 1.9 million years ago, the history of human society has been a history of environmental transformation, from hunter–gatherers through to the massive levels of urbanization that occurred in the twentieth century (see Chapters 1–3). Human society has tended to transform the surrounding environment for its own benefit, turning the environment into an increasing range of things as its technology improves. Over a hundred years ago Karl Marx tried to understand how humans interact with nature, suggesting that the critical driving force was that of *production.*

> Not only do the objective conditions change in the act of reproduction, e.g. the village becomes a town, the wilderness a cleared field etc., but the producers change, too, in that they bring out new qualities in themselves, develop themselves in production, transform themselves, develop new powers and ideas, new modes of intercourse, new needs and new language.
>
> (Karl Marx 1973, originally 1861, *Notebook V*)

One of the most basic arguments underpinning Marxism is the idea that social development unfolds through the transformation of nature. From the basic work of a peasant clearing a field to the industrial exploitation of oil resources, economic transformation *always* goes hand in hand with the transformation of nature and society.

Taking this argument to its extreme conclusion, Bruce McKibben (1999) argues that it is impossible to find a part of the planet that has not been affected by humans in some way – even the remotest inland Antarctic ice sheets are affected by massively elevated levels of CO_2 in the atmosphere because of industrial emissions. This observation leads him to argue that 'nature', in its traditional sense as something separate from and untouched by human society, has now ended!

But did nature ever really exist? Conservationists talk about 'restoring' landscapes that have been spoilt by human activities to their natural condition. Environmentalists hark back to the natural conditions that existed before humans altered the levels of CO_2 in the atmosphere and biologists talk about how many more species there would be in the absence of humans. But *when* was this pre-human condition? For example, the dominant vegetation in pre-industrial Britain was temperate forest, and what is generally considered to be Britain's 'natural' landscape is constituted by the species that typified this landscape, like oak trees. But if one goes back to the last glaciation some ten thousand years ago when human impact was less apparent, there was no forest, and no temperate species existed in Britain (Birks 1997). Quintessentially British species like oaks did not arrive until the end of the last ice age. Depending on what timescale is used they can be seen as either native to Britain, or an invasive species. When we go back in time no historical cut-off point is any more valid than another as the supposed *natural* state of the environment.

12.1.1 The social construction of nature

So, if nature does not exist as something separate, and never really did, then what is it? Many academics have argued that instead of being a pre-given reality, nature is actually an idea, or a *social construct,* which varies between different groups, places and times. The phrase social construction was coined by Peter Berger and Thomas Luckmann (1966). They argued that humans are essentially social animals, and that meanings and truths are established socially through habits and institutions. For example, things like laws are social constructions. They have no essential existence outside of the society in which people have decided to agree and act as if they exist – they are 'social facts'.

Social constructionism recognizes that individuals and groups actually participate in the creation of their realities through the way in which they perceive things and events. Perception is always subjective and tends to be coloured by previous experiences or social norms, leading to different interpretations between various groups. David Harvey gives a classic example of how this problem of perception can work in relation to nature:

> What's nature? What if I give you a chemical formula and say, how do you feel about your relationship to that? If I then take that chemical formula and represent it to you as a tree, then you relate to it in a different way. Now, what's nature? Is it all the molecules that make up

Plate 12.1 Carbon, tree, forest, wood.
(Photodisc, Inc.; P. Lunnon)

the tree, or is it the tree? The point is that if you see a tree, you'll react to it differently than you would if you saw a bunch of molecules, and if you see a tree in a habitat in a forest with a spotted owl sitting in it you would react very differently than if you just saw a tree.

Harvey (in Banrffalo, 1996)

Plate 12.1 shows four representations of the same 'thing'. It is easy to argue that each is natural, and yet each prompts very different reactions. Perception depends upon individual preferences and experience. For example, a trained chemist will recognize the carbon molecule as constitutive of wood, while a conservationist will recognize that the forest depicted is not just any forest, but a rainforest – the most biodiverse habitat on the planet. Indigenous inhabitants of that rainforest will react with a whole different set of emotions, although they may not recognize their home from an aerial photo. The pile of timber is a resource to be burned, while the solitary tree is not. The way in which each representation is perceived depends on the experience and knowledge that the viewer has, which in turn depends on the society and culture in which he or she lives. A key part of the social construction of nature involves exploring how ideas of nature vary from people to people, place to place, and period to period. To return to the example of trees, early US settlers were fearful of the impenetrable dark forests that dominated the Eastern seaboard, and believed that clearing trees for agriculture was 'God's work'. In stark contrast to the attitudes of today, clearing trees was actually seen as morally good.

The idea of nature is subjective – it means something slightly different to you than it does to me. This matters because claims over nature and how we should behave towards it confer power over landscapes and those people and things in them. As the American environmental historian William Cronon puts it, environmental discourses tend to 'appeal to nature as a stable external source of non-human values against which human actions can be judged' (1996: 26). Constructions of nature are thus *political* – anyone invoking the *authority* of nature implies that they are privileged to speak for nature. Geographers have sought to understand how different social constructions of nature benefit some while harming others (see Case study 12.1).

Thematic Case Study 12.1

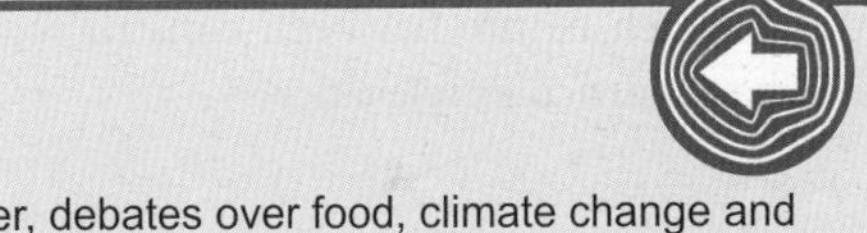

The radical critique of limits to growth

- The idea of natural limits comes from the writings of Thomas Malthus in the late eighteenth century. Malthus was a churchman who deplored the overcrowding and unsavoury conditions in which the newly created working classes of the British industrial revolution lived. In his *Principle of Population* he attributed this problem to a disjuncture between the geometric rate of population increase (1,2,4,8,16 . . .) and the arithmetic rate of food supply (1,2,3,4,5 . . .). Overstepping the limits of natural resources such as food would inevitably result in famine and death. Malthus deplored the moral depravity of the working classes and sought to control their behaviour through invoking the idea of nature in the form of natural limits. Of course Malthus has been proved wrong. Food supply has far out-stripped population growth with the advent of mechanized production and better-yielding varieties. The population has increased almost ten-fold since Malthus's time and a higher percentage of the population enjoy clean water than ever before.
- However, debates over food, climate change and oil still revolve around the idea of natural limits. But are these resources really 'natural'? For example, in what way is it 'natural' to burn oil? Oil did not become a resource until the invention of the internal combustion engine. It will cease to be the most important and fought-over global resource with the proliferation of hydrogen cells and biofuel over the twenty-first century. Oil is only important to individual human existence in the relatively recent era of industrialization, and even then, only if you are one of the fortunate few who either buy it or own it. Far from being a 'natural' resource, the demand for oil depends entirely on the existence of a socio-economic system that values and uses it.
- David Harvey (1974) argues that the real cause of scarcity is the unequal distribution of resources. For example, the richest 1 per cent of the world's adult population owned 40 per cent of global assets in the year 2000. By contrast, the poorest half of the world adult population owned only 1 per cent of global wealth (Davies et al. 2006). As Harvey notes, the earth could not support the current

→

population if everyone enjoyed a Western lifestyle. The idea that resource shortages are due to overstepping 'natural' limits of population conceals these grotesque imbalances. When we look at famines on a global scale, the problem is not food production, but the distribution of the food that we produce. Blaming problems on overpopulation shifts the blame to the poorest people, while protecting the monopoly that the rich nations have over resources. Radical geographers are interested in how constructions of nature serve different political interests. The notion of natural limits clearly serves the interests of those who control the lion's share of the world's wealth.

Q What other environmental debates are framed by the language of limits?

Q Why do people talk about natural limits in preference to social equality?

Q Is there any such thing as the 'basic natural needs' of a human being?

Social constructionists seek to understand two aspects of this process: first, how ideas of nature reproduce the cultural, political and spiritual beliefs of a society, and second, the ways in which ideas of nature are used to support specific political viewpoints. If certain types of behaviour or ways of organizing society can be shown to be 'natural', then it makes it hard to justify any other way of being. The authority to speak for nature becomes the authority to speak for society. Geography has traditionally placed itself at the interface between society and the environment, and the ways in which nature has been constructed and contested have particular relevance.

12.2 Cultural constructions of nature

The great enemy of truth is very often not the lie – deliberate, contrived and dishonest – but the myth – persistent, persuasive and unrealistic.

(J. F. Kennedy)

David Demeritt (1998) identifies the main use of social constructionism within geography as *refuting* established claims about nature. This is perhaps not surprising as social constructions can be understood as a form of 'myth' that may become widely accepted in the absence of any counter evidence. The majority of work in geography on the construction of nature has battled against a variety of myths concerning nature.

12.2.1 Wilderness

One of the most cherished notions of nature in Western societies is that of *wilderness.* Images of wilderness are used widely, to sell holidays, relax us and inspire us. Motivational posters showing 'man conquering the great outdoors' with captions like 'Success' and 'Risk' will be familiar to you if you have ever set foot in a gym. The idea of wilderness as a form of nature untouched by humans has a long history, and underpins many environmental movements. In the USA it has been argued that the idea of wilderness appealed to settlers, who, as they moved West across the country, preferred to believe that the lands they were claiming were devoid of humans and waiting to be inhabited. Exploding what he calls the 'Pristine Myth', Bill Denevan (1992) uses extensive archaeological and historical research to argue that native Amerindians actually managed their landscapes through cutting, burning, terracing and building. These forms of management created the landscapes that European settlers assumed were untouched. America was only a 'discovery' for the white European and, similarly, the plains of the USA only appeared to be uninhabited wildernesses to white European settlers. While the idea of wilderness is obviously related to the inability of the settler to 'see' the human impacts upon unfamiliar landscapes, it had a clear political effect, serving to marginalize the native population.

The idea of wilderness was fundamental to the birth of the modern environmental movement in North America. The first National Park was created by Congress in 1864 in Yosemite Valley. It became seen as a national symbol, representing the spiritual home of the modern nation in the aftermath of the civil war. In 1892 the so-called 'wilderness prophet' John Muir founded the Sierra Club in California, which went on to establish a national wilderness preservation system. Of course, national parks are exactly that – parks – and great effort was needed to preserve these areas as wildernesses. People had to be kept out, animals had to be kept in, views had to be opened up for tourists while simultaneously not compromising the untouched feel of the area. The landscape historian Simon Schama notes that Muir actually praised Yosemite for its resemblance to an artificial landscape garden (1995: 191).

The artificiality of wilderness is taken to its extreme in the designed landscapes of Frederick Olmsted, who is perhaps most famous for designing New York's Central Park. In 1887 Olmsted became involved in the preservation of another symbolic American landmark of natural vigour, Niagara Falls. As William Irwin notes (1996: 77), Olmsted 'believed that parks and nature retreats relieved the festering distress of the poor and working classes'. The area had become overrun with disorganized tourist developments and industry, and Olmsted sought to transform the area into a pure pastoral park for the visitor to enjoy the splendour of nature in peace. In its return to nature, however, the minutest details were planned, including widespread landscaping and the installation of viewing points to maximize the vistas over the falls. The landscape that remains around the falls represents an idealized version of how Olmsted thought the Falls *should* look, rather than a reproduction of the original landscape, which was renowned for hindering the efforts of visitors to see the falls.

Wilderness appeals to the American psyche because it is reminiscent of the birth of their country through settlement, and it remains a powerful and emotive cultural construction of nature in contemporary North America. However, while the idea of wilderness is aesthetic and moral on one level, the politics involved in valuing landscapes that are free of humans are still influential (see Case study 12.2).

Thematic Case Study 12.2

Contesting wilderness

- Clayoquot Sound is a small ocean inlet on Vancouver Island in British Columbia. The surrounding area contains 'old growth' temperate rainforest, one of the rarest habitats in the world. While such forests once covered large parts of Europe and Asia they now comprise 0.01 per cent of the earth's land area. In the early 1990s the logging firm MacMillan Bloedel applied for a logging licence in the area, but was vigorously opposed by the Western Canada Wilderness Society. The geographer Bruce Braun (2002) explored how the resulting dispute played out through a series of representations and counter-representations of the area.
- MacMillan Bloedel produced a brochure titled *Beyond the Cut*, which represented Clayoquot Sound as a natural resource that was needed by the Canadian people. Sticking to the discourse of forest as resource, they presented their credentials as responsible and experienced resource managers who would use scientific management techniques to create a sustainable industry and create much-needed jobs.
- By contrast, the Western Canada Wilderness Society produced a popular and glossily illustrated coffee-table book called *On the Wild Side*. In it they championed Clayoquot Sound as a pristine wilderness, evoking emotive spiritual discourses of the forest as a paradise unspoilt by humans.
- The representations of the forest as economic resource and ecological wilderness reproduced very familiar Western constructions of nature, corresponding to the differing political aims of the logging company and the environmental group. Braun's real insight, however, is that both representations fail to include the indigenous Indians who live in the forest. He argues that this erasure of the indigenous population is a common feature of both the wilderness and resource discourses of nature, and can be traced back to tendency of settlers to appropriate nature from the indigenous peoples. The importance of constructing nature as separate from humans (in this case uninhabited) allows powerful claims to be made over the area by both the logging company *and* the environmental group. The real political import of this process is that the people who live there are excluded and marginalized from their own home.

Q Who wins and who loses in this example?

Q Can you think of other examples when constructions of nature support economic exploitation?

Q How does the idea of wilderness relate to the emergence of environmental thought in the USA?

In Britain the excesses of industrialization and urbanization in the eighteenth and nineteenth centuries generated a cultural backlash that began to value environments that were typified by a lack of human impact. The Romantic movement which emerged at this time was typified by Wordsworth's love of the Lake District and an increasingly negative attitude towards cities. This was in part a reaction to the physical degradation of the environment associated with the Industrial Revolution, but it was also a moral reaction. In opposition to the squalor of the new industrial cities, the countryside became seen as the repository of moral purity. Interestingly, this represented a U-turn in British attitudes to wilder landscapes. In the sixteenth and seventeenth centuries mountainous and uninhabited regions were regarded with fear and disregard. We have reports of travellers drawing the curtains of their carriages to block out the 'Satanic' view as they travelled through mountainous regions. In the 1720s the great diarist Daniel Defoe calls the Lake District a 'barren and frightful place' on his tour of Great Britain. Less than one hundred years later Wordsworth claims of the Lake District 'Who comes not hither ne'er shall know how beautiful the world below'. So if you are one of those people who hate the outdoors and hiking up mountains, simply tell people you have an early Georgian attitude to nature.

As with the US environmental movement, the British Romantic movement established the idea that nature has a spiritual value. In the United Kingdom this happened primarily as a reaction to industrialization, while in the United States it is closely related to the pioneer/settler mentality. In each case the idea of nature has been carefully constructed by societies seeking to preserve a world free of human development.

12.2.2 Landscape

This new relation to nature was expressed through the genre of landscape painting that emerged as a way of seeing in the seventeenth and eighteenth centuries. Landscape takes a three-dimensional world and represents a single view of it in two dimensions, establishing a highly *visual* relation between the viewer and nature that emphasizes aesthetic detachment.

A classic example of how a landscape painting conveys social power is found in John Berger's (1990) analysis of Gainsborough's *Mr and Mrs Andrews*, which was completed c.1750 (Plate 12.2). The painting depicts an upper-class landowning couple in the foreground, with their estate stretching into the distance behind them. While the painting shows a highly cultivated patchwork of fields, there are no workers in this rural landscape. They have been effaced to represent the new balance of social power in the countryside after the Enclosures Acts removed peasants from the land. The landscape is familiar to us as an archetypically *English* landscape, Constable Country if you like, and the nature in it is archetypically English. There are other things going on in this representation too, such as the dominance of the male figure standing over his seated wife, and the symbol of the gun. The symbolism of masculine control over nature could not be more obvious. Landscape painting invented a quintessentially

Plate 12.2 *Mr and Mrs Andrews* by Thomas Gainsborough.
(*Mr and Mrs Andrews*, c.1748–9 (oil on canvas), Gainsborough, Thomas (1727-88)/National Gallery, London, UK/The Bridgeman Art Library)

English aesthetic of landscape, but it also established a new era of control, both of humans over nature and of the landed upper classes over the countryside in Britain.

Representations of landscape are central to national identities. The word 'nature' derives from the Latin *natura*, which comes in turn from *nasci*, to be born. Thus *nature* is linked to other words from the same root, such as nascent, innate, native, and nation. Just as wilderness plays an important symbolic role in the pioneer mentality of the USA, so the 'countryside' landscape is quintessentially English. Authors have explored the importance of the oak tree to British culture, as a sign of strength derived from their use to build the ships that allowed the British navy to rule the seas for 200 years. The historical geographer David Lowenthal (1994) argues that in France the legacy of peasant agriculture has left a national landscape of diverse smallholdings, which is cherished for the variety of cheese, wines and foods that it produces.

These constructions of nature also have profound political effects. Staunch French opposition to reform of the European Common Agricultural Policy, which would increase the exposure of its small producers to market forces, suddenly becomes intelligible as an attempt to preserve the French landscape and way of life rather than simply blunt economic protectionism. The relationship between landscape and national identity was carried to its ideological extreme by the Nazis, who were fond of being photographed in forest settings, and viewed the Black Forest of southern Germany as their spiritual home. They also attempted to eradicate non-native plants in exactly the same way that they attempted to exterminate the non-Aryan human population. As Anne Whiston Spirn notes (1997: 253–4) 'the use of "native" plants and "natural" gardens to represent the Nazi political agenda should dispel forever the illusion of innocence surrounding the words *nature, natural,* and *native* . . .'

The British association of nature with idealized rural landscapes also has impacts that are felt beyond the art gallery. One is the geographical split between the affairs of the city and the affairs of the country. The graffiti artist Banksy produced a series of works exploring the politics of this division, in which he placed typically 'urban' objects such as parking tickets and graffiti into traditional landscape paintings (see Plate 12.3). This juxtaposition disturbs unspoken discourses of the countryside as a somehow 'natural' place untroubled by the problems of cities, and in doing so highlights how exclusive this vision of the countryside actually is. As rural geographers have noted, the discourse of the rural idyll is not particularly helpful as it conceals real problems of poverty and deprivation in rural areas (see Chapter 11).

Plate 12.3 Painting by Banksy.
(Banksy)

Urban areas are often more biodiverse than massive swathes of agricultural land, and require vast infrastructures to control elements such as water and waste (Gandy 2002). In many ways these elements are *more* important in the city, but have simply been ignored until recently, as they do not *look* like nature. The political impacts of this town–country divide in the UK psyche should not be underestimated. The framework for sustainable development in the United Kingdom, which, lest we forget, is supposed to be integrated and holistic, remains split into urban and rural strategies.

12.3 Environmental myths

12.3.1 The power of science

What we observe is not nature itself, but nature exposed to our method of questioning.

(Heisenberg 1958)

Science adheres to a very detailed and rigorous set of rules for building knowledge that accurately reflects the natural world, but in practice science has not been immune to social constructions of nature either. As the biologist Richard Lewontin says (1993: 3), 'Scientists do not begin life as scientists, after all, but as social beings immersed in a family, a state, a productive structure, and they view nature through a lens that has been moulded by their social experience'. However sanitized lab work is made, experiments still require human interaction and judgements to work. Thus the findings and models of science are always influenced by cultural factors. Science is a particularly influential realm in which nature is constructed and contested, because it has the capacity to make very strong truth claims about how reality *is*, and hence how we should live and behave.

12.3.2 Ecology and politics

In the past thirty years, geographers have become increasingly interested in how dominant models of ecology rely on social constructions of nature. The field of political ecology recognizes that ecological knowledge is not always neutral, but reflects the specific aims of those involved in producing it. It can have profound political consequences upon how different landscapes are managed, and for whom. In the words of Paul Robbins (2004: 12), the goal of political ecology is to 'take the hatchet' to environmental myths, using both scientific and social scientific studies to expose the false assumptions and unsuitability of certain ecological models. Bill Deneven's work on the 'Pristine Myth' described in the previous section is an example of this, but the hatchet has also been taken to other scientific myths (see Case study 12.3).

These myths have a political effect. The assumption that environmental degradation is caused by mismanagement generally leads to the imposition of solutions that exclude local people (see Robbins 2004, Chapter 8, for a wealth of examples). This resonates with Bruce Braun's work on Clayoquot Sound discussed in the previous section, where representations of the forest as a wilderness actively excluded the indigenous population living there.

Ecology has been influenced by wider social and political ideas, and it is possible to identify clear traditions in ecological science that reflect the places and periods in which they developed. Directly parroting the language of US settlement, ecologists in the early twentieth century talked about the succession of plant 'communities' from *pioneers*, who come in first and settle an area, through to the stable climax forest. There is a circularity to this process. Models of ecology reflect the social context in which they are produced, but then having become established as fact become models for human society. Some authors have argued that the appeal of 'nature in equilibrium' goes back to the Judaeo-Christian myth of the Garden of Eden, when humans supposedly lived in perfect harmony with nature before sinning and falling from grace. This yearning for a return to innocence often finds expression in the kinds of stories that are told about indigenous peoples, such as Amazonian Indians, who are supposedly still at one with nature (Slater 1995). But the propensity of ecology to exclude humans has effects on conservation practice that are felt today.

12.4 Constructing human nature

There is a long tradition within the sciences of looking directly to the world of animal behaviour in order to establish how humans should behave. Again, this is a game of high stakes; 'the problem for political

Thematic Case Study 12.3

Desertification

Desertification became an important international issue in the 1970s when successive years of drought in the Sahel (the area bordering the southern edge of the Sahara) caused widespread famine. The United Nations responded with the Environment Programme's 1977 conference on desertification, and the subsequent Convention to Combat Desertification, with a focus on desertification as land degradation in arid, semi-arid and dry sub-humid areas. This conference popularized and publicized the term, but was based on little actual science. David Thomas and Nick Middleton (1994) identified four 'myths' of desertification in dryland areas:

1. It is a 'voracious process' affecting one-third of the world's land area
2. Drylands are fragile ecosystems
3. Desertification is the primary cause of human suffering and misery
4. The United Nations (UN) is central to its understanding and solution.

They situate the idea of desertification within a longer history of reports from Westerners in the Sahara concerning advancing deserts. For example, early accounts of desertification originated in the eighteenth and nineteenth centuries when scholars believed the Sahara to have been created by the Romans and Phoenicians through deforestation, overgrazing and over-cultivation. In the twentieth century, colonial land managers clearly replicated the idea that deserts advanced because of the mismanagement and over-exploitation of land. In 1935 E.P. Stebbing, a forester, identified the causes of degradation in British West Africa to be shortened agricultural fallow periods, shifting agriculture and overgrazing, and published his views widely on the 'encroaching Sahara'. It was concluded that the Sahara had grown, and was still growing, owing to poor land management, which had worsened under the colonial regime. The spectre of sand dunes encroaching upon fertile land remained an enthralling one, and a report in 1975 suggested that the Sahara was advancing at the astonishing rate of 5.5 km per year.

Thomas and Middleton question both the pace of this process and the causes. They suggest that biophysical processes have been the primary cause of Saharan advance over the twentieth century, because of the progressive desiccation (drying out) of North Africa since the end of the Pleistocene ice-age 10,000 years ago. The advances and retreats reflect climatic variations in rainfall over tens of years. They then attack the idea that humans are to blame for desertification, arguing that policies that reduce the use of dryland areas are ineffective. 'Expert' knowledges about dryland fragility undermine local adaptive strategies that have evolved over centuries. For example, strategies like lowering livestock densities damage the ability of local farmers to resist drought. The focus on mismanagement obscures both the biophysical causes of dryland degradation and specific social problems, such as firewood scarcity. As Batterbury and Warren note (2001), degradation is usually localized and ephemeral, while it is very hard to overgraze in a dynamic non-equilibrium system, dominated by annual grasses, where the external forces like drought are more powerful than animal numbers as influences on rangeland quality.

The myth of desertification is bound up with the failure of Western scientists to understand the dynamics of dryland areas, simply falling back on received stereotypes of overgrazing and advancing deserts. As Paul Robbins states (2004: 109), the notion of soil erosion was 'a social construction that helped to secure colonial power', and while the term is generally considered unhelpful amongst scientists, it perseveres within the post-colonial policy discourses of international agencies such as the UN and NGOs. Once these versions of nature become established they assume a life of their own, with funding streams for research, policies and the vested interests of those that work within the field, in keeping the concern alive. It is by circulating through networks of scientists, institutions, funding bodies, media and politicians that constructions of nature become accepted as fact.

Q Why do environmental myths like desertification persevere?

Q Who benefits and who loses in these kinds of debates?

Q Why is it hard to produce scientific answers to large scale environmental questions?

philosophers has always been to try to justify their particular view of human nature' (Lewontin 1993: 87). If science can show certain behaviour or modes of social organization to be 'natural', then it is assumed that they are unquestionably right. Think about how often you hear political commentators talk about things like war being 'a part of human nature', or capitalists justify the free market on the grounds that competition and the need to own things are 'in our nature'. It does not take much to spot the flaws in this logic. Animals do not cook, but no one would say cooking food is wrong because it is unnatural (apart from Raw Food advocates, of course). As James Weinrich says, 'When animals do something that we like, we say it is natural. When they do something that we don't like, we call it animalistic' (1982: in Bagemihl 1999: 77).

The attempt to draw parallels between animal and human behaviour is one of the most obvious ways in which nature acts as a mirror for the values of society. Donna Haraway's (1989) work on primatology (the study of apes) presents evidence that major research projects on ape behaviour tend to be biased by the researchers conducting the study. Male research teams tend to demonstrate a tendency to explain ape behaviour in terms of sexual competition between aggressive males to impregnate passive females. Female primatologists focus on different observations that require more communication and basic survival activities, offering very different perspectives on social behaviour than the currently accepted ones. Depending on who is doing the research, different aspects of behaviour will be measured and different inferences drawn from them. Gender and politics become tangled up in the process of scientific enquiry, so that the ape becomes the site of legitimization for what is and is not 'natural', and by proxy the moral framework for human social behaviour. A more recent debate that has hit the media involves natural homosexuality (see Case study 12.4).

Thematic Case Study 12.4

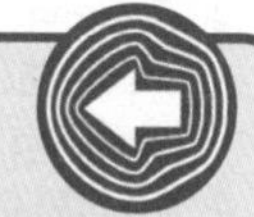

Gay animals

WHAT DOES A DOG HAVE TO DO ROUND HERE TO PROVE HE'S GAY????!!!!! (Ingilby 2002)

In 1999 a biologist called Bruce Bagemihl published a book called *Biological Exuberance: Animal Homosexuality and Natural Diversity* (Bagemihl 1999). Drawing upon a lifetime of research, Bagemihl works his way through the animal world, from mammals and insects to birds and cockroaches, describing a profusion of homosexuality, bisexuality, orgies, transvestism and transgenderism. The book includes field sketches of female bonobos engaging in cunnilingus and photographs of male whales entwining their penises in sexual communion, provoking a considerable degree of outrage. Being gay is seen as an exclusively human behaviour, something that by definition is not natural. This is a longstanding idea; the Church Council of Nablus in AD 1120 wrote the first law condemning homosexuality which was subsequently preserved in the Vatican library. By the time of the Renaissance this idea had found its way into the laws of many countries, and today many groups in society claim that homosexuality is wrong because it is 'unnatural'.

Bruce Bagemihl's book threw this illusion into doubt. In some colonies, as many as one in 10 pairs of penguins may be same-sex, while in bonobo chimpanzees the whole species can be said to be bisexual. If homosexuality is found throughout the animal kingdom then how can it possibly be unnatural? As Bagemihl says, 'What is remarkable about the entire debate about the naturalness of homosexuality is the frequent absence of any reference to concrete facts or accurate, comprehensive information about animal homosexuality' (ibid. p.78). Bagemihl also asks a more fundamental question: why has science tended to suppress this vast body of information? He argues that zoology is essentially a conservative profession, dominated by white male heterosexuals who impose social assumptions about homosexuality as perverse and unnatural on to the things it studies. He gives examples of how the term 'homosexual' is avoided in favour of terms like male-only social interactions, multifemale associations, unisexuality, isosexuality or intrasexuality. As one paper reported at the time,

> A female ape wraps her legs around another female, rubbing her own clitoris against her partner's while emitting screams of enjoyment. The researcher explains: It's a form of greeting behavior. Or reconciliation. Possibly food-exchange behavior. It's certainly not sex. Not lesbian sex. Not hot lesbian sex . . .
>
> (McCarthy 1999)

It is not only science that is to blame either. How common is it for parents to tell small children who enquire why two male dogs are mounting each other that they are just 'confused'? In the USA pet-owners have been acquitted for shooting their dogs on the grounds that they were exhibiting 'gay' behaviour. The topic of gay animals is laced with contradiction, which may explain why it is a topic so beloved of comedians, from Ricky Gervais to South Park's 'Big Gay Al's Big Gay Animal Sanctuary' and Sparky the Gay Dog (famously played by George Clooney).

Prudishness is only half the story though. Scientifically Bagemihl's work is controversial because it suggests that sex is not solely practised for reproduction. This seems to contradict the most important theory in biology – that of evolution. Researching animal homosexuality is not the road to success: applications for research grants that appear ignorant of the theory of evolution do not tend to stand a good chance of receiving funding. However, sex between animals – as between humans – is often a matter of enjoyment, rather than procreation, and this applies to animals of the same sex as well as opposite sexes. Non-heterosexual practices can also be beneficial. For example pairs of male birds may rear eggs 'donated' by a female, and two males can command a larger territory than a heterosexual pair, improving the chances of survival for chicks.

In 2006 the Oslo Natural History Museum ran an exhibition on gay animals, tellingly called 'Against Nature' (BBC 2006a). There has been considerable hostility to the exhibition, with organisers reporting criticism of the project, and being told by one opponent they would 'burn in hell'. An American commentator said it was an example of 'propaganda invading science'. More recently still, scientists in America have claimed that they can 'cure' gay sheep by manipulating hormone levels in their brains to make them fancy ewes (Oakeshott and Gourlay 2006). The implication that gayness might be 'curable' in humans has caused outrage. Competing constructions of nature are not only highly charged politically, but cut across science and religion, parts of modern society that are often seen as unrelated.

Q Why do scientists often resist work that shows the effects of culture on their work?

Q How does the 'gay animals' argument fit into the nature/nurture debate?

Q Is there any such thing as human nature?

12.4.1 Our place in nature

The biological sciences have also attempted to interrogate our genes to find out what makes humans distinctive from other organisms. The attempt to place humans *in* nature while simultaneously maintaining their position *above* nature is not a new one. The 'great chain of being' devised by medieval cosmologists represents a clear ordering of different things and beings in the world, with humans only lower than God. The trick of maintaining human uniqueness has become progressively harder since Darwin placed us squarely among the apes, but has generally been based upon the argument that while made of the same stuff, humans are more complex organisms than other animals.

While our attempts to benchmark human behaviour against 'natural' standards have tended to fail, even our own biology is not quite what we were hoping. The Human Genome Project (HGP) was a major international scientific collaboration to map the entire human genome, begun in 1990 and completed in 2003. The human genome comprises the entire set of human genes that determines the sequence of DNA. The project aimed to advance our understanding of human genes in order to support biotechnology research. The genomes of other organisms were mapped simultaneously in order to shed further light on the functioning of the human genome. Tori Holmberg (2005) has argued that the HGP represented a modern attempt to discover some inherent physical basis for what makes humanness distinct from nature. By counting the number of genes the HGP promised a potential verification of human distinctiveness and superiority.

As the HGP progressed the issue of how many genes human beings possessed became increasingly debated. Estimates varied wildly between 40,000 and 200,000. The subtext to this debate was the idea that individual genes controlled elements of an organism's phenotype (appearance and behaviour). A relatively large number of genes would support this argument, while smaller numbers would suggest that the relationship between genes and phenotype is more complex. When the preliminary results were published in 2001 it turned out that humans only have about 32,000 genes, far fewer than expected. The most disturbing element of this discovery was how close other organisms were, with the popular press jumping on the fact that the weed thale cress has 25,000 gene pairs, worms 19,000 and banana flies 13,600. Suddenly we were not as special as we had thought.

Within the scientific world, the discourse of human uniqueness, although momentarily shaken, was soon re-established through the idea that the genes interacted in complex ways to allow them to produce greater numbers of proteins than in other organisms. Thus the idea of natural human superiority was transposed from gene numbers to protein numbers. Similarly, the things being compared shifted from genes to base pairs. So while reporters could write that we share 99 per cent of our genes with apes, they could simultaneously state that this means that we do not share 3.5 million base pairs, thus upholding the distinction between human and animal. Just as science reveals our commonality with our animal relations, so our cultural beliefs require a shift of debate towards ground that reaffirms our uniqueness.

Plate 12.4 'The Blue Marble'.
(NASA)

12.5 Nature and the media

Did the planet betray us? Or did we betray the planet?

(Trailer for *An Inconvenient Truth*)

In each of the examples discussed so far, the idea of what is and is not natural forms a critical bone of contention. The way in which different positions are represented is important in understanding how certain ideas become dominant. For example, the idea of wilderness is inseparable from its representation in landscape painting and, more recently, photography. But how does this process work out in the mainstream media?

It is possible to identify two dominant discourses of nature that animate media discussions of environmental change: fragility and violence. The first is closely related to the emergence of the political movement of environmentalism, as a concern with the health of the planet as something separate that humankind depends upon for its survival. While the idea of caring for the planet seems quite normal to us now, this mode of thinking only emerged in the 1960s and 1970s. Denis Cosgrove (1994) argues that the emergence of an environmental mindset was shaped by the release of photographs of the earth taken during the Apollo Space programme in 1968 and 1972. These pictures of the earth from space were splashed across newspapers and television, and had a huge cultural impact. Never before had the planet upon which we live been viewed as a single object. The image of an orb hanging in space seemed to show the planet at its most vulnerable, as a finite, isolated island in a sea of nothingness (Plate 12.4). Books were written likening the earth to a 'spaceship', replete with life-support systems that require managing and conserving. The image of the earth became the leitmotif for Western environmentalism, the fragile blue image of the environment that transcends national borders and political divisions, on which human features are invisible.

The second discourse of nature beloved of the media is that of an uncontrollable, violent force. This version of nature is older, speaking to a pre-modern period in which human fortunes were largely determined by climate and disease. Press descriptions of floods and hurricanes speak of nature's wrath, and the unstoppable forces of nature. While the discourses of fragility and violence appear to be contradictory, they both reveal insecurities in the way in which society relates to the environment. Our original fear of nature's capricious power to destroy drove us to strive to control and dominate our surroundings, but our technological mastery has now made us the potential authors of our own destruction.

12.5.1 Natural disasters

Today, these versions of nature come together in media descriptions of natural disasters, which hold both a fascination with the frailty of human life, and highlight the potential for future ecological destruction. In the environmental era, the spectre of global disaster is constantly invoked, from the impending ice age predicted in the 1970s to the threat of global warming that emerged in the 1990s. In *The Ecology of Fear*, Mike Davis (1998) notes the fascination with disaster, claiming that the city of Los Angeles has been destroyed in film 145 times by directors and authors in the twentieth century, by everything ranging from earthquakes and nuclear weapons to Bermuda grass and the Devil. Crisis fascinates humans; it sells papers and makes people watch TV and films. The film *The Day After Tomorrow* suggests what may happen if the Gulf Stream shuts down because of the melting of the Canadian and Greenland ice caps. While there is evidence from Quaternary studies that this has possibly happened in the past, the film suggests that it could freeze the North Atlantic overnight. In reality the process would take at least fifty years. As the poster for the film shows (Plate 12.5) nature's capacity for destruction is used to sensationalize the film. Representations of crisis can also be politically useful, for example to international aid agencies as a way to secure further funding.

But what is 'natural' about a natural disaster? The Cambridge dictionary defines a disaster as 'an event causing great damage, injury or loss of life', which means that it cannot be a disaster unless there are people involved. By this definition a natural disaster would be a massively harmful event not caused by humans in any way. But there is undoubtedly a politics to natural disasters. The people who tend to suffer most are those who are the poorest and least able to respond. Increasingly disasters are seen as exacerbating pre-existing social and economic problems, rather than as the author of problems in their own right. The flooding of New Orleans in 2005 in the wake of Hurricane Katrina demonstrates this point well, with the majority of those left trapped in the city being amongst the poorest. Media representations of the disaster tended to fall back on a traditional discourse of destructive nature. A *New York Times* editorial on 30 August 2005 led with the headline 'Nature's Revenge', even though it went on to criticize the policies that exacerbated the disaster. Nature is often personified as a subjective force, in this case harbouring a whim for revenge. The quote from Al Gore's film on climate change, which opens this

Plate 12.5 Film poster for *The Day After Tomorrow*.
(Ronald Grant Archive)

section, asks whether we betrayed the planet, or the planet betrayed us. While again it is fair to ask whether it is possible for a lump of rock and some gases to betray anything, let alone 'us', the personification of nature he uses evokes both our power over nature (the capacity for us to betray the planet) and fear of nature (the possibility that it is going to betray us).

Again, the parallels with older religious discourses of God punishing humans are hard to ignore, and while we laugh at ancient polytheistic beliefs that the natural world was inhabited by a pantheon of gods, it is easy to forget that the US National Hurricane Center continues an old tradition of giving *names* to hurricanes. Such disasters are also commonly referred to as 'acts of God' by politicians and insurance companies. The BBC website offers some clues as to what nature may have been taking revenge for, claiming that the levees surrounding New Orleans were a 'snub to nature'. Statements like this beg the question of whether it is actually possible to offend water and wind, and if so, how water and wind would be able to seek revenge. Representations of these disasters as somehow being the 'fault' of nature distract attention from the real causes of human loss.

12.5.2 Environmental knowledge and the media

Many commentators have argued that global environmental change and the success of our efforts to cope with it will define the twenty-first century. Issues such as catastrophic climate change and natural disasters seem to be pervasive, invading our TV screens and filling newspaper column inches. And everyone seems concerned, from movie stars like Angelina Jolie to trendy politicians like David Cameron. Never has nature been so hot a topic.

The media is a critical sphere in which knowledges about nature are produced and consumed. The wildlife documentaries that are broadcast into our homes try to reproduce notions of Africa as a great wilderness, going to great lengths to capture scenes in which people and evidence of people are absent (Davies 2000). This is an increasingly difficult task in the face of tourist traffic and indigenous peoples that are found across many African savannah areas. Wildlife films have been edited for years to present only heterosexual animals. The Internet represents a huge source of scientific knowledge but is largely unregulated, producing

Spotlight Box 12.1

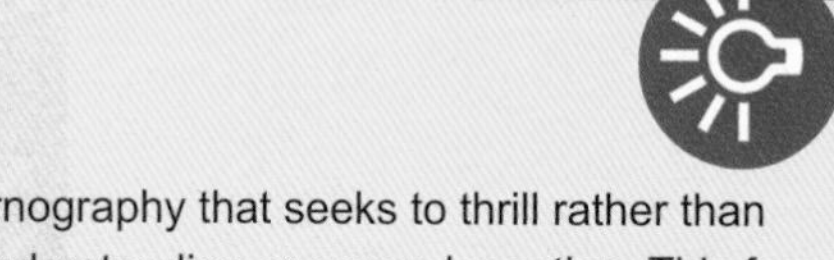

Climate porn

In August 2006 the Institute for Public Policy Research released a report claiming that the representations of climate change in the media are unhelpfully alarmist, focusing on thrilling headlines that prevent the public from understanding what they can do about the problem (Ereaut and Segnit 2006). The report is interesting because it shows that the government is increasingly aware that the media is critical in shaping how people understand environmental change. Based on research that analysed more than 600 newspaper articles, and 90 TV and radio excerpts, the report identified the dominant discourse driving the climate change debate as 'alarmist'. This discourse describes articles that focus on the overwhelming size of the problem, and the potentially disastrous effects of climate change. The alarmist discourse of climate change clearly draws upon our deeper cultural constructions of nature.

By focusing on the terrible unstoppable forces of nature, the alarmist discourse represents a form of climate pornography that seeks to thrill rather than increase understanding or engender action. This form of debate distracts us from what we can actually *do* to help tackle environmental change. The second discourse they identify is that of 'small actions', which typifies the current approaches of government and campaign groups, which emphasize the small actions that can be done to help combat climate change and save money in the process. The report suggests that because these two discourses often occur together, the small actions discourse is rendered impotent by the sheer scale of the alarmist claims.

Q Can you think of other ways in which humans personify nature?

Q Can the extinction of the dinosaurs be considered a disaster?

Q How important do you consider the media to be in tackling environmental change?

a range of conflicting claims about the state of nature. Popular environmental websites vary wildly in their estimates of biodiversity loss (Ashlin and Ladle 2006) (a salutary tale for students who rely solely on information from the Internet to write their coursework.)

The media has assumed increased importance as policy makers attempt to encourage people to shift towards more sustainable lifestyles. You may have heard of a recent UK government report branding the sensationalization of the climate debate a form of 'climate porn'; a scaremongering approach to environmental change that excites, but prevents meaningful action (Spotlight box 12.1). As the primary interface between the majority of the population and environmental issues, the way in which ideas of nature are communicated has a decisive impact on how people understand and respond to environmental challenges.

12.6 Conclusions

12.6.1 Nature: a dangerous idea

This chapter started by asking what nature is. Geographers have sought to answer this question in a range of ways, suggesting that nature is a socially constructed idea, which varies between place and period. Rather than lessen its importance, it is as an idea that the importance of nature truly becomes apparent. Even with the help of scientific methods and analysis it is hard to prevent preconceptions and assumptions about landscapes and nature from interfering with conclusions. Nature is used to dictate how we should behave, how we should live and what we should value. It allows the exploitation of certain people and validates the luxuries of others. It has been used to persecute homosexual behaviour while legitimizing genocide. Nature is indeed a dangerous idea.

At the beginning of the chapter it was argued that social constructionists seek to demonstrate two aspects of nature: first how ideas of nature do not represent some external 'truth' but, rather, come to reflect the cultural, political and spiritual beliefs of a society; and second, the ways in which ideas of nature are used to support specific political viewpoints. In light of the case studies, this picture has become more complex. This process is circular; as dominant cultural norms and preferences influence ideas about nature, these arguments about nature are then used to support dominant cultural norms. Geographers seek to reveal the circularity of this process in order to expose repressive ideologies and open up new political possibilities for existence.

12.6.2 The challenge

The idea of nature remains central to the challenges facing society in the twenty-first century. The growing priority attached to global environmental change by leading political figures means that debates concerning how we should live, who should make sacrifices and how change should be managed are becoming ever more apparent. The increasing influence of biotechnology as a major global industry raises questions concerning how far we should manipulate biology to our benefit, and whether or not knowledge about our genes will reveal some essential truths of human existence. The idea of nature looms large in all these areas, and will continue to do so. You will see people talking about nature on the programmes every day on TV, in newspapers and around you in everyday life. Think about what people mean when they use the word. What are they excluding? What kinds of arguments are they trying to support? Be critical of how nature is used; the power to refute dominant presentations of the world and our place in it is the power to create a different world.

Learning outcomes

Having read this chapter you should be able to:

- Understand nature as a social construction.
- Recognize that perceptions of nature vary across time and space.
- Identify basic discourses of nature like wilderness in cultural representations.
- Understand the political power of different representations of nature.
- Critique ideas and concepts such as the 'limits to growth' and desertification.
- Appreciate how science is used to make arguments about human 'nature'.
- Understand how the media uses ideas of nature to report environmental issues.

Further reading

Castree, N. (2005) ***Nature***, Routledge, London. Covers the major approaches to nature that geographers use.

Demeritt, D. (2002) What is the social construction of nature? ***Progress in Human Geography*, 26**: 767–90. Reviews the social construction of nature literature in human geography.

Hinchliffe, S. (2007) ***Space for Nature***, Sage, London. An introduction to current geographical thinking beyond nature.

Macnaghten, P. and Urry, J. (1998) ***Contested Natures***, Sage, London. Covers a range of approaches to the sociology of nature.

Robbins, P. (2004) ***Political Ecology***, Blackwell, Oxford. Accessible introduction to major work on the myths of environmental science, with lots of case studies.

Useful websites

www.sierraclub.org/john_muir_exhibit Interesting information and quotes concerning how wilderness became a valued idea in the USA.

www.banksy.co.uk/indoors/02.html Banksy's art tries to highlight cultural blind spots. It is no coincidence that much of his work has natural elements in it.

www.nhm.uio.no/againstnature/index.html The website of the Natural History Museum of Norway for the 'Against nature?' exhibition.

www.funtrivia.com/ubbthreads/showthreaded.php?Cat=0&Number=34605&page=6 Great blog about gay animals.

www.earthobservatory.nasa.gov/Study/Desertification NASA website with an interesting take on the desertification myth in the era of GIS.

www.genome.gov Information on the Human Genome Project. Lots of outlandish rhetoric about how the HGP will reveal the secrets of our inner universe.

www.youtube.com/watch?v=O97kS25uNzA The trailer for *An Inconvenient Truth*. Climate porn at its best! Notice the images and wording. Does it '*shake you to your core*'?

For annotated, clickable weblinks and useful tutorials full of practical advice on how to improve your study skills, visit this book's website at **www.pearsoned.co.uk/daniel**

GEOGRAPHY, CULTURE AND GLOBAL CHANGE

Chapter 13

Cheryl McEwan

Topics covered

- A definition of culture
- The 'cultural' and 'spatial turns' in the social sciences
- An evaluation of the extent of cultural globalization
- The impacts of globalization on local cultures
- An exploration of the concepts of multiculturalism and hybrid cultures
- Progressive ways for geographers to think about culture

Culture is a word on everybody's lips these days. Hardly a moment seems to pass when we do not hear on the radio or television, see in newspapers and magazines, or read in academic texts some account of the world of the cultural. Governments at all levels announce cultural policies and provide funding for cultural activities; intellectuals announce the need for cultural initiatives or bemoan the loss of traditional cultural values; famous cultural icons – musicians, artists, novelists – themselves culture makers, are increasingly sought out for their opinions on the state of the world. (Kahn 1995: ix)

13.1 What is culture?

This chapter explores some of the challenges posed by and for culture in the twenty-first century. First, however, it is necessary to define what is meant by **culture**. This itself is a complex and difficult task. By the 1950s, for example, there were over 150 different academic definitions of culture. As Mike Crang (1998: 1) argues, despite sounding like the most airy of concepts, culture 'can only be approached as embedded in real-life situations, in temporally and spatially specific ways'. Cultures are part of everyday life. They are systems of shared meanings often based around such things as religion, language and ethnicity that can exist on a number of different spatial scales (local, regional, national, global, among communities, groups or nations). They are embodied in the material and social world and are dynamic rather than static, transforming through processes of cultural mixing or transculturation (discussed below).

Cultures are also socially determined and defined and, therefore, not divorced from power relations. Dominant groups in society attempt to impose their ideas about culture and these are challenged by other groups, or **subcultures**. The latter might include various types of youth cultures, gang cultures, and different ethnicities or sexualities, where identities are organized around different sets of practices and operate in different spaces from dominant cultures (Crang 1998; Skelton and Valentine 1998). Culture makes the world meaningful and significant. As Phil Crang (1997: 5) argues, we should think of culture as a process we are all involved in rather than as a thing we all possess (see Spotlight box 13.1).

13.1.1 The 'cultural turn'

As Kahn suggests, culture has generated a great deal of interest in recent years, for academics, policy makers, and at the popular level. Geographers have turned their attention towards cultural explanations of global, national and local phenomena, exploring issues such as the cultural embeddedness of economic processes (e.g. Amin and Thrift 2004), the relationship between cultures, identities and consumption (see Chapter 18), and cultural constructions of social relations of gender, ethnicity and class that shape people's lives (e.g. Nelson and Seager 2004). However, the current popularity of culture is not simply a trend in academe, but is reflective of a broader cultural turn in (Western) society as a whole.

The world has changed fundamentally in the past two decades and these changes are deeply cultural in character. For example, enormous changes have occurred in 'advanced' economies since the early 1980s (the decline in manufacturing, the growth of services, the feminization of the workforce, increased flexibility – all characteristic of 'post-Fordism' as discussed in Chapter 3, p. 77). However, as Stuart Hall (1996: 233) argues, if 'post-Fordism' exists, it is as much a description of cultural as of economic change. Florida (2002) characterizes this as a shift from an industrial to a creative age, with 40 per cent of people in

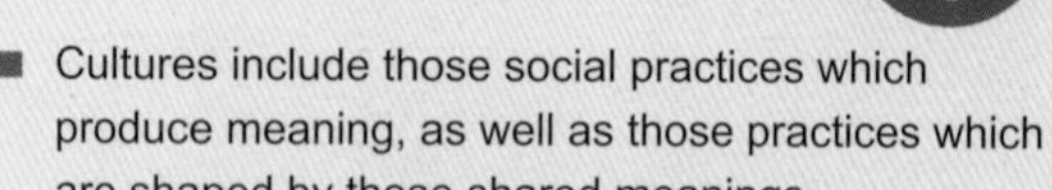

Spotlight Box 13.1

Cultures: a summary

- The systems of shared meanings which people who belong to the same community, group or nation use to help them interpret and make sense of the world, and to reproduce themselves.
- These systems of meanings include language, religion, ethnicity, custom and tradition, and ideas about 'place'.
- Cultures cannot be fixed, but shift and change historically through diffusion and transculturation.
- Culture is a process rather than a thing, but is embodied in the material and social world.
- Cultures include those social practices which produce meaning, as well as those practices which are shaped by those shared meanings.
- Cultures are one of the principal means by which identities are constructed; they give us a sense of 'who we are', 'where we belong' – a sense of our own identity and identity with others.
- Cultures are not divorced from power relations.

Source: Adapted from Hall (1995)

Plate 13.1 Image marketing is central to popular consumption: the world of commodities is profoundly cultural.
(© Benjamin Rondel/CORBIS)

the US and UK economies now working in 'creative' sectors of science, technology, culture, arts and entertainment, and creative economies burgeoning in cities as diverse as Shanghai and Bangalore. Modern consumption depends overwhelmingly on image (for example, the marketing of food and drink products and fashion clothing) (see Plate 13.1). Movements around the world of images, symbols, modes of thought and communications are unparalleled in terms of their volume, speed and complexity. As computer technology, video imagery and electronic music demonstrate, the material world of commodities and technologies is profoundly cultural. In addition, culture has become increasingly commodified; in other words culture is being translated into material goods that can be marketed and sold. For example, the annual Gay Games in New York has been transformed from a political event into a cultural commodity. The Games draw the support of large corporate sponsors eager to capture the 'pink dollars' of a supposedly affluent gay population. The mainstream press covers the games as a cultural, rather than a sporting or a political event. As Zukin (1995: 263) argues, culture now has great appeal – culture sells.

13.1.2 The 'spatial turn'

There has also been a 'spatial turn' in explanations in cultural and social theory. The world is changing fast, and the rate of change is probably greater than ever before. New technologies such as the Internet and satellite communications mean that the world is becoming more global and more interconnected. The increased speed of transport and communications, the increasing intersections between economies and cultures, the growth of international migration and the power of global financial markets are among the factors that have changed everyday lives in recent decades. There is no historical equivalent of the global reach and volume of 'cultural traffic' (Held et al. 1999) through contemporary telecommunications, broadcasting and transport technologies. The challenge for geographers is to find ways of understanding and interpreting these changes.

Culture can be said to operate at three spatial scales: local, national and global. Two main interpretations have dominated discussion. The first highlights the global aspects of change. At its simplest, this approach suggests that it is possible to identify processes of cultural homogenization – the idea that everywhere is becoming the same – dominated by the USA and most easily recognized in terms such as 'Coca-Colonization', 'McDonaldization' and 'Hollywoodization' (Cochrane 1995: 250) (see Chapter 18, pp. 380–87). This *cultural globalization* involves the movement of people, objects and images around the world through telecommunications, language, the media industries, radio and music, cinema, television and tourism. The second interpretation places emphasis on the local and the localization of people's everyday lives and experiences. Instead of homogenization, emphasis is placed on the diversity of culture, on the ways in which global icons such as Coca-Cola or McDonald's are reinterpreted locally so that they take on different meanings in different places. The emphasis here is on the *interconnectedness* of global and local processes. For example, although the same event can be witnessed simultaneously around the world (e.g. an incident broadcast in a CNN news report, or an international sporting event), this event will be interpreted differently in different places. Furthermore, locality does not necessarily refer to the opposite of globality. For example, some environmentalists imagine the world as a locality, a 'global village'. Cultural theorists have a growing interest in how increasing globalization, especially of cultural production and consumption, affects people's sense of identity and place at both local and national levels (McDowell 1994: 147). Thus a geographic or spatial perspective has become central to studies of culture more widely. These are some of the concerns that form the focus of this chapter. Subsequent sections explore in more detail ideas about a global culture, examine ways of rethinking local cultures, and explore progressive ways of thinking about cultures in contemporary contexts.

13.2 Towards a global culture?

13.2.1 Imagining a global culture

Processes of cultural globalization have a very long history and are not peculiar to contemporary times. Through global patterns of trade and migrations, and through the spread of religions and empires, people, objects and ideas have been circulating for centuries (see Chapters 1–3). However, contemporary globalization is distinctive in extent, form, rapidity of change, intensity and impact. Some commentators suggest that the idea of a global culture is becoming as meaningful as the idea of national or local cultures, with different places and cultural practices around the world converging and becoming ever similar. As Shurmer-Smith and Hannam (1994: 76) argue, a global culture might be the product of two very different processes:

1. The export or diffusion of supposedly 'superior' cultural traits (e.g. Western time-frames – the 24-hour day and the Gregorian calendar) and products (e.g. the motor car, television) from advanced countries, and their worldwide adoption ('westernization', 'Americanization', 'modernization'). This is believed to create global cultural convergence – people around the world are becoming increasingly similar in terms of consumption, lifestyle, behaviour and aspirations (see Case study 13.1). It can be perceived positively (as 'modernization' or 'development') or negatively (as 'cultural imperialism', where 'we' assume that others in the world should aspire to be like 'us').
2. The mixing, or hybridization, of cultures through greater interconnections and time–space compression (the shrinking of the world through transport links and technological innovation), leading to a new universal cultural practice. This challenges the notion of unidirectional 'westernization' and allows us to consider how western cultures have influenced and are also being influenced by this mixing of cultures. Flows of music, food, ideas, beliefs and literature continue to percolate from around the world into the cultures of the West.

In reality, both these processes are flawed explanations for what is happening today. If a global culture exists, it is far from a product of unidirectional 'westernization'. However, alternative ideas about cultures mixing to produce a universal global culture are also problematic. Cultures are mixing, but this does not necessarily mean we are all becoming the same.

13.2.2 Debunking global culture?

A different departure point for discussing global culture is that there is no such thing. Ideas about a singular global economy, politics and culture imply some sort of worldwide commonality that does not exist. First, the image of rampant cultural imperialism by the West, and especially the USA, is flawed since apparent cultural sameness is limited in scope, located only in the consumption of products and media images. The possibilities of this eroding centuries of local histories, languages, traditions and religions are far-fetched and

Thematic Case Study 13.1

The globalization of culture: some examples

Music

Music lends itself to globalization because it is one of the few popular modes of cultural expression that is not dependent on written or spoken language for its primary impact (Held et al. 1999: 352). The relative ease of cultural diffusion has led to the spread of many genres and major artists all over the world. The production, distribution and consumption of music has a particular geography. The global music industry is dominated by transnational corporations, with the United States and United Kingdom dominating domestically generated popular music. 'World music' is now a significant part of the marketing strategies of these corporations and exposes global audiences to local musical traditions from around the world. Migrations of people have also had cultural impacts on music, with increasingly 'hybridized' forms (see Case study 13.6).

Sport

Sports are forms of cultural expression that are becoming increasingly globalized, as well as increasingly commodified. Football/soccer is the most obvious example. The World Cup was held in Japan/South Korea in 2002 and will be held for the first time in Africa in 2010 in an effort to expand its global popularity. European leagues are increasingly internationalized, with some clubs fielding teams made up almost exclusively of overseas players. Clubs like Real Madrid and Manchester United are globally recognized icons owing in part to global television coverage. Similar trends can be observed in US Major League Baseball. The New York Yankees are a global icon; many major league players hail from countries such as Cuba, the Dominican Republic, Puerto Rico and Costa Rica; the sport is becoming increasingly globalized through television coverage.

Tourism

Tourism is one of the most obvious forms of globalization. The geography of tourism is skewed, since it is dominated by people of all classes from developed countries (North America, Western Europe and, increasingly, Japan and Australia). It can also be exploitative, particularly through the growth of international sex tourism and the dependency of some developing economies on the exploitation of women. However, it is a form of international cultural exchange that allows vast numbers of people to experience other cultures and places. It also locks specific places (tourist destinations) into wider international cultural patterns (Urry 1990).

people in different parts of the world respond to these images and products in different ways. Many millions of people do not have access even to a television. (In 2003, there were only 55 TV sets per 1,000 people in Africa, and 208 per 1,000 people in Asia, compared with 471 per 1,000 in Europe, and 754 per 1,000 in the USA (*CIA World Factbook* 2003). There is no single global culture and this is because of the unevenness of globalization.

Secondly, some theorists would argue that national cultures remain stronger than global cultures. This is borne out when we consider the many conflicts occurring throughout the world along the geopolitical fault-lines of national cultures (the ongoing conflicts between Pakistan and India are but one example of this). For the past 200 years, nation-states and national cultures have monopolized cultural power (state television is one example of a national institution influencing national cultures within national territorial boundaries). At the end of the twentieth century, this balance began to change, with international telecommunications and media corporations challenging the centrality and importance of national cultures. However, it could be argued that despite these changes a great deal of cultural life is still organized along national and territorial lines.

Thirdly, if culture is a system of shared meanings, then looking at the world today there are clearly many systems of shared meanings and many different cultures. People in different places use different techniques and technologies to reproduce culture, such as oral histories, literature or television and cinema. These techniques have different patterns of dispersion, penetration and scale. Therefore, some cultures are more likely to become globalized than others – those reproduced through television and cinema have a greater range and speed of dispersion than those reproduced through oral histories. However, this does not mean that globalized cultures completely erode localized cultures; the ways in which these different cultures intersect is important. Those 'things' (products, symbols, corporate entities) that have become global signifiers are clearly globalized (they are recognized the world over), but the ways in which people around the world make cultural responses to them are complex and multiple. Globalization of products and symbols does not necessarily equal westernization. For example, Japanese consumer goods do not sell on the back of exporting Japanese culture but on a market strategy based around the concept of *dochaku* ('glocalism'). This involves a global strategy not of imposing standardized products but of tailoring Japanese consumer products to specific local markets. These goods are, therefore, both globalized and localized. Consequently, how intersections between cultures are played out at local levels are of significance, and this suggests that imagining a universal global culture is quite problematic.

13.2.3 Rethinking global culture

Instead of imagining a global culture that is erasing local and national cultures, we can think of local, national and global as three important spatial levels at which culture operates. Those aspects of culture that operate at the global level are 'third cultures' (Featherstone 1995: 114). National institutions are no longer in complete control of cultural globalization and 'third cultures' (sets of practices, bodies of knowledge, conventions and lifestyles) have developed in ways that have become increasingly global and independent of nation-states. Phenomena such as patterns of consumption, technological diffusion and media empires are part of these third cultures, and transnational and multinational corporations are the institutions that make them global. In this sense, global cultures exist but only as third cultures, outside national and local cultures, yet intersecting on both these scales in different ways around the world.

Acknowledging that 'our' global view might be very different from that of people elsewhere, living in very different contexts, is also important. It is clear that multiple global cultural networks exist, such as those connecting the overseas Chinese with their homeland, or those linking Islamic groups around the world. These networks disrupt any notion of a singular global culture.

Power and inequality bring into question the idea of global cultures. As Massey (1993) argues, a **power geometry** exists, which gives people with different access to power different notions of what global means. New institutions (like global media corporations) for the production, transmission and reception of cultural products are creating infrastructures supporting cultural globalization, including electronic infrastructure (radio, television, music, telecommunications), linguistic infrastructure (the spread of bi- and multilingualism, particularly the dominance of English), and corporate infrastructure (producers and distribution networks). As we have seen, these new institutions often operate at scales beyond the nation-state, and they are sites of power in the production of culture. The ownership, control and use of these institutions remain uneven across and within countries (Held et al. 1999: 370), thus creating 'power geometries' that are centred overwhelmingly on the West. People have very different experiences of culture because of their different locations in the world and their relationship to these sites of power. Mapping this power geometry, identifying sites of power and revealing the marginalization of some peoples around the world by cultural globalization are increasingly significant. Of equal significance are forms of resistance (e.g. **culture jamming** (Dery 1993) – media hacking, information warfare, 'terror-art' and graffiti – which aims to invest advertisements, newscasts and other media artefacts with subversive meaning) and the ways in which marginalized peoples might be empowered by engaging with, and perhaps transforming, the new institutions driving cultural globalization.

In summary, global processes are occurring, but they do not produce a universal global culture, they are not distributed evenly around the world and are not uncontested. Global cultural processes are not simply a result of a unidirectional 'westernization', since culture flows transnationally. A number of different global cultures exist as 'third cultures' – in patterns of consumption, flows of knowledge, diffusion of technologies and media empires that operate beyond, but connecting with, the local and national scales.

13.3 Reinventing local cultures?

13.3.1 Locality and culture

It has often been assumed that there is a simple relationship between local place and local culture. Places were thought of as having a distinct physical, economic and cultural character; they were unique, with their own traditions and local cultures that made them different from other places. It is clear, however, that processes of globalization are also posing serious challenges to the meaning of place. Places and cultures are being restructured. According to Massey and Jess (1995a: 1), 'on the one hand, previous coherences are being disrupted, old notions of the local place are being interrupted by new connections with a world beyond'. The appearance of 7–11 stores in the rainforests of northern Thailand is one example of how even the remotest of places are becoming increasingly internationalized, in this case through tourism. 'On the other hand', Massey and Jess continue, 'new claims to the – usually exclusive – character of places, and who belongs there, are being made.' A dramatic example of this is the revival of place-based ethnic identities in former Yugoslavia and the post-Soviet states, and the conflicts that have resulted. Therefore, modern life is characterized both by decentralization and globalization of culture and by the resurgence of place-bound traditions. Following this, the impact of the new global context on local cultures has two, possibly contradictory, outcomes.

13.3.2 Negative sense of culture

Where global processes are perceived to pose a threat to local culture, there might be an attempt to return to some notion of the exclusivity of culture. At the extreme, this might take the form of exclusivist nationalism or even 'ethnic cleansing'. Reactions to the perceived threat to local cultures include nationalistic, ethnic and fundamentalist responses, which also entail a strong assertion of local cultures, such as reviving or inventing local traditions and ceremonies. These can create a level of local fragmentation, with a parochial, nostalgic, inward-looking sense of local attachment and cultural identity. In this sense, cultures are thought of as **bounded**, with very clear definitions of 'insiders' and 'outsiders' in the creation of a sense of belonging, and producing **geographies of exclusion** (Sibley 1995). For example, English rural areas are often conceptualized as the preserve of culture and identity. This idea is mobilized through the myth of the 'rural idyll', where rural cultures are thought of as timeless, unchanging and unaffected by global processes. This cosy vision of a peaceful countryside excludes many people who live in rural areas but do not fit this stereotype. Gypsies, travellers, environmental protesters, hunt saboteurs, people with alternative lifestyles and people from minority ethnic communities are deemed to be 'outsiders' and a threat to local cultures (Cloke and Little 1997). As Halfacree (1996) argues, the 'rural idyll' is a selective representation. It is exclusive in its class, race and status connotations, is profoundly conservative and demands conformity. It is based on a very inward-looking sense of place and culture (see Plate 13.2).

Conservative reactions to change can be thought of as a kind of cultural fundamentalism through which the process of cultural change is often bitterly contested. Gender plays an important role in this. Women are often considered as guardians of the borders of culture (Yuval-Davis 1997). They not only bear children for the collective, but also reproduce it culturally. In closed

Plate 13.2 Exclusion in the countryside: travellers and other minority groups are often excluded from shops and public houses, marking their position as 'outsiders'.
(Joanne O'Brien Photography)

cultures, the control of women's sexuality is seen as imperative to the maintenance of the purity of the cultural unit; women are discouraged from marrying outside their cultural and ethnic group. Ethnicity and culture, therefore, are seen to be one and the same. In addition, symbols of gender play an important role in articulating difference between cultural groups. Women's distinctive ways of dressing and behaving very often come to symbolize the group's cultural identity and its boundaries. Women are often the intergenerational transmitters of cultural traditions, customs, songs, cuisine and the 'mother' tongue, primarily through their role as mothers. This is especially true in minority situations where the school and the public sphere present different and dominant cultural models to that of the home. Recent controversies over the Islamic veil, for example, have seen countries across Europe wrestling with issues of religious freedom, civil rights, women's equality, secular traditions and escalating fears of terrorism. France banned Muslim head-scarves and other 'conspicuous' religious symbols at state schools in 2004; Turkey, a secular Islamic country, banned the wearing of head-scarves in all civic spaces in 2005; the Italian parliament approved anti-terrorist laws in July 2005, which make covering one's features in public – including through wearing the burqa – an offence; the Dutch cabinet backed a proposal in 2006 to ban the few dozen Muslim women who choose to wear the burqa from doing so in public places on grounds that it disturbs public order, citizens and safety. In the United Kingdom, debate about Islamic dress was opened in October 2006 when Member of Parliament Jack Straw stated that he asks constituents to remove their veils when attending his surgeries, and by the sacking of a teacher in November 2006 for refusing to remove her niqab in class.

Societies do not evolve smoothly from closed, bounded perceptions of culture to more open, dynamic notions. The question of cultural power, identity and resistance also needs to be considered. For many groups cultural survival is seen to depend on a closed idea of culture, with strongly marked boundaries separating it from 'others'. The controversies over Islamic dress and the continuing sectarianism in Northern Ireland are examples of different cultures, religions and national identities colliding with each other (see Case study 3.1). Elsewhere the mixing of cultures under the impact of globalization is often seen as threatening and as weakening the sense of cultural identity. Immigration is seen as a particular threat, creating a revival of ethnicity that cuts across the political spectrum. Examples include the 'little England' reaction to closer European integration and the rise of neo-fascism in Europe (characterized by anti-immigration and racism). Perceived threats to religious identities have also witnessed the strengthening of Islamic and Christian fundamentalism around the world. These phenomena are not all the same, but they do share a response to globalization that involves a closed, fixed, bounded and often place-specific definition of culture, and a strong resistance to changes heralded by cultural globalization.

13.3.3 Positive sense of culture

A more positive response to global processes would be to imagine cultures as fluid, ever-changing, unbounded, overlapping and outward-looking – akin to Massey's (1994: 151) 'progressive sense of place'. This involves people being more cosmopolitan (free of prejudice and tolerant of difference). Increasing interconnectedness means the boundaries of local cultures are seen to be more permeable, susceptible to change, and difficult to maintain than in the past. Rather than everywhere becoming the same, some nation-states have reconstituted their collective identities along pluralistic and multicultural lines, which take into account regional and ethnic differences and diversity. In Europe, this involves re-creation and invention of local, regional and sub-state or new 'national' cultures (for example, the cultural renaissance of the Scots, Basques and Catalans, or the cultural assertiveness of minority ethnic communities in cities such as London, Paris and Berlin). Thus, what can be perceived as destruction of local cultures by globalization might in fact be the means of creating new senses of locality. This still involves notions of local identity, but recognizes both the differences between cultures and their interconnectedness, taking account of the positive aspects of cultural mixing and increased cosmopolitanism. This new sense of local identity is based on notions of inclusion rather than exclusion.

This is not to say, of course, that all people within the same place will share the same culture and the same sense of locality. Within these more culturally pluralistic and cosmopolitan locales, different class factions, ages, genders, ethnicities and religious groupings mingle together in the same sites, consuming the same television programmes and products, but in highly uneven ways. These groups often possess different senses of affiliation to places and localities, possess different cultural identities and belong to different cultural groupings (Featherstone 1995: 97). A progressive sense of culture does not foresee the locale as a 'melting-pot', where everything becomes the same, but rather recognizes the different experiences of people, and that increasing interconnections might create new, dynamic and exciting cultural forms. An understanding of this is crucial to the creation of a progressive notion of place and culture, which recognizes cultures as fluid and open and interconnected, and accepts that older local cultures might decline as new ones emerge.

In summary, localities are important in maintaining cultural difference, but can also be sites of cultural mixing and transformation. Ideas about culture can be negative (bounded, fixed, inward-looking) or positive (progressive, dynamic, outward-looking). Bounded, fixed notions of culture can lead to localized resistance, racism, nationalism, and even 'ethnic cleansing'. Progressive ideas about culture involve the recognition of differences between cultures, the interconnectedness of cultures and their constant evolution.

13.4 Activity: thinking about 'own' and 'other' cultures

In the light of the above, this is a useful juncture at which to reflect upon how we think about our 'own' and 'other' cultures and how this might relate to notions of cultural superiority/inferiority. Consider first the newspaper report reproduced in Case study 13.2.

This recounts the remarks of a European political leader, which were quickly condemned around the world and subsequently retracted. It reveals the sensitivities of global politics shortly after the shocking events in the USA in September 2001 but also prompts us to think about how we see the world from the perspective of our own cultures and the implications of this. The equation of 'democracy', 'civilization', 'superiority' and 'human rights' with 'European'/'Western' cultures and the explicit assertion that these qualities do not exist in other (in this case, Islamic) cultures remain pertinent issues today.

13.4.1 Questions about cultures of democracy

Three questions are raised by the article in Case study 13.2:

1. What is the assumed relationship between the West (specifically Europe and the USA) and democratic cultures?
2. What assumptions are made about the superiority of European cultures and how they are exported around the world?
3. How might we question the supposed universalism of Western cultures and the assumption that it is appropriate to export/impose these beyond the West?

First, let us think about relationships between place and the assumed origins of democratic cultures. A crucial point is that the West is assumed to have a special place in the history of democracy and its associated qualities (e.g. 'freedom', 'civilization', 'human rights'). Western cultures are depicted as the cradle of human rights, progress, enlightened thought and philosophical reflection on creating the conditions for the improvement of people's lives. Democracy, then, is seen as an intrinsic part of Western cultures. The

Thematic Case Study 13.2

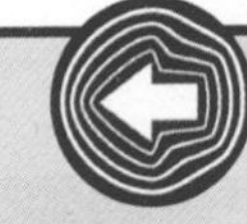

Berlusconi breaks ranks over Islam

Italy's prime minister, Silvio Berlusconi, yesterday went out of his way to stress what every other leader backing America's 'war on terrorism' has been desperate to deny – that the looming conflict is . . . a clash of civilisations. On three occasions during a visit to Berlin, Mr Berlusconi enthusiastically . . . boasted of the 'supremacy' and 'superiority' of western civilisation and called on Europe to recognise its 'common Christian roots' . . . He declared that he and his host [the German Chancellor] 'consider that the attacks on New York and Washington are attacks not only on the United States but on our civilisation, of which we are proud bearers, conscious of the supremacy of our culture, of its discoveries and inventions, which have brought us democratic institutions, respect for the human, civil, religious and political rights of our citizens, openness to diversity and tolerance of everything.'

Mr Berlusconi said: 'We should be conscious of . . . [our cultural] superiority . . . , which consists of a value system that has given people widespread prosperity in those countries that embrace it, and guarantees respect for human rights and religion.' Mr Berlusconi added: 'This respect certainly does not exist in the Islamic countries.'

Source: Hooper and Connolly (2001)

second question concerns ideas about the supposed superiority of Western cultures. Democracy is thought of as produced solely within Western cultures; it is not depicted as a product of cross-cultural encounters or influenced by ideas from other parts of the world. This assumption is rooted in ideas about the supposed superiority of Western cultures. The third question concerns the supposed universalism of Western cultures. Western democracy is depicted as a common step forward for humanity as a whole and as encapsulating progressive and civilized cultures. This assumption is based on the representation of non-Western cultures in negative terms and as less civilized. Western democracy is seen as the best way of improving people's lives and should be exported to other places assumed to be culturally inferior. How might we think differently about culture by looking at examples from other places?

13.4.2 Democratic cultures in other places

Now consider these three questions in terms of the two examples in Case studies 13.3 and 13.4. How do these examples challenge ideas of (i) democracy originating solely in Western cultures; (ii) the superiority of Western cultures; and (iii) the universal applicability of Western cultures?

These two examples do not suggest that non-Western cultures have 'better' cultures of democracy. Critics argue, for example, that *ubuntu* fails to take account of the current socio-political context of poor communities in South Africa and perpetuates gender inequalities; there are questions about the armed resistance of the Zapatistas within Mexico. However, they demonstrate that different cultural models exist; Western and indigenous cultures of democracy might

Regional Case Study 13.3

Ubuntu and democracy in South Africa

The relatively non-violent transition in South Africa from a totalitarian state to a multi-party democracy is not merely the result of the mimicry of Western democratic cultures, but also of the emergence of an ethos of solidarity and a commitment to peaceful coexistence amongst many South Africans in spite of their differences. *Ubuntu*, a cultural philosophy eroded for generations by the slave trade, Western missionary movements, colonialism and apartheid, is central to this. The South African government officially recognizes *ubuntu* as: 'The principle of caring for each other's well-being . . . and a spirit of mutual support . . . Each individual's humanity is ideally expressed through his or her relationship with others and theirs in turn through recognition of the individual's humanity. *Ubuntu* means that people are people through other people. It also acknowledges both the rights and the responsibilities of every citizen in promoting individual and societal well-being' (Anonymous 1996). *Ubuntu* philosophy is one of generating agreement or building consensus. South African democracy is not simply majority rule; the Constitution and legislation explicitly protect the rights of minority groups. *Ubuntu* differs from Western cultures of individual rights by seeing democracy as built upon interdependent relationships.

Regional Case Study 13.4

Zapatistas: democracy and indigenous rights in Chiapas, Mexico

Struggles for democratization in Mexico have occurred since the early twentieth century. New philosophies began to emerge in the 1980s and 1990s with the Zapatista uprising in the impoverished Chiapas region. Globalization and free trade agreements are seen as threats to the livelihood of large sectors of the Mexican population, especially peasants and workers. The Zapatista uprising aims to democratize and decentralize Mexican society, secure autonomy and self-determination for indigenous communities, and protect their cultural heritage. It aims to resist the negative effects of globalization and protect women's rights. The Zapatista conceptualization of democracy does not flow from Western political philosophy but rather emanates from older, indigenous Maya social organization, based in cultures of reciprocity, communal values and respectful dialogue.

(Slater 2003)

thus inform each other in a respectful dialogue. In this sense, understandings of different cultures are enriched rather than based on Western-centred and Western-biased assertions of cultural superiority evident in the newspaper report. Expanding our geographies of reference and learning about examples from beyond our own cultures encourages the production of less partial/more informed visions and we can also begin to think about how cultures elsewhere have shaped and continue to shape our own.

13.5 Multi- and hybrid cultures?

13.5.1 Hybridity

One of the major contemporary challenges concerns what we do with the concept of culture in the changing global scene, where nation-states are forced to tolerate greater diversity within their boundaries. Some want to see national identity as homogeneous and assimilatory – in other words, different cultures are subsumed into the dominant culture (the 'melting-pot' idea) (see Case study 13.5). Denmark, for example, is highly assimilationist and has one of the toughest policies on immigration in Europe. Others call for the acceptance of ethnic pluralism and the preservation of minority ethnic cultures as a legitimate part of the national project. This is the politics of **multiculturalism**, which instead of thinking of different cultures as being absorbed by dominant cultures, relies more on a notion of a cultural mosaic, or a 'patchwork quilt' of cultures. Each culture is recognized as different and distinct, but these differences are understood and valued. Sweden, for example, rejects Denmark's assimilationist model in favour of multiculturalism.

Multiculturalism might seem more progressive, but it can sometimes reinforce difference because culture is

Thematic Case Study 13.5

Bounded or hybrid national culture?

Nationalists around the world cling to a notion of bounded cultures that make them distinct from others. One effect of globalization has been resistance in the form of increased nationalism to what is perceived to be the erosion of national cultures. Increased mixing of cultures is seen to pose a threat to the survival of national cultures. Nationalists seek to preserve the symbols of nationhood, such as language, lifestyles and cultural forms in the face of what are perceived to be sweeping changes. However, cultures are not unchanging; a fundamental flaw in nationalist ideology is the adherence to a notion of static culture, and its reliance on a mythical history of the origin of the nation. For example, English nationalists define Englishness as distinct, which is used to justify anti-immigrationist ideas, anti-Europeanism and, in some cases, racism. But who are 'the English'? After the last ice age many communities settled in the British Isles from all over Europe. They lived and fought with each other and in a short space of time produced a mixed group of people who eventually called themselves English. The British Isles have been subject to waves of invasion and settlement (e.g. Celts, Vikings, Romans, Anglo-Saxons, Normans). England and the British Isles have always been hybrid; peoples and cultures have mixed and evolved together. Some English nationalists avoid thinking about this point by arguing that the final invasion (by William the Conqueror in 1066) marks the origin of England and Englishness (Anderson 1983). This myth is also flawed. William the Conqueror spoke no English. Whom did he conquer? He conquered 'the English'. For many nationalists the founding father of England is French! It is also no small irony that one of the symbols of Englishness, the monarchy, changed its official name to Windsor in 1917 from Saxe-Coburg-Gotha, thus hiding its German origins. Even protectionist policies towards language are flawed; as with culture more generally, language is always hybridized and evolving. Most sentences in 'English' contain words that derive from German, French, Spanish, Latin, Nordic and Celtic languages. Similar myths of origin and notions of bounded cultures exist elsewhere in the world. In some places they emanate from deeply contradictory ideas (anti-immigrationist views in the USA and Australia, for example); in other places they have led to conflict (recent events in Bosnia and Kosovo, for example). The idea of nationhood, based on fixed, bounded and unchanging cultures, is an ideological creation that masks profound cultural divisions of gender, race, class and religion within a nation-state, and ignores the fact that, in reality, all cultures are hybrid and dynamic.

seen as essentially connected to **race**, and racial difference as rooted in biological difference. (These ideas are no longer considered acceptable; anti-racists have demonstrated that 'race' is socially constructed and has little basis in biology – see Kobayashi and Peake 1994. We could just as easily have 'races' of blue-eyed and brown-eyed people.) Multiculturalism, therefore, still relies on a negative notion of bounded cultures. It might suggest tolerance, but often results in segregation and ghettoization. The United Kingdom and Netherlands, for example, have tended towards multiculturalism but have increasingly witnessed tensions surrounding the lack of integration of some Muslim communities. The murder of Dutch film-maker Theo Van Gogh in 2004 and the London bombings in July 2005 – both by apparent 'home-grown radical Islamists' – are seen by some to represent the failures of multiculturalism. Similarly, the 2006 riots in the suburbs of Paris were seen as a product of the deep alienation of poor, largely immigrant communities facing high levels of unemployment, discrimination in housing and jobs markets, and police harassment.

A similar example is the appearance of 'gay spaces' in cities such as San Francisco, London, Manchester and Cape Town, which might appear to signal greater tolerance and cultural diversity, but points to segregation rather than integration. Indeed, in the USA politically correct accommodation agencies now ask applicants to tick a box for 'gay' or 'straight' so that they can be housed suitably. As Reeves (1999) argues, 'Homophobia, overt and covert, is herding lesbians and gay men into ghettos – which are then celebrated as signs of a tolerant society.' It is OK to be gay in these segregated spaces; it is still not OK to be openly gay in other spaces of the city.

In contrast, a more progressive idea of culture (and its manifestations in ethnicity, gender and sexuality) might be developed through the concept of **hybridity**. Hybridity breaks down barriers, adhering to neither the 'melting-pot' nor 'mosaic' idea of cultural mixing, but rather seeing different cultures coming together and informing each other in different ways to produce something entirely new. This process has a long historical trajectory. Indeed, some argue that cultures have always been hybrid forms – they have never existed in isolation from other cultures, and thus have always been subject to change and influences from elsewhere (Werbner 1997: 15).

One of the most obvious places we can observe hybridity is in popular mass culture, an immediate example being popular music. Innovations in music have always involved the fusion of different styles to create new sounds and rhythms. Rock-and-roll, rhythm-and-blues and Latin jazz are obvious examples; we might also think of contemporary forms of music that fuse different styles, such as 'trip hop' and 'nu metal'. Authors such as Barthes (1972), Bourdieu (1984) and Bakhtin (1984) see popular hybridity as an exciting challenge to, or subversion of, dominant cultures and the exclusive lifestyles of dominant elites. Such popular mixings and inversions, like the subversive elements of youth cultures (see Hebdige 1979; Skelton and Valentine 1998), are hybrid in the sense that they bring together and mix languages and practices from different and normally separated domains. They have the potential to disrupt dominant cultures by their 'out-of-placeness'.

In many ways hybridity is related to the notion of **transculturation**. Transculturation describes one of the key cultural processes that operate between hitherto sharply differentiated cultures and peoples who are forced (usually by the processes of imperialism or globalization, and primarily through migration) to interact. This interaction often takes place in profoundly asymmetrical ways in terms of relative power between different groups. Despite the illusion of boundedness, cultures evolve historically through borrowings, appropriations, exchanges and inventions. Cross-fertilization of cultures is endemic to all movements of people throughout history (see Case study 13.5). However, for those who aspire to bounded notions of culture and refuse this idea of perpetual hybridity, cultural mixing is felt to be threatening and a deliberate challenge to social order. In reality there are no fixed cultures in modern nation-states, but some people cling to ideas of pure or impure cultures. For others, however, hybridity remains the site of revitalization, resistance and fun (see Case study 13.6).

13.5.2 Diaspora

Related to the idea of hybridity is the notion of **diaspora**. This term was originally used to refer to the dispersal of Jewish peoples, but is now used in reference to the long-term settlement of peoples in 'foreign' places that follows their scattering or dispersal from their original homeland. It refers to a modern condition where a sense of belonging is not derived from attachment to territory, and where different peoples mix together through the processes of migration (forced or free). European imperialism and associated processes of globalization have set many of these

Thematic Case Study 13.6

Hybridity/Diaspora – some examples

British Bhangra

British Bhangra developed out of traditional forms of folk dances and music of the people of the Punjab, which arrived in Britain in the late 1960s with the migration of people from South Asia and East Africa. It emerged in the mid-1980s among second-generation South-Asian British musicians, primarily in Birmingham and west London, combining the robust and energetic punctuated rhythms of double-sided drums (*dhol*), modern technology and urban sounds with Punjabi poetry, traditional Punjabi lyrics, acrobatic dance skills and celebrations of the Punjabi harvest festival (*bhaisakhi*). It played an important role in the proclaiming of a cultural identity for groups of youths defining themselves as South-Asian British. Although it appeals to the Punjab as a specific region of origin, Bhangra, with its fusion of instruments and musical styles from South Asia and urban Britain, has a large following among diverse groups of South-Asian British youths. Throughout the 1990s, it was characterized by ongoing processes of fusion with broader ranges of musical styles (ragga, reggae, soul, jazzfunk, rock, hip-hop, pop), both to reflect the lived experiences of South-Asian Britons, and as an attempt to 'cross over' into mainstream charts. It is the antecedent to newer and more fashionable forms of post-Bhangra and 'Asian Kool' which have been less marginalized by mainstream culture, and are part of an emergent and vibrant British-Asian urban youth culture.

Source: Adapted from Dudrah (2002)

Cuban Santería

Santería is an example of a syncretic (hybrid, mixed) religion. It is based on West African religions brought by slaves imported from what are now Nigeria and Benin to the Caribbean to work the sugar plantations. These religions were suppressed by the European plantation owners and in Cuba slaves were forced to convert to Catholicism. However, they were able to preserve some of their traditions by fusing together various West African beliefs and rituals and syncretizing these with elements from Catholicism. One factor enabling this process was that many of the *orishas* (primary gods) shared many of the same characteristics of Catholic saints. This enabled slaves to appear to be practising Catholicism while practising their own religions. This has evolved into what we know today as Santería, the Way of the Saints, whose traditions are transmitted orally from generation to generation. Despite suppression by Fidel Castro's Socialist Revolution since 1961, its influence is pervasive in Cuban life. Devotees are found in most households, Yoruba proverbs litter Cuban Spanish, and high priests (*babalawos*) offer guidance based on ancient systems of lore. Today, with less religious persecution, Santería is experiencing a rise in popularity and is part of an emerging Cuban youth culture. Similar syncretic religions are found in Haiti, Puerto Rico and other Caribbean and Latin American countries.

Source: Adapted from Betts (2002)

migrations in motion. Diasporas are classic **contact zones** (spaces in which two cultures come together and influence each other) where transculturation or hybridization takes place. Diasporic identities are at once local and global and based on transnational identifications encompassing both 'imagined' and 'encountered' communities (e.g. Irish-Americans belong to an imagined international community of people who have 'Irishness' in common, but whose identities are also informed by the communities in which they live in the USA). In other words, diasporas are a direct challenge to the idea that there is a simple relationship between place and culture. They transgress the boundaries of the nation-state and provide alternative resources for constructing identity and fashioning culture.

The concept of diaspora space allows us to think of 'culture as a site of travel' (Clifford 1992), which seriously problematizes the idea of a person being a 'native' or an 'insider'. Diaspora space is the point at which boundaries of inclusion and exclusion, of belonging and otherness, of 'us' and 'them', are contested. As Brah (1996: 209) argues, diaspora space is 'inhabited' not only by those who have migrated and their descendants, but equally by those who are constructed and represented as indigenous or 'native'. In the diaspora space called 'England', for example, African-Caribbean,

Irish, Asian, Jewish and other diasporas intersect among themselves as well as with the entity constructed as 'Englishness', thoroughly reinscribing it in the process. Like notions of hybridity, the concept of diaspora is important since it allows for the recognition of new political and cultural formations that continually challenge the marginalizing impulses of dominant cultures.

13.5.3 Selling hybridity and the commodification of culture

In today's world, culture sells. Hybrid cultures, in particular, sell. Cities are now constructing themselves as cosmopolitan, and hybridity has become a form of 'boosterism' – where city authorities create marketing images to attract investment in the form of business and tourism. Hybrid culture is perceived as creating economic advantage. With increasing deindustrialization in Europe and North America, cultural strategies have become key to the survival of cities. Examples include the international marketing of cultural/religious festivals such as Mardi Gras in Sydney or New Orleans, or the importance of 'Chinatowns' and other 'ethnic' districts to tourism in cities throughout the developed world (see Plate 13.3). Ironically, hybridity is in danger of becoming just another marketable commodity. For example, treating the political work of some British-Asian bands as marketable hybridity trivializes black political activity and leaves problems of class exploitation and racial oppression unresolved. As Hutnyk (1997: 134) suggests, to focus on hybridity while ignoring (or as an excuse for ignoring) the conditions in which this phenomenon exists (the commodity system, global economic inequality, inequitable political relations), is problematic in that it maintains the status quo. Hybridity and difference sell, but in the meantime the market remains intact, power relations remain unequal, and marginalized peoples remain marginalized. Moreover, as culture is subsumed into capitalism, those marginalized peoples who might be capable of oppositional politics are also subsumed under the rubric of hybridity.

The notion of hybridity, therefore, can be problematic. In some Latin American countries, cultural elites and nation-states have appropriated the hybrid mestizo (mixed) identity, making it dominant. This has been seen to be oppressive of 'Indian' populations, who have in turn been accused of ethnic essentialism (or emphasizing their racial difference) because of their desire to protect their cultures (Radcliffe and Westwood 1996). In the West, ideas of hybridity are currently popular with highly educated cultural elites, but ideas about culture, ethnicity and identity that develop in poverty-stricken underclass neighbourhoods are likely to be of a different nature (Friedman 1997: 83–4). Evidence of racial tensions in many North American and European cities points to the fact that class and local ghetto identities tend to prevail, with little room for the mixing pleaded for by cultural elites. The global, cultural hybrid, elite sphere is occupied by individuals who share a very different kind of experience of the world, connected to international politics, academia, the media and the arts. In the meantime, the world becomes more polarized in terms of wealth, and heads

Plate 13.3 Chinatown in San Francisco: 'ethnic' districts are promoted as major tourist attractions in many 'global' cities.

(© Dave G. Houser/Corbis)

towards increasing balkanization where regional, national and ethnic identities are perceived as bounded, threatened, and in need of protection. As Bhabha (1994) reminds us, hybridity is an insufficient means through which to create new forms of collective identity that can overcome ethnic, racial, religious and class-based antagonisms – it sounds nice in theory, but does not necessarily exist outside the realms of the privileged.

Stuart Hall (1996: 233) argues, however, that we should not view the current fashionability of hybridity in a wholly negative light. Even as cultures are increasingly commodified, we should not forget the potential for the democratization of culture in this process, the increased recognition of difference and the diversification of the social worlds in which women and men now operate. This pluralization of social and cultural life expands the identities available to ordinary people (at least in the industrialized world) in their everyday working, social, familial and sexual lives. For Bhabha (1994: 9), the interconnections of different cultural spaces and the overlapping of different cultural forms create vitality and hold out the possibility of a progressive notion of culture.

We thus need to think about the place and meaning of cultural hybridity in the context of growing global uncertainty, xenophobia (fear of foreigners) and racism. Why is cultural hybridity still experienced as an empowering, dangerous or transformative force? Why, on the one hand, is difference celebrated through a consumer market that offers a seemingly endless choice of identities, sub-cultures and styles yet, on the other hand, hybridity continues to threaten and shock? Conversely, why do borders, boundaries and 'pure' identities remain important, producing defensive and exclusionary actions and attitudes, and why are the latter so difficult to transcend? Is the sheer pace of change in cultural globalization producing these reactions?

To summarize, hybridity and diaspora are examples of more progressive ways of thinking about culture. It could be argued that all cultures are always already hybrid; they are never pure, have always evolved and changed through time and through contact with other cultures, and they continue to evolve. In today's world, hybridity is being commodified, which might make it less radical. However, despite this, it has the potential to democratize culture and to allow us to rethink culture in ways that are more tolerant of difference. Finally, cultural hybridity needs to be understood in the context of growing global uncertainty, xenophobia and racism.

13.6 Conclusion

In this chapter it is suggested that there are two apparently contradictory tendencies in thinking about cultures – the attempt to secure the purity of a culture by conceptualizing it as strong, fixed, bounded, permanent and homogeneous, and the hybridity of most cultures. Culture is thus a contested concept. A progressive way of thinking about culture is to reject the idea of boundedness and internal cohesion. In the modern world especially, culture is a meeting point where different influences, traditions and forces intersect. There is, therefore, a continual process of change in cultural practices and meanings. Globalization is undermining closed, fixed ideas of culture and leading to new ways of conceptualizing cultures (transculturation, contact zones, hybridity and diaspora). However, the fact that cultures are not fixed or homogeneous does not mean that we will stop thinking of them in this way. As Stuart Hall (1995: 188–9) argues, this is because some people need 'belongingness' and the security that closed conceptions of culture provide.

Despite this, recent years have witnessed a decentring of culture, with nation-states increasingly superseded by transnational institutions that are producing cultural globalization and greater cultural diversity. There has also been a shift in the awareness of the cultural capital of the West, and an understanding of the cultural dominance of developed countries. At the same time, there are now more voices talking back, reflecting the cultural assertiveness of marginalized groups and making us aware of new levels of diversity. Even though most people remain physically, ideologically and spiritually attached to a local or a national culture and a local place, complex cultural flows and networks ensure that it is becoming increasingly impossible for people to live in places that are completely isolated and disconnected culturally from the wider world. Thus,

> if there is a global culture it would be better to conceive of it not as a common culture, but as a field in which differences, power struggles and cultural prestige contests are played out. . . . Hence globalization makes us aware of the sheer volume, diversity and many-sidedness of culture.
>
> (Featherstone 1995: 14)

This points to a 'more positive evaluation by the West of otherness and differences' (ibid.: 89). For Massey and Jess (1995b: 134), globalization is not simply a threat to existing notions of culture, but a 'stimulus to a positive new response'.

People around the world have different cultures and systems of meaning, but we cannot avoid reading the world from within our own cultures and interpreting it through our own systems of meaning. Understandings of global culture for the majority of the readers of this book are filtered through the logic of the West (Spivak 1985). Western ideas and cultural forms are still considered superior and have become hegemonic, or dominant. Similarly, one's own cultural positioning (on the basis of gender, ethnicity, class, location, sexuality, stage in life cycle, ability) also influences understandings of local cultures. The same processes operate at local levels; dominant cultures marginalize others on the basis of ethnicity, sexuality, gender and religion. However, as we have seen, those dominant cultures also produce resistances that have the potential to create new ways of thinking about culture.

The challenge is to confront the limits of 'our' knowledge, to recognize other worlds, to acknowledge the legitimacy of other cultures, other identities and other ways of life. Accepting 'cultural translation' (Bhabha 1994) involves understanding the hybrid nature of culture, the influence of marginal cultures on dominant cultures, and that people in marginal cultural systems at local, national and international levels are also active in creating their own systems of meaning. They do not simply absorb ideas from, or become absorbed into, more dominant cultures. It is possible to develop cosmopolitanism in the twenty-first century that is global, sensitive to cultural difference, and dynamic.

Learning outcomes

Having read this chapter, you should be able to:

- Understand the complexities of culture; it is a process, rather than a thing, and subject to change over time.
- Demonstrate ways in which cultures operate at local, national and global levels.
- Discuss examples of global cultural processes and how these are filtered through localities to contest notions of a singular global culture.
- Discuss examples of negative (closed, bounded), or positive (progressive, hybridized) local cultures.
- Critique and utilize concepts such as hybridity and diaspora as a progressive way of thinking about culture.
- Reflect on your own cultures of knowledge, and the ways in which these condition your ideas about global, national and local cultures.

Further reading

Anderson, K., Domosh, M., Pile, S. and Thrift, N. (eds) (2003) ***Handbook of Cultural Geography***, Sage, London. This book offers an assessment of the key questions informing cultural geography and contains over 30 essays. It is an invaluable resource for students looking for an assessment of major issues and debates and the breadth, scope and vitality of contemporary cultural geography.

Cook, I., Crouch, D., Naylor, S. and Ryan, J. (eds) (2000) ***Cultural Turns/Geographical Turns***, Prentice Hall, London. Aimed at advanced undergraduates and postgraduates, this book addresses the impact, significance and characteristics of the 'cultural turn' in contemporary geography and the intersections between space and culture.

Crang, M. (1998) ***Cultural Geography***, Routledge, London. A useful introduction to the ways in which geographers think about culture, exploring the global and the local, cultures of production and consumption, and ideas about place, belonging and nationhood.

Duncan, J.S., Johnson, N.C. and Schein, R. (eds) (2004) ***A Companion to Cultural Geography***, Blackwell, London. A comprehensive introduction to cultural geography, with 32 chapters written by leading authorities in the field and covering debates about cultural theories, nature/culture, culture and identity, landscapes and colonial/postcolonial geographies.

Massey, D. and Jess, P. (eds) (1995) ***A Place in the World? Places, Cultures and Globalization***, Oxford University Press, Oxford. A comprehensive first-year text, it explores the conceptualization of place, and how this relates to identity, cultures and contestation.

Shurmer-Smith, P. (ed.) (2002) ***Doing Cultural Geography***, Sage, London. This is a good book for those students thinking of undertaking projects or writing dissertations in cultural geography. It explains the theory informing cultural geography and encourages students to engage directly with theory in practice.

Useful websites

www.culture.gov.uk The official website of the UK Government's Department for Culture, Media and Sport (DCMS). An animated site that includes information about the role of the DCMS, and government policy towards media and arts, heritage, libraries and museums, sport, the National Lottery and tourism. Also has links to other useful sites in each category.

www.alt.culture.com An encyclopaedia of modern culture spanning modern musical styles, film, cyberpunk and street fashion, extreme sports and political correctness. A useful site for those interested in global cultures of consumption.

www.adbusters.org/home An example of 'culture jamming'.

www.un.org/womenwatch The official United Nations Internet Gateway on the Advancement and Empowerment of Women, and part of the global phenomenon of cyber-feminism, which might be considered an example of a 'third culture'. Has useful links to other sites advancing women's rights through new technologies.

www.asiansociety.co.uk A comprehensive site on British-Asian culture, especially the Asian entertainments scene across the United Kingdom (Bhangra, TV, films and radio, etc.).

For an overview of the South African transformation, visit **www.nobel.se/peace/index.html** and its articles on President Mandela and Bishop Tutu. There are some interesting links to other sites on South African culture and history at **www.columbia.edu/cu/lweb/indiv/africa/cuvl/SAfrcult.html**, and **http://www.roughguides.com/website/travel/destination/content/default.aspx?titleid=35&xid=idh125970096_0843** contains a useful introductory summary of the diversity of South African cultures.

For an introduction to Cuba's hybrid cultures see **http:// folkcuba.com/** and **www.afrocubaweb.com**

For annotated, clickable weblinks and useful tutorials full of practical advice on how to improve your study skills, visit this book's website at **www.pearsoned.co.uk/daniels**

PRODUCTION, EXCHANGE AND CONSUMPTION

Section 4

Edited by Peter Daniels

The idea of going 'global' is often associated with the economic aspects of human geography. This section of the book demonstrates how 'the economic' shapes many aspects of human and organizational behaviour and, thereby, the spatial patterns of growth and development. Many economic processes and their geographical outcome are increasingly a function of links and dependencies that extend far beyond the immediately local or national; they involve diverse interactions with places and people distributed around the world. Many of the symptoms of this 'globalization' are most obvious in the sphere of consumption (Chapter 18); Coca-Cola (USA), Nokia (Finland), BMW (Germany) and Louis Vuitton (France) are brands that are almost universally known, if not equally accessible, worldwide. Such products and services are often slightly customized to conform with local tastes, cultures or regulations (often reflected, for example, in the way that they are advertised in different markets), but essentially they offer a uniform, high quality, experience or convey a distinctive fashion statement wherever they are consumed in the world (see Chapter 18).

Consumption is both a consequence, and a driver, of the globalization of production. The shift from production systems largely focused on national markets to production facilities of the same company distributed around the globe has been accelerating and increasingly involves sophisticated global production chains and networks (see Chapter 15). In part it reflects a search for new markets as national sales opportunities have become saturated; but it also reflects the demand created by a growing band of global consumers, boosted by significant economic growth in newly industrializing countries such as India and China, that can easily access common sources of information via the Internet, cable and satellite services that have, metaphorically, made the world smaller. Fordist methods of production helped both to fulfil and to create demand for high-volume, low-cost, standardized products at a time of steadily increasing incomes per capita during the first quarter of the twentieth century, but since then there has been a shift towards post-Fordist production systems with an emphasis on greater specialization, customization, different firms

→

joining together to undertake research while remaining competitors (strategic alliances) (see Chapter 15), even more fragmented divisions of labour as well as expertise (see Chapter 16) and horizontal rather than vertical integration of production. These have helped, on the one hand, to continue the trend towards globalization, with the first global shift led by manufacturing and the second global shift by service industries (see Chapter 16), and on the other, to reinstate the role of localities. The latter have been reasserting their role as the focus for production in conjunction with global-level influences. The massive production plants employing thousands of workers that were symbols of the Fordist era have been replaced by industrial districts made up of networks of firms, small and large, each contributing particular skills and knowledges relating to particular products. Together they form distinctive nodes of production such as the Third Italy (small-scale skilled production in the North East, Emilia, Tuscany and Marches), Bangalore (Silicon Valley of India), or the City of London.

These changes should not lead us to conclude that large firms are no longer important in the globalization of economies. On the contrary, they are even more influential in shaping the geographies of production and consumption or the distribution of spaces that are strongly engaged with the global and those that are much less engaged (see Chapters 14, 15 and 16). Typical are the transnational corporations (TNCs) that often have more financial resources at their disposal than many national governments. They therefore exercise significant influence on decisions relating to the economy, investment in infrastructure, or regulation by nation-states and by international organizations such as the Organization for Economic Cooperation and Development (OECD) or the World Trade Organization (WTO).

Consumption and production are mediated or enabled by the circulation of finance capital (Chapter 17). National governments, companies large and small, or individuals require finance to allow them to execute investment strategies in the case of companies or purchasing decisions in the case of individual consumers. Advances in information technology and political shifts have greatly increased the global circulation of finance and encouraged innovation in the financial instruments traded daily on stock, commodity, currency and other exchanges in business centres around the world. Yet again, however, as with the other themes explored in this section of the book, the flows of finance capital and access to it are spatially uneven. This operates to the significant disadvantage of marginalized nations, regions and groups within society while reinforcing the advantage of those individuals, companies and nation-states that already have very good access to finance capital.

This section demonstrates how the geography of the economy simultaneously incorporates global and local economic perspectives. They are inseparable even though there have been some important shifts in modes of production and mechanisms of consumption over time and across space. Some of the important causes and the consequences of the uneven economic development that results are highlighted in this section. An overview of all the contributions suggests that there are numerous important issues, questions and challenges that economic geographers can usefully address as the twenty-first century unfolds.

GEOGRAPHIES OF THE ECONOMY

Chapter 14

Peter Daniels

Topics covered

- The changing nature of economic geography
- The rise of the global economy
- Is globalization inevitable?
- Role of localities in a globalizing economy
- Digital economies in the twenty-first century

The economy is everywhere. Wherever we happen to live, work or play our daily existence invariably involves making decisions that have an economic basis. Yet, we tend to take for granted the ways in which society determines how the scarce resources at its disposal are used to provide for its material needs and to produce wealth. Few days pass without the news media reporting on the economy: the balance of exports over imports, the unemployment rate, consumer spending, announcements of factory closures, trends in house prices, the relocation of jobs to other countries. Separate sections on business and the economy in the 'quality' daily newspapers, specialist weekly business publications, or 24-hour satellite television channels devoted exclusively to news and information about economic affairs are numerous (Plate 14.1). In summary, the economy is a set of human activities and institutions linked together in the production, distribution, exchange and consumption of goods and services.

You may have noticed that 'economies' or 'economy' are preceded by various adjectives such as: global, local, market, command (or redistributive), capitalist,

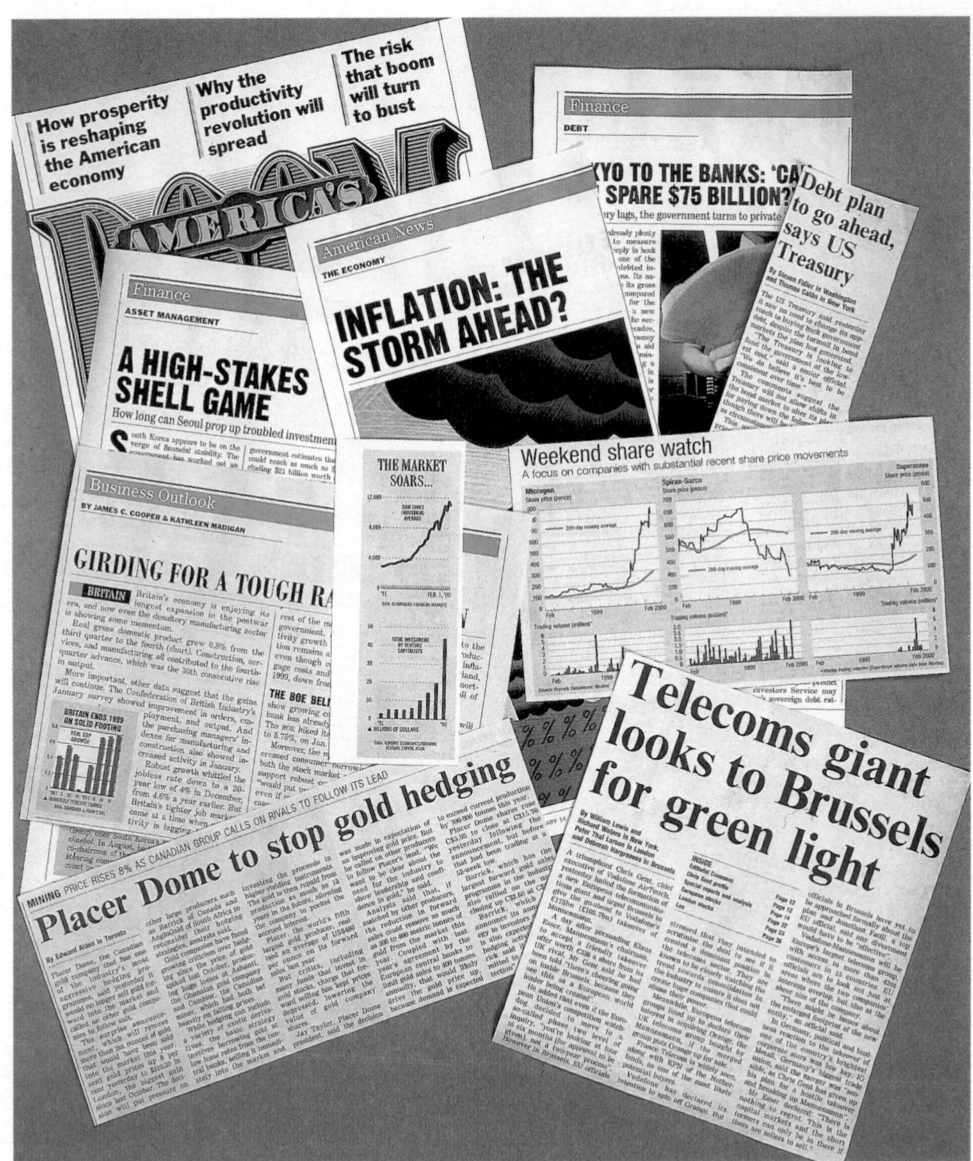

Plate 14.1 Economic and business information occupies an important place in daily newspapers and periodicals such as *The Economist* or *Business Week*.
(Alan Thomas)

informal, subsistence, mixed, Internet, information, new, or space economy. There are clearly many different kinds of 'economy' depending on the starting point for analysis; for example, an informal economy signifies how it is organized while a digital economy signifies something about the medium used to undertake economic transactions. Economic geographers are interested in the space economy; the way in which the organization of geographical space reflects the behaviour, values, and actions of actors (individuals, firms, institutions) that together make up the economy. Each actor operates from or occupies some unit of space: a field for a crop, a room used by a self-employed person working from home, a building of several thousand square metres for a corporate call centre, a site of several hundred hectares used for vehicle production in Korea or Japan, or an area of hundreds of square kilometers occupied by a sheep station in the Australian outback. The households that provide the labour required for production also require space in which to live and are, in turn, a major source for consumption goods and services that result from production; these are accessed via informal and formal transactions and trading. The challenge for the economic geographer is how to analyse and explain the geographical patterns of economic activity at different scales and how they change over time.

14.1 The changing nature of economic geography

The uneven distribution of economic activities is plain enough to see and can be explained by the way in which different parts of our planet are endowed with different resources, climates, or different levels of accessibility. But there is much to understand and to explain and geographers who are interested in the consequences of these spatial variations comprise the major sub-discipline of economic geography which 'is concerned with the spatial organization and distribution of economic activity, the use of the world's resources, and the distribution and expansion of the world economy' (Stutz and de Souza 1998: 41). While the distribution of economic activities is influenced by historical, political, social, and environmental factors the outcomes are a function of **spatial interaction**. It is not possible for all the actors in the economy to be located at one point in geographical space. In its early guise economic geography was termed commercial geography: a largely factual compilation of the circumstances under which different commodities were produced and exchanged around the world. There was some analysis of the factors that affected where particular commodities such as coal or iron ore were produced. Gradually the emphasis shifted towards explaining observed geographical variations in the location patterns of commodity production and trade. The transformation from commercial to economic geography was consolidated when explanations for the location of economic activity were derived from neoclassical economics (Case study 14.1). This is an early example of the way in which economic geography has used ways of thinking and modes of analysis used by other disciplines, in this case economics, to inform its own agenda. Neoclassical economics is the study of the allocation of scarce resources amongst alternative ends when there are several alternative outcomes. Interestingly, one of these is the maximization of social welfare which is based on the idea that the economic basis of society determines the form of its social institutions. We will see later that the social basis of the economy is very much part of contemporary analyses of geographies of the economy.

In the meantime, economic geography experienced a number of transformations during the second half of the twentieth century (see Scott 2004). During the 1950s and 1960s economic geographers led the way in the testing of theories and ideas, using rigorous quantitative methods and modelling. Analysis of economic crises such as the global oil shortage of 1973 used techniques derived from the political economy of regional inequalities or of manufacturing decline (Bluestone and Harrison 1982); economic geographers extensively used Marxist-based interpretations of economic change and its socio-economic consequences. In the 1980s the emphasis shifted again, this time towards empirical and theoretical analysis of some new and surprising economic growth in places previously regarded as peripheral or outside the mainstream. These included the proliferation of small-scale, skilled production units to form new industrial districts in North-east Italy, Emilia and Central Italy collectively known as the Third Italy. This period also saw a marked increase in the international division of labour, led by multinational corporations (Fröbel et al. 1980), which became but one of the issues symbolic of globalization and international economic integration. This has continued to attract the attention of economic geographers (Dicken 2003) and we will return to this theme later.

Thematic Case Study 14.1

Models of economic location

There has been much interest in economic geography in theories or models that help to explain the development of economic activities within a spatial context. ***Neoclassical economic theory*** was first used by Von Thunen in 1826 (Hall 1966) to model patterns of agricultural land use and later applied to industrial location (Weber 1929) and for explaining the distribution of services and of settlements (Christaller 1966). In each of these and numerous other cases the objective is to generalize about patterns of economic activity. However, such is the complexity of the real world that it is necessary to make some assumptions: decision makers behave in a sensible (or rational) fashion; they possess complete and correct knowledge (often referred to as perfect knowledge); everyone is attempting to maximize profits; competition is unconstrained; economic activity takes place on a uniform land surface. The key assumption is that distance is the main influence on decision-making by households and businesses with the resulting spatial patterns of economic activity explained by examining the relationship between distance and transport costs. The outcome of this approach is an ideal or optimal pattern of land use or industrial location.

The main criticisms are the unrealistic nature of the assumptions and the oversight of many other factors that have an impact on the geography of economic activity such as:

- changes in the technology of production and consumption
- the variety of ways in which businesses are organized or make decisions on where to locate
- the role of social and cultural factors
- the effects of international and national political and regulatory environments.

The limitations of neoclassical models encouraged the development of alternatives. The *behavioural model* also attempts to arrive at generalizations but the focus is shifted to the role of the individual as the principal explanation for spatial patterns. The motives, opinions, preferences and perceptions of the individuals making location decisions are incorporated and it is not unexpected that the outcomes are suboptimal (i.e. decision makers do not have perfect information and are not assumed to maximize profits). Examples of the behavioural approach include studies of consumer behaviour (Potter 1979), office location (Edwards 1983), and location change by manufacturing industries (Hayter and Watts 1983). The limitations of the behavioural approach are that it is too descriptive and only highlights variations in behaviour rather than why, for example, an entrepreneur starts a certain business at a particular location. Too much attention is given to factors specific to the firm or households, at the expense of more general processes operating in the wider environment.

The *structuralist approach* is a more holistic way to explain the location of economic activity and how it changes. It is based on the premise that behaviour is shaped or constrained by wider processes in the social, political and economic spheres. Notions of culture and of class rather than individual ideas determine the spatial structure of economic activity. The structuralist approach holds that firms work within a capitalist society in which production is essentially a social process structured by the relationship between capital and labour. It is not enough to explain patterns of economic activity by studying the various sectors such as manufacturing or services for their own sake; rather, it is necessary to examine the 'hidden' mechanisms or processes (e.g. social, political) that underpin economic patterns. These cannot be measured in the way typical of the neoclassical or behavioural approaches, so that much of the structuralist approach involves developing theories about, for example, the way in which changes in economic conditions affect the requirements for production that in turn change the requirements of an economic activity at a given location and that might ultimately cause it to move elsewhere (see Massey 1981).

The various modes of analysis used by economic geographers were increasingly seen as too deterministic. During the 1990s more inclusive, flexible interpretations of the geography of the economy began to surface. Debates about this **new economic geography (NEG)** continue, even down to the level of what exactly is 'new' (see Bryson et al. 1999), and engages economists as well as economic geographers. There is also a debate about whether NEG is really economic geography as opposed to geographical economics (Martin 1999a).

Economists claim that, methodologically, NEG belongs to them. Economic geographers claim that economists have finally acknowledged the role of space (geography) and moved it from the edges of their discipline into the mainstream of economic theory; away from the 'wonderland of no dimensions' (Shepherd and Barnes 2000: 3). In any event, NEG is also about recognizing the importance of the 'cultural' when interpreting the 'economic' (see Chapter 18 for a discussion of the links between culture and the consumption of goods and services). It is argued that lifestyles, beliefs, languages, ideas, imaginations and representations interact with the economic to produce culturalization of the economy rather than the economization of culture. Thus, goods and services incorporate cultural attributes in, for example, their design, marketing, packaging and potential benefits to users. Material possession of a good (car, camera, mobile phone) or consumption of a service (tourism, fast food, financial transaction) is only part of the experience (a real as well as an imagined event) of using or being seen to use or to consume. Consumption of goods and services involves beliefs about what it says about us as individuals or groups: social or job status, image or wealth. Economic geographers have previously recognized cultural factors when using terms such as 'socialist economy', 'Chinese family production networks', or the various 'corporate cultures' encountered in transnational corporations (TNCs); but only recently have they considered the economic and the cultural to be intertwined rather than worthy only of separate study.

We could conclude the overview of NEG here but it is worth loitering a little longer to consider some further issues. For example, is it too simplistic to talk of 'economic geography' rather than 'economic geographies'? If the economy is shaped, at least in part, by social relations (that are by their very nature complex) and exchanges that reflect multiple variations in the value associated with the production, consumption, or circulation of a good or service, then 'are the geographies constituted through peoples' struggle to construct circuits of value sustainable across space and time' (Lee 2006: 417). As Lee points out, if societies cannot do this in ways that allow them to make a living they can only materially reproduce themselves with great difficulty. In this sense, there is more than one economic geography.

NEG is therefore also about how to theorize economies and their geographies. The scope that this provides for a healthy, ongoing debate can be illustrated using a set of basic propositions that are assumed to be true when we conceptualize a capitalist economy (Hudson 2004). First, there should be a variety of concepts of a capitalist economy that reflect the diversity of the flows of people or knowledge, for example, in space and time that make up economies. Secondly, the concepts used should not be constructed on the basis that there are separate economies; rather, they should necessarily be treated as interrelated. Thirdly, economic behaviour and practice is undertaken by individuals or subjects possessing knowledge and skills, although not in the sense that everyone has complete knowledge and skills as in perfect competition (see Spotlight box 14.1). Therefore, and fourthly, economies are a social construct incorporating the full range from the informal habits of individuals to the formal institutions of the state. Fifthly, individual and collective behaviour in a capitalist economy is influenced by, and structures and institutions based on, long-term social relationships. Finally, capitalist economies are reproduced using various governance institutions that exist because capitalist economies are formed via social relations and practices that are competitive and not natural.

You may recall from the beginning of this chapter that there are many different kinds of economy; the capitalist economy is but one among many and you

Spotlight Box 14.1

Perfect competition

The intensity of competition among firms encourages efficiency and helps to keep prices low. The ultimate expression of this is *perfect competition*. In these purely theoretical circumstances the actions of any one individual buyer or seller have a very limited impact on

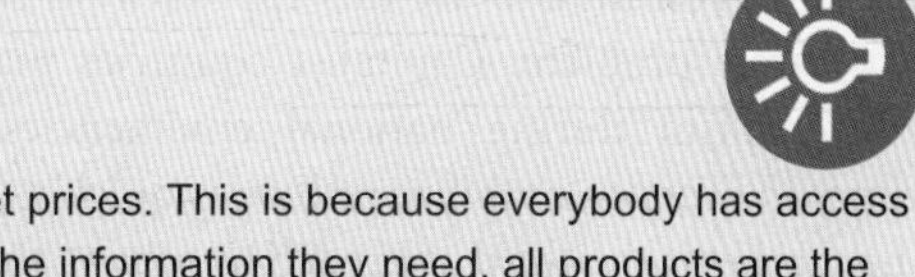

market prices. This is because everybody has access to all the information they need, all products are the same, and firms earn only the base minimum profit. If firms earn excess profits (more than the base minimum) other firms will enter the market. This will continue to happen until profits are driven down so that only normal profits are made.

should consider whether conceptualizations such as the one above are likely to be the same, or different, to those applied to an informal, digital, or knowledge economy?

14.2 What is the economic problem?

Each of us has a variety of needs and wants that, when they are combined, comprise demand for goods and services. The resources available to fulfil this demand are finite. The economic problem (Heilbronner 1972) is how to devise a system that combines the physical and human resources needed to produce and distribute the output of goods and services to attain given ends, such as the maximization of social welfare. Possible solutions are numerous but they can be distilled to just three mechanisms (Dicken and Lloyd 1990). The first of these is **tradition** in which the allocation of resources to production relies on a set of 'rules' based on convention or past practice, such as the handing down of land from father to eldest son. Given the contemporary interest in the relationship between the economic and the social, it is notable that traditional solutions to the economic problem incorporate socially determined norms.

A second mechanism for resolving the economic problem is a **command economy** in which a key objective is redistribution of wealth based on public ownership of the factors of production. Prior to its demise during the late 1980s countries such as the former Soviet Union, Poland, Hungary and East Germany used central political authority to prepare directives setting production targets for a defined period (say three or five years) for the types, and quantities of goods and services to be produced. In order to achieve the targets and to distribute the output the centralized direction of national economic development is combined with control over the allocation of human and physical resources. Some countries such as Cuba and North Korea continue to use this approach for solving the economic problem. Capitalist countries may use this mechanism during national emergencies (e.g. wartime) in order to mobilize resources quickly.

The market is the third approach to solving the economic problem. In contrast to command economies, market economies rely on decentralized decisions by consumers and by firms about the quantities of goods to produce, the prices to charge, and how, when and where exchange transactions take place. Generalizations about the modern market system and its outcomes are actually quite difficult because decentralized decision-making creates complexity; for example, professional economists can rarely agree about the significance of the latest unemployment figures, the impact of a change in bank interest rates, or the likely impact of a steep rise in crude oil prices on the stability of the global economy.

Economies that operate by voluntary exchange in a free market that is not in some way managed or controlled by a central authority do not in practice exist. Most developed nations are mixed economies in that they allow markets to drive most of their economic activities, using regulations and government intervention to ensure stability and efficient economic operation. Market/mixed economies are now in the majority, with fast emerging economies such as China making big strides in the same direction.

Within such **capitalist** economies most of the resources and the means of production are controlled by a relatively small proportion of individuals and firms who are seeking to improve their economic well-being through competition (see Spotlight box 14.1). If we assume that there is more than one producer of a given good or service, each is seeking to create the most favourable value (such as lower price, better quality, superior design, or efficient after-sales support) so as to retain or enhance its share of the market.

The principle of freedom of action may be paramount but, in practice, market economies incorporate regulation by national governments and/or international institutions. In the interests of equity, society needs to control the behaviour of those who own the means of production; the market is not necessarily the most effective and equitable way to allocate its rewards. Thus, most countries have a national organization that monitors and adjudicates company acquisitions and mergers in an attempt to avoid the development of monopolies, i.e. a market situation in which there is only one provider of a particular product or service. The lack of economic competition will allow a monopoly to sell a lower quantity of a product at a higher price than firms in a purely competitive market, leading to monopolistic profits. Because of the increasing integration of national economies and firms (see below in this chapter), the actions of individual national governments are complemented by groupings of nation-states such as the European Union (EU) and the North American Free Trade Area (NAFTA) or international institutions such as the World Trade Organization (WTO) that regulate and facilitate trade flows of goods and services as well as the movement of labour.

Spotlight Box 14.2

Transnational corporations

The definition of a TNC is that it:

- controls establishments/economic activities in at least one other country apart from its home country;
- possesses an ability as a result of its size or the ownership of particular knowledge or skills to move its operations and its resources quickly between international locations, i.e. it is relatively footloose;
- can exploit or take advantage of differences between countries, regions or cities around the globe in factor (land, labour, capital) and non-factor (information, knowledge, regulation) endowments;
- owns and controls overseas activities, although this is not a prerequisite for TNC status since there are many other ways, such as franchising, licensing or joint ventures, of achieving a presence in markets outside the home country.

You may well encounter a related term: multinational corporation (MNC). This suggests a firm that has premises or production plants in several countries. TNC is therefore a much more all-embracing term than MNC; there are many more TNCs than there are MNCs and they provide the basis for a more realistic assessment of the scale of international investment.

One of the key targets for cross-border control and regulation are transnational corporations (TNCs) which serve markets and operate production plants or distribution facilities that extend well beyond their home countries (see Spotlight box 14.2) (see also Chapter 16 in this volume). The distribution of TNCs varies by industry sector and by country; the majority are small relative to a limited number of very large global TNCs. But they generate significant flows of capital, knowledge, information, expertise, products, raw materials and components amongst their own establishments and between countries. There are few parts of the world that are not in some way affected by the activities and decisions of TNCs and, for some countries, it makes the difference between inclusion as opposed to exclusion from the economic mainstream.

14.3 The structure of the economy

So far we have considered the economy in more or less abstract terms, although we have made reference to various 'actors' such as individuals, households, companies, government departments, national governments, and so on. In order to help economic geographers to measure and interpret the processes and interactions that help to shape spatial patterns of economic activity, it is necessary to simplify the complexities of the real world into manageable categories or groupings.

One of the most common approaches to generalizing the structure of economies is to divide them into four broad economic sectors:

1. Activities engaged in the exploitation of natural resources, such as agriculture, fishing, mining and oil extraction, form the **primary sector**. Although still a significant share of activity in the less developed economies, this sector is declining overall. Some of the output from the primary sector has limited use and value until it has been transformed in some way to become part of usable goods.
2. Whether the transformation takes place near the source of the primary commodity or after transfer to a location some distance away it requires a **secondary sector** (or manufacturing). This sector is still expanding in some less developed economies and the emerging economies but contracting (employment or GDP share) in most developed countries. The outputs from the manufacturing sector may be immediately suitable for final use by consumers, or they may be components for incorporation in other final products.
3. The outputs of the secondary sector require distribution to the places and markets where they can be assembled, consumed, or purchased (Plate 14.2). The **tertiary sector**, which includes wholesale and retail trade, transportation, entertainment and personal services, fulfils this role. Improvements in transport and telecommunications and their integration following major advances in computing technology since the early 1980s have transformed

Plate 14.2 Large-scale distribution facilities at strategic locations with easy access to motorway networks serving national or global markets are a vital part of modern production.
(© Walter Hodges/CORBIS)

the operation and reach of firms in both the secondary and tertiary sector. The growth of international purchasing by firms and individuals via the Internet is symbolic of these changes.

4. A fourth, **quaternary**, sector has increasingly been identified as a separate grouping which includes banking, finance, business and professional services, the media, insurance, administration, education, and research and development. These intellectual services assemble, transmit and process the information, knowledge and expertise used by activities in the other three sectors to enable them to adjust effectively and efficiently to the changing geographic, economic, social and cultural parameters of doing business in the twenty-first century. The share of the tertiary (quaternary) sector in the total economy has expanded steadily for at least a hundred years and it now accounts for four out five jobs in countries such as the USA, Canada, Britain, Hong Kong or Australia.

Another useful distinction for tracking change in economic structures is that between white-collar and blue-collar occupations. The former are salaried professionals (such as doctors, lawyers or airline pilots) and employees engaged in clerical or administrative occupations performing tasks that are mentally rather than physically demanding. Blue-collar workers, on the other hand, tend to earn hourly wages for performing skilled or unskilled tasks in factories, construction, or technical installation work. An intermediate category, the pink-collar worker, is also sometimes used to distinguish women mainly engaged in white-collar occupations that do not require as much professional training or who perform tasks to which lower prestige is attached. In line with the increasing share of the tertiary and quaternary sectors, the proportion of white-collar workers tripled during the twentieth century to around 60 per cent of total occupations.

14.4 Dynamic economies

There are clearly several different ways of analysing the structure of economies and it is not possible to examine them all here. But one of the benefits flowing from such segmentations is that they enable economic geographers to measure the dynamics of change and whether, and how, this impacts at different spatial scales. These dynamics occur at all levels of analysis: from small rural localities to the largest metropolitan areas, from peripheral to core regions, from least developed to most developed economies. This is not new or surprising but during the closing decades of the twentieth century it was dominated by changes which will shape the economic agenda well into the present century (such as more flexible production systems, see Chapter 15; or the changing nature of work, see Chapter 16 in this volume).

Table 14.1 Distribution of employment, by sector (%), selected OECD countries, 1994 and 2004

Country	Primary sector 2004	Primary sector 1994	Diff.	Secondary sector 2004	Secondary sector 1994	Diff.	Tertiary sector 2004	Tertiary sector 1994	Diff.
Australia	3.7	5.1	–1.4	21.4	23.6	–2.2	74.9	71.3	3.6
Czech Republic	4.3	6.9	–2.6	39.4	42.7	–3.3	56.3	50.4	5.9
Germany	2.4	3.3	–0.9	31	37.7	–6.7	66.6	59	7.6
Greece	12.6	20.8	–8.2	22.5	23.6	–1.1	64.9	55.6	9.3
Japan	4.5	5.8	–1.3	28.4	34	–5.6	67.1	60.2	6.9
Korea	8.1	12.6	–4.5	27.5	33.6	–6.1	64.4	53.8	10.6
Poland	18	23.8	–5.8	28.8	31.8	–3	53.2	44.4	8.8
Sweden	2.1	3.5	–1.4	22.6	25.1	–2.5	75.2	71.4	3.8
United Kingdom	1.3	2.1	–0.8	22.3	27.6	–5.3	76.4	70.3	6.1
United States	1.6	2.9	–1.3	20	24	–4	78.4	73.1	5.3
G7	2.6	4	–1.4	24.1	28.8	–4.7	73.3	67.2	6.1
EU-15	3.7	5.3	–1.6	27	30.9	–3.9	69.3	63.7	5.6
OECD Total	**6.1**	**8.8**	**–2.7**	**24.9**	**28.4**	**–3.5**	**69**	**62.8**	**6.2**

Note: G7: Group of seven industrialized nations – Canada, France, Germany, Japan, Italy, United Kingdom, United States.

Source: *Organization for Economic Co-operation and Development in Figures 2005*

As we have already noted, the share of country employment in the secondary sector has been declining steadily; by 2004 it directly supported only one in four jobs in the OECD countries (Table 14.1). The primary sector, already a very small part of the economy in developed countries, is contracting further as farm productivity is enhanced by improved crop disease resistance and better fertilizers. Meanwhile, the tertiary and quaternary sectors have expanded and diversified as rising standards of living and disposable incomes have boosted demand for tourism, travel or private health services. The production of goods and services also now incorporates more knowledge-intensive services that are heavily weighted towards the employment of engineers and scientists (see Chapter 16). These activities are growing most rapidly in the newly industrialized countries that are 'catching up' with economies where services already provide more than 70 per cent of total employment (see Table 14.1).

There are numerous historical, political, social and institutional explanations for structural and functional changes in economies. Their relative importance will vary according to the geographical scale involved; local economic change will reflect the circumstances that are particular to that place, such as the depletion of a basic resource endowment or the influence of one major employer. As we move up the spatial scale the factors shaping economic change are more numerous; local or endogenous factors intertwine with external or exogenous influences. The ultimate expression of this is the way in which economic change is increasingly linked to **internationalization**: the spread of economic activities beyond national boundaries. This is not a new phenomenon, but since the late 1980s it has been taking place more quickly and has involved an ever wider range of economic activities. Internationalization should not be confused with **globalization**, which is a more contemporary phenomenon that involves integration across national boundaries of markets, finance, technologies and nation-states in a way not witnessed in the past. It has enabled nation-states, TNCs, as well as individuals, to extend their reach (markets, travel) across the globe faster, further, deeper, and at lower cost than could ever have been imagined, even ten years ago (Plate 14.3). Thus: 'These two themes are intimately connected in the sense that the pervasive internationalization, and growing globalization, of economic life ensure that changes originating in one part of the world are rapidly diffused to others.' (Dicken 1992: 1)

Globalization implies convergence of economic, social and cultural values across the localities, regions

Plate 14.3 Globalization has been accompanied by increasing geographical concentration of corporate control as symbolized by the density of office development and the skyline of Manhattan, New York City.
(T. Paul Daniels, Bromsgrove)

and nations that come together as the 'global economy'. Perhaps economic diversity is being transformed into economic uniformity? However, as all the other chapters in this section of the book show, we should beware of such a sweeping assumption (see also Dicken 2007). 'Globalization' is a convenient shorthand that should not divert us from the reality of persistent variations in the levels of participation in the global economic system of different parts of the world as revealed, for example, by unemployment rates (Figure. 14.1). Low-income countries are home to almost half of the world's population but produced only 3.4 per cent of world gross national income (GNI) in 2001, equivalent to just US$430 per capita. The high-income countries have one-sixth of the world's population but almost 81 per cent of total GNI in 2001, at a rate per capita that is 62 times greater than for low-income countries such as Ethiopia or Nicaragua where severe poverty (more than 50 per cent of the population living below US$1 a day) is a major challenge.

An analysis of the world distribution of household wealth in 2000 underlines the scale of global economic inequality (Davies et al. 2006). Global household wealth averaged US$20,500 per person, ranging from US$144,000 per capita for the USA to US$1,100 per capita in India. Some 90 per cent of wealth is concentrated in North America, Europe and selected high-income Asia-Pacific countries like Japan, New Zealand and South Korea. The USA and Canada have just 6 per cent of the world adult population but a staggering 34 per cent of household wealth (Figure. 14.2). On the other hand, China, which is widely regarded as a fast emerging economy, accounts for 23 per cent of world population but for less than 3 per cent of global household wealth. It is also more unequally distributed amongst countries than income; middle- and low-income nations generate a larger overall share of GNI than their share of wealth.

14.5 The rise of a global economy

Globalization may be far from smoothing economic inequalities but if we assume for the moment that 'Every country may not feel part of the global[ization] system, but every country is being globalized and shaped by it' (Friedman 1999), what are the symptoms? (Amin and Thrift 1994). The first is the pivotal role of *knowledge.* Because the creation and exchange of knowledge tends to be embodied in people (see also Chapter 16) rather than machines, national economic health relies more than ever on 'producing' an educated and skilled workforce. Secondly, *technology has become transnationalized,* especially amongst knowledge-intensive economic activities such as financial services or telecommunications. This does not necessarily mean better access to the factors of production; the increased complexity of the opportunities created by advanced technology means that only those individuals, companies and institutions with the resources to manage high technology can really take advantage.

Thirdly, *there has been a marked increase in the power of finance over production.* Finance capital now takes many forms and moves almost seamlessly and with great speed, especially between the world's stock, currency, commodity and futures exchanges located in 'global cities' such as Tokyo, New York, London and Hong Kong

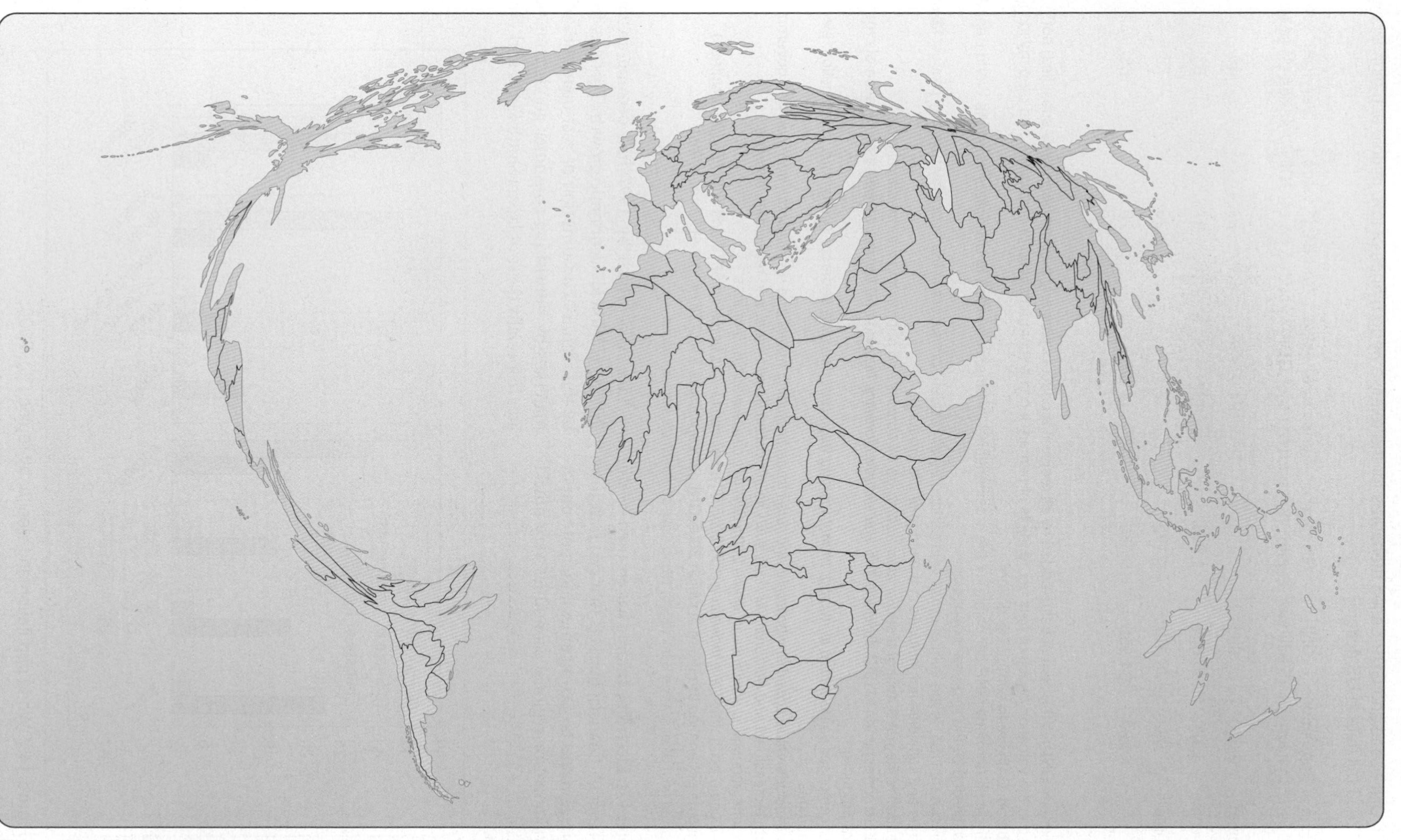

Figure 14.1 Cartogram of unemployment rates, by country, 2006.

Source: *The CIA World Factbook* (2006)

Table 14.2 GNI-based and GDP-based measures for the relative performance of low, middle and high income countries, 2003 (all data in US dollars)

Country groups	Population (millions)	GNI[1] ($ billions)	GNI per capita[2] ($)	Agriculture[5] value added[4] (% of GDP)[3]	Industry[6] value added (% of GDP)	Services[7] value added (% of GDP)
Low income[8]	2,270	995	439	24	27	49
Middle income[9]	3,020	5,854	1,938	10	37	53
High income[10]	997	28,119	28,195	2	26	72
World	6,288	34,959	5,589	4	28	68

Notes:

1. GNI (formerly GNP) is the sum of value added by all resident producers plus any product taxes (less subsidies) not included in the valuation of output plus net receipts of primary income (compensation of employees and property income) from abroad. Data are in current US dollars.
2. GNI per capita (formerly GNP per capita) is the gross national income, converted to US dollars using the World Bank Atlas method, divided by the mid-year population.
3. GDP (in current US$) is the sum of gross value added by all resident producers in the economy plus any product taxes and minus any subsidies not included in the value of the products.
4. Value added is the net output of a sector after adding up all outputs and subtracting intermediate inputs.
5. Agriculture corresponds to ISIC divisions 1–5 and includes forestry, hunting and fishing, as well as cultivation of crops and livestock production.
6. Industry comprises value added in mining, manufacturing (also reported as a separate sub-group), construction, electricity, water and gas.
7. Services include value added in wholesale and retail trade (including hotels and restaurants), transport and government, financial, professional and personal services such as education, health care, and real estate services.
8. Low-income economies are those in which 2005 GNI per capita was $875 or less.
9. Middle-income economies are those in which 2005 GNI per capita was between $876 and $10,725 (90 countries).
10. High-income economies are those in which 2005 GNI per capita was $10,726 or more (52 countries).

Source: Extracted from data in World Bank (2005) *World Development Indicators* database

(see Chapter 17). Electronic trading has ensured volatile and fast-changing financial markets that can transform the economic prospects of companies and, more importantly, of national or regional economies very quickly indeed.

Take Asia, which by 1997 had become an engine driving worldwide economic growth. This was fuelled by the region's consumption of vast quantities of raw materials by flourishing consumer goods conglomerates (especially those of Japan, South Korea and Taiwan)

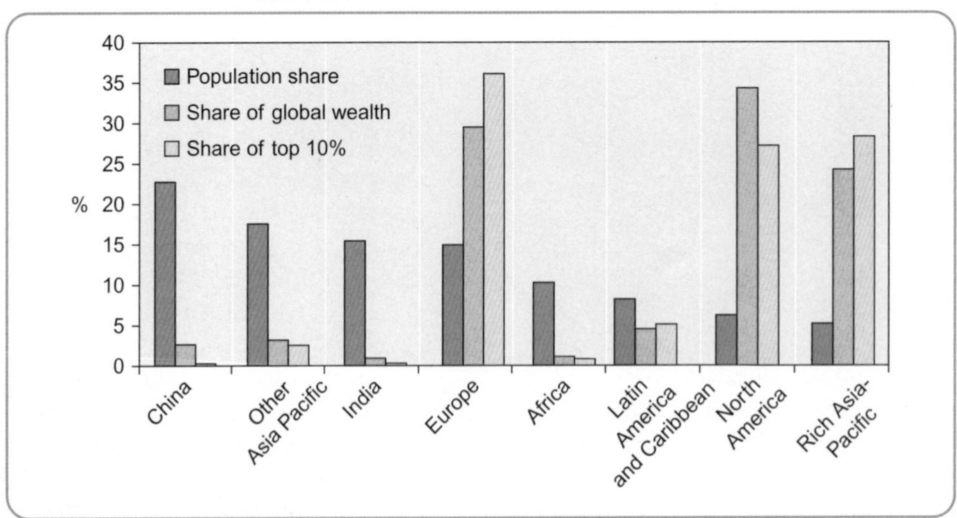

Figure 14.2 World population and wealth, by region (%), 2000.

Source: Davies et al. (2006), derived from figure showing population and wealth shares by region that exported merchandise such as electronic products to every corner of the world. At the end of 1997 global speculators began to think that Thailand's economy was much weaker than suggested by official financial indicators. This 'softening of investor sentiment' started a free fall in the value of the Thai currency (baht) and the government announced that it was closing almost all of the country's leading finance houses, mostly private banks, because they had almost instantly been made bankrupt. This triggered a domino effect in the form of loss of confidence in the economies of the other emerging markets in Asia. With the exception of Singapore, the large volume of finance capital that had previously flowed into the region, often as foreign direct investment (FDI), was rapidly diverted elsewhere.

That was just the start of the economic contagion. World prices for crude oil, aluminium and copper began to collapse as demand fell, transmitting the domino effect to the economies that supplied these commodities. One of the hardest hit was Russia which, since the early 1990s, had been refashioning its economy along market rather than centrally planned lines. With many of its factories not up to the task of producing internationally competitive goods, Russia was very dependent on the revenues from sales of crude oil combined with foreign loans to fund economic readjustment. Foreign banks and corporations, attracted by unrealistically high interest rates (70 per cent and more), were persuaded to purchase financial instruments such as government bonds believing that the International Monetary Fund (IMF) would help Russia in the event of any defaults, thus protecting their loans. In the event, the collapse of crude oil prices instigated by the Asian crisis greatly reduced Russian government revenues, making it difficult to pay interest on government bonds. When it found it necessary to devalue its currency (the rouble) in August 1998, it was not long before Russia defaulted on its government bonds.

Although the Russian economy was small by world standards, events there had further knock-on effects. With some of the world's largest banks reporting losses of several million US dollars on their Russian investments, they needed to plug the shortfalls by selling assets in countries that were more financially sound. Thus the economy of Brazil had been performing quite well during the mid-1990s, but although the government raised interest rates to 40 per cent, it could not prevent a flight of investment to safer havens such as the USA or Europe (United Kingdom, France, Germany).

A fourth symptom of globalization is the emergence of *global oligopolies*, that is, markets dominated by a small number of suppliers of a product or service (Case study 14.2). For a market sector in which there are very few buyers the term *oligopsony* has been coined (see www.oligopolywatch.com). This is the converse of an oligopoly; for example, just six companies in the USA own the majority of movie theatres so that a film distribution company has very few negotiating alternatives. The majority of books in the USA are sold by Borders and by Barnes and Noble which gives them major bargaining power when negotiating with publishers. There is now a sense in which corporate and economic survival requires 'going global'. The fifth symptom of globalization is that the power of individual nations to regulate their own economic development or to exercise a strong influence on the outcome, for example, of trade agreements has been diluted by *the rise of transnational institutions* such as the IMF that coordinate, steer and even regulate aspects of the world economy. The sixth symptom, *the 'wiring' of the world*, has stimulated more extensive flows of people and cultures and has 'deterritorialized signs, meanings and identities' (Amin and Thrift 1994: 4). Finally, these symptoms of globalization present themselves as *new (global) economic geographies*. These take various forms, such as the centralization of economic power in global cities (Sassen 1991), the creation of a 'borderless' global economy (Ohmae 1990), the formation of a 'space of flows' (Castells 1989), or the emergence of a new international division of labour (Dicken 1992) (see also Chapters 15 and 16, this volume).

14.6 Is globalization of the economy inevitable?

Economic globalization seems all-pervasive, but are we dazzled by the rhetoric? Two examples will suffice. First, labour markets have largely continued to function within national boundaries. Of course many millions of people move temporarily or permanently to jobs in countries other than their own (see Chapter 4). But the numbers involved do not stand comparison with the massive migration of labour from Europe to North America or to Australia and New Zealand in the late nineteenth century and the early part of the twentieth century. Between 1890 and 1914 an estimated 200,000 people a year left the United Kingdom alone for the United States, Canada and Australia. In Asia, Latin America or Europe, labour markets are essentially national and regional for, amongst others, linguistic, cultural and educational

Thematic Case Study 14.2

An oligopoly: Nestlé

Nestlé is a Swiss-based company founded in 1866 that is now the top-ranking food and beverage producer in the world with sales of over US$100 billion in 2007. It employs 265,000 and has factories and operations in almost every country in the world. Nestlé has achieved this position by acquiring other related companies and brands; in 2001 it purchased Ralston Purina, making it the leading pet food manufacturer in the world; it purchased the leading French bottled water producer (Perrier) in the late 1990s to become number one for that product. Nestlé is the leading producer of instant coffee (Nescafé was the first instant coffee on the market), and a major manufacturer of sweets and chocolates, and has increased its dominance of the ice-cream sector by merging the Nestlé Ice Cream Company with the leading US company (Dreyers) in 2003. In addition to food products, Nestlé is involved in cosmetics (it has a large stake in L'Oréal), nutritional supplements and eye care. Its long-term strategy is to continue expanding worldwide, especially into emerging markets such as China, India, Latin America and Russia where it has invested US$120 million in a new coffee-processing plant.

Other oligopolies include Coca-Cola, Pepsico, Cadbury Schweppes, Pearson, Interbrew and Gillette. For these leading oligopolies the key objective is to protect or to acquire world-class brands. There are currently some forty consumer brands worldwide that sell more than one billion dollars annually (Coke, for example, sells over US$15 billion). The companies with these exceptional products are growing and expanding internationally at more than 10 per cent per annum.

reasons. The European Union (EU) has actively dismantled the administrative and legal barriers to cross-border movements of labour or the ability of professional workers to practise outside their home countries; this has encouraged some migration of labour from the less prosperous (southern) regions and post-socialist member states to the more prosperous (northern) regions and there are also reverse movements of retirees and others.

A second example is that globalizing markets should equate with more uniform prices and product specifications that reflect the liberalization of trade rules or the pressure of competition on producers to lower their prices. Yet, significant price differentials persist for the same products (DVD recorders, digital camcorders, personal computers, cars, washing machines, books, etc.) available in various regional markets. This is most obvious when making purchases via the Internet; obstacles appear when consumers attempt to take advantage of cross-border price differences for identical products. Such transactions may simply be blocked, home-market taxes or customs charges will be levied, or product warranty terms and conditions will only be honoured at regional rather than global level. The obstacles are numerous and, while companies may also be culpable because they use inefficient distribution networks or charge premiums for minor product modifications in order to meet environmental or other regulations in particular markets, product price convergence is more about potential than reality.

A truly integrated, joined-up global economy may be some way off. More certain is that it will be part of the economic development agenda for the early decades of the present century. ICT will continue to set the pace, changing the way in which information and knowledge is transferred and accessed, modifying the nature of work and the skills required, as well as smoothing the friction of distance and time. The global redeployment of work (see Chapters 15 and 16) offers opportunities for both rich and poor countries. An American company specializing in display technology will have an R&D contract with a technology company in India which brings together the bits that will turn the US company's technology into a high-end television. The Indian company sources its parts for the television from companies in South Korea, Taiwan, the USA and Japan. After design and testing, final assembly will pass to a specialist electronics manufacturer. The finished television may be purchased with a credit card administered from an office in Kuala Lumpur, Malaysia with after-sales services provided from a call centre in Ireland. Satellite and other surface-based telecommunications media have made it much more difficult for governments to regulate such activities (using trade barriers as well as internal controls). This fuels the potential for further displacement of jobs and growth opportunities for less developed economies. It also creates increased uncertainty as TNCs can relatively easily move production from one country to another.

Further economic convergence will reflect changes in national and international political priorities, including the extent to which individual nations are prepared to give up some of their sovereignty over economic policies. Greater economic integration in Asia, for example, will only be hastened by a reduction in the cultural differences/values across the region. To some extent this is also true for the European Union which launched a new currency in 1999. The Euro (€) became legal tender (replacing the old national currencies) in 12 of the 15 EU member countries on 1 January 2002 (the number of member states is now 25). Only the United Kingdom, Sweden and Denmark remain outside the 'eurozone'. With Slovenia also adopting the Euro on 1 January 2007 the new currency now seriously rivals the US dollar, the Japanese yen and the UK pound sterling as a foreign currency reserve in the coffers of companies as well as governments such as Sweden, Russia and China.

14.7 Global convergence: the examples of tourism and foreign direct investment

14.7.1 International tourism flows

One advantage of transnational currencies like the Euro is that travellers no longer have to endure the costs of currency exchange within Europe as they move from one country to another. Companies and governments can also raise money across borders without the need to allow for the possible effects of exchange rate fluctuations. Combine these advantages with all the other drivers of globalization and it becomes apparent that *flows* of all kinds both enable and sustain the process. These flows vary in intensity, duration, volume, space and time but where they touch down determines how, and in what ways, places and regions either benefit or lose out from globalization.

We can illustrate this using international tourism for which the United Nations World Tourism Organization (UNWTO) divided country arrivals in 2005 into leisure, recreation holidays (402 million), business travel (125 million), and visiting friends and relatives, religious purposes, health treatment and so on (212 million). Apart from demand for land, air and sea transport services, such large-scale flows also represent tourism receipts at a large number of destinations. Whether travelling on business or for leisure, visitors spend on accommodation, food and drink, local transport, shopping, entertainment and so on. In many cases this creates much-needed employment and other economic development opportunities, often in those less developed regions that have been largely by passed by some of the other symbols of participation in the global space of flows (Table 14.3). Tourist destinations such as the Caribbean islands, Thailand, Bali, the Maldives or Fiji are benefiting from **comparative advantage** (Spotlight box 14.3) as well as the concept of **competitive advantage** which is the advantage that a country or a business has over its competitors because of the quality or superiority of its products or services which will persuade other countries or customers to buy from it rather than from competitors (Porter 1990). The latter is a more useful way of explaining how individual nations participate in the international tourist market in that it incorporates differences in values, cultures, histories and institutions as well as variations in endowments of the factors of production that make some tourist destinations more appealing than others.

14.7.2 Foreign direct investment

Another indicator of global integration is *foreign direct investment* (FDI). The development of many international tourist destinations relies on FDI, which is defined as the acquisition by an individual or enterprise resident in one country of assets (such as hotels, restaurants and clubs) located in another. World FDI inflows and outflows for developed and developing economies fluctuate from year to year in line with the performance of world, regional or individual national economies (Table 14.4). Overall, flows have been increasing in real terms since 1980, with the developed countries generating a net outflow of FDI in excess of US$1,271 billion in 2000, with developed countries accounting for almost 80 per cent of the total. Flows had almost doubled since 1998 but then declined to an estimated US$760 billion in 2001. FDI fluctuates year on year. A slowdown of the world economy in 2000–2001, for example, caused a decline in corporate cross-border mergers and acquisitions (M&As) which make up a large part of FDI activity. The value of worldwide M&As (domestic and foreign) during the first eight months of 2001 was half the value reported during the

Table 14.3 International tourist arrivals and receipts, world regions 1990 and 2005

	International tourist arrivals (million)				Tourist receipts (US$billion)	
Region	1990	2005	Share (%)	Change (%)	2005	Share (%)
Europe	266	442	54.8	66	348	51.2
Asia and the Pacific	56	155	19.3	176	139	20.4
Americas	93	134	16.6	44	145	21.2
Africa	15	37	4.6	146	22	3.2
Middle East	10	39	4.8	290	28	4.0
World	439	806	100	84	680	100

Source: United Nation World Tourism Organization (2006) *World Tourism Highlights, 2006 Edition*. UNWTO, Madrid

same period in 2000. The tragic events in September 2001 in New York were also followed by a cutback in FDI flows, but by 2006 M&A activity reached a record high with deals worth $US4 billion. FDI between developed countries is more important than FDI into developing countries. However, some of the biggest inflows in recent years, although still a relatively small proportion of the total, have been recorded by the newly emerging economies of Central and Eastern Europe and China.

TNCs contribute significantly to FDI activity. Although the number and size of developing country TNCs is now increasing, most FDI inflows result from the activities of developed country TNCs and over 60 per cent of their investments are made in other developed economies. Outflows from developing and transition economies are now increasing as TNCs from these countries evolve into major regional and even global players. For those countries able to attract it, FDI helps to secure access to capital, technologies and organizational expertise. The net effect is modernization of infrastructure, an increase in industrial capability and an improvement in the quality and breadth of much-needed financial, business and professional services. China, a vast potential market comprising more than one-fifth of the world's population, was closed to FDI

Spotlight Box 14.3

Comparative advantage

Comparative advantage explains the tendency for countries (or regions/localities within countries) to specialize in certain goods and/or services even if they have the ability to fulfil their needs from domestic production. As long as countries or regions specialize in those products or services in which they have comparative advantage, they will gain from trade. Advantages can stem from spatial variations in, for example, mineral or land resource endowments, from variations in the educational levels of the labour force, in access to markets, or differences in levels of technology. For comparative advantage to work effectively it is necessary to assume a system of free trade, hence the significance of the trade liberalization that has been high on the agenda of many countries since the 1980s. There is a nagging concern, however, that comparative advantage is not reflected in actual patterns of world trade. We would expect the biggest flows to be between the countries with the largest cost differences. It seems that consumer tastes and geographical proximity are actually more important than cost differences. This explains the fact that the vast majority of trade is between countries with relatively small cost differences, often involves similar rather than different goods (such as cars, electrical goods of all kinds, certain kinds of business services such as management consulting), and occurs on a 'nearest neighbour' basis. Well over half of EU goods and services trade takes place between the member states while Canada and Mexico are the major trading partners of the USA.

Table 14.4 FDI inflows and outflows, developed and developing economies, 1980–2004

	$US (million)	1980	1990	2000	2004
Developed economies	Inflows	46,629	172,067	1,134,293	380,022
	Outflows	50,407	225,965	1,092,747	637,360
	Balance (I-O)	−3,777	−53,898	41,546	−257,338
Developing economies	Inflows	8,455	35,736	253,179	233,227
	Outflows	3,336	12,701	143,226	83,190
		5,119	23,036	109,952	150,037
World	Inflows	55,108	207,878	1,396,539	648,146
	Outflows	53,743	238,681	1,239,149	730,257

Note: FDI inflows and outflows comprise capital provided (either directly or through other related enterprises) by a foreign direct investor to a FDI enterprise, or capital received by a foreign direct investor from a FDI enterprise. FDI includes the following components: equity capital, reinvested earnings and intra-company loans.

Source: United Nations Conference on Trade and Development (UNCTAD), Major FDI Indicators (extract from table at http://stats.unctad.org/FDI/TableViewer/tableView.aspx?ReportId=5 accessed 31 December 2006

until very recently but is now one of the leading recipients. Prior to the collapse during the late 1990s, almost two-thirds of the investment flows to developing countries went to Asia (excluding Japan). Some of the states of Latin America such as Brazil and Chile were also attracting significant FDI, along with Eastern Europe before the knock-on effects of the Asian financial crisis surfaced. During 2002 and 2003, investor confidence in these regions recovered but much of the African continent continues to be bypassed for any form of FDI, even though many of its countries are resource-rich. With the exception of South Africa, consumer-purchasing power is low and average incomes per capita are not anticipated to grow in a way that will raise the consumption of major consumer goods to a level that justifies the investment attention of most TNCs.

To summarize, we can observe that the globalization of economies has strengthened the role of market forces while, as a result of advances in telecommunications and transportation technology, it has eased some of the constraints on interaction imposed by space and time. With the real cost of international telecommunications declining steadily over the past fifteen years, global financial integration has strengthened and has been accompanied by diversified opportunities for new kinds of international trade, especially in services, and enabled easier transfer of the technology and innovation that encourages economic development and participation. The importance of non-government organizations, TNCs and regional trading blocs such as the European Union and NAFTA for shaping the economic geography of globalization has been greatly enhanced and will continue to increase.

14.8 Role of localities in a globalizing economy

While the geography of economies accommodates a global dimension, this is not at the expense of individual localities. Indeed 'it is the combination of national and intensely local conditions that fosters competitive advantage' (Porter 1990: 158). Localities are subdivisions within nations, such as cities or regions or places that often have a particular economic identity because of the kinds of activities that take place there. Rural districts or subdivisions that specialize in the production of very particular kinds of wine (France), dairy products (Denmark) or woollen goods (Wales), for example, are also included. Many are identified as named territorial units while others are industrial districts or agglomerations whose identities are derived from specific economic activities, such as Motorsport Valley (Pinch and Henry 1999) in southern England, Silicon Valley in California or the light industrial districts of the Third Italy. Although these are often highly specialized and self-contained localities with systems of governance and regulation that fit their particular needs, they are inextricably linked with the wider national and global economic system.

These local economies and 'new industrial spaces' thrive on the dynamism, innovation and 'untraded interdependencies' (knowledge and information that circulates through the transfer of key workers between firms or via social and other networks) that are made possible by proximity. Even TNCs, that are now often portrayed as being 'placeless' because their operations are so extensive and relatively mobile, started from businesses that were nurtured in a particular locality with its own economic and other characteristics to such an extent that the locally shaped attributes of such firms are carried through into their organization and transformation into TNCs (Dunning 1981).

An outstanding example of the synergy between local and global processes is the City of London. This one-square-mile (259 hectares) district and the areas that fringe it in central London provided employment for approximately one million workers in 2004, with 30 per cent of these crammed into the City. The special characteristic of the City in particular is that the majority of its business activities are knowledge-intensive services, including commercial banks and insurance companies, international banks, sophisticated private and corporate banking, firms operating in the foreign exchange, securities, commodities, shipping and derivatives markets, fund management, corporate finance, professional advisory services (legal, accountancy) that are associated with the City's financial-services complex, advertising firms and other highly specialized activities. Furthermore, a large proportion of these activities are owned or managed by non-UK enterprises to such an extent that it is sometimes said that the City has stronger links with Europe and the rest of the world than it has with the rest of the United Kingdom. In 2004 one-third of all global foreign currency transactions were undertaken in London; it was involved in 45 per cent of the total global market turnover in company equities (shares) that took place in exchanges other than the companies' domestic (or home market) exchanges (City of London Corporation 2005a).

These are just a few of the indicators of the unique localization of highly dynamic entrepreneurial, innovative and international economic activities in one very small space. There is clearly something about the environment in such localities that facilitates their competitiveness. A study by the City of London (2005b) shows that local factors such as the availability of skilled personnel, access to suppliers of professional services, or the relatively open regulatory environment places the City ahead of its global competitors (Tokyo, New York, Frankfurt) (Table 14.5). It is not just about economic advantages; quality of life, cosmopolitan cultures, linguistic diversity, and a fair and just business environment are some of social/cultural factors that help us to understand how such localities work. On the basis of its prominent international status, firms operating from the City of London are able to attract the most able foreign workers who meet its needs both for highly educated staff offering very specialized knowledge and skills as well as for more routine occupations in hotels and catering, transport services, or office servicing.

Localization as a counterpoint to globalization of the economy is also being sustained by national governments that are decentralizing resources and decision-making to local and regional governments (World Bank 1999). Spending by these sub-national institutions was 10 per cent of the national total for Mexico in the mid-1980s, rising to more than 30 per cent in the mid-1990s; equivalent figures for South Africa are 21 per cent and 48 per cent, and for China, 45 per cent and 58 per cent. The idea is to give localities greater autonomy over the institutions and instruments that they put in place, using the insights gained from detailed knowledge of local networks, to capitalize on their economic advantages or to create economic and social environments that will attract new investment. Such strategies still require some national coordination because some localities within a nation, especially cities, are usually better positioned to act as independent competitors for globally mobile investment or employment than more peripheral or rural localities. The association between globalization and localization has boosted urbanization, encouraged by the rise of service economies in ways that divert economic development and investment from marginal locations (European Commission 1999). Metropolitan areas are major drivers if national economic growth in developed and developing economies; they are the ultimate expression of the benefits of localization such as scale and agglomeration economies, access to diverse and lower cost information, and an environment conducive to innovation. They are also more easily promoted as locations for investment and good factor productivity. The City of London is emblematic of these effects. Promoting cities as locations for investment is likely to be more successful and can be done in ways that stimulate vibrant rural industrial sectors

Table 14.5 Rank of competitive factors, leading world financial centres, 2005

Factor of competitiveness	Rank	Average score	London rank[1]
Availability of skilled personnel	1	5.37	1
Regulatory environment	2	5.16	1
Access to international financial markets	3	5.08	1
Availability of business infrastructure	4	5.01	2
Access to customers	5	4.90	1
A fair and just business environment	6	4.67	1
Government responsiveness	7	4.61	1
Corporate tax regime	8	4.47	2
Operational costs	9	4.38	2
Access to suppliers of professional services	10	4.33	1=
Quality of life	11	4.30	3
Cultural and language	12	4.28	1
Quality/availability of commercial property	13	4.04	2
Personal tax regime	14	3.89	2

[1] Compared with three other leading financial centres: New York, Paris, Frankfurt.

Source: City of London Corporation (2005b)

such as Guangzhou, north of Hong Kong, or Zhejiang and southern Jiangsu adjacent to Shanghai (World Bank 1999) (Plate 14.4).

14.9 Digital economies in the twenty-first century?

In many respects the interaction between local and global economies is shaped by the so-called digital economy. It is estimated that there were 1.1 billion Internet users worldwide (approximately 17 per cent of total population) at the end of 2006, up tenfold from 100 million in 1997 (Table 14.6). E-commerce was already expanding at the end of the 1990s but it is now growing and diversifying even more rapidly, led by business-to-business and business-to-customer transactions by service sector activities such as telecommunications, information technology, publishing and the media, travel and tourism, retailing, transportation and professional services like management consultancy, industrial design and engineering. Although consumer sales attract the publicity (such as the phenomenal growth of eBay or Amazon), much of the expansion of e-commerce is between and within businesses. Global e-commerce sales were estimated to be worth US$2.3 billion in 2003, rising to US$13 billion in 2006. The potential for growth in direct cross-border sales to consumers is substantial as Internet access (see Table 14.6), user confidence, payment systems, Internet/Web security, and mechanisms for tracking transactions for levying taxes and duties, continue to improve.

Apart from the numerous legal, regulatory, security and other challenges it presents, the global digital economy is likely to be accompanied by new geographies of the economy. Just as the Industrial Revolution generated significant economic and social changes (both positive and negative) in the form of new jobs, new industries and new industrial regions, so will the digital economy stimulate its own revolution. Perhaps the Internet will finally eliminate the effects of friction of distance on economic interactions. But this will require a truly global network of telecommunications and computer infrastructure that is accessible to all. As the data in Table 14.6 show, some national economies are yet to be 'plugged in' to the Internet. To be excluded is to widen the gap that already exists between those countries 'inside' and those 'outside' the global digital economy. A networked readiness index (Figure 14.3) reveals wide differences in the geographical distribution (number, density and processing power, for example) of the

Plate 14.4 A street vendor walks past an old house marked with the Chinese character for 'demolish'. Parts of China, such as Shanghai or Guangzhou, are experiencing rapid economic transformation which is expected to continue well into the 21st century.
(AP Photo/Eugene Hoshiko)

computers, telecommunications networks and software required to participate in the global economy. However, Africa's digital economy deficit may be rectified much more rapidly than expected with the expansion of cellular (mobile) telephony. Cell phones not only offer voice services but also technologies that bypass the need for a computer to access the World Wide Web so that as early as 2001 Africa was the first world region where the number of mobile subscribers exceeded the number of fixed-line subscribers. Access to mobile phones opens up all kinds of economic and social benefits such as better access to information about crop prices, guidance and support from government institutions on how to run small businesses, accessing bank services without the need to travel (including money transfers – see Grameen Bank case study, Chapter 17, this volume), or keeping in touch with family and friends.

A second challenge posed by the digital economy is the demand for human resources. More jobs will be created by the digital economy than will be lost but they are generally in higher-skilled and better-paid occupations. In order to fill new digital-economy jobs, economies everywhere need to retrain existing workforces and to train future workers. History suggests that the response to this challenge will be uneven; comparative advantage will enable some economies and some localities to respond more quickly and effectively than others. Perhaps less certain is how this will modify existing spatial patterns of economic development. A digital economy, for example, enables a wide range of purchasing and banking transactions to be undertaken from home via the telephone, personal computers, televisions, or on the

Table 14.6 Internet usage, by world region, 2006[1]

Region	Internet usage (000s)	Penetration (% of total population)	Usage (% of world)	Population (% of world)	Usage growth 2005–2006 (%)
Africa	32,766	3.6	3.0	14.1	625.8
Asia	387,593	10.6	35.5	56.4	239.1
Europe	312,723	38.7	26.8	12.4	197.6
Middle East	19,382	10.2	1.8	2.9	490.1
North America	232,057	70.0	21.3	5.1	114.7
Latin America/ Caribbean	88,779	16.0	8.1	8.5	391.3
Oceania/Australia	18,430	54.3	1.7	0.5	141.9
World total	**1,091,731**	**16.8**	**100.0**	**100.0**	**202.4**

[1] 30 December 2006.

Source: http://www.internetworldstats.com/stats.htm (accessed 3 January 2007)

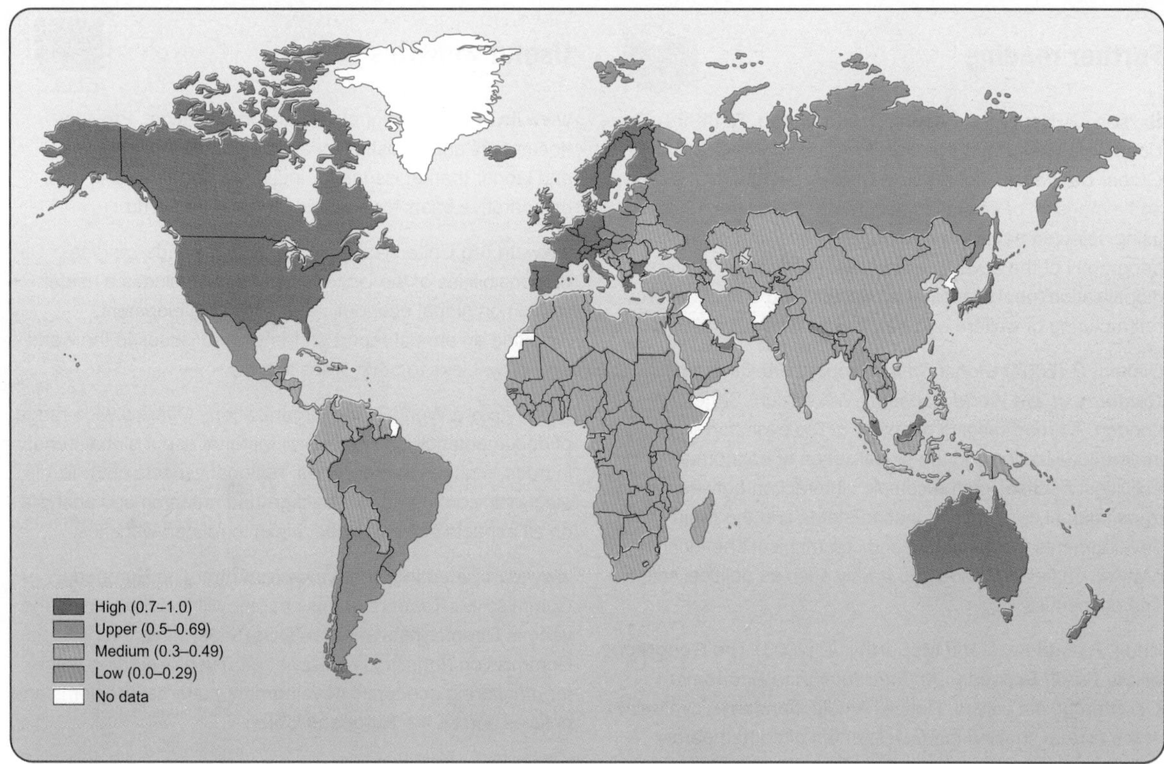

Figure 14.3 The Networked Readiness Index, 109 countries, 2005–2006.

move using mobile phones that provide access to Internet services. An international door-to-door delivery time of 24 hours or much less is now commonplace, depending on the type of good or service, whether it is required just in time, or the location of the supplier. Moreover, cross-border transactions and payments are as easy as those made to the local supermarket or bookstore. With shopping sites on the Internet presenting themselves as 'virtual malls' and encouraging users to place their selections in 'baskets' or 'carts', they are attempting to replicate offline retail outlets. More than 40 per cent of Internet users shopped online in the USA and in Western Europe in 2003, and by 2008 it has been estimated that 40 per cent of online buyers will purchase travel and 58 per cent will purchase books (http://cyberatlas.internet.com/markets/retailing). This, combined with trends such as increased working from home or teleworking, whether self-employed or as an employee of a TNC, points to the potential for significant changes.

The challenge for economic geographers is how to understand the ways in which the digital economy impacts on the structure and pattern of economic activities at different scales, on the structure of localities and their relationship with the global economy, or on the relative importance of nations and TNCs in regulating or influencing the outcome.

Learning outcomes

Having read this chapter, you should be aware of:

- The nature and changing concerns of economic geography during the last quarter of the twentieth century.
- The importance of economic restructuring and the shift from manufacturing to services in the modern economy.
- The significance of the globalization process in the structure and spatial development of the world economy.
- The major role performed by transnational corporations in trade, foreign direct investment and other indices of globalization.
- The challenges posed to economic geographers by the rapid growth of the digital economy and its potential for changing established patterns of economic activity.

Further reading

Bryson, J., Henry, N., Keeble, D. and Martin, R. (1999) ***The Economic Geography Reader: Producing and Consuming Global Capitalism***, John Wiley, Chichester. Charts the revival and expansion of economic geography during the 1990s using selected papers that explore the evolving economic geography of the advanced capitalist economies. Covers globalization, new spaces of production and consumption, restructuring of welfare and new landscapes of work.

Dicken, P. (2007) ***Global Shift: Mapping the Changing Contours of the World Economy***, 5th edition, Sage, London. A broad-ranging overview of the evolution of internationalization and the globalization of economic activities. Focuses on the complex interaction between transnational corporations, nation-states and the rapid developments in information and communications technology, drawing on fields outside geography such as political science and economics.

Knox, P., Agnew, J. and McCarthy, L. (2003) ***The Geography of the World Economy: An Introduction to Economic Geography***, 4th edition, Hodder Arnold, London. A synthesis of the factors shaping the development of contemporary economic patterns in market-oriented and centrally planned economies. Strong emphasis on interdependence of economic development at different spatial scales, from local through national to international.

Mackinnon, D. and Cumbers, A. (2007) ***Introduction to Economic Geography: Uneven Development, Globalization and Place***, Pearson, London. Explores the wide range of approaches and models that are debated about and used by economic geographers. Focuses on globalization, uneven development and place. Covers more conventional topics such as regional development and labour markets alongside an introduction to economic topics that are subject to rapid change such as consumption, information and communications technologies and tourist geographies.

Useful websites

www.ilo.org International Labour Organization. Provides documents and statistics on a wide variety of employment and labour market issues for individual countries and on a comparative basis for countries around the world.

www.un.org United Nations. Reflects the wide-ranging responsibilities of the United Nations but includes a useful section on global economic and social development, including an annual report on trends and issues in the world economy. Links to other websites.

www.wto.org World Trade Organization. Offers a wide range of documentation and statistical material about global trends in trade in goods and services, regional aspects of trade, electronic commerce and background research and analysis on all aspects of world trade. Links to related websites.

www.europa.eu.int/comm/indexren.htm The European Commission. Readers of this chapter will find the links to the various Directorates-General (DGs) sites within the Commission (Industry, Transport, etc.) useful starting points for monitoring economic development patterns, problems and policies across the European Union.

www.oecd.org Organization for Economic Cooperation and Development. Represents mainly the developed economies and provides free documents, summaries of OECD economic surveys and statistics.

www.unctad.org United Nations Conference on Trade and Development. Focuses in particular on the interests of less developed countries in relation to international trade and foreign direct investment. Publications, statistics and links to related websites.

For annotated, clickable weblinks and useful tutorials full of practical advice on how to improve your study skills, visit this book's website at **www.pearsoned.co.uk/daniels**

THE GEOGRAPHIES OF GLOBAL PRODUCTION NETWORKS

Chapter 15

Neil M. Coe

Topics covered:

- Defining and identifying production networks
- Spatial divisions of labour
- The governance of production networks
- Production networks in their institutional contexts
- Reshaping production networks through 'standards' and 'codes'

15.1 Engaging with global production networks

Imagine the scene. We could be in Manchester, Minneapolis, Munich, Mumbai or Melbourne, or many other places besides. A student – having just been to a lecture and en route to their evening job in a restaurant – stops for a drink in a café. After choosing from the extensive menu of coffee options, they take a window seat and boot up their laptop computer – the café has recently installed free wireless Internet for its customers. Essay assignment in hand, the student starts surfing the Web, looking for the reading materials and background information they will need. Every so often, they pause, take a sip of coffee – a decaf, skinny latte or some such like – and gaze out at the street . . .

On the one hand, what is being described here is a unique consumption event, an individual lost in their thoughts as they consume their coffee and use their laptop at a particular time and place (see Chapter 18). And yet, on the other hand, the two central non-human artefacts in this particular story – the coffee and the laptop – may have remarkably similar economic geographies and histories, wherever the event is taking place.

For example, the coffee bar in question might be a branch of Starbucks, the world's largest chain, serving over 30 million customers weekly in 12,000 cafés across 37 countries (as of November 2006). It is one of many other similar chains offering their own standardized take on the continental European coffee house experience: for example, Caffè Nero, Costa Coffee or Coffee Republic in the United Kingdom; Caribou Coffee in the United States; Barista in India; and Gloria Jean or Dome in Australia. There is a strong chance that the coffee beans were roasted and processed by one of just five American and European firms – Kraft (Maxwell House brand), Nestlé (Nescafé), Sara Lee (Douwe Egberts), Proctor and Gamble (Folgers) and Tchibo – that handle almost half of the world's coffee each year. Equally, the laptop used by our student may well be made by either Dell or Hewlett Packard (HP), who, in late 2006, were running neck and neck for leadership of the laptop industry with around 17.5 per cent global market share each. Indeed, just five companies – Dell, HP (both US), Acer (Taiwan: 12 per cent market share), Toshiba (Japan: 11 per cent) and Lenovo (China: 8 per cent) – account for two-thirds of laptop sales worldwide.

Why are these observations interesting? In short, because both the cup of coffee being consumed and the laptop being used are the end result of far-reaching global production networks that are in turn dominated by powerful corporate interests of various kinds. Almost all commodities that we consume have complicated histories and geographies and yet, as a system, capitalism seems to conceal these. The purchase of commodities such as coffee or computers with money serves to *disconnect* producers and consumers, meaning many consumers are unaware of the nature of the production system that has enabled those commodities to be available to them. This poses profound challenges

Plate 15.1 A student using her laptop in a coffee shop.
(Alex Segre/Alamy)

to both conscientious consumers actively curious about the history of the commodities they consume, and economic geographers who want to understand connections and interdependencies within the global economy. In reality, even drinking just one cup of coffee links the consumer – albeit unknowingly in most cases – to hundreds of thousands of workers involved directly or indirectly in its production through complex global webs of connections.

This chapter is therefore about how we, as economic geographers, can explore and understand the inherent variability of global production networks. In what follows, we will look in turn at four generic dimensions of all global production networks, namely their organization, geography, power relations or 'governance', and institutional context. We will use the contrasting examples of the coffee and laptop global production networks as illustrations of these various dimensions. In the final section of the chapter we will look at ongoing attempts to reshape production networks through the implementation of different kinds of standards and codes of conduct.

15.2 Production chains, production networks . . .

Every economic activity can be thought of as a production chain – a linked series of value-adding activities (see Figure 15.1). In very simple terms, material and non-material inputs are combined and transformed through some kind of production process, leading to a new good or service that needs to be delivered to the customer who then consumes it. Consumption, in turn, is not just a single act of purchase, but an ongoing process that may include maintenance, repair, waste disposal, recycling and the like. This basic model holds whether it is a physical good – such as a television or bicycle being produced – or a service such as a haircut or insurance policy, with the difference coming in the relative balance of tangible and intangible elements in the production chain (see Chapter 16). While simple chains can be thought of in linear terms, in reality, as Figure 15.1 shows, they are enmeshed within much wider *networks* of relationships involving a broad range

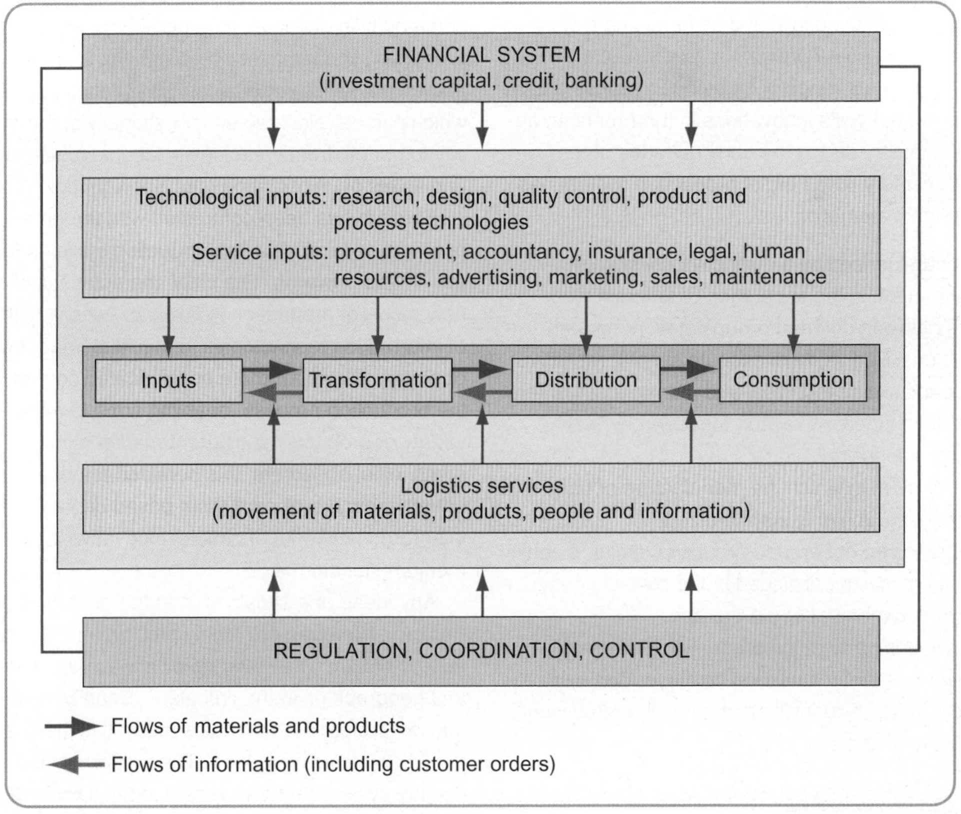

Figure 15.1 A generic production network.

Source: Dicken (2007), Figure 1.4c

of other functions that are necessary for economic activities to take place: for example, research and development, market research, technological inputs, logistics services, advertising, legal services, accounting, personnel management, software, security and so on. In turn, these inter-firm relations are embedded in broader financial and regulatory systems that may bring non-firm entities such as the state and its many institutions into view. In this chapter we use the term *production network* to refer to the full mesh of relationships that lies behind any economic activity.

Production networks vary greatly in their complexity. At one end of the scale is the single farmer who grows one particular kind of vegetable and transports it to a market for sale at harvest time. Conversely, a complicated manufactured product such as a car mobilizes a wide range of skills and technologies to bring together the thousands of components that comprise the finished product. The next step is to think about *who* undertakes the different activities in a given production network. Hypothetically, every single function might be carried out by one huge firm, which would coordinate the production network through its internal management hierarchy (a vertically integrated system). Equally, every single function might be undertaken by separate firms, meaning that the production network would take the form of a series of external, inter-firm relationships (a vertically disintegrated system). In reality, all production networks fall somewhere between these extremes. In a context of increased global competition, however, the dominant trend has been for firms in many sectors to focus on their core activity or 'competency', while seeking non-core inputs via external relationships. For some commentators, this trend is part of a broader shift from *Fordist* to more flexible *post-Fordist* production systems (see Spotlight box 15.1).

Spotlight Box 15.1

From Fordism to post-Fordism?

Some commentators have characterized recent changes in production systems as heralding a shift from *Fordism* to *post-Fordism*. In its narrowest sense, Fordism signifies a range of industrial practices emerging from Henry Ford's innovations in the American automobile industry during the early decades of the twentieth century. The Fordist production system has four important elements:

1. It is characterized by a distinctive division of labour – that is, the separation of different work tasks between different groups of workers – in which unskilled workers execute simple, repetitive tasks and skilled technical and managerial workers undertake functions related to research, quality control, finance, etc.
2. It is a system in which the manufacture of parts and components is highly standardized.
3. It is organized not around groups of similar machinery, but machines arranged in the correct sequence required to manufacture a product.
4. The various parts of the production process are linked together by a moving conveyor belt – the assembly line – to facilitate the quick and efficient fulfilment of tasks.

Together, these four attributes can provide *economies of scale* whereby the large volume of production reduces the cost of producing a single product. In turn, prices can then be reduced, leading to increased sales and the potential development of a *mass market* for the commodity in question. In geographical terms, Fordist production is often associated with the formation of *branch plants* for undertaking the production process which can be relocated either nationally or internationally to benefit from variations in labour wages, skills, and levels of unionization (see Spotlight box 15.2).

Fordism has been contrasted with the emergence of so-called 'post'-Fordist production systems from the mid-1970s onwards. The chief characteristic of this new mode of production is *flexibility*, whereby the use of information technologies in machines and their operation allows for more sophisticated control over the production process, enabling firms to make a wide variety of products for different market niches. For some initial observers, this heralded the rise of an entirely new era of capitalism based on vertically disintegrated networks of nimble, innovative, small and medium-sized firms.

Any ideas of a simple shift from Fordism to post-Fordism, however, are highly problematic. For one thing, Fordism was never dominant across all sectors and geographies in the first place. Equally, no one production system predominates today. It is more realistic to think in terms of, at least, three different kinds of production system that are evident today (Dicken 2007: see Table 15.1 for more details). First, *Fordism* is alive and well in certain sectors where simple scale economies remain crucial, e.g. agriculture and food processing, routinized services activities, and some electronic component industries. Secondly, in other instances, flexible production technologies have led to the resurgence of craft-based production driven by innovative, small independent firms. At the heart of this so-called *flexibly specialized* production system are skilled craft workers using flexible machinery to produce small volumes of customized goods such as shoes, jewellery, clothing and furniture, ceramics, and the like. Thirdly, there are forms of *flexible production* emanating originally from Japan – for example, through the auto industry – that are distinctive from both the Fordist and flexibly specialized systems. Key here is the combination of information technologies with flexible ways of organizing workers that allows the highly efficient production of large volumes of customized goods. Intriguingly for economic geographers, the flexible specialization and Japanese flexible production systems both seem to exhibit a newfound tendency for firms to agglomerate – or group together – in particular places in what are commonly known now as *clusters* (see Spotlight box 15.3).

These increasingly important external networks can take on many forms:

- *Markets*: some inputs – usually of low value and standardized – will simply be purchased by firms on the open market. In this case, there is no long-term relationship between the two parties and firms can readily switch between suppliers.
- *Subcontracting*: this involves firms buying inputs that have been made, under contract, to meet their own specific requirements. The stability of the relationship will vary according to the formality and length of the contract. Subcontracting may involve the entire manufacture of a particular good or services, known as *commercial* subcontracting, or it may take the form of a firm buying in particular inputs that it does not have the skills or capacity to produce cost-effectively 'in-house' – *industrial* subcontracting.
- *Strategic alliances/joint ventures*: this is where firms come together to create a new corporate entity in order to undertake a particular task, for example costly joint research. In many other respects the participating firms may still remain competitors.
- *Franchising and licensing*: here, firms allow a company (the franchisee or licensee) to sell their product or service in a particular territory under given terms and conditions, and in return for a set fee. Franchising is very common in the service sector (including fast food, retailing, coffee shops etc.), allowing rapid geographical expansion.

In sum, every production network constitutes a unique constellation of activities performed via complex combinations of internal and external network connections. The first step in understanding any production system, therefore, is to map out the key participants and the *nature* of the relationships that connect them.

The coffee production network is relatively straightforward and is represented schematically in Figure 15.2. Even so, coffee travels a long way and changes hands several times on the journey from bean to cup. The major participants in the network are depicted by boxes, and the transactions that move coffee in its different forms between participants are shown by arrows. Coffee flows up the network from the growers, who in effect begin the system, to the consumers who represent the end point. Coffee – which comes in two main types, Arabica and Robusta – is generally grown on small farms or estates in the developing world. Once basic processing has extracted the 'green' coffee beans from picked coffee cherries, they will pass in 60-kg bags from an exporter to a developed country importer, trader or broker, then on to a roaster or instant coffee manufacturer, and then finally to a consumer via either a supermarket shelf or café of some kind. These relationships are essentially 'arms-length' *market* connections, with prices being set by international commodity markets such as the New York Board of Trade (Arabica) and the London International Financial Futures and Options Exchange (Robusta). We will see later, however, how the roles of these various participants have evolved over time.

In contrast, laptop computers are the outcome of a much more complicated production network that brings together hundreds of different components into the finished product. The personal computer (PC) industry is therefore a complex network of firms involved in a wide range of different industry segments – from microprocessors and other electronic components

Table 15.1 Alternative post-Fordist production systems

Characteristic	Flexible specialization	Fordist mass production	Japanese flexible production
Technology	Simple, flexible tools/machinery; non-standardized components	Complex, rigid single-purpose machinery; standardized components; difficult to switch products	Highly flexible (modular) methods of production; relatively easy to switch products
Labour force	Mostly highly skilled workers	Narrowly skilled professionals 'conceptualize' the product; semi/unskilled workers 'execute' production in simple, repetitive, highly controlled sequences	Multi-skilled, flexible workers, with some responsibilities, operate in teams and switch between tasks
Supplier relationships	Very close contact between customer and supplier; suppliers in physical proximity	Arms-length supplier relationships; stocks held at assembly plant as buffer against disruption of supply	Very close supplier relationships in a tiered system; 'just-in-time' delivery of stocks requires 'close' supplier network
Production volume and variety	Low volume and wide (customized) variety	Very high volume of standardized products with minor 'tweaks'	Very high volume; total partially attained through the production of range of differentiated products

Source: adapted from Dicken 2007, Table 3.2.

to applications and systems software providers – and covering a wide range of activities: R&D and design, manufacturing, assembly, logistics, distribution, sales, marketing, service and support. How these various functions are split between different companies has changed over time. Historically in the computer industry, vertically integrated giants such as IBM, HP and Siemens operated in all the industry segments and carried out the key functions of product innovation, manufacturing and customer relations internally.

Since the advent of the PC, however, a much more complex 'tiered' network has evolved in which most companies concentrate on one particular market segment, for example, assembling PCs, or making circuit boards or disk drives (see Figure 15.3). The PC can now be described as a *modular* product, whereby ten to fifteen relatively self-contained sub-components (e.g. keyboard, monitor, hard drive etc.) are brought together and assembled, an attribute that facilitates the disintegration of the production network into separate firms. Branded PC companies now focus primarily on design and customer relations, 'outsourcing' the remainder of the production process to other firms. This system reflects the nature of the PC as a standardized product assembled from components that can be produced in a wide variety of locations by a broad range of firms. Only limited value is added by assembling a PC: in most cases PC firms add value through customer relationships, either directly through their own direct sales and service relationships (e.g. help desks, repairs, etc.) or indirectly through their branding, marketing and quality assurance practices. They also extract value from the network through coordinating the logistics operations that turn components into finished products on customer doorsteps. As Curry and Kenney (2004: 114) suggest, 'a PC assembler is, in many ways, more a logistics coordinator than a manufacturer'.

Most laptop assembly operations, therefore, are subcontracted to contract manufacturers (CM), and

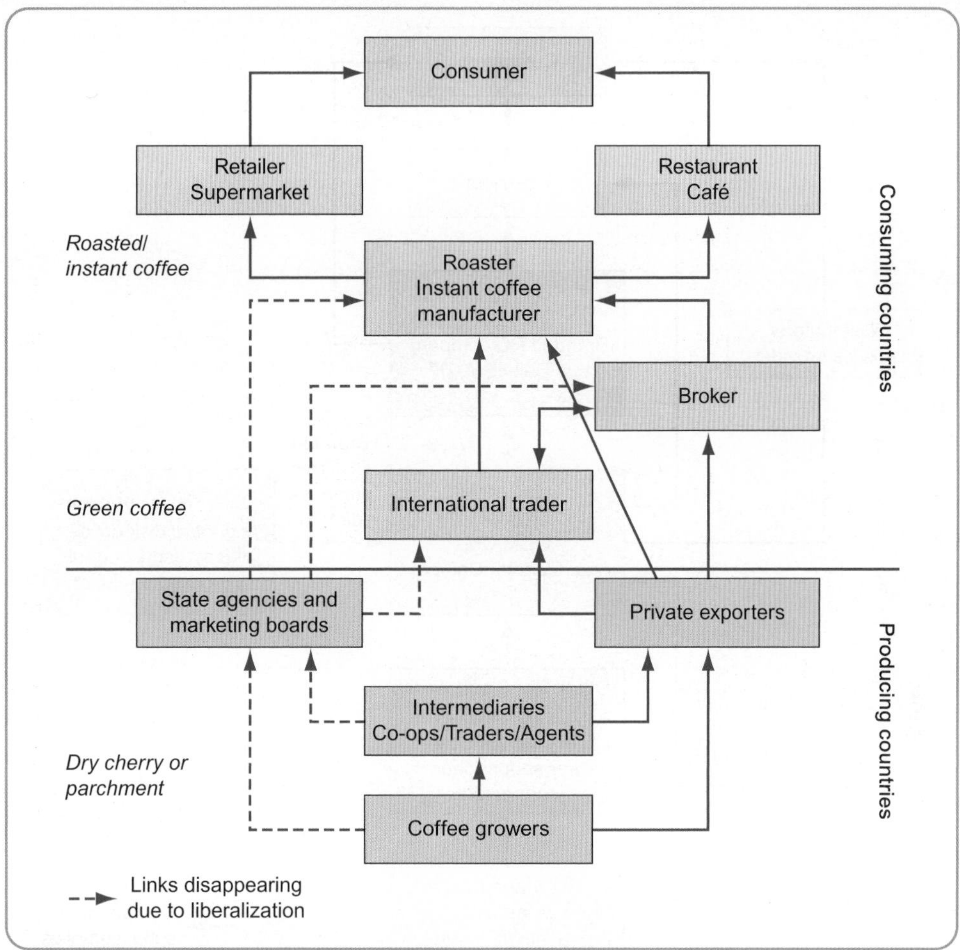

Figure 15.2 The coffee production network.

Source: adapted from Ponte (2002), Figure 1 and Talbot (2004), Figure 2.1

increasingly, to a particular form of subcontractor known as the Original Design Manufacturer (ODM) that also contribute to the design process. In the case of contract manufacturers, a branded PC firm is likely to employ its own, on-the-ground design and development teams throughout the production process to supervise subcontractors: this is the model preferred by Toshiba and Lenovo, for example. With an ODM, however, the PC maker may take primary responsibility for design, but it will then pass a product specification on to the ODM for final development and manufacturing; Dell and HP operate in this way and this is currently the most common model in the industry. In yet another model, smaller PC vendors without the scale to undertake their own design activity may simply purchase generic 'off-the-shelf' products from ODMs to be labelled and sold under their own name. Backwards relationships from branded PC firms to ODMs may also extend forwards to the customer through the delivery and servicing of laptops on behalf of PC vendors.

It is important to bear in mind that Figure 15.3 only provides a limited window on what is, in reality, a highly complex network connecting together hundreds of firms and tens of thousands of workers. For example, the hard disk drive industry (see 'Core Tier 3 supplier' box in Figure 15.3) is itself a significant sector involving several tiers of firms and assemblers spread across East and South-east Asia, with a particular concentration of final assembly and testing activity in Malaysia, Thailand and Singapore (Gourevitch et al. 2000). It is to these complex, on-the-ground geographies of production networks that we now turn.

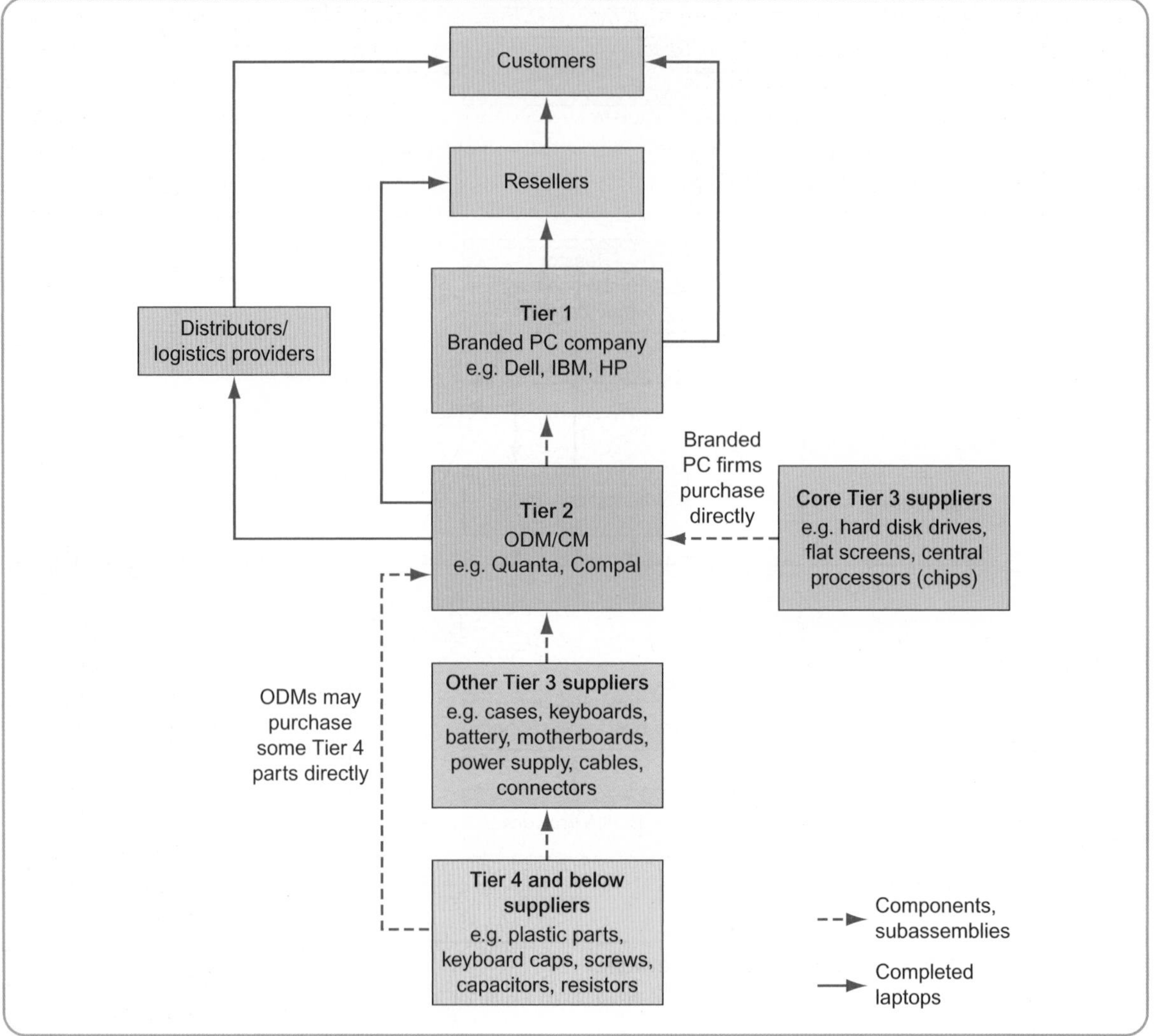

Figure 15.3 The laptop PC production network.

Source: adapted from Foster *et al.* (2006), Figure 2

15.3 Geographies of production networks: spatial divisions of labour

It goes without saying, of course, that the production networks described above do not exist on the head of a pin, but rather connect together, and indeed constitute, real places within the global economy. Every production network requires a *spatial division of labour*, which refers to the way in which certain elements of the production process are concentrated in particular places. The concept of the spatial division of labour is elucidated particularly well by Massey (1984) who explored how these spatial variations are both *created* and *exploited* by the ongoing restructuring of capitalist firms in their pursuit of profit. After examining the spatial restructuring of the UK economy in the 1960s and 1970s, her key argument was that under conditions of increased global competition, corporations were increasingly looking to separate spatially the *control* functions performed by managerial workers from the *execution* functions undertaken by manual workers. In its most simple form, low-skilled manual tasks were tending to locate in peripheral areas, while managerial and R&D tasks were concentrating in core cities and their surrounding regions. Massey argued that, over time, different *layers* of investment fan out across the economic landscape, redefining the nature of relationships between places as they gain or lose different kinds of activities.

Different spatial divisions of labour arise from variations in corporate structures which, at a simple level, can be divided into *cloning* and *part-process* structures. In the cloning structure, the production apparatus is simply replicated in different localities, with ultimate control residing at a headquarters, usually sited in the firm's initial place of origin. In the part-process format, there is a technical division of labour between branch plants, with components being made in one location and passed on to another for final assembly. The different stages of the production process will have varying requirements, and hence tend towards different kinds of location.

Massey's conceptual apparatus is still extremely powerful today, although her study – which was primarily about *intra-firm* and *intra-national* divisions of labour – needs to be extended in two important ways. First, as we have already noted, spatial divisions of labour can be constructed through combinations of intra-firm (internal), and increasingly, inter-firm (external) networks. Secondly, we need to apply these ideas at the *international* scale. One of the defining characteristics of the world economy over the past two to three decades has been the dramatic increase in the number of transnational corporations organizing their spatial divisions of labour at the international scale. It is now difficult to think of a production network that does not have at least some international elements, even if it is just seen in the sourcing of one or two inputs, or a limited export market for the final good/service. *Global* production networks, as we will call them here, have become one of the most important organizational features of the contemporary global economy. Spotlight box 15.2 considers

Spotlight Box 15.2

Conceptualizing global divisions of labour?

It is helpful to distinguish between three separate attempts to conceptualize international spatial divisions of labour, which broadly coincide with different phases of development of the global economy.

- First, it is possible to identify the traditional **International Division of Labour** (IDL) that took shape by the nineteenth century and prevailed largely unaltered until the 1950s. The IDL essentially depicted a *trading* system – shaped initially by global trading empires, and later by the rise of the USA as an economic power – in which the developing world or 'periphery' was largely relegated to providing raw materials and agricultural plantation products (e.g. coffee from Brazil, copper from Chile, gold and diamonds from South Africa) for the industrialized economies of the 'core' (Western Europe and the USA). High-value manufactured goods were exchanged between the industrialized countries, and some were exported back to developing countries.
- Second, from the 1960s onwards, a **New International Division of Labour** (NIDL) started to emerge in which European, North American and Japanese TNCs created labour-intensive export platforms in so-called 'newly industrializing economies' (especially in East Asia and, to a lesser extent, Latin America) in response to falling profit rates in the core countries. Crucially, the system depended on new technologies that allowed production fragmentation, thereby creating tasks that could use, often young and female, semi-skilled or unskilled workers in the periphery. Two kinds of technology were important: *process* technologies that allowed the subdivision of the manufacturing process into simple and self-contained tasks, and *transportation* technologies such as jet aircraft and containerized shipping that allowed the efficient shipment of both components and finished goods.
- Third, it has become clear that in recent decades a *New Global Division of Labour* has emerged that is far more complex than the system depicted in the NIDL model, which does not capture, for example: how the range of NIE economies has broadened and deepened considerably, particularly in Asia; increasing investment into NIEs, such as China and India, to access their domestic markets; significant outward investment by NIE transnational corporations e.g. from South Korean giants such as LG and Samsung; continued high levels of investment between developed economies and the emergence of complex international divisions of labour in service sectors. The contemporary global economy is clearly not characterized by one single type of IDL but rather many different forms. Rather than disappearing, the traditional IDL and the NIDL remain important in a range of industries (e.g. natural resources and clothing/toys, respectively) and have subsequently been overlain by, and interacted with, newer and more complex international divisions of labour.

different attempts to conceptualize the nature of international divisions of labour at different points in the global economy's evolution.

We can make three further arguments about the geographies of global production networks:

- First, their *geographical complexity* is increasing, enabled by a range of developments in transport, communication and process technologies. As we shall see shortly, the assembly of a laptop requires components manufactured at places all across East Asia.
- Second, the geographic configurations of global production networks are becoming *more dynamic* and liable to rapid change. This flexibility arises, first, from the use of certain 'space-shrinking' information and communication technologies and secondly, from organizational forms that enable the fast spatial switching of productive capacity (Dicken 2007). In particular, the increased use of subcontracting and strategic alliance relationships (noted above) allows firms to switch contracts between different firms and places without incurring the costs of moving production themselves.
- Third, we need to connect these ideas about the geographical extensiveness and complexity of global production networks with notions concerning the geographical *clustering* of economic activity. Not all the connections within the system can be 'stretched out' across the global economy; some interactions will need to take place within the same locality because of the sheer intensity of transactions or because of the importance of place-specific knowledge to the activity concerned. From this perspective, global production networks need to be seen as the organizational forms that connect clusters together.

Let us now move on to apply these ideas to our two products. The global map of coffee production and consumption is not very complicated. Growing coffee involves several million farmers, who ultimately support the consumption of an estimated 2 billion cups of coffee every day. The vast majority of coffee production takes place in developing countries while most of the consumption occurs in the wealthy markets of North America, Western Europe and East Asia. Although over 50 countries currently produce coffee, the top ten exporters account for around 80 per cent of global production (Figure 15.4). The geography of production is shaped by the ecology of the coffee plant which requires a consistently warm and wet climate, making it most

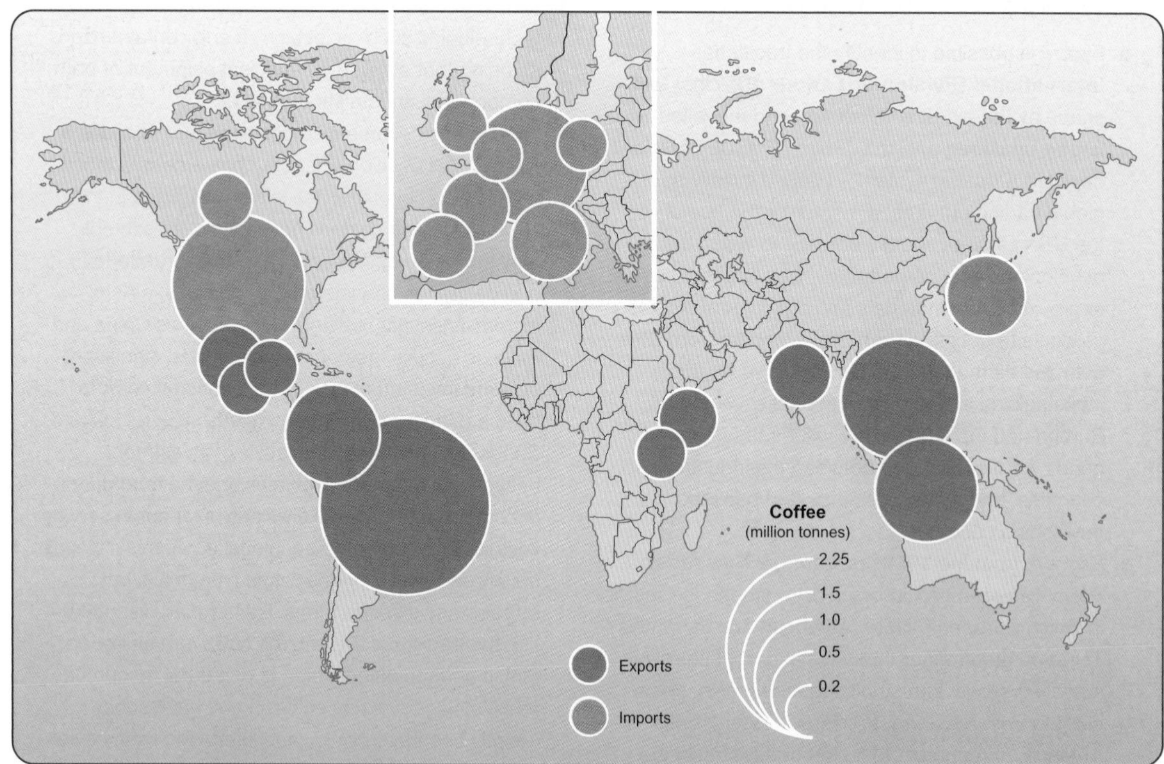

Figure 15.4 The top ten coffee producing (exports) and consuming (imports) countries, 2005.

Source: data from http://www.ico.org, accessed 9/1/07

suitable for growing in tropical highland areas. The production geography is not static, however. Most notably, exports of Robusta coffee grown in the central highlands of Vietnam have expanded dramatically over the last decade – rising from just 100,000 tonnes in 1990 to 1 million tonnes by 2005 – creating an over-supply in the global market and putting severe downward pressure on prices. As a result, coffee growing is an increasingly marginal practice for growers across Central and Latin America and Africa. The consumption of coffee has a long history linked to colonial expansion by Europeans, which served to both spread the taste for coffee, particularly from the sixteenth century onwards, and spread coffee cultivation from its supposed origins in present-day Ethiopia to the band of tropical countries shown in Figure 15.4. The key commodity markets and international traders and roasters (see Figure 15.2) are now all based in the USA and Western Europe.

Turning to the geography of the laptop production network, the majority of the world's laptops are produced by Taiwanese firms. Most branded manufacturers – such as Dell, HP, Apple and Toshiba – rely on Taiwanese ODM firms for manufacturing and product development. In 1995, Taiwanese companies accounted for 27 per cent of world laptop production, rising to 40 per cent of a world market of 15 million units in 1998. By 2004, the figure had reached 72 per cent, with the top ten Taiwanese manufacturers (see Table 15.2) producing 33 million of the 46 million laptops sold worldwide. Production is actually growing faster than a very rapidly growing PC market: in 2005, one in three PCs sold globally was a laptop, and this share is expected to keep rising. Each leading laptop vendor tends to contract with two or three Taiwanese firms, arrangements which in the cases of Dell and HP account for upwards of 90 per cent of their total global laptop production: for Japanese companies such as Toshiba and Sharp, the proportion tends to be lower (20–50 per cent). Taipei – and the nearby region of Hsinchu where many Taiwanese high-tech firms are based – is therefore a critical node in the laptop production network.

However, these geographies are also far from static. Taiwanese electronics firms have been moving production 'offshore' – to Southeast Asia, Europe and, most importantly, China – since the early 1990s. There are now two key PC clusters in China: the Shenzhen area of Guangdong province (in the south of the country), specializing in desktop machines, and the Shanghai/Suzhou/Yangtze River delta area, home to the laptop PC industry (see Figure 15.5). Before 2001, the Taiwanese government prohibited its laptop manufacturers from undertaking final assembly in China. When this restriction was lifted, the ODMs moved collectively, and incredibly rapidly, to the Shanghai area: in 2001 only 5 per cent of Taiwanese laptops were produced in China but by 2004 the figure had shot up to 80 per cent, and is soon expected to approach

Table 15.2 Top ten Taiwanese laptop PC manufacturers, 2004

Company	2004 volume (thousands)	Major customers
Quanta	11,100	Gateway, Dell, HP, IBM*, Apple, Sharp, Sony, Fujitsu-Siemens
Compal	7,700	Dell, HP, Fujitsu-Siemens, Toshiba, Acer
Wistron	3,200	IBM, Dell, Acer, Hitachi, Fujitsu-Siemens
Inventec	2,800	HP, Toshiba
Asus	2,700	Epson, Canon, Sony, Apple, Trigem
Uniwill	1,400	Fujitsu-Siemens, Samsung, clones
Mitac	1,400	Sharp, Fujitsu-Siemens, NEC
Arima	700	HP, NEC
FIC	600	NEC
ECS	500	Apple

*IBM's personal computer business was bought by the Chinese firm Lenovo during 2004.

Source: Dedrick and Kraemer (2006), Table 1 © 2006 IEEE

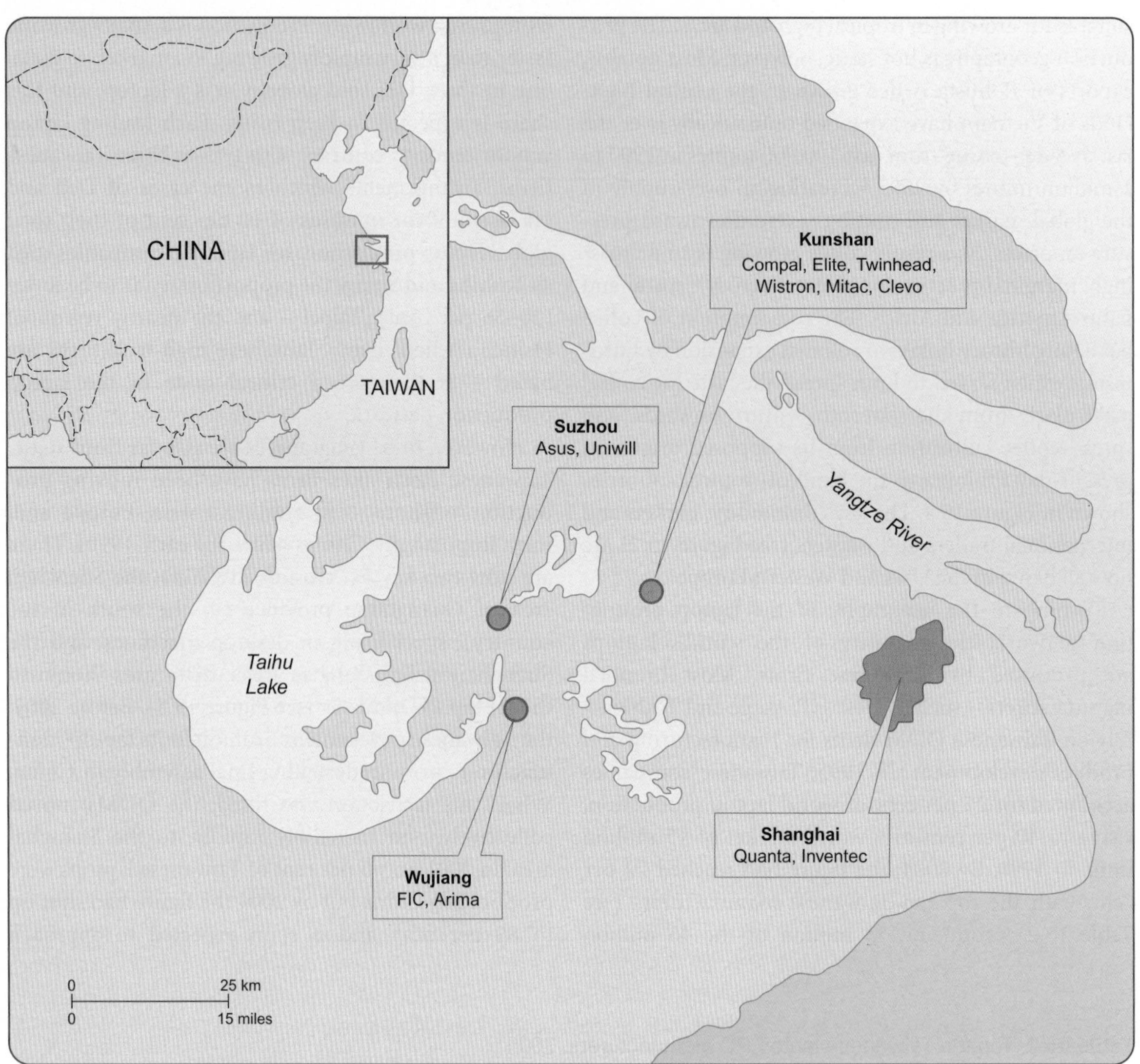

Figure 15.5 The laptop manufacturing cluster, Yangtze delta region, China.
Source: based on Yang (2006), Appendix 2

100 per cent. The lower costs of land, labour and facilities on the mainland allowed the firms to build giant factories, expand output and reduce costs through scale economies. By 2004, for example, Quanta (a Taiwanese firm) was manufacturing over 1 million laptops per month at its four plants in China. The offices of these firms in Taiwan remain responsible for product development, technology research, materials procurement, financial management and marketing (Saxenian 2006). In this way, a new *intra-firm* spatial division of labour is added to the system depicted in Figure 15.3.

Many Taiwanese component suppliers had already relocated to the region by 2001, and it is now estimated that nearly 90 per cent of the parts needed to make a laptop – excluding the highest-value components such as the central processor, hard drive and displays, which tend to be made by Japanese, Korean and US companies – are manufactured in the Suzhou area. In 2002, for example, the laptop keyboard supplier Sunrex built a factory in Wujiang to be close to Quanta and its other customers. Moving the entire supply chain to the Shanghai region in this way allowed the Taiwanese firms to overcome deficiencies in the Chinese logistics and distribution sectors. The laptop cluster in and around Shanghai is augmented by two other important types of firms: foreign electronics components manufacturers (e.g. Hitachi, AMD and Infineon) and transnational computer brand manufacturers (e.g. Toshiba, Samsung and Sony).

While the laptop global production network has a relatively concentrated geography around the key nodes of Taipei/Hsinchu and Shanghai, the PC industry generally has a somewhat more disparate geography. We can use the case of Dell as an example. There are three distinctive elements to Dell's business model. First, the company only sells direct to its customers, bypassing almost entirely distributors, resellers, and retailers (see Figure 15.3). Secondly, unlike some PC manufacturers, Dell chooses to undertake the final assembly of both desktop and laptop computers itself, allowing customers to specify the components included in a particular product line (a process that can be thought of as 'mass customization'). Thirdly, it has aggressively used the Internet to establish not only its system of direct sales, but also the procurement and assembly operations, thereby dramatically reducing its inventory of components. Figure 15.6 illustrates how Dell services its global market by using a series of six regional production clusters that bring together local and global suppliers and logistics centres (Fields 2004). Such clusters are an integral part of the global space economy (see Spotlight box 15.3).

15.4 The governance of production networks

So far we have exemplified the various actors within production networks, the functions that they undertake, and where they are located. The next step is to explain how such production networks function, and more specifically, the way in which some firms use their power over other firms to control or 'drive' the overall system. These 'lead' firms are able to define which other firms can join the network, the roles they perform, and the financial and technological conditions under which they perform those roles – despite not directly owning them. This shaping of global production networks by powerful corporations through inter-firm relations is known as **governance**.

Governance can take different forms. A useful starting point is the distinction between networks that are *producer-driven* and those that are *buyer-driven* (Gereffi 1994: see Figure 15.7 and Table 15.3). Producer-driven networks tend to be found in sectors where large industrial corporations play the central role in controlling the production system, for example in capital- and

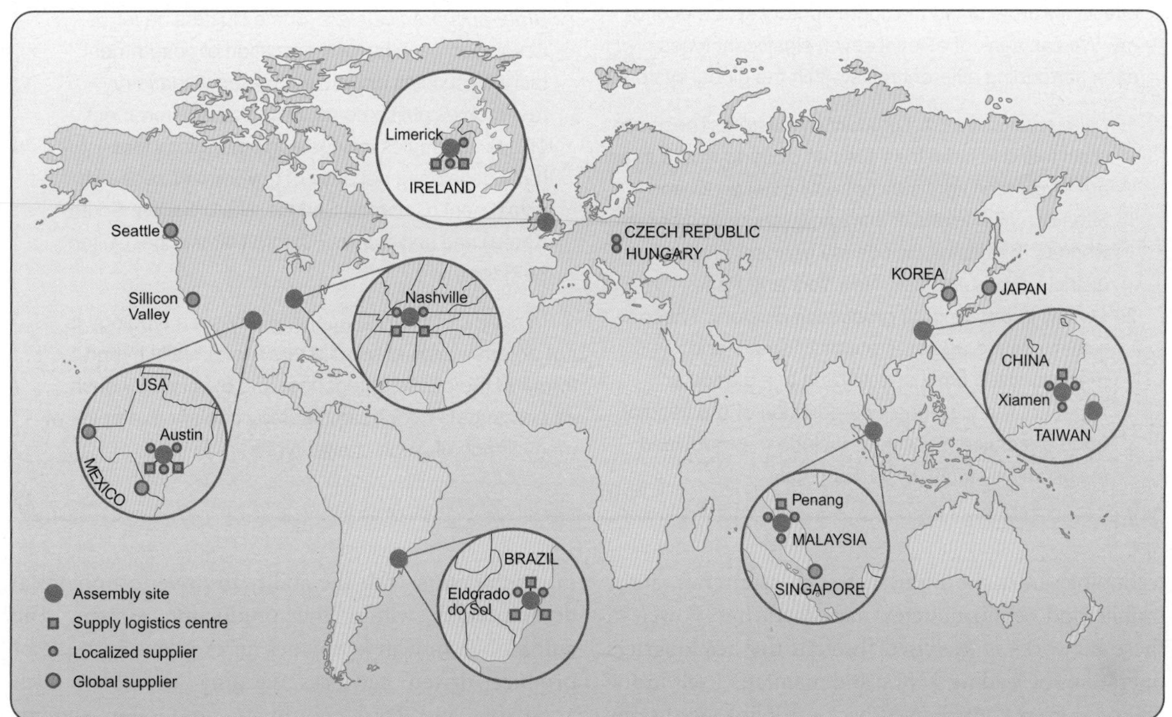

Figure 15.6 Dell's global production network, 2001.

Source: adapted from Fields (2004), Map 6.3

Spotlight Box 15.3

Types of clusters in the global economy?

There is a wide variety of different kinds of clusters in the global economy, created through different historical processes, and bound together by varying combinations of both *traded* and *untraded* interdependencies. Traded interdependencies are created by firms co-locating in a cluster alongside suppliers, partners and customers with which they have formal trading relationships. Proximity to these firms reduces the costs of transportation, communication, information exchange, and searching and scanning for potential customers. Untraded interdependencies are the less tangible benefits of being located in the same place, for example the emergence of a particular pool of specialized workers, or the development of local institutions such as business associations, chambers of commerce, government offices, and universities with particular attributes and capabilities. In particular, clusters can facilitate patterns of intense and ongoing face-to-face communication – and thereby transfers of important forms of intangible knowledge – between people working in the same or closely related industries.

Recognizing the diversity of clusters in the world economy is important if economic geographers are to effectively understand the contemporary space economy. We can think of at least seven significant types, each performing different roles within the global system:

1. *Labour intensive craft production clusters.* These are often found in industries such as clothing where firms are involved in tight subcontracting networks and often use high levels of immigrant labour and homeworkers. Examples include the garment production districts of Los Angeles, New York and Paris.
2. *Design intensive craft production clusters.* These refer to dense agglomerations of small- and medium-sized firms specializing in a particular aspect of the high-quality production of a particular good or service. Examples include the renowned towns and districts of the Third Italy.
3. *High-technology innovative clusters.* These clusters tend to have a large base of innovative small- and medium-sized firms and flexible, highly skilled labour markets in sectors such as computer software and biotechnology. Examples include Silicon Valley in the USA and Cambridge, United Kingdom.
4. *Flexible production hub-and-spoke clusters.* In these clusters, a single large firm, or small group of large firms, buys components from an extensive range of local suppliers to make products for markets external to the cluster. Examples include Boeing in Seattle, USA, and Toyota in Toyota City, Japan.
5. *Production satellite clusters.* These clusters represent congregations of externally owned production facilities. These range from relatively basic assembly activity, through to more advanced plants with research capacity. Examples are to be found across the export processing zones (EPZs) of the developing world – the Shanghai laptop cluster would fall into this category.
6. *Business service clusters.* Business services activities such as financial services, advertising, law, accountancy are concentrated in leading cities – such as New York, London and Tokyo – and their hinterlands.
7. *State-anchored clusters.* Some clusters have developed because of the location of government facilities such as universities, defence industry research establishments, prisons, or government offices. Examples include agglomerations that have developed because of government research investment (Colorado Springs, USA; Taejon, South Korea) and universities (Oxford/Cambridge, United Kingdom).

In reality, of course, applying any such typology is not an easy task. Many clusters are actually hybrid forms of the above categories, and a world city such as Los Angeles, for example, may contain examples of many, if not all, such cluster types.

technology-intensive industries such as aircraft, automobile and semiconductor manufacturing. Power in these networks is exercised through the headquarters operations of leading TNCs, and manifests itself in the ability to exert control over 'backward' linkages to raw material and component suppliers, and 'forward' linkages with distributors and retailers. High levels of profits are secured through the scale and volume of production in combination with the ability to drive technological developments within the production system. The automobile industry provides an excellent example of producer-driven networks. Leading assemblers such as Toyota and Ford coordinate production systems involving literally thousands of subsidiaries and tiers of subcontractor firms around the world, as well as extensive global networks of distributors and dealers.

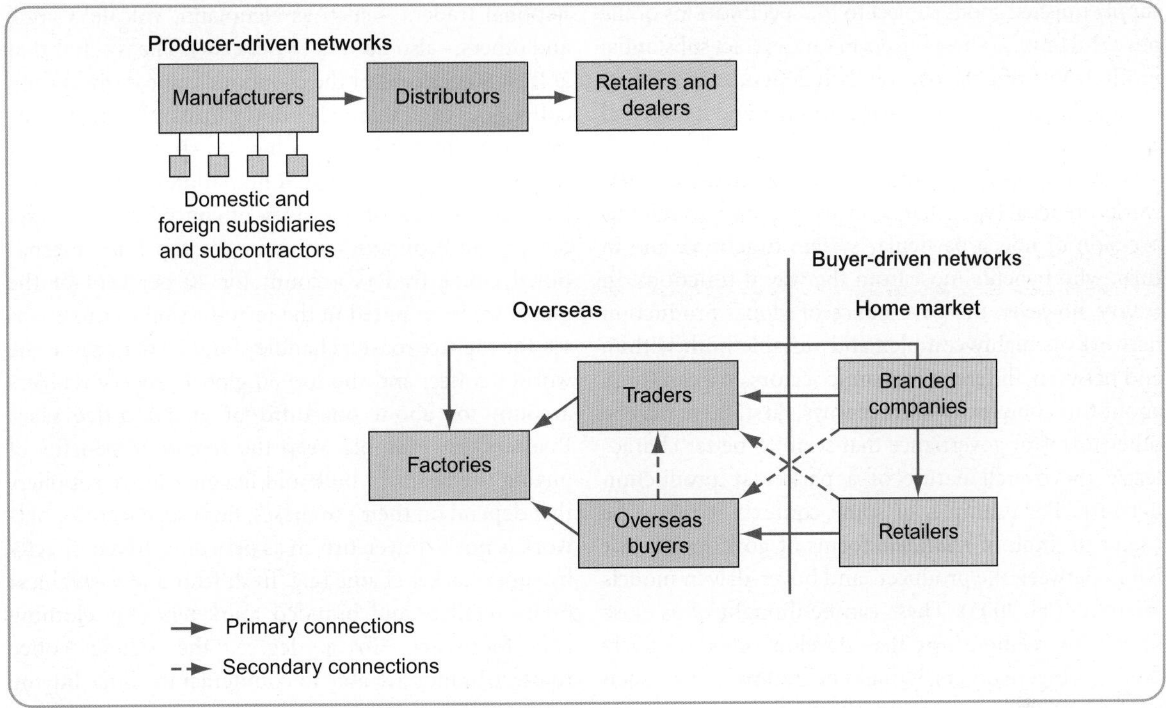

Figure 15.7 Producer- and buyer-driven production systems.

Source: *Commodity Chains and Global Capitalism,* Gereffi, G., Copyright © (1994) by Gary Gereffi and Miguel Korzeniewicz. Reproduced with permission of Greenwood Publishing Group, Inc., Westport, CT.

Table 15.3 Producer-driven and buyer-driven production networks compared

	Form of economic governance	
	Producer-driven	*Buyer*-driven
Controlling type of capital	Industrial	Commercial
Capital/technology intensity	High	Low
Labour characteristics	Skilled/high wage	Unskilled/low wage
Controlling firm	Manufacturer	Retailer
Production integration	Vertical/bureaucratic	Horizontal/networked
Control	Internalized/hierarchical	Externalized/market
Contracting/outsourcing	Moderate and increasing	High
Suppliers provide	Components	Finished goods
Examples	Automobiles, computers, aircraft, electrical machinery	Clothing, footwear, toys, consumer electronics

Source: adapted from Kessler (1998)

Buyer-driven networks, by comparison, are characteristic of industries where large retailers (e.g. Wal-Mart or Tesco) and brand-name merchandisers (e.g. Nike or Reebok) play the central role in establishing and shaping the network, with production activity itself often located in developing countries (see Figure 15.7). This form of global production network is common in labour-intensive consumer goods sectors, such as clothing, footwear, toys and handicrafts. Production is usually undertaken using tiered networks of subcontractors that supply finished goods subject to the specifications of the powerful buyers. These buyers in turn extract substantial profits from bringing together their design, sales, marketing and financial expertise with strong brand names and access to large consumer markets in developed countries.

The notions of producer- and buyer-driven networks are ideal types that allow us to get an overall impression of how a particular system functions, and in turn, who benefits most from the way it functions. In reality, however, the governance of global production networks is highly complex and variable both within, and between, different economic sectors. We can think about this complexity in three ways. First, there may be other forms of governance that seem to better characterize the overall nature of a particular production network. For example, in some contexts it might be useful to think of *relational* forms of governance that fall in between the producer- and buyer-driven models (Gereffi et al. 2005). These can be thought of as close inter-firm relationships that develop on a relatively even footing. In others, it might be trading firms – such as the Japanese giant trading firms, the so-called *soga shosha* – or intermediaries that drive the network. In the Internet era, for example, there is evidence that new forms of 'infomediaries' with extensive access to online consumer information (e.g. Google, AOL etc.) may play increasingly important roles in production networks (Gereffi 2001). Secondly, when we look at the detail of a particular production network, it is important to recognize that there may be a variety of governance types in operation between the various actors within the overall system. Thirdly, governance regimes are not static, but need to be seen in *dynamic* terms. The nature of a production network – and indeed the constituent relationships within it – may change over time because of a wide range of factors, both internal and external to the system. In particular, firms and/or particular regions may seek to improve their position within the global production network through pursuing *upgrading* strategies.

At first glance, comparing coffee and laptops would seem to provide a nice contrast between buyer- and producer-driven networks. The coffee network is 'clearly buyer-driven and coffee roasters play the lead role in determining the functional division of labour along the chain. In other words, roasters define the key terms of participation directly for their immediate suppliers and indirectly for other actors further upstream' (Gibbon and Ponte 2005: 83). Roasters capture about 30 per cent of the value that is added in the coffee production network, but retailers (22 per cent) and international traders – such as Neumann, Volcafe, Cargill and others – also do well (8 per cent). The result is that at least 60 per cent of the economic value derived from coffee goes to developed country firms, although some estimates put the figure as high as 80 per cent (Fitter and Kaplinsky 2001). The domination by these actors is in large part a simple story of concentrated market power and economies of scale: the top four international coffee traders account for 40 per cent of the global trade; as noted in the introduction to this chapter, the top five roasters handle about 50 per cent of the world's coffee; and the top 30 global grocery retailers account for about one-third of global coffee sales. Roasters, for example, reap the economic benefits of buying the beans in bulk and having captive suppliers that depend on their purchases. Even so, the coffee network is not *as* buyer-driven as networks driven directly by supermarket chains (e.g. fresh fruit and vegetables) or by retailers and branded marketers (e.g. clothing and footwear). To a degree, the strong coffee roasters/brands are able to counteract the huge buying power of the retailers, meaning that influence is shared. Ideal-type buyer- and producer-driven chains, then, should perhaps be seen as the ends of a spectrum rather than discrete categories.

Typically, as noted in Table 15.3, the computer industry has been seen as a capital-intensive producer-driven network dominated by vertically integrated giants such as IBM. However, the computer industry has changed beyond recognition over the past fifteen years as the PC has increasingly become a mass-market standardized product. The branded PC companies continue to drive the global production network, taking the decisions that drive the whole system and coordinating the activities of the other main players. But their role has changed over time. Leading manufacturers like Dell and HP have increasingly focused on final assembly, design and branding, with the vast majority of production outsourced to Asian subcontractors, thereby sharing the characteristics of the lead firms in buyer-driven systems such as clothing and shoe production. Over time, there has clearly been a shift in the governance characteristics of the PC industry.

However, it is possible to make another argument relevant to the PC sector. The branded PC marketplace has become increasingly competitive over recent years as the standardization process has progressively reduced profit margins for even the leading firms. The result has being ongoing consolidation and rationalization, with several large brands being swallowed up by competitors (e.g. Compaq by HP, IBM by Lenovo).

O'Riain (2004) has suggested that we might think of the PC network (and others like it) as falling into another governance category – the technology-driven network, in which it is control over technical standards which is critical for exercising power over the system. In that sense, it is the firms that set the dominant architectural standards for PCs in terms of software (Microsoft's operating systems) and hardware (Intel's computer chips) that benefit most from the production network and secure profits rates significantly higher than any other participants in the industry (40 and 31 per cent of revenues for Microsoft and Intel, respectively, in 2000). 'Power . . . resides, therefore, in what is getting made and mobilizing networks of support for those products rather than in how that product is itself produced – emphasizing the place of intellectual property holders in securing the rewards . . . within the system' (O'Riain 2004: 645).

Returning to coffee, we can start to appreciate the complexity and dynamism of global production networks. The coffee production network is not static in organizational and governance terms. It can no longer simply be understood as a sequence of market transactions. Three changes illustrate this point. First, the roasters are increasingly concentrating on their roasting, blending and branding activities, employing coffee traders to source coffee and manage their supply networks for them, along with undertaking the steaming of certain kinds of beans. These are particularly important roles in a coffee network which is increasingly concerned with coffee quality testing and assurance, and reflect the establishment of more long-standing, relational connections between roasters and traders. Secondly, international traders are vertically integrating their operations by buying export firms in producing countries in order to ensure secure and high-quality supplies. This serves to consolidate the position of traders within the system and potentially increase further the value captured by developed-country firms. In turn, the potential for developing countries' producers and processors to upgrade their activity is restricted by the fact that roasting and final processing ideally needs to be undertaken near final markets. Thirdly, growing consumer demand for new kinds of speciality, organic and fair-trade coffees has opened up some direct connections between producers and smaller roasting houses, potentially at least improving the financial returns for the coffee farmers connected to these new networks. We shall return to this issue in the last section of the chapter.

In terms of the laptop production network, at the same time as branded PC firms have retreated from manufacturing activity, Taiwanese manufacturers have evolved or upgraded to take on new roles and, as a result, now occupy much more significant positions within the production network. Many Taiwanese firms started out as simple assemblers of components for PCs. Over time, they moved – through what is known as *process* upgrading – into the manufacture and assembly of PCs sold under the brand names of other firms (i.e. they became OEMs – Original Equipment Manufacturers). The next step was to develop design capacity, as we have already noted, becoming ODMs (Original Design Manufacturers) through a step known as *product* upgrading. Next, some firms have taken things further and started to manufacture and sell PCs under their own brand, thereby becoming OBMs (Original Brand Manufacturers) through *functional* upgrading. Acer, for example, now a leading PC brand in its own right, started out as one of Taiwan's main electronics suppliers, producing various computer products for international buyers. Over time, it separated off its OEM activity, leaving it to focus on the marketing and R&D activities of its own brand laptops. In reality, however, these categories do not neatly map onto individual firms, some of which may take on multiple roles in the production network. Asus, for example, the number one PC brand in Taiwan itself, combines both OEM/ODM and OBM activity. In some cases, firms may be able to use their expertise to shift into entirely new production networks (e.g. from laptops to WAP phones) in a shift termed *chain* upgrading (Kaplinsky 2005).

Two further points should be noted here. First, there is no automatic or simple progression along this upgrading path. These shifts reflect changing *strategies* on the part of Taiwanese firms in response to changing market conditions. As PCs have become ever more standardized commodities, however, OEM and ODM producers have increasingly been squeezed for cost savings by branded PC firms. They have tended to respond in one or more of three ways: moving into original brand manufacturing themselves; making more components themselves as a way of cutting costs (i.e. vertically integrating their activity); or trying to cut costs through production relocation to China, as described earlier. Secondly, the successful ongoing upgrading of Taiwanese electronics firms is not only due to firm strategies and inter-firm networks. It also reflects a range of deliberate strategies pursued by the Taiwanese state to develop the industry since the early 1970s, for example: the establishment of the Hsinchu Science Park; direct investment in early key players in the sector including

UMC (United Microelectronics Corporation) and TSMC (Taiwan Semiconductor Manufacturing Corporation); and steering developments in the industry through its agency ERSO – the Electronics Research Service Organization. These reveal the importance of also considering the institutional context of production networks, a topic to which we now turn.

15.5 The institutional context of production networks

So far we have focused almost entirely on the production network as a mesh of intra- and inter-firm connections. Production networks are also shaped by a wide range of *extra*-firm relationships that, as we saw in Figure 15.1, are also an integral part of the system and may incorporate a wide range of non-firm entities (e.g. supranational organizations, the state, labour unions, business associations, etc.). Expressed slightly differently, every relationship in a production network is shaped by its *institutional context.*

We can unpack the complexity of these institutional contexts in two ways. First, institutional context is significant at various spatial *scales.* At the sub-national scale, local governments may seek to stimulate particular kinds of economic activity in their locality, for example by providing low-rent premises for small high-tech businesses. At the national scale, nation-states still wield a huge range of policy measures to try and promote, and steer, economic growth within their boundaries (Dicken 2007). Increasingly important in an era of globalization are measures designed to promote (or restrict) movements of traded products, investment and migrants across national boundaries. At the macro-regional scale, regional blocs such as the European Union or the North American Free Trade Agreement (NAFTA) have considerable influence on trade and investment flows within their jurisdiction. At the global scale, supra-national institutions like the World Trade Organization (WTO) and the International Monetary Fund (IMF) increasingly determine the regulatory frameworks for global financial and trading relationships. Even a relatively simple global production network will cross-cut and connect a wide range of institutional contexts. Secondly, it is important to distinguish between *formal* and *informal* institutional frameworks. The former relates to the rules and regulations that determine how economic activity is undertaken in particular places (e.g. trade policy, tax policy, environmental regulations and so on), while the latter refers to the rather less tangible, and often place-specific, ways of doing business that relate to the social, economic and political *cultures* of particular places.

Again, we can use the examples of coffee and laptop production to illustrate these arguments. First, the coffee production network demonstrates the importance of *formal* institutional frameworks at the national and global scales. From 1962 to 1989, the international trading of coffee was governed by a series of International Coffee Agreements (ICAs) managed by the International Coffee Organization (ICO), a supra-national organization made up of representatives of a wide range of coffee importing and exporting countries. The ICAs combined price bands and export quotas to provide a coffee trading system that was widely credited with both raising and stabilizing coffee prices. During this period, many exporting countries established coffee marketing boards: government institutions that controlled markets, monitored quality and acted as a link to exporters and international traders (see Figure 15.2). Such boards provided an important protection for farmers and growers from the vagaries of the international coffee market. In 1989, however, the ICA was not renewed in the face of rising production levels and low-cost competition from non-member exporting countries.

The ending of the ICA regime has dramatically altered the balance of power in the coffee chain, as the now market-based coffee trading system has led to lower and more volatile coffee prices. Its demise has served to concentrate power in the hands of consuming-country firms, and in particular the small group of roasters introduced earlier. The collapse of the ICA has meant that coffee-producing countries are no longer cooperating, but are in fact competing with each other, resulting in overproduction and a drop in prices which the roasters have been able to benefit massively from, a process exacerbated by the fact that consumer coffee prices have not dropped in a similar way, but have in fact risen significantly. At the same time, the national coffee marketing boards in the exporting countries either have been eliminated, or have retreated into a restricted overseeing role that has left them marginalized within the production network (see dotted lines in Figure 15.2). As a result of these changes to the interlinked international and national institutional contexts, millions of coffee farmers worldwide are exposed to price fluctuations on the global coffee market. On occasion, this can result in farmers receiving less for their coffee beans that it costs to grow them.

Plate 15.2 Coffee growers in Guatemala having their organic coffee beans weighed at the local co-operative.
(Sean Sprague/Panos Pictures)

What this story shows is how changing institutional frameworks can impact on all the other dimensions of a global production network. In terms of its basic structure, the demise of the ICA has led to the bypassing of a previously important actor, namely the coffee marketing boards. In terms of changing geographies, the rapid growth of exports from Vietnam was both a cause of the ICA's demise and at the heart of the subsequent overproduction. In governance terms, the post-ICA regime has enabled a further concentration of power in the hands of roasting firms.

The laptop industry is similarly affected by multi-scalar regimes of formal institutional relationships. For example, as we have seen, rules governing outward investment from Taiwan to China were crucial in determining the timing of shifts in laptop production. Here, however, we will use the laptop industry to open a window on a range of other less formal connections that are integral to the success of Taiwan's electronics industries. More specifically, there is now increasing recognition of the significance of migration flows of skilled engineers between Taiwan and California for driving the economic success of both high-tech regions. Hsu and Saxenian (2000) describe the development of a *transnational technical community* linking Hsinchu and Silicon Valley. For several decades now, Taiwanese migrants to Silicon Valley – in combination with those from India, China and other parts of Asia – have played an important part in the continued dynamism of the Californian region through their entrepreneurial activity and links back to Asian markets and supplier firms.

However, the more recent reversal of this standard 'brain drain' phenomenon has seen thousands of US-educated engineers returning to Taiwan – reaching a peak of 5,000 per year in the mid-1990s – and playing a pivotal role in transforming Taiwan into a high wage and skill economy. 'These returnees to Taiwan, many of whom had worked for at least a decade in the United States, brought with them not only technical skill but also organizational and managerial know-how, intimate knowledge of leading-edge IT markets, and networks of contacts in the United States technology sector' (Saxenian 2006: 149). Return migrants have taken on leading roles in Taiwan's technology sector: they are highly represented in the management cadre of leading firms, appear to be more entrepreneurial than non-migrants, and are active players in Taiwan's burgeoning venture capital markets. Not all are permanent returnees, however: some have instead become the 'new argonauts' of the global economy (Saxenian 2006), criss-crossing the Pacific on a weekly or monthly basis and facilitating connections between firms, suppliers, clients and investors in the two high-tech clusters. What these migrant flows serve to demonstrate is that global production networks in industries such as the PC sector are more than just formally regulated firm-to-firm connections – they are also constituted by broader personal and knowledge flows.

Plate 15.3 A worker at the Acer computer factory in Taiwan.
(Chris Stowers/Panos Pictures)

15.6 Reshaping global production networks?

We now have an appreciation of how to identify, analyse and explain global production networks. But what if various actors – and in particular, we as consumers – want to try to change how a production network is structured and operates? In an increasingly media-saturated and interconnected world, awareness has grown in consuming countries of the social, economic and environmental conditions under which commodities are produced, particularly in the developing world. This has seen a rise of so-called *ethical* consumption, whereby interventions surrounding the consumption stage are designed to improve conditions at points 'upstream' in the production network. There are a number of forms such intervention may take. Perhaps the simplest form of consumer campaign is to *boycott* – that is, not purchase – the products of a particular company. We can also think of *corporate campaigns* that seek to target highly visible corporations by mobilizing information regarding violations against workers or the environment, such as were targeted at the sweatshop operations of sportswear giants like Nike and Reebok in the late 1990s. Increasingly, however, the development of various kinds of *benchmarks* or *standards* – against which various end products and their production processes can be measured – are an increasingly important part of production network regulation within the global economy (see Figure 15.8).

The world of standards and codes has very rapidly become a broad and varied one. There are seven dimensions to this complexity:

1. They may be applied to different facets of the production system, for example environmental, social or labour conditions or rates of economic return.
2. They may take a variety of forms: a code of conduct, a label on a finished product, a tightly specified technical standard, a set of voluntary initiatives or a combination of some or all of these forms.
3. They may apply to a particular chain (e.g. beef), a sector (e.g. fresh meat), or be generic (e.g. all fresh foods).
4. They may be developed by firms, NGOs, trade unions or international organizations, and usually by a combination of some, or all, of these institutions.
5. The certification or accreditation of the standards – that is, judging whether they have been met – may be undertaken by public or private, and profit or not-for-profit organizations;
6. They will range from the voluntary (e.g. seeking 'Fair Trade' status for a product) to the mandatory (e.g. safety standards for plastic toys).

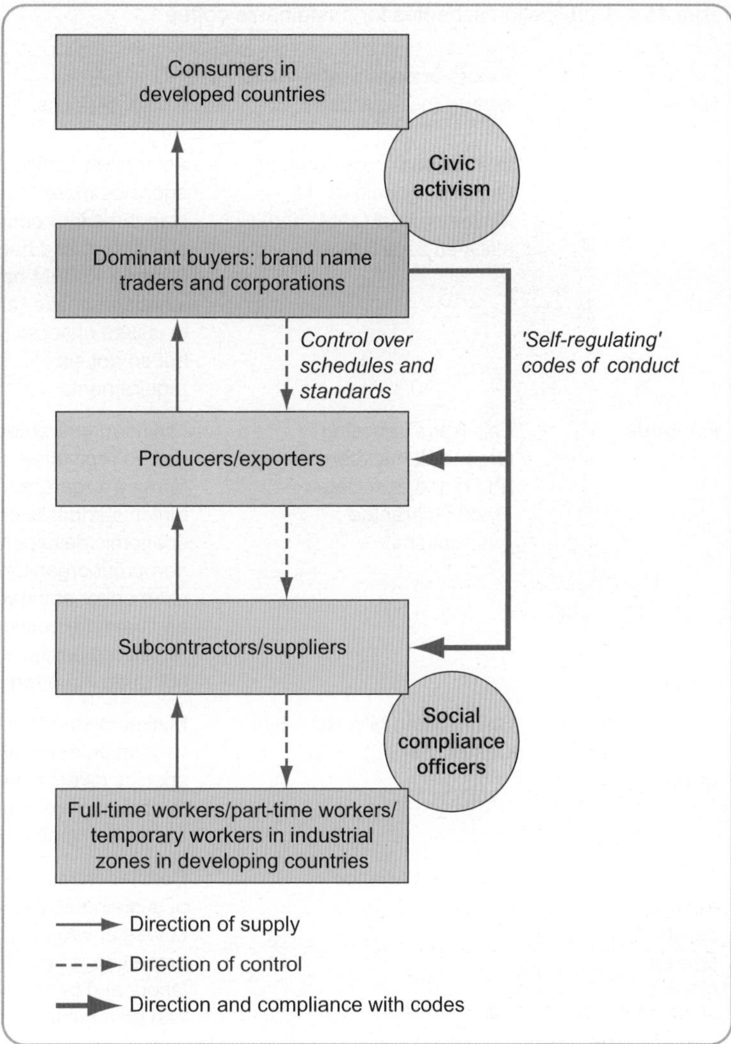

Figure 15.8 New modes of regulating production networks.

Source: adapted from Sum and Ngai (2005), Figure 1

7. They are inherently geographical, in terms of both the territory in which they apply (often the place of consumption) and the places in which the effects are felt (usually places of production).

In terms of our two case studies in this chapter, we will focus exclusively on coffee, as these kinds of initiatives are more prevalent and developed in buyer-driven sectors characterized by labour-intensive production in the developing world (e.g. shoes, garments, food stuffs, etc.). A number of schemes seeking to promote sustainable coffee production currently exist (Table 15.4). The five listed schemes clearly illustrate the degrees of variation just described. For example, they may apply to environmental or economic conditions (bird-friendly vs. Fair Trade, respectively); involve a label or be primarily a code of conduct (Fair Trade vs. Utz Kapeh); be global or regional in their coverage (organic certification vs. bird-friendly) and apply to all farms or just estates/cooperatives (organic vs. Utz Kapeh). Beyond these examples, there are other schemes initiated by private companies, for example, Starbuck's CAFE (Coffee and Farmer Equity) Practices, designed to 'to evaluate, recognize, and reward producers of high-quality sustainably grown coffee' or Green Mountain Coffee Roasters' Stewardship Programme. While only a small proportion of the world coffee trade currently falls under these schemes – just one per cent in 2003 – the level is rising rapidly, and leading roasters such as Sara Lee and Kraft are starting to buy significant volumes of certified coffee (Daviron and Ponte 2005).

Such schemes have certainly initiated improvements in the social, economic and environmental conditions of the coffee growers that are enrolled in them. Growers of certified Fair Trade coffee, for example, are

Table 15.4 Certification schemes for sustainable coffee

Name	Actors or organizations setting the standards	Characteristics	Geographic and farm-size coverage
Organic	International Federation of Organic Agriculture Movements (IFOAM) and affiliated associations	Accredited certification agencies monitor organic standards in production, processing and handling; formally, IFOAM basic standards make reference to issues of social justice, but do not set requirements	Global, but most organic coffee comes from Latin America, especially Mexico; all farms
Fair trade	Fair Trade Labelling Organizations International (FLO) and associated Fair Trade Guarantee Associations	Minimum guaranteed price paid to registered small farmers' organizations that match standards on socio-economic development; non-profit organizations set/monitor standards and mediate between registered producers and fair trade importers	Global; only smallholders
Bird-friendly coffee (shade-grown)	Smithsonian Migratory Bird Center (SMBC)	Minimum standards on vegetation cover and species diversity needed to obtain use of label; also covers soil management	Standard applies only to Latin American coffees so far; mainly estates
Rainforest Alliance-certified (shade-grown)	Rainforest Alliance	Certifies farms on the basis of sustainability standards; covers environmental protection, shade, basic labour and living conditions, and community relations	Latin American countries only; mostly estates but also some cooperatives
Utz Kapeh	Utz Kapeh Foundation	Code of conduct for growing sustainable coffee formulated on the basis of the 'good agricultural practices' of the European Retailer Group (EUREP); includes standards on environmental protection and management, and labour and living conditions	Mainly in Latin American countries, but growing also in Asia (India, Indonesia and Vietnam) and in Africa (Uganda and Zambia); mostly estates, but some cooperatives

Source: Daviron and Ponte (2005), Table 5.11

guaranteed a premium above the market price, and their product may fetch significantly more in certain market conditions. And yet we need to add some significant notes of caution, such as the fact that the vast majority of the world's coffee farmers are still labouring outside the jurisdiction of these schemes which are voluntary in terms of participation. Coffee roasters and retailers that benefit from association with such schemes may in reality only obtain a small proportion of their coffee from certified channels. And we need to ask who pays for, and who benefits from, the certification process? The rise of standards-based schemes has fuelled the emergence of a new category of commodity chain participant, the independent auditor – such as Scientific Certification Systems (SCS), for example – many of whom themselves are profit-seeking firms.

The costs of auditing long and complicated production networks are significant, and firms will vary in their ability and willingness to meet these costs. In many instances, buying firms will expect suppliers to meet the extra production costs. Farmers involved in certified organic coffee production in Oaxaca State, Mexico, for example, may pay between 10 and 30 per cent of their gross receipts to certify their produce. This serves as a significant barrier to entry and only the best organized and most well-funded farmers can turn organic production into a profitable enterprise (Mutersbaugh 2005). Together, these points suggest that we need to look carefully and critically at initiatives to alter the nature of global production networks; there is always the risk that, in certain contexts, they will serve to enhance the very injustices and inequalities they seek to diminish.

15.7 Conclusion

Returning to where we began, we now understand far more about the global economic interconnections and exchanges that enable the student in the café to drink a coffee and surf the Internet on their laptop computer. We could tell similar, yet at the same time profoundly different, stories about the mobile phone and iPod in their jacket pocket, about the jeans they are wearing, about the machine that brewed the coffee, the furniture in the café and so on. Whether the student in question is aware of the complex 'back stories' of the commodities around them is another matter: as economic geographers, however, it is beholden on us to unravel those stories and their implications for the people and places across the globe that they connect.

This chapter has provided us with a framework and a language for exploring and understanding these economic geographies. Sophisticated global production networks lie behind nearly all of the products and services that we consume on a daily business. These global production networks can be understood as the meshes of intra-, inter- and extra-firm relationships through which material and non-tangible inputs are transformed into consumable outputs of many kinds. They provide the organizational glue that connects together the disparate local clusters of economic activity that constitute the 'on the ground' reality of the global economy. As a result of recent shifts in the global economy associated with increased competition, technological change, and trade and foreign investment deregulation – for which globalization is often used as shorthand – global production networks in general have arguably become more geographically extensive, disintegrated and dominated by developed-country buyers. Ultimately, however, each global production network has its own unique organizational structure, geographical configuration, governance regime and institutional context. In this chapter we have used just two examples, namely the coffee and laptop computer production networks, to explore some of this inherent variability in the configuration of global production networks.

What this chapter has also demonstrated is the futility of seeking to understand the contemporary global economy and its workings without adopting a geographical perspective. Lead firms in global production networks both *use* uneven geographies – taking advantages in differences in labour costs and skill levels, for example – and in turn *reshape* those uneven geographies through their investment and disinvestment decisions. Today's global economy is undoubtedly more complicated and interdependent than ever before: it is the task of the economic geographer to unravel and explain this complexity.

Learning outcomes

After reading this chapter, you should understand:

- How all economic activities can be conceptualized as a form of global production network.
- That global production networks have distinctive, yet changeable, geographical forms.
- That global production networks exhibit a range of governance regimes according to the types of firms involved.
- That the range, complexity and efficiency of global production networks is heavily shaped by multi-scalar institutional contexts of different types.
- How attempts can be made to reconfigure production networks by introducing various kinds of standards and codes of conduct.
- The importance of a geographical perspective for understanding the organization of the contemporary global economy.

Further reading

Coe, N.M., Kelly, P.F. and Yeung, H. (2007) ***Economic Geography: A Contemporary Introduction***, Blackwell, Oxford. A clear, engaging and student-friendly introduction to contemporary economic geography.

Curry, J. and Kenney, M. (2004) The organizational and geographic configuration of the personal computer value chain, in M. Kenney and R. Florida (eds) ***Locating Global Advantage: Industry Dynamics in the International Economy***, Stanford University Press, Stanford, pp. 113–41. Provides an up-to-date and detailed account of the PC industry production network described in this chapter.

Dicken, P. (2007) ***Global Shift: Mapping the Changing Contours of the World Economy***, 5th edition, Sage, London. Now in its fifth edition, this remains far and away the best textbook on the structure, evolution and geography of the global economy.

Massey, D. (1984) ***Spatial Divisions of Labour***, Macmillan, London. This landmark book theorizes, and exemplifies, the concept of spatial divisions of labour. This is one of the most definitive accounts of why economic geography matters. A second edition was published in 1995 with a new concluding chapter.

Peck, J. and Yeung, H. (2003) (eds) ***Remaking the Global Economy***, Sage, London. A thought-provoking series of essays on the changing nature of the global economy.

Ponte, S. (2002) The 'latte revolution'? Regulation, markets and consumption in the global coffee chain, ***World Development***, 30, 1099–122. As with the Curry and Kenney reference above, a good source for more information on the coffee networks introduced in this chapter.

Useful websites

http://www.econgeog.org.uk The website of the UK's Economic Geography Research Group provides a window on current economic geography activity in the UK context.

http://www.geography.uconn.edu/aag-econ Does likewise for the US Economic Geography Specialty Group.

http://www.globalvaluechains.org This site contains a wealth of conceptual and empirical material on global production networks/value chains.

http://www.pcic.gsm.uci.edu The Personal Computer Industry Centre at University of California, Irvine, offers a wide range of up-to-date materials on the industry.

http://www.ico.org/index.asp The website of the International Coffee Organization provides a range of information on the coffee industry, and in particular, its evolving regulatory structures.

http://www.fairtrade.org.uk The United Kingdom's Fairtrade Foundation is one of the most well-known attempts to improve the economic returns offered to producers.

For annotated, clickable weblinks and useful tutorials full of practical advice on how to improve your study skills, visit this book's website at **www.pearsoned.co.uk/daniels**

SERVICE ECONOMIES, SPATIAL DIVISIONS OF EXPERTISE AND THE SECOND GLOBAL SHIFT

Chapter 16

John R. Bryson

Topics covered

- Defining services
- Services capitalism
- Service employment versus output
- The body, the personality market and emotional labour
- The division of labour
- Spatial and gender divisions of labour
- Spatial divisions of expertise
- The second global shift

Other chapters in this book explore the environmental implications and inequalities of wealth creation at a variety of spatial scales. This chapter is about the large and complex capitalist economic system and the ongoing shift from manufacturing to services that is being experienced by a range of economies. Put another way, it is a shift towards various forms of expertise- or knowledge-intensive employment. The focus of this chapter is on understanding some of the dynamics of expertise-intensive service work and employment but with the important caveat that the complexity of the modern economy makes it problematical to view service activities as a distinct and separate category from manufacturing. It follows that the other chapters in this section that include references to manufacturing, finance, knowledge-intensive services, and even consumption, must be seen as an integrated whole. Isolating finance or manufacturing processes from a range of service functions overlooks the multiple interconnections that form a complex and integrated production system. The flows of money and finance that are transferred around the world in the twinkling of an eye are manipulated and coordinated by service workers, or more specifically knowledge workers who convert information into knowledge that then informs investment decisions of all kinds.

This chapter begins by defining services and placing them within the context of the overall production system at large. The continued development of this system over time (and space) reflects the process of economic specialization. This is driven by continuous extension of the **division of labour** and the development of a new **spatial division of expertise**. The latter is a key concept that will be examined in more detail in the second part of this chapter which explores the ways in which service expertise is integrated into complex production systems. The chapter concludes with an introduction to some of the characteristics of services offshoring or the rise of the **second global shift**.

16.1 Defining services

In the economically developed world, the vast majority, often more than 75 per cent, of all jobs involve some form of service work (Bryson et al. 2004). Furthermore, in Europe, North America, Japan, Australia, New Zealand as well as parts of the developing world, around 90 per cent of new jobs are created in services. As a result the economies and societies of these countries appear to revolve predominantly around service activities and the experience of service work. It is generally accepted by scholars and policy makers that developed market economies are dominated by various forms of service work, ranging from extremely well-paid lawyers and merchant bankers to less well-paid hotel and retail workers (Case study 16.1). Services contribute to economic growth in a variety of ways; they are traded locally, regionally, nationally and internationally. They are also heavily wrapped within and around the production processes of manufactured goods as well as other services; they add value by smoothing the relationship between production and consumption, for example via market research, product design, development and testing and advertising (Spotlight box 16.1). Services can be

Plate 16.1 The Orwell pub – a former warehouse by the Leeds & Liverpool Canal, Wigan Pier, Greater Manchester, UK.
(Dorothy Burrows/Photofusion)

Regional Case Study 16.1

The changing economic structure of the West Midlands, United Kingdom

The West Midland's economy is still dominated by manufacturing, but this is under threat as the competitiveness of low-value added or labour-intensive manufacturing is undermined or value-added creation in other economic sectors outstrips that of manufacturing. As a consequence, over the past thirty years the region has been experiencing considerable economic restructuring and turbulence. This restructuring has caused considerable turbulence related to the rise of unemployment which is linked to the closure, relocation or downsizing of manufacturing facilities. Alternative employment opportunities have arisen as the service side of the economy has grown, for example, in retail distribution, education, business and professional services and the public sector (Figure 16.1).

Motor vehicles and other transport equipment comprise some one in seven of the region's manufacturing employment, but this ratio continues to decline. In 2002, 19.2 per cent of total employment in the West Midlands was in manufacturing compared to a national average for Great Britain of 13.4 per cent. The region's over-representation in manufacturing employment is based on male rather than female employment. Just under 30 per cent of male employees worked in manufacturing (Great Britain 19.3 per cent) compared with only 9.7 per cent for female workers (Great Britain 7.2 per cent).

The second most important sector in employment terms is financial and business services. In fact, the region has the largest financial and business services centre outside London and this is driving service-output growth in the region. By 2002, this sector accounted for 16.7 per cent of all employees compared with a British average of 19.6 per cent. Employment figures are a relatively poor indicator of the contribution an economic sector makes to a regional or national economy. A far better measure is gross value added (GVA) which measures the contribution to the economy of each individual producer, industry or sector. On a regional basis GVA needs to be used with care as the figure is calculated using workplace rather than place-of-residence data. This means that GVA does not take into consideration GVA that is created by individuals living in an area but who work in adjacent regions.

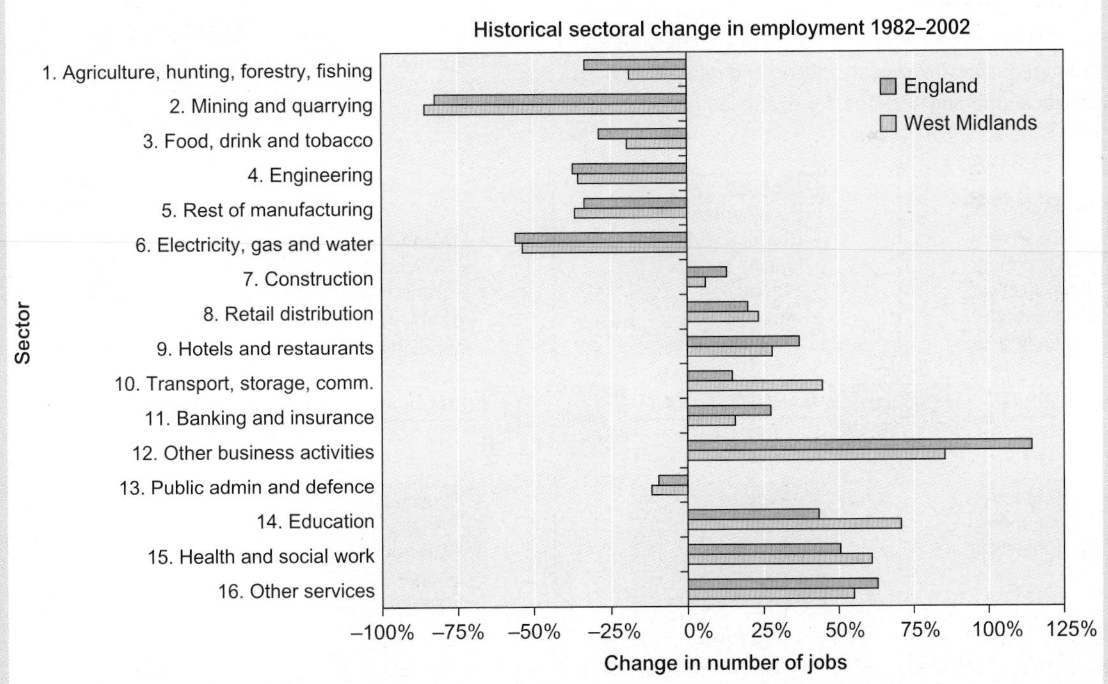

Figure 16.1 Historical change in sectoral employment in the West Midlands, UK, 1982–2002.
Source: Bryson and Taylor (2006)

Manufacturing's contribution to the region's GVA has been declining relatively rapidly in recent years. Between 1997 and 2002, manufacturing GVA fell from 29.3 per cent to 21.0 per cent while that of real estate, renting and business activities increased from 17.2 per cent to 20.8 per cent. The region is still the most industrialized region in the United Kingdom, with significant manufacturing activity in car and automotive components, engineering, metal processing and manufacture, ceramics, brewing and carpets, as well as becoming increasingly well known for specialist niche manufacturing.

Spotlight Box 16.1

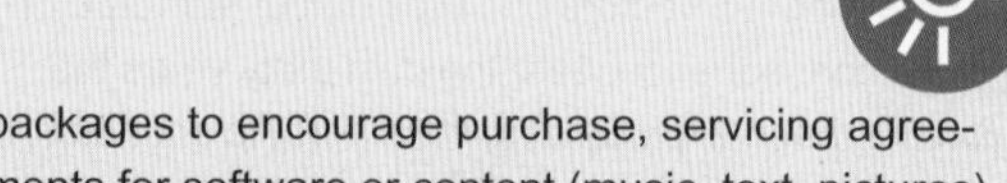

Services and the production process

Producing any product or service involves service expertise to be embedded in different parts of the production process – pre-production, during production and post-production or consumption (Figure 16.2). Pre-production involves understanding the design process, including market research and the ability to innovate. This may involve the design of a production process or of an actual product or service. The development of services may require the creation or modification of a process, for example the systems that support a financial services transaction or the check-in process at an airport. Services incorporated into production are concerned with the efficient management of the production process, for example queue management at leisure parks or call monitoring in call centres. Post-production involves marketing and related services or with supporting services, for example finance packages to encourage purchase, servicing agreements for software or content (music, text, pictures).

The production process can be divided into five parts with each part requiring different forms of service knowledge and expertise:

1. *Pre-manufacturing* – product development, research & development, design, product testing, market research, finance
2. *During manufacturing* – finance, quality control, stock control, purchasing, safety, management, continuity/contingency planning etc.
3. *Selling* – logistics, distribution networks, marketing, finance
4. *During product and system utilization* – maintenance, leasing, finance etc.
5. *After product and system utilization* – waste management, recycling etc.

Source: Bryson et. al. (2004)

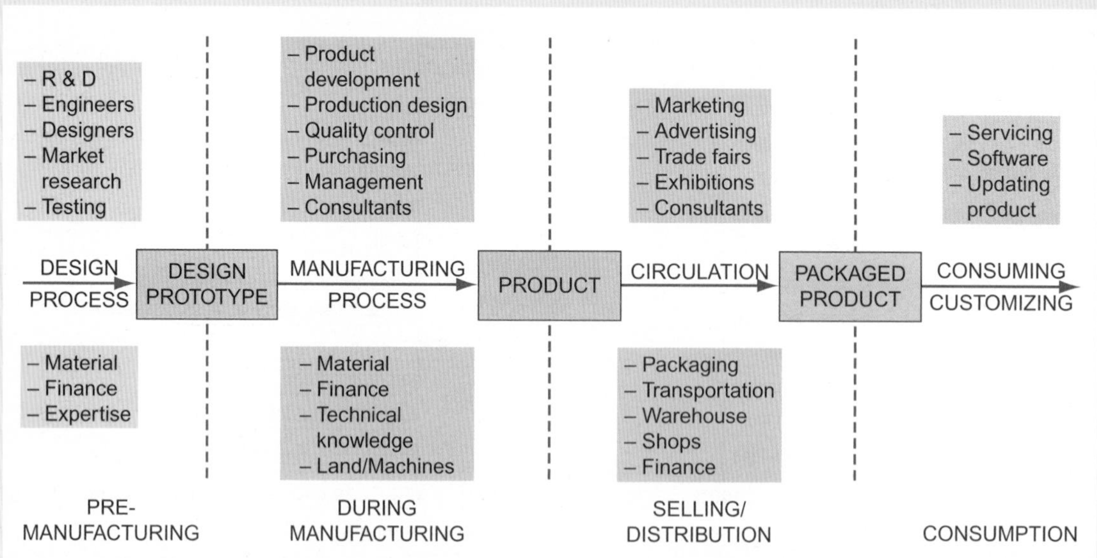

Figure 16.2 The production process.

Source: Bryson, Daniels and Warf (2004), p. 52

exported either directly via transfer across borders or direct representation of the provider in another country or indirectly through the incorporation of a service into a product or another service which is then exported.

During the 1970s economists tended to consider services as '"immaterial goods" or simply as "goods"' (Hill 1977: 315). But services are not goods; they exhibit a range of characteristics that fundamentally distinguish them from goods. Hill (1977) identified some of the primary differences between goods and services beginning with the observation that a diverse selection of services, the shipment of goods by a transport firm, vehicle repair, painting and decorating, the cleaning of a house by servants, hairdressing and dentistry, actually share a common set of characteristics. For Hill the two defining characteristics of services were the way in which they brought about:

1. A change or transformation in the condition of a good or person.
2. A change that is the result of the activity of another individual or firm.

These two defining principles are used by Hill to construct what is considered the classic definition of services: 'A service is defined as a change in the condition of a person, or of a good belonging to some economic unit, which is brought about as the result of the activity of some other economic unit, with the prior agreement of the former person or economic unit' (Hill 1977: 318). This definition highlights one of the primary problems with the category of services which is that in many instances the consumer often only observes or experiences the production or performance of the service. This simultaneous production and consumption, and even co-production of a service, creates confusion between the process of producing it and the final end process of consuming the service. This debate may appear to be quite complex but fundamental to the distinction between goods and services should be the realization that the delivery of a service does not result in the exchange of ownership of a material product. Thus, a service cannot be thought of as an immaterial good because nothing is exchanged. The output of many services is ephemeral or non-material, for example attendance at a lecture or viewing a live theatrical performance. It is often suggested that software is a material service since it can be stored on a computer, but this overlooks the fact that the software provider does not transfer complete ownership of the software to the user: it is owned under licence from the service provider and does not mirror the way in which a product such as a laptop computer or a mobile phone is physically transferred from the seller to the purchaser.

It is important not to become too distracted by attempts to arrive at a precise definition of services. Nevertheless, the classification of service activities is an important activity for those engaged in measuring the economy, for example national statistical agencies. It is far too easy to become preoccupied with systems of classification at the expense of understanding the changing dynamics of capitalism. In any event, such is the pace of economic change that classifications have to be continually modified to incorporate the development of new types of work or the on-going 'extended division of labour' (Sayer and Walker 1992).

All attempts to classify services must accommodate the complexity and diversity of the business activities involved (Illeris 2007). For our purposes a simple classification of service activities incorporates five different types:

1. Consumer services that provide services for final end-users, for example retailers, opticians, hotels and retail banks.
2. Producer and business services that provide intermediate inputs into the activities of private- and public-sector organizations.
3. Public services provided directly by the state or indirectly by the private sector and not-for-profit organizations.
4. Not-for-profit organizations working beyond the confines of the state.
5. Informal services or unpaid service work, that is usually predominantly undertaken by women, and which is a vital element of peoples' daily lives.

Each of these categories includes a heterogeneous collection of service functions and it is not proposed to cover them all in this chapter. Rather, the focus here is predominantly on knowledge-intensive services (second in the list above). This reflects the increasing importance of a group of distinctive activities that have exhibited dramatic growth rates in both the numbers of firms established and their contribution to employment creation. It incorporates key sectors such as legal services, accountancy, market research, management consultancy, design and technical consultancy. All these services make an important contribution to economic development. They contribute directly to the creation of added value; they contribute to a national economy's balance of payments through exports; and they have also experienced dramatic growth rates both in employment as well as new firm formation.

16.2 The body, services and emotional labour

The classic definition of services developed by Hill (1977) highlights the importance of people-based expertise in the creation of services. This implies that many services must be delivered by people rather than machines. Unlike manufacturing, it is difficult and in many instances impossible to replace service workers with machines and this means that productivity improvements are difficult for those services that rely on people-based expertise. This is an important point in that some service activities can be codified and replaced by relatively simple computer programs. A good example is an application for a mortgage that can be processed by a computer using simple rule-based logic. Each rule can have a simple 'yes or no' answer, for example, related to the value of the property or the income of the applicant(s) (Levy and Murnane 2004: 16–17). Many service jobs cannot be replaced by simple rule-based software as they require complex pattern recognition that may involve reading body language in service encounters.

The importance of face-to-face contact in the simultaneous production and consumption of a range of services plays an important role in differentiating the new world of service work from that of manufacturing. At the centre of interactive service relationships are three important elements: client interaction, an individual's reputation and embodied knowledge. In this service age, the workplace is increasingly conceptualized as a stage upon which employees must execute an aesthetically pleasing performance. Service employment is not simply the exchange of goods or services but is a complex skill in which presentation, communication and display are integral to success (Bryson and Wellington 2003: 60). The implication is that for some service jobs, appearance, the body and accent matter (McDowell 2007) and this implies that service economies have within them cohorts of people who do not possess the desired bodily attributes to meet the expectations of employers and perhaps customers (Bryson and Wellington 2003). McDowell argues that in the United Kingdom this group contains white working-class males who are finding it difficult to obtain well-paid service work.

The academic literature on service work is heavily influenced by Arlie Hochschild's (1983) important work on the commercialization of emotions in face-to-face service encounters. Hochschild was inspired by C. Wright Mills's (1959 [1951]) classic work, *White Collar*. Mills was the first sociologist to explore the complexity and diversity of service work in his analysis of the new American middle class. Mills developed the concept of the 'personality market' to describe the 'shift from skills with things to skills with people' (1959: 182); central to this shift is the enhanced importance of the psychological dimensions of service work. In this account of the new world of work, men and women 'are to be shaped' (1959: 183) and their personalities managed, by themselves and by others.

Horchschild developed Mills's work by undertaking theoretically grounded research into the commercialization of the body and feelings of male and female flight attendants and debt collectors. Central to this story is '**emotional labour**', a concept which describes the management of employees' feelings during social interaction in the work process (Hochschild 1983: 137). The best example is the emphasis placed on providing a 'service with a smile' to show customers that they are valued. Hochschild reveals that much face-to-face interactive service work (flight attendants, debt collectors, waitresses, secretaries, fast-food operations) involves having to present the 'right', managerially prescribed, emotional appearance or mask to the customer or client, and that this involves real labour. In these occupations workers are faced with the dilemma of how to identify with their work role without it becoming part of their identity.

Service employees have to depersonalize the work by 'surface acting' and 'deep acting'. In surface acting 'we deceive others about what we really feel, but we do not deceive ourselves' (Hochschild 1983: 33); the body not the soul is the main tool of the trade; the smile on the face of the worker is a false smile, but it is still a smile. In emotional labour a smile becomes attached to the feelings that a company wishes to project rather than being attached to its usual function – to show a personal feeling (Hochschild 1983: 127). In deep acting the 'act' is no longer an act but becomes part of the individual's persona. It is about persuading employees to be sincere, 'to go well beyond the smile that's just "painted on"' (Hochschild 1983: 33). Unprecedented efforts are being made by employers to control employees not simply in terms of what people say and do at work, but also how they feel and view themselves. In deep acting the disjunction between displayed emotions and private feeling is severe and potentially psychologically damaging. The danger is that deep acting becomes part of the worker's personality and is used beyond the workplace. If this occurs then the new 'emotional proletariat' (Macdonald and Sirianni 1996) might find it difficult to 'interpret and take appropriate action in response to bodily signals' (Shilling 1993: 119). Hochschild's account of male debt collectors highlights how the display of aggression required by debt collectors can spill over into personal relationships with wives and children. According to Bryson and Wellington (2003: 62), 'active involvement in "emotional labour" renders employees' appearance and personality a form of "adjudicated cultural capital" that can be recruited, managed, manipulated and utilised to buy the hearts and minds of consumers'. The commodification of image, not surprisingly, is highly visible in interactive business service occupations and, in particular, the 'professional managerial classes'.

The importance of the body and image in service economies has led to a new occupation, that of image consultancy or impression management. Image consultants are employed by individuals and firms to alter the surface appearance of the body. Image consultants provide seminars and one-to-one consultations directly related to the restructuring of professional employees' bodies. According to Wellington and Bryson (2001: 940):

> KPMG [the accountancy firm] employs image consultants on a monthly basis primarily to provide employees with a 'confidence boost', and also to provide staff with a 'bonding experience'. The accountants Coopers and Lybrand as well as Ernst and Young hire image consultants to provide seminars in personal presentation for their audit teams. These seminars examine ways of increasing credibility and projecting the right image when undertaking client audits and when pitching for new business. Price Waterhouse employed image consultants to instruct potential partners in dining etiquette and in the art of looking, acting and sounding like a partner of a major global accountancy company. Note the use of the terms act and art and the link to the literature on flight attendants.
>
> (Taylor and Tyler 2000)

The literature on emotional labour suggests that within the service economy employees have to develop skills in dealing directly with people. The implication is that extrovert personalities may have little difficulty in fitting into this new world of service work, but that introverts may experience some difficulties. It is important to remember that in many instances facing-based work that is heavily involved in emotional labour will be supported by back office or out-of-client-sight supporting labour. This suggests that a new division of labour is developing that links front and back office workers together.

16.3 Two common misconceptions about service economies

It is far too easy to assume that manufacturing no longer matters in economies that are dominated by service employment. Many service jobs are highly visible within the economy. A visit to a local shopping centre is saturated with service experiences. In contrast, manufacturing employment remains largely invisible as the production process is isolated from the consumption of the product. This separation of process from consumption is not evident in many service-based economic relationships. The dominance of service employment and the apparent recent shift from manufacturing has led to two common misconceptions about service employment and it is to these that we now turn our attention.

16.3.1 Services as old as the Industrial Revolution

It is a common mistake to assume that the transformation of economies towards services is a phenomenon of the twentieth century. The social sciences have paid too little attention to the role services played during the Industrial Revolution or earlier. A good example is the development of London in the nineteenth century as the command and control centre of the British Empire. A detailed analysis of London's fire office registers between the years 1775 and 1825 concluded by noting that the

> service industries made no less contribution to the British economy during the Industrial Revolution than manufacturing, and that nowhere was this more true than in London. Its service economy was on a very large scale, serving the nation as a whole as well as the capital . . . London's service industries underpinned both its own and the national manufacturing and commercial infrastructure and at the same time contributed to the new 'commercialisation of leisure'.
>
> (Barnett 1998: 183)

During the late eighteenth century London was already being transformed into an important world city; a process that has continued to the present and which has further enhanced its status as a global city. Deane and Cole (1962: 166, 175) calculated that in 1851 some 45.3 per cent of the United Kingdom's national income was derived from service activities (trade, transport, housing, the professions and the civil

Plate 16.2 Clerks at work in the offices of a London bus company.
(Topical Press Agency/Getty Images)

service). The structure of employment in the United Kingdom changed dramatically during the nineteenth century as a result of technological innovation and the increasing maturity and extension of the capitalist system. Growth occurred in occupations that facilitated the exchange of goods and services between producers and consumers. Between 1881 and 1901 the number of business clerks increased from 175,000 to 308,000; bank officials from 16,000 to 30,000; and insurance officials and clerks from 15,000 to 55,000 (Marsh 1977: 124). During the nineteenth century the expansion of international trade was restricted by financial problems until it was enabled and supported by the introduction of bill markets and banking facilities of the kind associated with the flow of tea and silk from China to Europe between 1860 and 1890 (Hyde 1973). As the United Kingdom and the United States were becoming industrialized societies they were simultaneously being transformed into service economies; the growth of manufacturing employment went hand in hand with the growth of service employment.

16.3.2 Productivity and services and the myth of service economies

Since the nineteenth century the employment structure of many countries has steadily shifted away from manufacturing to service employment. This shift should not be equated with the demise of manufacturing or the complete displacement of manufacturing with service work. Many national economies are still dominated by the development, design, manufacture and sale of goods; the absence of manufacturing within a national economy would present serious problems since everyday life as well as service work is supported by a complex array of manufactured products – toasters, cars, clocks, toothpaste, deodorant, laptops and toys. Some of these products need to be manufactured or customized locally, for example products that are difficult or impossible to transport (commercial air-conditioning systems) or production processes that require close contact between producers and consumers (Bryson et al. 2008).

It is possible to argue that manufacturing no longer needs to be undertaken within service-dominated economies as products can be traded for services. Nevertheless, many services support and are supported by manufacturing activities, for example the complete removal of manufacturing from the West Midlands (United Kingdom) would reduce the client base of local service companies by 25 per cent (Daniels and Bryson 2005: 3). It is also becoming apparent that a national economy that is over-reliant on products that are manufactured elsewhere is increasingly exposed to risk, as it is difficult to ensure that products meet national safety standards. In August 2007, for example, Mattel, the world's largest toy company, issued a product recall for 436,000 toys made in China that were painted with lead-based paint and 18.2 million toys that were designed with small magnets that could become detached. This type of product recall provides American toy manufacturers with a competitive advantage as they are able to market their toys as made in America (Rusten et al. 2007) and as '100 percent kid-safe' (Martin 2007).

After the Second World War manufacturing employment encountered gales of creative destruction (Schumpeter 1942) that eventually culminated in, what some commentators have termed, the crisis of Fordism (Gaffikin and Nickson 1984). This led to an ongoing reduction in manufacturing employment and a shift towards a diverse collection of service jobs. The difficulty is that scholars and policy makers tend to equate the decline of (often) high-profile manufacturing firms with a contraction in manufacturing overall. In practice the shift towards service employment does not mean that countries in Europe and North America, for example, have been transformed into service economies. It is essential that a distinction is made between alterations in the employment composition of an economy and changes in economic output (productivity). These two indicators of economic change are not necessarily related, or they are related in unexpected ways.

The type of productivity improvements that have been achieved by manufacturers through automation and process improvements are not necessarily achievable by service providers. The category 'services' includes an extremely diverse group of business activities: from very highly-paid professional occupations (lawyers, surgeons, bankers) to very poorly paid occupations (janitors, waiters). The common denominator is that, unlike manufacturing, many of these occupations rely on people-based skills or what is commonly termed 'embodied labour' (Bryson et al. 2004). In these circumstances the challenge of improving productivity is considerable; it is extremely high and increasing where there is interaction between work practices and new technologies (such as the introduction of people-facilitated as well as automated call centres) but either non-existent or extremely low in, for example medical services, teaching and cultural or creative industries. It is very difficult for a dance company to achieve productivity improvements; it involves putting on more performances, using fewer dancers, or increasing the speed of delivery to enable two performances to be given during the time previously devoted to a single performance. This is illustrative of the fact that many service occupations are labour-intensive and often involve face-to-face interactions between service providers and clients for which productivity improvements are difficult to achieve. As a result, overall productivity improvements in the service side of the economy have, as a general rule, lagged behind manufacturing. The implication is that this productivity differential partly explains the shift from manufacturing to service work.

The significance of this reasoning can be demonstrated with reference to UK manufacturing exports which amounted to £158 billion or 55 per cent of total exports in 2004 (Mahajan 2005: 23). Yet manufacturing directly employed only 16 per cent (3.3 million) of all workers, with a further 3 million jobs dependent on the sector. Although the gross added value created by manufacturing increased by 26 per cent over the period 1992 and 2003 (Figure 16.3), it actually increased at much faster rates in other sectors of the economy. Other sectors were therefore growing at a much faster rate than manufacturing even though in absolute rather than relative terms manufacturing's contribution to the overall economy increased over this time period. That said, it is important to recognize that a restructuring of economic activity is occurring; new types of economic activity such as computer services or information management are flourishing alongside the development of innovations in well-established services such as insurance, banking and retailing. Such restructuring arises from, first, the externalization of previously in-house services to independent providers and, second, the continuation of an extended division of labour. The latter arises from the growing complexity that accompanies many service innovations as service providers seek to refine product differentiation in the marketplace.

While most of the key manufacturing sectors have therefore declined in relative terms, and some in absolute terms (for example, footwear, knitted goods, leather goods, man-made fibres, wearing apparel), the service sector has expanded, whether measured by employment or as GVA. Between 1992 and 2003, gross value added increased in the United Kingdom by 354.3 per cent in computer services, 221.5 per cent in other business services and 225.6 per cent in market research and management consultancy (Figure 16.3 and Table 16.1). In 2003, computer services accounted for 2.9 per cent of total GVA (£28.7 billion). It is clear then that recent growth in the UK economy has been led by service industries which made up nine of the top ten fastest growing industries between 1992 and 2003. An inexorable decline in manufacturing employment has therefore been offset by employment gains in services.

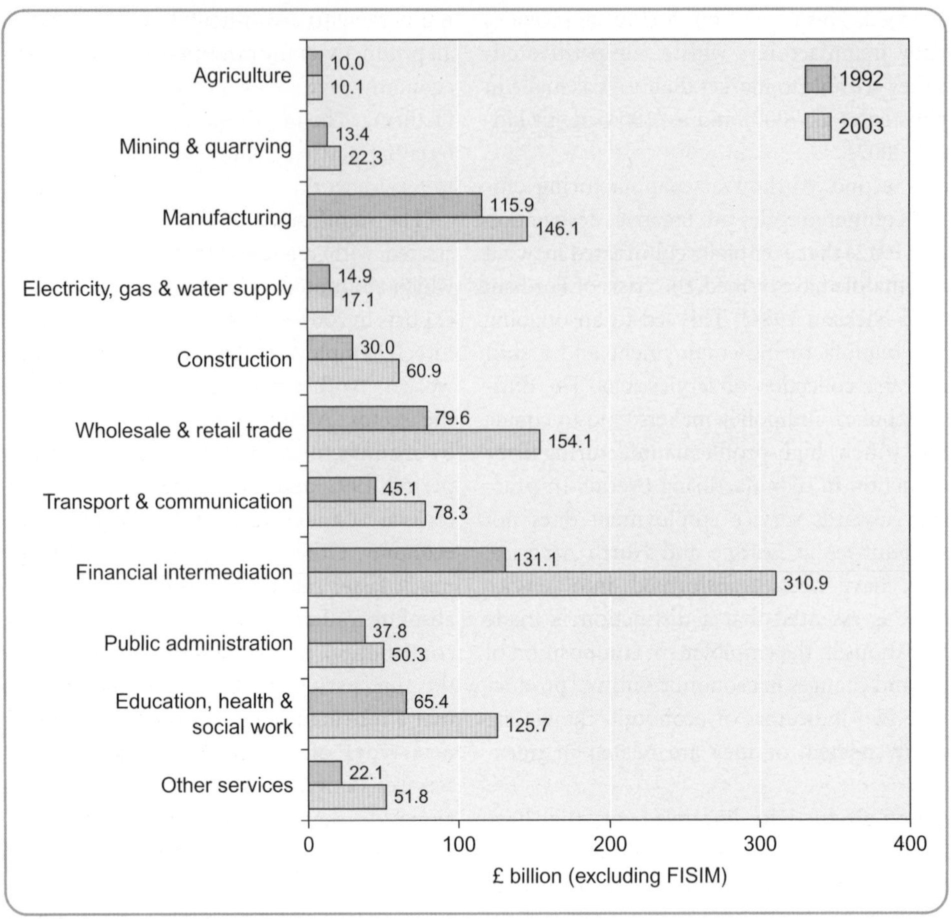

Figure 16.3 Comparison of gross value added in the UK at current basic prices by industry between 1992 and 2003.

Source: Mahajan, S. (2005) Input-Output Analysis: 2005, p. 23, Office for National Statistics, London

Table 16.1 The top ten fastest growing industry groups in the United Kingdom in terms of Gross Value Added (GVA) between 1992 and 2003 (weighted by per cent growth and £ million)

		Change 1992–2003	
I-O no.	I-O group name	%	£m
107	Computer services	354.3	22,401
114	Other business services	221.5	23,900
101	Insurance and pension funds	220.7	15,915
121	Recreational services	152.3	17,032
103	Owning and dealing in real estate	164.0	11,550
104	Letting of dwellings	105.4	39,814
111	Market research, management consultancy	225.6	7,813
88	Construction	103.1	30,911
92	Hotels, catering, pubs, etc.	117.4	17,623
91	Retail distribution	101.6	28,342

Source: Mahajan, S. (2005) Input-Output Analysis: 2005, p. 27, Office for National Statistics, London

However, it has been suggested earlier that the economy is integrated in complex ways and it is important to begin to try to identify and explore some of the interactions that occur between manufacturing and service activities. How does this affect our interpretation of employment gains in services at the expense of manufacturing? A manufacturing company can be dominated by service workers employed in activities such as marketing, advertising, management, accounting and purchasing. Take the task of classifying and counting service activities that occur inside manufacturing companies. On the one hand, we can suggest that part of the shift towards service employment reflects a transfer of service tasks that were previously undertaken within manufacturing firms to external service providers. On the other hand, we can suggest that 'although the objective of manufacturing industry is the production of commodities many people employed in manufacturing are not directly employed in the actual production process' (Crum and Gudgin 1977: 3). Questions follow about the wisdom of dividing the economy into service and manufacturing activities (Daniels and Bryson 2002). In 1971 every two production workers needed to be supported by rather more than one non-production worker (Crum and Gudgin 1977: 5); the term non-production is used to identify people within a manufacturing firm who are not directly involved in the actual production process (managers, designers, sales team, research scientists). Since Crum and Gudgin's research, there have been two major organizational changes to production processes. First, knowledge has become more important as well as being increasingly combined or embedded within products and services. Secondly, the production of goods and services has increasingly required the blending of different types of expertise that include processes involving shop-floor workers (manufacturing as well as services) and a complex array of service and management expertise. All these different types of expertise can be organized within a manufacturing company or, alternatively, by combining the activities provided by many separate and independently owned service firms.

16.4 Services and the spatial division of expertise

The division of labour is a key concept for understanding the organization of work under capitalism and the shift towards service activities. In 1776 Adam Smith published *The Wealth of Nations*, in which he established the foundations of economic theory. For Smith,

> the greatest improvement in the productive powers of labour, and the greatest part of the skill, dexterity, and judgement with which it is anywhere directed, or applied, seem to have been the effects of the division of labour.
>
> (Smith 1977: 109)

To illustrate the importance of the division of labour, Smith explores pin-making, suggesting that an individual without knowledge of the production process would be unable to make even one pin in a day. The production process of pin-making can be divided into eighteen distinct operations: from the drawing out of the wire to the making of the pin head. If one person performs all of these tasks, then they might make 20 pins in a day. If, however, individuals specialize in particular tasks, by introducing a division of labour, Smith shows how ten people could make 48,000 pins in a day (Bryson and Henry 2005). Adam Smith's division of labour is based on his observations of specialization within factories. There is no reason, however, why its principles should not be applied to the services production process, not least to the ways in which new service activities are identified and firms developed to create a market for the provision of new forms of services.

Adam Smith's theory makes no explicit reference to geography. The obvious extension of the division of labour is to incorporate geography into the process which leads to a spatial division of labour (Massey 1984). Once a production process had been subdivided into its component parts it is a comparatively simple step to develop warehouses or offices dedicated to a particular part of the production process, such as consignment assembly and packing for despatch or a marketing function. The spatial division of labour began with manufacturers establishing specialist units that were contiguously located, but soon developed into a more dispersed spatial division of labour. Different parts of the production process were located close to sources of raw materials, the market or cheap or skilled labour. Very quickly, some regions and countries came to specialize in particular types of economic activity. For example, an (international) division of labour soon developed between the industrial countries (core) producing manufactured goods and the non-industrialized countries (semi-/periphery) supplying raw materials and agricultural goods as well as a market for manufactured products. This spatial division of labour can occur *within* service or manufacturing firms (Thematic case study 16.2)

Thematic Case Study 16.2

ICT and the establishment of data-processing factories in the Caribbean and call centres in India

Developments in ICT have made the transfer of service jobs to low-cost locations possible (Bryson et al. 2004). What is occurring is an intriguing ongoing international division of labour, but with a difference. The difference is that branch plants in developing or less developed countries used only to be associated with the assembly of products designed by and for the developed world: now, many such branch plants process data and/or interact with customers located thousands of miles away.

Data processing factories, for example, in the Caribbean are linked via cable and satellite with service workers in Ireland, the Dominican Republic, Jamaica, Mauritius and the United States. To Freeman this development 'signals an intensification of transnational production and consumption – of labour, capital, goods, services and styles' (2000: 1). The Barbados Data Processing Centre, for example, employs 100 women who enter data from 300,000 ticket stubs for one airline's 2,000 daily flights, while on another floor women enter data from medical claims sent for processing by one of America's largest insurance companies (Freeman 2000: 1). This company employs nearly one thousand people, mostly women, who work eight-hour shifts and who are constantly monitored for productivity and accuracy.

Call-centre operations are labour-intensive, labour accounting for over 65 per cent of their running costs. In the 1990s, in the United Kingdom, American Express and British Airways (BA) transferred their customer services divisions from the United Kingdom to Delhi and then Bombay. BA was attracted by India's large pool of English-speaking graduates who could be employed on starting salaries of between £1,500 and £2,500. India currently dominates the market for the provision of English, speaking call centres. Every year India produces 2 million graduates, mostly taught through English, and who are desperate for well-paid employment in a country where the national average wage is £300. In the United Kingdom, call centres are staffed by students and temporary workers and are considered to be twenty-first century sweatshops with high staff turnover rates and low salaries (£12,000–15,000). In India, they are desirable places, regarded as 'hip and funky places to work, somewhere to hang out with like-minded, outward-looking young people' (Spillius 2003: 44). Six years ago the call-centre industry did not exist in India, but by 2003, 1,500 Indian call-centre providers employed 102,000 young people in the 'remote services' industry; the industry is growing by 70 per cent a year (Spillius 2003: 41). Over the next five to ten years, over 200,000 British call-centre jobs will be exported to South Africa, Malaysia, the Philippines and China, but India is currently becoming the world's back office.

Plate 16.3 A call centre where business is outsourced from Western companies in New Delhi in India.

(Fredrik Renander/Alamy)

(Bryson et al. 2004). One location (a major city such as London) is used for the headquarters (HQ), another for research and development (R&D), yet another (a peripheral region, South Wales, or a country, China) for the manufacturing branch plant, and sales and service centres will be distributed over the globe (Bryson and Rusten 2008).

Within geography, the division of labour concept is traditionally associated with deskilling, branch plants of manufacturing companies, and the ongoing development of an international division of labour, mostly associated with textiles and clothing, automotive, electronics and to a much lesser extent services (Dicken 2003). The services literature increasingly emphasizes the importance of emotional labour and embodied expertise in service relationships and production (Hochschild 1983; Fineman 2000; Warhurst et al. 2000; McDowell 2007). It follows that the concept of a spatial division of labour needs to accommodate the increasing centrality of embodied expertise in the economy (Wellington and Bryson 2001; Bryson and Wellington 2003). There has also been a shift in the nature of particular forms of work. Some forms of high-paid work, for example, are moving away from being delivered by labour that is controlled by capital and regulated by company law to a situation in which control, or more correctly power, in the 'employment' relationship is transferred to the employed expert (from the employer). Such experts are the most important asset 'owned' or managed by expertise-intensive firms; these walking, highly mobile resources leave their firms' offices each evening and may, or may not, return next morning. This contrasts with manufacturing where owners of capital exercise most of the power in the relationship with their employees. The balance has shifted with the development of expertise-intensive occupations whereby power transfers from employers to employees. Indicative of this 'new order' are the difficulties experienced by firms trying to manage business service professionals and the role played by bonus payments and golden handcuffs in staff retention (Alvesson 2000; Leicht and Fennell 1997).

The *spatial division of labour* has continued to evolve and is centered on the expertise that is located within or outside client companies. This is not to suggest that the concept of a spatial division of 'production' labour is still not important, but it is now conceptually useful to distinguish between corporate strategies that affect production workers and those that involve expertise or forms of creative work. A spatial division of expertise distinguishes between the *spatial organization of expertise* (non-production activities) (advanced BPS and management functions) and the *spatial organization of production* (Bryson and Rusten 2005, 2006, 2008). It reflects the distinction between production and non-production activities within manufacturing (Crum and Gudgin 1977). The concept of a spatial division of expertise shifts the focus of analysis to the places where expertise is produced and consumed; it also exists within and between international business service firms and manufacturing firms. The building blocks of this 'expertise' economic system are distributed in a complex mosaic that reflects the ways in which a range of firms (from micro- to large transnational firms) try to maximize profitability. The spatial division of expertise is informed by the social division of labour which gives considerable emphasis to relationships between people and to the joint supply and co-production of service knowledge/expertise.

The development of a spatial division of expertise has important implications for the geographies of service as well as manufacturing companies (Case study 16.3). Companies, and even governments, increasingly develop services (as well as products) using production processes designed to exploit differential comparative advantages. This is not an entirely new process; earlier international divisions of labour were based on core countries producing and exporting manufactured goods whilst peripheral countries exported raw materials. This was the first manufacturing division of labour or the first global shift (Dicken 2003). The evolution into a spatial division of expertise came later. The comparative advantage for services is derived from a different form of raw material, that is, a highly educated and expert labour force and, importantly, language skills. It is to geographies of the new spatial division of expertise that we now turn our attention.

16.5 The second global shift

Traditionally services were considered to be produced and consumed locally; they were regarded as having limited export potential. As we noted earlier in this chapter this reflects the way in which services were conceptualized in traditional economic theory. Today, services are exported and companies are increasingly adopting business models that involve combining service expertise located in different parts of the world or, in other words, by developing competitive advantage

Thematic Case Study 16.3

Boeing's new spatial division of expertise

It is becoming increasingly difficult to distinguish between manufacturing and service companies (Daniels and Bryson 2002). Many physical products contain embedded services, for example software, or are supported by service agreements. The increasing symbiosis that is developing between manufacturing and services is changing the geographies of production in complex ways, creating new business models that capitalize on service/manufacturing expertise and is also repositioning expertise as a key source of competitive advantage. There are many examples, but perhaps one of the most dramatic is found in the aerospace industry. Boeing, the American manufacturer, is currently building a new aircraft, the 787 Dreamliner, that will carry between 200 and 300 passengers with a range of 8,500 nautical miles (see Bryson and Rusten 2006, 2008). For Boeing the Dreamliner represents a new way of designing and manufacturing aeroplanes. An aeroplane is a complex product that is manufactured using an extended, complex supply chain. Before the Dreamliner programme, Boeing followed a build-to-print model. Companies in Boeing's supply chain were provided with detailed part specifications and they only became involved towards the end of the design and development process. In this production model, Boeing was responsible for the detailed design of the complete plane. For the Dreamliner, Boeing is responsible for the overall design of the plane, for systems integration and final assembly, but detailed design work has been transferred to a network of global partners. For the Dreamliner project, Boeing has created a new spatial division of expertise that is constructed around capitalizing on Boeing's internal expertise combined with that of its partners that are distributed around the globe. This expertise is a combination of research and development with industrial design inputs. This expertise has been developed in centres of excellence that support high-tech engineering activities.

Plate 16.4 The first Boeing 787 Dreamliner takes shape in the assembly plant in Everett, WA, USA.
(AP Photo/John Froschauer)

The new model represents a radical alteration in the way Boeing designs and manufactures aeroplanes; most of the manufacturing will be undertaken at factories owned and managed by Boeing's partners and the only major part of the Dreamliner's airframe being manufactured by Boeing is the vertical tail. Boeing's major partners include three major Japanese manufacturing companies (Kawasaki Heavy Industries, Mitsubishi Heavy Industries and Fuji Heavy Industries) and Alenia Aeronautica, an Italian company. For the first time Boeing has outsourced the most important part of an aircraft – the design and production of the wings. The three Japanese companies are responsible for the wings. Initially, the Japanese wing design team worked closely with Boeing's design team and was based at Boeing's Everett facility in Washington State (USA). The Japanese design team returned to Japan to concentrate on the development of the detailed designs for the wing while the Boeing design team shifted its focus to concentrate on systems integration. The Dreamliner combines Boeing's 'service' expertise with that of its partners. Once assembled in the Kawasaki plant the completed wing box will be transported by barge to Nagoya's new Centrair airport and from there flown to Charleston, South Carolina, in modified 747 cargo freighters (Figure 16.4). Charleston is the main fuselage hub in this spatial division of expertise; the fuselage sections manufactured in Japan, Italy and the USA will be assembled before being flown to Everett for a final assembly process that should take only three days. This is both a conventional spatial division of 'production' labour, with finished parts travelling from one supplier to another down the chain, and a spatial division of 'embodied' expertise. This is an expertise-rich production system in which conventional shop-floor manufacturing workers play a relatively minor role. A total of 3,600 engineers are directly employed by Boeing in

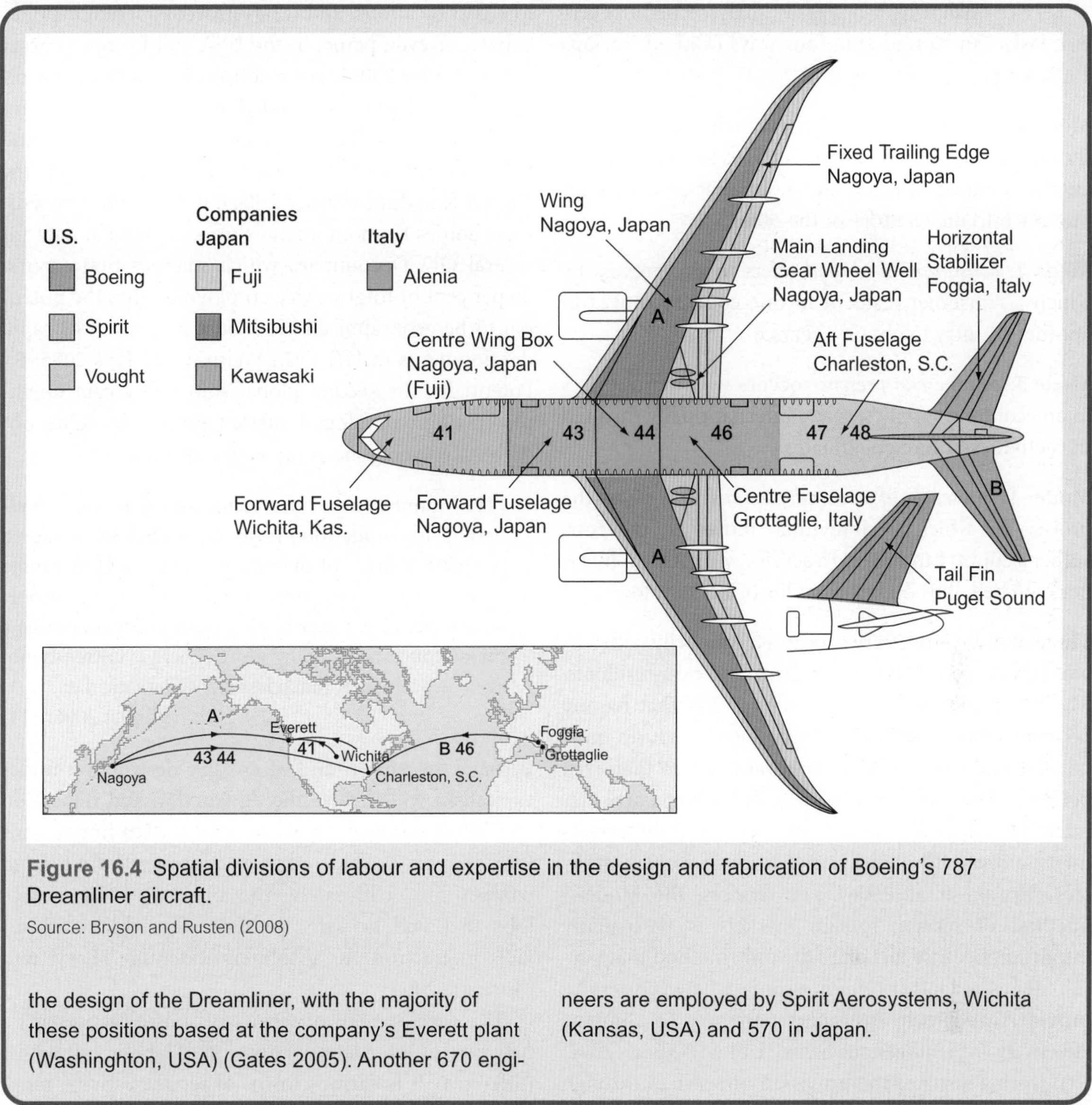

Figure 16.4 Spatial divisions of labour and expertise in the design and fabrication of Boeing's 787 Dreamliner aircraft.

Source: Bryson and Rusten (2008)

the design of the Dreamliner, with the majority of these positions based at the company's Everett plant (Washinghton, USA) (Gates 2005). Another 670 engineers are employed by Spirit Aerosystems, Wichita (Kansas, USA) and 570 in Japan.

through the development of an evolving spatial division of expertise. In the popular media or in political accounts this trend is known as 'service offshoring', a term developed in the Anglo-American context which should be used with considerable care. Services can indeed be 'offshored' but this is a less than helpful and even misleading term when applied to the transfer of work from, say, France to Eastern Europe.

Service offshoring occurs when firms shift production to foreign locations. The objective may be to reduce costs, to service a foreign market, to reduce exposure to country risk, or to access skilled labour. An additional factor influencing the location of offshore service centres is the requirement to provide a 24-hour service to customers or an extended service beyond standard working hours. The cost of providing such services can be high as late shift workers must be attracted by higher wages or extended holidays. Advanced call-routing and networking technologies enables companies to get around this by implementing a 'follow-the-sun' geographical policy. Companies can link two or more call centres together with each open from between 8 to 12 hours per day. Country risk is removed when a company is able to shift the provision of a function between facilities located in different countries.

Service offshoring is not easily analysed because service tasks can be traded in four ways (United Nations 2002: 1):

Mode 1: *cross-border supply* occurs when suppliers of services in one country supply services to consumers in another country without either supplier or consumer moving into the territory of the other.

Mode 2: *consumption abroad* refers to the process by which a consumer resident in one country moves to another country to obtain a service.

Mode 3: *commercial presence* occurs when enterprises in an economy supply services internationally through the activities of foreign affiliates.

Mode 4: *presence of natural persons* describes the process by which an individual moves to the consumer's country to provide a service, whether on his or her own behalf or on behalf of his or her employer.

Three of these modes are concerned primarily with service transactions between residents and non-residents. Mode 1 involves the provision of services that require no direct contact with customers but procedures must be developed to overcome cultural barriers that exist between countries. Recently, there has been a particular interest in Mode 3, whereby enterprises supply services internationally through the activities of foreign affiliates (Bryson et al. 2004). For services, the Mode 3 'method of serving foreign markets is particularly important because it is often the only method that permits the close and continuing contact between service providers and their customers necessary to compete effectively with indigenous firms' (United Nations 2002: 54). In this instance the provision of services through foreign direct investment represents a type of *captive offshoring* or offshoring without outsourcing. Trade in services must address cultural differences between countries that restrict the ability of service providers to export standardized services. Modes 3 and 4 enable service providers to localize provision to take into consideration local cultures and client expectations. Modes 1, 3 and 4 involve what is commonly termed 'service offshoring' or more correctly 'service global sourcing'. This is encapsulated by the concept of a '**second global shift**' (Bryson 2007). The first global shift involved the relocation of manufacturing employment to low-cost production locations while the second implicates services in this process.

The second global shift highlights the potential for transferring service work from developed economies to low-cost locations, and it has provoked a major policy debate, or even panic, in the USA and Europe (Parker 2004; Blinder 2006). For example, Forrester (a consultancy company) has estimated that 1.1 million Western European jobs will move offshore during 2005–15 and that two-thirds of these jobs will originate from the United Kingdom (Parker 2004). Perhaps the best estimate comes from an analysis of occupational data for several OECD countries which suggests that around 20 per cent of total service employment has the potential to be geographically footloose as a result of rapid developments in ICT (van Welsum and Reif 2005: 6). Potentially, the second global shift will create unemployment in developed market economies. One observer has even gone as far as to argue that

> [The] number of service-sector jobs that will be vulnerable to competition from abroad will likely exceed the total number of manufacturing jobs. Thus coping with foreign competition, currently a concern for only a minority of workers in rich countries will become a major concern for many more. There is currently not even a vocabulary, much less any systematic data.
>
> (Blinder 2006: 114)

Perhaps the next round of creative destruction under capitalism will involve the destruction and relocation of a whole series of service jobs that, according to convention, were considered to be protected from foreign competition. Ultimately, the only types of service jobs that will be safe are those for which face-to-face interaction is absolutely essential (Levy and Murnane 2004).

The development of service offshoring represents a new type of international division of labour, but with a difference. It is various forms of service activity, ranging from call-centre-based work to back-office administration that is being relocated to low-cost locations rather than manufacturing or assembly activities. This new form of trade involves low-value call-centre-type activities as well as high-value services such as legal work, accountancy, design, business analysis and equity research (Table 16.2). Data scanned in Britain or America is transmitted to a low-cost offshore location along high-speed fibre-optic cables and undersea telephone lines to be processed in back offices or used in call centres. A number of factors influence the decision to send a particular service activity offshore. First, it must be capable of some degree of standardization that does not require face-to-face interaction with consumers or clients. Secondly, the inputs and outputs required to deliver the service must be capable of being

Table 16.2 Jobs that have been sent offshore from the United Kingdom

Customer-facing – constant contact between producer and consumer	Back-office – business process outsourcing
Call centres	Business services
Customer services	Accountancy
Out of hours claims (call centre)	Legal services
Back office administration	Finance
Back office processing	Human relations
E commerce	IT provision
Credit control	Business systems
Internet services	Graphics and architectural services
Internet claims	Document production
IT helpdesk	Underwriting
Rail timetable enquiries	Technical lists
	Equity research
	Business analysts
	Legal secretaries
	Software development
	City analysts
	Legal secretarial
	Transcription services (voice and shorthand)
	Human relations
	IT provision

Source: Bryson (2007)

traded or transmitted with the assistance of ICT (OECD 2005b: 12). Thirdly, some service activities are not fixed in space and can be provided either as a form of foreign trade or by the temporary relocation of a service worker to a client's premises, for example, management consultancy or various forms of auditing. Fourthly, specialist services can be provided from central locations with consumers travelling to avail themselves of the service. In many cases such services would be provided within the confines of a nation-state, but some of these services are being consumed by a form of service-based travel, for example education (secondary and tertiary), plastic surgery and a whole range of other surgical procedures.

Unlike the first 'global shift', the geography of the second global shift is determined by the educational and language abilities of service workers located in foreign locations that may also perhaps, but not always, be lower-cost locations (Bryson 2007). For the English speaking world this means that potential suppliers must be able to provide English speaking employees in other countries, for example France, Norway or Sweden require a pool of staff with fluent French, Norwegian or Swedish. Language and culture plays a much more important part in this global shift than they did during the development of an international division of manufacturing labour. This means that countries with relatively localized languages may be protected from the global sourcing of services whilst countries with more widely spoken 'global' languages (such as English, Spanish, or French) will almost certainly participate in the second global shift. This means that the geography of the second global shift is also different from the first; it is more constrained by language as well as cultural nearness, that is, the ability of foreign service suppliers to relate to customers located in other countries (Figure 16.5).

The implication is that countries which developed extensive empires during the nineteenth century may have inadvertently laid the foundations for the emergence of foreign competitors who are able to exploit the benefits associated with their acquired or imposed non-native language. This is a form of 'post-colonial twist' in that the various trading and political empires encouraged the use of a common language and, in many cases, common legal and educational systems (Bryson 2007). It is therefore not surprising that countries such as the United Kingdom and the United States that are part of a globalized language grouping may be most at risk from companies choosing to deliver services from lower-cost locations either by establishing their own operations via a process of foreign direct investment or by subcontracting the activities to a third party and often foreign service provider. Whether this is a 'real' threat will depend on whether the availability of lower-cost services will actually enhance the competitiveness of other parts of the developed market economies, leading to future innovation and wealth creation opportunities. It is also important to remember that the second global shift does not just involve the transfer of jobs from high- to low-cost economies but it is also a two-way process in which service providers based in low-cost locations must also establish branch offices in developed market economies. Service offshoring is a complex process

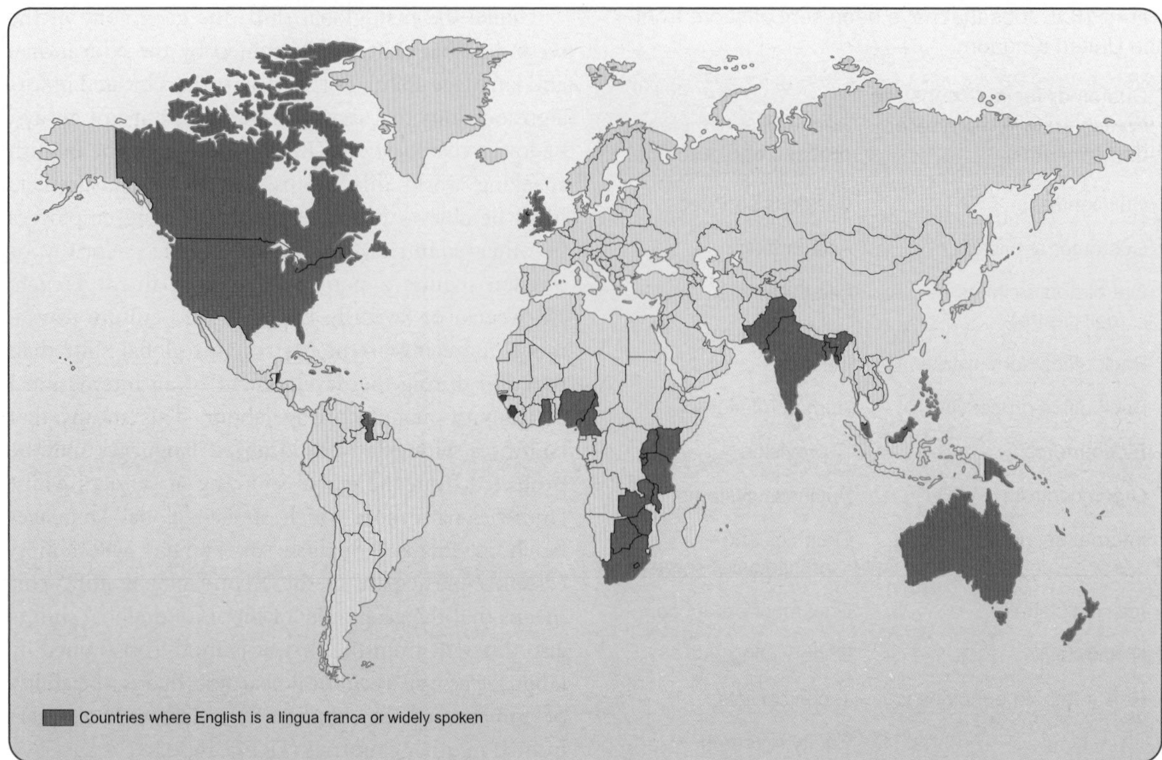

Figure 16.5 Language as a factor in the geography of outsourcing countries where English is a *lingua franca* or relatively widely spoken.

involving the development of firms that have the ability to provide service functions from a number of different locations. This type of transnational service firm can develop out of the activities of firms located, for example, in the United States or United Kingdom (developed market economies) or in India or China (developing or transition economies).

The global sourcing or offshoring of services does not have to entail the supply of services over large distances. It may occur as 'near-shoring' or the relocation or provision of services over short distances and often between locations on the same continental land mass, for example the 'nearshoring' of American services to Canada. Services can now be supplied on-shore, near-shore and offshore. Presented in this manner these may appear as simple alternatives but in many instances firms have developed *'blended delivery systems'* that capitalize on the place-based advantages of coupling or blending activities located in a variety of different locations: home country, near, far (Bryson 2007).

16.6 Conclusion

The history of economic development is one of increased economic specialization combined with the development of new and often complex geographies. All this represents and reflects the constant working through of new divisions of labour and new ways of maximizing wealth creation. The development of service-dominated economies represents a working through of the division of labour but it is worth remembering that it is important to differentiate between employment and output. All services must be supported by manufactured products and all manufactured products are supported by services. The implication is that we have an integrated economy in which the worlds of production (services and production functions), finance and consumption combine to produce local, national and international economies.

The developing spatial division of expertise and the related second global shift have the potential to rework the national and global dynamics of the capitalist economic system. Employment that was considered to be safe from foreign competition is suddenly exposed to business models that operate by blending together different forms of expertise related to differential comparative advantage. This may enhance productivity improvements, leading to new innovations, but it also has the potential 'to undermine the service-based solution to the employment crisis that is still being experienced by many developed regional economies' (Bryson 2007).

Learning outcomes

Having read this chapter, you should understand:

- Different ways in which services have been defined and conceptualized by economists and geographers.
- The concept of a 'personality market' or emotional labour highlights the importance of embodied labour or skills with people in the world of service work.
- Appreciate that as the United Kingdom and the United States became industrialized societies they were simultaneously transformed into service economies.
- Service economies should not just be identified by measuring employment, but should also take into consideration Gross Value Added (GVA).
- Economic production is a story of economic specialization and the continual reworking of the division of labour.
- A spatial division of labour occurs with the geographical separation of production tasks, for example, writing, printing, binding and distributing a book.
- The development of a concept of a spatial division of expertise draws attention to the development of new corporate strategies that are founded upon exploiting comparative advantage based on expertise.
- Different economic geographies can be explained by distinctive combinations and patterns of spatial divisions of labour and of expertise.
- Services were conventionally conceptualized as local untraded activities. The rise of the second global shift or service offshoring highlights one of the latest developments in the capitalist economic system.

Further reading

Bryson, J.R. (2007) A 'second' global shift? The offshoring or global sourcing of corporate services and the rise of distanciated emotional labour, ***Geografiska Annaler***, **89B**(1), 31–44.

Bryson, J.R. and Daniels, P.W. (eds) (2007) ***The Handbook of Service Industries***, Edward Elgar, Cheltenham.

Bryson, J.R., Daniels, P.W. and Warf, B. (2004) ***Service Worlds: People, Organizations, Technologies***, Routledge, London. The most recent account of the different divisions of labour amongst service activities and occupations as well as the role of ICT in altering the geographies of production systems.

Bryson, J.R. and Rusten, G. (forthcoming) Transnational corporations and spatial divisions of 'service' expertise as a competitive strategy: the example of 3M and Boeing, ***The Service Industries Journal***, **28**(3).

Levy, F. and Murnane, R.J. (2004) ***The New Division of Labour: How Computers are Creating the Next Job Market***, Princeton University Press, Princeton.

For annotated, clickable weblinks and useful tutorials full of practical advice on how to improve your study skills, visit this book's website at **www.pearsoned.co.uk/daniel**

THE GLOBAL FINANCIAL SYSTEM: WORLDS OF MONIES

Chapter 17

Jane Pollard

Topics covered

- Money in economic geography
- The social, cultural and political significance of money
- Geographies of monies
- Global monies, local monies and monetary networks
- Money, space, and power: geographies of inclusion and exclusion

Money is simultaneously everything and nothing, everywhere but nowhere in particular.
(David Harvey 1985b: 167)

This chapter explores the economic, social, cultural and political networks through which we establish value and manage exchange. Our attitudes to money and how we acquire and use it reveal not only its economic significance, but also its social, cultural, political and geographical qualities. As the quote from Harvey suggests, however, money can be a difficult subject. Novelists, film-makers and social theorists alike have grappled with its contradictory characteristics. Money can be everything and nothing, everywhere and nowhere; money can be the root of all evil and a source of independence and freedom. Even simple questions like 'What is money?' present some difficulties. Davies (1994: 29) suggests that money is 'anything that is widely used for making payments and accounting for debts and credits'. Historically, a wide range of items, including beads, shells, whales' teeth, cattle, salt, skins, tobacco, beer, gold and silver have been used as money. Such variation suggests that rather than asking what money is, we should perhaps ask: what functions does money perform? This question reveals the important economic functions of money and its status as a source of social power.

Spotlight Box 17.1

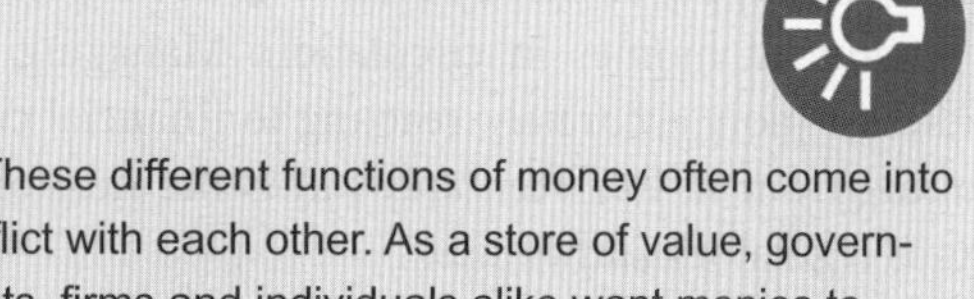

The functions of money

- Money is a *unit of account:* it is the base of economic accounting systems.
- Money is a *measure of value:* it is the commodity against which the values of other commodities can be calibrated.
- Money is a *store of value,* in that you can sell a good for a certain amount of money and then use that money to buy something else at a later date. Money thus allows for the separation of the sale and purchase of commodities over space and time *if* it is a reliable store of value. Preserving this ability of money to store value is one of the reasons why governments are so concerned about the phenomenon of inflation.
- Money is a *medium of exchange and circulation.*
- Money is a *means of payment and a standard for deferred payments.*

These different functions of money often come into conflict with each other. As a store of value, governments, firms and individuals alike want monies to maintain their value and hence their purchasing power, between different times and spaces. As a medium of circulation, however, it is important for money to be extended as credit.

Banks create credit by extending loans to individuals, firms and governments. During economic growth, there is temptation for banks and other lending institutions to create large volumes of credit. Any sudden loss of confidence in the quality of credit, as happened in the Wall Street Crash of 1929 (see Plate 17.1) or, more recently, in South-east Asia and Russia, can trigger financial panics and devaluation as depositors, investors and financial institutions flee from credit monies and seek out safer havens (see Figure 17.1 on page 372).

Plate 17.1 The Wall Street Crash 1929: the stock market collapse of October 1929 brought the 1920s boom to an end and led to widespread bank failures, bankruptcies and drastic reduction in the availability of credit.

We are used to thinking of money 'oiling the wheels' of commerce and also being a measure of worth or value. Yet these two facets of money can be contradictory. As a store of value, it is desirable for money to be a stable, fixed representation of value. As a medium of circulation and as a form of **capital** – money that is thrown into circulation with the intent of generating more money – it is desirable to have money freely available as credit, divorced from its moorings in the 'real economy'. As Davies (1994: 29) argues, the history (and geography) of the evolution of money reveals 'unceasing conflict' between borrowers, keen to expand the *quantity* of money in circulation, and lenders, keen to limit the supply of money and, at all costs, preserve the *quality* of monies in circulation. Managing this conflict or more accurately, reacting to financial crises resulting from it, has been and still is the mission of various regulatory authorities.

17.1 Money in economic geography

Many accounts of money and 'the global financial system' are written by economists or finance specialists, who pay little attention to its geographical anatomy. Is money something that we should think about geographically? As the Harvey quote at the start of the chapter suggests, money seems to be 'everywhere but nowhere in particular' and many contemporary debates about **globalization** stress the increasingly integrated, global machinations of the financial system in which 'virtual' monies can be moved around the globe at the touch of a button. One of the purposes of this chapter is to persuade you of the existence and importance – economically, culturally and politically – of different geographies of money and financial institutions that make up the so-called 'global financial system'.

However, the organization of monies into different financial markets can appear dauntingly complex. The 'global financial system' can sometimes seem like something 'out there', a fast-moving, volatile 'thing', running out of control, increasingly uncoupled from the 'real' economy and operating beyond the regulatory reach of any one nation-state (Langley 2003). This global sense of money and finance can be difficult to grasp and sometimes it is difficult to understand just how its machinations affect us. For example, how did the South-east Asia financial crisis of the late 1990s change the lives of people in Europe, Africa, the United States or elsewhere in the world (see also Chapter 14).

For all the difficulties, real and imagined, involved in understanding the workings of international, national and local finance, money and finance are now attracting a great deal of interest from geographers, other academics, governments and the popular press. This is partly explained by a number of recent financial and political crises, including those in Pacific and South-east Asia, Russia and Brazil, and growing unease about levels of household indebtedness in the United Kingdom, United States and elsewhere. These crises, growing indebtedness, and the emergence of complex new financial products like derivatives and hedge funds have fuelled growing unease over the volatility and instability in global financial markets and a concern that regulatory bodies and nation-states are increasingly unable to control international finance. Also, as Leyshon (1995a) argues, money and finance have assumed a growing social and cultural prominence since the 1980s, related to the growth of financial services employment and the growing media and popular attention devoted to the rhythms and practices of financial services in London, New York and elsewhere.

A growing body of work on money and finance has grappled with a number of challenges including:

- the economic, social and cultural aspects of money;
- the links between our everyday experiences of money and the workings of 'the global financial system';
- the changing geographies of money in these increasingly global times;
- the creation of progressive, non-exclusionary forms of financial institutions in the contemporary world.

In dealing with these issues, geographers are not talking about 'the global financial system' as though it were a machine, but rather seeking to understand 'global finance' as a social, cultural and political phenomenon. Revealing the *social construction* of 'global finance' is important because the workings of global finance make a real difference to the well-being of individuals and the choices/predicaments facing governments. The workings of the financial system shape who has access to credit, how, where and how much they pay for it and whether governments can afford welfare, education, health care and other services.

In the subsequent sections of this chapter, the above issues are explored in more detail. In the next section, we explore some of the different social and cultural aspects of money and how it is bound up with individual and group identities. The focus is broadened in the following section which considers the geographies of money that connect individuals and groups in distant places and at different times. By way of conclusion, the final section considers how some groups and regions are included or excluded from particular geographies of money, before exploring some examples of progressive, inclusive monetary geographies.

17.2 Everyday worlds of money

Although we may be used to thinking of money as primarily an economic phenomenon, geographers and other social scientists are now recognizing that money is saturated with cultural, social and political significance (Dodd 1994; Tickell 2003). In this section, we examine some of the social and cultural aspects of money in different societies. We focus initially on money and power, before moving on to consider money and identity, money as a transformer or leveller, and money as a mediator of our experience of social relations.

17.2.1 Money and power

Many of the social and cultural aspects of money are related to its role as 'the very incarnation of social power' (Harvey 1982: 245). Again, however, we confront the contradictory qualities of money. As a means of expressing social power money has both desirable and less desirable qualities. Leyshon and Thrift (1997) talk of two discourses about money, one of suspicion and one of liberation.

To understand the discourse of suspicion, we can turn to the work of Georg Simmel. For Simmel (1978: 277), money is 'the most perfect representation' of the tendency to reduce quality to quantity. As a store of value and unit of account, money transforms social relations – qualitatively different commodities and experiences – into an abstract quantity, namely their price. In contemporary capitalism, money has become *the* mediator and regulator of economic relations between individuals, a measure of wealth and a means of expressing social power. All manner of social, political and cultural issues – from health care, housing, education, leisure and sports, the environment, and so forth – are debated not in terms of what they are worth, but in terms of *how much* they are worth in the sense of what can be afforded. Money 'affects our very ideals of what is good and beautiful and true' (Mitchell 1937: 371). Money, in this sense, has corroded the importance of other meanings and measures of 'value'.

Yet there is also a discourse of liberation that accompanies money. First, money provides a degree of individuality, freedom, security and independence for those that hold it. This is especially true in societies like Britain where laws regarding private property buttress ideologies of liberty, equality and freedom. The idea that money is empowering and liberating is central to the desire to possess money, to work for it, to accept it as payment and to save and invest it. Secondly, as Simmel (1991: 20) argues, money 'creates an extremely strong bond among members of an economic circle'; people have far more connections to other people in modern, monetized societies than was the case in feudal society. Through the **division of labour** in capitalist societies (see Chapter 16, this volume), money links people together in offices, factories, homes and shopping malls in different parts of the world. Money may in some senses be corrosive of social bonds, but it also creates a community that can forge new connections and hold people together.

17.2.2 Money and identities

Thus far, the discussion about money and power has been rather abstract. To bring this discussion closer to home, let us think about the role of money in the construction of identity. If we think of identity not as something that is fixed or innate but as a 'capacity to keep a particular narrative going' (Giddens 1990: 54), then money can be an extremely important part of telling and maintaining stories about ourselves. In the remainder of this section, we consider four brief examples.

First, money affects how individuals identify their class positions, what accommodation they live in, what type of schools they attend, and whether they can afford to stay on and pursue higher education, and so forth. Being one of Britain's 'very rich' is associated with social and cultural networking practices which include particular (very expensive) rituals of conspicuous consumption, including balls and dinners and attendance at annual sporting events like Ascot and Henley. Such activities maintain relatively tight-knit networks, often linked by marriage. Within such moneyed networks, children can be sent to exclusive schools to mix with children from other wealthy families and be endowed with the social, if not financial, capital necessary to smooth their journey along their chosen future path.

Secondly, money stirs passions about nationhood, sovereignty and *national identities*. Consider the birth of the Euro (€) which became legal tender on

1 January 2002. Swyngedouw (2000: 71) has described the new currency as 'arguably the most radical political–economic and cultural transformation of space in Western Europe since the French Revolution'. Each Euro note has an image of an arch and a bridge, yet none of these are real bridges. Instead, they are all simulated bridges, symbolizing connection, interchange, communication; they hint of the economic, political and cultural aspirations of European integration (Pollard and Sidaway 2002).

Thirdly, just as money can be a powerful symbol of class or national identity, it is also integral to the performance of *gender identities*. By way of example, we can think of how central the role of 'the male breadwinner' is to the construction of many masculinities. Films like *Roger and Me* (USA) and *The Full Monty* (United Kingdom) illustrate only too well that loss of employment in the car, steel and mining industries is not simply an economic catastrophe for the men involved. For those whose masculinity is rooted in 'industrial work' and the status of a 'breadwinner', the loss of regular employment and earnings can entail wrenching changes in social status, self-esteem and relations with partners and children.

Fourthly, money is intricately linked to the *virtual financial identity* of individuals (and indeed countries), which are built up by credit-scoring companies that monitor credit histories and vet applications for credit. Those who seek credit cards, mortgages and other loans are the subject of 'credit checks' by major credit-scoring companies that assess a wide range of factors including employment status, income, expenditure, outstanding debts and repayment histories. Credit-scoring companies are not the only ones to construct virtual identities; many organizations construct virtual identities for their customers. For example, grocery retailers in the USA, Canada, Britain, Australia and elsewhere use what are variously referred to as 'club cards', 'reward cards' or 'loyalty cards' as a means of tracking their customers' buying habits. The aim is to use customer information to target marketing initiatives towards those socio-demographic groups which will generate the greatest profits for the lowest marketing-related expenses.

17.2.3 Money talks . . .

Money, as an incarnation of social power, can create new links between groups of people. As such, money can be a force for making people more tolerant of difference. For example, overseas investment in Birmingham, London, Manchester and other British cities is such that they embrace ethnic and racial diversity as part of their economic development strategies (Henry et al. 2002). In many British cities, 'Chinese Quarters' and other distinctive neighbourhoods associated with minority ethnicities are taking shape. Similarly, the 'pink pound' – the earning, investing and purchasing power of gay consumers – is courted by financial services providers, film and music media, travel firms, airlines, health and fitness firms, restaurants and bars. Surveys in Britain, the United States and Australia characterize gay consumers as groups that are relatively affluent, active, financially literate, and likely to travel overseas and eat out regularly (Anonymous 1999). Although this stereotype is not representative of gay couples on low incomes or with children, the economic significance of pink pounds and dollars – already acknowledged in cities like London, San Francisco, Toronto and Sydney – is being recognized in a growing number of places.

With its ability to forge new associations between individuals and groups, money is also, potentially, a leveller. Money can disrupt traditional cleavages along the lines of class, gender, ethnicity and sexuality such that singers, actors, footballers and other high earners can join the ranks of the financial elite regardless of their backgrounds. The realization that 'money talks', that it can be transformative, a bridge to a different life (McLuhan 1964) is integral to the appeal of the forms of gambling like lotteries and casino gambling that are now very big business in Europe, North and South America, Australia, Hong Kong and, more recently, China. Market researchers have estimated that by 2009, lotteries will account for 40 per cent of the global gambling market and generate US$7.9 billion in revenue (http://corporate.sms.ac/industryresources/mobile_gambling.htm).

17.2.4 Money as a mediator of social relations

The idea of money being a language that can connect and liberate different individuals and groups suggests that money can, in all sorts of ways, mediate our experience of social relations. By way of example, we can develop our previous discussion of gender identities and think about how money, its organization and uses are structured in ways that are gendered. Viviana Zelizer (1989), for example, suggests that how monies are exchanged within households is usually subject to a very different, and usually gendered, set of rules than those that govern the use of money in the market place. Zelizer (1989: 351) argues that social structures

> systematically constrain and shape (a) the *uses* of money, earmarking, for instance, certain monies for specified purposes; (b) the *users* of money, designating different people to handle specified monies; (c) the *allocation* system of each particular money; (d) the *control* of different monies; and (e) the *sources* of money, linking sources to specified uses.

Research in Britain has shown that men and women tend to be responsible for different forms of expenditure. Women are primarily responsible for expenditures on clothing, school-related expenses, food and charity donations and men responsible for expenditure on trips and holidays, repairs, redecorating, car maintenance and so forth (Pahl 1989).

Gendered assumptions about earnings and uses of money are also writ large in financial services industries. Financial spaces like the City of London and New York are renowned as elitist, masculinist environments (Lewis 1990; McDowell 1997) and many financial products target 'a male breadwinner' and are designed around gendered assumptions and stereotypes about work histories and sensitivity to risk. Pension products, for example, are designed to privilege those clients (most usually men) who work full-time and invest premiums regularly, and without interruption, over a 40-year period (Knights and Odih 1995). Similarly, Ravenhall (1999), citing a survey by *Money Magazine,* argues that male financial advisers treat men and women differently, with women being steered towards so-called 'widows and orphans investments', that is, relatively conservative products with lower risks, and potential returns, than other investments. In Bangladesh, by contrast, it is often women who are the clients of banks, who receive credit and organize themselves and others into village groups to oversee the repayment of debt. Later in this chapter we will examine the establishment of Grameen Bank in Bangladesh. Its founder, Mohammad Yunus, established the bank believing that conventional banks were anti-poor and anti-women and insisted on collateral before making loans. Ninety-four per cent of Grameen Bank's clients are women. Women are regarded as better borrowers because 'they had most to lose, were more careful, had longer vision and were more concerned with the family than men' (Caulkin 1998: 2). In different ways in different societies, then, money can mediate the experience of, in this example, gender relations. Households and financial institutions can operate to reinforce or challenge prevailing assumptions and stereotypes concerning women and money.

In summary, money has cultural, political, social and not just economic significance and this is reflected in the negative and positive discourses about the power of money. Money, as an 'incarnation of social power' is integral to the construction and performance of different identities. It is a 'leveller', a language for linking different individuals and groups, and it mediates our experiences of social relations.

Plate 17.2 Money changing lives? The possibility of a big win, however remote, helps to fuel various forms of gambling.
(AP Photo/Luke Palmisano)

17.3 Geographies of monies

How can we understand the links between some of our everyday experiences of money, described in section 17.2, and the workings of so-called 'global finance'? This section broadens the focus to consider the geographies of money that connect individuals and groups in different places and times. In an increasingly globalized era, some would argue that geography is becoming relatively less important. Below, we examine some of the arguments concerning the globalization of money and finance, before outlining some of the problems with this view. The section concludes by introducing the concept of *monetary networks* as a way of thinking about money and its 'global' and 'local' qualities.

17.3.1 Global monies?

Many geographers, particularly those working in a **political economy** tradition, have argued that capitalism seems to be speeding up and spreading out (Harvey 1989; Massey 1994). Capitalism is seen to be spreading out in that more people in different countries and regions are becoming bound up with the logics of contemporary capitalism. It is also speeding up in that the pace of life seems to be increasing in different parts of the globe. In the second section, we noted the ability of money to transform qualitatively different commodities and experiences into an abstract quantity, a price. This has led many theorists to argue that money is a vehicle for the *homogenization of space*, for making different spaces more similar. While there are ongoing debates about the extent to which capitalism is generating a *global culture* (see Chapter 18, this volume), it is difficult to argue with the contention that financial markets in different countries have, since the mid-1970s, become more interconnected (Harvey 1989; Martin 1994) and homogeneous.

When we talk about the *globalization of finance*, however, we are talking about more than just the growth of international financial transactions, or the growing presence of multinational companies in domestic financial markets. Globalization implies

> a strong degree of *integration* between the different national and multinational parts . . . the emergence of truly transnational banks and financial companies . . . that integrate their activities and transactions across different national markets. And above all, [globalization] refers to the increasing freedom of movement, transfer and tradability of monies and finance capital across the globe, in effect integrating national markets into a new supranational system.
>
> (Martin 1994: 256)

The emergence of this *supranational* system coincides with the increasing difficulties faced by *nationally* based regulatory authorities like the Bank of England or the US Federal Reserve. The 'discourse of suspicion' we mentioned earlier rears its head in many popular representations of 'global finance':

> [T]he world is now ruled by a global financial casino staffed by faceless bankers and hedge fund speculators who operate with a herd mentality in the shadowy world of global finance. Each day they move more than two trillion dollars around the world in search of quick profits and safe havens, sending exchange rates and stock markets into wild gyrations wholly unrelated to any underlying economic reality. With abandon they make and break national economies, buy and sell corporations and hold the most powerful politicians hostage to their interests.
>
> (Korten 1998: 4 © Guardian News & Media Ltd 1998)

Whether or not you agree with Korten's view, it is true that the degree of integration of the financial system profoundly shapes and connects the lives of people thousands of miles away from each other. In this sense, Leyshon (1995b) talks of money being able to 'shrink' space and time, bringing some (but not other) parts of the globe relatively closer together through the working of financial markets. A good example of this is considered in Case study 17.1, which describes how the fates of property developers in Bangkok (Thailand) ultimately affected the fortunes of a bartender, Graham Jones, in Whitley Bay (United Kingdom). The fates of individuals and companies in Britain and Thailand are connected through the workings of foreign exchange markets that, in turn, link national currencies to the competitiveness of different nation-states.

This example highlights only some of the interconnections between changes in the value of the Thai baht and how these affected businesses across Asia and Europe. We could extend the example by thinking through how job losses in any other region/locality would also have been felt by those not in paid work. For those in households dependent on the wages of someone made redundant, there would be further belt-tightening or the search for alternative sources of income. For the unemployed, job losses in the area might mean more competition for any new jobs that are created, and so forth.

Regional Case Study 17.1

From Bangkok to Whitley Bay . . .

1 Devaluation in Thailand, July 1997

In 1997, several property companies collapsed in Thailand; property prices and the stock market started to fall. Currency speculators, already nervous about slowing growth in the region, started to sell the baht (Thai currency) as they expected the currency to be devalued. In July, the baht was devalued. As a result, Thai exports became cheaper and, to stay competitive, Indonesia, Malaysia, South Korea and the Philippines allowed their currencies to fall sharply. Some Korean firms who needed foreign currency to pay off loans started dumping cheap microchips, forcing microchip prices down from US$10 a unit to US$1.50.

2 Factory closure in North Tyneside, United Kingdom, 31 July 1997

Rapidly falling semiconductor prices meant losses of £350 million for German electronics company Siemens. Siemens semiconductors were produced at plants in Tyneside (United Kingdom), Taiwan, Germany, France and the USA. Managers in Munich announced plans to close their £1.1 billion plant in North Tyneside that employed 1,100 people and had opened only in 1994.

3 The cleaning company

The Siemens factory provided 10 per cent of business in the north-east for Mitie, a cleaning company. One-third of the workforce of 90 people were facing redundancy and their boss had his salary bonus cut 10 per cent, in line with the loss of work from Siemens.

4 The hotel

Siemens and its contractors, like Mitie, were the single biggest sources of business for the Stakis Hotel. The hotel responded to the closure by switching their market focus. As the flow of German executives and their UK contractors slowed and then ceased, the hotel sought to attract more families.

5 The taxi firm

Foxhunters, a taxi firm, had a contract with Siemens that generated more than a dozen runs a day, usually to the Stakis Hotel, the airport or the university. Since August, Siemens business had dried up and 85 drivers were chasing work for 50. Despite working longer hours, drivers' takings were down by between £100 and £200 a week. Drivers started to economize by bringing in their own lunches and by cutting down their trips to the local pub.

6 The local pub

Takings at Cameron's had fallen by £600 a week since August. The landlord, who blamed the Siemens shutdown for the reduced trade, cut his opening hours. Attempts to drum up more trade by price reductions had little effect. The landlord was hoping for more business around Christmas. He and his partner had less money for spending on their leisure activities, which included visiting places like Whitley Bay.

7 Graham Jones, bartender in Whitley Bay, October 1998

Graham was fired from his job at a pub in Whitley Bay; there was not enough business to occupy two bartenders in the public bar.

Source: Adapted from Carroll (1998)

17.3.2 Forces for globalization

How has this globalization of money come about? There are a range of factors to consider here. First, through the 1980s and 1990s, different governments and international institutions, like the **International Monetary Fund (IMF)**, have pursued neoliberal, 'free market' policies and encouraged the deregulation of financial markets (by eliminating exchange and capital controls) and the liberalization of flows of capital across national borders.

Secondly, there have been advances in telecommunications and computing technologies that can be summarized under three headings: computers, chips and satellites (Strange 1999). Computers have transformed payments systems. For many hundreds of years, payments and transfers of money were completed in cash – coins and notes – and written down in ledgers. Coins and notes were superseded by cheques which in turn have been superseded by electronic monies that can be moved around the world at the speed of light. Semiconductor chips in computers mean that consumers in many countries can use credit and debit cards to pay for goods. In some countries, computer chips are now embedded in plastic 'smart' cards to allow the use of 'digital money', or, more accurately, electronic representations of currencies, to pay for groceries and other goods. Finally, there are the systems of communication, using earth-orbiting satellites, that are integral to the operation of computing systems, e-mail, the Internet and other forms of communication that are used by banks, governments and other financial players. These innovations have made 24-hour trading possible as stock markets in different countries are linked by computer.

The growth of the Internet through the 1990s has been closely linked to the rise of discourses about a 'new economy' (see Aglietta and Breton 2001; Thrift 2005). The Internet has not only altered how business can be done in financial markets – think, for example, about the rise of online banking in North America and parts of Europe – but also changed the cast of characters involved in undertaking that business. New communications and software technologies have spawned the proliferation of financial websites, information and intermediaries able to provide financial information to firms and consumers. Online providers, without the costs of maintaining the bricks and mortar associated with a high-street presence, provide insurance, banking, pension and other financial products to firms and consumers with access to Internet technologies.

Thirdly, and closely related to developments in technology, there have been innovations in tradable financial products that have made it easier and faster to move money around the globe. Derivatives are one example of this kind of innovation (Tickell 2003). Derivatives are contracts that specify rights/obligations based on (hence 'derived' from) the performance of some other currency, commodity or service. In the 1970s, financial derivatives were created to allow financial managers to deal with currency risk, but since that time they have become increasingly sophisticated and extended to more markets. Since 1997, for example, energy suppliers, transport agencies, construction companies, wine-bar owners and other firms exposed to weather risk (the possibility that weather could have an adverse impact on their profits and cash flow) have been able to purchase weather derivatives contracts to protect themselves against this risk (Pollard et al. 2007). With this extension, however, is increased speculative trading of derivative contracts, some high-profile losses (for example, Barings Bank in London) and growing concern on the part of regulatory authorities (Strange 1999; Tickell 2003).

Innovations in computing and software technologies and in financial instruments are fundamental to arguments about the growing *financialization* of the economy; this is an argument that asserts that the speculative accumulation of capital has become an end in itself in contemporary capitalism, that the financial system has come to feed on itself, so to speak, rather than supporting firms and industries and other elements of the 'real' economy (Strange 1999). Round-the-clock trading of financial instruments in different places and time zones opens up opportunities for speculation and arbitrage, the ability to profit from small differences in price when the same financial product is being traded on more than one market. One indicator of how growth in the international financial system is outpacing the growth of the 'real' economy is provided by foreign-exchange trading data. According to data from the Bank for International Settlements (www.bis.org) the daily turnover of foreign exchange trading in 1973 was \$10–\$20 billion, roughly twice the amount necessitated by world trade. By 2004, daily trade in foreign exchange averaged \$1.9 trillion, roughly *ninety-five* times that necessitated by world trade. For individual consumers with access to the appropriate technology, there are now online financial bookmakers encouraging clients to enjoy spread-betting (and possibly tax-free profits) on price movements of stocks, stock indices, currencies, interest rates, commodities and even house prices.

These, then, are some of the ways in which money has become more global since the 1970s. But what motivates such changes? What motivates the implementation of 'free market' policies, the development of new technologies and financial instruments like derivatives? For those working in a *political economy tradition*, like David Harvey (1989), the motivation for these changes is the search for profit. Capitalism is fundamentally about the accumulation of surplus. Competition drives capitalists to seek out new markets, new products and to reduce the *turnover time of capital* (see Spotlight box 17.2). In different parts of the globe, money is the language through which the imperatives of capitalism are being communicated.

So, there is a very strong economic rationale for the globalization of money. And it is difficult to argue

Spotlight Box 17.2

The circulation of money as capital

In capitalist societies, the circulation of money as capital is as follows. From left to right in the equation, money (M) is invested by producers to purchase commodities (C), namely labour power (LP) and the means of production (MP), say pieces of wood and wood-cutting machinery. Labour power and the means of production are combined in production (P) to make more commodities (C′), in this example, let us say chairs, which are then sold for more money (M′) than was originally invested (M).

$$M \rightarrow C \rightarrow \frac{\{LP\}}{\{MP\}} \rightarrow P \rightarrow C' \rightarrow M' \rightarrow \ldots$$

The purpose of production in capitalist societies is to produce profit and accumulate capital. The *turnover time of capital* is the amount of time it takes for money to complete this circuit. The shorter the turnover time, the more often money can be lent out and the more profit can be made. Producers therefore have a very strong incentive to, where possible, reduce the turnover time of capital; time *is* money.

with the contention that money has become more globalized since the 1970s, that it has, increasingly, connected people in distant places and homogenized financial space. For Richard O'Brien (1992), the growth of the international financial system is tantamount to 'the end of geography', the notion that geography is becoming relatively less important because money, in its different forms, is able to overcome the friction of distance and link distant places together. Others argue that this view is too simple, that geography remains critical to our understanding of global finance. In the next part of this section, we will consider these views in more detail.

17.3.3 Debunking 'global' monies

A rather different departure point for talking about the 'globalization of finance' is to argue that the financial system is not really global at all, that it is more like a *web of connections* between different financial systems, some of which are bound together more tightly than others. This argument has developed because the idea of a 'global financial system' is problematic for several reasons.

The first problem stems from the simple observation that 'global finance' is largely the province of North America, Europe and parts of Asia, most notably Japan, Hong Kong and Singapore; it is an idea centred on the experiences of the West (see Table 17.1). To continue with one of the examples mentioned in the previous section, consider the geography of Internet banking. In 2004, there were an estimated 122 million users of online banking. More than 47 per cent (58 million) of these users lived in Western Europe and another 18 per cent (22.8 million) lived in the USA. Beyond Western Europe, the USA and the Asia Pacific region, there were just 6.1 million users (http://www.epaynews.com/statistics/bankstats.html). Many parts of Africa, Asia and Latin America are not only excluded in 'the global financial system', they are also being actively excluded as banks refuse to lend money until outstanding debts have been repaid. There is, what Massey (1993) describes as, a *power geometry* at work when some commentators describe the financial system as 'global'. For some financial workers in New York, London and Tokyo, international finance may be regarded as 'global' in the sense that all countries and regions of the globe deemed creditworthy and capable of producing profits have been included; sub-Saharan Africa, however, ceases to exist in such a conception of 'the global'.

Secondly, and resulting from this geographical concentration of the management of 'global finance' in North America, Europe and parts of Asia, some currencies circulate more widely and have greater spatial reach than others. The Japanese yen, the Euro and most especially the US dollar are very useful in international markets because they are accepted as forms of payment, unlike, for example, Indian rupees. Historically, the country that occupies a dominant economic and political position has underwritten the soundness of the international financial system and had its currency accepted internationally as the currency in which commodity prices are quoted and payments made. Before the Second World War, Britain and the pound sterling

Table 17.1 Average daily turnover of equities trading 2005 (millions of $US)

Exchange	Average daily turnover	Region % of world total	Exchange	Average daily turnover	Region % of world total
North America		50	Spanish Exchange (BME)	6,141.6	
Amex	2,394.1		Swiss Exchange	3817.7	
Bermuda	0.3		Vienna (Austria)	187.4	
Mexican Exchange	222.3		Warsaw (Poland)	121.2	
NASDAQ	40,026.7		**Middle East**		0.2
NYSE	56,052.6		Tehran (Iran)	32.3	
Toronto (TSX Group)	3,587.6		Tel Aviv (Israel)	199.9	
South America		0.4	**Africa**		0.4
Buenos Aires (Argentina)	27.2		JSE South Africa	803.9	
Lima (Peru)	10.6		**Asia/Pacific**		17
Santigo (Chile)	75.2		Australian	2,668.2	
São Paulo (Brazil)	663.8		Colombo (Sri Lanka)	4.6	
Europe		31	Hong Kong	1,879.6	
Athens (Greece)	260.5		Jakarta (Indonesia)	171.3	
Budapest (Hungary)	95.5		Korea	4,862.1	
Deutsche Bourse (Germany)	7,452.5		Malaysia	208.9	
Euronext (Amsterdam, Brussels, Lisbon, Paris)	11,308.2		Mumbai (India)	633.4	
Irish	266.5		National SE India	1,253.7	
Ljubljana (Slovenia)	5.4		New Zealand	83.2	
Istanbul (Turkey)	790.8		Osaka (Japan)	883.4	
Italian exchange	5,053.4		Philippines	28.4	
London (UK)	22,530.6		Shanghai (China)	985.6	
Luxembourg	1.3		Shenzhen SE (China)	637.4	
Malta	0.6		Singapore	465.8	
OMX[1]	3,760.6		Taiwan	2,370	
Oslo (Norway)	926.4		Thailand	390.4	
			Tokyo (Japan)	18,292.7	

[1]OMX includes exchanges in Copenhagen, Stockholm, Helsinki, Riga, Tallinn and Vilnius.

Source: Adapted from World Federation of Exchanges at: www.iasplus.com/stats (accessed 4 December 2006)

fulfilled this role; after the Bretton Woods conference in 1944, the US dollar became the key international currency. More recently, as the economic dominance of the USA has declined, the yen and the Euro have become relatively more important in international markets.

So, some currencies, like the US dollar, are truly international while some others are national. There are also over 2,000 *local currencies* in operation around the globe that facilitate exchange only within very specific spatial and social contexts. For example, Local Exchange Trading Systems (LETS) are associations whose members list their services needed/offered in a directory and trade with each other in a local unit of currency, for example 'bobbins' in Manchester and 'solents' in Southampton (Williams 1996). Since their establishment in Canada in 1983, LETS schemes have spread to Europe, Australia, New Zealand and North America, allowing members not only access to credit, but also the chance to engage in productive activity to earn such credit (Williams 1996). Some local currency schemes are devised in times of hardship to allow local people to trade goods and services when they are unemployed and have little money. Other schemes are motivated by ecological concerns, the desire for community development and social cohesion, and the desire to construct alternative local economic geographies as a form of resistance against the global spread of capitalism (Lee 2000).

Thirdly, as Martin (1999b: 6) argues, for all the talk of 'global finance', there remain different *geographical circuits of money* that form the 'wiring' of an economy, along which 'currents' of wealth, consumption and power are conveyed. He identifies four geographies of money: locational, institutional, regulatory and public. The *locational geography* refers to the location of different financial institutions and markets. Financial institutions and specialized functions (like foreign exchange markets) tend to be agglomerated in large urban centres, with London, New York and Tokyo sitting at the top of the global hierarchy. Different countries also have different *institutional geographies,* in that they organize their financial institutions and markets in distinct ways. In Britain, for instance, banking is highly concentrated amongst a few large players; we are used to thinking of the 'big four' retail banks – NatWest, Lloyds-TSB, HSBC and Barclays – and relatively new banks like the HBOS or Woolwich that have converted from building societies. In the USA, by contrast, there were 7450 federally insured commercial banks in 2006 (http://www.fdic.gov/bank/statistical/stats/2006sep/industry.pdf). Meanwhile, in parts of Asia and West and South Africa, banks and other community-based financial institutions like Rotating Savings and Credit Associations (ROSCAs) (see Spotlight box 17.3) are important in organizing and funding economic activity. ROSCAs take different forms in different regions and vary with the class, gender and ethnicity of their members (Ardener and Burman 1995).

These varied institutional geographies are products, in turn, of contrasting *regulatory geographies.* There are a wide range of supranational, national and regional regulatory agencies that govern the workings of financial institutions and markets. To return to the example of how the Asian financial crisis cost Graham Jones his job in Whitley Bay, a key player in how the crisis unfolded was the International Monetary Fund (IMF). The Chief Executive of Siemens argued that Korea was able to sustain a 'suicidal pricing strategy' for its memory chips (a strategy that led to losses of £350 million for Siemens) because it was receiving financial aid from the IMF (Milner 1998). To give another example, as Martin (1999b: 9) observes,

> [M]oney has had a habit of seeking out geographical discontinuities and gaps in these regulatory spaces, escaping to places where the movement of financial assets is less constrained, where official scrutiny into financial dealing and affairs is minimal.

Offshore financial centres like the Bahamas, Cayman Islands, Jersey and Guernsey are attractive because of their low tax rates and minimal regulation.

Finally, when Martin (1999b) discusses *public geographies,* he refers to the role of states in distributing monies across regions in the form of goods and services, infrastructure, health, education and so forth, and in transferring monies in the form of various social and welfare programmes.

Spotlight Box 17.3

Rotating savings and credit associations (ROSCAs)

A ROSCA is 'an association formed upon a core of participants who agree to make regular contributions to a fund which is given, in whole or in part, to each contributor in turn' (Ardener 1995: 1). ROSCAs are known by different names, depending on their form and scale, the social classes of their members and their location. In South India they are known as *kuris, chitties* or *chit funds*, in Cameroon as *njangis* or *tontines.*

ROSCAs were well developed in China, India, Vietnam and parts of West Africa by the end of the nineteenth century. Variants of ROSCAs also existed in Scotland and parts of northern England. Members, usually ranging in number from a few to several hundred, make regular contributions, in cash or in kind, to a fund. The fund, or part of it, is then given to each member in turn, depending on age, kinship seniority, or by rules established by the organizer.

In addition to encouraging regular savings and providing small-scale capital and credit for their members, many ROSCAs have strong moral and social dimensions. Trust, social solidarity and responsibility are emphasized and members have an interest in ensuring that no member defaults on regular payment. Some ROSCAs are women-only and their potential to empower women, by giving them greater control over income and credit, has attracted the attention of anthropologists, sociologists and feminists (see Ardener and Burman 1995).

What becomes clear from this discussion is that the world remains made up of a patchwork of different financial spaces or systems, and although there has been considerable integration of financial practices and processes across such systems in recent years, we are a long way from the seamless global financial space described by some commentators.

(Leyshon 1996: 62–3)

17.4 From 'global monies' to monetary networks

The phrase 'global finance' therefore seems to be something of a misnomer. While it is true that different monies and markets have become more interconnected since the 1970s, 'the global financial system' seems to comprise some elements that are international in their reach, juxtaposed with other local and national currencies, regulations and institutional geographies. One way of juggling with this complexity is to think about the idea of *monetary networks.*

Recall that in the introduction to this chapter 'the global financial system' was described as a social, cultural and political *network of relations* in which individuals have different capacities to participate. When we talk about 'networks' what do we mean? Dodd (1994: xxiv) argues that monetary networks are, above all, *networks of information* that have five abstract properties (see Spotlight box 17.4). These networks are formed by the production, interpretation and circulation of information between actors through time and space. Thus we can think of foreign exchange traders in London, Tokyo, New York and elsewhere constituting a network of relations through which foreign exchange trading is performed, organized, regulated and understood.

How do networks work? Nigel Thrift (1996, 1998) has argued that the forms of information that constitute monetary networks are always ingrained in *practices.* To give an example, we can return to the workings of global telecommunications networks discussed earlier. Telecommunication networks rely on the workings of thousands of workers and machines. Similarly, the Internet consists of thousands of miles of fibre-optic cables, most usually laid alongside existing transport infrastructure, and computing hardware and software managed and sustained by programmers, engineers, users and machines (Bray and Kilian 1999). What we think of as the 'global financial system' is really an assemblage of consumers, workers, computers, software, telephones, office buildings, bits of paper, financial reports etc. There is no escaping the *materiality* (i.e. the bits of cable, bank branches, telephones, desks, etc.) and *practices* (using computers, paying for goods and so forth) that constitute these networks.

When we talk of 'the global financial system', then, we are really talking about actor networks that combine human beings, technologies and documents which come together to sustain, in particular spaces and times, the flows of information that constitute different monetary networks. Note the use of the plural.

Spotlight Box 17.4

Properties of monetary networks

- Networks contain a *standardized accounting system* into which each monetary form within the network is divisible; any goods, priced in terms of this standardized accounting system, are exchangeable.
- Networks rely on *information regarding expectations of the future;* 'money is accepted as payment almost solely on the assumption that it can be re-used later on' (Dodd 1994: xxiv).
- Networks rely on *information regarding their spatial extent;* institutional frameworks limit the territories in which specific monetary forms may be used.
- Networks are based on *legal information,* usually in the form of rules that govern forms of contractual relations of network members.
- Networks presuppose *knowledges of the behaviour and expectations of others.* In order to trust money, those holding it need to anticipate, and trust, that it can be used and reused within the network.

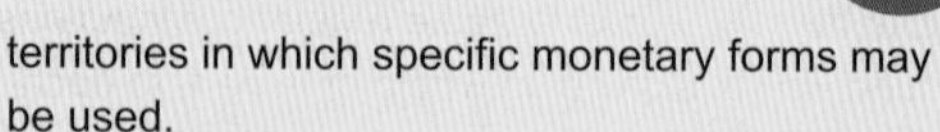

Source: Dodd (1994: xxiv)

Rather than thinking of one 'global financial system' we can think instead of there being a *multitude of monetary networks that overlap and intersect.* This is an important point, because different monetary networks, as we have seen, have very different geographies. Some monetary networks, like those of the foreign exchange markets, have the capacity to connect people and machines in London, New York and Tokyo at the touch of a button. Other monetary networks, like LETS schemes in Manchester, are much more compact and connect people over much shorter distances. Rather than thinking about networks as being either 'global' or 'local', Bruno Latour (1993: 122) argues that we should think of networks as being 'more or less long and more or less connected'.

Monetary networks differ not only in the *length of connections* they are capable of sustaining, but also in the *speed of connections* they are capable of sustaining. We have talked already of the computing networks that facilitate foreign-exchange trading and allow banks to move monies around the world in seconds. One of the attractions of ROSCAs (Spotlight box 17.3) for men and women in Ghana and elsewhere is the speed at which news of hardship or an emergency can be spread, and the order of rotation of the fund changed, to help out a member in trouble. ROSCAs may not have quite the speed of connection of foreign-exchange trading (assuming, of course, that all the computers are working), but they can react with a speed rarely matched by local banks (Ardener 1995).

In summary, financial markets have become increasingly globalized since the 1970s, although truly 'global' markets are concentrated in a select number of major financial centres. There remain distinct regional, national and supranational geographies of money, such that the 'global financial system' can be conceptualized as a web of intersecting networks, some fast, some slow, some long, some short. These financial networks connect different geographies of employment, technology, regulation, government policies, leisure trends and so forth. Such a conceptualization allows us to understand how, over a period of fifteen months, the fate of property developers in Thailand affected Graham Jones in Whitley Bay. This approach also reinforces the point that when we talk about 'global finance', we are not talking about a machine operating 'out there', but a socially, economically, culturally and politically constructed network of relations with which we engage, in some way, shape or form, on a daily basis.

17.5 Conclusion: money, space and power

From the preceding discussion, it should be clear that many monetary networks are very uneven and exclusive. Different monetary networks are not open to all who would like to participate and there are some very distinctive *geographies of financial inclusion and exclusion* that have already been mentioned. The benefits of some of the most extensive, fastest monetary networks are, for the most part, extended only to those areas where individuals, companies and governments can produce a profit for the major lenders in North America, Japan and Europe. Access to such networks is carefully screened and countries, firms, banks, public agencies and individuals are credit-rated. On the basis of their credit rating, decisions are made by major financial institutions on whether to offer credit or buy debt and, if so, for how long, and at what rate of interest.

So, who does this screening and what are the geographies of this credit scoring? The big three international agencies are Moody's Investment Services, Standard and Poor, and Fitch Ratings, all US firms. Not surprisingly, only the major capitalist economies are assigned high ratings when it comes to government-issued debt in the form of bonds (Figure 17.1). Standard and Poor's so-called 'Investment grade' scales rate types of debt on a scale from 'AAA' ('Superior') through 'A' ('Good') to 'BBB'. Below this are the 'Speculative' grades which range from 'BB' ('Questionable') through to 'D' ('Default').

This is not to say that only the major capitalist economies are able to secure credit. In the late 1970s and early 1980s, total lending to Africa, Asia and most especially Latin America increased sharply. Yet rising interest rates, falling commodity prices and a rapid appreciation of the US dollar left debtor countries unable to cope with repayments. Affairs came to a head in August 1982, when Mexico suspended its debt repayments, followed by Brazil, Argentina, Peru, Venezuela, Ecuador and more than twenty other countries by the end of 1983 (Porter and Sheppard 1998). The response of banks to the debt crisis was to reduce drastically their lending to these countries, to effectively exclude them from international bank lending networks.

More recently, investors in North America and Western Europe have been encouraged by fund managers and financial media to invest in 'emerging markets', which include parts of the Russian Federation and former Soviet Republics and other parts of Asia,

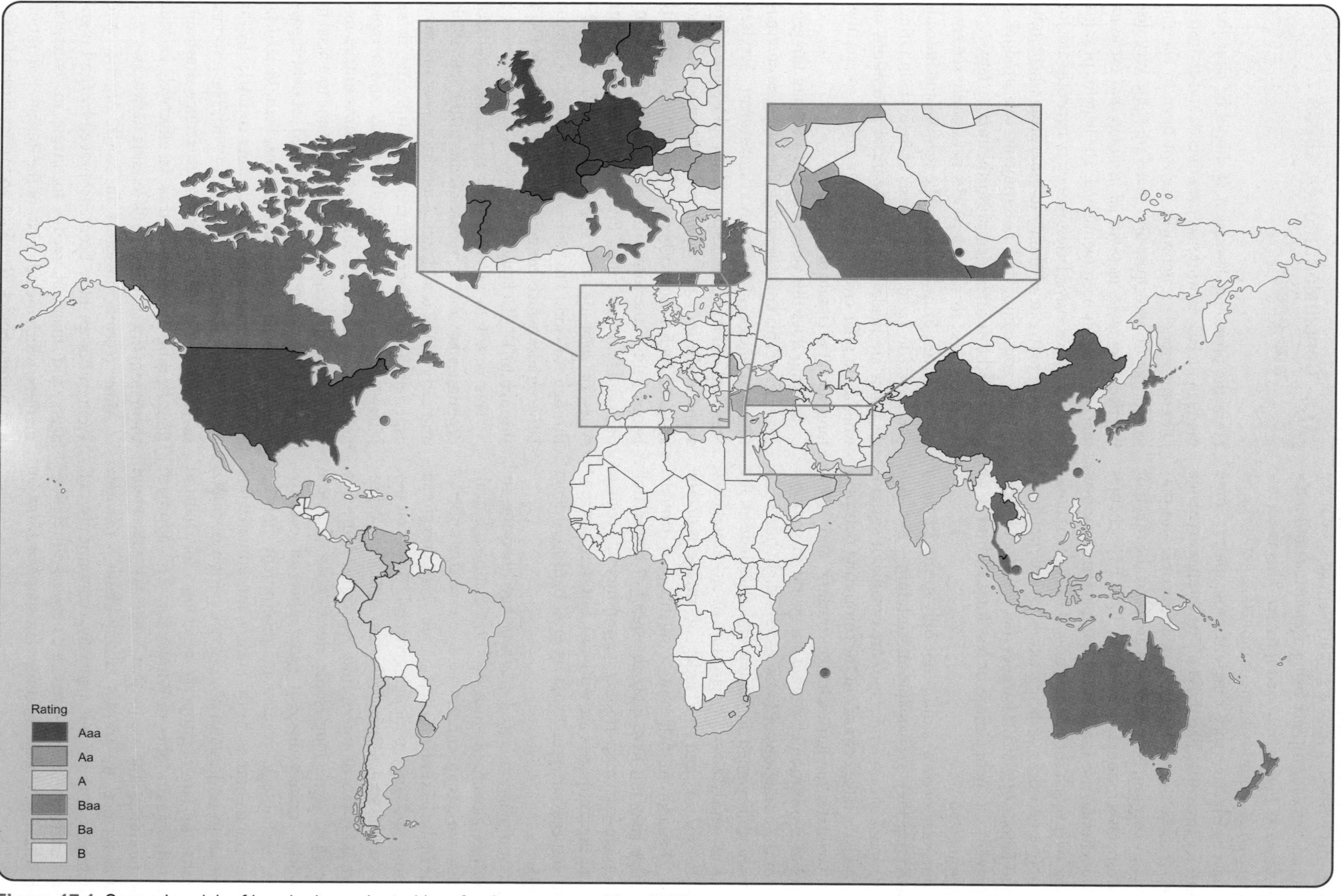

Figure 17.1 Sovereign risk of bonds denominated in a foreign currency, May 1996.
Source: Adapted from Laulajainen (1998: 56)

Latin America and Africa. The lure for investors is the prospect, real or imagined, of higher rates of return than those that can be secured in what are represented as more conservative, less volatile, confines of Western European and North American equity markets (see Sidaway and Pryke 2000). Again, the US-based credit-ratings agencies are key players in the construction of these financial markets. Research in emerging economies demonstrates that a downgrade in, say, a country's bond rating not only affects the price of its bonds but also affects stock market returns and 'spillover' to neighbouring states as international investors anticipate further downgrades in the region (Kaminsky and Schmukler 2002).

The concentration of wealth and power in the major capitalist economies, most especially the USA, is such that they hold sway, economically and politically, over the networks through which international lending is conducted and its terms negotiated. Even supranational agencies like the World Bank and the IMF rely on funds from major capitalist countries that exert considerable influence over where the funds go and how they are spent. If South Africa and Russia and other post-socialist economies want to be included in the monetary networks of the World Bank and the IMF, then instituting reform packages that mimic the political and economic organization of major capitalist economies is the price of admission. Such **structural adjustment** packages have spread widely through east and west Africa, Latin America and Asia since 1982 (Porter and Sheppard 1998). For those that do not toe the line, life can be made very difficult. Money, as we saw in section 17.2, is a very powerful tool of social and political control.

Such geographies of inclusion and exclusion operate within countries too as banks, private investors and other lenders seek out the most prosperous growth regions. For example, in Los Angeles major retail banks have closed bank branches in low-income neighbourhoods like South Central Los Angeles and opened new branches in prosperous areas like Beverly Hills (Pollard 1996); this pattern has been repeated in many other US cities. The financial institutions that tend to move into South Central Los Angeles are so-called 'alternative' financial institutions like pawnbrokers and cheque-cashing establishments, which often charge very high fees for basic banking services. In similar fashion, the distribution of venture capital investments in the United Kingdom has traditionally favoured firms located in the prosperous south-east of England (Mason and Harrison 2002).

Financial exclusion tends to reinforce existing patterns of wealth, poverty and inequality. Those individuals and nations that already have wealth are deemed the most creditworthy and are offered the cheapest forms of credit to expand businesses, pay off debts, buy new goods and so forth. And, as discussed in section 17.3.4, those consumers that have the financial, cultural and technical capacity to access the Internet will be the beneficiaries of new and sometimes cheaper financial products available online. For low-income groups and countries reeling from crippling debt repayment schedules, credit is difficult to find and usually very expensive. For those groups suffering varying degrees of financial exclusion, what alternatives are available?

17.5.1 Some responses and alternatives to financial exclusion?

Although some alternative monetary networks, like those of doorstep lenders, may be expensive to use, there are plenty of examples of other monetary networks that are designed to be inexpensive and inclusive. One very famous example is the Grameen Bank, formed in 1983, which now employs over 14,000 people in 35,000 villages in Bangladesh (Case study 17.2). The successes of the Grameen Bank and other micro-credit institutions in Indonesia, Bolivia, Kenya and elsewhere have attracted a great deal of attention and spawned innumerable imitations in Africa, Asia, Latin America and elsewhere. In recognition of their impact in helping the poor, Yunus and Grameen Bank were awarded the 2006 Nobel Peace Prize. The prospect of linking access to credit for the poorest with job creation has made these so called micro-finance schemes increasingly popular in Britain, the United States and Western Europe, and not just in those regions of the United Kingdom and the United States that are experiencing financial exclusion.

It is also important to note that some 'alternative' financial networks may be preferred by individuals and households who do not want to use financial institutions like traditional retail banks and instead want to participate in networks that encourage broader social, religious, environmental and other assessments of 'value'. In the last fifteen years, for example, Islamic banking and finance (IBF) has grown very rapidly cross the globe. IBF practices are rooted in the rules and norms of Islam which prohibit the payment of *riba* (interest) (as it is viewed as exploitative and unfair) and investments deemed excessively risky or uncertain (*gharar*). Islamic institutions thus shun investments

Regional Case Study 17.2

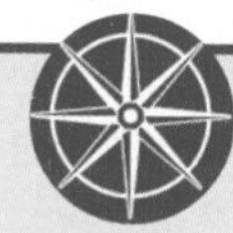

Grameen Bank

The basic principles of Grameen Banking are as follows:

- the bank lends only to the poorest landless villagers in Bangladesh
- the bank lends, overwhelmingly, to women; 94 per cent of its clients are women
- loans are extended without borrowers needing collateral
- the borrower, not the bank, decides what the loan will be used for
- borrowers pay sufficient interest to keep the bank self-sustaining.

The average loan is for about US$100. Rather than insisting on collateral or security, the bank makes sure that women wanting a loan join a village group, attend weekly meetings and assume responsibility for the loans of other group members. Thus, there is strong peer pressure for individual women to maintain payments, otherwise the group is unable to secure further loans. A total of 97 per cent of Grameen Bank loans are repaid.

The success of Grameen Bank has spawned imitations in 58 countries in many parts of Africa, Asia and Latin and North America. Grameen has now branched out into fisheries, mobile telephones, energy and the Internet.

Sources: Bornstein (1996), Caulkin (1998), www.grameen.com/bank/GBGlance.htm

into, for example, gambling, alcohol and prostitution (Henry and Wilson 2004).

The growing popularity of various micro-credit schemes, like the LETS and ROSCA schemes is based not only on their economic successes, but also on their status as 'alternative' and, potentially, progressive schemes. One of the interesting features of micro-credit schemes, like the Grameen Bank, is their often strong emphasis on creating social bonds and a sense of shared responsibility between members. Another feature of many of these micro-finance experiments is that it is often women who make up the bulk of the membership. As Goetz and Gupta (1996) point out, however, access to credit does not necessarily equate to control over money and, more generally, empowerment for women in environments where they frequently face extreme constraints on their lives. As we mentioned in section 17.2, how monies are exchanged within households is usually subject to a rather different set of rules than those governing the use of money in the market place (Zelizer 1989).

To conclude, these examples of monetary networks illustrate only too well that money, how it is obtained, how it is used, how it is managed and by whom, is an economic, social, political, cultural and *geographical* phenomenon. Although many accounts of 'the global financial system' pay little regard to the geographies of money and finance, this chapter has argued that a host of monetary networks – some relatively slow, some relatively fast, some longer than others – make up the 'global financial system'. These monetary networks intersect most densely through institutions and practices in North America, Japan and Western Europe. This geopolitical reality profoundly shapes patterns of uneven development and structures which regions, industrial sectors, governments and individuals are deemed credit-worthy and on what terms. 'Global finance' is thus a socially constructed, geographically rooted network of relations with which we engage, in some way, shape or form, on a daily basis.

Learning outcomes

Having read this chapter, you should be aware that:

- Money has geographical, economic, cultural, social and political significance and mediates our experiences of social relations.
- The 'global financial system' is a socially constructed web of intersecting networks operating at different speeds and over different distances.
- There are distinct regional, national and supra-national networks of money.

- There are many geographies of financial inclusion and exclusion.
- There are a range of financial networks which may be more inclusive and politically progressive than others.

Further reading

Leyshon, A. and Thrift, N. (1997) ***Money Space: Geographies of Monetary Transformation***, Routledge, London. This is a compilation of Leyshon and Thrift's essays on money.

Martin, R.L. (ed.) (1999) ***Money and the Space Economy***, Wiley, Chichester. A wide-ranging collection of essays covering geographies of banking, financial centres, money and the local economy, and money and the state.

McDowell, L. (1997) ***Capital Culture: Gender at Work in the City***, Blackwell, Oxford. This is a more specialist book that examines the masculine cultures and practices of parts of the City of London.

Sassen, S. (2001). ***The Global City: New York, London, Tokyo.*** Princeton University Press, Princeton, NJ. This book looks at the growth and internationalization of the financial system and its effects on the economic base and social structure of its three 'command centres', London, New York and Tokyo.

Useful websites

www.ft.com The *Financial Times* website contains useful links to financial and other firm, sector and country data, as well as breaking news stories.

http://money.cnn.com The US-based CNN financial website, like the *Financial Times* site, contains numerous links to business, financial and market news, by country, sector and firm.

www.worldbank.org The **World Bank** website is home to a host of information on World Bank publications, research, data and related organizations.

www.bis.org/statistics This site, maintained by the Bank for International Settlement, provides sources of banking data and statistics.

www.imf.org The home page of the International Monetary Fund. This site provides a wealth of information about the IMF and its mission in addition to country information.

www.grameen-info.org The website of the Bangladesh-based Grameen Bank. It has information on the mission of Grameen Bank and how it works, together with material on other micro-credit schemes.

http://www.ecb.int/home/html/index.en.html A site dedicated to European central banking issues.

www.creativeinvest.com The site of Creative Investment Research, a US-based firm specializing in 'socially responsible' investment and providing many links on women and minority-owned financial institutions.

http://www.cdfa.org.uk The site of the CDFA, the trade association for Community Development Finance Institutions (CDFIs) which are designed to be sustainable, independent financial institutions that provide capital and support for individuals/organizations in disadvantaged communities or under-served markets.

For annotated, clickable weblinks and useful tutorials full of practical advice on how to improve your study skills, visit this book's website at **www.pearsoned.co.uk/daniels**

CONSUMPTION AND ITS GEOGRAPHIES

Chapter 18

Philip Crang

Topics covered

- Why consumption matters
- Global geographies of consumption
- Local geographies of consumption
- Consumption, knowledge and geographies of (dis)connection

18.1 Consumption matters

Watching television, eating a burger, wearing designer-label clothing, going clubbing, drinking a Starbucks cappuccino, shopping, skateboarding – all of these are examples of consumption. All will get a mention in this chapter. At first sight they seem trivial, mundane, unimportant things. They may seem unworthy of academic scrutiny, especially when set alongside more obviously important issues like global corporate geographies, geographies of poverty, or new forms of global monetary regulation. Not only that but these instances of everyday consumption may also not seem terribly *geographical*. In this chapter I argue that such snap judgements would be wrong. In fact I will suggest precisely the opposite: that these everyday practices of consumption have great significance, economically and culturally, and that they are fundamentally geographical, implicated in productions of the 'global', 'local' and non-academic or 'lay' geographical knowledges. In sum, I argue that in significant part we both produce and understand our geographies through our everyday, mundane activities as consumers.

Plate 18.1 Man relaxes in the company of Ronald McDonald and Starbucks in Shanghai, China. This chapter argues that consumption is implicated in both global cultural flows and local fashionings of urban space.

(© CLARO CORTES IV/Reuters/Corbis)

There are a number of reasons why we are not predisposed to think of consumption as important in this way. Technically, consumption can be defined as the utilization of the products of human labour. On that basis, it seems necessarily to occur after and follow from those activities of production. Let's hypothesise that I eat a burger in McDonald's. I couldn't do this if burgers weren't being made there, if McDonald's hadn't developed its transnational corporate structure so there is a McDonald's outlet where I live, if I didn't work to earn money to spend, and if complicated financial systems weren't established to enable such commercial transactions as me buying a burger to take place. But we must not mistake the chronology of the life of an individual burger and its consumption – burger made first, burger eaten afterwards; money obtained first, spent afterwards – with issues of import or causality. Yes, consumption depends on production. But, obviously, production also depends on consumption. The nature of these relations are complex and open to both change and dispute, but in outline there is now widespread agreement that consumption is neither the inevitable, final stage of a unidirectional economic production chain nor simply a by-product of monetary circulation.

Specifically, in the form of 'retail capital', consumption in fact plays an increasingly central role within many economies. It accounts for a growing proportion of paid employment and often exercises great powers over manufacturing through the organizational power of retailers over their suppliers (for work in this vein see Wrigley and Lowe 1996). Moreover, in parts of the economy one can also see how economic production has been reconfigured away from just material manufacture and towards the fashioning of 'signs and spaces' (Lash and Urry 1994) devoted to marketing and sales. Nike would be exemplary. According to Robert Goldman and Stephen Papson (1998) Nike do not so much make sportswear as make and sell the brand Nike, constituted through the sign and typography of the Nike name, the logo of the 'swoosh', and philosophies such as 'Just do it'. The sportswear is the vehicle for that brand. So, Nike sub-contract manufacture. They invest far more in advertising and promotion (around

10 per cent of annual revenues) than in capital expenditure (around 3 to 5 per cent) (ibid.: 13). And that capital expenditure is largely devoted to distribution infrastructure, administrative and design infrastructure, and display or retail infrastructure (such as 'Niketown' retail outlets) rather than material production. Nike's business and its corporate geographies are constructed to be market-led and market-leading. They are consumption focused. This is a wider trend, with discourses of consumption – of thinking and acting in ways that are consumer focused – coming to increasing prominence in both commercial and public organizations (see du Gay 1996).

More generally, every day on news bulletins all over the world 'consumer confidence', or its lack, is cited as the cause of the 'productive' sectors of an economy booming or contracting. Consumers, or more often 'the consumer' or 'the market', are constantly being invoked as the reasons why actions elsewhere in the economy must be undertaken. Bananas must be a certain size because 'the consumer' wants them to be; hence Caribbean growers are passed over by UK food retailers in preference to those in South America. Costs must be cut in banking as 'the consumer' is unwilling to choose institutions that do not offer the best deals; so branches close and jobs are lost. Supermarkets switch food suppliers, or at least threaten to and thereby exert their power, claiming their concern with matching the quality standards that 'the consumer' demands. In the words of Danny Miller, this figure of 'the consumer' has become a sort of 'global dictator': 'Today, real power lies unequivocally with us, that is the waged consumers of the First World' (1995: 10).

This also suggests that any squeamishness about studying consumption simply because so many in the world are not affluent consumers is profoundly misplaced. For Miller, 'The acknowledgement of consumption need not detract from the critique of inequality and exploitation, but this critique is foundering precisely because the enormous consequences and attractions of consumption are left out of the analysis' (ibid.: 21). A fundamental challenge for the twenty-first century is to

Plate 18.2 'BMW-Alpina: the car with flair for people with flair'. Advertising, commodity auras and the production of false needs: you may want to own a BMW convertible, but do you really need to?
(Vintage Ad Gallery)

fashion economic forms that fulfil those attractions of consumption, but deal too with its consequences. But what about these attractions of consumption? Here we come to its social and cultural importance. For consumption is rarely, in fact almost never, simply a matter of meeting material needs. Rather, consumption plays a central role in our wider social and cultural lives and those wider contours of society and culture inflect our consumption practices. Consumption is centrally important to economies and their geographies, but it is not purely economic.

Just what social and cultural processes are implicated in our consumption practices, then? To simplify somewhat, one can identify two main sets of ideas on this. On the one hand there are those who emphasize how consumption is socially and culturally produced, how we as individual consumers are, to put it crudely, made to consume by being immersed in a wider consumer culture. Here, for example, the philosopher Herbert Marcuse wrote about how in capitalist societies there is a social production of 'false needs' (Marcuse 1964), a sense of wanting and requiring far more than is materially necessary (Plate 18.2). More specifically, numerous studies have debated the role of advertising in stimulating consumption, creating desires for products that would otherwise not exist (for a classic study in this vein see Packard 1977, where the role of adverts on our subconscious is emphasized; for more recent and subtle interventions on advertising and its role in mediating production and consumption see Nixon 1997 and Jackson and Taylor 1996). In this portrait, then, it is the producers who dominate consumer culture. Indeed consumer culture is a way of ensuring that mass production is accompanied by the appropriate levels of mass consumption. All in all, consumers are portrayed as rather sad and pacified figures:

> irrational slave[s] to trivial, materialistic desires who can be manipulated into childish mass conformity by calculating mass producers. This consumer is a cultural dupe or dope, the mug seduced by advertising, the fashion victim . . . yuppies who would sell their birthright for a mass of designer labels. Ostensibly exercising free choice, this consumer actually offends against all the aspirations of modern Western citizens to be free, rational, autonomous and self-defining.
>
> (Slater 1997: 33)

On the other hand, it is possible to see consumer culture as less passive, as less controlled, as less subordinate. Here the emphasis is not on consumers being made to consume, but on consumers *using* things for their own ends. In this perspective, consumption is less the death of the commodity, the end of an economic chain that began with the commodity's production, and more its resurrection. In this spirit, some stress the ways in which (many) consumers creatively rework the products being sold to them, giving them new meanings in the process, and using them as launch pads for their own symbolic and practical creativity (see, for example, Willis 1990). A classic example would be the Italian motor scooter (see Hebdige 1987). Having been originally aimed in the post-war years at a market of young, urban women in Italy, in the United Kingdom the Lambretta and Vespa scooters became a central identifier of the 'Mod' subculture. Initially attractive to Mods for its Italianicity and design aesthetics, by the late 1960s the scooter was increasingly symbolically recast and materially customized to act as a subcultural icon opposed to the British, brute masculinity of 'Rockers' and their motorbikes (Plate 18.3). More

Plate 18.3 Originally designed to be a stylish mode of travel for Italian women, the Italian motor scooter became a crucial component of the British Mod subculture. The meaning of things we consume are not simply controlled by those who design them, but are remade through their consumption.
(Robert Stiggins/Express/Getty Images)

generally, there have been a range of approaches that have emphasized how consumption involves the use of commodities within at least partially autonomous cultural fields: within existing status structures and patterns of class distinction, so that consumption is regulated by socially structured and inherited tastes (Bourdieu 1984); within the private dream worlds of consumers, so that consumption is about fulfilling our fantasies and desires (Campbell 1987); or within domestic relations of love and care, so that consumption is about concretizing our relationships to our nearest and dearest (cooking and sharing a meal, buying treats and gifts, helping someone choose their clothes) (Miller 1998).

In all these different approaches, then, to understand consumption one has to understand the social and cultural relations within which it is incorporated. One cannot simply read off our worlds of consumption from the corporate strategies of producers. Furthermore, what are identified here are the more positive social possibilities of consumption, the feelings of agency and self-autonomy it can engender. It could be argued, for example, that in comparison to many people's worlds of work, worlds of consumption can seem comparatively fertile spaces for creativity, self-fulfilment and, indeed, liberation (Miller 1995). It is through consumption that we can make the world meaningful to us and make our worlds.

Of particular interest to geographers is the geographical character of the worlds produced through consumption. In the remainder of this chapter, we examine this geographical character through three interrelated issues: the production of the global through consumption; the production of the local through consumption; and the geographical knowledges produced through consumption. First then, our focus will be on the character of the global geographies consumption produces. Some argue that consumption has increasingly corroded cultures and their meaningful geographies of place and locality, ripping apart the historical sedimentations that mark places and people as different from each other, and smoothing them over into more homogenized markets that better suit the supra-local operations of capital. So, are our worlds of consumption increasingly all the same? Does contemporary consumption promote a global space characterized by the end of geography? Reviewing examples from East Asia, Russia, Israel, the Congo and London, I will suggest not, arguing that consumption forges different, localized experiences of the global and incorporates and produces cultural and geographical difference rather than eliminating it.

Secondly, the discussion then shifts scales to think about the production of the local through consumption more directly. Not only do we make the world global in diverse forms through consumption, we also make local environments through consumption. Taking the examples of shopping centres, nightclubs and skateboarding in urban public space, we will be looking at the roles played by such places in the very process of consumption itself, asking just what it is that is going on in them, and what this tells us about the character of consumer cultures more generally. My argument will be that in part such local places represent strategies of commercial interests designed to promote and regulate consumption. However, they also involve people making and remaking places into forms and through practices that accord to a different logic than that of commerce and economic rationality.

Finally, I will turn directly to the connections between local and global geographies of consumption. Consumer worlds are always simultaneously local and global. For example, the food I eat, the television I watch, the music I listen to in my very local, domestic spaces are sourced from all over the world. The everyday acts of consumption I undertake in my kitchen and living room are dependent upon huge networks of provision. The last part of this chapter therefore reflects on these networks and what I and other consumers know about them. In part through a critical consideration of ideas of a (geo)ethical consumption, it considers how consumption enacts connections between human beings and places that are also, at the same, often disconnected from each other in terms of knowledgable contact. One of the characteristics of our world of consumption is its connecting of the local and the global, the immediately present to the absent, and this raises difficult geographical questions, both intellectually and personally. These, it is suggested, are appropriately important issues for consideration by human geography.

18.2 Global geographies of consumption

18.2.1 Consumption, cultural imperialism and the end of geography

Consumption is at the heart of some of the most significant intellectual debates of our age, being seen as central to nothing less than the cultural fate of the world. These debates centre on the extent to which modern consumption both produces and reflects a profound erosion of cultural traditions and differences, homogenizing the world into increasingly similar landscapes, people and social systems as it spreads from its heartlands in North America and Europe. Dick Peet, for example, argues that global capitalism and its consumer culture comprises 'a powerful culture which overwhelms local and regional experience . . . breaking down the old geography of society and culture' (Peet 1989: 156). Sack concurs that through consumption there is a 'trend toward a global economy and culture, which seems to require that places all over the world contain similar or functionally related activities and that geographical differences or variations that interfere with these interactions be reduced' (Sack 1992: 96).

More prosaically, it is hard to miss the frequent commentaries on the worldwide spread of products and brands such as McDonald's or Coca-Cola, to the extent that they have entered academic language – in the form of theses of the 'McDonaldization of society' (Ritzer 2000) or of 'Coca-colonization' – as bywords for a consumerist homogenization (Plate 18.4). Travellers' tales too are replete with mournful stories of how one can trek to the farthest ends of the earth, only to find people doing exactly what the traveller could have seen at home: in the title of Iyer's account from the 'not-so-far East', having 'video nights in Kathmandu' (1989). Some of these issues of cultural homogenization are discussed in Chapter 13. This section will consider them in more detail, focusing upon the issue of consumption.

For many, what modern consumption enacts is a form of 'cultural imperialism' (see Tomlinson 1991 for an excellent review). This thesis takes a number of forms. For some it is evidenced by the 'dumping' of First World, and especially American, consumer products on to Third World markets: 'authentic, traditional and local culture in many parts of the world is being battered out of existence by the indiscriminate dumping of large quantities of slick, commercial and media products, mainly from the United States' (Tunstall 1977: 57). What concerns critics here is both the dominance of First World and American multinational companies (Mattelart 1979) and the predominance of products that promote First World and American values (Dorfman and Mattelart 1975) (for a general review of these emphases see Fejes 1981). Also raised are concerns about the consequent loss of indigenous, 'authentic' local cultures, so that the 'real' culture of an area and a people are lost.

For others, the issue is not just the conquest of some national and regional cultures by others, but also the ways in which consumerism destroys all culture as we have traditionally understood it. The suggestion here, then, is that what consumption represents is the contemporary dominance of a very peculiar cultural form – consumer culture – which in many respects is barely cultural at all. Consumption, some suggest, is all about the political economy of market shares, niches and profit, and as such fundamentally opposed to the intrinsic values of cultural meaning and communication. For others, such as George Ritzer in his recent account of what he calls 'the globalization of nothing' (2004), consumption thereby tends to produce empty forms and spaces, devoid of ties to particular times and places (Spotlight box 18.1). What matters here is partly the loss of cultural diversity – the making of a duller

Plate 18.4 Coca-colonization?
(Stefan Boness/Panos Pictures)

Spotlight Box 18.1

Consumerism as the end of culture?

'Culture' was one of the crucial terms through which anti-liberal forces counted the cost of modernity. '*Consumer culture*', in this perspective, is merely an ersatz, artificial, mass-manufactured and pretty poor substitute for the world we have lost in post-traditional society. In fact, it is the antithesis and enemy of culture. In it individual choice and desire triumph over abiding social values and obligations; the whims of the present take precedence over the truth embodied in history, tradition and continuity; needs, values and goods are manufactured and calculated in relation to profit rather than arising organically from authentic individual or communal life. Above all, consumerism represents the triumph of economic value over all other kinds and sources of social worth. Everything can be bought and sold. Everything has its price. 'Consumer culture', therefore, is a contradiction in terms for much of modern Western Thought.

(Slater 1997: 63)

The social world, particularly in the realm of consumption, is increasingly characterized by nothing. In this case, 'nothing' refers to a social form that is generally centrally conceived, controlled and comparatively devoid of distinctive substantive content . . . we are witnessing the globalization of nothing.

(Ritzer 2004: 3, 2)

world – but perhaps more importantly the loss of cultural alternatives. Uniformity is doubly problematic as it entails conformity. This results in a form of culture that is inadequate and unfulfilling for many of those it is imposed upon:

> A cultural system which would be adequate for the poorest people in that system would mean a set of instrumental, symbolic and social relations that would help them to survive in meeting such fundamental needs as food, clothing, housing, medical treatment and education. Such needs are not met if they are identified with the consumption of Kentucky Fried Chicken, Coca-Cola, Aspro, or Peter-Stuyvesant cigarettes.
>
> (Hamelink 1983: 15)

What is questioned here, then, are the values that underpin consumer culture, and the ability of consumption to satisfy the 'real' as opposed to the 'false' needs of consumers. Clearly these are fundamental issues. However, the accounts related above are partial and rest upon some problematic assumptions: that the authenticity of a culture is dependent upon its resistance to external forces and artefacts; that non-Western consumers are duped or coerced into having consumption tastes that are inappropriate to their real needs, just as the mass of Western consumers have been too (on this basis it seems that only the critics of consumer culture know what it is any of us 'really' need); and that consumers passively consume Western goods, and make of them exactly the same the world over. We shall now turn to some of these criticisms of the cultural imperialism approach to consumption, through the lens of work that emphasizes processes not of 'homogenization' but of 'indigenization'.

18.2.2 Consumption and cultural indigenization

Let us start with someone who may seem an unlikely proponent of this idea of indigenization. Marshall McLuhan wrote one of the most famous social science books of the post-war period, as he sought to understand the character of a contemporary society so saturated by the mass media (McLuhan 1964). His argument about the increasing linkages between different places, and phrases expressing this such as the notion of the 'global village', have passed into common parlance, echoed in numerous advertising campaigns promoting the unifying qualities of new communication technologies. However, his argument always went beyond suggesting that this increasing interconnection between people and places – or in the much-cited language of the Marxist geographer Harvey (1989), this 'time–space compression' of the world – meant simple homogenization. He elaborates on this in a later paper:

> The question arises, then, whether the same figure, say Coca-Cola, can be considered as 'uniform' when it is set in interplay with totally divergent *grounds* from China to Peru? . . . The 'meaning' of anything . . . is the way in which it relates to the user . . . [For example] it might be suggested that the use of American stereotypes in the United Kingdom yields a widely different

> experience and 'meaning' from the use of the same stereotypes or *figures* in their American *ground*. The interplay between the *figure* and the *ground* will produce a wide range of different effects for the users.
>
> (McLuhan 1975: 44; emphases in original)

McLuhan argues, then, that consumer culture does not float above the world in some abstract space of capital or modernity. It lands and takes root in particular places and times. In so doing it not only impacts on its surroundings; it is also shaped by its locations.

Especially within anthropology a number of studies have developed this theme by exploring how globally distributed facets of consumer culture – whether particular goods, styles or cultural orientations – are locally incorporated. Take, for example, one of the exemplars of homogenizing American consumer culture: McDonald's. On an average day over 52 million customers are served at one of 31,000 McDonald's restaurants in more than 100 countries. Over one-third of these restaurants are in the United States, but based on recent figures the United Kingdom has over 1,000 outlets, Brazil over 500, China over 800, Thailand 90. Given their date, and notwithstanding more recent commercial problems, these figures are almost certainly substantial underestimates; a recurrent promotional corporate statistic is that a new McDonald's restaurant opens somewhere in the world every three hours. Not only this, of course, but McDonald's are famed for their uniformity; the same decor, the same basic menu (with very small variations, such as McSpaghetti in the Philippines or the Maharajah Mac in India), and the same service style the world over.

And yet, McDonald's may not be just the force for cultural homogenization that this suggests. This was certainly Watson's (1997a) conclusion, based on studies of the cultural geographies of McDonald's in East Asia (Case study 18.1). These studies found that McDonald's has been localized, indigenized and incorporated into traditional cultural forms and practices, even as it has played a part in the ongoing dynamism of those traditions. And exactly how this has happened varies across East Asia. In Beijing, McDonald's has lost its American role as a place of fast and cheap food. Instead, Yan suggests that it has become a middle-class consumption place, somewhere for a special family outing, somewhere 'customers linger . . . for hours, relaxing, chatting, reading, enjoying the music' (1997b: 72). McDonald's here is seen as American, but Americana means something stylish, exotic and foreign, and as such actually results in the meanings and experiences of McDonald's in Beijing being very un-American (Plate 18.5). In contrast, in Japan, whilst there is a similar leisurely use of McDonald's, it is not a place of exotic social prestige, but a youth hangout, a place where someone in a business suit would be out of place (Ohnuki-Tierney 1997). In Hong Kong, McDonald's was likewise marketed to the youth market, and adopted as such initially, but here its American origins are of little contemporary consequence (Watson 1997b). Whilst initially patronized in the 1970s by adolescents seeking to escape products associated both with

Regional Case Study 18.1

Cultural geographies of McDonald's in East Asia

In the introduction to a collection of findings on the cultural geographies of McDonald's in East Asia, Watson poses the following questions:

> Does the spread of fast food undermine the integrity of indigenous cuisines? Are food chains helping to create a homogeneous, global culture better suited to the needs of a capitalist world order? . . . But isn't another scenario possible? Have people in East Asia conspired to change McDonald's, modifying this seemingly monolithic institution to fit local conditions? . . . [Perhaps] the interaction process works both ways.
>
> (Watson 1997a: 5–6)

His conclusion is that there is indeed a two-way interaction:

> McDonald's has effected small but influential changes in East Asian dietary patterns. Until the introduction of McDonald's, for example, Japanese consumers rarely, if ever, ate with their hands . . . this is now an acceptable mode of dining . . . [However,] East Asian consumers have quietly, and in some cases stubbornly, transformed their neighborhood McDonald's into local institutions. In the United States fast food may indeed imply fast consumption, but this is certainly not the case . . . [in] Beijing, Seoul, and Taipei . . . [where] McDonald's restaurants are treated as leisure centers, where people can retreat from the stresses of urban life.
>
> (ibid.: 6–7)

Plate 18.5 McDonald's in Beijing: American fast food becomes exotic, stylish Americana.
(National Geographic/Getty Images)

China and with Hong Kong's perceived provincialism, by the 1990s McDonald's has become routinely local, just another mundane part of the Hong Kong landscape: '[t]oday, McDonald's restaurants in Hong Kong are packed – wall to wall – with people of all ages, few of whom are seeking an American cultural experience. The chain has become a local institution in the sense that it has blended into the urban landscape . . . McDonald's is not perceived as an exotic or alien institution' (Watson 1997b: 87, 107).

Hence the meanings and practices of consuming McDonald's – often cast as an icon of global homogenization – vary from place to place, from culture to culture (for a parallel account on an American global TV product, the series *Dallas*, see Case study 18.2). Melissa Caldwell's work (2004) on McDonald's and consumerism in Moscow develops the significance of these variations. In Russia too McDonald's has been indigenized or 'domesticated', and not just at the level of everyday practice. Counter-intuitively, McDonald's – which as outsiders we might read as an American export to Russia – has come to be understood by consumers in Moscow as an 'authentically Russian product', to the extent that eating there is seen as responding to 'nationalist-oriented consumer campaigns' (ibid.: 5). Whilst initially perceived and promoted as an exotic American import, more recently McDonald's has become Russian, explicitly marketed as 'Nash Makdonalds' (which Caldwell translates as 'our McDonald's'). Since the mid-1990s Russian consumer culture more generally has evidenced a patriotic enthusiasm for goods understood as 'Nash'. This Nash ideology entangles notions of national space with those of everyday familiarity and trust. McDonald's has thus been transformed from American exotica to both mundane familiarity and thence to imagined Russianness.

Parallel accounts of domestication have been produced through research on a range of other global products: for example, in the context of a study of modern Trinidad, Danny Miller writes about how both the American television soap opera *The Young and the Restless* and 'Coke' are authentically Trinidadian (Miller 1992, 1997). Particularly telling is Jonathan Friedman's study of the subcultural research of European designer fashions in the Congo (Case study 18.3). Here, not only does *haute couture* come to mean something different to what it means in Paris or Milan, it comes to mean it through relations of consumption, identity and personhood that are different. Rather than one consumer culture spreading globally, this suggests many consumer cultures, their forms and meanings being locally specific. At a broader level this marks a conceptual shift away from seeing the modern world as a singular universal entity, and towards recognizing the many 'specific', 'prismatic', 'comparative' or 'global' modernities being forged through consumption practices worldwide (see for example Breckenridge 1995 on South Asia; Featherstone et al. 1995 generally; and Miller 1994 on Trinidad).

Thus, rather than necessarily destroying local, indigenous cultures, the consumption of non-local products may be irrelevant, indeed even helpful to, the ongoing production of authentic local cultures. Miller argues that in a world of globalized consumption 'authenticity has increasingly to be judged a posteriori

Regional Case Study 18.2

Media products and cultural indigenization: the case of *Dallas*

For proponents of the cultural imperialism thesis, television programmes are primary instruments for the imposition of Western, American, consumerist values on their global audiences. *Dallas* was the most watched programme in the world in the early 1980s. Set, unsurprisingly, in Dallas, Texas, it was a weekly soap opera following the fortunes, relationships and fabulous squabbles of an extended oil baron family, the Ewings. Among many other places, *Dallas* was very popular in Israel where two media researchers, Liebes and Katz, were investigating what was made of it by a variety of Israeli audiences. What they found was that, far from the programme simply imposing a world-view on its audience, different groups of people understood the programme in very different ways and enjoyed it for very different reasons.

> Theorists of cultural imperialism assume that . . . [meaning] is prepackaged in Los Angeles, shipped out to the global village, and unwrapped in innocent minds. We wanted to see for ourselves . . . In the environs of Jerusalem, we watched the weekly episode in the homes of Arabs, veteran settlers from Morocco, recent arrivals from Russia and second-generation Israelis in kubbutzim – *Dallas* fans all . . . Each cultural group found its own way to 'negotiate' with the program – different types of reading . . . We found only very few innocent minds, and a variety of 'villages'.
>
> (Liebes and Katz 1993: xi)

The recent immigrants from the Soviet Union denied watching it (even when they did) or dismissed it as American, capitalist propaganda. They saw it in ideological terms. Middle-class kibbutzniks had a more psychological take on the series, concentrating on the personalities of the protagonists and enjoying the task of understanding their motivations. The interviewees from North African and Arabic backgrounds emphasized the convolutions of the plot, framing the programme as an epic family saga. Different people with different backgrounds, and different 'readings' of the same show.

not a priori, according to the local consequences not local origins' (Miller 1992: 181). It matters less where things come from than what they are used for. Indeed, it may be that the non-local origins of products may be positively beneficial, giving 'the periphery access to a wider cultural inventory' and providing new raw materials for local cultural and economic entrepreneurs (Hannerz 1992: 241; for specific cases, see Tranberg-Hansen 1994; Bloeman 2004 and Durham 2004 on the economy of *Salaula* or Western 'second-hand clothes' in Zambia, and Edensor and Kothari 2006 on the acquisition and sale of locally produced branded goods in a market in Mauritius). Moreover, when it does matter where things come from, when their origins are part of what they are valued for, then those origins are understood in locally specific, imaginative ways. America, for example, is mythologized the world over, but in a range of different forms (see, for example, Webster 1988).

18.2.3 Consuming difference

This last point – on the mythologies or 'imaginative geographies' (Driver 2005) attached to the things and places we consume – leads on to a further set of qualifications of the cultural imperialist thesis. These concern the assumption that the goal of those promoting consumer culture is simply to eradicate cultural and geographical differences, viewing them as unhelpful encumbrances to the smooth operation of the world of consumption. In fact, one can see cultural and geographical difference being actively embraced, even produced, within consumer culture. Global arrays of cultural and geographical difference are produced and sold within consumer culture.

Exemplary here, it is often suggested, are the themed environments of consumption spaces such as theme parks and shopping centres. In Disneyland and Disney World, for example, the amusements and other consumption opportunities are not just randomly arrayed, but organized into distinct lands (for example, Frontierland themed on the American West, or Adventureland themed on landscapes of European colonialism and imperialism) (see Gottdiener 1982; Wilson 1992; Bryman 1995). This echoes much earlier consumer festivals and fairs such as the Midway attractions at the 1893 Chicago exhibition (Rubin 1979) or Luna Park at Coney Island, New York (Kasson 1978), and is also a pattern to be found in many large shopping centres, most famously the huge West Edmonton Mall in Canada (Crawford 1992; see also Goss 1999a). Indeed,

Regional Case Study 18.3

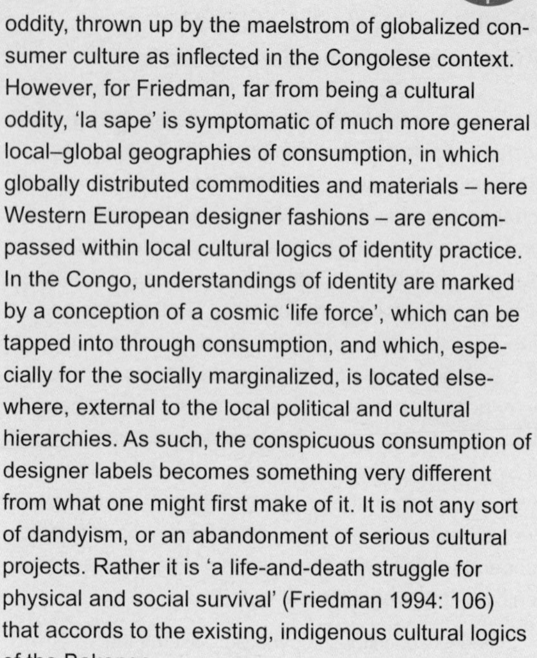

Western goods and cultural indigenization: the case of 'Les Sapeurs'

'Les Sapeurs' are a group of consumers in what was the People's Republic of the Congo (see Friedman 1994). They are men of the politically marginalized Bakongo ethnicity, drawn especially from the poor in large cities such as Brazzaville and Pointe Noire, who aim to move through a graded and spatialized consumption process of 'Western' fashions. Beginning with the conspicuous wearing of imported ready-to-wear items, Sapeurs look to progress to a stay in Paris ('*l'aventure*') accumulating *haute couture* from French and Italian design houses whilst usually living in extreme poverty, before returning to Brazzaville's 'Sape clubs' to perform competitively in '*la danse des griffes*', a catwalk display of one's dressed self that is augmented by the sewing of a host of designer labels to jacket lapels. All of which, especially at a first hearing, might be rather too easily dismissed as an extreme case of fashion victimhood, or as a bizarre oddity, thrown up by the maelstrom of globalized consumer culture as inflected in the Congolese context. However, for Friedman, far from being a cultural oddity, 'la sape' is symptomatic of much more general local–global geographies of consumption, in which globally distributed commodities and materials – here Western European designer fashions – are encompassed within local cultural logics of identity practice. In the Congo, understandings of identity are marked by a conception of a cosmic 'life force', which can be tapped into through consumption, and which, especially for the socially marginalized, is located elsewhere, external to the local political and cultural hierarchies. As such, the conspicuous consumption of designer labels becomes something very different from what one might first make of it. It is not any sort of dandyism, or an abandonment of serious cultural projects. Rather it is 'a life-and-death struggle for physical and social survival' (Friedman 1994: 106) that accords to the existing, indigenous cultural logics of the Bakongo.

Gottdiener (1997) goes so far as to posit that consumer culture has produced a general 'theming of America', so that theme parks become models for what is happening across the urban landscapes outside them. Sack (1992: 98) also points to the production of 'pseudo-places' in consumer culture. Harvey (1989) sees the same general trend towards highly packaged, and indeed simulated, experiences of cultural and geographical difference, such that microcosms of the whole world are encountered in particular consumption spaces, from the television screen, to the High Street, to the shopping centre and theme park.

Rather than eradicating cultural and geographical difference – and thus producing the end of geography – the world of consumption (re)produces geographies, framing certain local places of consumption as global centres. What then becomes the issue is not the end of geography, but what sort of new geographies are taking its place, and the kinds of understandings of people and place being circulated through consumer culture.

For instance, these kinds of questions have framed a number of pieces of research on the global arrays of food now being sold and consumed in the United Kingdom, as well as elsewhere (see, for example, Cook and Crang 1996; Cook et al. 1999, 2000). This explicit globalization of food provision is captured in an appreciative review of the diversity of restaurants in London, penned by the listings magazine, *Time Out*:

> The world on a plate. From Afghani ashak to Zimbabwean zaza, London offers an unrivalled selection of foreign flavours and cuisines. Give your tongue a holiday and treat yourself to the best meals in the world – all without setting foot outside our fair capital.
>
> (Author: Caroline Stacey, issue 1304 *Time Out*, 16–23 August 1995)

A central concern in research thinking about such developments has been just what sorts of geographical understandings and knowledges are being produced through this facet of consumer culture; what conceptualizations of people, place and culture are being mobilized. Take, for example, the following anecdote from the tabloid British newspaper, the *Daily Star*. Under the heading 'korma out for a meal', this item reported on the popularity of 'Indian' restaurants and take-aways amongst British consumers ('curry lovers have sparked an eating out boom in Britain'). At the same time, however, its notion of Britishness was pointedly narrow as, in the very next sentence, it reported that 'just 26 per cent [of those dining out] chose British restaurants' (*Daily Star*, 5 December 1995). Indian restaurants are a popular British pastime, but not British themselves, it seems. The logic here is all too clear. Indianness is something (non-Indian, white) British people may enjoy consuming, but that does not make it, or the British-Asian (in fact primarily British-Bangladeshi or British-Pakistani, not British-Indian) restaurateurs and waiting staff who embody it, culturally British. Difference is appreciated and enjoyed, at least when it comes to food, but it also distances those with whom it is associated. Here, an ethnicized geographical difference is incorporated into a very white British way of seeing the world, in which everything non-white becomes a resource to 'liven up the dull dish that is mainstream white culture' (hooks 1992: 21).

Of course, Britain's culinary multiculturalism, and its fashionings of 'Indian' food, are much more complicated than this anecdote allows: 'Indian' food is constructed and consumed in many forms in the United Kingdom – from the 'exotic', to the 'authentic', to the 'mundanely British' curry house, to the modern, contemporary 'Asian Kool' upmarket restaurant – each enacting different consumer tastes, market strategies and 'multicultural imaginaries' (Cook et al. 1999; see also Cook et al. 2008; Heldke 2003; Duruz 2005). More generally, the politics of cross-cultural consumption are not fixed and are open to intervention (generally see Jackson 1999; for examples about British-Asian fashion see Dwyer and Crang 2002; Dwyer and Jackson 2003 and Jackson et al. 2007). But what is common is how through consumption ideas of the global and the local, the foreign and the domestic, the there and the here, of difference, are not eradicated but actively produced (see also Lury 2000).

Plate 18.6 Eating into Britishness: by what process do there 'Indian' restaurants become 'British'?

(Photography courtesy of G.P. Dowling)

18.3 Local geographies of consumption

18.3.1 Shopping and place

I have been arguing, then, that consumption is not productive of global homogeneity but instead involves various interrelations between the global and the local (Crang and Jackson 2000). We now turn to the local geographies of consumption more directly through attention to the sites where consumption takes and makes place. We start in a shopping centre. Shopping centres (or malls in North America) are routinely cast as the cathedrals of our consumerist age, the symbols of all we value and hold dear. But what goes on inside a shopping centre, and what does this tell us about the kinds of places and 'public' space being produced in the world of consumption?

One response would be that in shopping centres we are manipulated into behaving in certain ways, and especially into buying things, by the power of the shopping centre as a place. Shopping centres are in part designed according to a 'merchandise plan' (Maitland 1985: 8) that tries to maximize the exposure of consumers to goods. In older shopping centres and North American malls this plan tended to be understood in quite mechanical terms. Shopping centres were seen as functional 'machines for shopping'. 'Generator' stores are used to pull us to the shopping centre in the first place, and once we are there 'magnet' stores every 200 metres or so make sure we do not just pop in and out, but explore every part of it. Here, then, 'the public mall [is defined] as essentially the passive outcome of a merchandising plan, a channel for the manipulation of pedestrian flows' (Maitland 1985: 10). There are well-known tricks to make us behave as the shopping centre management and its shops want us to: escalators arranged so we have to walk past shop fronts or merchandise when going between floors; hard seats so we do not linger too long without getting up and seeing some more potential purchases; no water fountains so we have to buy expensive drinks (see Goss 1993b).

In more recent shopping centres, however, the manipulation may be more subtle. Increasingly, retail planners and developers see their job as providing spectacular places that people will want to spend time in. This is because research suggests that the longer people stay, the more on average they spend, and because as competition between centres hots up, each one needs more than just the usual shops to attract us in the first

Regional Case Study 18.4

The shopping mall experience: Woodfield Mall, Schaumburg, Ilinois

Leaving the comfort of her air-conditioned car, the shopper hurries across the sun-broiled or rain-swept tarmac and thankfully passes through doors into an air-conditioned side mall. Drawn down this relatively dark, single-level tunnel, lined with secondary units, by the pool of light ahead, she reaches one of the squares ahead located at the end of each arm of the main mall system. She discovers this to be a two-storey space . . . On one side of

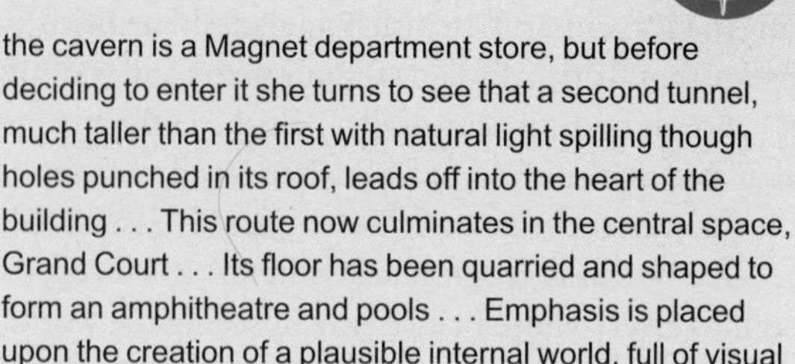

the cavern is a Magnet department store, but before deciding to enter it she turns to see that a second tunnel, much taller than the first with natural light spilling though holes punched in its roof, leads off into the heart of the building . . . This route now culminates in the central space, Grand Court . . . Its floor has been quarried and shaped to form an amphitheatre and pools . . . Emphasis is placed upon the creation of a plausible internal world, full of visual interest and variety, in which one is tempted not simply to pass through, but to stay.

(Maitland 1985: 26, 28)

place – everything from fountains to funfairs (see, for example, Goss 1999b). Here, then, the manipulation is less physical and more emotional. The hope is to excite, inspire, relax and please (Case study 18.4). Of course, these are promotional intentions rather than shoppers' experiences. As Chaney (1990) puts it in his review of the MetroCentre in Gateshead, north-east England, the designer's utopian impulses to produce a place of perfection often result in rather underwhelming, 'subtopian' forms. Nonetheless, there is much here to be taken seriously. There is the attempt to produce a fabricated space in which the individual consumer can be made to feel like consuming (see Goss 1999a). There is the emphasis on creating an internal, closed-off, privately owned but partially public environment, divorced from the harsh exterior world, not only climatically but also socially, through the operation of security systems that ensure the absence of anyone who might threaten this consumer paradise or disrupt the pleasure of the shopper. Thus, these privately owned and managed spaces rework the older public spaces of the street or market:

> They reclaim, for the middle-class imagination, 'The Street' – an idealized social space free, by virtue of private property, planning and strict control, from the inconvenience of the weather and the danger and pollution of the automobile, but most important from the terror of crime associated with today's urban environment.
>
> (Goss 1993b: 24)

Such shopping centres recognize that there is much more to going shopping than buying things (Glennie and Thrift 1992). Shopping is in part about experiencing an urban space, seeing and being seen by other shoppers. It is a social activity. But they seek to manage this social experience. The street gets recast as a purified space of leisurely consumption, cleansed of nuisances that inhabit the 'real' streets, like 'street people'.

However, this fashioning of the mall can be disturbed. Retail spaces can be reclaimed by those who use them. Leading the charge are a mixed assortment of senior citizens and adolescents. Take the case of Bert, 78 years old, and a regular visitor to a shopping mall in New England (USA):

> I come here at quarter to eleven and I leave at twenty minutes past one . . . I do this five days a week, except I missed one day this year . . . And I have my lunch here at noontime . . . I move every half-hour. I go from here down to where the clock is, I go down in front of Woolworth's, then I come back again and go up there and take my bus by the front of JC Penney.
>
> (Quoted in Lewis 1990: 126)

For Bert, the mall is not a place to shop, but somewhere sheltered to go outside his house, a chance to meet up with friends. For the mall management there is the option of evicting these low-spending visitors, but ejecting old people forcibly on to the pavement is hardly good public relations. Instead, and as part of wider moves to see malls as the new civic centres, Bert's mall has actually established its own walking club for the elderly, with over 300 members, for whom the doors open at 6 a.m., rather than 9 a.m. when shops open.

Adolescents also use the mall as a meeting place, somewhere to hang out. They also spend very little. 'Suburban kids come to malls to look around, meet and make friends, and hang out – because there is

Plate 18.7 The new Bullring shopping centre, Birmingham, United Kingdom was opened on 4 September 2003. The round tower (known as the Rotunda) that was part of the 'old' Bullring has been retained and redeveloped.
(Images of Birmingham/alamy)

nowhere else to go' (Lewis 1990: 130). Or as Ed and Tammy, self-confessed 'mall rat' and 'mall bunny' respectively, put it:

> '[Gesturing around himself] I met all these people here. I've met lots of other people, too. One place where you can always find someone . . .'
>
> (Ed; quoted in Lewis 1990: 130)

> 'I used to come here every Saturday from 11 am to 9.30 pm and just walk around with my friends, like Gina here, and just walk around and check out the guys.'
>
> (Tammy; quoted in Lewis 1990: 130)

Unlike the elderly, the mall is less indulgent of these teenagers. There are constant skirmishes with the security staff over how long ten people can share one Coke at the food court, and some mall rats and bunnies do get banned, but they resort to getting new hair cuts or clothes in attempts to get back in.

So what are we to make of our trip to the mall? What does it tell us about the local geographies of consumption? Well, it certainly tells us that place matters to consumption. But how? Here the answer is more ambivalent. In part what we see is a pseudo-public space symbolizing and enacting claims for consumption to be all there is to civic life, making citizens consumers in a very concrete way. Yet one also sees the small-scale, trivial resistances that people enact, their refusal to be just consumers in the sense of purchasers, to use this space for their own ends. And that too has a wider symbolism, signalling the possibility that consumption is never just about satisfying those selling to us, but also about fulfilling our own needs for social affiliation and participation. Two more examples extend this argument.

18.3.2 The geographies of skateboarding and dancing

As geographers have recognized for some time now, shopping centres are not the only sort of consumption space to explore (Crewe and Gregson 1998). Indeed the kinds of semi-illicit place consumption undertaken by the elderly and the adolescent in malls can be extended to a wider account of the consumption of urban space. Writing from the discipline of architecture, Iain Borden (2001) pursues this possibility in a study of skateboarding, space and the city. Paying particular attention to 'street skating', he argues that skateboarders remake urban spaces designed for other purposes: sometimes explicitly symbolic monumental space (e.g. town halls, national theatres, historical monuments, tourist attractions); but more often everyday spaces of neglect, the left-overs of rational, economically focused planning (e.g. mini-roundabouts, the spaces under bridges and urban highways, mini-malls). Thus spaces of public, official symbolism and of arid architectural functionality become the stages for energetic practices of place consumption. Handrails become tools for 'ollie nose grinds'; cement banks opportunities for 'shredding'. Skaters consume the city with their boards, recomposing it in the process. Skateboarding enacts a way of using the city, of consuming the city, that precisely emphasizes use over commercial or other values (see Case study 18.5).

Skateboarding is the consumption of urban space through skilled performances of embodied competencies. A parallel case, though often taking place in more clearly demarcated commercial arenas, would be practices of conviviality and togetherness. Clubs (in the sense of nightclubs) are exemplary. In London alone there are over 500 club nights on offer every week. In Britain the industry has a turnover of £1.8 billion per annum, and it is estimated that 42 per cent of the population now visit a club once a year, 42 per cent of 15–19-year-olds at least once a month (Mintel 2004, 2006).

Ben Malbon (1997, 1999) argues that clubs are spaces of play. This should really not need saying. It is patently obvious that people go to clubs to have fun. It is also patently obvious that this fun is achieved through a variety of means. Through enjoyment of the music; through an enjoyment of being part of a crowd of people with whom, temporarily, one can feel at one; through the feelings of competency and bodily expressiveness that dancing can give you; through the sharing of a special place and time with friends; through behaving in ways one cannot in other social situations, for example at work. Perversely, this fun needs to be taken seriously. We need to understand the playfulness of many consumption practices. We need to analyse the complicated micro-spatial practices through which play is enacted; the ways in which one gets to feel, and enjoy being, part of the clubbing crowd; the ways in which one dances so as to lose oneself in the music. We need to grasp the importance of place to playful consumption; the ways in which one has to be in the right place, with the right music and the right crowd, to have these feelings of exhilaration, communality and vitality; the way that 'the emphasis is placed on the near and the affectual: that which unites one to a place, a place that is experienced among others' (Maffesoli 1996: 128). And we need to appreciate the wider significance of such playful consumption for our productions of local experience, such that consumer culture is not just made up of atomistic, avaristic individual consumers, but also facilitates temporary, some would say

Thematic Case Study 18.5

Skateboarding and the consumption of urban space

Skateboarding responds meaningfully to the city . . . saying and living the city on its own terms.

(Borden 2001: 208)

A curb is an obstacle until you grind across it. A wall is but ledge until you drop off it. A cement bank is a useless slab of concrete until you shred it.

('In the streets today', *Thrasher*, vol. 1, no. 1, January 1981, p.16; quoted by Borden 2001: 192)

Skaters create their own fun on the periphery of mass culture. Sewers, streets, malls, curbs and a million other concrete constructions have been put to new uses.

('Lowboy, Skate and Destroy', *Thrasher*, vol. 2, no. 11, December 1982, p. 25; quoted by Borden 2001: 191)

Skateboarding shows that pre-existing uses of space are not the only possible ones, that architecture can instead be productive of things, and consumed by activities, which are not explicitly commodified.

(Borden 2001: 247)

'neo-tribal' gatherings, through which people come to feel part of something bigger than themselves (see Maffesoli 1996: 72–103).

Clubbing is, then, a place-specific practice of consumption. Whereas in the shopping centre, senses of communal belonging tend to be forged by those occupying the space for purposes beyond those for which it was commercially designed – such as the elderly and the adolescent – in clubs that communal belonging is the product that people are paying to consume. In many ways, clubs are therefore rather unusual places of consumption, characterized by an exceptional emotional and bodily intensity. However, they also point to wider qualities of our local places of consumption: how we use them for experiences of 'public' life and senses of social belonging; and how those experiences of public life and social belonging are produced through practices and places of consumption.

Skateboarding and clubbing represent, then, how consumption is about far more than the purchase and end or 'death' of a manufactured commodity. Consumption is about the (re)making of the spaces of everyday life. Commodities are the props and tools that we use to manufacture these local geographies.

18.4 Making connections

So far this chapter has explored some of the debates over the impacts of modern consumption on both global geographies (of cultural difference) and local geographies (of public space and life). Especially with regard to the latter, it has been deliberately optimistic. By way of conclusion, though, I want to think about an emergent concern with reconnecting the local contexts of everyday consumption with the global networks of commodity flows and production that make them possible. This reconnection has emerged as a fundamental intellectual, political and personal challenge in the field of consumption studies (for a wonderful, short evocation of some of the issues involved see Cook et al. 2000). Miller, for example, emphasizes how the 'study of consumption should . . . be increasingly articulated with, and not become an opposition to, the study of the mechanisms by which goods are produced and distributed' (1995: 17). Sack arrives at a similar conclusion on the basis of his determination to develop a moral framework for analysing consumption. Consumers are, he says, profoundly ignorant of the geographies of provision that bring goods to them: 'A shop that sells Colombian coffee does not reveal the social structure that produces the coffee, the economic impact of coffee production on the Colombian economy, or the way coffee growing affects the Colombian environment' (Sack 1992: 200). Such ignorance, he argues, makes it impossible for consumers to behave with a sense of responsibility to those they are in fact connected to through the webs of commodity flows. In this, modern consumption is deeply amoral. Crucially, then, to think about reconnecting practices and places of consumption and production demands that we somehow articulate play and morality, pleasure and exploitation, creativity and responsibility.

But what sorts of articulations can one make between worlds of consumption and production? One perspective is to see consumers as blinded by a veil that is draped over the things we consume, through a process of what Marx called 'commodity fetishism'. In this portrait the true history or biography of a product, and its basis in the work of others, is obscured to consumers. Instead, one is confronted by false advertising imagery that seeks to associate the product with other fantasy worlds. The task then becomes to puncture these fantasies with some sharp shafts of reality. Hartwick points to the startling incongruities and inequalities that such a puncturing can bring to light:

> Michael Jordan, basketball player extraordinaire, receives US$20 million a year to endorse Nike; Philip Knight, cofounder and chief executive officer of Nike, is worth US$5.4 billion; Indonesian workers are paid US$2.40 a day and Vietnamese workers are paid US$10.00 a week to make the sneakers . . . The Walt Disney Company is publicly criticized for its treatment of Haitian workers making Pocahontas shirts for 28 cents an hour; Disney's chief executive officer makes US$78,000 a day.
>
> (Hartwick 1998: 423)

Smith's account of Starbucks Coffee shops makes similar points. He highlights how unfairly traded coffee sees only 25 per cent of the retail price returning to the producing country, with even less going to those who actually cultivate the crop (the smallholders, waged workers and seasonal pickers). He contrasts the stylish coffee-shop culture Starbucks epitomizes with the ways in which that coffee is sourced. For the poorly paid women who pick the coffee beans, their working day begins somewhere between three and four in the morning. Getting back home at six in the evening, they still have evening meals to prepare, water to fetch from the river, a family to feed and clear up after. Few live in

Plate 18.8 'Which one is it you're [looking for]?' 'Ethiopian': from a scene in the documentary film *Black Gold* in which Tadesse Meskela, General Manager of the Oromia Coffee Farmers Co-operative Union in Ethiopia asks a UK supermarket employee where 'his' coffee is on the shelf [see www.blackgoldmovie.com/video/hires_trailer.mov].

(www.blackgoldmovie.com/dvd)

houses with electricity or running water. Along with malnutrition, a major health hazard are the pesticides used on the coffee plants (M.D. Smith 1996: 512–13).

So one way to reconnect worlds of consumption and production is to acquaint consumers with the harsh realities of commodity production. It is hard to criticize such an approach. It is a very simple message, but it is always worth getting it across. However, it does have its problems. Rather than articulating worlds of consumption and production together, it can tend to dismiss worlds of consumption, and their positive possibilities, altogether. Consumption just becomes inherently bad. That is neither true, as hopefully the discussions earlier in this chapter have demonstrated, nor terribly attractive in terms of what it suggests one should do. Particularly if one's own worlds of production, as a worker, are thoroughly unrewarding – admittedly not usually the case for academic critics of consumption, but true for many others – then giving up all the pleasures of consumption is pretty hard to stomach. Certainly the last two decades in Western politics suggest that failing to recognize the genuine senses of autonomy and creativity that consumption can facilitate is a sure way to alienate those very same workers in whose name one wishes to re-centre the world of production (Miller 2001).

There are, though, more subtle variations on this theme. So-called 'ethical consumerism', in particular, is a fascinating phenomenon (see also Freidberg 2003). Here, consumption is not disavowed, but it is channelled through concerns with the wider provisioning systems, and their human and non-human participants, of which the consumer is a part. Green consumerism, fair trade, vegetarianism and other animal welfare concerns are the main strands. Empirical research on ethical consumers in London challenges any sense of ethical consumption as joyless moral prescription (Bedford 1999), although it does contain some elements of self-denial, and most definitely of hard work. Instead it emphasizes the positive pleasures gained through such consumer movements: the feelings of achievement, the senses of empowerment. However, in its full-blown forms, ethical consumerism is not only hard work and expensive, but requires the adoption of consumption lifestyles that are very difficult to maintain unless one is immersed in wider friendship and family networks of like-minded people. Consumption is, after all, thoroughly social. Moreover, the levels of knowledge required to understand the incredibly complex biographies of products are more amenable to academic study than they are to someone who has something else to do. Even the most committed ethical consumers therefore rely on shorthand rationales, judgements and knowledges such as 'Nestlé is bad, Brooke Bond is good'; 'meat is murder, milk is allowed'; and so on.

What this suggests is that rather than simply replacing a superficial and false advertising fetish with the real history of commodities, it is more realistic to see ethical forms of consumption as reworking the fetish or advertising form, and reworking the pleasures of consumption rather than disavowing them in a blanket condemnation of consumerism (Cook and Crang 1996; Cook et al. 2004; Goodman 2004). Thus, for example, if we consider the huge growth in the market for organic produce in Britain and other Western countries at present what we see are products that are branded as in some way ethical, in the United Kingdom with a Soil Association approved designation and logo. What is more, one of the attractions of organic food is that it can easily be seen as benefiting not only the environment but also those to whom it is being fed (in terms of 'healthiness'). The 'organic' in organic food becomes understood as more generally 'good', drawing on vague, positive associations of the 'natural': good for the environment, good for my family, good for me, good to eat. Of course, not all ethics have such an easy semantic slippage. In particular, the Fair Trade logo has struggled to make the same market impact. Connecting care for one's nearest and dearest – a very strong existing consumer ethic – to caring for the distant strangers involved in commodity production and distribution has proved trickier (see Barnett et al. 2005; Clarke et al. 2007).

More broadly, what such an analysis suggests is that consumption is not only a way in which we make geographical entities such as the global and local: it is also an arena for the production, circulation and reception of geographical knowledges. It is tempting, especially perhaps for geography academics and students, to respond by looking to export the factual knowledge of our subject to this everyday arena, to correct the ignorances that we all operate under as consumers. But *that*, I have been suggesting, is both too easy – offering a very simplistic account of both geographical knowledge and the politics of consumption – and impossibly difficult, given the complex histories and geographies of the multitude of things we consume (Cook et al. 2007). The geographies of consumption offer no such easy resolution. We consumers are both global dictators and local freedom fighters. We all have to continue to think about and develop practically the connections between these two roles.

Learning outcomes

After reading this chapter, you should have:

- An understanding of the importance of consumption to modern economies and cultures.
- An understanding of the global geographical impacts of modern consumption.
- A sensitivity to the continuing importance of cultural and geographical difference within contemporary consumer cultures.
- An understanding of the character of public space fashioned within the worlds of consumption.
- An understanding of connections that exist between worlds of consumption and production.
- An ability to think about the geographies of your own everyday consumption practices.

Further reading

Bell, D. and Valentine, G. (1997) ***Consuming Geographies: We Are Where We Eat***, Routledge, London. A lively exploration of the geographies at stake in food consumption, at every scale from the body to the home, the city, the nation and the world.

Cook, I. and Crang, P. (1996) The world on a plate: culinary culture, displacement and geographical knowledges, ***Journal of Material Culture***, **1**(2), 131–53. Modesty prevents me from saying how wonderful this article is, but I can say that it offers an account of the global array of foods found in the United Kingdom that emphasizes the global, local and knowledgeable geographies of consumption.

Cook I. (2004) Follow the thing: papaya. ***Antipode***, **36**(4), 642–64. An engaging but jarringly written description of lives connected through the travels of this tropic fruit from its Jamaican 'production' to its UK 'consumption', this shows the complexity of 'defetishisation' research, and is written to provoke discussion.

Goss, J. (1993) The magic of the mall: form and function in the retail built environment, ***Annals of the Association of American Geographers***, **83**, 18–47. A brilliant, if somewhat sceptical, analysis of shopping malls and the consumer worlds they embody.

Hartwick, E. (1998) Geographies of consumption: a commodity-chain approach, ***Environment and Planning D: Society and Space***, **16**, 423–37. A pithy comparison of the advertising imagery used to sell gold and the ways in which that gold is produced. Argues for a knowledgable and ethical connection of consumers and producers. Simplistic but powerful.

Malpass, A., Cloke, P., Barnett, C., and Clarke, N. (2007) Fairtrade urbanism: the politics of place beyond place in the Bristol Fairtrade City campaign, ***International Journal of Urban and Regional Research***, **31**(3):231–49. An important paper challenging the assumption that 'ethical consumption' is just a matter of personal choice. Through influencing public procurement policies in favour of fair-trade principles, the Bristol campaign ensured the consumption of fair trade products in restaurants, cafes and canteens across the city.

Miller, D. (1992) The young and the restless in Trinidad: a case study of the local and the global in media consumption, in Silverstone, R. and Hirsch, E. (eds) ***Consuming Technologies: Media and Information in Domestic Spaces,*** Routledge, London, 163–82. A typically insightful and exuberant analysis from perhaps the best contemporary writer on consumption. Here he analyses the consumption of an American soap opera in Trinidad. His conclusion is that paradoxically this soap, despite its American origins, is authentically Trinidadian.

For annotated, clickable weblinks and useful tutorials full of practical advice on how to improve your study skills, visit this book's website at **www.pearsoned.co.uk/daniels**

POLITICAL GEOGRAPHIES: TERRITORIALITY, STATES AND GEOPOLITICS

Section 5

Edited by James Sidaway

A few years ago, in April 2005, Russian President Vladimir Putin declared that the collapse of the Soviet Union (which took place at the start of the 1990s) had been the greatest geopolitical catastrophe of the twentieth century. Putin made reference to the millions of Russians who found themselves living outside Russian territory in one of the fourteen other states that emerged after the collapse of the vast Soviet Union. But what did Putin mean by 'geopolitical'? The final chapter in this section will explore how geopolitics has been defined in different times and places: including ongoing debates in Russia, as well as through the prior Cold War confrontation between the Soviet Union and the USA, all building on the moment when geopolitics was first established as a way of understanding (and indeed seeking to influence) world politics, in early twentieth-century Europe.

As Chapter 21 details, one significant codifier of geopolitics was the Chilean General Augusto Pinochet, one-time President of Chile (1973–90). Pinochet had directed the Chilean armed forces in their violent overthrow of the democratically elected Chilean government on 11 September 1973, and he continued to preside over Chile for more than 15 years afterwards, before finally retiring in 1990, thus enabling his country to begin to return to democratic rule. Eight years into his retirement, Pinochet paid a visit to London for private meetings and medical treatment. Whilst in hospital he was arrested by British police. The arrest took place on behalf of the Spanish courts who had charged Pinochet with crimes against Spanish citizens in Chile who were murdered by the brutal military regime that Pinochet once headed. His detention in London reminded the world of Pinochet's politics and the crimes committed by his right-wing regime. It also raised complex questions about sovereignty. Could a former head of state be prosecuted in one country for alleged crimes committed in another? Did this mean that Spain and the United Kingdom had the right to violate Chilean 'sovereignty'? Or is 'sovereignty' and the right of the 'nation-state' an excuse for the otherwise inexcusable? Which law should apply? Spanish, Chilean, English, European, International? What about crimes committed abroad or in their own countries by the

Spanish and British armed forces in the past? What geographical *scale* of justice should apply? For his supporters in Chile, Pinochet had 'saved' the country from Soviet/Cuban-led communism. He had after all (with decisive support from the USA) overthrown a leftist government, and many of those imprisoned and murdered by his regime were seen as 'communists'. Many other Chileans celebrated Pinochet's arrest. Those who felt themselves or their families and friends to have been victims of Pinochet's regime saw the possibility of some justice. The former British Prime Minister Margaret Thatcher reminded whoever cared to listen to her that her friend General Pinochet had assisted Britain during its 1982 war with Argentina to recapture the Falkland Islands (another struggle over sovereignty and territory). In the end, Pinochet was released and returned to retirement in Chile. He died peacefully from heart failure in 2006, having faced further charges (but never a trial) of responsibility for torture and murder during his years in power. The Chilean story and career of General Pinochet always had *global* dimensions. These incorporate a variety of domains, scales and modes of political analysis, imagination and action, a variety of 'political geographies'.

This section will enrich understandings of what can be meant by geopolitics and political geographies and their myriad intersections. By the time they finish the section, readers will have an understanding both of some of the ideas that Pinochet drew upon to justify and codify his rule, and of powers (and sometimes the fragility) of states, nations and sovereignty. We aim to whet appetites for the complex topics of political geographies, to encourage critical thinking about often controversial, contested and complex issues and topics.

Those who have reached this section after reading the rest of the chapters (congratulations!) should already be aware that many elements of geography touch on political issues. Such 'politics' may not be confined to those things that are usually designated by the term: governments, elections (or the lack of them), political parties, and so on. Rather, readers will have noticed that debates about culture, economic change, history, changing gender relations and many other things described in this text are 'political'. They are political in the sense of being about power, albeit at all kinds of levels and in lots of different ways. So, in an important way, *all* geographies are political geographies. Though focused on territory, nation-state and geopolitics, the chapters that follow aim to bring that home. Beginning at the household, local and urban scales, Chapter 19 examines 'territoriality'; the ways that human individuals and human social groups claim or are assigned to particular areas: in short, the human territorial strategies which regulate, demarcate and divide social and political spaces, and their uses. Whilst this is perhaps most clearly expressed in the form of the political geography of nations and states and their boundaries (the focus of Chapter 20), Chapter 19 stresses that territoriality operates in myriad ways and at local scales: for example, in the manner in which particular areas within many cities become associated with different ethnicities (as in the history of ghettos and 'quarters') or, more widely, in the ways that some spaces are seen as public and others private. Readers will therefore (hopefully) find many connections with earlier sections of the book here. Chapter 20 introduces the historical and geographical variability of nations and states. It stresses that neither the nation nor the state is to be taken at face value. Or in other words, it examines how nation-states are complex symbolic systems that crucially depend upon particular visions and associations of territory, place and space. In the same vein, Chapter 21 will extend understandings of geopolitics in an interlinked, but conflict-ridden world. The order of the chapters follows a broad scalar logic (starting from small-scale and local through to planetary) – from household and local (19), through 'national' (20) to 'global' (21). In earlier editions of this text, we organized this section differently and there also was an additional chapter on citizenship (which is still worth well reading as a follow-up to issues that follow: track down an earlier edition of this text and read Low 2005). We have had to leave that chapter out of this new edition (along with some other changes to stop the book becoming too big!), so it should be clear that the chapter order here is as much for convenience in organizing the ideas in this section (and helping readers to work through them) rather than reflecting neat divisions in the world. For, in practice, as will become clearer when these chapters are compared with each other and others earlier in the book, scales are inter-tangled in complex ways. Indeed, as you deepen your reading in human geography, it is likely that you will encounter many more arguments about the roles and complexity of scale, for these are themes that cut across political, economic, social and cultural (and indeed physical) geography and are the subject of much wider debate and reflection in the discipline (if you want to sample these debates now, try: Marston 2000; Mansfield 2005 or – more closely linked to many of the arguments in the following chapters – Jones and Fowler 2007). The last one of these (Jones and Fowler 2007) contains an extensive discussion of the literatures on nations and nationalism from a geographical perspective (using Welsh nationalism as a case study) and is therefore especially rewarding to those who wish to read more about some of the key themes in this section. But, as the section also sets out to show, there are many other political geographies that merit exploration.

TERRITORY, SPACE AND SOCIETY

Chapter 19

David Storey

Topics covered

- Territorial strategies and the concept of territoriality
- Social processes and spatial relations
- Geographies of class, race, gender, sexuality
- Private property and personal space
- Contested spaces and geographies of resistance

Human spatial relations are the results of influence and power. Territoriality is the primary spatial form power takes.
(Sack 1986: 26)

19.1 Territoriality

We come up against territorial behaviour on a daily basis when we are confronted with signs saying 'authorized personnel only', 'keep out', 'no tresspassing', 'strictly no admittance' and so on. These and other manifestations lead people to sometimes assume that humans have a natural tendency to behave in a territorial manner: to claim space and to prevent others from encroaching on 'our' territory. However, in his landmark book *Human Territoriality: Its Theory and History*, published in 1986, Robert Sack rejects determinist views of human territoriality as a basic instinct and argues instead that it is a geographic and political strategy. For Sack, territoriality is 'the attempt by an individual or group to affect, influence, or control people, phenomena, and relationships, by delimiting and asserting control over a geographic area' (Sack 1986: 19). He draws attention to the means through which territorial strategies may be used to achieve particular ends. In essence the control of geographic space can be used to assert or to maintain power, or, importantly, to resist the power of a dominant group.

> Territoriality, as a component of power, is not only a means of creating and maintaining order, but is a device to create and maintain much of the geographic context through which we experience the world and give it meaning.
>
> (Sack 1986: 219)

In this way territoriality, which is deeply embedded in social relations, can be viewed as a process linking space and society. Rather than being natural entities, territories result from social practices and processes (Delaney 2005). They are produced under particular conditions and serve specific ends.

Sack (1986) argues that territoriality involves a classification by area whereby geographic space is apportioned. However, territories are more than mere spatial containers, they also communicate important ideas relating to authority, power and rights (Sassen 2006). A territory is 'a bounded social space that inscribes a certain social meaning onto defined segments of the material world. A simple territory marks a differentiation between an "inside" and an "outside"' (Delaney 2005: 14). These geographic spaces convey messages of political power and control which are communicated through various means, most notably through the creation and maintenance of *boundaries*. These and other mechanisms facilitate control over space and of those within it; they serve as divisions between those inside and those outside. Territoriality also tends to reify power so that it appears to reside in the territory itself rather than in those who control it. Attention is thereby deflected away from the power relationships, ideologies and processes underpinning the maintenance of territories and their boundaries. In this way, as Delaney argues, 'territory does much of our thinking for us and closes off or obscures questions of power and meaning, ideology and legitimacy, authority and obligation' (2005: 18). Ultimately territoriality can be seen as 'a primary geographic expression of social power' (Sack 1986: 5). Once created, territories can become the spatial containers in which people are socialized through various social practices and discourses. As Paasi (2003) suggests, a number of important dimensions of social life and social power are brought together in territory. There is a material component such as land, there is a functional element associated with control or attempts to control space and there is also a symbolic component associated with people's social identity. People identify with territories in such a way that they can be seen 'to satisfy both the material requirements of life and the emotional requirements of belonging' (Penrose 2002: 282).

> Territoriality . . . forms the backcloth to human spatial relations and conceptions of space . . . People do not just interact in space and move through space like billiard balls. Rather human interaction, movement, and contact are also matters of transmitting energy and information in order to affect, influence, and control the ideas and actions of others and their access to resources. Human spatial relations are the results of influence and power. Territoriality is the primary spatial form power takes.
>
> (Sack 1986: 26)

The creation of territories and the utilization of territorial strategies can be observed at a variety of spatial scales. It is also obvious that human territories and their boundaries may be contested, modified and destroyed (Paasi 2003). While some of the most obvious (and contested) expressions of territoriality are manifested at the level of the state (see Chapter 20), many more micro-level examples of territorial control and territorial strategies may be observed. These may be less obvious and may often seem more vaguely defined with less clear-cut boundaries but these 'informal' territories can convey quite clear meanings to those concerned (Delaney 2005). This chapter explores some of these examples, demonstrating how particular social practices are mapped on to space thereby marking and configuring territories.

The examples are arranged under five main headings: class-based divisions, racialized spaces, gendered spaces, sexuality and space, and private property and personal space. While the various topics discussed in this chapter are examined in relatively discrete sections, it should be abundantly clear that many of the issues raised are interrelated (and also overlap with earlier chapters in the book and others in this section). People have more than one single identity: gender, sexuality and ethnicity cross-cut each other in a system of overlapping identities. Ethnic groups and classes are not immutable but are social constructs. In other words, they vary in time and space and are not simply natural categories, but human (that is social) products. As such we need to be aware of the dangers of seeing them as rigidly defined.

Running through these various examples are issues related to power and exclusion. They raise questions about identity, rights and belonging thus touching on ideas of citizenship and the rights of individuals and groups (see Storey 2003). Traditional notions of citizenship are centred on the relationship between the individual and the state with a focus on rights and duties. More recently, ideas of citizenship have broadened to consider aspects of citizenship below the level of the state (local issues) and beyond the state (responsibility to distant others). At the same time issues of group (as distinct from individual) rights have been a focus of attention. While people may appear to enjoy full citizenship rights, as we shall see, particular territorializations and the use of territorial strategies can result in some groups being excluded from participating fully within society. In effect some people may be made to feel like second-class citizens as a consequence of the creation and maintenance of informal territorial boundaries. This is one reason why various social movements and groupings have arisen to assert group identities and rights (Low 2005). Thus ideas of citizenship, and with them, ideas of identity and belonging, are deeply contested. But what is also contested here is geographic space and who belongs where. Who is 'allowed' to be in particular spaces and who is barred or discouraged from being there?

19.2 Class-based divisions

As other chapters have pointed out, geographers and others involved in the study of urban areas have long drawn attention to spatial divisions in cities, linked to residential patterns, economic activities and land use (see in particular Chapter 10). Most cities have distinct residential neighbourhoods, colloquially defined as 'rich' or 'poor', 'working class' or 'middle class'. Socio-economic differentiation may well be the most important cleavage within the urban landscape (Knox and Pinch 2000). The idea of people living on the 'wrong side of the tracks' and the attendant ideas of parts of the city as dangerous, threatening and to be avoided has a long history and reflects a territorial awareness rooted in class divisions. These zones (seen by some as 'no-go areas') are often viewed as separate spaces inhabited by many people who are marginalized not just in social and economic terms, but also spatially. Morley (2000) points to the *banlieues* of Paris which are both physically separate and socially distant from the heart of the city, cut off by a ring road (the *périphérique*) from the centre and relatively isolated from each other. Urban riots in the poorer *banlieues* on the edge of Paris in 2005 spread rapidly to other French cities. Young people there, including many from families of North African migrants, reacted to incidents of police violence in a wave of protests that rapidly spread. In these areas of high unemployment younger residents gave violent expression to their feelings of social and spatial alienation.

Segregation along class lines is effected through various mechanisms such as the housing market, effectively determining who can afford to live where. This appears at its most formalized in US cities where the process of municipal incorporation means that better-off territorially defined urban areas can effectively secede from the larger city of which they are a part. In doing so they enjoy a degree of fiscal autonomy which means that residents do not have to support services, such as public transport, for poorer areas outside their own municipality (Johnston 1984; Knox and Pinch 2000). In this way, through a combination of land-use zoning and territorially based funding of services, effective barriers prevent those deemed 'undesirable' from moving in. Economic 'apartheid' results as an area opts out of the broader urban environment and so residents shed themselves of any sense of collective responsibility for the poor. The incorporated municipality can decide on certain local regulations such as excluding industrial developments. It also has the power to enforce minimum lot sizes and prohibit such things as mobile homes and vagrancy. In this way it can effectively exclude poorer residents, thereby insulating the relatively affluent residents from potential declines in property prices. This is a clear example of the political manipulation of space with power mediated through a territorial process, which may have serious racial as well as class connotations (see next section). These broader functional divisions of suburbia both reflect and reproduce social divisions.

Other more subtle, but perhaps equally effective, processes contribute to territorially based residential segregation (as discussed in Chapters 5, 9 and 10). As previously noted, gentrification is where parts of the urban area experience regeneration or renewal resulting in more affluent residents moving in and displacing the original predominantly working-class inhabitants. Driven in part by economic considerations and in part through consumer choice, it serves to reinforce economic divisions within society and thereby perpetuate the idea that some households do not belong in particular places (see N. Smith 1996). As Short (1989) suggests, the built environment reflects the needs or perceived needs of different household types and social categories. It is also argued that the role of 'urban gatekeepers' (such as estate agents) may play a key role here in altering (or endeavouring to maintain) the social composition of particular areas (Knox and Pinch 2000). Gentrification reflects broader socio-economic processes and the resultant residential territorialization can be seen as an expression of the financial power of home-owners and the power of finance capital. The regeneration of many dockland areas and older industrial zones in cities in Europe and North America reflects this transformation from manufacturing and working-class residential spaces into service sector (particularly financial services) zones with a resident middle-class population. This, as Short (1989) points out, reflects more than a simple change in land use, it also reflects changes in the meaning of place. Dockland and waterfront areas, like other 'regenerated' urban zones, have been transformed into different places, with quite different uses and symbolic meanings. In South Wales for example, the regenerated Cardiff Bay area has been transformed from a place associated initially with the export of coal and subsequently with industrial decline and decay. Now, with its mix of residential and commercial developments, it is promoted as a symbol of a modern forward-looking city (see http://www.cardiffbay.co.uk/) (Plate 19.1).

Residential segregation appears at its most territorially visible through the creation of gated communities which are becoming increasingly common in the United States, Europe and elsewhere. Almost twenty years ago, Davis (1990) drew attention to what he termed 'Fortress LA' where security guards patrol the perimeter of walled residential zones in an effort to exclude what are seen as 'undesirables'. This phenomenon is repeated in many other cities, with private security firms patrolling more affluent urban areas in order to exclude those seen as not belonging there, thereby maintaining the 'undefiled' nature of the neighbourhood. The increasing prevalence of apartment blocks and other new housing developments surrounded by security fences appears driven by fears of security as well as by a desire to live in 'exclusive' developments seen as being more prestigious through having limits on public access regulated by intercoms and associated 'screening' devices (Plate 19.2). These exclusionary devices could be said to be components in contemporary landscapes of power (see Jones et al. 2004).

These territorial strategies work in ways which ensure a particular residential mix and may well serve to link together both racial and class divisions. Conversely, many working-class housing estates, seen as being inhabited by an underclass, are often perceived as unsafe 'no-go' areas. As Hubbard suggests in Chapter 10, such stigmatizing of place (and people) in itself becomes part

Plate 19.1 Regenerating place: Cardiff Bay, Wales. (Author)

Plate 19.2 Closed space: private gated residential development, Isle of Dogs, England.
(Photofusion Picture Library/Alamy)

of the problem, serving to reinforce class divisions and to reproduce various forms of social exclusion. Effectively this results in a form of ghettoization with a whole series of negative consequences for both people and place flowing directly from this. As Winchester et al. suggest, the 'inhabitants come to wear the myths of that place' (2003: 176). Morley (2000) suggests social groups separate out from each other and, indeed, are encouraged to do so. He further suggests that these tendencies have been effectively reinforced in recent years through geodemographics and the use of postcode data and associated marketing strategies of companies who are keen to identify particular types of consumer, and link these to geographic areas. In this way residential homogeneity is both reflected and reproduced.

The mixture of subtle and visible processes of residential segregation are mirrored in other arenas. Taken alongside the privatization and 'gating' of residential zones, the proliferation of covered shopping malls is seen by critics as the destruction of shared urban street space and its replacement with privatized, more exclusionary spaces of consumption (M. Davis 1995). Where once streets were open to a broad public, there are now privatized spaces whose owners (invariably resorting to security firms who take on some police powers) can evict those seen to behave inappropriately or who simply look 'out of place'. In the United Kingdom, recent debate and associated social panic over young people wearing 'hoodies' reflects particular hegemonic ideas about how teenagers and young adults should behave in socially acceptable ways in particular places. In 2005 a shopping mall in Kent banned young people from wearing 'hoodies' and baseball caps as part of a crackdown on what was deemed to be antisocial behaviour. The centre manager was reported as saying: 'We're very concerned that some of our guests don't feel at all comfortable in what really is a family environment' (see http://news.bbc.co.uk/1/hi/england/kent/4534903.stm). The reference to 'guests' clearly indicates that the centre is considered private space into which people are 'invited' rather than having any automatic right to be there. These 'secure' shopping centres, office blocks and apartment buildings, complete with gates and intercom systems, exemplify a trend towards socio-spatial design whereby territorial strategies associated with crime prevention effectively exclude those not wanted. Some people are effectively barred from certain areas – a policy of territorial containment enforced through increased surveillance and architectural design features (Cozens et al. 1999).

While there are many negative consequences of these forms of territorialization, the apparent consignment of the poor to certain areas can also of course provide the spatial framework for forms of resistance. It may facilitate the election (at local level at least) of political representatives for residents of those areas. Thus they are given a voice that might otherwise be denied them. It may facilitate the mobilization of people in support of, or in opposition to, issues of direct concern to them such as transport provision, banking facilities, the provision (or non-provision) of a range of services and the preservation of open spaces (see Jones et al. 2004). In this way a territorial strategy can be utilized in order to defend the interests of those who, if more spatially scattered, would be unable to do so. Community groups in working-class areas

Thematic Case Study 19.1

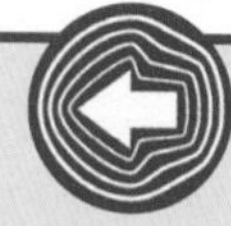

Marking territory

Within contested political spaces the names given to areas or to streets itself reflects power relations or challenges to existing power. This can be seen in a variety of ways. At an official level, streets have often been named in honour of royalty or leading political or military figures or as a reminder of important battles. Significant political change is often accompanied by name changes. Thus, the collapse of communism and the disintegration of the Soviet Union saw Leningrad revert to its earlier name St Petersburg, while in Montenegro, Titograd (named after the former Yugoslav communist leader) became Podgorica. Elsewhere in eastern Europe street names were changed to reflect the changing political order (Jones et al. 2004). The creation of an independent Ireland in the 1920s saw Dublin's principal thoroughfare change from Sackville Street to O'Connell Street in honour of the nineteenth-century Irish nationalist Daniel O'Connell. The defeat of the United States and the unification of Vietnam in the 1970s saw Saigon renamed Ho Chi Minh City in honour of the former leader of Vietnamese nationalism. In conflict zones the unofficial changing of such things as street names are important symbolic ways in which territory is reclaimed, and the phenomenon was well known in black townships during the apartheid era in South Africa and also in Northern Ireland. Unofficial Irish language street signs were often erected in republican parts of Belfast, for example, and one such street, Beechmount Avenue, was unofficially 're-named' as RPG (rocket propelled grenade) Avenue. In these ways the authority of the dominant political or ethno-nationalist group is either asserted or challenged.

A different form of street marking is often indulged in by graffiti artists and by members of urban gangs, the activities of which have received considerable media attention in recent years, particularly in the USA. In places such as South Central Los Angeles, a combination of high unemployment, deprivation and political powerlessness creates a situation where territoriality may be a means of expressing power, using the only resource available – the streets and neighbourhoods in which they live. These gangs tend to lay down territorial markers to indicate to others their 'ownership' of particular places. Graffiti on walls, bridges and buildings is one very visual method of claiming space. Markers are quite literally placed on the landscape to signal control of territory or 'turf ownership'. Work in Philadelphia indicated that graffiti became denser closer to the core of that gang's territory. Ley and Cybriwsky (1974) were able to demarcate reasonably accurately the spatial extent of gang control in the city. In this way, aspects of popular culture are translated into a territorial frame. This claiming of space may be a means by which marginalized youth make their claim to existence; through planting their mark on territory, that territory becomes theirs (see Scott and Brown 1993).

in European cities or in the *barrios* and *favelas* of many Latin American cities exemplify this tendency. Resistance may also take more overtly subversive forms such as street gangs laying claim to their 'turf', the re-naming of streets and the placing of territorial markers (see Case study 19.1). Ultimately the contestation of space through various forms reflects resistance and an assertion of rights. They indicate the ways in which the meanings of place are disputed and, hence, they could be said to reflect contested ideas of community.

19.3 Racialized spaces

Just as class is mapped on to the urban fabric, so too is ethnicity (see Chapters 10 and 13). In this way we can speak of **racialized spaces**. Race and ethnic categories are, however, social constructions rather than simple biological realities. While race can be questioned as a problematic and often dubious form of social classification derived from past notions of hierarchy and domination (notably slavery and colonialism), there is no doubt that racism or 'race thinking' is a very real social phenomenon. Although the idea of a 'scientific' categorization of humanity into different 'races' has long ago been challenged and is now thoroughly discredited, its legacy persists. Kobayashi and Peake argue that race is socially constructed but that 'racialization' is 'the process by which racialized groups are identified, given stereotypical characteristics, and coerced into specific living conditions, often involving social/spatial segregation and always constituting racialized places' (2000: 293). While issues linked to 'race' are clearly social phenomena, they are often manifested spatially. Amongst the most rigid examples of racialized space was that devised under the apartheid system in South Africa: a territorial system that enhanced and entrenched the political, economic and social power of a minority white population over non-white populations. Both nationally and at the more localized level of individual urban areas, space was divided on racial lines. Non-white people were 'placed' in locations not of their own choosing in order to entrench minority white power. In this way, there was a legal transposition of inequality on to geographical space. This spatial arrangement was designed to ensure greater degrees of control over the majority black population and is a classic example of the utilization of a territorial strategy to attain political objectives. At very localized levels, there was a racialization of space with buses, public toilets and other amenities reflecting this divide. A racial ideology was mapped on to the South African landscape. Although apartheid ended in the early 1990s, after a long struggle for democratic rights for all in South Africa, its legacy means that a division of space based on the racial and class lines reinforced during the apartheid era has left enduring marks on the social landscape.

Such extreme racializations have by no means been confined to South Africa, however. The generally negative stereotyping of gypsies in much of Europe has led to considerable discrimination, with gypsies seen as an undesirable 'other', as a consequence of which they are effectively de-territorialized; they are seen not to belong anywhere and active attempts are made to exclude them from certain spaces (Fonseca 1996). The phenomenon of ethnic cleansing in parts of the former Yugoslavia is a striking example of an attempt to 'purify' territory of those deemed to belong to other ethno-cultural groups (Case study 19.2).

Some groups may find themselves confined to very marginal spaces. The tribes of native Americans who

Thematic Case Study 19.2

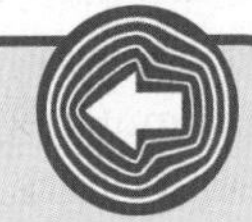

'Ethnic' territorial cleansing

Following the disintegration of Yugoslavia in the early 1990s, violent conflicts erupted in Croatia, Bosnia-Herzegovina and, later, Kosovo. These ethno-national conflicts were characterized by attempts by various groups to eradicate other ethnic groups from 'their' territory. This strategy was built on an essentialist version of defining ethno-national identity and, quite literally, clearing the territory of those possessing a supposedly 'different' identity. Armed movements claiming to be representing Serbs, Croats and Bosnians (predominantly Muslims) tried to carve out spaces which they could call their own in the ruins of the former federal state. Viewed from the outside it is easy to argue that such a strategy is both dangerous and simplistic. However, for those directly involved the control of territory was seen as an essential element in the conflict. The struggle over Bosnian territory was particularly violent. Despite the area's complex multicultural history reductionist interpretations of identity led to attempts to assert territorial control through the elimination of 'others'. Nationalist rhetoric, and the associated desire to control particular portions of territory in the name of a specific group, hardened divisions which were of relatively minor significance only a few years previously when Yugoslavia was a federal (and communist) republic under one-party rule. Once the conflict was underway however, it became impossible to reverse, as the conflict drew upon and *reinforced* group identities and competing territorial claims. In order to achieve peace, areas were mapped and information gathered on the various ethno-national groupings living in different localities. In this way territory was designated 'Serb' 'Muslim' 'Croat', and so on, with lines dividing towns and cities into different zones. After a number of failed alternatives, the Dayton Agreement of 1995 divided Bosnia-Herzegovina into two autonomous units: a Muslim-Croat Federation and a Bosnian Serb Republic (*Republika Srpska*). While this solution was lauded in many circles, critics have argued that it is inherently unstable and that it also reinforces ethnic divisions rather than rising above them. The partition of territories and 'transfer' of populations has occurred in many other situations of territorial, ethno-cultural, national and state conflict. Greece and Turkey in the early 1920s, India and Pakistan in 1947, Palestine and Israel in 1948 and Cyprus (whose northern part was occupied by Turkey in 1974) are all examples. Understanding them requires critical attention not only to the details of each case, but also to the concepts (territoriality, the political geography of nations and states and geopolitics) and background that are considered *both* in this *and* the subsequent chapter in this section.

(See Campbell 1999; Jeffrey 2006; Storey 2002; Gagnon 2006; Robinson and Pobric 2006.)

were assigned to 'reservations' provide a classic example of people whose territory was effectively taken from them and who were then allocated spatially restricted zones. Elsewhere, overt racial discrimination may be illegal but people may well experience exclusion through the implementation of regulations excluding the poor (referred to in the previous section). In most US cities, for example, the elites and middle class are disproportionally white, contributing to race/class segregation (Crump 2004). These patterns of exclusion and inclusion and attendant territorialities reflect the complex intersections of race, class and ideology.

In considering the evident spatial concentrations of ethnic groups in urban areas, it might be argued that individuals choose to locate in such areas for a variety of reasons. In brief there are a combination of 'positive' and 'negative' factors; for some there are attractions such as 'being amongst one's own', while others may feel driven to seek sanctuary from a racist, hostile society. Following Knox and Pinch (2000), key reasons may be summarized as follows. Clustering affords *defence* against attack by the majority group. It also provides a degree of *mutual support* and bolsters a sense of belonging and community. This in turn may be a useful means through which group cultural norms and heritage may be preserved. Finally, clustering produces *spaces of resistance* whereby external threats, whether to cultural norms or of physical attack, may be reduced.

None of the above should detract from the fact that such residential clustering may be more a function of necessity rather than free and unconstrained choice. It should be borne in mind that the degree of choice available to many may be extremely limited. The idea that people may choose to cluster is to ignore the fact that quite often no easy alternatives are available. Discriminatory ideologies of race work to exclude people from particular areas, thereby translating social exclusion into geographical exclusion. As Wacquant (1995) argues, 'ghetto' areas are institutionally produced; they are far from being simply the product of choice by those who come to inhabit them.

It should also be borne in mind that such clustering itself further contributes to future rounds of marginalization and exclusion so that the spatial divisions resulting from economic and social marginality have a tendency to lock people into a socio-spatial milieu which impedes economic advance or social mobility (Smith 2005). This further demonstrates the idea that territorial practices serve to reproduce particular social outcomes in a complex relationship whereby society does not simply impact on space, but spatial arrangements in turn impact on society.

In considering so-called ghetto areas, there is a tendency to see them in quite negative terms. The ghetto is a territorial entity and it is one that evokes many negative connotations; the term is often seen as synonymous with 'slum', the juxtaposition leading to a stigmatizing of its residents. In such ways the hegemony of the dominant group is maintained and the 'other' (in this case, ethnic minorities) remains geographically, as well as socially and economically, marginalized. In Western cities, 'ghettos' and other 'inner city' areas are frequently seen as the home of a so-called 'underclass' experiencing relative and absolute deprivation and disadvantage. Places such as South Central Los Angeles, for example, are represented as dangerous areas to avoid. In this way the 'ghetto' is portrayed *as a problem* rather than a place whose residents *experience* problems. It is also depicted in a somewhat monolithic way that ignores its internal social, economic and cultural diversity.

Such areas can be seen as territorial manifestations of social inequality reflecting the uneven distribution of power in society. Of course, like any other territorial entity, the ghetto can be used as a means of mobilizing residents, of providing the territorial frame with which people identify and within which they can operate with a view to improving their own conditions. Successful residents' groups often emerge and may become sufficiently well organized to be able to engage in lobbying, in forms of self-help, and so on. Such forms of community action may well make a positive difference to the lives of ghetto residents. In this way, it can be argued that the ghetto, just as with any other territorial formation, can take on its own sense of identity and can become a mechanism for the expression of group interests.

Places associated with visible ethnic minority groups can seek and develop positive connotations. Chinatowns (and latterly 'Banglatown', centred on Brick Lane in London, which is predominantly inhabited by people of Bangladeshi origin) in Western cities for example, have often become sites on tourist and gastronomic circuits. In the case of Brick Lane, an area with a long history of Jewish and Irish immigration, has latterly come to be seen as epitomizing a certain cosmopolitanism, which can be mobilized as a resource (Dwyer 2005). However, while there may be many potentially positive outcomes to this territorial construction, it may also be contested. Residents may not agree on the way in which the area is represented or portrayed. For example, there was a hostile reaction from some local residents to the fictional portrayal of the area by Monica Ali in her novel *Brick Lane* (2003).

Another outcome has been the commercialization of the area and its 'culture', with tourist guidebooks promoting its restaurants. The area also hosts an annual festival celebrating cultural diversity (see http://www.bricklanefestival.com/). While there are clearly positive dimensions to this, such strategies also run the risk of essentializing the identity of both places and their inhabitants. There is a risk of the portrayal of such 'ethnic spaces' as centres of exotica, to be exploited for commercial purposes.

More worryingly, the reputation of such areas makes them relatively easy targets for those hostile to ethnic minorities. In 1999 a nail-bomb attack was carried out in Brick Lane, an apparent racist assault. Thus, the promoting of a multicultural or **diasporic space** (Dwyer 2005) is met with resistance from racist movements (such as the British National Party) and others with hostile attitudes towards immigration and multiculturalism. Areas like Banglatown become contested spaces and the subject of struggle over their identity and meaning.

But so too are many predominantly white areas. We should also be mindful of the fact that much discussion surrounding issues of 'race' and ethnicity in Western societies tends to assume 'whiteness' as the norm. As McGuinness (2000) argues, much progressive research itself falls into this trap with a focus on non-white groups, tending to deflect attention away from white ethnicity. One consequence of the pursuit of this 'new exoticism', as McGuinness terms it, is that relatively little attention is given to 'white spaces'. Ideas of white flight to the suburbs (in response to the evolution of 'black' ghettoes) and the creation of 'white' territories are themselves elements in the racialization of space. Similarly the construction of rural Britain as a relatively 'white space' reflects deeply embedded ideas associated with belonging, rurality and with national identity. Such constructions can have serious implications for those who do not (or are seen not to) 'fit in' with the dominant assumptions and ethos. Current attention on immigration in the United Kingdom indicates a heightened concern with a supposed 'invasion' of asylum seekers, 'illegals' and 'hordes' of eastern Europeans. While much of this is inaccurate and misleading, the nature of the comments suggests an idea of the United Kingdom as a territorial entity that should increasingly seal itself off from invasions from 'outside', bolstered by concerns about terrorism. This racialization of space requires further investigation in order to broaden and deepen our understanding of the connections between geography and ethnicity (Bonnett and Nayak 2003).

This section has demonstrated the ways in which racist and exclusionary ideologies are transposed on to space. It has also indicated ways in which those racist constructions are opposed. Just as particular power relations are refracted through a territorial frame, so those relations are contested through territorial strategies. The spaces to which people are consigned may provide the means through which they contest their marginalization. A territorial base may serve as a means through which an ethnic identity or a class identity is reinforced and reshaped, in part at least, in opposition to other identities. Other categorizations, such as religious affiliation, may also provide the basis for parallel territorialities. (Case study 19.3 provides an example of segregation based on religious affiliation.) Of course we need to be mindful that such identities may be deeply contested and are far from monolithic, though there may often be attempts by some 'inside' or 'outside' to portray them as such (Yuval-Davis 2003).

Regional Case Study 19.3

Religious-territorial segregation in Northern Ireland

Linked to the idea of racialized spatial divisions, religious differences can also result in territorial separation within urban areas. One of the best-known examples is that which occurs in Northern Ireland. The region's long-standing political problems and conflict between unionists (predominantly Protestant) who wish the region to remain part of the United Kingdom and Irish republicans (predominantly Catholic) whose aim is a united Ireland (see Case study 3.1 and Spotlight box 20.1) is reflected in the social geography of Belfast and other large urban centres with high levels of residential segregation along religious lines. However, it should be borne in mind that this reading of the conflict can be slightly misleading; not all Catholics are nationalist and similarly not all Protestants are unionist. Nevertheless, it is apparent that there is considerable religious

→

segregation even if this cannot, and should not, be seen to correspond completely to people's wider political beliefs. Once again, this spatial segregation highlights the ways in which political relationships are mapped on to space.

Religious segregation has been reinforced through population movements, both voluntary and forced. In recent decades 'minority' families have periodically been burnt out of their homes in areas in which 'the other side' formed an overall majority. Equally, perceived influxes of the 'other' community have occasionally led to a drift away from certain areas by the 'other side'. This further reinforces pre-existing divisions, with housing estates becoming predominantly (and in some cases almost exclusively) Catholic or Protestant. As with other forms of ghettoization, this can be seen to be partly through choice and partly through force of circumstance.

Despite a peace process, ongoing since the mid-1990s, parts of Belfast are divided by walls – euphemistically known as 'peace lines' – literally dividing roads in the area and designed to prevent confrontations (Plate 19.3). Parts of the city are seen as out-of-bounds to one side because it is the other's territory. Once again, this highlights the territorial nature of political conflict and emphasizes the manner in which territory very easily comes to be seen in terms of 'ours' and 'theirs'. The territoriality of Belfast is such that a person's address is quite likely to reflect their place in a religious as well as a geographical sense.

Just as international borders have flags and other territorial markers to indicate their location, so also these segregated residential areas come complete with their own sets of boundary signifiers. Red, white and blue kerbstones indicate loyalist areas, while Irish flags are highly visible symbols in nationalist areas. Wall murals reflecting political allegiances adorn walls on both sides of the religious divide. These markers serve both to reassure residents and to send out a clear message to the 'other side' that their presence is not welcome. Paramilitary organizations on both sides have tended to exert a degree of control over 'their' areas and present themselves as protectors of the local population. Once again territoriality allows a mobilizing process to take place. Support is garnered for both political parties and paramilitary organizations.

A dispute surrounding the Holy Cross School in the Ardoyne area of North Belfast in 2001 can be seen as a microcosm of the dispute. In order to attend their local primary school Catholic schoolchildren had to walk past houses in a small Protestant enclave in this otherwise almost exclusively Catholic area. For several months, Protestants demonstrated along the road in an attempt to prevent the schoolchildren using this route through 'their' territory.

For more information on the Northern Ireland conflict see http://cain.ulst.ac.uk/. See also *Political Geography* **17**(2), 1998 (special issue: Space, place and politics in Northern Ireland) and Shirlow and Murtagh 2006.

Plate 19.3 Divided space: 'peace wall' Belfast, Northern Ireland. (Author)

19.4 Gendered spaces and the public–private divide

The growth of what became known as the 'feminist movement' from the 1960s onwards, building on earlier attempts to achieve equality for women, has been influential in gaining recognition in many societies for the unequal status of women and men in all dimensions of life; in the home, in (paid) workplaces, in the broader political and social arena. Feminist writers and activists have been instrumental in attempting to explain how patriarchal systems of power have tended to reinforce male dominance and how women have often been marginalized (Dixon and Jones 2006; England 2006). Feminist geographers have highlighted the 'geography' of discrimination against women, particularly in drawing attention to the manner in which space and place are heavily gendered and in challenging the relationships between gender divisions and spatial divisions (see, for example, McDowell 1999). Critical attention has been focused on divisions between the public and private domain and its spatial corollary of a separation between what is seen as public space and private space. It is argued that patriarchal systems of power have led to a division between predominantly 'male' public and mainly 'female' private space, resulting in social practices whereby certain activities and certain spaces are seen as male preserves. This duality reflects broader distinctions centred on the binary divide between masculinity and femininity (see Pratt 2005).

Issues of gender are mapped on to space in various ways. In its most simple form this is reflected in the idea that 'a woman's place is in the home'. The home has tended to be seen as a space of reproduction juxtaposed to the workplace as a space of production (Laurie et al. 1999). The implications of gender are seen to be as important as other political, social and economic factors in the structuring of spaces and places. Underpinning this are ideas that distinguish between sex as a biological fact and sex as gender, which refers to the socially constructed roles of both male and female identities. In emphasizing the role of social conditioning, the argument is that as individuals we are not biologically predetermined to be more suited to some roles rather than to others.

One reason for the relative absence of women in particular places is overt discrimination or active discouragement in the sense of certain activities or pursuits not being deemed suitable for women (see Case study 19.4). Historically, women who transgressed these boundaries were often portrayed in a negative light, an idea reflective of notions of 'good' and 'bad' women. Women out alone at night might be seen as not conforming to what is expected of them. A crucial aspect of the relationship between women and place centres on the perception of some specific places as 'unsafe'. Many women do not feel safe in certain public places, most notably darkened streets. As Valentine (1989) investigates, women transfer a fear of male violence into a fear of certain spaces, which has profound implications for the ways in which men and women negotiate their way through urban areas (Fell 1991). Clearly, the various strands of feminist thought and practice have resulted in significant advances with regard to equal rights for women. While this can be seen within the arena of legislation in many countries associated with equal pay and related issues, it is also reflected in terms of spaces. Thus, the heightened visibility of women in public space reflects the changing status of women. Phenomena such as 'reclaim the night' marches demonstrate the overt use of a spatial strategy to make a political and human point. While particular groups may find themselves excluded from certain spaces, those spaces can also be reclaimed (as an example see http://www.isis.aust.com/rtn/).

Historically, this gender division of labour has tended to confine women to the private realm, leaving men to inhabit (much of) the public domain. This view of women as playing a subordinate role has in the past been reflected in discriminatory attitudes and practices, particularly in relation to women in the paid workforce, with active discouragement through lower wages, if not actual exclusion, from many jobs. These views are predicated on the undesirability of women going out to work. It can be argued that this ascribing of women's role, through delimiting the spaces in which women were encouraged to appear, is another spatial expression of power. In other words, the confining of women to domestic space, and their exclusion from male territories, was the key element in male control. With increasing female participation in the workforce and a raft of equal opportunities legislation in many countries, such a generalization may appear to have lost some of its validity. Nevertheless, the division between a (largely) male public sphere and a (largely) female private sphere still has considerable resonance in many societies (although the extent of this is itself immensely geographically variable across the world).

Where women enter the workforce, they may still encounter territorial divisions in the workplace. Thus,

Thematic Case Study 19.4

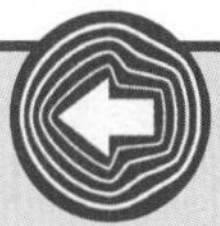

Space, sport and gender

In the United Kingdom and many other European countries, football stadia are usually markedly territorial, divided into areas reserved for 'home' and visiting ('away') supporters. However, there are also other (gender) divisions associated with this and many other sports, which often give these territorialities a distinctively marked masculine format. The geographer Doreen Massey (1994) recounts how, as a teenager, she used to get a bus into central Manchester on Saturdays and how she was struck by the fact that the numerous football pitches she passed en route were spaces inhabited totally by males. In this way, she argues, there is a clear gendering of space whereby some spaces come to be seen as the preserve of one or other gender. To take Massey's example one step further, it can be argued that professional football stadia in Britain were, at least until quite recently, almost exclusively male preserves. With a resurgence of interest in football in the 1990s, this phenomenon has altered, with an increase in the numbers of female supporters and clubs encouraging family attendance. The relative growth in popularity of women's football is another dimension to this. In the United Kingdom major women's international matches and some club games are now broadcast live on television complete with studio pundits airing their views, thereby mimicking coverage of the men's game. While these are undoubtedly marked changes, gendered ideas and practices about leisure activities and, hence, separate spaces for men and women, are still common.

Such social practices are built upon ideas of *what* is or is not acceptable behaviour for men and women to engage in (built on socially or culturally constructed notions of masculinity and femininity) and *where*. Stereotypes of women spending leisure time shopping while men attend sporting events, or watch them on television at home or in pubs, have a self-perpetuating quality. Media interest surrounding the shopping habits of the so-called 'WAGS' (wives and girlfriends of English international football players) provides further reinforcement of this. Despite this, these stereotypes are sometimes challenged or subverted, sometimes in an ironic manner. Irish female football supporters at a European championship match in Portugal were reported to display a banner saying 'Our husbands think we are shopping in Dublin!' (cited in Corry 2006).

Spain (1992) documents the 'closed door' jobs of managers (mainly men) and the 'open floor' jobs of manual workers (who may be predominantly women in certain countries/regions/sectors). Employers may locate in particular localities (or countries) in order to take advantage of what they see as an available (and exploitable) workforce based on prevailing wage levels or skills and assumptions about gender roles. Hanson and Pratt (1995), for example, show how some firms located in particular parts of Worcester, Massachusetts in order to avail of what they saw as desirable characteristics based on stereotypical assumptions associated with gender and ethnicity. These firms (like many elsewhere) presented some jobs (on assembly lines, for example) as more 'suitable' for women. In 2007 Iran's arrest of 15 British sailors, one of whom was female, in what it claimed were Iranian territorial waters led to media commentary on the appropriateness of a woman, particularly a mother (as in this instance) being engaged in military activity in a war zone, rather than being at 'home' playing the key role in bringing up her children. Social processes reproduce attitudes that tend to 'naturalize' a gendered division of labour in which women perform certain functions which are acted out in specific spaces, e.g. 'home-making' and child-rearing in domestic space. Socially constructed gendered difference is inherently also spatialized.

If women have sometimes been seen as 'belonging' in or closely associated with the home, it is not the case that the space of the home is undifferentiated. Even within the home, territorial divisions take place, the most obvious being the notion of the kitchen as a female preserve. The relative neglect of such territorial behaviour in all its manifestations is commented on by Sibley:

> In geography, interest in residential patterns wanes at the garden gate, as if the private province of the home, as distinct from the larger public spaces constituting residential areas, were beyond the scope of a subject concerned with maps of places.
>
> (Sibley 1995: 92)

There has, however, been a growing interest in the domestic sphere. Daphne Spain (1992) provides many examples of sex segregation in the home from different cultural contexts and from different time periods.

Many communities and societies often had a clear spatial division between men's and women's spaces and roles. Within most Western societies, deep-seated ideas about woman's role as homemaker, cook, cleaner, child-rearer, and so on, mean that women have often been historically presumed to 'belong' in some rooms and spaces more than others. While such ideas have been challenged (and in many cases transformed), they sometimes endure. All these reflect territorial expressions of power whereby the designation or apportionment of space within the domestic sphere reflects the relative status or roles of the individuals concerned. Spain therefore reminds us that 'houses are shaped not just by materials and tools, but by ideas, values and norms' (1992: 111).

19.5 Sexuality and space

The idea of places territorialized by particular groups on the basis of sexual orientation has been widely discussed. At its most elementary level this has seen the mapping of gay and lesbian 'zones' in selected cities. It is fair to point out that such spaces are not as easy to identify as, say, areas inhabited predominantly by a particular ethnic group. It is equally obvious that, in the main, these are not strictly demarcated areas. Rather, they are zones where gay people may feel more at ease through being accepted rather than rejected, scorned or ignored (or worse) by their neighbours. There have been criticisms of such straightforward mapping of 'gay territories' in that it may be interpreted as a mapping of what some see as deviant behaviour, that it essentializes sexuality and that it reinforces a gay/straight dichotomy (Knopp 1995; T. Davis 1995). Nevertheless, the fact that those who identify themselves as gay and lesbian do, in some instances, become associated with particular (usually urban) spaces, suggests that another form of territorial behaviour may be evident. A concentration of visibly gay restaurants, bars and clubs results in the creation of what Castells calls 'a space of freedom' (1997: 213). Areas such as the Castro District of San Francisco (Plate 19.4) and the more spatially confined 'gay village' in Manchester serve as important examples.

The construction of such zones may arise for reasons similar to those associated with ghettos and other forms of segregated space. Castells (1997) has argued that there are two key factors: protection and visibility. The first of these is fairly obvious. The idea of 'strength

Plate 19.4 Gay space: rainbow flag, Castro District, San Francisco, USA. (Author)

in numbers' may make people feel safer from homophobic 'gaybashers'. Ironically, however, crowds at gay venues may sometimes offer targets for racists and homophopes. In 1999 a gay pub in Soho, central London, was the target for a bomb, resulting in three deaths and over eighty people injured. The second reason, that of visibility, relates to the need for those who identify themselves as gay or lesbian to assert an identity within a culture that is predominantly straight (heterosexual) and in which a straight discourse dominates, a society in which other sexualities are often still seen by many as deviant or abnormal. Gay neighbourhoods become a means of asserting identities. Harry Britt, a one-time key figure amongst San Francisco's gay community, once commented that 'when gays are spatially scattered, they are not gay, because they are invisible' (cited in Castells 1997: 213). Rothenberg (1995) examined the development of lesbian spaces in cities in relation to gentrified housing. In this way the transformation of place becomes associated with the assertion of an identity. The relations between class, capital, sexual identity and place become embodiments and manifestations of territoriality.

The significance of San Francisco's gay area was reflected in that community's ability to gain political representation. In obtaining power over territory they also gained political representation and San Francisco has become recognized as something of a 'gay capital' of the United States with a somewhat more liberal attitude. In this way, the designation of 'gay territories' plays a crucial role in raising awareness of gay people and issues and also provides a means by which some degree of power and self-confidence can be attained. Thus celebratory events such as 'gay pride' marches can be seen as an assertion of citizenship rights through staking a claim to public space. The examples of gay and lesbian spaces suggest another important point, that of the temporality of territory. The longevity of these spaces may be quite brief as the 'scene' moves to somewhere else. Lesbian spaces may be very short-lived in time, whether caused by the transience of lesbian bars/clubs or the even more short-term phenomena of lesbian or gay evenings (Valentine 1995).

Of course such places and such events may often be 'co-opted' to present an 'acceptable' image of the group concerned. The evolution of 'gay spaces' has often been associated with an economic imperative. The importance of the 'pink pound (dollar, or euro)' in aiding urban regeneration has frequently added a strong commercial angle to these developments. Some activists have expressed disquiet over the appropriation of such events and their dislocation from their original social and cultural roots and from their original territorial base. The fact that Manchester's 'gay village' and San Francisco's Castro District are firmly on the tourist trails of their respective cities may be lauded as an acceptance of identities previously scorned but it can also be seen as a commercialization of that identity which may, to some extent, serve to further ghettoize it. It might be argued that it is acceptable to be openly gay in an area so designated but in other spaces and societies the pressure to keep this identity hidden may well persist (see http://www.sfguide.com/sights/neighborhoods/castro.htm). While gay spaces may allow for more open expressions of sexual identity, homophobic assaults on sexual minorities reflect contested place meanings (Sumartojo 2004). Openly gay behaviour may be accepted or tolerated in some places but may remain decidedly unacceptable elsewhere. Here of course there are profound links between the small-scale (often neighbourhood or urban) territorialities and the wider policies of the city or state concerned. Many states still criminalize gay sexuality and almost everywhere sexual mores and regulations are highly territorialized.

19.6 Private property and personal space

We can recognize two important social tendencies that bolster territoriality: the wish by people to have space of their own and the wish by others to exclude people from certain spaces. We have already seen ample evidence of the latter in this chapter. At its most elementary level, the assertion of territoriality is reflected in claims to private property. Thus, people desire to mark their own home, to adorn it in their chosen style (influenced of course by social trends, technologies and fashions) and, in various ways, to mark it out as theirs. Homeowners are generally keen to stamp their personality on their home through the ways in which they choose to decorate it, alterations to layout, choice of colour schemes, furnishings and so on. This personalizing of space is further manifested through such things as the display of paintings, posters or photographs and the collection and arrangement of ornaments. The geographer Jean Gottman suggested that people 'always partitioned the space around them carefully to set themselves apart from their neighbours' (1973: 1). This manifestation is commonly interpreted as being symptomatic of our inherently territorial nature. This emphasis on the centrality of the home also has a broader cultural and political significance. Symbolic connections are often made between the domestic home and the nation whereby images of the former are seen to give material meaning to the latter. The home is seen in some ways to be at the heart of the nation. In times of war, for example, people have been encouraged to fight for the 'homeland' and the defence of 'hearth and home'.

Private property is regarded by many as an outcome of human territorial behaviour. It represents a claim to space that is reinforced by the legal system of many countries. However, as Alland points out, it might well be the case that 'private property is the child of culture and develops into a major preoccupation only with the evolution of complex society' (1972: 64). It follows that we need to be careful to avoid the trap of translating a need for personal space into an ideological claim for the sanctity of private property. The centrality of the family home, encapsulated in such phrases as 'home, sweet home', glosses over the fact that the privacy which many of us associate with the home is comparatively recent and is specific to some societies. Where 'domestic' space is limited (or for those who find it constrains them), life may be lived in the street (or perhaps the mall or car), much more evidently than the 'behind four walls' lifestyle many take as 'natural'. We need to be mindful of social, ethnic and geographic differences in the ways in which the home is conceived. In the United Kingdom, the notion of private home space was initially quite a middle-class idea which has since permeated the rest of society (Morley 2000).

Hegemonic ideas of the home within Western societies, exemplified by such notions as the home being an 'Englishman's castle' or associated with 'The American Dream', can be argued to have led to an ignoring of internal tensions and, in particular, a consideration of the different positions, roles and experiences of men, women and younger people within this domestic space, as discussed earlier (McDowell 1999). While the home is commonly depicted as a refuge from the outside world it may also be a site for domestic violence and fear (Pain 1997). Similarly teenagers may view the home as somewhere to escape from. In any event, the emphasis on the domestic idyll may have highly exclusionary consequences (Delaney 2005).

Even within buildings territorial behaviour can be recognized. As we have seen, the idea of the kitchen as a 'woman's place' is one example of this. Spain (1992) has collated details of how the domestic home in many different cultural contexts is often spatially divided, not just in terms of gender but also in terms of age, with certain spaces being designated for women or for children. The banning of children from some rooms and the proprietorial attitude towards one's own room in a house are other examples of this. In the home, space is even being claimed at the level of 'my chair', 'my place at the table' and so on. There are also distinctions amongst those allowed in, with differential access for close family and friends on the one hand and more casual acquaintances on the other. Even then, friends may be welcomed into the living room but are less likely to be invited into the more 'private' spaces such as bedrooms (Morley 2000). In any consideration of the home, we need to be mindful that it is not a straightforward and unambiguous entity. While for some it conjures up feelings of comfort and security, for others it may be a place of discomfort, alienation and tension (Blunt and Dowling 2006). For some the home may come to feel like a prison – quite literally so for those sentenced to home imprisonment of the type long imposed on political activist Aung San Suu Kyi in Burma or the many others subject to degrees of curfew, control and house arrest.

Equally, in workplaces some areas and rooms can only be entered by staff of a certain level and are out of bounds to more junior staff. These can be interpreted as managerial strategies designed to ensure a particular outcome; staff know their 'place' and can be more effectively controlled. Hanson and Pratt (1995) reveal how companies reproduce social segregation through spatial practices within the workplace whereby different sets of workers inhabit different parts of the factory and rarely, if ever, meet. Thus, office staff may be located downstairs in 'cubicles' separated by room dividers, with sales staff and management upstairs in individual or shared offices while production staff are located in an entirely separate part of the building. Socializing between workers tends to reflect their spatial segregation even to the extent of each department having separate annual parties. Work hierarchies are reflected in the spatial arrangements of the workplace. These practices have clear outcomes. They may render it difficult for workers to organize through physically keeping them separate and through engendering a sense of difference between different sections of the workforce.

At a more elementary level, there are echoes of this in how people treat their workspace, whether it is an office, a workstation within an office, or commercial patch. In part, of course, this relates to notions of comfort and familiarity and the desire to protect our own personal space. The psychological need for space and privacy does not, however, take from the fact that what is happening is an attempt to wield power through a territorial mechanism.

Taking the idea of territory down to its most elementary level, the desire for personal space can be seen as a form of territorial behaviour. Humans like to have a pocket of space around them that is 'theirs' and they resent others 'invading' their space (unless invited!). This can be interpreted as a territorial claim to a portion of geographic space. While this might be taken as reflecting a natural tendency, it is worth noting that the amount of space needed appears to vary from one society to another, a fact noted long ago by Hall (1959). Nurture, culture, power and politics all need to be considered where the complexities and diversities of human territorialities are concerned.

19.7 Conclusion

The key argument of this chapter is that social practices are reflected in struggles and territorial claims over the use and control of space. Territorial strategies are utilized in conflicts concerned with social power and identity. This may be to do with maintaining power or with resisting the imposition of power by a dominant group. Forms of exclusion can be consolidated and reinforced through territorial practices, yet they can also be resisted through similar means. The examples provided are evidence of the way in which social relations are expressed through spatial patterns and they highlight the ways in which this geography helps in turn to shape social relations. Social phenomena such as racial or gendered identities invariably embody a territorial component. Territorial strategies are often used to control and police those who are defined as 'out of (their) place'. In this way particular ideologies are transposed on to space. People are confronted with wider practices through their use of space or through the ways in which they are allowed to use space. Power relationships take on a spatial dimension, even at the most mundane and everyday level. Class, racial, religious or gender divides are given material form through spatial divisions. The examples used here demonstrate the spatialization of wider ideas and they show how people are kept 'in their place' whether through legal means (apartheid), administrative practices (urban incorporation in the United States) or surveillance strategies (related to fear of danger, 'terrorism' and crime). Social boundaries are being communicated through space. As Harvey has suggested, 'the assignment of place within a socio-spatial structure indicates distinctive roles, capacities for action, and access to power within the social order' (1990: 419). This also means that some groups of people are effectively treated as 'second-class (or indeed as non-) citizens' denied full rights in the society in which they find themselves. Social and spatial exclusion has the effect of denying full access to those rights often thought of as inalienable for all people everywhere. In this way many territorial strategies are invariably discriminatory and exclusionary and can be used to deny people effective participation in 'society': the latter being based on a categorization of those who 'belong' (by virtue of citizenship and income levels that enable a range of choices and possibilities to participate). They raise serious questions about the nature of citizenship and demonstrate the uneven ways, both socially and spatially, in which citizenship rights are distributed (see Low 2005). In all societies, those who are convicted (or even in some cases accused) of criminal offences may also be excluded by territorial confinement (this is a key function of prisons) or controls on their movement. The range of these territorial strategies will vary according to the crime and legal jurisdiction. But in the USA, the prison population has quadrupled in recent decades, so much so that a new extended geography of prisons and incarnation can be critically mapped (Gilmore 2007). In the new geopolitical context of the US-led war on terror (see Chapter 21), the USA has also taken the lead in making new spaces of confinement and interrogation (such as the prison camp at Guantánamo Bay, Cuba) resorting to complex territorial strategies in so doing – placing the prison *outside* the territorial United States to allow the indefinite detention without trial of its inmates, but arguing nonetheless that the inmates be subject to the judgment of (and interrogation by) American military authorities re their status and future. The transportation of men and boys captured in Afghanistan, Iraq and other countries to Guantánamo renders the relationships between territory, extraterritoriality (being beyond territory), sovereignty and movement even more complex and blurred (Elden 2005; Gregory 2006).

Moreover, just as dominant ideologies can be reinforced through territorial practices, they can also be resisted. Territorial strategies are useful mechanisms in the assertion of identity. Spatial concentrations within particular geographic areas make visible people and issues that might otherwise remain unseen. They can be used to draw attention to exclusionary practices and to assert the right to be equal citizens. In doing so, this demonstrates the 'positive' and 'negative' dimensions to territoriality; it can be both a force for oppression and also one for liberation. Particular strategies can be used to assert an identity and territorially transgressive acts can be employed to reclaim space and, hence, to assert basic rights.

While many people do not necessarily freely choose their 'place', they may, nevertheless, identify with their immediate neighbourhood or locality. This sense of identity can in turn be converted into forms of action aimed at obtaining particular outcomes. The formation of community or residence groups reflects feelings of belonging or attachment to a particular place. It follows that notions of territory are connected with ideas of social power. The claiming of space is a political act whether it occurs in the 'public' or 'private' arena (and the categorization and demarcation of these areas is a key expression of territoriality). It is important to be aware of 'how relations of power and discipline are inscribed into the apparently innocent spatiality of social life' (Soja 1989: 6). In other words, *power* permeates society and this is something which is marked in and expressed through a range of human geographies. And as the next chapter details, the nation-state has come to be a key container of power and expression of human territoriality. As such, nation-states are frequently contested, not only by other states, but also by those acting in the name of other loyalties and other scales and ideologies of belonging.

Learning outcomes

Having read this chapter you should understand:

- How social issues linked to class, racism, ethnicity, identity, gender and sexuality are mapped on to geographic space and become the basis for territorial strategies.
- Some of the ways in which social inequalities and differences are manifested spatially.
- How territorial practices are used to exert or to resist control.
- How territorial strategies can be mechanisms for groups to promote their identities, rights and claims to meaningful citizenship.
- The ways in which many aspects of everyday life are reflected in spatial arrangements and territoriality.

Further reading

Delaney, D. (2005) ***Territory: A Short Introduction***, Blackwell, Malden.

Storey, D. (2001) ***Territory: The Claiming of Space***, Prentice Hall, Harlow.

These both build on Sack's work and deal with territory and territoriality in social and political contexts. Storey's book contains many more case studies of territoriality than can be considered here. Delaney's book also offers a good survey (and some critique) of Sack's ideas and lots of case studies. Ideally, read both as a follow-up to this chapter.

For some reflections on how territorial strategies delimit and shape violence and terror, see:

Gregory, D. and Pred, A. (2006) (eds) ***Violent Geographies: Fear, Terror, and Political Violence***, Routledge, London.

Panelli, R. (2004) ***Social Geographies: From Difference to Action***, Sage, London.

Valentine, G. (2001) ***Social Geographies: Space and Society***, Prentice Hall, Harlow.

The following two very readable social geography texts deal with many of the concerns of this chapter. Read them with a view to the complex operation and manifestations of territoriality.

Sack, R. (1986) ***Human Territoriality: Its Theory and History***, Cambridge University Press, Cambridge. As outlined in this chapter, Sack's book has become a classic geographic text on territory and territorial behaviour, setting it firmly within a 'social' rather than 'natural' framework.

For an essay revisiting Sack's book more than two decades after its original publication, see Classics in human geography revisited, ***Progress in Human Geography***, 24(1) 91–9. And for two contrasting cases (one on territoriality in a school playground, another on parades), both drawing on Sack, see:

Thompson, S. (2005) 'Territorialising' the school playground: deconstructing the geography of playtime, ***Children's Geographies***, 3(1), 63–78.

O'Reilly, K. and Crutcher, M. E. (2006) Parallel politics: the spatial power of New Orleans Labor Day parades, ***Social and Cultural Geography***, 7(2), 245–65.

For a study of intersections between identity, 'race' and forms of territorial exclusion in the English countryside, see:

Holloway, S.L. (2005) Articulating Otherness? White rural residents talk about Gypsy-Travellers, ***Transactions of the Institute of British Geographers***, 30, 351–67. Worth comparing with:

Sibley, D. (1995) ***Geographies of Exclusion: Society and Difference in the West***, Routledge, London.

Knox, P. and Pinch, S. (2000) ***Urban Social Geography. An Introduction***, 4th edition, Prentice Hall, Harlow. A useful urban geography text touching on many of the issues (such as gentrification and segregation) dealt with here. As with the social geography texts listed above, look for manifestations of territoriality in the case studies, and details on urban social segregation and change.

Blacksell, M. (2005) ***Political Geography***, Routledge, London.

Jones, M., Jones, R. and Woods, M. (2004) ***An Introduction to Political Geography: Space, Place and Politics***, Routledge, London.

The above two are both reader-friendly introductions to the intersections between politics and geography, providing very good examples of some of the themes covered here. Blacksell's book is a good starting point, the other one takes the arguments to a deeper level of analysis. Both texts also contain lots of material (for example, on the territorial state and geopolitics) relevant to other chapters in this section.

Gagnon, V. P. (2006) ***The Myth of Ethnic War: Serbia and Croatia in the 1990s***, Cornell University Press, Ithaca. An examination of the political mobilization of ethnicity, identity and territory in a conflict situation and a good way to follow up in detail some of the themes touched on in Case study 19.2.

Nelson, L. and Seager, J. (2004) ***A Companion to Feminist Geography***, Blackwell, Oxford. A collection which illustrates the diverse contributions feminist geographic thought and practice have made to human geography. A good way to follow up some of the ideas in this chapter is Eleonore Kofman's chapter on feminist political geographies.

Blunt, A. and Dowling, R. (2006) ***Home***, Routledge, London. A useful examination of the complex natures and meanings of home.

Bell, D. and Valentine, G. (eds) (1995) ***Mapping Desire: Geographies of Sexualities***, Routledge, London. This was an agenda-setting collection of articles on aspects of sexuality and space: the main focus was on the making of lesbian and gay spaces and 'territories' in Western cities. For a more recent case study of contest over political strategies to claim and mark a gay urban territory, see:

Sibalis, M. (2004) Urban space and homosexuality: the example of the Marais, Paris 'Gay Ghetto', ***Urban Studies* 41**(9): 1739–58.

Finally, the law is a key aspect in regulating territorial behaviours and strategies (and related and overlapping economic, social and political geographies). See how via any of these:

Blomley, N.K. (1994) ***Law, Space, and the Geographies of Power***, Guilford Press, New York.

Blomley, N.K. (2004) ***Unsettling the City: Urban Land and the Politics of Property***, Routledge, New York.

Blomley, N.K., Delaney, D. and Ford, R. (eds) (2001) ***The Legal Geographies Reader: Law, Power, and Space***, Blackwell, Oxford.

For annotated, clickable weblinks and useful tutorials full of practical advice on how to improve your study skills, visit this book's website at **www.pearsoned.co.uk/daniel**

THE PLACE OF THE NATION-STATE

Chapter 20

Carl Grundy-Warr and
James Sidaway

Topics covered

- The ubiquity and geographical diversity of nations and states
- Interpretations of nationalism as a territorial project
- Relationships between nations, states and territory

The territory of a nation is not just a profane part of the earth's surface. It is a constitutive element of nationhood which generates plenty of other concepts and practices directly related to it: for example, the concept of integrity and sovereignty; border control, conflict, invasion and war. It defines and has some control over many other national affairs, such as the national economy, products, industry, trade, education, administration, culture and so on. Unarguably, the territory of a nation is the most concrete feature of a nation for the management of nationhood as a whole. For a theoretical geographer, it is the territoriality of a nation . . . For people of a nation, it is a part of SELF, a collective self. It is a nation's *geo-body* . . . Geographically speaking, the geo-body of a nation occupies a certain portion of the earth's surface which can be easily identified. It seems to be concrete to the eyes and having a long history as if it were natural, and independent from technology or any cultural and social construction. Unfortunately, that is not the case . . . the geo-body of a nation is merely the effect of a modern geographical knowledge and its technology of representation, a map. The geo-body, the territoriality of a nation as well as its attributes such as sovereignty and boundary, are not only political but also cultural constructs. (Winichakul 1996: 67)

20.1 Historical and geographical variability of states

Although it is also other things, notably a provider of services, a system of regulations, ideologies, legal regulations and police powers ('law and order'), flows of capital (budgets, taxation and government spending) backed up by the threat of discipline and violence (for example, armed forces), the state can be interpreted as a form of *community*. As earlier chapters (in particular those in Section 1) have shown, forms of human community and their attendant territorialities (see Chapter 19) have been extremely variable historically and geographically. This variation is enormous, from the claims of spokespersons for a state (it could almost be any state) that it is a natural embodiment of a 'nation' with the right to a select segment of the world (a territory with boundaries) and an ancient history (so even archaeological remains of past societies somehow come to be classified as 'national monuments', like Stonehenge in England, Machu Picchu in Peru, the Acropolis in Greece, Angkor Wat in Cambodia, the Pyramids in Egypt and so on) to nomadic communities or complex tribal or dynastic ones with little or no sense of themselves as belonging to or requiring a national state, to those who would reject national and state identity in the name of what are seen as other or more worthy causes (and attendant territorial visions), for example a religious identification, local community or world revolution.

The contemporary system of states, in which all of the land surface of the earth (with the partial exception of Antarctica as is detailed in Chapter 21) is divided into state units, whose outlines become familiar to us from maps and globes, is after all fairly new. In the early twentieth century, the borders between many of today's states were only vaguely defined, and more recently large areas of the world were ruled by colonial empires or dynastic realms (the Austro-Hungarian in Central Europe, Ottoman in the Balkans and Eastern Mediterranean, and Ch'ing empire in China, for example). Alternatively, the concept of nation-statehood was simply not in the political vocabulary. In the latter cases, ethnically, linguistically or religious specific groups owed allegiance to the imperial order rather than to a defined nation. That is, loyalty would be *foremost* to the empire and any sense of national or ethnic identity would be a local or 'private' matter.

Such imperial visions are no longer dominant and today few formal colonies and no large-scale dynastic empires remain (see Chapter 3, pp. 72–76). Therefore the territories that were once ruled as part of, for example, a Japanese, British, French, Ottoman (Turkish), Portuguese or Russian empire are today mostly divided into self-avowed and recognized *sovereign* states. They possess the same apparatus of statehood (leaders, flags, capital cities, administrations, postage stamps, seats at the United Nations and so on) that the former imperial powers have. Moreover, the global map of states continues to change. In recent decades, some have disappeared as separate states (like the former East Germany and South Yemen), whilst others have split into component parts (like the former USSR and Yugoslavia). All this reinforces the point that states, like other communities, particularly the 'nations' with

Plate 20.1 The state as a system of organized violence: armed forces.
(GrahamGrieves/TRIP)

which they are associated, are not to be taken at face value. The claims made for and on behalf of them deserve critical examination. However, this is not an easy task. For, as Benjamin Akzin (1964) pointed out nearly half a century ago, to discuss nations, states and nationalism is to enter a terminological maze in which one easily and soon becomes lost. This chapter will indicate some pathways into the maze. It begins with an account of 'nations' and **nationalism** before returning to the relationship of these to states.

20.2 Nations as 'imagined' political communities

One of the most influential and suggestive critical studies of nations and nationalism is a book by Benedict Anderson (1983) entitled *Imagined Communities: Reflections on the Origin and Spread of Nationalism.* He begins with a reminder of the ubiquity of nations and nationalism:

> Almost every year the United Nations admits new members. And many 'old nations' once thought fully consolidated find themselves challenged by 'sub'-nationalisms within their borders – nationalisms which naturally dream of shedding this subness one happy day. The reality is quite plain: the 'end of the era of nationalism', so long prophesied, is not remotely in sight. Indeed, nationness is the most universally legitimate value in the political life of our time.
>
> (Anderson 1983: 3)

Although enormously variable between, say, Nepali, Israeli, Singaporean, Nicaraguan, Vietnamese, Eritrean, American, Greek, Turkish and Irish versions, the ideology of nationalism holds that everyone will have a primary identity with a particular 'nation'. Such communities should be able to express themselves in a state; that is, they should enjoy what is called 'sovereignty' within certain geographical boundaries. 'Sovereignty', which is a term of long vintage and was previously associated with royal dynasties (the sovereign monarch), shifts to the 'people' of a 'nation', and even if a royal figurehead is retained, she or he will have to become in some way a 'national' symbol. However, it is important to remember that territorial sovereignty as depicted on the world political map of today is of relatively recent origin (see section 20.3), that more differentiated forms of sovereignty and other territorialities have existed in the past, and that sovereignty is continually being challenged in various ways.

Examples of quite different conceptions of sovereignty in the past are abundant in many parts of the world, from medieval Europe to most parts of the pre-colonial world. The labyrinthine world of medieval times in Europe was at some levels intensely 'local', involving much smaller communities and political units than today, although these were usually a 'part of a complex hierarchy of political or cultural entities, such as the Church of Rome, the Hanseatic League, or the dynastic Habsburg Empire' (J. Anderson 1986: 115). Sovereignty was not rigidly territorial as it mostly is with modern nation-states (see Chapters 1 and 2). As James Anderson (1995: 70) explains:

> Political *sovereignty* in medieval Europe was shared between a wide variety of secular and religious institutions and different levels of authority – feudal knights and barons, kings and princes, guilds and cities, bishops, abbots, the papacy – rather than being based on territory *per se* as in modern times. Indeed the territories of medieval European states were often discontinuous, with ill-defined and fluid frontier zones rather than precise or fixed borders. Then the term 'nation' meant something very different and *non*-political, generally referring simply to people born in the same locality. Furthermore, the different levels of overlapping sovereignty typically constituted *nested* hierarchies, for example parish, bishopric, archbishopric for spiritual matters; manor, lordship, barony, duchy, kingdom for secular matters. People were members of higher-level collectivities not directly but only by virtue of their membership of lower-level bodies.

In many parts of the pre-colonial world, sovereignty was not based upon fixed boundaries and territorial control per se. For instance, much of North Africa and the Middle East had very different forms of political sovereignty in pre-modern times. According to George Joffé (1987: 27):

> political authority was expressed through communal links and was of varying intensity, depending on a series of factors involving, *inter alia*, tradition, geographic location and political relationships. The underlying consideration, however, was common throughout the region and involved a concept of political sovereignty that derived from Islamic practice. The essential condition was that ruler and ruled were bound together through a conditional social contract in which the ruler could expect loyalty in return for enduring the conditions in civil society for the correct practice of Islam.

Not surprisingly, many of the colonially inspired geometrical boundaries that define the modern states of the Middle East and North Africa have limited relation to pre-colonial political landscapes. This is also true of other parts of the world. In many pre-colonial Asian states, for example, the emphasis of sovereignty was not on the territorial limits of control 'but on pomp, ceremony and the sacred architecture of the symbolic centre' (Clarke 1996: 217). In the classical Indianized states of South and South-east Asia, sovereignty was often focused on rulers who claimed divinity, and further eastwards, emperors held the 'mandate of Heaven', and the mandarins' right to exercise their authority was derived from their being 'superior men' (Sino-Vietnamese, *quan-tu*; Chinese *chun-tzu*) 'who acted according to Confucian ideals' (Keyes 1995: 195). But it would be too simplistic to think of sovereignty purely in terms of emperors, kings, queens, chiefs and so on, as ruler–ruled/state–society relations in pre-modern societies were often complex, hierarchical, shifting and not based on strict territoriality. O.W. Wolters (1982: 16–7) described the scheme of power relations in South-east Asia as *mandala* (a Sanskrit word that defies easy translation, but which refers to a political apparatus that was without fixed boundaries, but which rested on the authority of a central court):

> [The] *mandala* represented a particular and often unstable political situation in a vaguely definable geographical area without fixed boundaries and where smaller centers tended to look in all directions for security. Mandalas would expand and contract in concertina-like fashion. Each one contained several tributary rulers, some of whom would repudiate their vassal status when the opportunity arose and try to build up their own networks of vassals.

This system of tributary relationships carried its own forms of obligations, sanctions and allegiance. Mandalas created complex geographies of power, 'a polycentric landscape–seascape' (Friend 2003: 18), including smaller chiefdoms paying tribute to more than one 'overlord' at the same time. Initially, the multiple sovereignties of the region were very confusing to the European imperial powers in the region who were eagerly trying to carve out their own spheres of unambiguous control. As Theodore Friend (2003: 21) notes in relation to the making of Indonesia:

> The Netherlands required centuries to unify Indonesia in their own fashion: first for mercantilist advancement of trade and then for nineteenth-century motives of geographic empire . . . How could so few succeed over so many? The answer: because only a handful of mandalas had to be overcome, each caring little about the others or knowing nothing of them. The Dutch brought a layer of assiduous modernity to political vacuums strung throughout a vast archipelago. Geographically disconnected and culturally discordant but now administratively centralized, the Netherlands East Indies was for the length of one human generation the first comprehensive empire that region had ever known.

Elsewhere in South-east Asia, there were frequent clashes in conceptions of space and power relations between indigenous polities and the colonial intruders. Thongchai Winichakul (1995: 70) recounts a typical conceptual clash between the British and the Siamese Court. The British in 1845 were seeking to delimit precise boundaries of 'uniform rule' from Chiangmai to the Kra Isthmus, and so they sent an intimidating letter to the Court of Bangkok, advising them to 'issue strict orders along their frontiers so that all subordinate authorities may clearly understand the line of boundary'. In response, the Court's reply confounded the British, for instead of a boundary line the Siamese only recognized a zone of relatively fluid control incorporating 'teak forests, mountains upon mountains, muddy ponds where there were three pagodas . . . and so forth'. As Thongchai explains, the Court saw the outer margins of sovereignty as more a matter for the local authorities to decide, and the Siamese Kingdom had discontinuous boundaries, including some forests and mountains where sovereignty was ambiguous, and some tributaries where sovereignty was shared. It was only by the late nineteenth century that the Siamese rulers began to adopt the concept of territorial sovereignty in the same sense as the British and French were doing in mainland South-east Asia.

Although significant vestiges of such territorial structures remain, today they have mostly been displaced by nation-states with fixed (though sometimes disputed) boundaries. Through the twentieth century, the ideology of nationalism (and the associated idea of **nation-states**) became one of the dominant and most widespread influences on politics across the world, arguably becoming a key (or the key) manifestation of modern territoriality. Occasionally nationalisms recognize that the state itself may be multinational (as in British nationalism, which contains English, Scottish and other affiliations), but in so doing, the wish is usually expressed that somehow a more inclusive national identity will evolve or has evolved which coincides with the boundaries of the state. Britain, China, Switzerland, South Africa, Nigeria and the United States of America are all cases where different versions of the claim and goal of an inclusive 'national identity' that supposedly unites disparate 'sub-nations' or communities have been asserted. Often, the assertion of a particular dominant nationalism in a territory has required the suppression of or conflict with other national claims on the same territory (the emergence of the state of Israel is a clear example of this; and in turn the project of a national home for Jewish people in the land of Palestine is in part a reaction to the genocidal extremes associated with German and other European nationalisms earlier in the twentieth century). As a result many 'national' communities that assert a claim to statehood are denied this (see Case study 20.1, for example). Others remain contested. Burma, for example, has witnessed over five decades of protracted ethnonationalist struggles, particularly between successive military regimes holding power in Rangoon and the predominantly ethnic-Burman heartlands, and various movements in the provinces that are seeking either greater political autonomy within a federal structure or complete independence from the fragile 'Union of Myanmar' (as Burma is named by the ruling military junta) (see Case study 20.2). Similarly a number of 'multinational' communist countries, notably Yugoslavia and the Soviet Union, sought to regulate nationalism by assigning citizens to one of a number of constituent national identities. This sometimes involved inventing nationalities to rationalize and simplify more complex tribal and religious identities, whilst asserting that these should be subservient to an overarching sense of Soviet or Yugoslav identity which coincided with the boundaries of the USSR or of Yugoslavia. So, for example, in the USSR, people could be declared on their identity documents as having one of a number of officially recognized nationalities (Uzbek, Latvian or Russian, for example), but they would also and supposedly above all be citizens of the USSR. In due course, it was in part because of local nationalist challenges to wider Soviet and Yugoslav affiliations that the USSR and Yugoslavia collapsed in the early 1990s.

Many of those self-identified nations without their own state (the Kurds are an example, see Case study 20.1; the Karens, Karenni, Shan and Kachin are other examples, see Case study 20.2, as are Palestinians, Basques and Tibetans) claim either the right to one or at least a high level of self-rule or autonomy. And frequently, smaller recognized national identities within a multinational state (such as Scotland in the United Kingdom or Quebec in Canada) become the basis for claims that they should enjoy full statehood. In many cases too, either the central state or some other community with another affiliation resists. There are numerous examples of this, including the complex case of Northern Ireland, where most Catholics (Republicans) would wish to see the province united with the rest of the Irish Republic (which itself successfully broke away from the British colonial empire earlier in the twentieth century). Most Protestants (Unionists), who claim descent from settler populations from England and Scotland, wish to remain part of a 'United Kingdom' (see Case study 3.1).

The Irish case is just one of dozens of situations where conflicting nationalist and confessional (religious–cultural) logics collide, often with violent consequences. There is clearly something very powerful going on, whereby nationalist visions are linked with particular territories and conceptions of state. Yet, as Anderson (1983: 3) recognizes:

> But if the facts [of the existence of many and sometimes conflicting nationalisms] are clear, their explanation remains a matter of long-standing dispute. Nation, nationality, nationalism – all have proved notoriously difficult to define, let alone analyse. In contrast to the immense influence that nationalism has exerted on the

Regional Case Study 20.1

Nations without states: the Kurdish case

Gerard Chailand (1993: 4) noted that: 'the Kurdish people have the unfortunate distinction of being probably the only community of over 15 million persons which has not achieved some form of national statehood, despite a struggle extending back several decades.'

The lands predominantly inhabited by Kurds are in fact divided between four states (see Figure 20.1): Iran, Iraq, Turkey and Syria (there are also smaller Kurdish communities in the former Soviet Union – Armenia, Azerbaijan and Georgia – and significant Kurdish migrant populations in Europe, especially in Germany).

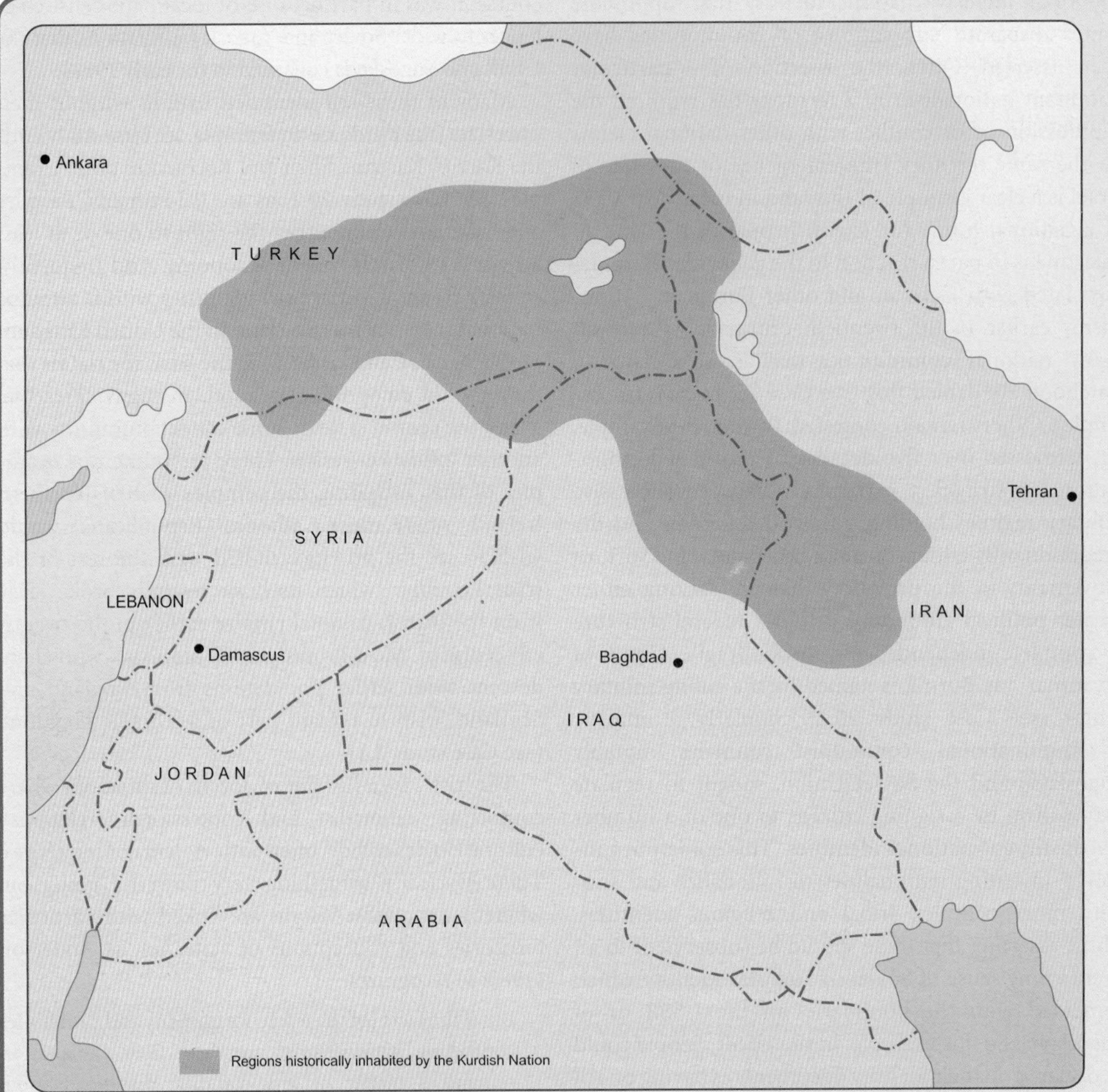

Figure 20.1 Regions historically inhabited by the Kurds.
Source: After Chailand (1993: ix)

Despite early twentieth-century promises from the then imperial powers (including Britain and France), in the end it suited no great power to establish an independent Kurdistan. As in much of Asia (and beyond), the borders were determined in large part by foreign imperialists, taking little or no account of the wishes and rights of local peoples. The rise of Turkish, Iranian and Arab nationalisms – themselves all partly responses to imperial domination – left little space for the Kurds.

Yet frequent attempts on the part of the states in the region to eradicate Kurdish identity and nationalism have not been able to quash either the widespread sense of Kurdishness or the often violent struggle in pursuit of a Kurdish nation-state. After the invasion of Iraq by the USA and its allies in 2003 and overthrow of Saddam Hussein's centralized dictatorship there, the northern part of the country – which is predominantly Kurdish – became an autonomous region. However, neither the occupying power, nor Iraq's neighbours have wanted to see this become the basis of a separate Kurdish state.

modern world, plausible theory about it is conspicuously meagre. Hugh Seton-Watson, author of by far the best and most comprehensive English-language text on nationalism, and heir to a vast tradition of liberal historiography [theories of history writing] and social science, sadly observes: 'Thus I am *driven* to the conclusion that no "scientific definition has existed and exists." '

Readers may wish to 'prove' Seton-Watson's observation for themselves, by trying to come up with a universally valid definition of a nation. Faced with this task, students will often work through a long list of characteristics ascribed to nationality. But it seems that exceptions can always be found. Language is a favoured criterion, but then many languages are spoken by more than one 'nation' (English is an example) and some 'nation states' (Switzerland, Belgium or Mozambique, for example) contain substantial communities speaking different languages. Religion is sometimes chosen, but the same objections apply. Ethnicity and 'race' turn out to be problematic criteria, favoured by racists of all stripes, and often part of the basis for national identities, but susceptible to the obvious and undeniable point that everywhere is much too mixed up historically and genetically for such categorizations to be watertight. Besides, some nationalisms have come to celebrate their multiracial and multicultural composition, as in the 'melting pot' United States or the 'rainbow nation' of South Africa. 'Culture' usually crops up as a criterion. Yet as Chapter 13 has indicated, cultures are always (though of course to varying degrees) heterogeneous and contested. Think, for example, of age, gender, class and other variations and the coexistence of multiple subcultures and identities (which, as Chapter 19 has shown, are often related to distinctive 'local' territorialities) that characterize every supposedly 'national society'.

All this leads Anderson to declare that nations are in a sense 'imagined communities'. This imaginary status is not to deny that they are not in a sense real to those who feel they belong to them. Indeed, Anderson (1983: 6) feels that:

> In fact, all communities larger than primordial villages of face-to-face contact (and perhaps even those) are

Regional Case Study 20.2

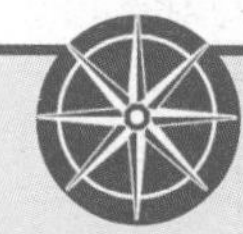

Burma: a national geo-body or a mosaic of ethnic homelands?

According to Chao Tzang Yawnghwe, a son of the first President of the Union of Burma, who took an active part in the Shan resistance movement in the mid-twentieth century: 'I feel that the greatest flaw in current works dealing with post-1948 Burma is the confusion over the term "nation-building" in general, and more specifically, its connotations within the internationally political perimeter known as Burma, which, in reality is a composite of many homelands' (Yawnghwe 1987: ix).

From a geographical perspective we could say that Burma is composed of the broad plain lying on both sides of the Irrawaddy River that flows into the sea at the Gulf of Martaban and the Bay of Bengal; and surrounding this plain there are extensive border regions populated by different ethnic groups. In each of these 'ethnic minority' areas (in Burma as a whole) the 'majority' Burman people are hardly present. It is these border states that Yawnghwe (1987) refers to as the ancestral 'homelands' of different ethnic groups, such as the Karen, Karenni (Kayah), Kachin, Shan or Tai, Chin, Arakan, and others (see Figure 20.2).

During the nineteenth century, the British colonial rulers sought to try to manage the complex political and ethnic mosaic that came to be defined as Burma, by dividing the colonial administration into 'Burma Proper' (covering much of Lower Burma of the Irrawaddy) and the so-called 'Frontier Areas' (encompassing the Kachin, Chin, Karenni uplands, the Karen Salween district, and the Shan plateau). In fact, the 'Frontier Areas' included almost 50 per cent of the total land area and about 16 per cent of the total population of Burma. However, it was only 'Burma Proper' that was to be under a highly centralized and direct form of bureaucratic rule, whilst the 'Frontier Areas' were thinly administered with forms of indirect rule that left some of the tiny chiefdoms with more political autonomy. By imposing different degrees of administration on the lowland and hill people 'colonial modernity began the institutionalization of ethnic categories' (Lang 2002: 31), which was to become intensely politicized in the years following Burma's independence in 1948.

In the extensive border regions (Figure 20.2), there have been many battles for control over territory, resources and people between the Burma Army or *tatmadaw* and the 'sub-state' nationalist movements,

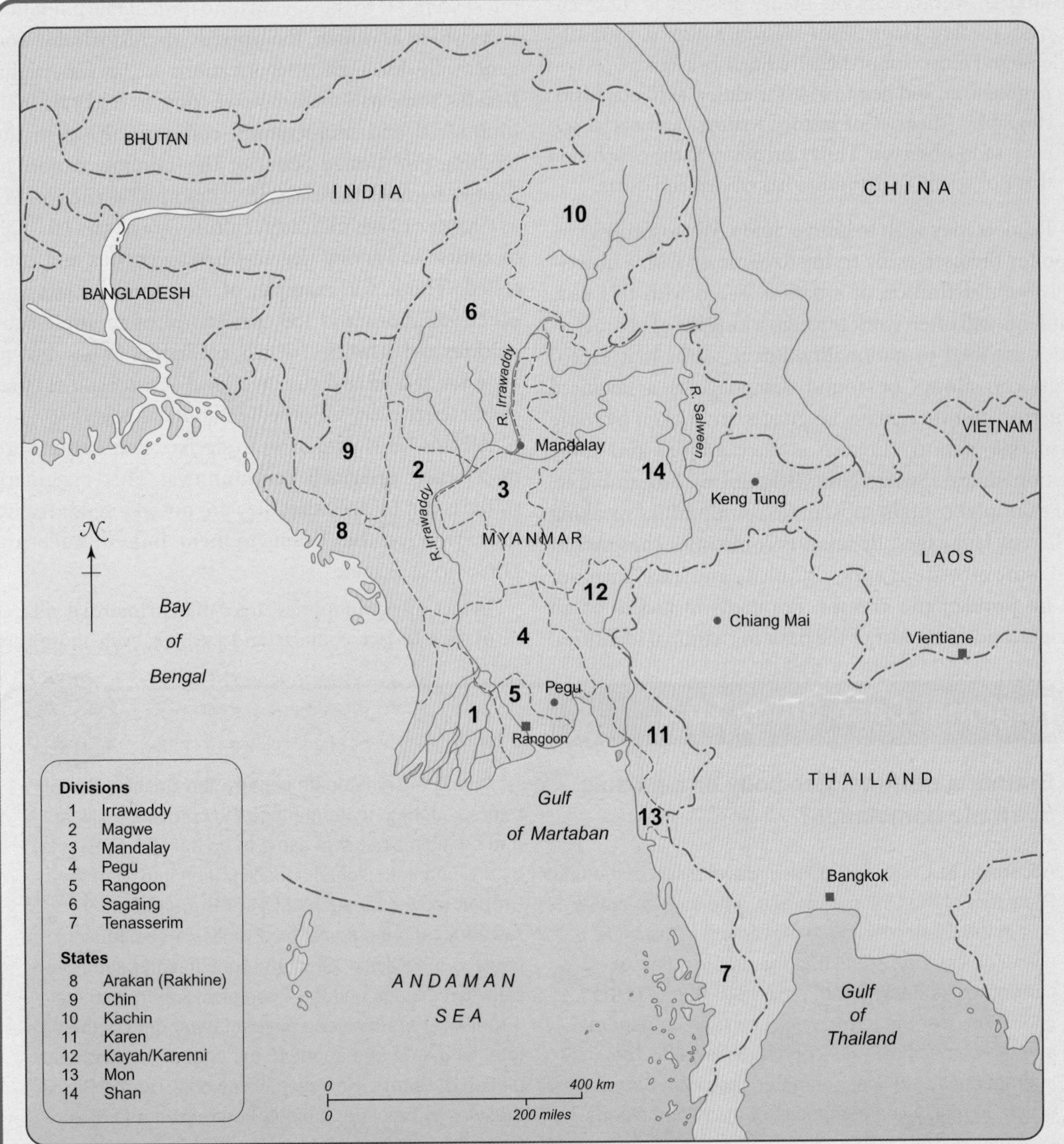

Figure 20.2 Burma: national geo-body or mosaic of ethnic homelands?

such as the Karen National Union (KNU), the Karenni National Progressive Party (KNPP), various Shan factions, and the Kachin Independence Organization (KIO), to name a few. Each of the main ethno-national groups maintained armies and for much of the period since independence they controlled large parts of what they called their 'national homelands'. Ethnic groups with sizeable populations straddling international boundaries, such as the Kachin people across the Burma–China border, have also been able to develop strong social and cultural connections beyond these homelands. Recent years have witnessed ongoing conflict in many of these troubled border regions of Burma. Countermoves by the Burmese army have led to huge numbers being forcibly displaced and many have fled to Thailand or live in constant fear of the Burma Army (Lang 2002; Grundy-Warr and Wong 2002).

imagined. Communities are to be distinguished, not by their falsity/genuineness, but by the style in which they are imagined.

The style in which nations are imagined (and therefore the way that they become to seem real by those who belong to them) is, according to Anderson (1983: 7), as 'limited', 'sovereign' and as 'communities':

> The nation is *imagined* as limited because even the largest of them, encompassing perhaps a billion living human beings, has finite, if elastic boundaries, beyond which lie other nations. . . . The most messianic nationalists do not dream of a day when all the members of the human race will join their nation in the way that it was possible, in certain epochs for, say, Christians to dream of a wholly Christian planet. . . .
>
> It is imagined as *sovereign* because the concept was born in an age in which Enlightenment and Revolution were destroying the legitimacy of the divinely-ordained, hierarchical dynastic realm . . . nations dream of being free . . . The gage and emblem of this freedom is the sovereign state.
>
> Finally, it is imagined as *community*, because regardless of the actual inequality and exploitation that may prevail in each, the nation is always conceived as a deep, horizontal comradeship. Ultimately it is this fraternity that makes it possible, over the past two centuries, for so many millions of people, not so much to kill, as willingly to die for such limited imaginings.

Plate 20.2 'Nation' imagined as race.
(Jim West/Alamy)

Within all this is a profound territorial link between the nation and the state. The state claims to be a sovereign expression of the nation – bound to it and to a particular place. That is, it has territorial limits. The nation-state has a *geography*, which is charted, demarcated, mapped and represented to the 'national population' in their school atlas and geography lessons. Such a system of 'national' geographical representation is always combined with an historical vision, a grand narrative of 'national history', often assuming that the nation is ancient, even primeval.

Yet Anderson and other critical accounts of nationalism stress that it is very much a modern ideology. For what makes mass nationalism possible are certain socio-economic and technical transformations, notably the arrival of media and national educational systems. Schooling, newspapers, and later radio and television all help to promote and popularize the idea that people belong to and share in the nation. Whilst others have emphasized the longer historical roots of many nationalisms in pre-modern ethnic affiliations (for example, Smith 1988), for Anderson and most other critical observers, what is striking is nationalism's relative modernity. Even where nationalists imagine the nation as ancient, such an imagination is itself more often than not predominantly a nineteenth- and twentieth-century phenomenon. At a pinch, nationalism might be traced to the seventeenth century. But the ideology of the nation-state was not anywhere very evident before then. So whilst someone might talk about a thousand years (or more) of, for example, English, Korean or British or Indian national history, they conveniently forget that English or British, Korean and Indian nationalisms are relatively modern concepts, much less than a thousand years old.

The tendency for nationalists to 'reinvent the past', to pick out selective moments from the past, or to manipulate history, is one of the main arguments of historian Eric J. Hobsbawm (1996: 255), who argues:

> For historians are to nationalism what poppy-growers in Pakistan are to heroin addicts: we supply the essential

raw material for the market. Nations without a past are contradictions in terms. What makes a nation *is* the past; what justifies one nation against others is the past and historians are the people who produce it.

Whilst history is undoubtedly significant, the nationalist brew would be incomplete without ancestral connections to 'homeland' and to particular places. In other words, *territory* is central to nationalism, and this is often why whenever there are divergent nationalist claims to one piece of land, extreme violence often follows. Indeed, the terrible forms of so-called 'ethnic cleansing' witnessed in the former Yugoslavia in the 1990s are illustrations of recent nationalist extremism and the significance of historical imaginations concerning territorial–political and cultural identity.

Nationalist narratives and claims are not always linked to such violent actions. At times they may be rather more rhetorical or theatrical. Graham E. Clarke (1996: 231–2) raises a particular Nepalese example:

> [I]n the 1960s in Kathmandu the sole national newspaper carried articles, no doubt read tongue in cheek by some educated Nepalese, arguing that since Lumbini, the birthplace of the historical Buddha (Gautama), was located some few five miles north of the current southern border, that Buddha was *therefore* Nepalese and not Indian.

In a sense, the nationality claims to Buddha are no more or less absurd than the Macedonian and Greek counter-claims about the 'national' identity of a distinctly pre-nation-state historical figure – Alexander the Great. As Loring Danforth (1995: 171) writes:

> At the level of popular discourse both Greeks and Macedonians make extensive use of Alexander the Great as a powerful symbol of the historical and cultural continuity, which, from their perspectives, links them with their glorious ancestors, the ancient Macedonians. Demonstrations held throughout the world in the early 1990s by Greeks and Macedonians in opposition to or in support of international recognition of the Republic of Macedonia were led by men dressed as Alexander the Great, wearing a crested helmet, breast plate, and greaves and holding a shield and a spear. Each group was outraged that the other group had attempted to appropriate its own famous ancestor for such obviously political purposes.

All of this relates to the important issue of how we deal with things like space and identity in history. As Morris-Suzuki (1996: 42) writes:

> The nation . . . casts a long shadow backwards on our vision of the past, and channels our perceptions into a particular spatial framework. In my bookcase, I have a volume on the history of Thailand since the tenth century, which, considering the repeated political and cultural realignments within the space we now label 'Thailand', seems only a little more bizarre than its neighbour on the shelf, a history of the Soviet Union from palaeolithic times to World War II.

Claims to nationhood frequently involve the blending of 'tangibles' and 'intangibles' in a unique brew that may contain such volatile ingredients as blood, language, race and religion. A fundamental goal of nationhood is to generate a strong *sense* of belonging associated with a particular territory. In practice it may prove difficult to distinguish between 'the nation' and other human collectivities, but as Walker Connor (1994: 93) puts it, 'what ultimately matters is not *what is* but *what people believe is*'.

There has been a strong tendency to see nations and nationalisms as essentially being derived from American and European prototypes. It is important to challenge the idea that nationhood is a nationalist project originating in the West that was simply implanted on to the rest of the world without significant transformations in the various cultural, ethnic and religious contexts it touched. African, Middle Eastern or Asian variants of nationalism or constructions of statehood are far more than simply colourful (or failed) replicas of the European–American model(s). To be sure, imported idioms and motifs from Europe and America have been important; for example, the official state language in a fair number of former colonies is that of the old colonial power. And European (and less often United States) imperialism has provided an important backdrop to the trajectory of nationalisms in Africa and Asia, not least because the anti-colonial nationalists themselves were often educated in Western institutions and adopted concepts such as self-determination, national liberation and territorial sovereignty in their political struggles against the colonial powers. Yet whilst 'globalizing' capitalism and especially imperialism are part of the picture, African, Middle Eastern and Asian nationalisms are also (like those of Europe and the Americas) rooted in *local* historical trajectories. In many cases these are of much longer vintage, often based on sophisticated cultural, religious and ethnic tapestries, some of the main patterns of which have subsequently been appropriated or (re)invented as pre-colonial 'national' histories.

Thus, we should perhaps think in terms of multiple histories and geographies of nations and states, and not just of particular European or American modular forms. The influential Bengali intellectual Partha Chatterjee (1993) has persuasively argued against the imagining of Indian national identity through the lens of the colonial power and stresses the 'essential' inner or spiritual domains of culture that were never colonized, never European. Furthermore, there are different coexisting national voices or 'fragments' – among women, peasants, elite, castes, outcasts, and so on – each with its separate discourse. As Stein Tønnesson and Hans Antlöv (1996: 32) put it:

> When the national idea entered Asia it could not be implemented without mediation, hence transformation, by indigenous agencies in particular settings. There were existing and alternative ideas with which European-style nationalism interacted and intermixed . . . the nationalist ideas were invested with local qualities, meanings and nuances which could not be found in Europe. People had their own views on what constituted a legitimate social order, and such views could not be ignored by the modernizing leaders of the anti-colonial struggles.

As well as different cultural 'forms of the nation', we may also consider different gendered perspectives of nationalism and nation-states. Although Chatterjee's (1993) arguments are quite specific to the South Asian case, they do reveal the coexistence of different voices, men and women, and this relates to the point about nationalism and constructions of nation-statehood invoking different conceptions of 'manhood' and 'womanhood' more widely. Although there are immense historical and geographical variations of this, there is a general tendency for women to be seen as particularly important transmitters of 'national culture' (for example, in the idea of a 'mother-tongue' language) (Yuval-Davis 1997) and as somehow the embodiment of the nation. McClintock (1994: 352) thus notes that:

> Nations are contested systems of cultural representation that limit and legitimize people's access to the resources of the nation-state, but despite many nationalists' ideological investment in the idea of popular *unity*, nations have historically amounted to the sanctioned institutionalization of gender *difference*. No nation in the world grants women and men the same access to the rights and resources of the nation-state.

Moreover, particular ideas of the 'family' are central to most imagined communities:

> Nations are frequently figured through the iconography of familial and domestic space. The term nation derives from [the Latin] *natio*: to be born. We speak of nations as 'motherlands' and 'fatherlands'. Foreigners 'adopt' countries that are not their native homes and are naturalized into the national 'family'. We talk of the 'Family of Nations', of 'homelands' and 'native' lands. In Britain, immigration matters are dealt with at the Home Office; in the United States, the president and his wife are called the First Family. Winnie Mandela was, until her recent fall from grace, honoured as South Africa's 'Mother of the Nation'. In this way, despite their myriad differences, nations are symbolically figured as domestic genealogies.

In several European nationalisms (including 'British') one significance of this is that the sexist notion of women (usually grouped with children) supposedly as *naturally* inferior to men, or as requiring the protection of *naturally* more powerful men, frequently provided the prior backdrop to depicting hierarchies (between dominating and dominated classes, for example) *within* the nation as natural, like those of a family. A similar conception provided part of the racist justification for overseas colonies. Peoples subject to colonial domination were often represented in such racist terms as being equivalent to 'families' of black 'children' ruled by a 'benevolent' white father.

20.3 Constructing boundaries

States come in diverse shapes and sizes. But they all have boundaries (which, if they have a coastline, also extend into the sea and are governed by international protocols: see Stienberg 2001). The extract from Winichakul (1996) quoted at the start of this chapter noted how the 'territoriality' of the modern nation-state is of a bounded space: 'a certain portion of the earth's surface which can be easily identified'. Pre-modern dynastic realms and empires could make do with loose boundaries. But modern nation-states have felt it necessary to *demarcate* their boundaries, to iron out perceived irrationalities and anomalies, sometimes by going to war. In short, they have sought to nationalize and unify 'their' space and 'their' (national) populations and render them into known, surveyed and defensible sovereign territory and communities. As Chapter 21 explores, the rise of geopolitics in the late nineteenth and early twentieth centuries represents a particular expression of just such a rationality, but it has wider forms and deeper roots. The nineteenth-century invention and refinement of statistics (*state-istics* = the science of the state) and demography (*demos* = people, *graphy* = writing) and the establishment of geography (*geo* = earth, *graphy* = writing) as a university discipline and field of research are all part of this regime of knowledge and power where people are surveyed and made into subjects of nation-states. For states with land boundaries the occupation and official demarcation of the frontiers is an important part of this process.

Consider the example of the border that is often recognized as the *oldest*, more or less stable, still-existing border in the world, that between Spain and Portugal. The mutual recognition of the border between Spain and Portugal is usually traced in modern Spanish and (particularly) Portuguese history to the Treaty of Alcañices signed on 12 September 1297 (in fact, the date on the Treaty is 12 September 1335, but amongst other things, our conventions for counting years have altered since). The Treaty has acquired something of the status of a foundational text. In other words, it is interpreted as a kind of proof of the ancient historical basis of the nations concerned.

Yet the treaty was *not* signed in the name of states or 'nations'. Named after the Templar castle in which it was sealed and witnessed, the text of the treaty of Alcañices begins with the words 'In the name of God. amen.' and is signed in the presence of the Templars and other holy orders by those who describe themselves as:

> by the grace of God the King of Portugal and the Algarve and by the grace of God the King of Castille and Leon,

and has as its subject:

> towns, castles and lands, town boundaries, divisions and orderings . . . [disputes over which] have caused many wars, homicides and excesses, as a consequence of which the lands of both [kingdoms] have been looted, burnt and ruined, weighing heavily on God . . . [and] because of our sins, risking the danger of losing them and them falling into the hands of our enemies in the faith and most gravely [causing] the violation of God's will and injury to the holy church of Rome and Christendom.
>
> (Cited in Martíñez 1997: 15; our translation)

Moreover, the actual demarcation of the Portuguese–Spanish borderline had to wait until the nineteenth and twentieth centuries. Two formal delimitation treaties set out an agreed frontier and established mechanisms for its physical demarcation on the ground. But gone are 'amens' and references to 'the will of God'. Instead, the treaties (the first of 1864 for the northern half of the border and the second of 1926 for the southern half of the border) make reference to the need to impose order and well-defined demarcations, to 'eliminate the anomalous situation in which, in the shadow of ancient feudal traditions' (Tratado de Limites 1866: 1, our translation), some frontier areas (including a number of villages) were recognized as *shared* or common lands with usage rights by communities of both states/'nationalities'. Following the treaties, such areas were divided and the border, where not demarcated by a river, was marked on the ground by rectangular boundary stones every few hundred metres (although it took a couple of decades to put in place all the boundary markers and resolve local conflicts and differences over the 'fine-scale' division of lands). In other words, the Portuguese–Spanish case indicates how the epoch of modern nationalism (i.e. the nineteenth and twentieth centuries) saw what James Anderson (1995: 71) calls:

> a territorialisation of politics, with a sharpening of differences at the borders of states and of nations between 'internal' and 'external', 'belonging' and 'not belonging', 'us' and 'them'.

By the same logic, or an extreme extension of it, minorities who do not 'fit in' sometimes have to be moved, murdered or deported – or, in the euphemisms of our times, 'cleansed' (see Sibley 1995). This has been associated with, for example, Nazi racial territorial visions of a Europe cleansed of Jews and Gypsies and an expanded racially pure Germany. The Nazi Holocaust represents an extreme case. But 'cleansing' or 'purification' of national space, amounting to the expulsion or murder of people defined as 'foreign' or 'alien', has been widespread in the twentieth century. There are examples from all continents: the creation of modern Turkey whose Armenian population were slaughtered in the second decade of the twentieth century (and which still has not resolved the status of its Kurdish minority); the creation of Israel in the 1940s, which was accompanied by the flight of indigenous Palestinian Arabs from territories deemed part of the new Jewish national state; the system of reservations for native Americans in the USA and Canada; the violent exchange of Muslim, Hindu and Sikh populations which accompanied the 1947 partition of India; the 1974 partition of Cyprus; the wars of the 1990s in the former Yugoslavia (see Case study 19.2); and the partition of Ireland (see Spotlight box 20.1). These logics have also produced calls for the modification of boundaries, sometimes resulting in violence and war.

Spotlight Box 20.1

Irredentism and secessionism

Irredentism is defined by Mark Blacksell (1994: 299) as: 'The assertion by the government of a country that a minority living across the border in a neighbouring country belongs to it historically and culturally, and the mounting of a propaganda campaign, or even a declaration of war, to effect that claim.' Whilst Blacksell focuses on official irredentism, sometimes irredentist parties and organizations promote the claim without it necessarily becoming their government policy. Often the irredentist claim includes areas where populations are mixed, making for potentially violent conflict. The collapse of Yugoslavia and the Soviet Union in the early 1990s were accompanied by many irredentist claims. But there are dozens of other examples around the world.

Often closely related to irredentism is the phenomenon of secessionism, whereby a minority in one state seeks the transfer of a section of the state's territory to another new or existing state. Although it has also had its own 'local' cultural, religious and political dynamics (see Case study 3.1 and Case study 19.3), the conflict in and over Northern Ireland also exemplifies irredentism and secessionism. The Irish Republic was born of a secession from the British Empire (whilst at the same time being seen as the first of many national liberations for colonized peoples) in the 1920s. The Irish constitution long claimed sovereignty over the whole of Ireland, thus making an irredentist claim on that part which remained under British jurisdiction, whilst nationalists in what they often called the north of Ireland (refusing the term Northern Ireland which signified a separate jurisdiction) sought secession and a unified Ireland.

Just over a decade ago, James Anderson (1997: 216) suggested that: 'Northern Ireland is the residue of failures in nation- and state-building, whether viewed from either a British or an Irish perspective. The disputed labels for territory – repeated at a local level in, for example, Derry/Londonderry (or, ironically, "Stroke City") – reflect the rival national identities and suggest that the ideal of the nation-state is unachievable.'

Today, Northern Ireland remains part of the United Kingdom, but with an autonomous power-sharing government comprising both nationalist and unionist forces. And the different terms used to describe this territory signify different visions of its role and status: 'the province' (which could be a province of the United Kingdom and/or of the Irish Republic and/or part of Ulster; a place apart), the north of the island of Ireland or Northern Ireland.

No continent has been without examples of such conflict and violence and it is probably impossible to find a state which has not at some time expelled people or murdered them in the name of some nation or other. Yet each nationalism must also be unique, establishing itself as different from others (even if the apparatuses of flags and anthems are superficially similar), constructing a sense of self against others who are defined as outside the imagined community. We will return to borders and borderlands in section 20.5.

20.4 Nation-states as symbolic systems

As we have seen, the state is the bureaucratic expression of nationalism. This is not to say that a widespread sense of nationalism inevitably precedes the state. Indeed, what is often termed the state apparatus (everything from regulations governing schools, to tax inspectors and politicians) seeks to foster national subjectivities out of the frequently ambivalent and disparate array of identities contained by its boundaries. If, as has been argued by many, nationalism is akin to religion, then the state becomes its symbolic structure. When examining the superficially quite different cases of Australia and Sri Lanka, Bruce Kapferer (1988) claims that nationalism is itself a religion, owing to the fact that, as with most religious-like beliefs, nationalism demands the recognition (his word is 'reification', which means something more than this) of an immanent and all-encompassing entity (the nation). The nation fulfils the role of a sacred cause, something greater than any individual and something which may be worth dying for. The nation is the God or deity of the religion of nationalism and the state is its theology or temple. Moreover, as Herzfeld (1992: 37) notes:

> Every bureaucratic action affirms the theology of the state. Just as nationalism can be viewed as religion, bureaucratic actions are its most commonplace rituals. There are other such everyday rituals: Hegel saw the

Plate 20.3 Representing the limits of the 'nation-state': boundary patrols.

(David R. Frazier Photolibrary, Inc./Alamy)

reading of the morning newspaper as the secular replacement of prayer.

There are indeed many public rituals of nationality and statehood: coronations and remembrance days, military parades, national holidays, national prowess (or the lack of it) at football or the Olympics, swearing in of governments, state funerals. (Football in the United Kingdom is also revealing of the complexity of British nationalisms, and the survival of separate Scottish, English, Welsh and Northern Irish squads is a testimony to the limits to 'Britishness' and the survival or 'revival' of other national affiliations.) In all these rituals, the nation is reaffirmed and the state performed and made to seem omnipresent, historical and real. Dramatic cultural or political events (as well as notable national sporting occasions, such as soccer World Cup or Olympic successes) covered by the media reinforce senses of national community. The events of 11 September 2001 were thus narrated in the United States as an 'attack on America'. It is also said, for example, that virtually every adult *American* alive at the time can recall where they were on hearing the news of the assassination of US President John F. Kennedy in 1963. Whilst his funeral, like that of Diana Spencer in 1997, became a global media spectacle, it was represented and felt most acutely as a kind of national loss. 'Goodbye *England's* Rose . . .' as Elton John put it at Diana's funeral. (It should be noted that there was some controversy over the lyrics of Elton John's tribute song, in so far as Diana was not only the Princess of *Wales*, but also a former member of the avowedly *British* monarchy. Rather like the case of football, this bears out the complexities and unresolved tensions of British nationality.)

But the nation-state also demands (like all effective religions) personal commitment and more minor ritualistic acts, many of which rest on a geographical imagination of inside and outside, belonging and otherness (see Taussig 1997). The anthropologist Michael Herzfeld (1992: 109) examines this theme and is worth quoting at some length:

> Nationalist ideologies are systems of classification. Most of them are very clear about what it takes to be an insider. That, at least, is the theory. In practice, however, divergent interpretations give the lie to such essentialist claims, as to take one prominent and current example, in the debate currently waging in Israel about the definition of a Jew. Such taxonomic [naming] exercises . . . are central to the very existence of the nation-state. All other bureaucratic classifications are ultimately calibrated to the state's ability to distinguish between insiders and outsiders. Thus . . . one can see in bureaucratic encounters a ritualistic enactment of the fundamental principles upon which the very apparatus of state rests. Seen in these terms, arguments about the number on a lost driver's license or an applicant's entitlement to social security do not simply challenge or reinforce the power of particular functionaries of state. They rehearse the logic of the state itself.

Not only that, but the whole exercise of state power gets taken for granted as the natural order of things. Only when many of those activities which are ascribed to the state are no longer carried out (in situations of war, for example) or when a person finds themselves on the wrong side of a state-sanctioned category (the wrong side of the boundary, the wrong side of the law) is the power (a power over life and death and thousands of lesser things) and the universality of the symbolic order of 'the state' revealed. The state claims the monopoly over these things. And for others to exercise judgment and to punish or to kill is 'to take the law into

Plate 20.4 Stansted Airport, UK – passport control signage.
(BAA Aviation Photo Library)

their own hands'. These things are reserved for the territorial state. But what if there is no state? Or as cultural anthropologist Bernhard Helander (2005: 193) asks: 'Who needs a State?' There are parts of the world where any semblance of effective state sovereignty has become so scattered, fragmented or *in*effective in practice that other forms of rule and competing authority structures take over. For instance, Helander wrote about the complex power situations in Somalia affecting every single aspect of daily life, where 'any opinion expressed or action performed must always be positioned in the political landscape' of rival clans, warlords, and issues of ethnic, religious and cultural identity. In the contexts of atomistic and antagonistic power struggles we need to appreciate the local power configurations that enable communities to continue with everyday life in the absence of a meaningful state apparatus. Helander was quite critical of international organizations that tend to act too rashly in countries such as Somalia, often distributing massive sums of inappropriate aid according to the 'shifting agendas of foreign donors' and frequently without great familiarity with the nuances of the 'local' human, cultural and political landscape. Such perspectives would clearly be even more applicable in situations of *failed international interventions* creating uncertain and extremely dangerous regime changes, unpopular occupations, and an almost complete absence of any form of security for large segments of the population. This happened in much of Iraq after the US-led invasion to depose the government of Saddam Hussein in 2003. Arguably this was also the situation in Afghanistan in the 1980s, after the Soviet Union invaded to prop up a communist government in the capital city who faced rebellion in the more conservative countryside. Afghanistan then became a focus of Cold War geopolitical conflict (as will be detailed in Chapter 21), laying the basis for many more years of confrontation, subsequent interventions (by the United States and its allies in the early twenty-first century) and enduring state failure. In such situations, we may actually have to ask, 'What kind of State? Where and who are the authorities?'

20.5 Sovereigntyscapes: 'shadows', 'borderlands' and 'transnationalisms'

We have seen that the geographies of 'nations' and states do not always coincide (consider again the Kurdish case, described in Case study 20.1). Thus sovereignty is rarely without contradictions or challenges from other states, potential states and nations. But how else might we think about the challenges to sovereignty and the historical and geographical complexity of nations? One way might be to think of political landscapes as 'soveriegntyscapes' (Sidaway 2003a; Sidaway et al. 2005) containing varieties of countervailing tendencies, fragmented state sovereignty, ambiguous forms of sovereign control, and situations (such as Somalia, as noted above) where state authority may have collapsed or been replaced by various contending de facto forms of authority. Even when we consider the

Plate 20.5 War-making and with it a sense of shared struggle can be productive of national identity.
(© Bill Ross/CORBIS)

state as strong and central, we may view sovereignty as being 'graduated', whereby, as Aiwha Ong (1999: 217) observed, 'states make different subject populations, privileging one gender over the other, and in certain kinds of human skills, talents and ethnicities; it thus subjects different sectors of the population to different regimes of valuation and control.'

In addition to highly uneven and complex 'sovereigntyscapes' there are economic, political, cultural and social forces that could be considered to be largely non-state 'shadow powers' that are able to operate across national, regional and international borders. These may have parasitic relations to conventional sovereignty. Thus Carolyn Nordstrom (2000, 2004) has examined examples of 'shadow powers and networks' within numerous 'war torn' countries. The notion of 'shadow' here enables us to think of 'extra-state' transactions that transcend territorial boundaries, but are not 'officially' sanctioned by states, yet may involve a wide amalgam of actors and agencies, including state functionaries within complex transactional processes and extended networks. Undoubtedly, such 'shadows' are often associated purely with powerful international crime syndicates such as the Italian mafia, Russian 'Mafiya', Japanese *yakuza*, Chinese *Snakeheads*, and so on (Lintner 2002). In practice, 'shadow networks' may also incorporate a great many people who are 'deeply immersed in society and civil life' (Nordstrom 2000: 46). As she notes,

> 'The relations of power and exchange I am concerned with here cross various divides between legal, quasi-legal, gray markets and downright illegal activities', but equally they may involve perfectly legal operators and exchanges that 'take place outside the formal state institutions'. (Nordstrom 2000: 36)

Nordstrom's analysis of 'shadow networks' is particularly relevant to the growing literature on the complex political economies of environmental resources within situations of conflict (Peluso and Watts 2001; Le Billon 2001), but it is important to remember that in the contexts of 'war torn' societies, such as Angola, Sudan, Congo, Afghanistan and Burma, 'the dangerously criminal, the illicit and the informally mundane cannot, in actual practice, be always or easily disaggregated' (Nordstrom 2000: 42). The 'shadow networks' are not simply chaotic and disorderly but often have complex 'hierarchies of authority, rules of conduct, ways of punishing transgression and codes of behaviour'. Furthermore, these networks are often far more significant economically and politically to the functioning of such societies than the malfunctioning apparatus of the official 'sovereign states'. In such contexts, the official state-centred political map appears to be little more than an international juridical nicety without real de facto force (Jackson 1990). More than this, the research of Nordstrom and others (Strange 1996; Lintner 2002) leads us to consider 'non', 'extra' and 'quasi' state forms of spatial power alignments and inter-territorialities. They lead us to (re)consider questions such as *who* wields authority in the junctures and 'borders' of state/non-state transactions? *Who* and *what* constitutes the state when 'shadow networks' dominate the political economy? *Why* and *how* do various communities, civil society groups and individuals become entangled in 'shadow networks'? Certainly, these questions relate to those situations of state 'failure' or 'collapse', which turns out to be much more complex than being a mere absence of authority.

A variation on the theme of 'shadow networks' is to consider myriad cross-border connections between people that somehow circumvent or subvert state-centred rules and regulations. In some contexts, such as the European Union, cross-border links are actively fostered and promoted (along with a common market and internal freedom of capital and personal mobility) with the aim of fostering a supranational (beyond nation) community (Kramsch and Hooper 2004). In other cases (such as the US–Mexican border) or places where there is large scale 'illegal' movement of people and goods, state power is defied or subverted. In this regard, Abraham and van Schendel (2005) draw upon a critical distinction between what states define as 'legal/illegal' and what ordinary people perceive as 'licit/illicit'. Thus many transnational movements of people, commodities and ideas are illegal because they defy the norms and rules of formal political authority, but they are quite acceptable, 'licit', in the eyes of the participants in these transactions and flows. Here the authors were not so concerned with the flows associated with big syndicates, but with the many 'micro-processes' and transactions that form 'everyday transnationality' in borderland spaces (van Schendel 2005: 55). Such transactions may involve kinship and family networks that were partitioned by superimposed boundaries, but revived subsequently in new forms. We should try to avoid overly dualistic perspectives of 'trans-border' or 'transnational' flows, such as 'domestic/foreign', 'internal/external', 'legal/illegal', because by doing so, we would be perceiving flows entirely from rigid state-centred and fixed territorial positions. Even if we just consider state practices such as the enormous efforts of the US federal authorities in trying to keep 'illegal aliens' 'out' (Plate 20.3), there are many more countervailing actions at official, quasi-official, unofficial levels, trying to make sure that as many people 'get in' because of their enormous contributions to economic sectors (Andreas 2000). As van Schendel (2005: 61) observes:

> The challenge is to look at territorial states, transborder territorial arrangements, and transnational flows as *complementary* elements in processes of global reterritorialization. The current drift of these processes appears to be toward making institutions at the national level less central and toward strengthening direct links between localized arenas (e.g. borderland networks) and supranational ones.

Why is such work of relevance to ideas about 'nation' and the 'nation-state'? It is precisely the very everydayness of many cross-border interactions, movements and connections that raises big questions about the impression of solidity of spatially fixed notions of nationhood. For instance, Case study 20.2 examines the complex spatial mosaic of 'ethnic homelands' within the geo-body of Burma. If we consider the fact that all of Burma's boundaries are relatively porous with multiple 'leaks' over which the state has varying degrees of knowledge, influence or control over, and where there are several active ethnic political organizations with either fragile ceasefires with or ongoing resistance to the regime, then we should perhaps not conceive of Burma as having solid boundaries at all. Despite attempts to fortify and police it, the southern border of the United States and the maritime and land frontiers of many southern and eastern member states of the European Union exhibit similar porosity. We may think of all these border spaces rather like a 'frontier' with multiple ethnic and commercial interactions and sometimes with interlocking identities (Leach 1960). Alternatively we may prefer to examine the myriad social, economic, cultural, political and everyday transnational processes that have successfully kept *other than 'Burman'* 'national' identities, such as that of the Kachin people, alive both *within* the territory and *in spite of* it across state borders (Dean 2005) – or indeed foster other transnational identities, such as Hispanics in the United States or Maghrebis in Europe.

20.6 Conclusions: the place of the nation-state?

There have never been so many states as there are today. In the nineteenth century a wave of new states (countries such as Argentina, Mexico, Bolivia) emerged in the Americas in revolt against Spanish and Portuguese empires. These American prototypes and the USA itself (which dates from 1776) provided an example to nationalists elsewhere in the colonial world. But it was not until the twentieth century that most of Africa and Asia could escape direct colonial domination. This often required violent 'national-liberation' struggles against entrenched resistance from the colonial powers and white settler populations. In the decades between about 1945 and 1975, dozens of new 'sovereign' states emerged in place of the old colonial map. And if, for example, the Tamil nationalists of Sri Lanka, the Karen, Karenni and Shan of Burma, the Palestinian and Kurdish nationalists of the 'Middle East' or the Basque, Scottish and Corsican nationalists in Western Europe get their way, there will be even more.

Yet, for many years the *demise* or decline of the state has been discerned or predicted, and such claims have of late become even more common in discourses about 'globalization'. These usually argue that the growing scale and power of transnational flows, particularly of capital (but also of people, ideas and religious affiliations, technologies and so on) is *subverting* the capacity of the state and *weakening* national identities. The nation-state is often described as being 'hollowed-out' or 'eroded'. In this view, the state no longer has the power to command, for example, the society and economy inside its boundaries that was once attributed to it. And such a 'hollowing out' is sometimes seen to point the way to a post-national world (or some kind of *shared* global culture in which national cultures are replaced by a more hybrid, but common global mixture).

Some writers such as Arjun Appadurai have argued that there are new 'postnational cartographies' emerging. An illustration of this is given by Appadurai (2003: 342–3):

> The case of Khalistan is particularly interesting. Khalistan is the name of the imagined home that some Sikhs in India (and throughout the world) have given to the place that they would like to think of as their own national space, outside of the territorial control of the Indian state. Khalistan is not simply a separatist, diasporic nationalism . . . Rather, Sikhs who imagine Khalistan are using spatial discourses and practices to construct a new, postnational cartography in which ethos and demos are unevenly spread across the world and the map of nationalities cross-cuts existing national boundaries and intersects with other translocal formations. This topos of Sikh 'national' identity is in fact a topos of 'community' (*qom*), which contests many national maps (including those of India, Pakistan, England, and Canada) . . .

Appadurai (2003: 243) also mentions 'translocal affiliations' that are 'global or globalizing', such as Islamic, Christian and Hindu fundamentalisms, as well as 'racial and diasporic' ones that 'cross-cut existing national boundaries' and intersect 'with other translocal formations'. The hyper-mobility of people, refugees, exiles, migrants and others, plus old and new identities, associations, networks and affiliations that are other than 'national', mean that the idea of the 'nation-state' is being challenged from all directions. Yuval-Davis (1997, reprinted in Brenner et al. 2002: 322) has pointed to the lack of congruence between nations and states, arguing that there is often a lack of 'overlap between the boundaries of state citizens and "the nation"', which requires us to have a much more 'multi-layered' notion of people's citizenship needs, 'because people's membership in communities and polities is dynamic and multiple'. And, as we have seen, in recent years, the sovereignty of many states has also been questioned through imperial 'interventions' in the name of humanitarian intervention or the 'war on terror'.

Moreover, others object that transnational forces of 'globalization' are really nothing new and that capitalism in particular has shown itself to be able to coexist with states and nations and in symbiotic relationship with them. In other words, they reinforce each other. The same goes for many organized religious movements that coexist or support nation-states. (This is not to say that many movements with religious inspirations do not see themselves as anti-statist, even radically and violently so, as in the case of some fundamentalist Christian militias in the USA and Islamicist movements in Asia. But in so doing of course they ascribe to the state a certain power and centrality – thereby making it more real. A parallel to help readers understand is the atheist statement: 'There is no God', which cites something called God in order to deny its existence!) Moreover, 'globalization' is not only massively uneven (as we have seen in earlier chapters) but as likely to produce local backlashes as a universal culture of, say, the *same* fast food, drinks and soap operas and political orientations everywhere in the world. That is, the technologies of capitalism (particularly media) provide the preconditions for strengthening, rather than undermining, imagined communities of nationalism, while states still act to fine-tune the regulatory frameworks for continued capital accumulation (resorting where needed to force to suppress opposition). States still enact laws about business, trade unions, property rights and so on. Everywhere, buying a property or land or setting up a legal business requires some kind of registration with the state. No amount of globalization has ended this. However, instead of further entering the large and complex debates about the impacts of 'globalization' on nations and states, let us return to the conception of the state as a symbolic system, and as a complex of *representations*. The historian-philosopher and activist Michael Foucault (1979: 29) argued in one of his most famous essays that:

> We all know the fascination which the love, or horror, of the state exercises today; we know how much attention is paid to the genesis of the state, its history, its advance, its power and abuses, etc. . . . But the state, no more probably today than at any other time in its history, does not have this unity, this individuality, this rigorous functionality, nor, to speak frankly, this importance; maybe after all the state is no more than a composite reality and a mythicised abstraction, whose importance is a lot more limited than many of us think.

Hence, to recognize the state as, in part at least, a symbolic system is also to recognize that, as Rose and Miller (1992: 172) argue: '"the state" itself emerges as an historically variable linguistic device for conceptualising and articulating ways of ruling'. In other words, things like the 'nation' and the 'state' are *made* real mainly in certain words, texts (including maps) and deeds, that is, in language and action. The nation-state is an historically specific way of governance that links land, nation, population and polity. Think again of the staging of those *national* sporting, political and cultural occasions, which bring the nation home, and the more mundane or bureaucratic acts of state, such as the display of maps in schools and public buildings or the action of showing a passport or filling in a

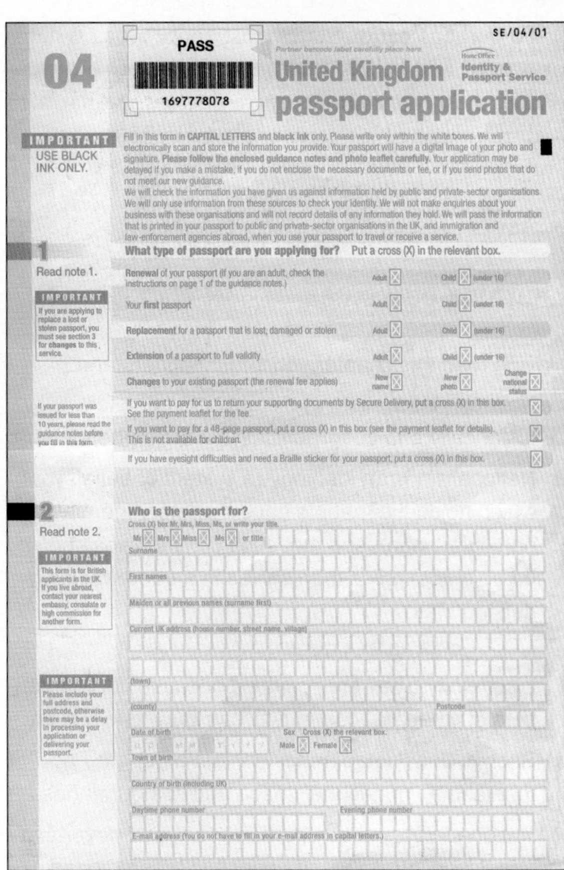

SE/04/01

04

PASS

1697778078

Partner barcode label carefully place here

Home Office Identity & Passport Service

United Kingdom passport application

IMPORTANT USE BLACK INK ONLY.

Fill in this form in **CAPITAL LETTERS** and **black ink** only. Please write only within the white boxes. We will electronically scan and store the information you provide. Your passport will have a digital image of your photo and signature. **Please follow the enclosed guidance notes and photo leaflet carefully.** Your application may be delayed if you make a mistake, if you do not enclose the necessary documents or fee, or if you send photos that do not meet our new guidance.

We will check the information you have given us against information held by public and private-sector organisations. We will only use information from these sources to check your identity. We will not make enquiries about your business with these organisations and will not record details of any information they hold. We will pass the information that is printed in your passport to public and private-sector organisations in the UK, and immigration and law-enforcement agencies abroad, when you use your passport to travel or receive a service.

1 Read note 1.

What type of passport are you applying for? Put a cross (X) in the relevant box.

IMPORTANT If you are applying to replace a lost or stolen passport, you must see section 3 for **changes** to this service.

Renewal of your passport (If you are an adult, check the instructions on page 1 of the guidance notes.) Adult ☒ Child ☒ (under 16)

Your **first** passport Adult ☒ Child ☒ (under 16)

Replacement for a passport that is lost, damaged or stolen Adult ☒ Child ☒ (under 16)

Extension of a passport to full validity Adult ☒ Child ☒ (under 16)

Changes to your existing passport (the renewal fee applies) New name ☒ New photo ☒ Change national status ☒

If your passport was issued for less than 10 years, please read the guidance notes before you fill in this form.

If you want to pay for us to return your supporting documents by Secure Delivery, put a cross (X) in this box. See the payment leaflet for the fee. ☒

If you want to pay for a 48-page passport, put a cross (X) in this box (see the payment leaflet for details). This is not available for children. ☒

If you have eyesight difficulties and need a Braille sticker for your passport, put a cross (X) in this box. ☒

2 Read note 2.

Who is the passport for?

Cross (X) box Mr, Mrs, Miss, Ms, or write your title.

Mr ☒ Mrs ☒ Miss ☒ Ms ☒ or title

IMPORTANT This form is for British applicants in the UK. If you live abroad, contact your nearest embassy, consulate or high commission for another form.

Surname

First names

Maiden or all previous names (surname first)

Current UK address (house number, street name, village)

(town)

(county) Postcode

IMPORTANT Please include your full address and postcode, otherwise there may be a delay in processing your application or delivering your passport.

Date of birth D D M M Y Y Y Y Sex Cross (X) the relevant box. Male ☒ Female ☒

Town of birth

Country of birth (including UK)

Daytime phone number Evening phone number

E-mail address (You do not have to fill in your e-mail address in capital letters.)

Plate 20.6 Identifying citizenship.

form with your *national* (insurance, social security, registration or identity) number. The political geographer Joe Painter (2006: 753) has called this 'prosaic stateness', 'the mundane *practices* through which something which we label "the state" becomes present in everyday life'. According to Painter (2006: 753) therefore:

> Behind each of these registration numbers, licenses and certificates are yet more documents and records held in state archives tracking employment, earnings, criminal convictions, academic performance, visits to doctors and hospitals, ownership of vehicles and landed property and numerous other features of individuals' 'private' lives. If weighed down by anxieties about the scope of the state's knowledge of us, we repair to the local pub for a drink, we will find that the state decides when and where the pub can open, the possible sizes of our serving of beer, how much of its price goes in tax . . . how our drinks are labeled . . . hygiene required in the pub kitchen and the minimum wages paid to the staff.

So perhaps nation-states exist above all as systems of actions and beliefs – an 'imagi*nation*' if you like – which must be continually re-enacted, re-narrated and re-imagined as territorial sovereign spaces in order to seem important and real to us. Just as they have been here. And as they will be next time you go to the pub, or perhaps even to a café, library or classroom instead.

Learning outcomes

Having read this chapter you should understand:

- Nations, nationalism and states are complex historically and geographically variable phenomena.
- Nationalism is an ideology (a system of beliefs) which holds that people have a primary identity to a particular nation and that such communities should be able to express themselves in a *geographically* defined state.
- Nations can be understood as a kind of imagined community. They are imagined because not all members of a national community can know each other.
- Such national imaginaries contain a geography, a mental map of national space and its boundaries.
- Nation-states embody particular ways of governing population, territory and economy: these modes of governance are variable in space and time (the role of the state has and continues to change) – and new modes and scales (such as the supranational project of the European Union) of governance have emerged.
- National imaginaries are also gendered, for example in the idea of mothers as key transmitters of 'national culture' to the next generation.
- States can be understood as complex symbolic systems.

Further reading

Agnew, J., Mitchell, K. and Toal, G. (eds) (2003) ***A Companion to Political Geography***, Blackwell, Oxford.

Cox, K., Low, M. and Robinson, J. (eds) (2008) ***The Sage Handbook of Political Geography***, Sage, London.

Both these edited collections contain several chapters relevant to the material covered here – and to other chapters in this section.

Anderson, B. (1983) ***Imagined Communities: Reflections on the Origin and Spread of Nationalism***, Verso, London. A revised (third) edition (2006) is available. Always readable and very worthwhile. It is worth comparing Anderson (who stresses the modernity of nations and nationalism) with another theorist who argues that many nations do have much deeper historical roots:

Smith, A. (1988) ***The Ethnic Origin of Nations***, Blackwell, Oxford.

Anderson, J., Brook, C. and Cochrane, A. (eds) (1995) ***A Global World? Re-ordering Political Space***, Oxford University Press and the Open University, Oxford. Chapter 2 on 'The exaggerated death of the nation-state' remains a good place to follow up debates about the impact of globalization mentioned above in section 20.6.

McDowell, L. and Sharp, J.P. (eds) (1997) ***Space, Gender and Knowledge: Feminist Readings***, Arnold, London. Section 7 on 'Gender, nation and international relations' contains extracts of some challenging articles and lots of suggestions for further reading.

McNeill, D. (2004) ***New Europe: Imagined Spaces***, Arnold, London. Contains material on European nation-states and their relationships to alternative scales of politics (such as the European Union and the municipalities/cities). A good way to follow this up is to read the set of papers in: ***Journal of Comparative European Politics***, 4 (2/3), (July/September 2006) Special Issue: Rethinking European Spaces: Territory, Borders, Governance.

Seton-Watson, H. (1977) ***Nations and States: An Enquiry into the Origins of Nations and the Politics of Nationalism***, Westview Press, Boulder, Colorado. There are hundreds of case studies of particular nationalisms. Select a nation and you will usually find many to choose from: Seton-Watson reviewed them in the mid-1970s. Although long out of print, this book can often still be found in libraries (or second-hand). If you cannot track down a copy of Seton-Watson, consult:

Delanty, G. and Kumar, K. (eds) (2006) ***Handbook of Nations and Nationalism***, Sage, London.

Taylor, P.J. and Flint, C. (2000) ***Political Geography: World-Economy, Nation State and Locality***, Prentice Hall, Harlow. Now in its fourth edition, this text may be consulted for further ideas on most political geography topics, including a treatment of nations and nationalism.

Yuval-Davis, N. (1997) ***Gender and Nation***, Sage, London. An accessible and rewarding survey of its theme.

Finally, some different (but equally fascinating) studies of (mostly European) state boundaries and the (re)making of national identities around them:

Agnew, J. (2007) No borders, no nation: making Greece in Macedonia, ***Annals of the Association of American Geographers***, **97**(2): 398–422.

Ballinger, P. (2003) ***History in Exile: Memory and Identity at the Borders of the Balkans***, Princeton University Press, Princeton, NJ.

Berdahl, D. (1998) ***Where the World Ended: Re-unification and Identity in the German Borderland***, University of California Press, Berkeley.

Kramsch, O. and **Hooper, B.** (2004) (eds) ***Cross-Border Governance in the European Union***, Routledge, London.

Megoran, N. (2006) For ethnography in political geography: experiencing and re-imagining Ferghana Valley boundary closures, ***Political Geography***, **26**(10): 622–40.

Paasi, A. (1996) ***Territories, Boundaries and Consciousness: The Changing Geographies of the Finnish–Russian Frontier***, Wiley, Chichester.

Pelkmans, M. (2006) ***Defending the Border: Identity, Religion and Modernity in the Republic of Georgia***, Cornell University Press, Ithaca and London.

It is worth comparing any of these with a short but readable book on minorities, forced migration and boundary-making in the Balkans in the aftermath of the First World War (when the modern state of Turkey emerged from the collapsing Ottoman Empire) and/or a study of the aftermath of the partition of India (into the states of India and Pakistan) as the British Indian Empire ended in 1947:

Clark, B., (2007) ***Twice a Stranger: How Mass Expulsion Forged Modern Greece and Turkey***, Granta, London.

Tan, T. Y. and **Kudaisya, G.** (2002) ***The Aftermath of Partition in South Asia***, Routledge, London.

For annotated, clickable weblinks and useful tutorials full of practical advice on how to improve your study skills, visit this book's website at **www.pearsoned.co.uk/daniel**

GEOPOLITICAL TRADITIONS

Chapter 21

James Sidaway

Topics covered

- Origins and history of geopolitics, critical perspectives and 'popular geopolitics'
- Diversity and dissemination of geopolitical discourses
- Changing 'World Orders'

In the fifteen years since the collapse of the Soviet Union in 1991, the geopolitical space occupied by the former Soviet Union and often dubbed the 'post-Soviet space' has undergone far-reaching political, economic and social change. (Nitkin 2007: 1)

21.1 Introducing the idea of a geopolitical tradition

It is evident that Russia and other countries that were part of the Soviet Union between its establishment in 1917 and dissolution in 1991 have undergone rapid changes since the end of the communist state. However, what does it mean to write of them, as in the quote above, as a 'geopolitical space'? One answer (or rather set of answers) is offered by O'Loughlin et al. (2005: 333) in a study of geopolitical ideas in contemporary Russia. They examine the range of debates taking place in Russia about what its role in the world should be:

> All across the post-Soviet space, fledgling bureaucratic institutions and administrative structures are striving to consolidate and legitimate themselves as genuine state institutions representing a transcendent and transparent national identity. Nowhere has this process been more difficult than in Russia, the home of the institutions most closely associated with Soviet ideology. Russia lost an empire virtually overnight and has been striving to find a place and identity for itself since that time. This crisis of national identity in Russia is not yet over and is closely related to the clarification of the role of Russia as the inheritor of the Soviet tradition of great power geopolitics as well as the dominant power on the Eurasian landmass.

In this context, Russian writing about geopolitics has proliferated, some of it in ultra nationalist and right-wing (racist and anti-Semitic) vein (Ingram 2001). One enduring topic in this new Russian geopolitics is about Russia as both a European and an Asian power (its putatively unique role and space), another is the strategic significance of energy supplies and the pipelines that transport oil and gas to consumers. In a world once again hungry for cheap oil and gas, Russia is both a source of and a space through which oil and gas pipelines traverse. In recent years, the projection of American power in Europe (with former Soviet allies and territories in the Baltic and Eastern Europe becoming members of the US-led NATO alliance and/or of the European Union and others leaning towards the West) and Central Asia (following the American-led occupation of Afghanistan since 2001) has led to resurgence of references to the geopolitics of energy and pipelines (indicated with maps such as that in Figure 21.1).

Yet whilst the term **geopolitics** can and does refer to many things, and is in vogue, it is important for students of geography (a subject with which geopolitics has often been associated) to understand that geopolitics is commonly associated with particular ways of writing (and thinking) about space, states and the relations between them. Often this takes the form of mapping, emphasizing the strategic importance of particular places (as in Figure 21.1).

Geopolitics in this sense often sees itself as a tradition: that is, something conscious of its unfolding historical development and with a sense of important founders (and certain key texts written by them). Many of those who write about this tradition or see themselves as working within or extending it usually trace its origins to the late nineteenth-century writings of a conservative Swedish politician Rudolf Kjellén. Kjellén is reputedly the first person to have used the term 'geopolitics' in published writings (see Holdar 1992). But beyond this idea of a founding moment when the geopolitical 'tradition' begins with the first use of the term by Kjellén, things start to get complicated. The 'tradition' divides, fractures, multiplies and finds itself translated into many languages and cropping up in everything from the writings and speeches of American politicians to texts written by Brazilian generals and Russian journalists. All these reinvent and rework the 'tradition' as they go along. As the introduction to a critical collection on 'rethinking geopolitics' explained a decade ago:

> the word 'geopolitics' has had a long and varied history in the twentieth century, moving well beyond its original meaning in Kjellén's work. . . . Coming up with a specific definition of geopolitics is notoriously difficult, for the meaning of concepts like geopolitics tends to change as historical periods and structures of world order change.
>
> (Ó Tuathail 1998: 1)

In other words, exactly what is meant by the term 'geopolitics' has changed in different historical and geographical contexts. However, a good way to understand something about this idea of a tradition of geopolitics is to look at some of the people, places and ideas associated with one of its most significant appearances, in the ideology of the right-wing dictatorships that ruled many South American countries in the 1960s, 1970s and 1980s. Although today they are nearly all democracies, between the mid-1960s and the mid-1980s many South American countries were dominated by their armies. In most cases they had come to power through military coups and became more concerned with internal security (patrolling the towns and country as a kind of police force and running the wider

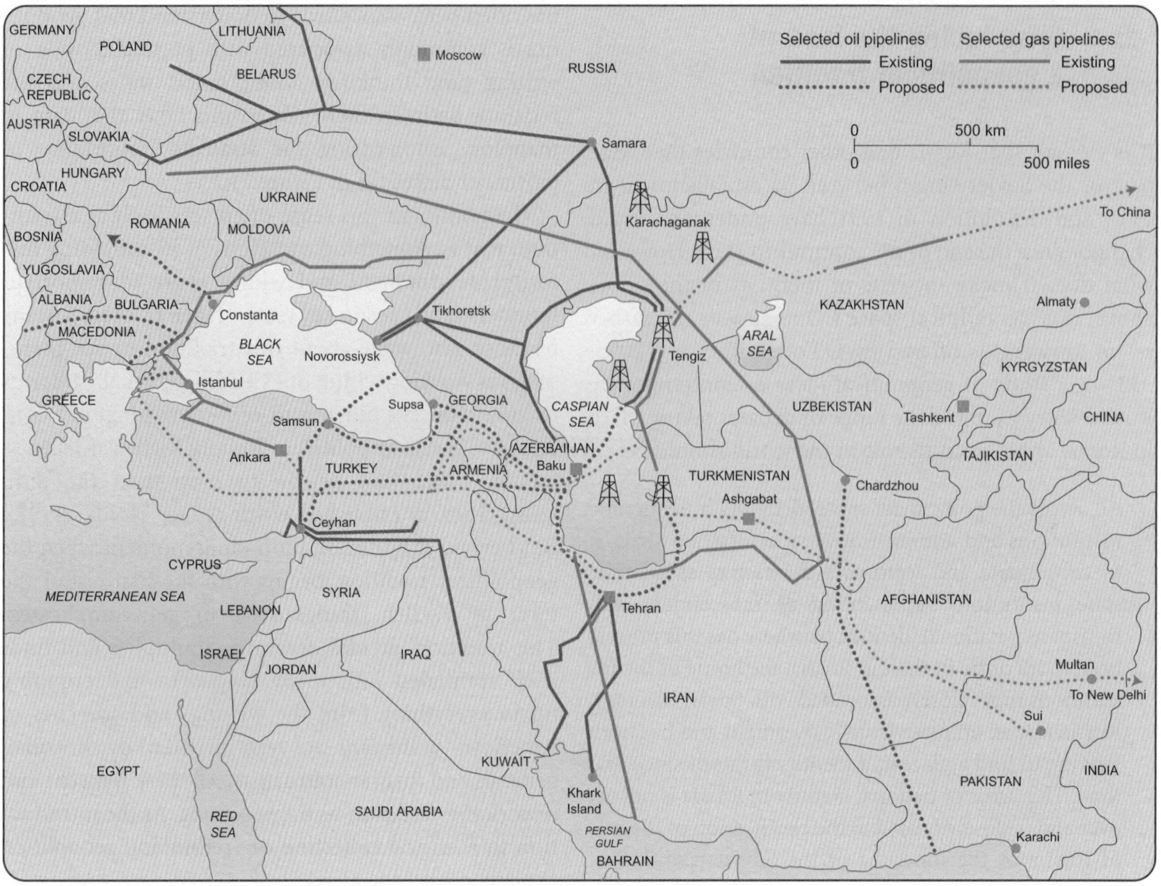

Figure 21.1 Central Asian and Transcaucasian oil and gas pipelines.

Source: Adapted from Dekmejian and Simonian (2003)

government) than with fighting or preparing to fight wars with other countries. In virtually all cases the military regimes received a significant degree of support from the United States, which had often trained the military elites in counter-insurgency and provided weapons and logistical support. For in the USA, Central and South America and the Caribbean were presented as its legitimate sphere of influence, America's 'backyard'. The USA had first declared this to be the case back in 1823 – when it was termed the 'Monroe Doctrine' (so named after the fifth US President James Monroe, 1817–25) (Figure 21.2). However, the full significance of the Monroe Doctrine would only be fulfilled once the USA achieved superpower status after 1945. Alarmed by the presence of pro-Soviet 'communist' forces in Cuba following the successful revolution there in the late 1950s which overthrew a corrupt US-backed regime, the United States was determined to allow no more radical governments to emerge in the region. US political and economic interests would be protected – by whatever means necessary.

Figure 21.2 The Monroe doctrine.

Source: Minneapolis Journal, 1912

Plate 21.1 General Pinochet: geopolitics in action.
(SIPA Press/Rex Features)

Chile is a good example, both of the US role and of the active presence of the geopolitical tradition. On 11 September 1973, the elected left-wing government of Chile (headed by the communist President Salvador Allende) was overthrown in a brutally violent coup led by the head of the armed forces General Augusto Pinochet (Plate 21.1). The left-wing regime that Pinochet overthrew had already faced American economic and political power and sanctions, and the CIA backed the coup. Promising to impose order on a country whose politics had long been democratic but was wracked by left–right divisions, Pinochet and his generals ruled Chile with an iron fist. All opposition was crushed and thousands of people were rounded up by the armed forces. Many 'disappeared' into secret military jails. Hundreds of these prisoners were tortured and murdered. The use of torture became a routine instrument of state. For the next seventeen years, Chile was ruled by Pinochet and his cronies who imposed a new 'neoliberal' economic and political model (of the form that is described in Chapter 8) and murdered any significant opposition. Pinochet finally stepped down as head of state in 1990, when democracy was restored, but not before he had entrenched himself as a senator (in the parliament) for life. He also guaranteed that the armed forces (which he continued to head) would continue to have a significant role in Chile's state structures and that they could not be prosecuted for any crimes committed during the coup and brutal years of military rule.

The ideology of the military regime in Chile (like those elsewhere in South America) was intensely nationalist. More importantly it was a particular conservative form of Chilean nationalism. The nation was held to be sacred and the military were rescuing this 'sacred body' from communists, subversives and so on (i.e. anyone who opposed the military vision of ultra-nationalism and order and the neoliberal economic and social model which was now imposed). It is in this nationalism and conception of the nation as a kind of sacred body that geopolitics enters the picture. A few years before the coup, Pinochet had published an army textbook on geopolitics. Intended mainly for use in Chilean military academies, the 1968 textbook indicated that Chile's future dictator took the subject deadly seriously. For Pinochet, geopolitics was a science of the state, a set of knowledges and programmes to perfect the art of statecraft, that is to strengthen the state in a continent and wider world in which it was held to be in competition with others. The starting point and the heart of Pinochet's textbook and of the wider bleak tradition of geopolitics that it presents is an organic theory of the state.

21.2 The organic theory of the state

This idea is at the core of most South American geopolitical writings and of the wider geopolitical tradition. In summary, it holds that the state or country (and the sense of nation that goes with it) is best understood as being like a living being. Like any living organism, it therefore needs space to grow and it will be in competition with other living beings. An idea that the state needs 'living space' (a certain amount of resources and land) for the nation to thrive is particularly evident in South American geopolitics. A certain reading of Charles Darwin's idea of a struggle for the 'survival of the fittest' is therefore transferred to the realm of states. Drawing upon conservative German writings of the nineteenth and early twentieth centuries (which had been codified by the German academic Friedrich Ratzel and were later elaborated by another conservative German geographer and general Karl Haushofer) and with this organic notion of the state in mind, geopolitics claims to identify certain laws that govern state behaviour and that, once identified, can be a guide for those charged with furthering and protecting the 'national interest' (or at least a certain definition of the latter).

It is not difficult to see how this idea could appeal to South American military dictators, for it gives the military, equipped with the supposedly scientific study of geopolitics, a special mission. In other words it legitimizes their rule. Moreover, the organic idea was extended further to legitimize the extermination of those whom the military defined as enemies of the state. People labelled as communist or subversive, indeed all those who oppose the military dictatorship, can be compared with a disease or cancer which threatens the lifeblood of the state-organism and is best 'cut out', that is, eliminated. The otherwise unthinkable (the murder of thousands of people in the name of the wider welfare of the nation-state) is made to seem natural and a good thing, since it serves the longer-term interest of the health and power of the 'living being' that is the state (see Hepple 1992 for an exploration of this). With the 'cancer' of subversion and disloyalty eliminated, the 'state-body' can go on to grow and thrive.

21.3 Brazilian national integration

Aside from Chile, one of the most important expressions of geopolitics in South America has been in Brazil. The military in Brazil overthrew the democratic government in 1964 and stayed in power for the next 20 years. Again the Brazilian military were supported by the USA, whose ambassador in Brazil termed the military coup 'the single most decisive victory of freedom in the mid-twentieth century' (cited in Chomsky 1998: 7). Although never as despotic as the regime led by Pinochet (or similar military governments in Argentina and Uruguay), the long years of military rule gave the Brazilian generals a chance to elaborate and impose a geopolitical vision on the country. One of the most evident aspects of Brazilian geopolitics is the idea that state security requires a measure of national integration. The vast scale of Brazil, the difficulty of travel across its Amazon 'heartland' (the world's largest tropical rainforest) and the fact that it shares borders with every other South American country except Chile gave the geopolitics of the Brazilian generals a special sense of its national mission and an obsession with the potential for the country to be a great power (known in Portuguese as *Grandeza*). The associated sense of the urgency of the integration of Brazil required the extension of a network of highways across Amazonia and the settlement of its lands by farmers and ranchers. It is this geopolitically inspired vision, combined with a highly corrupt system of patronage and favours to those close to the regime, that underlies the enclosure and division of Amazonia into private lands (some larger than European countries such as Belgium) and the accompanying transformation or destruction of the tropical rainforest (see Hecht and Cockburn 1989) (Figure 21.3).

Although the Brazilian military have been back in the barracks for more than two decades and Brazil is again a lively democracy, the long-term consequences of this geopolitically motivated (and economically profitable) strategy have been disastrous for the many of the indigenous peoples of Amazonia. They have found themselves forced off land that was traditionally theirs, sometimes murdered or attacked by ranchers, the state and settlers and disoriented by the arrival of a frontier culture of violence, destruction and consumption. Poor peasants, particularly from the impoverished north-east of Brazil, have also moved into the Amazon region in search of land and freedom. They are fleeing from the oppression and landlessness they themselves face in the north-east of Brazil, where the vast majority of the land is still in the hands of an elite class. Of course, in the geopolitical visions of the Brazilian generals, this population movement was seen as overwhelmingly positive, reinforcing the Brazilian population of the Amazon and redistributing what they regard as marginal 'surplus' people in the process (in much the same way that the colonization of the 'wild west' was seen as the advance of civilization and America's 'manifest destiny' in the United States). At the same time this has defused some of the political pressures for land reform in north-east Brazil. In all this we can see how Brazilian geopolitics was implicated in a complex of social and environmental transformations. The destruction of the Amazonian ecosystem is therefore not simply the result of 'population pressure' or abstract forces of development, rather it is to be understood as a social process and therefore as the consequence of particular political choices and (im)balances of power.

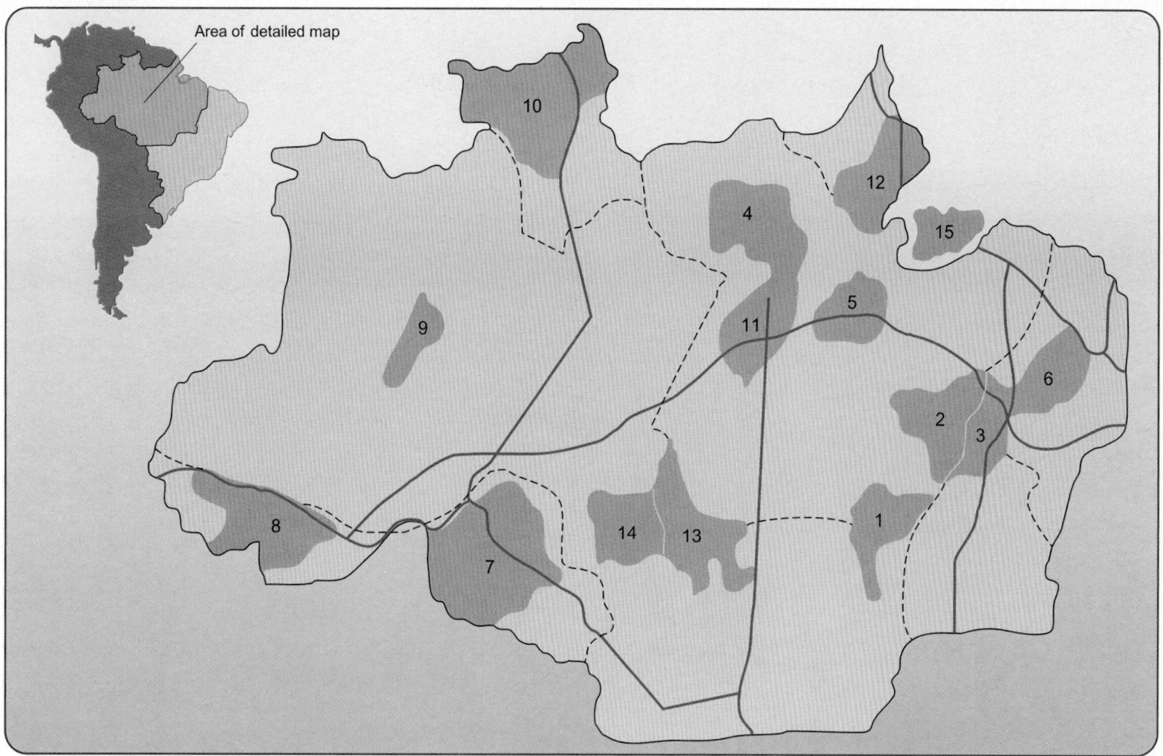

Figure 21.3 National integration in Brazilian geopolitics: Amazon development poles. The shaded areas represent the 15 'development poles' where financial resources would be concentrated and concessions offered to mining, settlement industrial, Lumber and agricultural development resources.

Source: Hecht and Cockburn (1989: 127)

21.4 Antarctic obsessions

An interest, at times an obsession, with the last unexploited continent of Antarctica is evident in much South American geopolitics. Antarctica is also prominent in geopolitical writings from Argentina, Chile and Uruguay and also appears in Peruvian, Brazilian and Ecuadorian geopolitical texts (see Dodds 1997). To understand what this amounts to and what forms it takes necessitates some understanding of Antarctica's exceptional political geography. Antarctica has a unique territorial status, in so far as it has no recognized state on its territory. Everywhere else in the world forms part of the patchwork of states that we learn to be familiar with and take for granted. Antarctica is a stark reminder that there is nothing natural or inevitable about this. The absence of recognized states in Antarctica reflects the inhospitability and remoteness of the continent, the only one without an indigenous human population. Although the plankton-rich waters around Antarctica were exploited for seal and whale hunting in the nineteenth century, it was not until the twentieth century that exploration of the interior began. Even today, Antarctica has only a non-permanent population (at any time) of a few hundred scientists. This presence, plus the relative proximity of the continent to South America and the possibility of exploitable mineral resources, has led to a series of territorial claims. Although Argentinean writers had already represented Antarctica as an extension of the southern Argentine area of Patagonia, the British made the first formal claim in 1908. In turn, parts of this claim were 'granted' to Australia and New Zealand. France and Norway made claims in the 1930s and 1940s, followed shortly by Argentina and Chile. After 1945, the USSR and the USA also established a wide network of bases, although without staking formal claims to territory. The claims made by Argentina, Chile and the United Kingdom overlapped – and foreshadowing the British–Argentine conflict over the Falklands/Malvinas, British and Argentine forces exchanged fire in Antarctica in the late 1940s and early 1950s. In the context of this potential for conflict and growing possibility of Cold War confrontation (on the Cold War, see section 21.7), a United Nations Treaty in 1959 agreed that all claims would be (forgive the pun) 'frozen' for at least 30 years, and the continent reserved for scientific (not commercial or

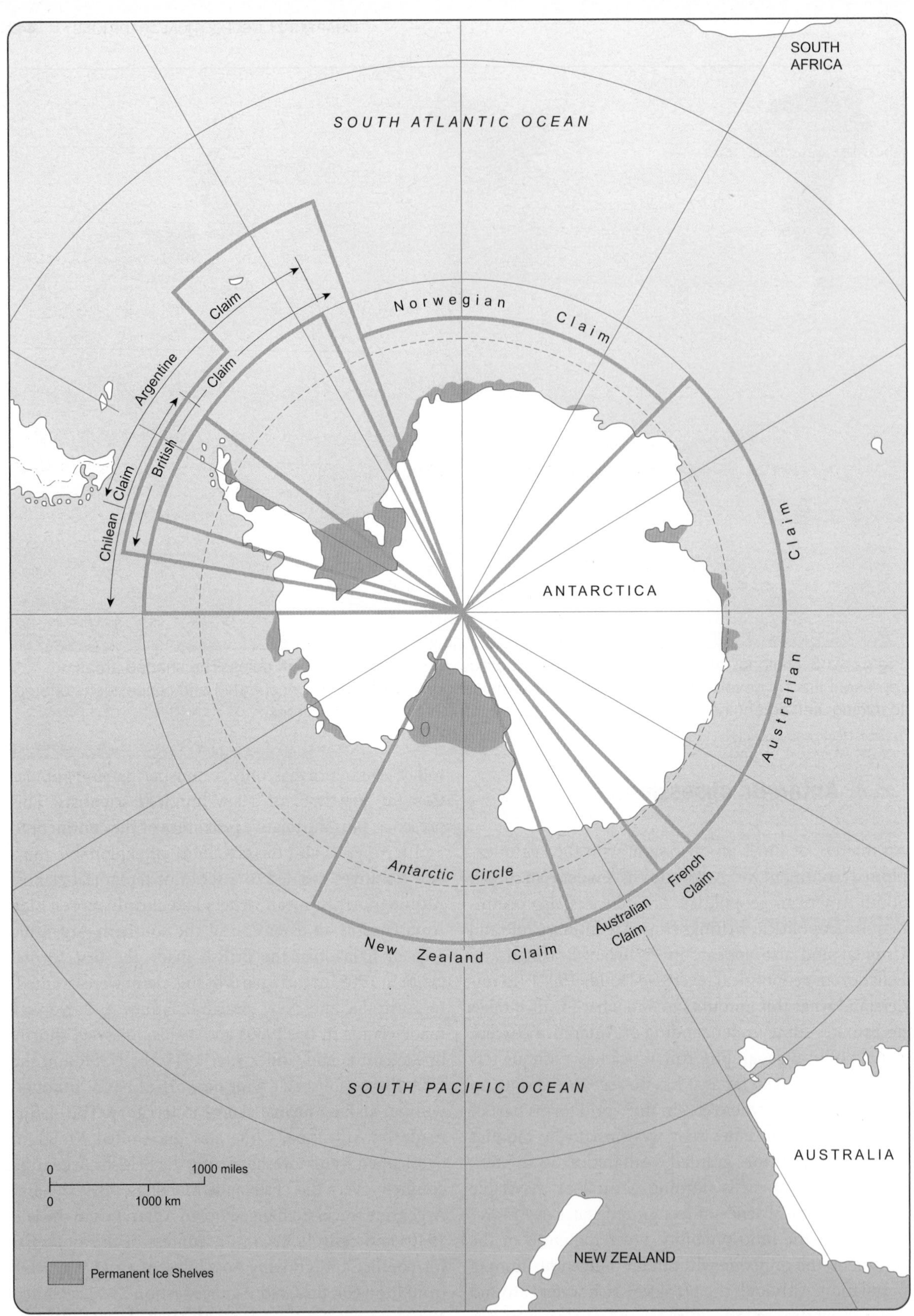

Figure 21.4 Claims on Antarctica.

Source: Adapted from M.I. Glassner, *Political Geography*, John Wiley & Sons Inc., 1993, p. 498. Copyright © 1993 John Wiley & Sons Inc. This material is used by permission of John Wiley & Sons Inc.

military) use. The Treaty was extended in 1991. But prior to then, it was not clear what the future status of the continent would be. Even today, the Treaty merely defers the issue of claims.

Argentina is the South American country with the largest claim (Figure 21.4). The Argentine claim has also become central to geopolitical discourse there. As Child (1985: 140–1) explains in a study of geopolitics and conflict in South America:

> For Argentine geopolitical writers, the subject of Antarctica is not only linked to tricontinental Argentina [a power in the South American continent and the South Atlantic], but also to . . . national sovereignty, patriotism, and pride. This is a particularly touchy combination after the humiliating defeat of the Malvinas/Falklands conflict. The Argentine National Antarctic Directorate has professors of Antarctic geopolitics on its staff. Through the media, maps, and postage stamps and the centralized educational system, Argentines are constantly taught and reminded that there is an Argentine Antarctic just as much as there are Argentine Malvinas. The need to assert Argentine rights in the Argentine Sea, islands, and Antarctica is linked to dreams and national projects of Argentine greatness.

21.5 Heartland

Although Antarctica and the South Atlantic are significant components of Argentine (along with Brazilian and Chilean) geopolitics, the former also looks beyond other Argentine frontiers. In particular, Argentine geopolitical writers together with some Brazilians and Chileans have taken a particular interest in the security of Bolivia. They have scripted Bolivia (a relatively impoverished land-locked mountainous country, which has borders with Argentina, Brazil, Chile, Paraguay and Peru) and proximate areas of its neighbours as a key strategic continental **heartland** (see Kelly 1997). Control of Bolivia, in this vision, would be a vital key to a relative dominance in the South American continent. That Bolivia has a strong revolutionary tradition and was for many years characterized by chronic political instability has reinforced the tendency of the other South American countries to meddle in Bolivian politics. Indeed, during the years (1976–82) when Argentina was last ruled by a geopolitically obsessed military junta, the Argentine armed forces were actively involved in supporting a Bolivian military government. This activity took the form of the kinds of brutal suppression and frequent murder of those (trade union leaders, dissidents, opposition members and leaders) who opposed the military government and its economic and social strategies. The idea of Bolivia as a 'heartland', control of which would be a kind of magic (geopolitical) key to domination of South America, links back to one of the best-known genres in classical (European) geopolitical thought. For although the South American countries have seen some of the most significant expressions of geopolitical discourse of modern times, geopolitics originates in Europe and it is to some examples of European geopolitics, including the idea of heartland, that we now turn.

The designation of a heartland was first made by the British geographer (and strongly pro-imperialist conservative politician) Halford Mackinder. In what has since become a widely cited (if less often read) article first published in 1904 following its presentation to the Royal Geographical Society (RGS), Mackinder argued that the age of (European) geographical exploration was drawing to a close. This meant that there were hardly any unknown 'blank' spaces left on European maps of the world. According to Mackinder, the consequence of this closing of the map, this end of the centuries-long task of exploration and discovery, was that political events in one part of the world would invariably affect all others, to a much greater extent than hitherto. There would be no more frontiers for Europeans to explore and conquer. Instead, the great powers would now invariably collide against each other. Mackinder called this end of European exploration 'the post-Columbian age' and the closing of frontiers, the emergence of a 'closed political system':

> From the present time forth, in the post-Columbian age, we shall again have to deal with a closed political system, and none the less that it will be one of world wide scope. Every explosion of social forces, instead of being dissipated in a surrounding circuit of unknown space and barbaric chaos, will be sharply re-echoed from the far side of the globe, and weak elements in the political and economic organism of the world will be shattered in consequence. There is a vast difference of effect in the fall of a shell into an earthwork and its fall amid the closed spaces and rigid structures of a great building or ship. Probably some half-consciousness of this fact is at last diverting much of the attention of statesmen in all parts of the world from territorial expansion to the struggle for relative efficiency.
>
> (Mackinder 1904: 422)

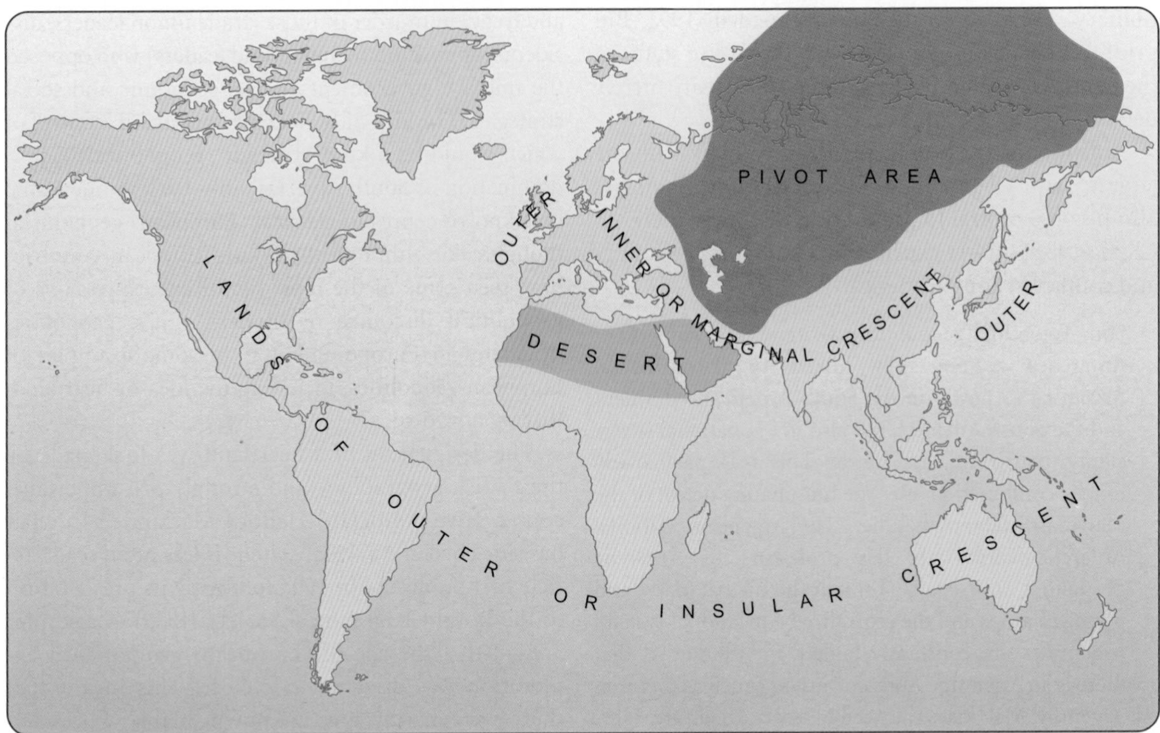

Figure 21.5 'The Natural Seats of Power' according to Mackinder (1904).

Given that this was the case, Mackinder claimed to identify the places of greatest world-strategic significance, control of which would give any great power a key to world power. In his 1904 paper, he termed this the 'pivot area'. With Mackinder's address to the RGS and his subsequent article came a series of maps, the most frequently reprinted one of which claims to describe 'The Natural Seats of Power' (Figure 21.5). As Ó Tuathail's (1996a: 25) critical account of Mackinder argues:

> Mackinder's January 25 04, address to the Royal Geographical Society, 'The Geographical Pivot of History', is generally considered to be a defining moment in the history of geopolitics, a text to which histories of geopolitics invariably point.

Mackinder himself did not use the term 'geopolitics'. But his 1904 paper and its maps were later appropriated into other geopolitical traditions. Indeed, the map of 'The Natural Seats of Power':

> is perhaps the most famous map in the geopolitical tradition. . . . The map is meant to create the illusion of the unveiled geographical determinants of power. It is the global vision, the post-Columbian made possible, the pure panorama produced by the penetrative gaze of the global voyeur. Yet, interestingly, the map is not purely visual but is marked with text. Huge swathes of territory are stamped with a definitive positionality and function: 'Pivot Area', 'Inner or Marginal Crescent', and 'Lands of Outer or Insular Crescent'. These are macrogeographical identities around which Mackinder spatializes history and reduces it to formulaic equations and timeless east/west, land/sea conflicts. 'The social movements of all times', he concluded, 'have played around essentially the same physical features.'
>
> (Ó Tuathail 1996a: 31–3)

A certain geographical determinism can be detected here, in which the physical environment ('essentially the same physical features' in Mackinder's words) provide the fixed, unchanging backdrop to and therefore something of the explanation of human history. This view, which ignores the social construction of resources (see Chapter 5) and other features of nature, is common to virtually all geopolitics (something which it shares with a great deal of twentieth-century geography). One of the key formulas in Mackinder's notion of geographical determinism is his identification of a 'pivot area' or 'heartland zone' in east-central Europe, control of which would be a kind of magic key to world domination. Reworking the ideas of his 1904 article after the

First World War, Mackinder refined this notion and penned a formula that encapsulated it:

> Who rules East Europe commands the Heartland.
> Who rules the Heartland commands the World-Island.
> Who rules the World-Island commands the world.
> (Mackinder 1919: 4)

According to this (simplistic) formula, 'rule' of the eastern portion of Europe offered the strategic path to that of the African–Asian–European continents (which together constitute what Mackinder terms the 'World-Island') and hence a dominant position on the world scene.

In addition to the general backdrop of the emergence of the sense of closed political space, or in Mackinder's terms the completion of the age of exploration (which he called, after its best-known figure, Christopher Columbus, the Columbian age), it is important to recognize that Mackinder, whilst claiming to offer a universal and objective model (with 'laws', for example the formula about 'Heartland') is in fact highly subjective. Heartland is another organic metaphor, invoking the idea that territory has a heart, like a living being. Moreover, Mackinder's writings reflect not only the moment at which they were penned, but also the place. Mackinder, writing from the vantage point of imperial Britain, is concerned to identify threats and dangers to British power. At the time when Mackinder presented his paper to the RGS, Britain was the pre-eminent world power. It still seemed that way to Mackinder in 1919 when he wrote about the 'Heartland'. But British imperial politicians, like Mackinder, were aware of the growing power of America, Germany and Russia. In fact, potential British imperial competition with the latter in Asia provided a key context to Mackinder's work. As Peter Taylor (1994: 404) explained:

> Behind every general model there is a specific case from which it is derived. For the heartland model this is particularly easy to identify. Throughout the second half of the nineteenth century Britain and Russia had been rivals in much of Asia. While Britain was consolidating its hold on India and the route to India, Russia had been expanding eastwards and southwards producing many zones of potential conflict from Turkey through Persia and Afghanistan to Tibet. But instead of war this became an arena of bluff and counter-bluff, known as the 'Great Game'. . . . Mackinder's presentation to an audience at the Royal Geographical Society would not have seemed so original as it appears to us reading this paper today. . . . Put simply, the heartland model is a codification and globalization of the Great Game: it brings a relatively obscure imperial contest on to centre stage.

Not only does this envisage the world in a particular way, as a 'stage', but it sees only select key actors as the significant figures at play. These are the European powers (plus Russia). Other peoples and places are merely the backdrop for action by White Men. The taken-for-granted racism of Mackinder's model, in which only Europeans make history, is also that of European imperialism and that of the bulk of wider European geographical and historical writings of the time (see Chapter 3).

Yet, although Mackinder's 1904 paper is very much a product of its time and Mackinder's own conservative world-view, it has proven durable and has been integrated into rather different contexts, which saved it from the relative obscurity that it deserves as a turn-of the-century imperialist text. The transfer of the discourse of 'Heartland' to Bolivia by South American codifiers of geopolitics has already been noted. In addition, 'Heartland' was appropriated by German geopolitics in the 1930s and 1940s and formed part of the backdrop to Cold War American strategy from the late 1940s through to the last decade of the twentieth century. Taylor (1994: 405) notes:

> First in the inter-war years the heartland theory became an integral part of German geopolitics. It fitted the needs of those who advocated lebensraum [living-space], the policy of expanding into eastern Europe [and enslaving and murdering its indigenous populations], coupled with accommodating the U.S.S.R., as a grand continental policy for making Germany a great power again. Second with the onset of the Cold war after World War II, the heartland theory got another lease of life as the geostrategic basis of nuclear deterrence theory. The west's nuclear arsenal was originally justified in part as compensation for the U.S.S.R.'s 'natural' strategic advantage as the heartland power.

The next two sub-sections of this chapter will examine aspects of Nazi, Fascist and Cold War geopolitics in greater detail.

21.6 Nazi and Fascist geopolitics

The formal tradition of writing about space and power under the title of 'geopolitics' also found fertile contexts in Italy, Portugal, Spain and Japan. Influenced and supported by Nazi Germany, all these countries (together with Hungary and Romania) saw the rise and victory of ultra-nationalist or Fascist governments (often through violent struggle or full-scale civil war with democratic or communist forces) (see Chapter 3, pp. 66–71). In each case geopolitical debates were crucially negotiated through other cultural and political debates about race, nationalism, the colonial pasts and futures, supposed national 'missions' and destinies and the European and global political contexts. The German case has become the best-known. In Germany, organic notions of the state had already been popularized by conservative nineteenth-century academics. Moreover, Germany was characterized by extreme political and economic turbulence in the decades following its defeat in the 1914–18 World War. This combination provided a fertile environment for the elaboration and circulation of a distinctive geopolitical tradition. In Ó Tuathail's (1996a: 141) words:

> After the shock of military defeat and the humiliation of the dictated peace of Versailles, the Weimar Republic proved to be fertile ground for the growth of a distinct German geopolitics. Geopolitical writings, in the words of one critic, 'shot up like mushrooms after a summer rain'.

The main features of these writings (which they shared with a wider German nationalism, later codified in Nazism) were a critique of the established 'World Order', and of the injustices imposed on Germany by the victors. German claims were often presented graphically in maps that were widely circulated (see Herb 1989: 97). Like the variants of the geopolitical tradition that were developed amongst right-wing and military circles in Italy, Portugal, Spain and Japan, German geopolitics also asserted an imperial destiny. Indeed, as Agnew and Corbridge (1995: 58–9) explain: 'The Nazi geopoliticians of the 1930s came up with formalized schemes for combining imperial and colonized peoples within what they called "Pan-Regions".'

In this vision, notions of racial hierarchy were blended with conceptions of state 'vitality' to justify territorial expansion of the Axis powers (see O'Loughlin and van der Wusten 1990) (Figure 21.6). In Europe, related conceptions of the need for an expanded German living-space were used as justification for the mass murder of occupied peoples and those who did not fit into the grotesque plans of 'racial/territorial' purity. The practical expression of these was the construction of a system of racial 'purification' and mass

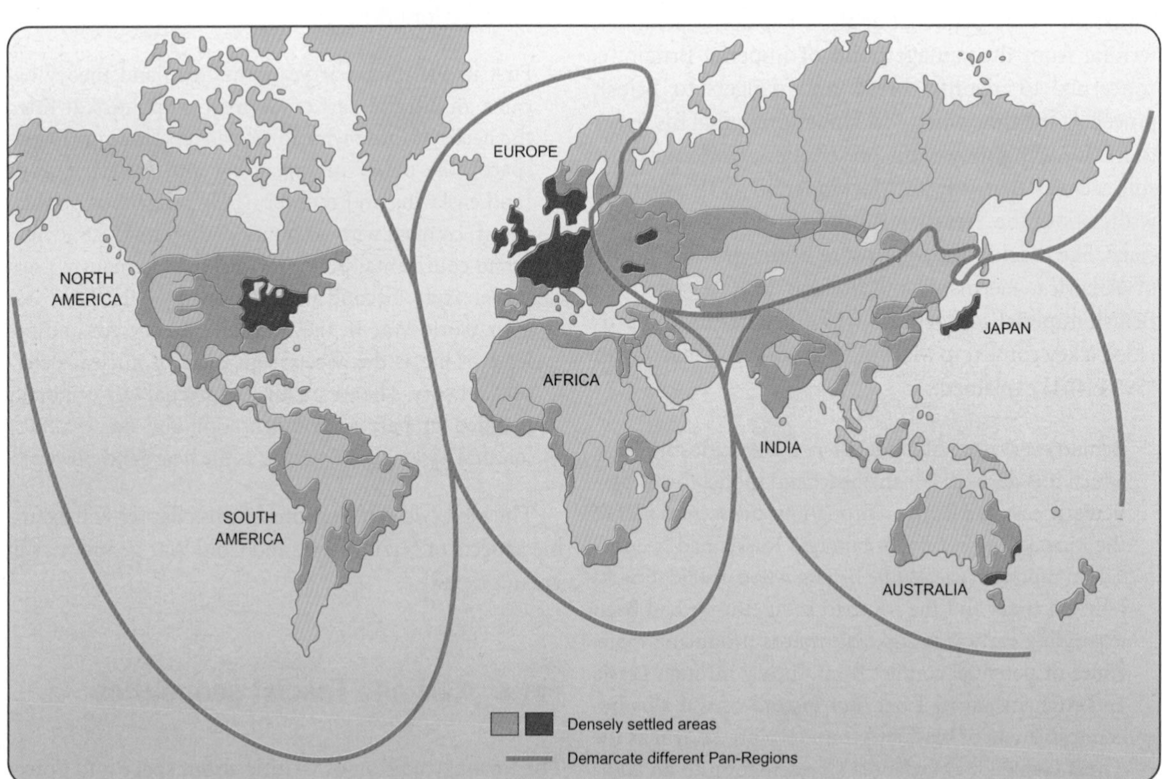

Figure 21.6 'Pan-Regions' as envisaged in Nazi geopolitics.

Source: O'Loughlin and Van der Wusten (1990) Political geography and the pan regions, *Geographical Review*, **80**, adapted with permission of the American Geographical Society.

extermination. At least 6 million Jewish people were murdered in concentration camps together with millions of others: disabled people, gays and lesbians, gypsies and political opponents. Historical debates about the role and relative significance of German geopolitics within the Holocaust and within broader Nazi ideology and strategy continue (see Heske 1986: 87; Bassin 1987; Ó Tuathail 1996a; Natter 2003). Paul Gilroy (2000: 39), however, reiterates the connections between racial, early ecological and geopolitical thinking:

> connected in profound ways to the notions of Lebensraum (living-space) that figured in but were not created by the racist population policies and agricultural and scientific planning of the Nazi period . . . [and to] the geo-organic, biopolitical and governmental theories of the German geographers Friedrich Ratzel and Karl Haushofer and the early-twentieth century geopolitican Rudolf Kjellén. These writers supplied important conceptual resources to Nazi racial science, helping it to conceptualize the state as an organism and to specify the necessary connections between the nation and its dwelling area.

In the United States in the 1940s, German geopolitics became the subject of lurid tales and depictions, cropping up in media, military and government 'explanations' of Nazi danger. Popular magazines, such as *Readers Digest*, would inform Americans of the 'scientists behind Hitler' describing German geopolitics as *the* key to Nazi strategy. And whilst the relative significance of the German geopolitical tradition in the wider genocidal ultra-nationalism of German Fascism was certainly overstated in such accounts, we should see geopolitics as a particular expression of *wider* academic and intellectual involvement and complicity in authoritarian state power, war-making and genocide. Much more widely, beyond the geopolitical tradition per se, academic geography was deeply implicated in these activities. Exploring this, Natter (2003: 188) notes how

> [T]he work of disciplinary historians of geography has demonstrated the extent to which the demarcation of geography seems inseparable from the history of war, imperialism and quests for national identity. . . . Geopolitics, thus, would mark a particular, but in no way separable (and hence containable) geopolitical deployment of geo-power.

An example of this wider complicity was the way in which the models of 'Central Place Theory' (an abstract model of the ideal spatial distribution of towns), which were developed in Germany by Walter Christaller in the 1930s, were elaborated with the express purpose of providing planners with a model of German settlement to impose on conquered territories of Eastern Europe, once the indigenous Jewish and Gypsy people had been gassed to death and the majority Slavic population enslaved (see Rössler 1989).

The defeat of the Axis powers in 1945 (culminating in the use of American atomic weapons against the civilian populations of Hiroshima and Nagasaki) and the lurid wartime depictions of Nazi geopolitics in the United States dealt something of a blow to the formal tradition of geopolitics. In the 1990s however, right-wing geopolitics returned – in Russia (as was noted at the beginning of this chapter), where references to Nazi-era concepts (such as *Groβraum*, 'large spaces') and associated racist and anti-Semitic ideologies have become influential within some contemporary geopolitical thought. Moreover, references to geopolitics had continued through the 1950s to the 1970s in both Spain and Portugal (see Sidaway 1999, 2000; Sidaway and Power 2005), where Fascist regimes remained in power to the mid-1970s, as well as in Turkey (especially when it too was under military rule through part of the 1960s, again in the early 1970s and early 1980s. On the Turkish example, Pinar Bilgin (2007: 753) argues that:

> Constructed through texts authored by military geopoliticians, and disseminated through a variety of institutions including compulsory military service (with access to all males 18+ years of age), the National Security Academy (proving in service training to high level civil servants and journalists), and the compulsory high-school course 'National Security', Turkey's geopolitical discourse has allowed the military to play a central role in shaping domestic political processes but also make this role seem 'normal'.

In addition, as this chapter has detailed, geopolitics has been influential in recent decades in a number of South American countries, especially in the 1970s' and early 1980s' epoch of military rule through much of the continent. These regimes (and plenty of others) were integrated into the US-led anti-communist network of allies. In a policy that became known as containment, the USA aimed to encircle and block the potential expansion of Soviet power and influence beyond the immediate borders of the USSR and the pro-Soviet states installed by the USSR in the Eastern European territories that it had occupied during its Second World War battle for survival against Nazi Germany. In the allied countries (and as an expression of this strategy of containment), the functions of geopolitics,

Plate 21.2 The gateway to the Auschwitz concentration camp. The Nazi regime used concentration camps to execute their policy of racial/territorial 'purification'.
(© Ira Nowinski/CORBIS)

in particular its 'strategic vision' and claim of 'scientific' objectivity, continued to operate or were displaced into other disciplines and branches of knowledge, including geography and the expanding subject of International Relations. This displacement of strategic knowledge was particularly evident in the United States, which by 1945 was the greatest power that the world had ever seen, constituting over half of the world economy and (until the recovery of the Soviet Union and its own development of atomic weapons in 1949) possessing a virtually unrivalled military capacity.

21.7 Cold War geopolitics and the logics of containment

> **Like a fast-moving glacier, the Cold War cut across the global terrain in the 1940s, scraping and grinding solid bases, carving and levelling valleys, changing the topography. For nearly half a century the Cold War competition between the United States and the Soviet Union dominated the international system. . . . Like all retreating glaciers, the receding Cold War left deposits of debris and exposed an altered, scarred landscape that leaders in the 1990s struggled to measure and fathom.**
>
> (Paterson 1992: 2)

The Cold War is used as a shorthand description of the conflict (which to many seemed to be a conflict for world domination) between the 'communist' East, led by the USSR, and the 'capitalist' West, led by the USA from around 1949 to around 1989. This was not simply a rift between two great powers, but a complex ideological conflict, which often appeared to be about different ways of life and contrasting social systems (the bureaucratic 'command economies' based on predominantly state ownership of the means of production, with a ruling class drawn from the bureaucracy; and the capitalist economies based on predominantly private ownership, with a ruling class drawn from the owners of capital). It was also a conflict in which *direct* military confrontation between the two great powers was avoided. Hence the metaphor of Cold, that is,

COPYRIGHT, 1942, BY ANDREAS DORPALEN
PRINTED IN THE UNITED STATES OF AMERICA
BY THE FERRIS PRINTING COMPANY, NEW YORK
ALL RIGHTS RESERVED

FOR
ROSE-MARIE

Dr. K. Haushofer

THE WORLD OF
GENERAL HAUSHOFER

Geopolitics in Action

BY ANDREAS DORPALEN

With an Introduction by
Colonel Herman Beukema, U.S.A.

FARRAR & RINEHART, INC.
New York • Toronto

Plate 21.3 German geopolitics: as represented in the USA.

'poor, frozen or frosty' relationships, but short of all-out 'hot' war, involving direct exchanges of missiles, bombs and so on. In part this avoidance of direct conflict arose because both powers were conscious of the enormous stakes, armed with nuclear, chemical and biological weapons as well as vast arsenals of increasingly high-tech 'conventional' armaments. By the 1960s, they were each capable of destroying virtually all human life several times over. Moreover, neither could attack the other without being sure that the enemy would still possess enough undamaged nuclear weapons to retaliate, a 'balance' of terror known, appropriately enough, as MAD (Mutually Assured Destruction). Instead, conflict took the form of the continual preparation for war, plus proxy wars in what became known as the (originally not pro-West, not pro-East, but contested) 'Third World' and many other political, economic and cultural forms of competition. That is, it was a conflict conducted by *every* means *except* direct military confrontation between the superpowers.

Of course there was no shortage of indirect confrontation and preparation for war. Europe was divided into two armed camps and split down the middle between Soviet and American zones of influence by a fortified military and ideological frontier that became known as the 'Iron Curtain'. Whilst Europe was characterized by an uneasy stability and balance of terror, by the 1970s the term 'geopolitics' had been revived (or rediscovered) by US national security advisers Henry Kissinger and Zbigniew Brzezinski (see Hepple 1986; Sidaway 1998) to refer to the strategic vision deemed necessary to circumvent the sense of growing Soviet power, particularly in the **Third World**, where a wave of successful revolutions had brought left-wing, pro-Soviet governments to power. As has already been noted, one of these was in Cuba, but by the end of the 1970s, there were also pro-Soviet governments in the countries of Indochina (Vietnam, Laos and Kampuchea) where the US had been unable to defeat communist insurgencies despite a massive military effort, in the African countries of Angola and Mozambique, in the Arabian country of South Yemen and in Afghanistan as well as in the small Central American state of Nicaragua. Communist China and North Korea added to American concerns. By the early 1980s, American concern about the security of oil supplies from the Persian Gulf region had escalated to the extent that the then US President Carter declared that it was of vital importance to the USA (see Spotlight box 21.1). In this world of superpower competition, geopolitics again found its moment and expression in the visions of the national security advisers and American generals.

But to focus on formal geopolitics would be too narrow. For the relative absence of explicit references to geopolitics in American strategic culture during the 1950s and 1960s did not mean that the kind of

(a)

(b)

Plate 21.4 (left) Containment in action: the aftermath of a napalm bombing on a Vietnamese village by US-backed (anti-communist) South Vietnamese forces. (right) 'Containment' explained: US Secretary of State for Defense briefing journalists during the Vietnam war.

(© Bettmann/CORBIS)

Spotlight Box 21.1

The Carter Doctrine: contexts and consequences

In late 1979, the USSR invaded Afghanistan. For the Soviets, this meant supporting a communist government in a neighbouring country, who had come to power just a couple of years before (in a coup, after several years of political instability in the country) and were facing rebellion and resistance in the Afghan countryside (already being fostered by the Pakistani and US intelligence agencies). For strategists in Moscow, the invasion would secure the USSR's southern flank, especially from the growing perceived threats from a new Islamic radicalism: the USSR had millions of Muslim citizens in its Central Asian territories, and feared that the Islamic revolution in Iran and the insurgency in Afghanistan might influence them and threaten Soviet power and stability. However, for the USA – at least to those in influential positions, like US National Security Advisor Zbigniew Brzezinski – this marked a new phase of Soviet expansion and a failure of containment. In tandem with the Islamic revolution in Iran which overthrew the Shah (who was an American ally and whose rise to power in the 1950s was orchestrated by the USA and Britain) was the second jump in oil prices in a decade, and American concern over the security of oil supplies from the Persian Gulf region grew. Brzezinski and the American media had already starting talking about the region as 'an arc [or crescent] of crisis' (see Sidaway 1998) (see Plate 21.5, which shows how the influential *Time* magazine represented these 'geopolitical dangers'). And soon after the Soviets invaded Afghanistan the US media started to describe it as the first Soviet step towards the Middle East, and President Carter announced that any threat by an outside force to the Persian Gulf was a matter of direct American strategic interest and would result in a military response. This

Plate 21.5 'The Crescent of Crisis' as depicted on the front cover of *Time* Magazine 15 January 1979.
(*Time* Magazine)

→

commitment became known as the Carter Doctrine, taking its place amongst a succession of other US Presidential geopolitical doctrines: from Monroe (see the main text, page 440), through Truman in 1947, who set out the overall commitment to containment, to (after Carter) the Reagan Doctrine in the 1980s (which sought to roll back Soviet influence in places like Afghanistan and elsewhere in the Third World where pro-Soviet regimes had come to power, rather than simply contain the USSR and its allies). Under the Reagan Doctrine, vast quantities of weapons were channelled to rebel forces fighting the pro-Soviet government in Afghanistan. Amongst those who took part in the anti-Soviet struggle in Afghanistan and through which the armaments were channelled was Osama Bin Laden – later to turn against the United States, whom he came to regard as intent on dominating the Muslim world (Coll 2004).

mapping of places as pacified/dangerous, ours/theirs, threatening/stable, strategic/secure, and the intersection of a geographical imagination with a military/strategic one that characterize the 'commonsense' understanding of the geopolitical tradition was absent. As has been noted, such understandings proliferated in other academic realms, practices, knowledges and disciplines. As well as cropping up in the subject of International Relations, this proliferation of quasi-geopolitics could be found in academic geography where, under the title of 'political geography', American geographers would proclaim their difference from the 'disreputable' tradition of geopolitics, and then proceed to engage in more or less the same kinds of strategic mapping and policy advocacy (see Smith 1984, 2003). More widely, American leaders and policy makers declared that the Soviet Union had to be *encircled* and *contained*. The metaphor of disease (containment) was mixed with that of 'dominoes' – if one country (say Vietnam or Cuba) had 'fallen' to Soviet control or influence, then it could (like a chain reaction) 'infect' other proximate ones. The irony of this (which is equally present in the formal geopolitical tradition) is that in the name of security and strategy, the real complexity of *human* geographies in the places (such as Vietnam or Chile, or Afghanistan, Nicaragua or the Persian Gulf) that are deemed strategic is sometimes obscured or erased. Forget about the complex details of the people, culture and society. What matters is the 'strategic value' of the place, or the political identity of its government as an ideological and strategic friend or foe. At times this could take quite extreme forms, as with US National Security Advisor Robert McNamara (previously head of Ford, and later head of the World Bank):

> Robert McNamara was, of course, the leading specimen of *homo mathematicus* – i.e., men who behave and believe other men [and, we might suppose, women too] behave primarily in response to 'hard data', usually numbers (infiltration rates, 'kill ratios', bomb tonnage). Like the classic private eye on television, they are always looking for 'the facts', but usually the wrong ones. They miss reality, for they never get close enough or related enough to another society to do more than count things in it. If you relate to a country as a military target you do not need to know anything about it except details as are easily supplied by reconnaissance satellites, spy ships, secret agents, etcetera. You need never know who the victims of your attack were. Your task is merely to assess the results of what you have done and this is done by counting bodies, destroyed factories, enemy soldiers.
>
> (Barnet 1973: 119)

Such logic is seemingly far removed from the prominence of maps in the formal geopolitical tradition. Yet we can see that, in *both* the kind of thought present in McNamara's mind and in classic geopolitics, are moments when the myriad complexity of the world is reduced to a simple black and white representation, be it numerical or cartographic:

> By gathering, codifying, and disciplining the heterogeneity of the world's geography into the categories of Western thought, a decidable, measured and homogeneous world of geographical objects, attributes, and patterns is made visible, produced.
>
> (Ó Tuathail 1996a: 53)

These logics are both *colonial* and *colonizing*. They are colonial, in so far as they are rooted in the fact-gathering epoch of colonial exploration and share the arrogance of its classifications of peoples and places, which are then turned into a hierarchy of civilized to uncivilized, with the former (who are seen in racist terms) as the 'natural' leaders, or most developed, free or 'advanced'. They are also colonizing, in so far as they expand into and affect other seemingly distant fields of knowledge and social or cultural activities. Everything from scientific debates to art and cinema becomes caught up in geopolitics, and registers its spread, its dissemination.

21.8 Cold War geopolitics in art and culture and 'popular geopolitics'

The art world and, in particular, the course of modern art in the second half of the twentieth century became intertwined with Cold War geopolitics. More generally, art and politics have frequently been intertwined. The Nazis (and their leader Adolf Hitler, who was a failed artist of minimal talents) detested modern art, condemning it as decadent. For them a gross parody of classical art, sculpture and architecture was all that was fit for the German Reich. Other dictators, such as Joseph Stalin, the leader of the Soviet Union from the mid-1920s till his death in 1953, were also wary of modern art and favoured a garish realist style, depicting heroic smiling workers and peasants. Meanwhile, and especially in the years between the two World Wars, Paris became the centre of production of modern art. But this was disrupted by the devastation of the Second World War and the rise of American economic, cultural and political powers. As Sylvester (1996: 405) explains:

> In the wake of World War II and the catapulting of the United States into superpowerdom and the Cold War, there was parallel activity in US art circles. Media figures, such as Howard Devree and Alden Jewell, advocated a position for US art in the world that would mirror the international ascendancy of the US political economy. Devree maintained in the *New York Times* that the United States was the most powerful country and therefore needed to create art that was strong and virile to replace the art of Paris.

But this was not only words from influential figures. The American Central Intelligence Agency (CIA) stepped in to fund and promote American modern art, bankrolling major exhibitions and tours, to demonstrate the creative and in Devree's words, the 'virile' artistic culture of the United States. This assertion of 'virility' reveals too the masculinity of much American Cold War language and policy, differentiating the thrusting, powerful American society (later personified by people such as President John F. Kennedy) from a Europe portrayed as weak and effeminate and from the more brutal, less cultured men in charge of the Soviet Union. This strategy was more subtle than the vulgar anti-communism that pervaded Hollywood at the time. It reinforced more general arguments about the USA as a bastion of 'free

Plate 21.6 Cold War in cinema: James Bond (007) saves the West from communists and criminals.
(© 2005 Topfoto)

expression' and seemed to tally with the relative dynamism of the post-war American economy and the sense that the leadership of the Western World (or, as American commentators preferred to call it, the 'Free World') had decisively passed in military, political, economic and cultural terms to the USA. As Cockcroft (1974: 129) explains:

> CIA and MOMA [the New York based Museum of Modern Art, which was funded by the multi-millionaire Rockefeller family] cultural projects could provide the well-funded and more persuasive arguments and exhibits needed to sell the rest of the world on the benefits of life and art under capitalism.

Of course, the Cold War was waged too in everything from science fiction comics and action movies and thrillers (in which good Americans, like Rambo, and the occasional heroic Brit, like James Bond, faced and defeated totalitarian, alien and frequently communist enemies) to Chess (where 'Soviet Man' could demonstrate intellectual superiority and training) and Olympic sports (in which the Soviet, Chinese, East Germans and Bulgarians invested enormous resources in training and drug-enhanced muscles, to 'prove' the superiority of the communist system in the Olympic stadiums). And the 'Space Race' to build and launch artificial satellites and then to put humans into space and reach the moon became a military, technical and a cultural expression of the Cold War – full of moments of national superpower pride, media spectacles and the basis of tales of heroic and daring American and Soviet men in space (Carter 1988). At the same time, the East–West Cold War intersected in complex ways with struggles against colonial rule, for 'development' of the 'Third World (so called because it was neither First-Western nor Second-Eastern) and against white supremacy (in the US South and South Africa, for example) (Borstelmann 2001; Parker 2006).

The important thing to understand here is that whilst the sense of a formal geopolitical tradition may be seen as a self-conscious and fairly direct expression of states, space, power and the relations between them, the latter also take a vast variety of other forms. At the same time, the backdrop of a geopolitical conflict (such as the Cold War) can provide the basis for the making of a coherent sense of a particular way of life, for bonding a country, emphasizing a shared conformist identity and a common enemy (see Tyler May 1989; Campbell 1992). We might thus write of a 'popular geopolitics' to designate how geopolitical concepts are reflected in and take on meaning through popular culture: from art and literature, through film to comics, sport, computer games (Power 2007) or the Internet (Dodds 2006) and more traditional media like magazines, newspapers, radio and television. Perhaps, however, a more appropriate critical term is 'banal geopolitics' or 'banal militarism' (see Kuus 2007; Sidaway 2003b). These terms reflect on how the *everyday* forms of talk about geopolitical themes (such as the 'war on terror since 9/11' in the USA) and warfare have become taken for granted as the primary way to understand and respond to violent acts, like the bombings in European cities or the aircraft hijackings and deliberate crashes of 11 September 2001 in the USA. Such acts, however, have some deep and complex roots in prior geopolitical orders and in actions, decisions and reactions made through the twentieth century (and arguably before, in the form of more than a century of Western imperialism and reactions to it).

21.9 New World order to the Long War

The notion of the Cold War faded with the collapse of the USSR and associated communist allies between 1989 and the early 1990s, leaving the United States as the sole effective global military superpower. Yet this Western and American 'victory' quickly produced a certain sense of disorientation. This had been evident before, as the so-called 'bipolar' world of the early years of Cold War confrontation in which Washington and Moscow were the main political, economic and ideological points of orientation became progressively more 'multipolar' from the 1960s onwards. That is, as Western Europe and Japan recovered from wartime devastation and as Communist China split on ideological grounds with the USSR, the sense that the world was divided into just two superpower points of orientation lessened. But the sensation of geopolitical complexity was to grow precipitously once one of the established 'poles' (Soviet communism and pro-Soviet regimes in Central and Eastern Europe) collapsed at the start of the 1990s. Having a clearly demarcated and identifiable enemy – as both sides had in the Cold War – offered contrasting but apparently solid ideas of identity, purpose and common cause that has faded in a post-Cold War world. If the Cold War embodied a relatively coherent geopolitical map (East against West, with the allegiance of the Third World as one of the prizes), the post-Cold War world is characterized by a diversity

Thematic Case Study 21.1

The continuing Cold War?

The Cold War in Asia was always a very complex mixture of ideological, territorial and national disputes. An extra dimension was added by the fact that China became a nuclear power in the early 1960s at the same time that the Chinese communists split with the USSR on the grounds of ideological differences.

Moreover, whilst it is now almost two decades since proclamations of the 'end of the Cold War' began to multiply in Europe, in the Asia-Pacific region remnants of the global Cold War confrontation remain in place. Despite substantial changes and accommodations, China, North Korea and Vietnam are still one-party 'communist' political systems. In addition, a series of divisions and territorial disputes whose history is deeply rooted in the Cold War remain unresolved. Most notably, the Korean peninsula remains divided between mutually antagonistic regimes and both the People's Republic of China and the offshore island of Taiwan have claimed to be the legitimate government of the whole of China and remain fundamentally opposed.

Arguably, therefore, there is a sense in which the narratives about 'the end of the Cold War' reflect the Euro-Atlantic experience rather than being globally valid. The 'end' may have happened in Europe and the Soviet Union may have ceased to exist years ago, but Cold War legacies and logics are active in East Asia (see Hara 1999) and beyond. However, as the main text details (see sections 21.9 and 21.10), the extent to which new geopolitical narratives, notably 'the war on terror,' have replaced the Cold War as organizing frames for Western strategy is much debated – and contested.

Plate 21.7 South Korean soldiers patrol their side of the demilitarized zone between North and South Korea.
(© Michel Setboun/Corbis)

of maps and scripts. A variety of interpretations and geopolitical 'models' now attempt to offer explanation and impose meaning on contemporary events.

Amongst the most influential of these has been the discourse of **New World Order**, articulated by then US President George Bush (senior) in the early 1990s. In Bush's vision, what is called the 'international community' (the member states of the United Nations, led by the USA, but with financial and military support from key allies such as Germany, the United Kingdom and Japan) would act as a kind of global police force, intervening where and when they felt necessary to maintain or restore 'order'. Both the 1990–1 Gulf War and the 1999 conflict with Yugoslavia were justified and conducted (in part) in the name of building such a 'New World Order'. But it was his son, President George W. Bush, who was left to articulate a new American strategy (which became the Bush Doctrine, of pre-emptive attack on states that may – it was claimed – threaten the United States or harbour terrorists) in the aftermath of 9/11. And even this was, arguably, partially within a long American geopolitical tradition of overthrowing overseas regimes during and before the Cold War, which has its roots in the rise of American power in the Pacific and Americas in the late nineteenth century and has some deeper roots in the colonization of the continent (Kinzer 2006). Even well before 9/11 and the Bush administration's strategic reactions to it, critics had pointed out that, as Booth (1999: 49) put it, the New World Order 'means the New World [gives out the] Orders'. That is, the United States was seeking global military hegemony and with the Soviet resistance out of the way, there seemed little, save its own 'public opinion' (or perhaps, economic limits?) to stop it. The 'Monroe Doctrine' mentioned in section 21.1 is extended to virtually the entire world. Critics point out that this suits Western and especially American commercial interests. The United States' global military posture in the aftermath of the drama of 11 September 2001, in particular the extension of US military power deeper into what were once the heartlands of the Soviet Union in Central Asia, is one key aspect of this. US strategy in the Middle East, Africa, Asia and the Americas points to a wider American imperial geopolitics, which extends the strategies and doctrines developed in the twentieth century and finessed in the Cold War (Smith 2003).

Others have stressed that such apparent American military superiority and rhetoric counts for less than might appear at first sight, and that what continues to be evident in the post-Cold War world is enhanced economic (and in some ways, cultural) competition between blocs or great powers. In this vision, American power is potentially contested by an integrating European Union and by Pacific powers, notably China. But the terms of the 'contest' are less in conventional military and strategic terms than in competition for markets, productivity and profit. A world of 'geopolitical' competition is, it was argued, being partially displaced by 'geoeconomic' competition (for an example, see Luttwak 1990). In the USA (and beyond), this was also sometimes interpreted as an impending 'supposed clash of civilizations' (e.g. Huntington 1993), between that of the 'West' and, for example, societies oriented to alternative belief systems such as (Chinese) 'Confucianism' or **Islam**. Although superficially quite different, notions of geoeconomic competition and a clash of civilizations both betray a sense that the relative military dominance of the USA at the end of the Cold War may not in itself be enough to preserve America's sense of leadership, power and world-historical destiny in the context of a complex 'multipolar' world. It remains too early to tell how enduring the reassertion of American military power that unfolded under the Presidency of George W. Bush will be. Certainly it has faced multiple challenges, as attempts to implement visions of 'A New American Century [of leadership]' (http://www.newamericancentury.org/) have mutated into an open commitment to what the US Department of Defense, in the form of the latest National Security Strategy documents, has called (and codified in the February 2006 Defense Review as) 'The Long War'. Superficially this appears as a new Cold War, with the 'global war on terror' (GWoT) now taking the centre stage once occupied by communism. However, as Buzan (2006: 1102) argues, despite the prominence of the global war on terror in these texts, it does not define contemporary US grand strategy overall:

> US grand strategy is much wider, involving more traditional concerns about rising powers, global energy supply, the spread of military technology and the enlargement of the democratic/capitalist sphere. US military expenditure remains largely aimed at meeting traditional challenges from other states, with only a small part specifically allocated for the GWoT. The significance of the GWoT is much more political . . . the main significance of the GWoT is as a political framing that might justify and legitimize US primacy, leadership and unilateralism, both to Americans and to the rest of the world.

21.10 Conclusions: postmodern geopolitics?

Despite early post-Cold War proclamations about the great prospects for a more peaceful and secure New World Order, the world soon came to appear to American strategists (and many others) as more 'disordered' and unpredictable. 'Enemy regimes' (at various times in the last few years signifying Afghanistan, Iran, Iraq, Libya, North Korea, Sudan and Yugoslavia) have been targeted as the United States continues to declare that certain regimes must be contained, punished or changed. Yet the USA is more conscious that the enemy is also configured as a loose and amorphous network (the form attributed to al-Qaeda, who are held responsible for the events of 9/11 and host of other terrorist attacks). At the same time, high-tech weapons spread, disseminating greater destructive capacity into areas of local confrontation and conflict, such as South and East Asia and the Middle East. The USA seeks to retain the high-tech military advantage in surveillance and remotely targeted weapons and anti-missile systems, for example, but faces low-tech, but frequently deadly responses (such as small roadside bombs) in Iraq and Afghanistan. Manufacturers in the West, China and the former communist countries have found plenty of lucrative markets for weapons of all kinds, from fighter planes to machine guns. Such exports very often end up in relatively impoverished societies, including many in Africa, the Balkans and Asia that are awash with the destructive debris of the Cold War. In parts of central Africa and the Horn of Africa large-scale (though low-intensity) violent conflict continues: often reusing cold-war guns and fuelled by competition (as in Congo) for mineral resources. And in East Asia a number of disputes and differences rooted in the Cold War endure (see Case study 21.1). With other significant geopolitical divides (for example, the rival claims made by India and Pakistan over the territory of Kashmir, and the Arab–Israeli conflict) that were sometimes overshadowed by the Cold War but have endured beyond it, the reality and future prospects for larger-scale wars and conflicts persists, and with them the spectre of nuclear or other (chemical or biological) 'weapons of mass destruction'. Commenting on this, Barkawi and Laffey (2002: 125–6) note how

> [t]he end of the Cold War arguably made nuclear war *more* likely, especially given that Soviet weapons, nuclear materials and technical personnel are far from being concentrated under imperial control and indeed may even be available for purchase on the open market. The buyers may well be non-state actors such as al Qaeda who, on the evidence of 11 September, would be far more willing to use nuclear weapons of mass destruction than the leadership of a state with a vulnerable homeland. If India and Pakistan, among other possibilities, indicate that inter-state and even nuclear war cannot be assigned to the dustbin of history, al Qaeda and the 'War on Terror' are indicative of new forms of international and globalised war . . .

Moreover, for the Western powers and for the USA in particular, no longer is the main 'enemy headquarters' so clearly demarcated and fixed behind a stable 'Iron Curtain' – as the USSR and its communist allies were. Following the hijacking of airliners and their use in attacks on highly symbolic economic and military targets in the United States on 11 September 2001 ('9/11'), this has been felt even more deeply. Although there is no absence of traditional forms of geopolitical discourse, which define a clear boundary of 'our' space against 'them', some have argued that, today, Western strategy operates in the frame of a new kind of 'postmodern geopolitics'. For example, Gearóid Ó Tuathail (1998: 31–2) examined US foreign policy debates in the 1990s and noted how they argue that:

> In contrast to the transcendent containment imperative and fixed posture of Cold War strategy, what is required [according to US strategists] in response to persistent territorial concerns and proliferating deterritorialized threats is a geo-strategic doctrine premised on flexibility and speed. The 1995 National Military Strategy of the United States is subtitled 'a strategy of flexible and selective containment'. . . . Threats are described as widespread and uncertain, the possibility of conflicts probable, but their geographic sites are too often unpredictable. This document describes the current strategic landscape as characterized by four principal dangers that the US military must address: 'regional instability; the proliferation of weapons of mass destruction; transnational dangers such as drug trafficking and terrorism; and the dangers to democracy and reform in the former Soviet Union, Eastern Europe and elsewhere'. . . . What is remarkable about these threats is that none has a fixed spatial location; regional instability refers not only to the Middle East but also to Europe and Africa; proliferation and transnational dangers are global; even dangers to reform, which is the only danger explicitly linked to certain places, are potentially ubiquitous, as the 'elsewhere' indicates. Military strategy still has to negotiate territory and place but it has also, in interesting ways, become untethered from place and territory. Anywhere on the globe is now a potential battlefield. The document concedes as much, noting that 'global interdependence and transparency coupled with our worldwide security interests, make it difficult to ignore troubling developments almost anywhere on earth'.

Since then, under the auspices of what came to be defined by George W. Bush's administration as 'The Long War', the commitment to open-ended conflict anywhere was reiterated:

> adjusting the U.S. global military force posture, making long overdue adjustments to U.S. basing by moving away from a static defense in obsolete Cold War garrisons, and placing emphasis on the ability to surge quickly to trouble spots across the globe.
>
> (http://www.defenselink.mil/qdr/report/Report20060203.pdf,v)

Critics have argued that the level of connections and reconfiguring of power amount to a new form of global empire (Hardt and Negri 2000), although the extent to which this was different from the imperialisms of the past (such as those described in

Chapter 2) is being debated. Whilst the global maps and scripts of geopolitical alliances (friends and foes) and actions have changed quite dramatically since the Cold War declined, there are also some important continuities and many legacies. The shocking attacks of 11 September 2001 have multiple causes and consequences. However, amongst the causes are links back to the superpower confrontation of the Cold War (see also Case study 21.1). Thus as Halliday (2002: 36) notes:

> the rise of [Islamic] fundamentalist groups [such as *al-Qaida* which claimed responsibility for the 9/11 attacks] is not subsequent to, but an integral result of the Cold War itself.
>
> The Cold War indeed contributed to this crisis and in particular to the destruction of Afghanistan from 1978 onwards, but in a way that should give comfort for few. One can here suggest a 'two dustbins theory' of Cold War legacy: if the Soviet system has left a mass of uncontrolled nuclear, chemical and biological weapons and unresolved ethnic problems, the West bequeathed a bevy of murderous gangs, from National Union for the Total Independence of Angola (UNITA) [which waged a violent struggle against the pro-Moscow regime in that African country in the 1970s and 1980s and later continued to wage war in Angola to control resources like diamonds] and Cuban exiles in the Caribbean and Miami to the *mujahidin* of Afghanistan, who are now on the rampage [both of which were earlier funded and armed by the United States and its allies].

Not all such non-state and anti-statist movements are of the right, however, and Routledge (2003) is among those who have celebrated the prospects of a radical 'geopolitics from below' and a counter- or anti-geopolitics from those movements (such as the Zapatistas of Mexico and the *Movimento Sem Terra* or Landless Peasants Movement of Brazil and other indigenous and social movements critical of 'globalization') that seek progressive alternatives. According to Routledge (2003: 236) therefore:

> myriad alternative stories can be recounted which frame history from the perspective of those who have engaged in resistance to the state and the practices of geopolitics.

Yet if it began with the modern age of geopolitics, announced by Kjellén and Mackinder, it might be argued that the twenty-first century has begun less in a proliferation of anti-geopolitics and more as an age of 'postmodern geopolitics'. The apparent simplicity of the great power and Cold War confrontations, let alone demarcated and fixed 'heartlands', have been replaced by the sense of a world of proliferating uncertainty and 'threats'. Scares, dangers and anxieties about terrorism since '9/11' have greatly reinforced this. The remaining superpower also faces long-term economic challenges (especially as the global role of the dollar diminishes) and must continue to reckon with more traditional state-centred powers, such as Russia and a rising China (even if they no longer pose the ideological completion of the Cold War years). There is no easy or predictable end to this uncertain situation.

At least with all this complexity, and the preceding points in mind, we might well argue that a critical examination of 'geopolitics' should not confine itself to the tradition that goes under that name (also see Tesfahuney 1998; Sharp 1999; Nagel 2002). To do so would be to miss much wider issues about how broader sets of ideas about space, power, order, society and the international scene are combined into common-sense notions about 'us' and 'them' and the whole geographical order of what is presented as the good life.

Learning outcomes

Having read this chapter, you should understand:

- The term 'geopolitics' has been associated with a wide variety of texts and contexts.
- However, geopolitics is frequently associated with the sense of a self-conscious tradition of writing about and mapping the relations between states, geography and power.
- This is usually traced to the late nineteenth-century writings of Rudolf Kjellén.
- One of the key features of this tradition is the idea that the state resembles a living organism which requires living space and will compete with others (the organic theory of the state).
- The oppressive military regimes in South America (notably Argentina, Brazil and Chile) during the 1960s and 1970s elaborated and applied a variety of the geopolitical tradition.
- However, it also proliferated in a number of European countries and Japan, particularly between the two world wars.

- The rise and eclipse of the Cold War have produced wide and deep geopolitical transformations.
- Since this proliferation of geopolitics was most evident in Fascist states, the defeat of the Axis powers in 1945 signified its relative decline as a formal tradition, although it continued in Fascist Spain and Portugal and in South American military circles.
- Beyond the self-conscious tradition of geopolitics, similar forms of thinking about territory, states and power have proliferated in many disciplines.
- This has been particularly evident during the Cold War, when broader geopolitical discourses suffused many aspects of art, science, culture and daily life.
- Since the decline of the Cold War, there have been a proliferation of geopolitical visions of the 'New World Order', amongst them a more complex cartography of perceptions of threats and dangers. However, the ability of any new single geopolitical 'big picture' (such as 'the war on terror') to inherit the mantle of the Cold War is radically open to question and contested.

Further reading

Agnew, J. (2003) ***Geopolitics: Revisioning World Politics***, 2nd edition, Routledge, London and New York. A historical and contemporary guide which places geopolitics in comparative historical perspective.

Agnew, J. and Corbridge, S. (1995) ***Mastering Space: Hegemony, Territory and International Political Economy***, Routledge, London and New York. Written in the aftermath of the Cold War and as debates proliferated about post-Cold War geopolitical and geoeconomic trends, Part II (Chapters 5, 6 and 7) remains useful on aspects of and critical reflections on the 'New World Order'.

Agnew, J., Mitchell, K. and Toal, G. (eds) (2003) ***A Companion to Political Geography***, Blackwell, Oxford. Includes some useful accounts of geopolitical traditions, amongst other aspects of past and present political geographies.

Campbell, D. (1992) ***Writing Security: United States Foreign Policy and the Politics of Identity***, University of Minnesota Press, Minneapolis. A challenging but rewarding account of how foreign policy and the Cold War have been productive of America's sense of self-identity.

Dodds, K. (2004) ***Global Geopolitics: A Critical Introduction***, Prentice Hall, Harlow.
Dodds, K. (2007) ***Geopolitics: A Very Short Introduction***, Oxford University Press, Oxford.
Flint, C. (2006) ***Introduction to Geopolitics***, Routledge, London.
Three accessible introductions: all good starting points.

Halliday, F. (2002) ***Two Hours that Shook the World. September 11, 2001: Causes and Consequences***, Saqi Books, London. Amongst the most nuanced of the hundreds of books that have appeared on this subject.

Heffernan, M. (1998) ***The Meaning of Europe: Geography and Geopolitics***, Arnold, London.
McCormick, J. (2006) ***The European Superpower***, Palgrave Macmillan, London.
The first book is a readable account of European geopolitics and the background to European integration, the second book claims that the European Union is a new form of 'superpower' (in geoeconomic rather than conventional geopolitical terms).

Hyndman, J. (2003) Beyond either/or: a feminist analysis of September 11th, ***ACME: An International E-Journal for Critical Geographies***, **2**(1), 1–13 (downloadable from www.acme-journal.org). Indicates how feminist approaches might problematize and counter dominant geopolitical interpretations.

O'Loughlin, J. (ed.) (1994) ***Dictionary of Geopolitics***, Greenwood Press, Westport, CT. A useful source.

Ó Tuathail, G. (1996) An anti-geopolitical eye: Maggie O'Kane in Bosnia, 1992–3, ***Gender, Place and Culture***, **3**(2), 177–85. Shows how commitment can produce alternative narratives amounting to an 'anti-geopolitics'. Suggestive and worth reflection.

Ó Tuathail, G. and Dalby, S. (eds) (1998) ***Rethinking Geopolitics***, Routledge, London and New York. A collection of critical reflections on Cold War, popular and post-Cold War geopolitical narratives.

Ó Tuathail, G., Dalby, S. and Routledge, P. (eds) (2006) ***The Geopolitics Reader (second edition)***, Routledge, London and New York. A collection of previously published short texts (readings) on different aspects of geopolitical discourse, including extracts from Mackinder, Haushofer and Kissinger and debates about the New World Order. Detailed editorial introductions contextualize the readings.

Three suggestive papers on aspects of popular geopolitics (cartoons and newspapers), worth comparing and contrasting:

Debrix, F. (2007) Tabloid Imperialism: American geopolitical anxieties and the war on terror, ***Geography Compass***, **1**(4), 932–945.

Dittmer, J. (2005) Capitan America's Empire: reflections on identity, popular culture and post-9/11 geopolitics, ***Annals of the Association of American Geographers,*** **95**(3), 626–43.

Falah, G-W, **Flint C.** and **Mamadouh, V.** (2006) Just war and extraterritoriality: the popular geopolitics of the United States war on Iraq as reflected in newspapers of the Arab world, ***Annals of the Association of American Geographers,*** **96**(1), 142–64.

There are many variants of 'popular geopolitics' in circulation and five minutes on the Internet will yield dozens of (mostly US authored and hosted) religious 'end times' sites on the New World Order as well as 9/11 conspiracy theories. Read them critically with a view to understanding the basis on which each claims truth and the consequences of and basis of their *geopolitical* assumptions.

Finally, a century after its original publication, Mackinder's (1904) paper was reprinted in ***The Geographical Journal*** (Volume 170, Number 4, December 2004), along with a set of commentaries and reflections.

For annotated, clickable weblinks and useful tutorials full of practical advice on how to improve your study skills, visit this book's website at **www.pearsoned.co.uk/daniel**

Glossary

A

AAA referring in this book to 'Anglo/American/Australasian' tradition in geography, which is sometimes also referred to (erroneously) as the Western tradition, where the latter should include the geography of scholars in Europe.

Abjection the process by which we seek to repress that which we regard as unclean, improper, impure or dangerous. In geographic terms, abject material tends to be located in marginal spaces which become repositories for those things which are regarded as threatening the social body.

Actor-network human actors in interaction through non-human intermediaries and power relations.

Agglomeration the concentration of productive activities in a particular region.

Agribusiness large-scale, capital-intensive, agricultural businesses incorporating supply, production and processing capacities.

Agri-food system the highly integrated system of agricultural production in the developed world which involves both upstream (e.g. suppliers) and downstream (e.g. processing) industries.

Alternative food networks a reaction to conventional (e.g. supermarket-dominated) food chains, in which local and/or organic foods are marketed through alternative outlets such as farmers' markets, box schemes and home deliveries.

Anthropocentric a way of understanding the value of the environment which prioritizes humans as having the most value. In other words the environment is only useful as a resource for humans. In this approach humans do not have any responsibility to ensure the environmental sustainability of our actions.

Apartheid the policy of spatial separation on racial grounds employed in South Africa under National Party rule between the late 1940s and early 1990s.

Appropriationism the replacement of agricultural inputs with industrial alternatives. It forms a key process in the industrialization of the agri-food system.

B

Balance of payments (BOP) the difference in value between a country's inward and outward payments for goods, services and other transactions.

Balkanization progressive subdivision of a region into small political units.

Bands societies of hunter–gatherers typically numbering up to 500 individuals. See Spotlight box 1.1.

Billion thousand million.

Biodiversity the variability among living organisms from all sources including terrestrial, marine and other aquatic ecosystems and the ecological complexes of which they are part; this includes diversity within species, between species and of ecosystems.

Biomass plant and animal residue burnt to produce heat.

Biotechnology a branch of technology concerned with the industrial production of living organisms and their biological processes.

Birth rate number of babies born per thousand population per year; known as the crude birth rate.

Bottom-up development economic and social changes brought about by activities of individuals and social groups in society rather than by the state and its agents.

Bounded applied to the behaviour of decision makers whose access to information, for example, is constrained by financial resources or time. They also have a limited capacity to process the information that they are able to obtain and they will also be constrained (bounded) by the environment within which their behaviour is taking place. Bounded cultures may be regarded as reactions to perceived threats to local cultures, entailing a strong assertion of the latter. These can create a level of local fragmentation, with a parochial, nostalgic, inward-looking sense of local attachment and cultural identity. Bounded cultures generally involve very clear definitions of 'insiders' and 'outsiders' in the creation of a sense of belonging.

Brundtland Report published in 1987, it adopted the position that it was possible to pursue economic growth without compromising the environment and introduced the first widely used definition of sustainable development: 'development which meets the needs of the present without compromising the ability of future generations to meet their own needs'.

Buyer-driven commodity chain a chain in which the nature of production (e.g. salmon farming) is shaped or driven by powerful 'downstream' actors such as distributors and retailers.

C

CAP Common Agricultural Policy.

Capital money put into circulation or invested with the intent of generating more money. See Spotlight box 15.2.

Capitalism an historically specific economic system in which production and distribution are designed to accumulate capital and create profit. The system is characterized by the separation of those who own the means of production and those who work for them.

Chiefdom social formation based on societies characterized by the internal and unequal differentiation of both power and wealth and organized around the principle of kinship. See Spotlight box 1.3.

City-states urban regions having political jurisdiction and control over a specific territory. A form of social organization associated with some of the earliest states.

Civilization refers to an advanced form of social development characterized by such things as urban life, commercial activity, writing systems and philosophical thought.

Class social distinctions between groups of people linked to their material conditions and social status.

Clean development mechanism (CDM) a mechanism whereby industrialized countries can earn credits towards their own greenhouse gas reduction targets by investing in emission-reducing projects in the developing world. Part of the aim is to aid economic growth in developing countries without a commensurate increase in greenhouse gas emissions. At the same time this assistance is to the industrialized countries; such assistance is often cheaper than achieving their own emission reduction.

Climate change a change in climate that is attributed directly or indirectly to human activity that alters the composition of the global atmosphere and which is in addition to natural climate variability observed over comparable time periods. Includes temperature rise, sea-level rise, precipitation changes, droughts and floods.

Cluster a localized concentration of similar and interlinked economic activities.

Cold War a period extending from the end of the 1950s to the late 1980s during which two ideologically opposed blocs emerged within the world system, both with nuclear capabilities. The first was headed principally by the USA while the other was dominated by the Soviet Union.

Command economy an economic system characterized by the state-led central planning of economic activity combined with the simultaneous suppression of market-type relations.

Commodity a good or service produced through the use of waged labour and sold in exchange for money.

Commodity fetishism the process whereby the material and social origins of commodities are obscured from consumers.

Communism a political theory attributed to the works of Karl Marx and Friedrich Engels. Communism is characterized by the common ownership of the means of production. The foundations for full communism were supposed to be laid during a transitional period, known as socialism. The Soviet Union was the first state to be ruled by a communist government, from 1917 to 1991.

Communitarianism a moral and political philosophy based on the extension and preservation of community. It is the opposite of liberalism which is often regarded as insufficiently sensitive to the social/community sources of self-worth.

Community food scheme a site of resistance to the industrialization of agriculture, in which the local community retains control and where consumers get seasonal, fresh food at a price that supports farmers who use sustainable practices.

Comparative advantage in economic geography, refers to an advantage held by a nation or region in the production of particular set of goods or services. See Spotlight box 13.3.

Competitive advantage exists when a firm can deliver the same benefits as competitors but at lower cost (cost advantage). A firm may also deliver benefits that are greater than those of competing products (differentiation advantage).

Conditional reserve those (mineral or ore) deposits that have already been discovered but that are not economic to work at present-day price levels with currently available extraction and production technologies.

Contact zone the space in which transculturation takes place – where two different cultures meet and inform each other, often in highly asymmetrical ways.

Containment the Western strategy of encircling the Soviet Union and its allies during the Cold War.

Conventionalization also referred to as 'mainstreaming', this occurs when larger scale businesses (e.g. agribusinesses, supermarkets) increasingly control the production of 'alternative' products such as fair trade and organic through the adoption of conventional patterns of marketing and distribution.

Core according to Wallerstein, the core refers to those regions of the capitalist world economy characterized by the predominance of core processes associated with relatively high wages, advanced technology and diversified production. See Spotlight box 2.1.

Cosmology study of the universe as an ordered whole, and of the general laws or principles that govern it.

Counterurbanization population increases in rural areas beyond the commuting range of major urban areas.

Crude birth rate number of births in one year per 1,000 population.

Crude death rate number of deaths in one year per 1,000 population.

Cultural capital the possession of taste, style or attitude that can be converted, in many instances, into financial capital. In the post-industrial economy, where ideas are a global currency, cultural capital is a lucrative commodity.

Culture a system of shared meanings often based around such things as religion, language and ethnicity that can exist on a number of different spatial scales (local, regional, national, global, among communities, groups, or nations). Cultures are embodied in the material and social world, and are dynamic rather than static, transforming through processes of cultural mixing or transculturation. See Spotlight box 13.1.

Culture jamming tactics such as media hacking, information warfare, 'terror-art' and graffiti, which aim to invest advertisements, newscasts and other media artefacts produced by powerful agencies and organizations with subversive meaning.

D

DDT a chemical pesticide banned by many countries in the 1970s as a pollutant which is a probable human carcinogen and causes damage to internal organs.

Death rate number of deaths per year per thousand population; known as crude death rate.

Deep ecology An environmental approach which asks us to consider humans not only as part of nature but of *equal* value to other non-human entities. This is a holistic vision whereby all human and non-human entities are interconnected and interdependent. A holistic perspective enables us to understand that if we upset one element, it will have an impact on all other elements.

Defensive localism a process where consumers purchase local foods to support local farmers and the local economy, irrespective of their quality and whether they are produced organically or conventionally.

Deforestation the removal of forests from an area.

Deindustrialization refers to a relative decline in industrial employment. It may also refer to an absolute decline in industrial output as well as employment.

Demarginalization the process whereby a marginal or stigmatized space becomes 'normalized', and its population incorporated into the mainstream.

Demographic transition model traces the shift from high birth and death rates to low ones.

Dependency a viewpoint or theory of development which argues that global inequality is explained by the patterns of exploitation of the periphery by the capitalist core, established during the colonial period and perpetuated by neo-colonial economic relations in recent times. The dependency theorists' recommendation for poorer countries is to de-link from the global economy.

Depopulation the reduction of population in an area through out-migration or a reduction in the birth rate below the death rate.

Diaspora literally, the scattering of a population; originally used to refer to the dispersal of Jews in AD 70, now used to refer to other population dispersals, voluntary and non-voluntary. Evokes a sense of exile and homelessness.

Diaspora space the spaces inhabited not only by those who have migrated and their descendants, but also by those who are conceptualized as indigenous or 'native'. Similar to contact zone – a meeting point of different cultures.

Discourse social and cultural theorists understand discourse to be the dominant meanings that are attached to a linguistic term or utterance.

Disorganized capitalism a form of capitalism identifiable since the mid-1970s. It is characterized by a relative decline in the importance of extractive and manufacturing industries, together with a relative increase in the importance of services. There is also an increasing tendency for the production process to be dominated by small-scale and flexible forms of organization. See Spotlight box 3.3.

Displaced persons refugees who have no obvious homeland.

Division of Labour the separation of tasks in the production process and their allocation to different groups of workers.

Doubling time time taken for a population to double in size.

E

Earth Summit the UN Conference on Environment and Development held in Rio de Janeiro in 1992. A total of 178 countries negotiated a global strategy centrally concerned with sustainable development.

Ecocatastrophism the prediction of impending environmental disasters, often as a result of human actions.

Ecocentrism a way of understanding the value of the environment which critiques the priorities of anthropocentrism and suggests that non-human entities have intrinsic value.

Ecological democratization an approach that suggests that the only way to achieve environmentally sustainable practices is to value participation and justice in environmental decision-making. Thus this approach requires extensive citizen participation and the development of democratic institutions to tackle environmental issues at all scales.

Ecological modernization an environmental management approach that argues that economic growth does not need to be slowed to ensure environmental protection. It does not undermine the limits to growth thesis entirely, but suggests that we can 'reorient' economic growth and use technological solutions for environmental problems, thus overcoming the environmental impact of growth. It is a very weak interpretation of sustainable development that allows for a reformist response where the current dominance of free-market capitalism is not challenged.

Economies of scale occur when mass production of a good (product) results in a lower average cost for each item. Economies of scale occur within a firm, such as using expensive equipment more intensively, or outside the firm as a result of its location, such as the availability of a local pool of skilled labour or good transport links.

Edge city (exopolis) a term referring to an area with city-like functions usually arising on the edge of an already urbanized area or conurbation and heavily dependent on fast communications systems. Edge cities are regarded as symptomatic of the most recent phase of urbanization.

Emerging markets refers to those markets that are perceived to have a substantial growth potential. The term is often used in relation to countries of the former Soviet Union, Asia, Latin America and Africa.

Emotional labour workers are expected to display certain emotions, often in line with the goals of their employer or organization, as part of their job.

Empire an extended territorial political unit or state built up, often by force, under a supreme authority. Empires usually involve rule over alien or subject peoples.

Energy mix the balance between various sources of energy in primary energy consumption.

Energy ratio the relationship between energy consumption and economic growth in an economy:

$$\text{Energy ratio} = \frac{\text{Rate of change in energy consumption}}{\text{Rate of change in economic growth (GDP)}}$$

A value greater than 1 indicates that the amount of energy required to create an additional unit of GDP is increasing; a value less than 1 suggests the reverse.

Enlightenment a philosophy or movement that emerged during the eighteenth century and was based on the idea of social progress through the application of reason.

Environmentalism a broad term incorporating the concerns and actions in aid of the protection and preservation of the environment.

Epidemiological transition switch in the predominant causes of death noted in the movement from a pre-transition to a post-transition society, primarily from infections to degenerative diseases.

Ethnicity refers to the process through which groups are recognized as possessing a distinct collective cultural identity.

EU European Union.

Exchange the process of interchange of goods and services between individuals, groups and/or organizations, whether involving money or not. Can also refer more broadly to social interactions.

Export processing zone a small closely defined area which possesses favourable trading and investment conditions created by a government to attract export-orientated industries.

F

Factors of production refers to those elements necessary for the effective functioning of the production process and typically includes land, labour and capital.

Fair trade attempts to overcome the injustices of free trade by guaranteeing producers a fair price and thus improving their lifestyles. Coffee, tea, bananas and chocolate are among the major fair trade products.

Fascism a term used particularly to describe the nationalistic and totalitarian regimes of Benito Mussolini (Italy, 1922–45), Adolf Hitler (Germany, 1933–45) and Francisco Franco (Spain, 1939–75).

Fast food a quick and accessible way to eat food; epitomized by McDonald's, it is usually eaten out of the home and out of our hands.

Fertility rates the number of like births per thousand women of child-bearing age.

Feudalism a hierarchical social and political system common in Europe during the medieval period. The majority of the population were engaged in subsistence agriculture while simultaneously having an obligation to fulfil certain duties for the landholder. At the same time the landholder owed various obligations (fealty) to his overlord.

Financial exclusion the processes that prevent disadvantaged social groups from gaining access to the financial system.

Flexible specialization new methods of production such as multi-purpose machines and small-batch production. The term is frequently associated with agglomerations of firms, or industrial districts.

Food chain the route traced by particular foodstuffs from 'farm to fork'. Food chains involve a number of production, processing, distribution and consumption nodes, and the connecting links between them.

Food desert a place, usually in inner cities or remote rural areas, where access to fresh, affordable and healthy food is poor.

Food miles a term used to describe the distance that food travels from the point of production to the place of consumption. As well as distance, the mode of transport (e.g. air vs ferry freight) is an important consideration.

Food regimes distinct relationships discerned between patterns of international food production and consumption and the developing capitalist system.

Fordism a regime of accumulation involving mass production and consumption. Named after Henry Ford (1863–1947), Fordism is known for a differentiated division of labour, assembly-line production and affordable mass-produced consumer goods.

Fourth World according to Castells (1998) a term referring to those regions effectively excluded from participation in the 'new' world economy. Importantly, it is of relevance not only to regions and countries of the recognized Third World but also to regions within the developed world. Also sometimes used to refer to indigenous or native peoples.

G

Gated communities residential developments protected by a range of mechanisms such as security gates, walls, private security guards and intercom systems.

Gemeinschaft a form of community said to be common in traditional societies (as distinct from industrial societies) and associated with notions of stability and informal personal contact.

Gender refers to socially constructed ideas of difference between men and women.

Gender division of labour a division of labour constructed around gender in which particular tasks and occupations are deemed to be male or female activities.

Gendered space the ways in which certain spaces are seen to be occupied exclusively or predominantly by either males or females.

Genetic modification (GM) human manipulation of genetic material (plant, animal and human) to create altered organisms.

Gentrification the process by which middle- and upper-class incomers displace established working-class communities. Often associated with new investment in the built environment, gentrification may be small-scale and incremental (i.e. instigated by individual incomers), or be associated with major redevelopment and regeneration schemes.

Geographies of exclusion the spatial processes by which a powerful grouping consciously seeks to distance itself from other less powerful groupings.

Geopolitics a term that has been used to refer to many things, including a tradition of representing space, states and the relations between them; also emphasizing the strategic importance of particular places.

Gesellschaft a form of association common in urban-based industrial societies (as distinct from traditional societies) and associated with non-permanent and utilitarian social relationships.

Ghetto refers to very high concentrations of people drawn from a particular ethnic or cultural background living in specific parts of an urban area. The term is now commonly associated with notions of deprivation, unemployment and social exclusion.

Global cities those cities performing a dominant role in the world economy and characterized by specialized service-type functions such as financial markets.

Global production networks the extensive webs of intra-, inter- and extra-firm connections through which commodities are produced, distributed, sold and consumed.

Global production system refers to the dominant form of production identifiable at a global scale. It has evolved over many centuries into its current form, referred to as post-Fordism.

Global triad this term describes how the world economy is essentially organized around a tripolar, macro-regional structure whose three pillars are North America, Europe, and East and South-east Asia.

Global warming an increase in the temperature of the Earth's surface caused by trapping infrared radiation in carbon dioxide, increased amounts of which are produced by burning fossil fuels.

Globalization a contested term relating to transformation of spatial relations that involves a change in the relationship between space, economy and society.

Glocalization the interaction between the particular character of places or regions and the more general processes of change represented by globalization.

GM See **Genetic modification**.

Governance the way in which power operates through the relationships between different organizations.

Gross domestic product (GDP) the value left after removing the profits from overseas investments and those profits from the economy that go to foreign investors.

Gross national product (GNP) a broad measure of an economy's performance; it is the value of the final output of goods and services produced by the residents of an economy plus primary income from non-residential sources.

H

Heartland identified by the British geographer Halford Mackinder (1904) as the zone in East-Central Europe, control of which would be a key to world domination. The term has since been appropriated by Latin American geopolitics.

Hegemony term derived from the work of Antonio Gramsci which refers to the ability of a dominant group to exert or maintain control through a combination of overt and subtle mechanisms.

Heritage according to UNESCO, 'heritage is our legacy from the past, what we live with today, and what we pass on to future generations. Our cultural and natural heritage are both irreplaceable sources of life and inspiration'. In human geography the focus is often on rural and urban heritage landscapes, frequently regarded as playing a formative role in the maintenance of national identity.

High-value foods includes such foods as fruit, vegetables, poultry and shellfish. World trade in high-value foods has increased markedly during the last two decades.

Horizontal integration occurs when two companies, within the same industry and at the same stage of production, merge.

Hybridity refers to groups as a mixture of local and non-local influences; their character and cultural attributes are a product of contact with the world beyond a local place. See Case study 12.5.

Hypothetical resources those resources that might be expected to be found in the future in areas that have only been partially surveyed and developed.

I

Imaginary geographies the ideas and representations that divide the world into spaces with particular meanings and associations. These exist on different scales (e.g. the imaginaries that divide the world into a developed core and less developed peripheries, or the imagined divide between the deprived inner city and the affluent suburbs).

Imperialism a relationship of political, and/or economic, and/or cultural domination and subordination between geographical areas.

Indigenous peoples those peoples native to a particular territory that was later colonized, particularly by Europeans.

Industrial revolution a term that is often taken to refer to the marked transformation of productive forces, initially within the British economic system, between the mid-eighteenth and mid-nineteenth centuries. It resulted in the movement of Britain from a largely rural-based economy to one that was dominated by manufacturing and industrial production. Such a transformation resulted in substantial social, political as well as economic changes.

Industrialization of agriculture (see also agribusiness) a process whereby methods commonly associated with the manufacturing industry are increasingly incorporated into farming, e.g. specialization of labour, assembly-line production systems.

Infant mortality rate (IMR) number of deaths of persons aged under 1 per 1,000 live births.

Informal economy those parts of the economy which operate beyond official recognition and outside formal systems of control and, often, of remuneration.

Inherent value the value something has for someone, but not as means to a further end.

Instrumental value the value which something has for someone as a means to an end.

International division of labour a term referring to the tendency for particular countries and regions of the globe to specialize in particular types of economic activity.

(International Monetary Fund) IMF international financial institution that originated from the 1944 Bretton Woods Conference. Its main roles include regulating international monetary exchange and controlling fluctuations in exchange rates in order to alleviate balance of payments problems.

Internationalization a process whereby products and services are adapted to specific local languages and cultures.

Intrinsic value simply the value something has. No appeal needs to be made to those for whom it has value.

Islam a monotheistic religion founded by the prophet Muhammad in the seventh century AD. Today there are two predominant divisions within Islam, the Sunni and Shi'i.

Islamophobia a prejudice against those who identify as Muslim, fuelled by post-9/11 discourses of Islamic fundamentalism and terrorist threat. Taking a variety of forms in the urban West, this can be interpreted as merely the latest inflection of a long-standing Orientalist discourse that contrasts Western democracy with Arab 'barbarism'.

K

Kondratieff cycle a term used to describe the cycles of boom and bust evident within the capitalist system since the mid-eighteenth century. Named after the Russian scholar N.D. Kondratieff. See Table 2.1.

Kyoto Protocol the agreed outcome of a meeting of 160 nations in Kyoto, Japan, in 1997 whereby developed nations agreed to limit their greenhouse gas emissions, relative to the levels emitted in 1990.

L

Legitimacy with regard to nation-states, a term meaning that the majority of people accept the rule of law of the governing political organizations.

Less developed countries (LDCs) countries at a disadvantage in today's global competitive environment because their comparative advantage in cheap labour or natural resource endowments has become subordinated to knowledge-based factors. LDCs suffer from poor productive capacities and competitiveness.

Liberalism a political philosophy distinguished by the importance it attaches to the civil and political rights of individuals. May also refer to belief in the importance of the free market.

Life expectancy (at birth) average number of years of life expected on the basis of age-specific mortality schedules for the specified year.

Limits to growth the belief that there are natural limits to possible growth which if exceeded will lead to environmental catastrophe.

Local Agenda 21 (LA21) the implementation by local administrations of sustainable development practices as defined in Agenda 21, an outcome of the Earth Summit, Rio de Janeiro, 1992.

Locality a place or region of sub-national spatial scale.

Low impact development (LID) a form of low impact living, a deep green vision where humans minimize their environmental impact in all aspects of their daily lives. LID is a radical form of housing and livelihood that works in harmony with the landscape and natural world.

M

Market-based states modern states where the market is the dominant means by which land, labour, capital and goods are exchanged and has a major influence over social and political organization.

Marxism a form of socialism and mode of analysis derived from the teachings of Karl Marx (1818–83). Marxism regards capitalism as an inherently unjust system with the capitalists (those who own the means of production) exploiting the proletariat (those who must sell their labour in order to live). It aims to replace capitalism with a fairer system, socialism maturing into communism.

Merchant capitalism refers to an early phase of capitalist industrial development dominant in the larger urban regions of Europe from the late fifteenth century. Merchants were the principal actors engaged in both the provision of capital and the movement and trade of goods (predominantly bulky staples such as grain and manufactured goods).

Mergers occur when two firms agree to form a new company.

Modernism a term typically associated with the twentieth-century reaction against realism and romanticism within the arts. More generally, it is often used to refer to a twentieth-century belief in the virtues of science, technology and the planned management of social change.

Modernity refers to a period extending from the late sixteenth and early seventeenth centuries (in the case of Europe) to the mid to late twentieth century characterized by the growth and strengthening of a specific set of social practices and ways of doing things. It is often associated with capitalism and notions such as progress.

Monopoly in theory, exists in an industry when one firm produces all the output of a market; in practice varies between countries. In the United Kingdom, for example, any one firm that has 25 per cent of the market is considered to hold a monopoly.

Monopsony in theory exists in an economy when one firm or individual purchases all of the output in a given sector. An agribusiness company can gain a geographical monopsony over a given area through the use of contracts that prevent sale of produce to other parties.

Moral panic a term describing periodic episodes of concern about the threat of a particular group to the nation-state. Moral panics are normally fuelled by sensationalist media reporting, and are generally diffused by the state through policies which aim to counteract this imagined threat.

More developed countries (MDCs) countries with significant competitive advantages in today's globalizing economy. They have well-developed, increasingly knowledge-based and strongly interconnected manufacturing and service sectors that provide a significant proportion of employment and contribute to significant national and individual wealth. Indices such as literacy levels, incomes and quality of life are high and these countries exercise considerable political influence at the global scale. Examples are the United Kingdom, the United States, Germany and France.

Multiculturalism refers to a belief or policy that endorses the principle of cultural diversity and supports the right of different cultural and ethnic groups to retain distinctive cultural identities. It has often been criticized for being too symbolic and not politically radical enough in challenging racism.

Multiple deprivation usually refers to a situation in which an individual or group suffers a series of disadvantages, including poverty, poor health, exposure to criminality, inadequate housing and so on. Geographically often used to denote those denied easy spatial access to a range of services, such as housing, health care, education and also transport.

N

Nationalism the ideology and sentiment of belonging to a 'nation' and the claim that the 'nation' should be expressed in a 'state'.

Nation-state and state a symbolic system of institutions claiming sovereignty over a bounded territory.

Natural increase surplus of births over deaths in a population.

Neoliberalism an economic doctrine promoting market-led growth, deregulation and the privatization of state-owned enterprises.

Neo-Malthusian the belief that environmental problems are a consequence of population growth, following the arguments set out by Thomas Malthus in the late eighteenth century.

New economic geography (NEG) an economic geography that recognizes the importance of culture as an influence on economic processes and outcomes. In this way it draws attention to the culturalization of the economy in contrast to the economization of culture.

New industrial districts (NIDs) areas that specialize in a particular industry because of external economies which result from specialization.

New international division of labour (NIDL) the global shift of economic activity that occurs when the process of production is no longer constructed primarily around national economies.

New World Order the notion of a Western-led post-Cold War structure of global power, which sanctions intervention.

Newly industrialized countries (NICs) countries where there has been a relatively recent and significant shift away from primary activities towards manufacturing production. In some cases the proportion of manufacturing production is similar to that of the United Kingdom or the United States. Examples are South Korea and Mexico.

NGO non-governmental organization.

Niche markets refer to the existence of consumer groups with identifiable tastes and lifestyles. A good example would be speciality food sales, often purchased by consumers as a symbol of social status or cultural capital.

Non-governmental organization (NGO) an organization formed by members of the public and one that has no government connections.

Non-renewable (stock) resources those resources, mainly mineral, that have taken millions of years to form. Their availability is therefore finite as there is no possibility of their stock being replenished on a timescale of relevance to human society.

O

Organized capitalism a form of capitalism that reached its zenith during the period 1945 to 1973. It was characterized by the dominance of extractive and manufacturing industries, large-scale manufacture, mass-production and a significant level of state involvement. See Spotlight box 3.1.

Overpopulation used to suggest that the finite resources of a particular area will run out if the population expands beyond a given point. Similar to the idea of a carrying capacity, that there is a limit beyond which environmental degradation occurs.

P

Palaeolithic the stage in the development of human society when people obtained their food by hunting, fishing and gathering wild plants, as opposed to engaging in settled agriculture. Also referred to as the Early Stone Age.

Participatory democracy a form of governance that encourages involvement of all people in political decision-making, as opposed to representative democracy.

Pastoral nomadism a form of social organization that is based on livestock husbandry for largely subsistence purposes. Pastoral nomads are characterized by a high level of mobility which allows them to search continually for new pastures in order to maintain their herds of animals.

Patriarchy system of gendered power relations through which men exercise power over women.

Peasant the term has had a number of different interpretations over the years and in different parts of the world. In general it usually refers to those individuals whose livelihood is largely dependent on the land and on rural subsistence-type activities. The term 'peasant' is usually reserved for those living in organized states, thus distinguishing them from band and tribal members, and so on.

Periphery according to Wallerstein, refers to those regions of the world capitalist economy characterized by low wages, simple technology and limited production.

Permaculture an environmental approach inspired by nature's patterns to create sustainable human habitats. It is based around three core principles – earth care (working with nature and designing systems that draw upon natural systems for inspiration), people care (looking after ourselves on a community and individual level through cooperation and mutual support) and fair shares (ensuring we only consume our share of the earth's resources).

Personal space the apparent desire by humans to have a pocket of space around them and into which they tend to resent others intruding.

Political economy an approach to social study that emphasizes the political/social construction and consequences of economic activity.

Post-Fordism production system comprises a mix of different ways of organizing production at a number of spatial scales. Common to all these types of production is some form of flexible production.

Postmodernism a philosophy that holds that the traits associated with twentieth-century modernism, such as belief in the possibility of managing social change according to sets of agreed principles, are now in retreat in the face of increasing individualism, pluralism and eclecticism.

Post-productivist transition a term used to describe the movement away from productivist agricultural systems. This new phase of agricultural production is characterized by extensive and diversified patterns of farming and the growing importance of non-agricultural activities in the countryside, for example, recreation.

Poverty the condition of possessing an income insufficient to maintain a minimal standard of living. Definitions of poverty are culturally specific, and thus relative to the social norms and expectations endemic to a given nation-state. However, the condition of absolute poverty (i.e. lacking the income to maintain a minimum diet) is acknowledged worldwide.

Power geometry the ways in which different social groups/individuals are placed in relation to the forces of globalization, enabling some to benefit and others to be disadvantaged.

Prehistoric societies societies that have left no written records.

Primary energy the energy in the basic fuels or energy sources used, e.g. the energy in the fuel fed into conventional power stations.

Primary sector comprises economic activities that exploit naturally occurring resources.

Proto-industrialization refers to the early phase of capitalist industrial development in Europe. Characteristics of this period include a significant level of rural-based industrial activity and a low level of technological application in the production process.

Proven (proved) reserves those deposits of a resource that have already been discovered and are known to be economically extractable under current demand, price and technological conditions.

Q

Quality of life a composite measure that reflects individual preferences that include, for example, education, health, entertainment, living environment.

Quality turn the idea that local/alternative foods are of higher quality than products produced under conventional farming systems; in this way, quality is inextricably link to locality.

Quaternary sector economic activities, mainly business enterprises, involved with the processing and exchange of information.

R

Race refers to the division of human beings into supposedly recognizable groups based predominantly on physical characteristics.

Racialized space the ways in which certain spaces are seen to be occupied exclusively or predominantly by a particular 'racial' or ethnic group.

Racism practices and attitudes that display dislike or antagonism towards people seen as belonging to particular ethnic groups. Social significance is attached to culturally constructed ideas of difference.

Ranked society society in which there is an unequal division of status and power between its members, where such divisions are based primarily on such factors as family and inherited social position.

Regulated states social formations organized on the basis of territory. The regulated, or pre-modern, state was not dominated by the market system with political, social and religious considerations usually being of greater importance.

Relocalization (of food) refers to the renewed interest in foods of local and regional provenance, in which the link between 'product' and 'place' is emphasized and often formalized through a system of quality assurance or quality labels. Such locally distinctive quality food products are sold in regional and national markets.

Renewable energy energy sources such as winds, waves and tides that are naturally replenished and cannot be used up.

Renewable (flow) resources those resources that are naturally renewed within a sufficiently short time-span to be of use to human society. The continued availability of such resources is increasingly dependent upon effective management.

Rent gap a term describing the difference between the current ground rental for a building and its potential ground rent. As such, gentrifiers seek to exploit the rent gap by investing in property and real estate in areas where they perceive rentals to be below their potential market value.

Replacement rate number of babies that an average woman should have to replace her generation, given prevailing levels of mortality (2.1 per woman in post-transition societies).

Representative democracy a form of governance that limits involvement of people in political decision-making to representatives often voted into position of power by the broader public.

Reserves/production ratio (R/P) if the reserves remaining at the end of any year are divided by the production in that year, the result is the length of time those remaining reserves would last if production were to continue at that level.

Residential segregation the ways in which, most obviously in urban areas, housing patterns can be observed where people live in areas divided along class or ethnic lines. In some instances this is conscious policy, in others it results from the interaction of social and economic processes.

Resource a substance in the physical environment that has value or usefulness to human beings and is economically feasible and socially acceptable to use.

Romantic movement artistic, literary and philosophical movement that originated in the eighteenth century as a backlash against industrialization.

Rural services social goods that comprise the fabric of non-productive life in rural areas (e.g. education, health, transport, shops).

Rurality functionalist, critical political economy and social representation approach for distinguishing between rural and urban economy and society.

Rust Belt a region of the north-eastern USA that suffered substantial industrial decline, especially after the Second World War.

S

Second global shift offshoring of service functions; the first global shift involved the offshoring of manufacturing functions.

Second industrial revolution a term sometimes used to account for the profound technological and accompanying social changes that affected industrial capitalism from the late nineteenth century. During this period Britain's industrial might was challenged by Germany and the USA.

Secondary sector comprises economic activities that transform primary sector outputs into useable goods.

Semi-periphery according to Wallerstein, refers to those regions of the world capitalist economy which, while exploiting the periphery, are themselves exploited by the core countries. Furthermore, they are characterized by the importance of both core and peripheral processes.

Sense of place the feelings, emotions and attachments to a locality by residents (past or present), which may be articulated in art, literature, music or histories, or may become part of individual or group memory.

Services work done as an occupation or business for other individuals or businesses that brings about a change in the condition of a person or of a good belonging to some economic unit which does not produce or modify physical goods.

Sexuality refers to social differences linked to sexual identity and behaviour.

Short food supply chains (SFSCs) where the number of nodes in the food chain is reduced to as few as possible. An ideal SFSC involves the direct marketing of food between a producer and the final consumer. The key feature is that foods reach the final consumer 'embedded' with value-laden information about the production, provenance and distinctiveness of the product.

Simulacrum a copy without an original. An example might be Disneyland's 'Main Street', which represents an ideal American high street, but is not modelled on an original.

Slow food promoted as an anti- 'fast food' culture, slow food aims to decelerate the food consumption experience and celebrate the cultural connections surrounding local cuisines and traditional products. The slow-food movement aims to embed food in the local territory and culture.

Social construct a social concept or idea (such as race, class, gender or age) that is institutionalized and normalized within a culture to the extent that people behave as if it were a 'real' or a pre-social given.

Social constructionism a sociological theory of knowledge which claims that meanings and truths are established socially through habitual use and institutionalization. In opposition to realism, which regards truths as pre-given and objective, social constructionism recognises that individuals and groups participate in the creation of their realities through the way in which they perceive things and events.

Social embeddedness the idea that economic behaviour is embedded in, and mediated by, a complex and extensive web of social relations. Trust and regard are examples of such social relations.

Social exclusion the various ways in which people are excluded (economically, politically, socially, culturally) from the accepted norms within a society.

Social movement comprised of individuals, groups and organizations united by a common purpose or goal.

Social representation lay and academic discourses that describe an event or context.

Spatial divisions of labour the concentration of particular sectors or production tasks in specific geographic areas.

Spatial interaction a term used to indicate interdependence between geographical areas. Covers the movement of people, goods, information and money between places.

Spatial relations the ways in which people are connected across geographic space through economic, social, cultural or political processes.

Speculative resources those resources that might be found in unexplored areas which are thought to have favourable geological conditions.

State a political unit having recognizable control (claiming supreme power) over a given territory. Unlike earlier social formations like bands and tribes, states have always based their power on their ability to control a specified territory and its inhabitants.

Stigma a term describing the condition of possessing a 'spoiled' or discredited identity.

Stratified society society within which there is an unequal division of material wealth between its members.

Structural adjustment programmes loans designed to foster structural adjustment in LDCs by promoting market-led growth and a reduction in government intervention in the economy.

Structuralism a theory, evolved in Latin America in the 1940s and 1950s, which argues that the relative position of a given economy vis-à-vis the global economy is a function of the nature of the manner of insertion of that economy historically. Changing the fortunes of an economy therefore requires state intervention to alter its structure.

Studentification an increase in the student population in a residential area, principally associated with student renting in the private sector, though also associated with newly built off-campus accommodation developed by private investors. A widely noted phenomenon in those countries where students in higher education tend to live away from home, it is sometimes associated with gentrification, but has markedly different social and environmental outcomes.

Subculture a subdivision of a dominant culture or an enclave within it with a distinct integrated network of behaviour, beliefs and attitudes.

Substitutionism refers to the increased use of non-agricultural raw materials characterized by the replacement of food and fibre with industrial alternatives. It forms a key process in the industrialization of the agri-food system.

Subsumption a process whereby an agribusiness TNC increases its control over farming by offering contracts to farmers to provide 'raw materials' for its food-manufacturing activities.

Suburbanization refers to the movement of middle- and skilled working-class people into residential areas located some distance away from their paid employment.

Sun Belt refers to the major growth areas of the southern and western parts of the USA during recent years in contrast to the contracting and declining industrial base of the north-east. Can also be used in other parts of the developed world to describe dynamic regions, for example, the M4 corridor in England.

Sustainable development a vague yet highly influential term popularized by the publication of the Brundtland Report in 1987. At its simplest, it requires contemporary societal development to be undertaken bearing in mind the needs and aspirations of future generations. It has been criticized for failing to represent any real challenge to the prevailing Western development model.

T

Takeover occurs when one company purchases a controlling interest in a second company against the wishes of the latter's directors.

Territoriality a term used to describe an expression of ownership and control by an individual, group or state over a particular area of land in order to achieve particular ends.

Territory a recognizable region (area of land or sea) occupied and controlled by an individual, group or state.

Tertiary sector comprises economic activities engaged in enabling the exchange and consumption of goods and services (often referred to as the service sector).

Third World a rather vague term used to describe those regions of the world in which levels of development, as understood by such measures as GDP, are significantly below those of the economically more advanced regions. The term is increasingly seen as an inadequate description of the prevailing world situation since it disguises a significant amount of internal differentiation.

TNCs see transnational corporations.

Top-down development economic and social changes brought about by activities of the state and its agencies.

Total fertility rate (TFR) in general terms equal to the average number of live births per woman within a population.

Traceability in economic geography refers to the ability to determine the origin of agricultural foodstuffs.

Trade bloc a group of nation-states that act together in order to further their influence over world trade, for example, the European Union (EU) or the North American Free Trade Area (NAFTA).

Transculturation the ways in which subordinated or marginal groups select and invent from dominant cultures; although such groups cannot control what emanates from the dominant culture, they do determine to varying extents what they absorb into their own and what they use it for.

Transgression a term describing actions that breach social expectations of what is appropriate in a particular place. Deliberate transgression may thus constitute an act of resistance.

Transnational corporations (TNCs) major business organizations that have the power to coordinate and control operations in more than one country. They are the primary agents of globalization in the agri-food sector. See Spotlight box 14.2.

Transnationalism multiple ties and interactions linking people or institutions across the borders of nation-states and measured, for example, as flows of capital, people, information and images.

Tribe a type of social formation usually said to be stimulated by the development of agriculture. Tribes tend to have a higher population density than bands and are also characterized by an ideology of common descent or ancestry. See Spotlight box 1.2.

U

Underclass a term referring to poorer, more marginalized groups in society who are seen to experience multiple deprivation.

Urban revanchism literally, the process of 'revenge' by which middle-class citizens take back the city from the marginal groups whom they portray as taking it from them. In practice, this is promoted by urban policies promoting law and order on the city streets and excluding those who do not fit in with the middle-class consumer ambience vital to the success of urban gentrification.

Urban village a residential sector of a city (often inner-city) believed to possess social characteristics typical of a village-type community.

Urbanization refers to the increasing importance of the urban relative to the total population. It is initially stimulated by the movement of people from rural to urban areas.

V

Value-adding activity sequential steps in a production process that enhance the saleable value of a commodity.

Vertical disintegration a process whereby segments of the production process (usually in a vertically integrated business) are subcontracted out to smaller-scale producers.

Vertical integration firms at different stages of the production chain merge together.

Vital trends changes in fertility and mortality.

W

World Bank international financial institution established in 1944. Its main role is to provide development funds to LDCs in the form of loans and technical assistance.

World systems theory Immanuel Wallerstein's conceptualization of the changing nature of the world socioeconomic system into three distinct historical categories.

WTO World Trade Organization.

Bibliography

A

Abraham, I. and van Schendel, W. (2005) *Illicit Flows and Criminal Things, States, Borders, and the Other Side of Globalization*, Indiana University Press, Bloomington and Indianapolis.

Abrahamsen, R. (2000) *Disciplining Democracy: Development Discourse and Good Governance in Africa*, Zed Books, London.

ActionAid (1995) *Listening to Smaller Voices: Children in an Environment of Change*, ActionAid, Chard, Somerset.

Aglietta, M. and Breton, R. (2001) Financial systems, corporate control and capital accumulation, *Economy and Society*, **30**, 433–66.

Agnew, J. (2003) *Geopolitics: Revisioning World Politics*, 2nd edition, Routledge, London and New York.

Agnew, J. (2007) No borders, no nation: making Greece in Macedonia, *Annals of the Association of American Geographers*, **97**(2), 398–422.

Agnew, J. and Corbridge, S. (1995) *Mastering Space: Hegemony, Territory and International Political Economy*, Routledge, London.

Agnew, J., Livingstone, D. and Rogers, A. (eds) (1996) *Human Geography: An Essential Anthology*, Blackwell, Oxford.

Agnew, J., Mitchell, K. and Toal, G. (eds) (2003) *A Companion to Political Geography*, Blackwell, Oxford.

Ahluwalia, P. (2001) *Politics and Post-colonial Theory: African Inflections*, Routledge, London.

Ake, C. (1996) *Democracy and Development in Africa*, Brooks, Washington.

Akzin, B. (1964) *State and Nation*, Hutchinson, London.

Ali, M. (2003) *Brick Lane*, Doubleday, London.

Alland, A. (Jr) (1972) *The Human Imperative*, Columbia University Press, New York.

Allen, J. (1995) Global worlds, in Allen, J. and Massey, D. (eds) *Geographical Worlds*, Oxford University Press and Open University, Oxford, 105–44.

Allen, J. (1999) Cities of power and influence: settled formations, in Allen, J., Massey, D. and Pryke, M. (eds) *Unsettling Cities*, Routledge, London, 182–227.

Allen, P., FitzSimmons, M., Goodman, M. and Warner, K. (2003) Shifting plates in the agrifood landscape: the tectonics of alternative food initiatives in California, *Journal of Rural Studies*, **19**, 61–75.

Allen, T. and Thomas, A. (2000) *Poverty and Development into the Twenty-first Century*, Oxford University Press and Open University, Oxford.

Alvarez-Rivadulla, M.J. (2007) Golden ghettos: gated communities and class segregation in Montevideo, Uruguay, *Environment and Planning A*, **39**, 47–63.

Alvesson, M. (2000) Social identity and the problem of loyalty in knowledge-intensive companies, *Journal of Management Studies*, **37**(8), 1101–23.

Amin, A. (ed.) (1994) *Post-Fordism: A Reader*, Blackwell, Oxford.

Amin, A. and Graham, S. (1997) The ordinary city, *Transactions of the Institute of British Geographers*, NS 22, 411–29.

Amin, A. and Graham, S. (1999) Cities of connection and disconnection, in Allen, J., Massey, D. and Pryke, M. (eds) *Unsettling Cities*, Routledge, London, 7–47.

Amin, A. and Thrift, N. (1994) Living in the global, in Amin, A. and Thrift, N. (eds) *Globalization, Institutions and Regional Development in Europe*, Oxford University Press, Oxford, 1–22.

Amin, A. and Thrift, N. (2002) *Cities: Reimagining the Urban*, Polity Press, Cambridge.

Amin, A. and Thrift, N. (2004) *The Blackwell Cultural Economy Reader*, Blackwell, Oxford.

Anderson, B. (1983) *Imagined Communities: Reflections on the Origins and Spread of Nationalism*, Verso, London.

Anderson, J. (1986) Nationalism and geography, in Anderson, J. (ed) *The Rise of the Modern State*, Harvester, Brighton, 113–26.

Anderson, J. (1995) The exaggerated death of the nation-state, in Anderson, J., Brook, C. and Cochrane, A. (eds) (1995) *A Global World? Re-ordering Political Space*, Oxford University Press and the Open University, Oxford, 65–112.

Anderson, J. (1997) Territorial sovereignty and political identity: national problems, transnational solutions? in Graham, C. (ed.) *In Search of Ireland: A Cultural Geography*, Routledge, London, 215–36.

Anderson, J., Brook, C. and Cochrane, A. (eds) (1995) *A Global World? Re-ordering Political Space*, Oxford University Press, Oxford.

Anderson, K. (1991) *Vancouver's Chinatown: Racial Discourse in Canada 1875–1980*, McGill-Queens University Press, Montreal.

Anderson, K. (2000) 'The beast within': race, humanity and animality, *Environment and Planning D: Society and Space*, **18**, 301–20.

Anderson, K., Domosh, M., Pile, S. and Thrift, N. (eds) (2003) *Handbook of Cultural Geography*, Sage, London.

Andreas, P. (2000) *Border Games: Policing the US–Mexico Divide*, Cornell University Press, Ithaca.

Andreasson, S. (2005) Orientalism and African Development Studies: the 'reductive repetition' motif in theories of African underdevelopment, *Third World Quarterly*, **26**(6), 971–86.

Anon. (2001) *Text painted on a wall at Carters Road Community*, Margaret River, Western Australia.

Anonymous (1996) Ubuntu and democracy in South Africa, *Government Gazette*, Pretoria, No. 16943, 18.

Anonymous (1999) South Africa's middle-class pink currency, *Johannesburg Mail and Guardian*, 18 June, available at: www.globalarchive.ft.com

Appadurai, A. (2003) Sovereignty without territoriality: notes for a postnational geography, in Low, S.M. and Lawrence-Zúñiga, D. (eds) *The Anthropology of Space and Place*, Blackwell, Oxford, 337–49.

Applebaum, A. (2003) *Gulag: A History*, Allen Lane, London.

Ardener, S. (1995) Women making money go round: ROSCAs revisited, in Ardener, S. and Burman, S. (eds) *Money Go-Rounds: The Importance of Rotating Savings and Credit Associations for Women*, Berg, Oxford.

Ardener, S. and Burman, S. (eds) (1995) *Money Go-Rounds: The Importance of Rotating Savings and Credit Associations for Women*, Berg, Oxford.

Argent, N. (2002) From pillar to post? In search of the post-productivist countryside in Australia, *Australian Geographer*, **33**, 97–114.

Ashlin, A. and Ladle, R. (2006) Environmental science adrift in the blogosphere, *Science*, **312**, 201.

Atkin, S. and Valentine, G. (eds) (2006) *Approaches to Human Geography*, Sage, London and Thousand Oaks, New Delhi.

Atkins, P. and Bowler, I. (2001) *Food in Society: Economy, Culture and Geography*, Arnold, London.

Atkinson, A. (1991) *Principles of Political Ecology*, Belhaven, London.

Atkinson, D. and Cosgrove, D. (1998) Urban rhetoric and embodied identities: city, nation and empire at the Vittorio Emanuele II monument in Rome, 1870–1945, *Annals of the Association of American Geographers*, **88**, 28–49.

Auty, R. (1979) World within worlds, *Area*, **11**, 232–5.

Auty, R. (1993) *Sustaining Development in Mineral Economies: The Resource Case Thesis*, Routledge, London.

Auty, R. (2001) *Resource Abundance and Economic Development*, Oxford University Press, Oxford.

B

Bagemihl, B. (1999) *Biological Exuberance: Animal Homosexuality and Natural Diversity*, Martin's Press, New York.

Baker, B. (1998) The class of 1990: how have the autocratic leaders of sub-Saharan Africa fared under democratisation, *Third World Quarterly*, **19**, 115–27.

Bakhtin, M. (1984) *Rabelais and his World* (trans. Helen Iswolsky), Indiana University Press, Bloomington, IN.

Ballinger, P. (2003) *History in Exile: Memory and Identity at the Borders of the Balkans*, Princeton University Press, Princeton, NJ.

Banrffalo, R. (1996) *Interview with David Harvey: The politics of social justice*, available at www.uky.edu/AS/SocTheo/DisClosure/Harveydi.htm Accessed 10/10.

Barkawi, T. and Laffey, M. (2002) Retrieving the imperial: empire and international relations, *Millennium*, **31**, 109–27.

Barnes, T. and Farish, M. (2006) Between regions: science, militarism, and American geography from World War to Cold War, *Annals of the Association of American Geographers*, **96**(4), 858–63.

Barnet, R.J. (1973) *Roots of War: The Men and Institutions behind US Foreign Policy*, Penguin, Harmondsworth.

Barnett, C., Cloke, P., Clarke, N. and Malpass, A. (2005) Consuming ethics: articulating the subjects and spaces of ethical consumption, *Antipode*, **37**(1), 23–45.

Barnett, D. (1998) *London: Hub of the Industrial Revolution: A Revisionary History: 1775–1825*, Tauris Academic Studies, London.

Barrett, H.R., Ilbery, B.W., Browne, A.W. and Binns, T. (1999) Globalisation and the changing networks of food supply: the importation of fresh horticultural produce from Kenya into the UK, *Transactions of the Institute of British Geographers*, **24**, 159–74.

Barthes, R. (1972) *Mythologies*, Paladin, London.

Bassin, M. (1987) Race contra space: the conflict between German Geopolitik and National Socialism, *Political Geography Quarterly*, **6**, 115–34.

Bater, J.H. (1986) Some recent perspectives on the Soviet city, *Urban Geography*, **7**, 93–102.

Batterbury, S. and Warren, A. (2001) Desertification, in Smelser, N. and Baltes, P. (eds), *International Encyclopædia of the Social and Behavioral Sciences*, Elsevier Press, Amsterdam, 3526–29.

Bayley, C. (1989) *Atlas of the British Empire*, Hamlyn, London.

BBC *see* British Broadcasting Corporation.

Beaverstock, J.V., Hubbard, P. and Short, J.R. (2004) Getting away with it? Exposing the geographies of the super-rich, *Geoforum*, **35**, 401–7.

Bedford, T. (1999) Ethical consumerism, consumption, identity and ethics, unpublished PhD thesis, University of London, London.

Bell, D. and Valentine, G. (eds) (1995) *Mapping Desire: Geographies of Sexualities*, Routledge, London.

Bell, D. and Valentine, G. (1997) *Consuming Geographies: We are Where we Eat*, Routledge, London.

Bell, M. (1994) Images, myths and alternative geographies of the Third World, in Gregory, D., Martin, R. and Smith, G. (eds) *Human Geography: Society, Space and Social Science*, Macmillan, London, 174–99.

Benko, G. and Strohmayer, U. (2004) *Human Geography: A History for the Twenty-First Century*, Arnold, London.

Benton, L. (1995) Will the real/reel Los Angeles please stand up? *Urban Geography*, **16**, 144–64.

Berdahl, D. (1998) *Where the World Ended: Re-unification and Identity in the German Borderland*, University of California Press, Berkeley.

Beresford, M.W. (1988) *East End, West End: The Face of Leeds during Urbanisation 1684–1842*, Thoresby Society, Leeds.

Berger, J. (1990) *Ways of Seeing*, Penguin, London.

Berger, M.T. (1994) The end of the 'Third World'?, *Third World Quarterly*, **15**, 257–75.

Berger, M.T. (2001) The post-cold war predicament: a conclusion, *Third World Quarterly*, **22**, 1079–85.

Berger, P. and Luckmann, T. (1966) *The Social Construction of Reality: A Treatise in the Sociology of Knowledge*, Anchor Books, Garden City, New York.

Betjeman, J. (1937) *Continual Dew*, John Murray, London.

Betts, V. (2002) Geographies of youth, religion and identity in (post-)socialist Cuba, unpublished PhD thesis, University of Birmingham.

Bhabha, H.K. (1994) *The Location of Culture*, Routledge, London.

Bilgin, P. (2007) Only strong states can survive in Turkey's geography: the uses of geopolitical truths in Turkey, *Political Geography*, **26**(7): 740–56.

Bingham, N., Blowers, A. and Belshaw, C. (2003) *Contested Environments*, Open University Press, Milton Keynes.

Birks, H. (1997) Environmental change in Britain: a long-term paleaoecological perspective, in Mackay, A. and Murlis, J. (eds) *Britain's Natural Environment: A State of the Nation Review*, Ensis, London, 23–8.

Blacksell, M. (1994) Irredentism, in Johnson, R.J., Gregory, D. and Smith, D.M. (eds) *The Dictionary of Human Geography*, 3rd edition, Blackwell, Oxford, 299.

Blacksell, M. (2005) *Political Geography*, Routledge, London.

Blaut, J.M. (1993) *The Colonizer's Model of the World: Geographical Diffusionism and Eurocentric History*, Guilford Press, New York and London.

Blinder, A.S. (2006) Offshoring: the next industrial revolution, *Foreign Affairs*, Mar/Apr, 113–28.

Bloeman, S. (2004) *T-shirt Travels*, accessed 24 September 2007, www.pbs.org/independentlens/tshirttravels/film.html

Blomley, N.K. (1994) *Law, Space, and the Geographies of Power*, Guilford Press, New York.

Blomley, N.K. (2004) *Unsettling the City: Urban Land and the Politics of Property*, Routledge, New York.

Blomley, N.K., Delaney, D. and Ford, R. (eds) (2001) *The Legal Geographies Reader: Law, Power, and Space*, Blackwell, Oxford.

Bluestone, B. and Harrison, B. (1982) *The Deindustrialization of America*, Basic Books, New York.

Blunden, J. (1995) Sustainable resources?, in Sarre, P. and Blunden, J. (eds) *An Overcrowded World: Population, Resources and Environment*, Oxford University Press and Open University, Oxford, 161–213.

Blunt, A. and Dowling, R. (2006) *Home*, Routledge, London.

Blunt, A. and Wills, J. (2000) *Dissident Geographies: An Introduction to Radical Ideas and Practice*, Prentice Hall, Harlow.

Bobek, H. (1962) The main stages in socio-economic evolution from a geographical point of view, in Wagner, P.L. and Mikesell, M.W. (eds) *Readings in Cultural Geography*, University of Chicago Press, Chicago, 218–47.

Bonnett, A. (2003) Geography as the world discipline: connecting popular and academic geographical imaginations, *Area*, **35**, 55–63.

Bonnett, A. (2008) *What is Geography?* Sage, London and Thousand Oaks, New Delhi.

Bonnett, A. and Nayak, A. (2003) Cultural geographies of racialization – the territory of race, in Anderson, K., Domosh, M., Pile, S. and Thrift, N. (eds) *Handbook of Cultural Geography*, Sage, London, 300–12.

Booth, K. (1999) Cold wars of the mind, in Booth, K. (ed.) *Statecraft and Security: The Cold War and Beyond*, Cambridge University Press, Cambridge, 29–55.

Borden, I. (2001) *Skateboarding, Space and the City: Architecture and the Body*, Berg, Oxford.

Borgelt, C., Ganssauge, K. and Keckstein, V. (1987) *Mietshaus im Wandel, Wohnungen der behutsamen Stadterneuerung*, S.T.E.R.N, Berlin.

Bornstein, D. (1996) *The Price of a Dream: The Story of the Grameen Bank and the Idea that is Helping the Poor to Change their Lives*, University of Chicago Press, Chicago.

Borstelmann, T. (2001) *The Cold War and the Color Line: American Race Relations in the Global Arena*, Harvard University Press, Cambridge, MA.

Boserüp, E. (1990) *Economic and Demographic Relationships in Development*, Johns Hopkins University Press, Baltimore.

Bourdieu, P. (1984) *Distinction*, Routledge & Kegan Paul, London.

BP (2007) *Statistical Review of World Energy*, available at: www.bp.com

Brah, A. (1996) *Cartographies of Diaspora: Contesting Identities*, Routledge, London.

Braidotti, R., Charkiewicz, E., Hausler, S. and Wieringa, S. (eds) (1994) *Women, Environment and Sustainable Development: Towards a Theoretical Synthesis*, Zed, London.

Brandt, W. (1980) *North–South: A Programme for Survival*, Pan, London.

Braun, B. (2002) *The Intemperate Rainforest: Nature, Culture and Power on Canada's West Coast*, University of Minnesota Press, Minneapolis.

Bray, K. and Kilian, T. (1999) God does not pay for it: the political economy of cyberspace. Paper presented at the 95th Annual Meeting of the Association of American Geographers, Honolulu.

Breckenridge, C.A. (ed.) (1995) *Consuming Modernity: Public Culture in a South Asian World*, University of Minnesota Press, Minneapolis.

Brenner, N., Jessop, B., Jones, M. and Macleod, G. (eds) (2002) *State/Space: A Reader*, Blackwell, Oxford.

Bridge, G. and Watson, S. (eds) (2000) *A Companion to the City*, Blackwell, Oxford.

British Broadcasting Corporation (BBC) (1999) *GM Foods: An Evaluation.* Online, available at: www.bbc.co.uk/science, accessed 12 December 2002.

British Broadcasting Corporation (BBC) (2005a) *The Changing Face of Poverty.* http://news.bbc.co.uk/1/hi/business/4070112.stm, accessed 23 May 2007.

British Broadcasting Corporation (BBC) (2005b) *Hurricane Aid Efforts Gather Pace.* http://news.bbc.co.uk/1/hi/world/americas/4210646.stm, accessed 23 May 2007.

British Broadcasting Corporation (BBC) (2006a) *Oslo Gay Animal Show Draws Crowds*, http://news.bbc.co.uk/1/hi/world/europe/6066606.stm, accessed 20/11.

British Broadcasting Corporation (BBC) (2006b) *Zero Carbon Homes Plan Unveiled*, http://news.bbc.co.uk/go/pr/fr/-/1/hi/sci/tech/6176229.stm, accessed 13 December 2006.

Brooks, R. (2005) *Our homegrown Third World*, http://www.commondreams.org/views05/0907-24.htm, accessed 23 May 2007.

Broome, R. (1994) *Aboriginal Australians: Black Responses to White Dominance, 1788–1994*, 2nd edition, Allen & Unwin, St Leonards, NSW.

Brown, L. (2001) *Eco-economy: Building an Economy for the Earth*, Norton, New York.

Bryman, A. (1995) *Disney and his Worlds*, Routledge, London.

Bryson, J.R. (2000) Spreading the message: management consultants and the shaping of economic geographies in time and space, in Bryson, J.R., Daniels, P.W., Henry, N.D. and Pollard, J.S. (eds) *Knowledge Space, Economy*, Routledge, London, 157–75.

Bryson, J.R. (2007) A 'second' global shift? The offshoring or global sourcing of corporate services and the rise of distanciated emotional labour, *Geografiska Annaler*, **89B**(1), 31–44.

Bryson, J.R. and Daniels, P.W. (eds) (2007) *The Handbook of Service Industries*, Edward Elgar, Cheltenham.

Bryson, J.R. and Henry, N.D. (2005) The global production system: from Fordism to post Fordism, in Daniels, P.W. et al. *Human Geography: Issues for the 21st Century*, Prentice Hall, London, 313–36.

Bryson, J.R. and Rusten, G. (2005) Spatial divisions of expertise: knowledge intensive business service firms and regional development in Norway, *The Service Industries Journal*, **25**(8) 959–77.

Bryson, J.R. and Rusten G. (2006) Spatial divisions of expertise and transnational 'service' firms: aerospace and management consultancy, in Harrington, J.W. and Daniels, P.W. (eds) *Knowledge-based Services, Internationalisation and Regional Development*, Aldershot, Ashgate, 79–100.

Bryson, J.R. and Rusten, G. (2008) Transnational corporations and spatial divisions of 'service' expertise as a competitive strategy: the example of 3M and Boeing, *The Service Industries Journal*, **28**(3) forthcoming.

Bryson, J.R. and Taylor, P. (2006) *The Functioning Economic Geography of the West Midlands Region, Advantage West Midlands*, Birmingham, West Midlands Regional Observatory.

Bryson J.R. and Wellington, C. (2003) Image consultancy in the United Kingdom: recipe knowledge and recreational employment, *The Service Industries Journal*, **23**(1) 59–76.

Bryson, J., Daniels, P.W. and Henry, N. (1996) From widgets to where? A region in economic transition, in Gerrard, A.J. and Slater, T.R. (eds) *Managing a Conurbation: Birmingham and its Region*, Brewin Books, Birmingham, 56–168.

Bryson, J.R., Daniels, P.W. and Warf, B. (2004) *Service Worlds: People, Organizations, Technologies*, Routledge, London.

Bryson, J.R., Taylor, M. and Cooper, R. (2008) Competing by design, specialisation and customization: manufacturing locks in the West Midlands (UK), *Geografiska Annaler*, forthcoming.

Bryson, J., Henry, N., Keeble, D. and Martin, R. (eds)(1999) *The Economic Geography Reader: Producing and Consuming Global Capitalism*, John Wiley, Chichester.

Buckingham-Hatfield, S. (2000) *Gender and Environment*, London, Routledge.

Bullard, R.D. (1990) *Dumping in Dixie: Race, Class and Environmental Quality*, Westview Press, Boulder, Co.

Bundy, D. and Gotur, M. (2002) *Education and HIV/AIDS: A Window of Hope*, World Bank, Washington.

Bunyard, P. and Morgan-Grenville, F. (eds.) (1987) *The Green Alternative*, Methuen, London.

Burgess, J., Bedford, T., Hobson, K., Davies, G. and Harrison, C. (2003) (Un)sustainable consumption, in Berkhout, F., Leach, M. and Scoones, I. (eds) *Negotiating Environmental Change: New perspectives from Social Science*, Edward Elgar, Cheltenham, 261–91.

Bush, G. (2007) *President Bush Delivers State of the Union Address*, 23 January 2007, http://www.whitehouse.gov/news/releases/2007/01/20070123-2.html, accessed 25 May 2007.

Butlin, R.A. (1993) *Historical Geography, Through the Gates of Time and Space*, Edward Arnold, London.

Buzan, B. (2006) Will the 'global war on terrorism' be the new Cold War? *International Affairs*, **82**(6) 1101–18.

C

Caldeira, T. (1996) Building up walls: the new pattern of social segregation in São Paulo, *International Social Science Journal*, **147**, 55–65.

Caldeira, T. (2001) *City of Walls: Crime, Segregation and Citizenship in São Paulo*, University of California Press, Berkeley.

Caldwell, M.L. (2004) Domesticating the French fry: McDonald's and consumerism in Moscow, *Journal of Consumer Culture*, **4**, 5–26.

Callicott, J.B. and Rocha, F.J.R. (eds) (1996) *Earth Summit Ethics: Toward a Reconstructive Postmodern Philosophy of Environmental Education*, State University of New York Press, New York.

Calvocoressi, P. (1991) *World Politics since 1945*, Longman, London.

Campbell, B. (1993) *Goliath: Britain's Dangerous Places*, Methuen, London.

Campbell, C. (1987) *The Romantic Ethic and the Spirit of Modern Consumerism*, Blackwell, Oxford.

Campbell, D. (1992) *Writing Security: United States Foreign Policy and the Politics of Identity*, University of Minnesota Press, Minneapolis, MN.

Campbell, D. (1999) Apartheid cartography: the political anthropology and spatial effects of international diplomacy in Bosnia, *Political Geography*, **18**(4) 395–435.

Carapico, S. (1985) Yemeni agriculture in transition, in Beaumont, P. and McLachlan, K. (eds) *Agricultural Development in the Middle East*, John Wiley & Sons, Chichester, 241–54.

Carneiro, A. de M. (1990) *Descrição da Fortaleza de Sofala e das mais da Índia*, Fundação Oriente, Lisbon.

Carroll, R. (1998) The chill east wind at your doorstep, *The Guardian*, 28 October, 2–3.

Carroll, R. (2000) Adopt a sheep on the internet, *The Guardian*, 7 November, 17.

Carson, R. (1962) *Silent Spring*, Hamish Hamilton, London.

Carter, D. (1988) *The Final Frontier: The Rise and Fall of the American Rocket State*, Verso, London and New York.

Carter, N. (2001) *The Politics of the Environment: Ideas, Activism, Policy*, Cambridge University Press, Cambridge.

Caspian Revenue Watch (2003) *Caspian Oil Windfalls: Who Will Benefit?*, The Open Society Institute, New York. Available at: www.revenuewatch.org/reports

Castells, M. (1989) *The Informational City: Information Technology, Economic Restructuring and the Urban and Regional Process*, Blackwell, Oxford.

Castells, M. (1996) *The Information Age: Economy, Society and Culture*, vol. 1, *The Rise of the Network Society*, Blackwell, Oxford.

Castells, M. (1997) *The Power of Identity*, Blackwell, Malden, MA.

Castells, M. (1998) *The Information Age: Economy, Society and Culture*, vol. 3, *End of Millennium*, Blackwell, Oxford.

Castles, S. and Miller, M.J. (2003) *The Age of Migration: International Population Movements in the Modern World*, Palgrave Macmillan, Basingstoke.

Castree, N. (2003) A post-environmental ethics? *Ethics, Place and Environment*, **6**(1) 3–12.

Castree, N. (2005) *Nature*, Routledge, London.

Caulkin, S. (1998) Credit where credit's overdue, *Observer*, 15 November, 2.

Central Intelligence Agency (CIA) (2003) *The World Factbook*, CIA, Washington, DC.

Central Intelligence Agency (CIA) (2005) *The World Factbook: Uganda*, CIA, Washington, DC.

Central Intelligence Agency (2006) *The World Factbook* (access online at https://www.cia.gov/library/publications/the-world-factbook/index.html).

Chaliand, G. (1993) Introduction, in Chaliand, G. (ed.) *A People without a Country: The Kurds and Kurdistan*, 2nd edition, Zed, London, 1–10.

Chaliand, G. and Rageau, P. (1985) *Strategic Atlas: World Geopolitics*, Penguin, London.

Chambers, R. (1982) *Rural Development: Putting the Last First*, Longman, London.

Chandler, J. (ed.) (1993) *John Leland's Itinerary*, Sutton, Stroud.

Chaney, D. (1990) Subtopia in Gateshead, the MetroCentre as a cultural form, *Theory, Culture and Society*, **7**, 49–68.

Charlesworth, A., Stenning, A., Guzik, R. and Paszkowski, M. (2006) 'Out of place' in Auschwitz? Contested development in post-war and post-socialist Ośięcim, *Ethics, Place and Environment*, **9**(2) 149–72.

Chatterjee, P. (1993) *The Nation and its Fragments: Colonial and Postcolonial Histories*, Princeton University Press, Princeton, NJ.

Chatterton, P. (1999) University students and city centres: the formation of exclusive geographies, *Geoforum*, **30**, 117–33.

Child, J.C. (1985) *Geopolitics and Conflict in South America: Quarrels among Neighbours*, Praeger, New York.

Chomsky, N. (1998) Power in the global arena, *New Left Review*, **230**, 3–27.

Chossudovsky, M. (1997) *The Globalisation of Poverty*, Zed Books, London.

Christaller, W. (1966) *Central Places in Southern Germany* (trans. by C.W. Baskin), Prentice Hall, Englewood Cliffs, NJ.

Chung, R. (1970) Space–time diffusion of the demographic transition model: the twentieth century patterns, in Demko, G.J., Rose, H.M. and Schnell, G.A. (eds) *Population Geography: A Reader*, McGraw-Hill, New York.

CIA *see* Central Intelligence Agency

City of London Corporation (2005a) *London's Place in the UK Economy, 2005–2006*, London, City of London Corporation.

City of London Corporation (2005b) *The Competitive Position of London as a Global Financial Centre*, London, City of London Corporation.

Clark, B. (2007) *Twice a Stranger: How Mass Expulsion Forged Modern Greece and Turkey*, Granta, London.

Clarke, G.E. (1996) Blood, territory and national identity in Himalayan states, in Tønnesson, S. and Antlöv, H. (eds) *Asian Forms of the Nation*, Curzon, Richmond, 205–36.

Clarke, N., Barnett, C., Cloke, P. and Malpass, A. (2007) Globalising the consumer: doing politics in an ethical register, *Geoforum*, **26**(3), 231–49.

Claval, P. (1998) *An Introduction to Regional Geography*, Blackwell, Oxford.

Clifford, J. (1992) Travelling cultures, in Grossberg, L., Nelson, C. and Treichler, P. (eds) *Cultural Studies*, Routledge, London.

Clifford, N. and Valentine, G. (eds) (2003) *Key Methods in Geography*, Sage, London.

Cloke, P. and Johnston, R. (2005) *Spaces of Geographic Thought: Deconstructing Human Geography's Binaries*, Sage, London.

Cloke, P. and Little, J. (eds) (1997) *Contested Countryside Cultures: Otherness, Marginalisation and Rurality*, Routledge, London.

Cloke, P., Crang, P. and Goodwin, M. (eds) (2004) *Envisioning Human Geographies*, Arnold, London.

Cloke, P., Crang, P. and Goodwin, M. (eds) (2005) *Introducing Human Geographies*, Arnold, London.

Cloke, P., Crang, P., Goodwin, M., Painter, J. and Philo, C. (2002) *Practicing Human Geography*, Sage, London.

Cloke, P.J. (2000) Rural, in Johnston, R.J., Gregory, D., Pratt, G. and Watts, M. (eds) *The Dictionary of Human Geography*, Blackwell, Oxford, 718.

Cloke, P.J. (2005a) Conceptualising rurality, in Cloke, P., Marsden, T. and Mooney, P. (eds) *Handbook of Rural Studies*, Sage, London.

Cloke, P.J. (2005b) The country, in Cloke, P.J., Crang, P. and Goodwin, M. (eds), *Introducing Human Geographies*, 2nd edition, Arnold, London.

Cloke, P.J., Marsden, T. and Mooney, P. (eds) (2005) *Handbook of Rural Studies*, Sage, London.

Cochrane, A. (1995) Global worlds and worlds of difference, in Anderson, J., Brook, C. and Cochrane, A. (eds) *A Global World*, Oxford University Press, Oxford, 249–80.

Cockcroft, E. (1974) Abstract expressionism, weapon of the Cold War, *Artform*, **12**, 39–41.

Coe, N.M., Kelly, P.F. and Yeung, H. (2007) *Economic Geography: A Contemporary Introduction*, Blackwell, Oxford.

Cohen, I. (1997) *Green Fire*, Harper Collins, Australia.

Coll, S. (2004) *Ghost Wars: The Secret History of the CIA, Afghanistan and Bin Laden from Soviet Invasion to September 10, 2001*, Penguin, New York and London.

Commission of the European Communities (2007) *An Energy Policy for Europe*, Commission of the European Communities, Brussels.

Cone, C. and Myhre, A. (2000) Community supported agriculture: a sustainable alternative to industrial agriculture? *Human Organisation*, **59**, 187–97.

Connelly, J. and Smith, G. (2003) *Politics and the Environment: From Theory to Practice*, 2nd edition, Routledge, London.

Connor, W. (1994) *Ethnonationalism: The Quest for Understanding*, Princeton University Press, Princeton, New Jersey.

Conzen, M.P. (1990) *The Making of the American Landscape*, Unwin Hyman, Boston.

Cook, I. (2004) Follow the thing: papaya, *Antipode*, **36**(4), 642–64.

Cook, I. (2006) Geographies of food: following, *Progress in Human Geography*, **30**(5), 655–66.

Cook, I. et al., (2000) Social sculpture and connective aesthetics: Shelley Sacks' 'exchange values', *Ecumene: A Journal of Cultural Geographies*, **7**, 337–44.

Cook, I. et al. (2008) Geographies of food: mixing, *Progress In Human Geography*, **30**(6), 1113–26, forthcoming.

Cook, I. and Crang, P. (1996) The world on a plate: culinary culture, displacement and geographical knowledges, *Journal of Material Culture*, **1**(2), 131–53.

Cook, I., Crang, P. and Thorpe, M. (1999) Eating into Britishness: multicultural imaginaries and the identity politics of food, in Roseneil, S. and Seymour, J. (eds) *Practising Identities*, Macmillan, London, 223–48.

Cook, I., Crang, P. and Thorpe, M. (2000) Regions to be cheerful, culinary authenticity and its geographies, in Cook, I., Crouch, D., Naylor, S. and Ryan, J. (eds) *Cultural Turns/Geographical Turns*, Longman, London, 109–39.

Cook, I., Crang, P. and Thorpe, M. (2004) Tropics of consumption: getting with the fetish of 'exotic' fruit?, in Hughes, A. and Reimer, S. (eds) *Geographies of Commodities*, Routledge, London.

Cook, I., Crouch, D., Naylor, S. and Ryan, J. (eds) (2000) *Cultural Turns/Geographical Turns*, Prentice Hall, London.

Cook, I. Evans, J., Griffiths, H., Morris, B., Wrathmell, S. et al. (2007) 'It's more than just what it is': defetishising commodities, changing pedagogies, mobilising change, *Geoforum*, **38**(6), 1113–1126.

Cooke, B. and Kothari, U. (2001) *Participation: The New Tyranny?* Zed, London.

Cooper, D.E. (2001) Arne Naess, in Palmer, J.A. (ed.) *Fifty Key Thinkers on the Environment*, Routledge, London, 211–16.

Corbridge, S. (ed.) (1995) *Development Studies: A Reader*, Edward Arnold, London.

Corry, E. (2006) *The Nation Holds its Breath: Great Irish Soccer Quotations*, Hodder Headline Ireland, Dublin.

Cosgrove, D. (1994) Contested global visions, *Annals of the Association of American Geographers*, **84**, 270–94.

Cosgrove, D. (2001) *Apollo's Eye: A Cartographic Genealogy of the Earth in the Western Imagination*, Johns Hopkins University Press, Baltimore.

Cowen, M.P. and Shenton, R.W. (1996) *Doctrines of Development*, Routledge, London.

Cox, K., Low, M., and Robinson, J. (eds) (2007) *The Sage Handbook of Political Geography*, Sage, London.

Cozens, P.M., Hillier, D. and Prescott, G. (1999) Crime and the design of new-build housing, *Town and Country Planning*, **68**(7) 231–33.

Crang, M. (1998) *Cultural Geography*, Routledge, London.

Crang, P. (1997) Cultural turns and the (re)constitution of economic geography, in Lee, R. and Wills, J. (eds) *Geographies of Economies*, Arnold, London, 3–15.

Crang, P. and Jackson, P. (2000) Consuming geographies, in Morley, D. and Robins, K. (eds) *British Cultural Studies*, Oxford University Press, Oxford.

Crawford, M. (1992) The world in a shopping mall, in Sorkin, M. (ed.) *Variations on a Theme Park: The New American City and the End of Public Space*, Noonday, New York.

Cresswell, T. (1996) *In Place/Out of Place: Geography, Ideology and Transgression*, Minneapolis, University of Minnesota Press.

Cresswell, T. (2004) *Place: A Short Introduction*, Blackwell, Oxford.

Crewe, L. and Gregson, N. (1998) Tales of the unexpected: exploring car boot sales as marginal spaces of consumption, *Transactions of the Institute of British Geographers*, **23**, 39–53.

Crone, P. (1986) The tribe and the state, in Hall, J.A. (ed.) *States in History*, Basil Blackwell, Oxford, 48–77.

Crone, P. (1989) *Pre-Industrial Societies*, Blackwell, Oxford.

Cronon, W. (1991) *Nature's Metropolis: Chicago and the Great West*, W.W. Norton, New York.

Cronon, W. (1996) *Uncommon ground: toward reinventing nature*, Norton, New York.

Crouch, D.P., Garr, D.J. and Mundigo, A.I. (1982) *Spanish City Planning in North America*, MIT Press, Cambridge, MA.

Crum, R.E. and Gudgin, G. (1977) *Non-production Activities in UK Manufacturing Industry*, Commission of the European Communities, Brussels.

Crump, J.R. (2004) Producing and enforcing the geography of hate: race, housing segregation, and housing-related hate crimes in the United States, in Flint, C. (ed.) *Spaces of Hate. Geographies of Discrimination and Intolerance in the USA*, Routledge, New York, 227–44.

Crush, J. (1995) Imagining development, in Crush, J. (ed.) *Power of Development*, Routledge, London, 1–26.

Cummins, S. and MacIntyre, S. (1999) Food deserts, evidence and assumptions in health policy-making, *British Medical Journal*, **325**, 436–8.

Curry, J. and Kenney, M. (2004) The organizational and geographic configuration of the personal computer value chain, in Kenney, M. and Florida, R. (eds) *Locating Global Advantage: Industry Dynamics in the International Economy*, Stanford University Press, Stanford, 113–41.

Cybriwsky, R. and Ford, L. (2001) Jakarta, *Cities*, **18**, 199–210.

D

Daily Star (1995) Korma out for a meal, *Daily Star*, 5 December.

Danforth, L.M. (1995) *The Macedonia Conflict: Ethnic Nationalism in a Transnational World*, Princeton University Press, Princeton, New Jersey.

Daniels, P.W. (1975) *Office Location: An Urban and Regional Study*, G. Bell, London.

Daniels, P.W. and Bryson, J.R. (2002) Manufacturing services and servicing manufacturing: changing forms of production in advanced capitalist economies, *Urban Studies*, **39**(5–6), 977–91.

Daniels, P.W. and Bryson J.R. (2005) Sustaining business and professional services in a second city region: the case of Birmingham, UK, *The Service Industries Journal*, **25**(4), 505–24.

Daniels, S. and Lee, R. (eds) (1996) *Exploring Human Geography: A Reader*, Arnold, London.

Dankelman, I. and Davidson, J. (1988) *Women and Environment in the Third World*, Earthscan, London.

Davies, G. (1994) *A History of Money: From Ancient Times to the Present Day*, University of Wales Press, Cardiff.

Davies, G. (2000) Narrating the natural history unit, *Geoforum*, **31**, 539–51.

Davies, J., Sandström, S., Shorrocks, A. and Wolff, E. (2006) *World distribution of household wealth*. London: Foreign Press Association and United Nations Secretariat, presentation 5 December.

Daviron, B. and Ponte, S. (2005) *The Coffee Paradox: Global Markets, Commodity Trade and the Elusive Promise of Development*, Zed Books, London.

Davis, J. and Goldberg, R. (1957) *A Concept of Agribusiness*, Harvard University Press, Boston, MA.

Davis, M. (1990) *City of Quartz: Excavating the Future in Los Angeles*, Verso, London.

Davis, M. (1995) Fortress Los Angeles: The militarization of urban space, in Kasinitz, P. (ed.) *Metropolis: Centre and Symbol of our Times*, Macmillan, Basingstoke, 355–68.

Davis, M. (1998) *Ecology of Fear: Los Angeles and the Imagination of Disaster*, Metropolitan Books, New York.

Davis, T. (1995) The diversity of queer politics and the redefinition of sexual identity and community in urban spaces, in Bell, D. and Valentine, G. (eds) *Mapping Desire: Geographies of Sexualities*, Routledge, London, 284–303.

de Ru, N. (2004) Hollocore, in Koolhas, R., Brown, S. and Link, J. (eds) *Content*, Taschen, Cologne, 336–49.

Dean, K. (2005) Spaces and territorialities on the Sino-Burmese boundary: China, Burma and the Kachin, *Political Geography*, **24**, 808–30.

Deane, P. and Cole, W. (1962) *British Economic Growth 1688–1959*, Cambridge University Press, Cambridge.

Dear, M. (2000) *The Postmodern Urban Condition*, Malden, MA, Blackwell.

Dear, M. (ed.) (2002) *From Chicago to L.A.: Making Sense of Urban Theory*, Sage, Thousand Oaks, CA.

Dear, M. and Flusty, S. (2002) The resistable rise of the LA school, in Dear, M. (ed.) (2002*) From Chicago to L.A.: Making Sense of Urban Theory*, Sage, Thousand Oaks, CA, 3–16.

Debrix, F. (2007) Tabloid imperialism: American geopolitical anxieties and the war on terror, *Geography Compass*, **1**(4), 932–945.

Dedrick, J. and Kraemer, K.L. (2006) Is production pulling knowledge work to China? A study of the notebook PC industry, *Computer*, **39**, 36–42.

Dekmejian, R.H. and Simonian, H.H. (2003) *Troubled Waters: The Geopolitics of the Caspian Region*, I B Tauris, London.

Delaney, D. (2005) *Territory: A Short Introduction*, Blackwell, Malden.

Delanty, G. and Kumar, K. (eds) (2006) *Handbook of Nations and Nationalism*, Sage, London.

DeLind, L. (2003) Considerably more than vegetables, a lot less community: the dilemmas of community supported agriculture, in Adams, J. (ed.) *Fighting for the Farm: Rural America Transformed*, University of Pennsylvania Press, Philadelphia, 192–206.

Demeritt, D. (1998) Science, social constructivism and nature, in Castree, N. and Braun, B. (eds) *Remaking Reality: Nature at the Millennium*, Routledge, London, 173–93.

Demeritt, D. (2002) What is the social construction of nature? *Progress in Human Geography*, **26**, 767–90.

Denevan, W. (1992) The pristine myth: the landscapes of the Americas in 1492, *Annals of the Association of American Geographers*, **82**, 369–85.

Dennis, R. (1984) *English Industrial Cities of the Nineteenth Century*, Cambridge University Press, Cambridge.

Department for International Development (DFID) (2005) *Country Profile: Uganda*, copy available at: http://www.dfid.gov.uk/countries/africa/uganda.asp, accessed 23 May 2007.

Dery, M. (1993) *Culture Jamming: Hacking, Slashing and Sniping in the Empire of Signs*, Open Magazine Pamphlet Series, Open Magazine.

Desai, V. and Potter, R. (eds)(2001) *The Arnold Companion to Development Studies*, Arnold, London.

Desforges, L. (1998) 'Checking out the planet': global representations/local identities and youth travel, in Skelton, T. and Valentine, G. (eds) *Cool Places: Geographies of Youth Cultures*, Routledge, London, 174–91.

Devall, B. and Sessions, G. (1985) *Deep Ecology: Living as if Nature Mattered*, Peregrine Smith Books, Salt Lake City, UT.

DFID *see* Department for International Development.

Dicken, P. (1992) *Global Shift: The Internationalization of Economic Activity*, Paul Chapman, London.

Dicken, P. (2003) *Global Shift: Reshaping the Global Economic Map in the 21st Century*, 4th edition, Sage, London.

Dicken, P. (2007) *Global Shift: Mapping the Changing Contours of the World Economy*, 5th edition, Sage, London.

Dicken, P. and Lloyd, P. E. (1990) *Location in Space: Theoretical Perspectives in Economic Geography*, Harper & Row, New York.

Dickenson, J., Gould, B., Clarke, C., Mather, S., Prothero, M., Siddle, D., Smith, C. and Thomas-Hope, E. (1996) *A Geography of the Third World*, Routledge, London.

Dikeç, M. (2006a) Two decades of French urban policy: from social development of neighborhoods to the republican penal state, *Antipode*, **38**(1), 59–81.

Dikeç, M. (2006b) The badlands of the republic: revolts, the French state and questions of the banlieues *Environment and Planning D*, **24**, 159–63.

Dikeç, M. (2007) *Badlands of the Republic: Space, Politics, and French Urban Policy*, Oxford, Blackwell, RGS/IBG Book Series.

Dittmer, J. (2005) Capitan America's empire: reflections on identity, popular culture and post-9/11 geopolitics, *Annals of the Association of American Geographers*, **95**(3), 626–43.

Dixon, D.P. and Jones, J.P. (2006) Feminist geographies of difference, relation and construction, in Aitken, S. and Valentine, V. (eds) (2006) *Approaches to Human Geography*, Sage, London, 42–56.

Dobb, M. (1946; 1963) *Studies in the Development of Capitalism*, Routledge & Kegan Paul, London.

Dobson, A. (2001) *Green Political Thought*, 3rd edition, Routledge, London.

Dodd, N. (1994) *The Sociology of Money: Economics, Reason and Contemporary Society*, Continuum Publishing Company, New York.

Dodds, K. (1997) *Geopolitics in Antarctica: Views from the Southern Ocean Rim*, Wiley, Chichester.

Dodds, K. (2004) *Global Geopolitics: A Critical Introduction*, Prentice Hall, Harlow.

Dodds, K. (2006) Popular geopolitics and audience dispositions: James Bond and the Internet Movie Database, *Transactions of the Institute of British Geographers*, **31**(2), 116–30.

Dodds, K. (2007) *Geopolitics: A Very Short Introduction*, Oxford University Press, Oxford.

Dodgshon, R.A. (1987) *The European Past: Social Evolution and Spatial Order*, Macmillan, Basingstoke.

Dodgshon, R.A. (1998) *Society in Time and Space: A Geographical Perspective on Change*, Cambridge University Press, Cambridge.

Dodgshon, R.A. and Butlin, R.A. (eds) (1990) *An Historical Geography of England and Wales*, 2nd edition, Academic Press, London.

Doherty, B. (2002) *Ideas and Actions in the Green Movement*, Routledge, London.

Dolan, C. and Humphrey, J. (2002) Changing governance patterns in the trade in fresh vegetables between Africa and the United Kingdom, *Environment and Planning A*, **36**, 491–509.

Dorfman, A. and Mattelart, A. (1975) *How to Read Donald Duck: Imperialist Ideology in the Disney Comic*, International General Editions, New York.

Dorling, D. (2004) *The Human Geography of the UK*, London, Sage.

Dorling, D. and Fairbairn, D. (1997) *Mapping: Ways of Representing the World*, Prentice Hall, Harlow.

Dorling, D. and Rees, P. (2003) A nation still dividing: the British census and social polarisation 1971–2001, *Environment and Planning A*, **35**, 1287–313.

Dorling, D. and Shaw, M. (2000) Life chances and lifestyles, in V. Gardiner and H. Matthews (eds) *The Changing Geography of the UK*, London, Routledge, 230–60.

Doyle, T. (2000) *Green Power: The Environment Movement in Australia*. University of New South Wales Press, Sydney.

Doyle, T. (2005) *Environmental Movements in Majority and Minority Worlds: A Global Perspective*, Rutgers University Press, London.

Doyle, T. and McEachern, D. (2001) *Environment and Politics*, 2nd edition, Routledge, London.

Driver, F. (2005) Imaginative geographies, in Cloke, P., Crang, P. and Goodwin, M. (eds) *Introducing Human Geographies*, 2nd edition, Arnold, London, 144–55.

Dryzek, J.S. and Schlosberg, D. (eds) (2004) *Debating the Earth: The Environmental Politics Reader*, Oxford, University Press, Oxford.

DTI (Department of Trade and Industry) (2007) Meeting the Energy Challenge. A White Paper on Energy, May 2007. London: HMSO.

du Gay, P. (1996) *Consumption and Identity at Work*, Sage, London.

Dudrah, R.K. (2002) Drum n dhol: British Bhangra music and diasporic South Asian identity formation, *European Journal of Cultural Studies*, **5**(3), 363–83.

Duncan, J.S., Johnson, N.C. and Schein, R. (eds) (2004) *A Companion to Cultural Geography*, Blackwell, London.

Dunn, K.M. (2001) Representations of Islam in the politics of mosque development in Sydney, *Tijdschrift voor Economische en Sociale Geografi*, **92**, 291–308.

Dunning, J.H. (1981) *International Production and the Multinational Enterprise*, Allen & Unwin, London.

DuPuis, E.M. and Goodman, D. (2005) Shall we go 'home' to eat?: towards a reflexive politics of localism, *Journal of Rural Studies*, **21**, 359–71.

Durham, M. (2004) Clothes line, *The Guardian G2 supplement*, 25 February, 2–4, accessed 24 September 2007: www.guardian.co.uk/g2/story/0,,1155254,00.html

Duruz, J. (2005) Eating at the borders: culinary journeys, *Environment & Planning D: Society & Space*, **23**, 51–69.

Dwyer, C. (2005) Diasporas, in Cloke, P., Crang, P. and Goodwin, M. (eds) *Introducing Human Geographies*, 2nd edition, Hodder Arnold, London, 495–508.

Dwyer, C. and Crang, P. (2002) Fashioning ethnicities: the commercial spaces of multiculture, *Ethnicities*, **2**, 410–30.

Dwyer, C. and Jackson, P. (2003) Commodifying difference: selling EASTern fashion, *Environment and Planning D: Society and Space*, **21**, 269–91.

E

Earthscan (1998) *Internally Displaced People: A Global Survey*, Earthscan, Guildford.

Edensor, T. and Kothari, U. (2006) Extending networks and mediating brands: stallholder strategies in a Mauritian market, *Transactions of the Institute of British Geographers*, **31**(3), 323–36.

Edwards, L.E. (1983) Towards a process model of office-location decision making, *Environment and Planning A*, **15**, 1327–42.

Edwards, M. and Hulme, D. (1992) *Making a Difference: NGOs and Development in a Changing World*, Earthscan, Guildford.

EIA (Energy Information Administration) (2007) International Energy Outlook 2007. Washington DC: US Department of Energy.

Elden, S. (2005) Territorial integrity and the war on terror. *Environment and Planning A*, **37**, 2083–104.

Elliott, D. (2003) *Energy, Environment and Society*, 2nd edition, Routledge, London.

Engardio, P. (2001) Smart bombs, so why no smart aid?, *Business Week*, 24 December, 58.

Engels, F. (1845) *The Condition of the Working Class in England*, 1987 edition, Penguin, London and New York.

England, K. (2006) Producing feminist geographies: theory, methodologies and research strategies, in Aitken, S. and Valentine, V. (eds) (2006) *Approaches to Human Geography*, Sage, London, 286–97.

Ereaut, G. and Segnit, N. (2006) Warm Words: How are we telling the climate story and can we tell it better? http://www.ippr.org.uk/members/download.asp?f=%2Fecomm%2Ffiles%2Fwarm%5Fwords%2Epdf, accessed 11/11.

Escobar, A. (1995) *Encountering Development*, Princeton University Press, Princeton, NJ.

Esposito, J.L. (ed.) (1995) *The Oxford Encyclopedia of Islam in the Modern World*, Oxford University Press, New York.

Esteva, G. (1992) Development, in Sachs, W. (ed.) *The Development Dictionary*, Zed, London.

EU (2005) *Sustainable Financing of Social Policies in the European Union*, Commission Staff Working Document. SEC (2005) 1774, Brussels.

European Commission (1999) *Sustainable Urban Development in the European Union: A Framework for Action*, Office for Official Publications of the European Communities, Luxembourg.

Eyben, R., Moncrieffe, J. and Knowles, C. (2006) *The Power of Labelling in Development Practice*, Institute of Development Studies, IDS Policy Briefing 28, April.

F

Fairbairn, D. and Dorling, D. (1997) *Mapping and map-making: new approaches to the teaching of cartography*, Proceedings of the 18th International Cartography Conference, Stockholm, 23–27 June, 4, 1955–1962.

Fairlie, S. (1996) *Low Impact Development: Planning and People in a Sustainable Countryside*, Jon Carpenter, Oxfordshire.

Falah, G-W., Flint, C. and Mamadouh, V. (2006) Just war and extraterritoriality: the popular geopolitics of the United States war on Iraq as reflected in newspapers of the Arab world, *Annals of the Association of American Geographers*, **96**(1) 142–64.

Featherstone, M. (1995) *Undoing Culture: Globalization, Postmodernism and Identity*, Sage, London.

Featherstone, M., Lash, S. and Robertson, R. (eds) (1995) *Global Modernities*, Sage, London.

Fejes, F. (1981) Media imperialism, an assessment, *Media, Culture and Society*, **3**, 281–9.

Fell, A. (1991) Penthesilea, perhaps, in Fisher, M. and Owen, U. (eds) *Whose Cities?* Penguin, London, 73–84.

Ferguson, J. (1990) *The Anti-Politics Machine: 'Development', Depoliticization and Bureaucratic Power in Lesotho*, Cambridge University Press, Cambridge.

Fieldhouse, D.K. (1999) *The West and the Third World*, Blackwell, Oxford.

Fields, G. (2004) *Territories of Profit*, Stanford Business Books, Stanford.

Fineman, S. (ed) (2000) *Emotion in Organization* Sage, London.

Fitter, R. and Kaplinsky, R. (2001) Who gains from product rents as the coffee market becomes more differentiated? A value chain analysis, *IDS Bulletin Paper*, IDS Sussex.

FitzSimmons, M. (1986) The new industrial agriculture: the regional differentiation of specialty crop production, *Economic Geography*, **62**, 334–53.

FitzSimmons, M. (1997) Restructuring, industry and regional dynamics, in Goodman, D. and Watts, M. (eds) *Globalising Food*, Routledge, London, 158–68.

Flint, C. (2006) *Introduction to Geopolitics*, Routledge, London.

Florida, R. (2002) *The Rise of the Creative Class and How It's Transforming Work, Leisure and Everyday Life*, Basic Books, New York.

Flowerdew, R. and Martin, D. (1997) *Methods in Human Geography: A Guide for Students Doing a Research Project*, Longman, Harlow.

Foley, R. (1995) *Humans Before Humanity: An Evolutionary Perspective*, Blackwell, Oxford.

Fonseca, I. (1996) *Bury Me Standing*, Vintage, London.

Forest, B. (1995) West Hollywood as symbol: the significance of place in the construction of a gay identity *Environment and Planning D: Society and Space*, **13**, 133–57.

Foster, R.F. (ed.) (2001) *The Oxford History of Ireland*, Oxford University Press, Oxford.

Foster, W., Cheng, Z., Dedrick, J. and Kraemer, K.L. (2006) Technology and organizational factors in the notebook industry supply chain, *The Personal Computer Industry Center publication*, UC Irvine (see http://www.pcic.gsm.uci.edu/pubs.asp)

Fotheringham, S.A., Brunsdon, C. and Charlton, M. (2000) *Quantitative Geography: Perspectives on Spatial Data Analysis*, Sage, London and Thousand Oaks, New Delhi.

Foucault, M. (1979) On governmentality, *Ideology and Consciousness*, **6**, 5–29.

Frank, A. (1966) The development of underdevelopment, *Monthly Review*, **18**(4), 17–31.

Frank, A.G. (1969) *Capitalism and Underdevelopment in Latin America*, Monthly Review Press, New York.

Frank, A.G. (1997) The Cold War and me, *Bulletin of Concerned Asian Scholars*, **29**, available at http://csf.colorado.edu/bcas/symmpos/syfrank.htm, accessed 8 January 2003.

Freeman, C. (2000) *High Tech and High Heels in the Global Economy*, Duke, Durham.

Freeman, M. (1986) Transport, in Langton J. and Morris, R.J. (eds) *Atlas of Industrializing Britain, 1780–1914*, Methuen, London and New York.

Freidberg, S. (2003) Cleaning up down South: supermarkets, ethical trade and African horticulture, *Social and Cultural Geography*, **4**, 27–44.

Friedberg, S. (2004) *French Beans and Food Scares: Culture and Commerce in an Anxious Age*, Oxford University Press, Oxford.

Friedman, J. (1994) Globalization and localization, in Friedman, J. (ed.) *Cultural Identity and Global Process*, Sage, London, 102–16.

Friedman, J. (1997) Global crises, the struggle for cultural identity and intellectual porkbarrelling: cosmopolitans versus locals, ethnics and nationals in an era of de-hegemonisation, in Werbner, P. and Modood, T. (eds) *Debating Cultural Hybridity. Multi-Cultural Identities and the Politics of Anti-Racism*, Zed, London, 70–89.

Friedman, J. and Rowlands, M.J. (1977) Notes towards an epigenetic model of the evolution of civilization, in Friedman, J. and Rowlands, M.J. (eds) *The Evolution of Social Systems*, Duckworth, London, 201–76.

Friedman, T. (1999) It's a small world, *Sunday Times*, 28 March, Section 5.

Friedmann, H. and McMichael, P. (1989) Agriculture and the state system: the rise and decline of national agricultures, 1870 to the present, *Sociologia Ruralis*, **29**, 93–117.

Friedmann, J. (1966) *Regional Development Policy: A Case Study of Venezuela*, MIT Press, Cambridge, MA.

Friedmann, J. (1986) The world city hypothesis, *Development and Change*, **17**(1), 69–83.

Friedmann, J. (1992) *Empowerment: The Politics of Alternative Development*, Blackwell, Oxford.

Friend, T. (2003) *Indonesian Destinies*, The Belknap Press of Harvard University Press, Cambridge, MA.

Friends of the Earth flyer (2006) What kind of world do you want? *Friends of the Earth Limited*, June 2006.

Fröbel, F., Heinrichs, J. and Kreye, O. (1980) *The New International Division of Labour*, Cambridge University Press, Cambridge.

Fukuyama, F. (1992) *The End of History and the Last Man*, Penguin, London.

Furtado, C. (1964) *Development and Underdevelopment*, University of California Press, Berkeley.

Fyfe, N. (2004) Zero Tolerance, maximum surveillance? Deviance, difference and crime control in the late modern city, in Lees, L. (ed.) *The Emancipatory City: Paradoxes and Possibilities*, Sage, London.

G

Gad, G. and Holdsworth, D.W. (1987) Corporate capitalism and the emergence of the high-rise office building, *Urban Geography*, **8**, 212–31.

Gaffikin, F. and Nickson, A. (1984) *Jobs Crisis and the Multi-Nationals*, Birmingham Trade Union Group for World Development, Birmingham.

Gagnon, V.P. (2006) *The Myth of Ethnic War: Serbia and Croatia in the 1990s*, Cornell University Press, Ithaca.

Gandy, M. (2002) *Concrete and Clay: Re-working Nature in New York City*, MIT Press, New York.

Gans, H. (1968) *People and Problems: Essays on Urban Problems and Solutions*, Basic Books, New York.

Garner, R. (2000) *Environmental Politics: Britain, Europe and the Global Environment*, 2nd edition, Macmillan, London.

Garreau, J. (1991) *Edge City: Life on the New Frontier*, Doubleday, New York.

Gary, I. and Karl, T.L. (2003) *Bottom of the Barrel: Africa's Oil Boom and the Poor*, Catholic Relief Services. Available at: www.catholicrelief.org/africanoil.cfm

Gates, D. (2005) Boeing 787: parts from around world will be swiftly integrated, *Seattle Times*, 11 September.

Geobusiness Magazine (2001) *The World Bank*, Geobusiness, Geneva.

Geographical Journal, The, **170**(4), December 2004.

Gereffi, G. (1994) The organisation of buyer-driven global commodity chains: how US retailers shape overseas production networks, in Gereffi, G. and Korzeniewicz, M. (eds) *Commodity Chains and Global Development*, Praeger, Westport, 95–122.

Gereffi, G. (2001) Shifting governance structures in global commodity chains, with special reference to the Internet, *American Behavioural Scientist*, **44**, 1616–37.

Gereffi, G., Humphrey, J. and Sturgeon, T. (2005) The governance of global value chains, *Review of International Political Economy*, **12**, 78–104.

Ghazvinian, J. (2007) *Untapped: The Scramble for Africa's Oil*, Harcourt, London.

Gibbon, P. and Ponte, S. (2005) *Trading Down: Africa, Value Chains, and the Global Economy*, Temple University Press, Philadelphia.

Giddens, A. (1990) *The Consequences of Modernity*, Polity Press, Cambridge.

Giddens, A. (1991) *The Constitution of Society*, Polity Press, Cambridge.

Giddens, A. (2001) *Sociology*, Polity Press, Cambridge.

Giles, C. and Goodall, I.H. (1992) *Yorkshire Textile Mills, 1770–1930*, HMSO, London.

Gilmore, R.W. (2007) *Golden Gulag: Prisons, Surplus, Crisis and Opposition in Globalizing California*, University of California Press, Berkeley.

Gilroy, P. (1993) *The Black Atlantic: Modernity and Double Consciousness*, Verso, London.

Gilroy, P. (2000) *Between Camps: Nations, Cultures and the Allure of Race*, Penguin, London.

Glassner, M.I. (1993) *Political Geography*, John Wiley, New York.

Gleeson, B. (1998) The social space of disability in colonial Melbourne, in Fyfe, N. (ed.) *Images of the Street*, Routledge, London, 93–110.

Glennie, P. and Thrift, N. (1992) Modernity, urbanism and modern consumption, *Environment and Planning D: Society and Space*, **10**, 423–43.

Godlewska, A. and Smith, N. (eds) (1994) *Geography and Empire*, Blackwell, Oxford.

Goetz, A.M. and Gupta, R.S. (1996) Who takes the credit? Gender, power, and control over loan use in rural credit programs in Bangladesh, *World Development*, **24**, 45–63.

Goldman, R. and Papson, S. (1998) *Nike Culture: The Sign of the Swoosh*, Sage, London.

Goodman, D. (2003) The quality 'turn' and alternative food practices: reflections and agenda, *Journal of Rural Studies*, **19**, 1–7.

Goodman, D. and DuPuis, E.M. (2002) Knowing food and growing food: beyond the production–consumption debate in the sociology of agriculture, *Sociologia Ruralis*, **42**, 5–22.

Goodman, D. and Watts, M. (eds) (1997) *Globalising Food*, Routledge, London.

Goodman, D., Sorj, B. and Wilkinson, J. (1987) *From Farming to Biotechnology: A Theory of Agro-industrial Development*, Blackwell, Oxford.

Goodman, M.K. (2004) Reading fair trade: political ecological imaginary and the moral economy of fair trade foods, *Political Geography*, **23**, 891–915.

Goodwin, M. (1995) Poverty in the city, you can raise your voice, but who is listening?, in Philo, C. (ed.) *Off the Map: The Social Geography of Poverty in the UK*, Child Poverty Action Group, London, 134–54.

Gore, A. (2006) *An Inconvenient Truth: The Planetary Emergency of Global Warming and What We Can Do About It*, Bloomsbury Publishing, London.

Goskomstat (2003) *Rossiiskii Statisticheskii Ezhegodnik (Russian Statistical Yearbook)*, State Committee of the Russian Federation for Statistics, Moscow.

Goskomstat (2006) *Rossiiskii Statisticheskii Ezhegodnik (Russian Statistical Yearbook)*, State Committee of the Russian Federation for Statistics, Moscow.

Goss, J. (1993b) The magic of the mall: form and function in the retail built environment, *Annals of the Association of American Geographers*, **83**, 18–47.

Goss, J. (1999a) Consumption, in Cloke, P., Crang, P. and Goodwin, M. (eds) *Introducing Human Geographies*, Arnold, London, 114–21.

Goss, J. (1999b) Once upon a time in the commodity world, an unofficial guide to the Mall of America, *Annals of the Association of American Geographers*, **89**, 45–75.

Gottdiener, M. (1982) Disneyland, a utopian urban space, *Urban Life*, **11**, 139–62.

Gottdiener, M. (1997) *The Theming of America: Dreams, Visions and Commercial Spaces*, Westview Press, Boulder, CO.

Gottman, J. (1973) *The Significance of Territory*, University Press of Virginia, Charlottesville.

Gould, P. and Pitts, F.R. (eds) (2002) *Geographical Voices: Fourteen Autobiographical Essays*, Syracuse University Press, Syracuse, NY.

Gourevitch, P., Bohn, R. and McKendrick, D. (2000) Globalization of production, *World Development*, **28**, 301–17.

Graham, B.J., Ashworth, G.J. and Tunbridge, J.E. (2000) *A Geography of Heritage: Power, Culture and Economy*, Arnold, London.

Graham, S. (ed.) (2004) *Cities, War, and Terrorism: Towards an Urban Geopolitics*, Blackwell, Oxford.

Graham, S. and Marvin, S. (1996) *Telecommunications and the City: Electronic Spaces, Urban Places*, Routledge, London.

Graham, S. and Marvin, S. (2001) *Splintering Urbanism: Networked Infrastructures, Technological Mobilities and the Urban Condition*, Routledge, London.

Green, A. (1997) Income and wealth, in Pacione, M. (ed.) *Britain's Cities: Geographies of Division in Urban Britain*, Routledge, London, 78–102.

Green, A. and Owen, D. (2006) *The Geography of Poor Skills and Access to Work*, Joseph Rowntree Foundation, York.

Green, F. and Sutcliffe, B. (1987) *The Profit System: The Economics of Capitalism*, Penguin, London.

Gregory, D. (1982) *Regional Transformation and Industrial Revolution: A Geography of the West Yorkshire Woollen Industry*, Macmillan, London.

Gregory, D. (1994) *Geographical Imaginations*, Blackwell, Oxford.

Gregory, D. (2006) The black flag: Guantánamo Bay and the space of exception, *Geografiska Annaler*, **88B**(4), 405–27.

Gregory, D. and Pred, A. (2006) (eds) *Violent Geographies: Fear, Terror, and Political Violence*, Routledge, London.

Grenville, K. (2006) *The Secret River*, Canongate Press, London.

Griffiths, I.L. (1994) *The Atlas of African Affairs*, 2nd edition, Routledge, London.

Groves, P.A. (1987) The northeast and regional integration, 1800–1860, in Mitchell, R.D. and Groves, P.A. (eds) *North America: The Historical Geography of a Changing Continent*, Hutchinson, London.

Grundy-Warr, C. and Wong, S.Y. (2002) Geographies of displacement: the Karenni and the Shan across the Myanmar–Thailand border, *Singapore Journal of Tropical Geography*, **23**, 93–122.

Guthman, J. (2004) Back to the land: the paradox of organic food standards, *Environment and Planning A*, **36**, 511–28.

Gutkind, E.A. (1964) *Urban Development in Central Europe*, Macmillan, London and New York.

Gwynne, R.N., Klak, T. and Shaw, D.J.B. (2003) *Alternative Capitalisms: Geographies of Emerging Regions*, Arnold, London.

H

Hagen, J. (2004). The most German of towns: creating an ideal Nazi community in Rothenburg ob der Tauber, *Annals of the Association of American Geographers*, **94**(1), 207–27.

Halfacree, K. (1996) Out of place in the country: travellers and the 'rural idyll', *Antipode*, **18**, 42–72.

Hall, C., Tharakhan, P., Hallock, J., Cleveland, C. and Jefferson, M. (2003) Hydrocarbons and the evolution of human culture, *Nature*, **426**, 318–22.

Hall, C.M. (2005) *Tourism: Rethinking the Social Science of Mobility*, Prentice Hall, Harlow.

Hall, E.T. (1959) *The Silent Language*, Doubleday, Garden City, NY.

Hall, P. (ed.) (1966) *Von Thünen's Isolated State*, Pergamon, Oxford.

Hall, P. (2002) *Cities of Tomorrow: an Intellectual History of Urban Planning and Design in the Twentieth Century*, 3rd edition, Blackwell, Oxford.

Hall, P. and Pain, K. (2006) *The Polycentric Metropolis: Learning from Mega-city Regions in Europe*, Earthscan, London.

Hall, P. and Pfeiffer, U. (2000) *Urban Future 21: A Global Agenda for 21st Century Cities*, Spon, London.

Hall, R. (1995) Households, families and fertility, in Hall, R. and White, P. (eds) *Europe's Population: Towards the Next Century*, UCL Press, London.

Hall, S. (1992) The West and the rest: discourse and power, in Hall, S. and Gieben, B. (eds) *Formations of Modernity*, Open University Press, Milton Keynes, 275–331.

Hall, S. (1995) New cultures for old, in Massey, D. and Jess, P. (eds) *A Place in the World? Places, Cultures and Globalization*, Oxford University Press, Oxford, 175–213.

Hall, S. (1996) The meaning of new times, in Morley, D. and Chen, K.-H. (eds) *Stuart Hall: Critical Dialogues in Cultural Studies*, Routledge, London, 223–37.

Halliday, F. (2002) *Two Hours that Shook the World. September 11, 2001: Causes and Consequences*, Saqi Books, London.

Hamelink, C. (1983) *Cultural Autonomy in Global Communications*, Longman, New York.

Hamnett, C. (2003) *Unequal City: London in the Global Arena*, Routledge, London.

Hannerz, U. (1992) The global ecumene, in Hannerz, U. (ed.) *Cultural Complexity: Studies in the Social Organization of Meaning*, Columbia University Press, New York, 217–67.

Hanrou, H. and Obrist, H.-U. (1999) Cities on the move, in Bradley, F. (ed.) *Cities on the Move: Urban Chaos and Global Change: East Asian Art, Architecture and Film Now*, Hayward Gallery, London, 10–15.

Hanson, S. and Pratt, G. (1995) *Gender, Work and Space*, Routledge, London.

Hara, K. (1999) Rethinking the 'Cold War' in the Asia-Pacific, *Pacific Review*, **12**, 515–36.

Haraway, D. (1989) *Primate Visions: Gender, Race and Nature in the World of Modern Science*, Routledge, New York.

Hardt, M. and Negri, A. (2000) *Empire*, Harvard University Press, Cambridge, MA.

Harper, S. (2005) *Ageing Societies*, Hodder Arnold, London.

Harris, D.R. (1989) An evolutionary continuum of plant–animal interaction, in Harris, D.R. and Hillman, G.C. (eds) *Foraging and Farming: The Evolution of Plant Exploitation*, Unwin Hyman, London, 11–26.

Harris, D.R. (1996a) Introduction: themes and concepts in the study of early agriculture, in Harris, D.R. (ed.) *The Origins and Spread of Agriculture and Pastoralism in Eurasia*, UCL Press, London, 1–9.

Harris, D.R. (1996b) The origins and spread of agriculture and pastoralism in Eurasia: an overview, in Harris, D.R. (ed.) *The Origins and Spread of Agriculture and Pastoralism in Eurasia*, UCL Press, London, 552–70.

Harris, F. (ed.) (2004) *Global Environmental Issues*, John Wiley and Sons, Chichester.

Hart, G. (2001) Development critiques in the 1990s: culs-de-sac and promising paths, *Progress in Human Geography*, **25**, 649–58.

Hartwick, E. (1998) Geographies of consumption: a commodity-chain approach, *Environment and Planning D: Society and Space*, **16**, 423–37.

Harvey, D. (1974) Population, resources and the ideology of science, *Economic Geography*, **50**, 256–77.

Harvey, D. (1982) *The Limits to Capital*, Blackwell, Oxford.

Harvey, D. (1985a) *The Urbanization of Capital: Studies in the History and Theory of Capitalist Urbanization*, Blackwell, Oxford.

Harvey, D. (1985b) Money, time, space and the city, in Harvey, D. (ed.) *The Urban Experience*, Johns Hopkins University Press, Baltimore, MD.

Harvey, D. (1989) *The Condition of Postmodernity: An Enquiry into the Origins of Cultural Change*, Blackwell, Oxford.

Harvey, D. (1990) Between space and time: reflections on the geographical imagination, *Annals of the Association of American Geographers*, **80**, 418–34.

Harvey, D. (2003) *The New Imperialism*, Oxford: Oxford University Press.

Haskell, T.L. (1985a) Capitalism and the origins of the humanitarian sensibility, part 1, *American Historical Review*, **90**, 339–61.

Haskell, T.L. (1985b) Capitalism and the origins of the humanitarian sensibility, part 2, *American Historical Review*, **90**, 547–66.

Hastings, A. and Dean, J. (2003) Challenging images: tackling stigma through estate regeneration, *Policy and Politics*, **31**, 171–84.

Hayter, R. and Watts, H.D. (1983) The geography of enterprise: a re-appraisal, *Progress in Human Geography*, **7**, 157–81.

Hebdige, D. (1979) *Subculture: The Meaning of Style*, Methuen, London.

Hebdige, D. (1987) Object as image: the Italian scooter cycle, in *Hiding in the Light: On Images and Things*, Comedia/Routledge, London, 77–115.

Hecht, S. and Cockburn, A. (1989) *The Fate of the Forest: Developers, Destroyers and Defenders of the Amazon*, Verso, London and New York.

Heffernan, M. (1998) *The Meaning of Europe: Geography and Geopolitics*, Arnold, London.

Heilbronner, R.L. (1972) *The Economic Problem*, 3rd edition, Prentice Hall, Englewood Cliffs, NJ.

Heisenberg, W. (1958) *Physics and Philosophy: The Revolution in Modern Science*, Harper, New York.

Helander, B. (2005) Who needs a state? Civilians, security and social services in north-east Somalia, in Richards, P. (ed.) *No Peace, No War: An Anthropology of Contemporary Armed Conflict*, Ohio University Press, Athens, 193–202.

Held, D., McGrew, D., Goldblatt, D. and Perraton, J. (1999) *Global Transformations: Politics, Economics and Culture*, Polity, Cambridge.

Heldke, L. (2003) *Exotic Appetites: Ruminations of a Food Adventurer*, Routledge, New York.

Helm, D. (ed.) (2007) *The New Energy Paradigm*, OUP, Oxford.

Hendrickson, M.K. and Heffernan, W.D. (2002) Opening spaces through relocalization: locating potential resistance in the weaknesses of the global food system, *Sociologia Ruralis*, **42**, 347–69.

Henry, C.M. and Wilson, R. (eds) (2004) *The Politics of Islamic Finance*, Edinburgh University Press, Edinburgh.

Henry, N., McEwan C. and Pollard J. S. (2002) Globalization from below: Birmingham – postcolonial workshop of the world?, *Area*, **34**(2) 118–27.

Hepple, L. (1986) The revival of geopolitics, *Political Geography Quarterly*, **5**, 521–36.

Hepple, L. (1992) Metaphor, geopolitical discourse and the military in South America, in Barnes, T.J. and Duncan, J.S. (eds) *Writing Worlds*, Routledge, London, 136–54.

Herb, G.H. (1989) Persuasive cartography in Geopolitik and National Socialism, *Political Geography Quarterly*, **8**, 289–303.

Herf, J. (1984) *Reactionary Modernism: Technology, Culture and Politics in Weimar and the Third Reich*, Cambridge University Press, Cambridge.

Herzfeld, M. (1992) *The Social Production of Indifference: Exploring the Symbolic Roots of Western Bureaucracy*, University of Chicago Press, Chicago.

Heske, H. (1986) German geographical research in the Nazi period, *Political Geography Quarterly*, **5**, 267–81.

Hettne, B. (1995) *Development Theory and the Three Worlds*, 2nd edition, Longman, London.

Hildyard, N. (1997) *The World Bank and the State: A Recipe for Change?*, N. Hildyard, Bretton Woods Project, London.

Hill, T.P. (1977) On goods and services, *Review of Income and Wealth*, **23**, 315–38.

Hinchliffe, S. (2007) *Space for Nature*, Sage, London.

Hinchliffe, S. and Belshaw, C. (2003) Who cares? Values, power and action in environmental contests, in Hinchliffe, S., Blowers, A. and Freeland, J. *Understanding Environmental Issues*, John Wiley and Sons, Chichester, 89–126.

Hindess, B. and Hirst, P. (1975) *Pre-Capitalist Modes of Production*, Routledge & Kegan Paul, London.

Hinrichs, C.C. (2000) Embeddedness and local food systems: notes on two types of direct agricultural market, *Journal of Rural Studies*, **16**, 295–303.

Hinrichs, C.C. (2003) The practice and politics of food system localization, *Journal of Rural Studies,* **19**, 33–45.

Hirschmann, A.O. (1958) *The Strategy of Economic Development,* Yale University Press, New Haven, CT.

Hobsbawm, E. (1995), *The Age of Extremes: The Short Twentieth Century, 1914–1991,* Abacus, London.

Hobsbawm, E.J. (1996) Ethnicity and nationalism in Europe today, in Balakrishnan, G. (ed.) *Mapping the Nation,* Verso, London, 255–66.

Hochschild, A.R. (1983) *The Managed Heart: Commercialization of Human Feeling,* University of California Press, London.

Hoggart, K., Lees, L. and Davies, A. (2002) *Researching Human Geography,* Arnold, London.

Holdar, S. (1992) The ideal state and the power of geography: the life-work of Rudolf Kjellen, *Political Geography,* **11**, 307–23.

Holloway, L. (2002) Virtual vegetables and adopted sheep: ethical relation, authenticity and Internet-mediated food production technologies, *Area,* **34**, 70–81.

Holloway, L. and Hubbard, P. (2001*) People and Place: The Extraordinary Geographies of Everyday Life,* Prentice Hall, London.

Holloway, S.L. (2005) Articulating otherness? White rural residents talk about gypsy-travellers, *Transactions of the Institute of British Geographers,* **30**, 351–67.

Holloway, S.L., Rice, S.P. and Valentine, G. (eds) (2003) *Key Concepts in Geography,* Sage, London.

Holmberg, T. (2005) Questioning 'the number of the beast': Constructions of humanness in a Human Genome Project (HGP) narrative, *Science as Culture,* **14**, 23–37.

Holt-Jensen, A. (1999) *Geography: History and Concepts,* 3rd edition, Sage, London and Thousand Oaks CA and New Delhi.

Homburger, E. (1994) *The Historical Atlas of New York City,* Henry Holt, New York.

Hoogevelt, A. (1997) *Globalisation and Postcolonialism,* Macmillan, London.

hooks, b. (1984) *Feminist Theory: From Margins to Centre,* South End Press, Boston.

hooks, b. (1992) Eating the other, desire and resistance, in hooks, b. (ed.) *Black Looks: Race and Representation,* South End Press, Boston, MA, 21–39.

Hooper, J. and Connolly, K. (2001) Berlusconi breaks ranks over Islam, *Guardian,* 27 September.

Hosking, G. (1992) *A History of the Soviet Union, 1917–1991,* final edition, Fontana Press, London.

House of Commons (2005) Hansard written answers for 18 October 2005 (pt 17). Available at: http://www.publications.parliament.uk/pa/cm200506/cmhansrd/vo051018/text/51018w17.htm

Hsu, J-Y., and Saxenian, A. (2000) The limits of guanxi capitalism: transnational collaboration between Taiwan and the USA, *Environment and Planning A,* **32**, 1991–2005.

Hubbard, P. (1999) *Sex and the City: Geographies of Prostitution in the Urban West,* London, Ashgate.

Hubbard, P. (2008) Regulating studentification through the housing market: a Loughborough case study, *Environment and Planning A,* forthcoming.

Hubbard, P., Kitchin, R., Bartley, B. and Fuller, D. (2002) *Thinking Geographically: Space, Theory and Contemporary Human Geography,* Continuum, London.

Hudson, R. (2004) Conceptualising geographies and their economies: spaces, flows and circuits, *Progress in Human Geography,* **28**(4) 447–71.

Hudson, R. (2005) *Economic Geographies: Circuits, Flows and Spaces,* Sage, London.

Hughes, A. (2000) Retailers, knowledges and changing commodity networks: the case of the cut flower trade, *Geoforum,* **31**, 175–90.

Hughes, A. (2005) Corporate strategy and the management of ethical trade: the case of the UK food and clothing retailers, *Environment and Planning A,* **37**, 7, 1145–63.

Hughes A. (2007) Supermarkets and the ethical/fair trade movement: making space for alternatives in mainstream economies?, in Burch, D. and Lawrence, G. (ed.) *Supermarkets and Agri-food Supply Chains,* Edward Elgar, Cheltenham.

Hughes, A. and Reimer, S. (eds) (2004) *Geographies of Commodity Chains,* Routledge, London.

Hugo, G. (2007) Population geography, *Progress in Human Geography,* **31**(1), 77–88.

Hulme, M. (2004) A change in the weather? Coming to terms with climate change, in Harris, F (ed.) *Global Environmental Issues,* John Wiley and Sons, Chichester, 21–44.

Huntington, S.P. (1993) The clash of civilizations? *Foreign Affairs,* **72**, 22–49.

Hutnyk, J. (1997) Adorno at Womad: South Asian crossovers and the limits of hybridity-talk, in Werbner, P. and Modood, T. (eds) *Debating Cultural Hybridity: Multi-cultural Identities and the Politics of Anti-racism,* Zed, London, 106–36.

Hutton, D. and Connors, L. (1999) *A History of the Australian Environment Movement,* Cambridge University Press, Melbourne.

Hyde, F.E. (1973), *Far Eastern Trade: 1860–1914,* Adam and Charles Black, London.

Hyndman, J. (2003) Beyond either/or: a feminist analysis of September 11th, *ACME: An International E-Journal for Critical Geographies,* **2**, 1–13, www.acme-journal.org

I

IBT *see* International Broadcasting Trust

IEA (International Energy Agency) (2006) *World Energy Outlook 2006*, OECD/IEA, Paris.

Ilbery, B. and Kneafsey, M. (1998) Product and place: promoting quality products and services in the lagging rural regions of the European Union, *European Urban and Regional Studies*, **5**, 329–41.

Ilbery, B. and Kneafsey, M. (2000a) Producer constructions of quality in regional speciality food production: a case study from south west England, *Journal of Rural Studies*, **16**, 217–30.

Ilbery, B. and Kneafsey, M. (2000b) Registering regional speciality food and drink products in the United Kingdom: the case of PDOs and PGIs, *Area*, **32**, 317–25.

Ilbery, B. and Maye, D. (2005) Alternative (shorter) food supply chains and specialist livestock products in the Scottish–English borders, *Environment and Planning A*, **37**, 823–44.

Ilbery, B., Morris, C., Buller, H., Maye, D. and Kneafsey, M. (2005) Product, process and place: an examination of food marketing and labelling schemes in Europe and North America, *European Urban and Regional Studies*, **20**, 331–44.

Illeris, S. (2007) The nature of services, in Bryson, J.R. and Daniels, P.W. (eds) *The Handbook of Service Industries*, Edward Elgar, Cheltenham.

Ingilby (2002) Are animals gay? http://www.funtrivia.com/ubbthreads/showthreaded.php?Cat=0&Number=34605&page=6, accessed 11/01.

Ingram, A. (2001) Alexander Dugin: geopolitics and neofascism in post-Soviet Russia, *Political Geography*, **20**(8), 1029–51.

International Broadcasting Trust (IBT) (1998) *The Bank, the President and the Pearl of Africa*, IBT/Oxfam, Oxford.

International Communications Union (2006) http://www.internetworldstats.com/stats.htm, accessed 3 January 2007.

International Council for Local Environmental Initiatives (2002) *Second Local Agenda 21 Survey*. Published as UN Dept of Economic and Social Affairs, Background Paper 15 for the 2nd Preparatory Conference for the World Summit on Sustainable Development, DESA/DSD/PC2/BP15, 29pp. Available at: www.iclei.org/rioplusten/finalrdocument.pdf

IPCC (2007) Working Group III Contribution to the Intergovernmental Panel on Climate Change Fourth Assessment Report, *Climate Change 2007: Mitigation of Climate Change*, IPCC Secretariat, Geneva.

Irwin, W. (1996) *The New Niagara: Tourism, Technology, and the Landscape of Niagara Falls, 1776–1917*, Pennsylvania State University, Pennsylvania.

Iyer, P. (1989) *Video Nights in Kathmandu: Reports from the Not-so-far East*, Black Swan, London.

J

Jackson, P. (1999) Commodity culture, the traffic in things, *Transactions of the Institute of British Geographers*, NS, **24**, 95–108.

Jackson, P. (2004) Local consumption cultures in a globalizing world, *Transactions of the Institute of British Geographers NS*, **29**, 165–78.

Jackson, P. and Taylor, J. (1996) Geography and the cultural politics of advertising, *Progress in Human Geography*, **19**, 356–71.

Jackson, P., Dwyer, C. and Thomas, N. (2007) Consuming transnational fashion in London and Mumbai, *Geoforum*, **38**(5), 908–24.

Jackson, P., Ward, N. and Russell, P. (2006) Mobilising the commodity chain concept in the politics of food and farming, *Journal of Rural Studies*, **22**, 129–41.

Jackson, R.H. (1990) *Quasi-States: Sovereignty, International Relations and the Third World*, Cambridge University Press, Cambridge.

Jacobs, J. (1961) *The Death and Life of Great American Cities*, Penguin, Harmondsworth.

Jacobs, J. (1984) *Cities and the Wealth of Nations*, Random House, New York.

Japan Institute for Social and Economic Affairs (1996) *Japan 1997: An International Comparison*, Keizai Koho Center, Tokyo.

Jeffrey, A. (2006) Building state capacity in post-conflict Bosnia and Herzegovina: the case of Brčko District, *Political Geography*, **25**(2), 203–27.

Joffé, G. (1987) Frontiers in North Africa, in Blake, G.H. and Schofield, R.N. (eds) *Boundaries and State Territory in the Middle East and North Africa*, MENAS Press, Berkhamsted.

John Muir Trust Flyer (2006) *Love the Wild? Help keep it Wild*, John Muir Trust flyer, 2006.

Johnson, N. (2003) *Ireland, the Great War and the Geography of Remembrance*, Cambridge University Press, Cambridge.

Johnston, R. (2007) Commentary: on duplicitous battleground conspiracies, *Transactions of the Institute of British Geographers*, **32**, 435–8.

Johnston, R.J. (1984) *Residential Segregation, the State and Constitutional Conflict in American Urban Areas*, Academic Press, London.

Johnston, R.J. and Sidaway J.D. (2004) *Geography and Geographers: Anglo-American Human Geography since 1945*, 6th edition, Arnold, London.

Johnston, R.J., Gregory, D., Pratt, G. and Watts, M. (eds) (2000) *The Dictionary of Human Geography*, 4th edition, Blackwell, Oxford.

Johnston, R.J., Gregory, D., Pratt, G. and Watts, M. (eds) (2008) *The Dictionary of Human Geography*, 5th edition, Blackwell, Oxford.

Jones, G. and Hollier, G. (1997) *Resources, Society and Environmental Management*, Paul Chapman, London.

Jones, H. (1990) *Population Geography*, 2nd edition, Paul Chapman, London.

Jones III, J.P., Woodward, K., and Marston, A.A. (2007) Reply: situating flatness, *Transactions of the Institute of British Geographers NS*, **32**, 264–76.

Jones, M., Jones, R. and Woods, M. (2004) *An Introduction to Political Geography: Space, Place and Politics*, Routledge, London.

Jones, P.S. (2000) Why is it alright to do development 'over there' but not 'here'? Changing vocabularies and common strategies across the 'First' and 'Third Worlds', *Area*, **32**, 237–41.

Jones, R. and Fowler, C. (2007) Placing and scaling the nation, *Environment and Planning D: Society and Space*, **25**, 332–54.

Journal of Comparative European Politics, Volume 4, Issue 2/3 (July/September 2006), Special Issue: Rethinking European Spaces: Territory, Borders, Governance.

Jurgens, U. and Gnad, M. (2002) Gated communities in South Africa: experiences from Johannesburg, *Environment and Planning B: Planning and Design*, **29**, 337–53.

K

Kabeer, N. (1994) *Reversed Realities: Gender Hierarchies in Development Thought*, Verso, London.

Kahn, J.S. (1995) *Culture, Multiculture, Postculture*, Sage, London.

Kaminsky, G. and Schmukler, S. (2002) Emerging markets instability: do sovereign ratings affect country risk and stock returns?, *World Bank Economic Review*, **16**, 171–95.

Kapferer, B. (1988) *Legends of People, Myths of State: Violence, Intolerance and Political Culture in Sri Lanka and Australia*, Smithsonian Institution Press, Washington, DC.

Kaplinsky, R. (2005) *Globalization, Poverty and Inequality*, Polity Press, Cambridge.

Kasson, J.F. (1978) *Amusing the Million: Coney Island at the Turn of the Century*, Hill & Wang, New York.

Kay, C. (2001) Reflections on rural violence in Latin America, *Third World Quarterly*, **22**, 5.

Kaye, H.J. (1984) *The British Marxist Historians*, Polity Press, Cambridge.

Kelly, P. (1984) *Fighting for Hope*, Chatto and Windus, London.

Kelly, P. (1997) *Checkerboards and Shatterbelts: The Geopolitics of South America*, University of Texas Press, Austin.

Kennedy, J.F. (1962) Address to Yale University, New Haven, Connecticut.

Kershaw, I. (2000) *The Nazi Dictatorship: Problems and Perspectives of Interpretation*, 4th edition, Arnold, London.

Kessler, J. and Appelbaum, R. (1998) The growing power of retailers in producer-driven commodity chains: a 'retail revolution' in the US automobile industry?, unpublished manuscript, Department of Sociology, University of California at Santa Barbara, USA.

Keyes, C.F. (1995) *The Golden Peninsula: Culture and Adaptation in Mainland Southeast Asia*, University of Hawaii Press, Honolulu.

Kinzer, S. (2006) *Overthrow: America's Century of Regime Change from Hawaii to Iraq*, Times Books, New York.

Kipling, R. (1912) *Rudyard Kipling's Verse: Definitive Edition*, Hodder & Stoughton, London.

Kirwan, J. (2004) Alternative strategies in the UK agro-food system: interrogating the alterity of farmers' markets, *Sociologia Ruralis*, **44**(4), 395–415.

Kirwan, J. (2006) The interpersonal world of direct marketing: examining conventions of quality at UK farmers' markets, *Journal of Rural Studies*, **22**, 301–12.

Klare, M. (2004) *Blood and Oil: How America's Thirst for Petrol is Killing Us*, London: Penguin.

Kleveman, L. (2003) *The New Great Game: Blood and Oil in Central Asia*, Atlantic Books, London.

Knights, D. and Odih, P. (1995) It's about time! The significance of gendered time for financial services consumption, *Time and Society*, **4**, 205–31.

Knopp, L. (1995) Sexuality and urban space: a framework for analysis, in Bell, D. and Knox, P. and Pinch, S. (2000) *Urban Social Geography: An Introduction*, 4th edition, Prentice Hall, Harlow.

Knox, P. and Agnew, J. (1994) *The Geography of the World Economy*, 2nd edition, Edward Arnold, London.

Knox, P. and Agnew, J. (2003) *The Geography of the World Economy*, 4th edition, Edward Arnold, London.

Knox, P. and Marston, S.A. (2001) *Human Geography: Places and Regions in Global Context*, 2nd edition, Prentice Hall, Upper Saddle River, NJ.

Knox, P. and Marston, S.A. (2004) *Human Geography: Places and Regions in Global Context*, 3rd edition, Prentice Hall, Upper Saddle River, NJ.

Knox, P. and Marston, S.A. (2006) *Human Geography: Places and Regions in Global Context*, 4th edition, Prentice Hall, Upper Saddle River, NJ.

Knox, P. and Pinch, S. (2000) *Urban Social Geography: An Introduction*, 4th edition, Prentice Hall, Harlow.

Knox, P., Agnew, J. and McCarthy, L. (2003) *The Geography of the World Economy: An Introduction to Economic Geography*, 4th edition, Hodder Arnold, London.

Kobayashi, A. and Peake, L. (1994) Unnatural discourse: 'race' and gender in geography, *Gender, Place and Culture*, **1**, 225–44.

Kobayashi, A. and Peake, L. (2000) Racism out of place: thoughts on whiteness and an antiracist geography in the new millennium, *Annals of the Association of American Geographers*, **90**, 392–403.

Koczberski, L. (1998) Women in development: a critical analysis, *Third World Quarterly*, **19**, 395–409.

Kondratieff, N.D. (1925) The major economic cycles, *Voprosy Konjunktury*, **6**, 28–79; an English translation is to be found in *Lloyds Bank Review* (1978), **129**.

Kong, L. (2007) The promises and prospects of geography in higher education, *Journal of Geography in Higher Education*, **31**(1), 13–7.

Korten, D.C. (1998) Your mortal enemy, *Guardian*, 21 October, 4–5.

Koter, M. (1990) The morphological evolution of a nineteenth-century city centre: Lódz, Poland, 1825–1973, in Slater, T.R. (ed.) *The Built Form of Western Cities*, Leicester University Press, Leicester and London.

Kothari, U. and Minogue, M. (eds) (2002) *Development Theory and Practice: Critical Perspectives*, Macmillan, London.

Kramsch, O. and Hooper, B. (2004) (eds) *Cross-Border Governance in the European Union*, Routledge, London.

Kuus, M. (2007) 'Love, peace and Nato': Imperial subject-making in Central Europe, *Antipode*, **39**(2), 269–90.

L

Laïdi, Z. (1988) Introduction: What use is the Soviet Union?, in Laïdi, Z. (ed.), *The Third World and the Soviet Union*, Zed Books, London, 1–23.

Landes, D.S. (1969) *The Unbound Prometheus*, Cambridge University Press, Cambridge.

Lang, H.J. (2002) *Fear and Sanctuary: Burmese Refugees in Thailand*, Southeast Asia Program Publications, Cornell University, New York.

Langley, P. (2003) *The Everyday Life of Global Finance*, International Political Economy Group Working Paper No. 5, University of Manchester, Manchester.

Langton, J. (1984) The industrial revolution and the regional geography of England, *Transactions of the Institute of British Geographers*, NS, **9**, 145–67.

Langton, J. and Morris, R.J. (eds) (1986) *Atlas of Industrializing Britain, 1780–1914*, Methuen, London and New York.

Langton, M. (1998) *Burning Questions: Emerging Environmental Issues for Indigenous Peoples in Northern Australia*, Centre for Indigenous Natural and Cultural Resource Management, Northern Territory University, Darwin.

Lash, S. and Urry, J. (1987) *The End of Organized Capitalism*, Polity Press, Cambridge.

Lash, S. and Urry, J. (1994) *Economies of Signs and Space*, Sage, London.

Latour, B. (1993) *We Have Never Been Modern*, Harvester Wheatsheaf, Brighton.

Laulajainen, R. (1998) *Financial Geography*, School of Economics and Commercial Law, Göteborg University.

Laurence, J. (1997) The poor of Britain are going hungry, *The Independent*, 11.6.97 (London), 7.

Laurie, N., Dwyer, C., Holloway, S. and Smith, F. (1999) *Geographies of New Femininities*, Longman, Harlow.

Lawton, R. (1989) Introduction: aspects of the development and role of great cities in the Western world in the nineteenth century, in Lawton, R. (ed.) *The Rise and Fall of Great Cities*, Belhaven, London.

Laxton, P. (1986) Textiles, in Langton, J. and Morris, R.J. (eds) *Atlas of Industrializing Britain, 1780–1914*, Methuen, London and New York, 106–13.

Le Billon, P. (2001) The political ecology of war: Natural resources and armed conflicts, *Political Geography*, **20**, 561–84.

Le Billon, P. (2007) Geographies of war: perspectives on 'resource wars', *Geography Compass*, **1**(2), 163–82.

Le Galès, P. (2002) *European Cities: Social Conflicts and Governance*, Oxford University Press, Oxford.

Le Heron, R.B. (1993) *Globalised Agriculture*, Pergamon Press, London.

Le Heron, R., Murphy, L., Forer, P. and Goldstone, M. (eds) (1999) *Explorations in Human Geography: Encountering Place*, Oxford University Press, Auckland.

Leach, E.R. (1960) The frontiers of 'Burma', *Comparative Studies in Society and History*, **3**, 49–68.

Lee, R. (2000) Radical and postmodern? Power, social relations and regimes of truth in the social construction of alternative economic geographies, *Environment and Planning A*, **32**, 991–1009.

Lee, R. (2006) The ordinary economy: tangled up in values and geography, *Transactions of the Institute of British Geographers*, **NS31**, 413–32.

Lee, R.B. (1968) What hunters do for a living, or how to make out on scarce resources, in Lee, R.B. and DeVore, I. (eds) *Man the Hunter*, Aldine Press, Chicago, 30–48.

Leicht, K.T. and Fennell, M.L. (1997) The changing organizational context of professional work, *Annual Review of Sociology*, **23**, 215–31.

Lemon, J.T. (1987) Colonial America in the eighteenth century, in Mitchell, R.D. and Groves, P.A. (eds) *North*

America: The Historical Geography of a Changing Continent, Hutchinson, London.

Lester, A. (1999) in Graham, B. and Nash, C. (eds) *Modern Historical Geographies*, Prentice Hall, Harlow.

Lester, A. (2002) Obtaining the 'due observance of justice': the geographies of colonial humanitarianism, *Environment and Planning D: Society and Space*, **20**, 277–93.

Levy, F. and Murnane, R.J. (2004) *The New Division of Labour: How Computers are Creating the Next Job Market*, Princeton University Press, Princeton.

Lewis, D., Rodgers, D. and Woolcock, M. (2005) *The Fiction of Development: Knowledge, Authority and Representation*, London School of Economics, Working Paper 05-61, Development Studies Institute, LSE, London.

Lewis, G. (1990) Community through exclusion and illusion: the creating of social worlds in an American shopping mall, *Journal of Popular Culture*, **24**, 121–36.

Lewis, M. (1990) *Liar's Poker: Rising Through the Wreckage on Wall Street*, Penguin Books, London.

Lewontin, R. (1993) *Biology as ideology: the doctrine of DNA*, Harper Perennial, New York.

Ley, D. and Cybriwsky, R. (1974) Urban graffiti as territorial markers, *Annals of the Association of American Geographers*, **64**(4), 491–505.

Leyshon, A. (1995a) Geographies of money and finance, *Progress in Human Geography*, **19**, 531–43.

Leyshon, A. (1995b) Annihilating space? The speed up of communications, in Allen, J. and Hamnett, C. (eds) *A Shrinking World?*, Oxford University Press, Oxford.

Leyshon, A. (1996) Dissolving difference? Money, disembedding and the creation of 'global financial space', in Daniels, P.W. and Lever, W.F. (eds) *The Global Economy in Transition*, Longman, London.

Leyshon, A. and Thrift, N. (1994) Access to financial services and financial infrastructural withdrawal: problems and policies, *Area*, **26**, 268–75.

Leyshon, A. and Thrift, N. (1997) *Money Space: Geographies of Monetary Transformation*, Routledge, London.

Liebes, T. and Katz, E. (1993) *The Export of Meaning: Cross-Cultural Readings of Dallas*, 2nd edition, Polity Press, Cambridge.

Lintner, B. (2002) *Blood Brothers. Crime, Business and Politics in Asia*, Allen and Unwin, Crows Nest NSW.

Little, J. and Leyshon, M. (2003) Embodied rural geographies: developing research agendas, *Progress in Human Geography*, **27**(3), 257–72.

Livingstone, D. (1992) *The Geographical Tradition: Issues in the History of a Contested Enterprise*, Blackwell, Oxford.

Low, M. (2005) States, citizenship and collective action, in Daniels, P., Bradshaw, M., Shaw, D. and Sidaway, J. (eds) *An Introduction to Human Geography. Issues for the 21st Century*, 2nd edition, Prentice Hall, Harlow, 442–66.

Lowenthal, D. (1994) European and English landscapes as national symbols, in Hooson, D. (ed.) *Geography and National Identity*, Blackwell, Oxford, 15–38.

Lury, C. (2000) The united colors of diversity: essential and inessential culture, in Franklin, S., Lury, C. and Stacey, J. (eds) *Global Nature, Global Culture*, Sage, London, 146–87.

Luttwak, E. (1990) From geopolitics to geoeconomics, *National Interest*, **20**, 17–24.

M

Macdonald, C.L. and Sirianni, C. (1996) *Working in the Service Economy*, Temple University Press, Philadelphia.

Machel, G. (1996) *UN Study on the Impact of Armed Conflict on Children*, UNICEF, online, available at: www.unicef.org/graca/women/html

Mackinder, H. (1904) The geographical pivot of history, *Geographical Journal*, **23**, 421–44.

Mackinder, H. (1919) *Democratic Ideals and Reality: A Study in the Politics of Reconstruction*, Constable, London.

Mackinnon, D. and Cumbers, A. (2007) *Introduction to Economic Geography: Uneven Development, Globalization and Place*, Pearson, London.

Macnaghten, P. and Urry, J. (1998) *Contested Natures*, Sage, London.

Maffesoli, M. (1996) *The Time of the Tribes: The Decline of Individualism in Mass Society*, translated by Don Smith, Sage, London.

Mahajan, S. (ed.) (2005) *Input–Output Analysis: 2005*, Office for National Statistics, London.

Mahon, R. (1987) From Fordism to ?: new technology, labour markets and unions, *Economic and Industrial Democracy*, **8**, 5–60.

Maisels, C.K. (1993) *The Emergence of Civilization*, Routledge, London.

Maitland, B. (1985) *Shopping Malls, Planning and Design*, Construction Press, London.

Malbon, B. (1997) Clubbing, consumption, identity and the spatial practices of every-night life, in Skelton, T. and Valentine, G. (eds) *Cool Places*, Routledge, London, 266–86.

Malbon, B. (1999) *Clubbing*, Routledge, London.

Malpass, A., Cloke, P., Barnett, C. and Clarke, N. (2007) Fairtrade urbanism: the politics of place beyond place in the Bristol Fairtrade City campaign, *International Journal of Urban and Regional Research*, **31**(3), 239–49.

Mansfield, B. (2005) Beyond rescaling: reintegrating the 'national' as a dimension of scalar relations, *Progress in Human Geography*, **29**(4), 458–73.

Marchand, M.H. and Parpart, J.L. (1995) *Feminism, Postmodernism, Development*, Routledge, London.

Marcuse, H. (1964) *One-Dimensional Man*, Abacus, London.

Marcuse, P. (2000) The language of globalization, *Monthly Review*, **52**(1–4), available at: www.monthlyreview.org/700marc.html

Marquez, P. (2005) *Dying Too Young: Addressing Premature Mortality and Ill Health Due to Non-Communicable Diseases and Injuries in the Russian Federation*, The World Bank, Europe and Central Asia Human Development Department.

Marsden, T., Banks, J. and Bristow, G. (2000) Food supply chain approaches: exploring their role in rural development, *Sociologia Ruralis*, **40**, 424–38.

Marsh, D.C. (1977) *The Changing Structure of England and Wales: 1871–1961*, Routledge and Kegan Paul, London.

Marsh, G.P. (1869) *Man and Nature; or, Physical Geography as Modified by Human Action*. C. Scribner & Co., New York.

Marston, S. (2000) The social construction of scale, *Progress in Human Geography*, **24**, 219–42.

Marston, S.A., Jones III, J.P. and Woodward, K. (2005) Human geography without scale, *Transactions of the Institute of British Geographers*, **NS 30**, 416–32.

Martin, A. (2007) Trouble in China is good news for American Toy Manufacturers, *New York Times*, 15 August.

Martin, R.L. (1994) Stateless monies, global financial integration and national economic autonomy: the end of geography?, in Corbridge, S., Martin, R. and Thrift, N. (eds) *Money, Power and Space*, Blackwell, Oxford, 253–78.

Martin, R.L. (1999a) The new 'geographical turn' in economics: some critical reflections, *Cambridge Journal of Economics*, **23**, 65–91.

Martin, R.L. (1999b) The new economic geography of money, in Martin, R.L. (ed.) *Money and the Space Economy*, Wiley, Chichester, 3–27.

Martin, R.L. (ed.) (1999c) *Money and the Space Economy*, Wiley, Chichester.

Martíñez, M. (1997) *Olivenza y el tratado de Alcanices*, Ayuntamiento de Olivenza.

Marx, K. (1867) *Das Kapital*; translated into English as *Capital*, 3 vols, International Publishers, New York (1967) and Penguin, London (1976).

Marx, K. (1973) *Grundrisse: Foundations of the Critique of Political Economy*, Penguin, London.

Mason, C. and R.T. Harrison (2002) The geography of venture capital investments in the UK, *Transactions of the Institute of British Geographers*, **27**, 427–51.

Massey, D. (1981) The geography of industrial change, in Potter, D., Anderson, J., Clarke, J., Coombes, P., Hall, S., Harris, L., Holloway, C. and Walton, T. (eds) *Society and the Social Sciences*, Open University Press, Milton Keynes, 302–13.

Massey, D. (1984) *Spatial Division of Labour: Social Structures and the Geography of Production*, Macmillan, London.

Massey, D. (1993) Power-geometry and a progressive sense of place, in Bird, J., Curtis, B., Putnam, T., Robertson, G. and Tickner, L. (eds) *Mapping the Futures: Local Cultures, Global Change*, Routledge, London, 59–69.

Massey, D. (1994) *Space, Place and Gender*, Polity, Cambridge.

Massey, D. (1996) Space/power, identity/difference, tensions in the city, in Merrifield, A. and Swyngedouw, E. (eds) *The Urbanisation of Injustice*, Blackwell, London, 190–205.

Massey, D. (1999) Cities in the world, in Massey, D., Allen, J. and Pile, S. (eds) *City Worlds*, Routledge in association with the Open University, London, 99–175.

Massey, D. and Jess, P. (eds) (1995a), *A Place in the World? Places, Cultures and Globalization*, Oxford University Press, Oxford.

Massey, D. and Jess, P. (1995b) The contestation of place, in Massey, D. and Jess, P. (eds) *A Place in the World? Places, Cultures and Globalization*, Oxford University Press, Oxford, 133–74.

Massey, D., Allen, J. and Pile, S. (eds) (1999) *City Worlds*, Routledge, London.

Mather, A.S. and Chapman, K. (1995) *Environmental Resources*, Longman, London.

Mattelart, A. (1979) *Multinational Corporations and the Control of Culture*, Harvester Press, Brighton.

Maxwell, S. (2002) More aid? – Yes – and use it to reshape aid architecture, ODI opinions (3), available at: www.odi.org.uk/opinions/

Maye, D. and Ilbery, B. (2006) Regional economies of local food production: tracing food chain links between 'specialist' producers and intermediaries in the Scottish–English borders, *European Urban and Regional Studies*, **13**, 337–54.

Maye, D., Holloway, L. and Kneafsey, M. (eds) (2007) *Alternative Food Geographies: Representation and Practice*, Elsevier, Oxford.

McCarthy, C., Rodriguez, A., Buendia, E., Meacham, S., David, S., Godina, H., Supriya, K. and Wilson-Brown, C. (1997) Danger in the safety zone: notes on race, resentment and the discourse of crime violence and suburban security, *Cultural Studies*, **11**, 274–95.

McCarthy, J. (2005) Rural geography: multifunctional rural geographies: reactionary or radical? *Progress in Human Geography*, **29**, 773–82.

McCarthy, J. (2006) Rural geography: alternative rural economies: the search for alterity in forests, fisheries, food, and fair trade, *Progress in Human Geography*, **30**(6), 803–11.

McCarthy, S. (1999) The fabulous kingdom of gay animals, available at http://www.salon.com/it/feature/1999/03/cov_15featurea.html, accessed 18/12/06.

McClintock, A. (1994) *Imperial Leather: Race, Gender and Sexuality in the Colonial Context*, Routledge, London.

McCormick, J. (1989) *The Global Environmental Movement*, Belhaven Press, London.

McCormick, J. (1991) *British Politics and the Environment*, Earthscan, London.

McCormick, J. (2006) *The European Superpower*, Palgrave Macmillan, London.

McDowell, L. (1994) The transformation of cultural geography, in Gregory, D., Martin, R. and Smith, G. (eds) *Human Geography, Society, Space and Social Science*, Macmillan, London, 146–73.

McDowell, L. (1997) *Capital Culture: Gender at Work in the City*, Blackwell, Oxford.

McDowell, L. (1999) *Gender, Identity and Place: Understanding Feminist Geographies*, Polity Press, Cambridge.

McDowell, L. (2007) Gender divisions of labour: sex, gender, sexuality and embodiment in the service sector, in Bryson, J.R. and Daniels, P.W. (eds) *The Handbook of Services*, Edward Elgar, Cheltenham, 395–408.

McDowell, L. and Sharp, J.P. (eds) (1997) *Space, Gender and Knowledge: Feminist Readings*, Arnold, London.

McGreal, C. (2002) Thabo Mbeki's catastrophe, *Prospect*, **72**, available at: www.prospect-magazine.co.uk/article_details.php?id=4995

McGuinness, M. (2000) Geography matters? Whiteness and contemporary geography, *Area*, **32**, 225–30.

McKibben, B. (1999) *The End of Nature*, Random House, New York.

McLaughlin, E. and Muncie, J. (eds) (1999) Walled cities: surveillance, regulation and segregation, in Pile, S., Brook, C. and Mooney, G. (eds) *Unruly Cities? Order/Disorder*, Routledge, London.

McLeod, G. and Ward, K. (2002) Spaces of utopia and dystopia: landscaping the contemporary city, *Geografiska Annaler*, **84B**(3–4), 153–70.

McLuhan, M. (1964) *Understanding Media: Extensions of Man*, Routledge & Kegan Paul, London.

McLuhan, M. (1975) The implications of cultural uniformity, in Bigsby, C.W.E. (ed.) *Superculture, American Popular Culture and Europe*, Bowling Green University Popular Press, Bowling Green, OH, 43–56.

McNeill, D. (2004) *New Europe: Imagined Spaces*, Arnold, London.

Meadows, D.H., Meadows, D.I., Randers, J. and Behrens III, W.W. (1972) *The Limits to Growth: A Report to the Club of Rome*, Universe Books, New York.

Megoran, N. (2006) For ethnography in political geography: experiencing and re-imagining Ferghana Valley boundary closures, *Political Geography*, **26**(10), 622–40.

Meinig, D.W. (1986) *The Shaping of America. A Geographical Perspective on 500 Years of History: Atlantic America, 1492–1800*, Yale University Press, New Haven and London.

Merchant, C. (1980) *The Death of Nature: Women, Ecology and the Scientific Revolution*, Harper Collins, San Francisco.

Meyer, D.R. (1990) The new industrial order, in Conzen, M.P. (ed.) *The Making of the American Landscape*, Unwin Hyman, London and Winchester, MA.

Middleton, N. (2003) *The Global Casino: An Introduction to Environmental Issues*, 3rd edition, Arnold, London.

Miles, D. (2006) *The Tribes of Britain*, Phoenix, London.

Miller, D. (1992) The young and the restless in Trinidad: a case study of the local and the global in mass consumption, in Silverstone, R. and Hirsch, E. (eds) *Consuming Technologies: Media and Information in Domestic Spaces*, Routledge, London, 163–82.

Miller, D. (1994) *Modernity, an Ethnographic Approach: Dualism and Mass Consumption in Trinidad*, Berg, Oxford.

Miller, D. (1995) Consumption as the vanguard of history, a polemic by way of an introduction, in Miller, D. (ed.) *Acknowledging Consumption: A Review of New Studies*, Routledge, London, 1–57.

Miller, D. (1997) Coca-Cola, a black sweet drink from Trinidad, in Miller, D. (ed.) *Material Cultures*, UCL Press, London.

Miller, D. (1998) *A Theory of Shopping*, Polity, Cambridge.

Miller, D. (2001) The poverty of morality, *Journal of Consumer Culture*, **1**(2), 225–43.

Millington, A.C., Mutiso, S.K., Kirkby, J. and O'Keefe, P. (1989) African soil erosion: nature undone and the limitations of technology, *Land Degradation and Rehabilitation*, **1**, 279–90.

Millstone, E. and Lang, T. (2003) *The Atlas of Food*, Earthscan, London.

Milner, M. (1998) IMF puts UK jobs at risk, *Guardian*, 17 July, 23.

Ministry of Defence (MOD) (2006) *Defense Spending*, copy available from: http://www.mod.uk/DefenceInternet/AboutDefence/Organisation/KeyFactsAboutDefence/DefenceSpending.htm, accessed 23 May 2007.

Mintel (2004) *Nightclubs*, Mintel Market Intelligence, London.

Mintel (2006) *Nightclubs*, Mintel Market Intelligence, London.

Mitchell, D. (1996) Public space and the city, *Urban Geography*, **17**, 127–31.

Mitchell, D. (2003) *The Right to the City: Social Justice and the Fight for Public Space*, Guilford, New York.

Mitchell, J.B. (ed.) (1962) *Great Britain: Geographical Essays*, Cambridge University Press, Cambridge.

Mitchell, R.D. and Groves, P.A. (eds) (1987) *North America: The Historical Geography of a Changing Continent*, Hutchinson, London.

Mitchell, W.C. (1937) *The Backward Art of Spending Money and Other Essays*, McGraw-Hill, New York.

MOD *see* Ministry of Defence

Mohan, J. (2000) Geographies of welfare and social exclusion, *Progress in Human Geography*, **24**, 291–300.

Mohanty, C.T. (1991) Under Western eyes: feminist scholarship and colonial discourses, in Mohanty, C.T., Russo, A. and Torres, L. (eds) *Third World Women and the Politics of Feminism*, Indiana University Press, Bloomington, 51–80.

Momsen, J. (1991) *Women and Development in the Third World*, Routledge, London.

Moore, R.I. (ed.) (1981) *The Hamlyn Historical Atlas*, Hamlyn, London and New York.

Moran, D. (2004). Exile in the Soviet forest: 'special settlers' in northern Perm Oblast, *Journal of Historical Geography*, **30**, 395–413.

Morley, D. (2000) *Home Territories: Media, Mobility and Identity*, Routledge, London.

Morris-Suzuki, T. (1996) The frontiers of Japanese identity, in Tønnesson, S. and Antlöv, H. (eds) *Asian Forms of the Nation*, Curzon, Richmond, 41–66.

Moseley, W., Lanegran, D. and Pandit, K. (eds) (2007) *The Introductory Reader in Human Geography*, Blackwell, Oxford.

Moser, C. (1993) *Gender, Planning and Development: Theory, Practice and Training*, Routledge, London.

Mosse, G.L. (1975) *The Nationalization of the Masses*, H. Fertig, New York.

Mountjoy, A.B. (1976) Worlds without end, *Third World Quarterly*, **2**, 753–7.

Munslow, B. and Ekoko, F. (1995) Is democracy necessary for sustainable development?, *Democratisation*, **2**, 158–78.

Murdoch, J. and Miele, M. (2004) Culinary networks and cultural connections: a conventions approach, in Amin, A. and Thrift, N. (eds) *Cultural Economy Reader*, Blackwell, London, 231–48.

Murray, W.E. (2001) The second wave of globalisation and agrarian change in the Pacific Islands, *Journal of Rural Studies*, **17**(2), 135–48.

Murray, W.E. (2005) *Geographies of Globalization*, Routledge/Taylor and Francis, London and New York.

Murray, W.E. (2006) Neo-feudalism in Latin America? Globalisation, agribusiness, and land re-concentration in Chile, *Journal of Peasant Studies*, **33**, 646–77.

Murray, W.E. and Terry, J.P. (2004) Niue's place in the Pacific, in Terry, J.P. and Murray, W.E. (eds), *Niue Island: Geography on the Rock of Polynesia*, INSULA, UNESCO, Paris.

Mutersbaugh, T. (2005) Fighting standards with standards: harmonization, rents, and social accountability in certified agrofood networks, *Environment and Planning A*, **37**, 2033–51.

N

Naess, A. (1973) The shallow and the deep: long-range ecological movement, *Inquiry*, **16**, 95–100.

Nagel, C. (2002) Geopolitics by another name: immigration and the politics of assimilation, *Political Geography*, **21**, 971–87.

National Research Council (2000) *Beyond Six Billion: Forecasting the World's Population*, National Academy Press, Washington, DC.

National Statistics Online (NSO) (2004) *Family Expenditure Survey HMSO*, London.

Natter, W. (2003) Geopolitics in Germany 1919–1945, in Agnew, J., Mitchell, K. and Toal, G. (eds) *A Companion to Political Geography*, Blackwell, Oxford, 197–203.

Nayek, A. (2003) Last of the real Geordies? White masculinities and the subcultural response to deindustrialization, *Environment and Planning D: Society and Space*, **21**, 7–25.

Nelson, L. and Seager, J. (2004) *A Companion to Feminist Geography*, Blackwell, Oxford.

New Internationalist (1999) *The A–Z of World Development*, New Internationalist, London.

Newman, O. (1972) *Defensible Space: People and Design in the Violent City*, Architectural Press, London.

Nitkin, A. (2007) The end of 'post-Soviet space': the changing geopolitical orientations of the newly independent states, *Russia and Eurasia Briefing Paper 07/01*, Chatam House, London.

Nixon, S. (1997) Circulating culture, in du Gay P (ed.) *Production of Culture/Cultures of Production*, Sage, London, 177–234.

Nordstrom, C. (2000) Shadows and sovereigns, *Theory, Culture and Society*, **17**, 35–54.

Nordstrom, C. (2004) *Shadows of War: Violence, Power, and International Profiteering in the Twenty-first Century*, University of California Press, Berkeley.

Nostrand, R.L. (1987) The Spanish borderlands, in Mitchell, R.D. and Groves, P.A. (eds) *North America: The Historical Geography of a Changing Continent*, Hutchinson, London.

Nove, A. (1987) *The Soviet Economic System*, 3rd edition, Allen & Unwin, London.

O

Oakeshott, I. and Gourlay, C. (2006) Science told: hands off gay sheep, *Sunday Times*, London.

O'Brien, R. (1992) *Global Financial Integration: The End of Geography*, Royal Institute of International Affairs, Pinter Publishers, London.

O'Connor, A. (1976) Third World or one world, *Area*, **8**, 269–71.

Odell, P.R. (1989) Draining the world of energy, in Johnston, R.J and Taylor, P.J. (eds) *A World in Crisis?*, 2nd edition, Blackwell, Oxford.

OECD *see* Organization for Economic Co-operation Development

Ogborn, M. (1999) Historical geographies of globalisation, *c.*1500–1800, in Graham, B. and Nash, C. (eds) *Modern Historical Geographies*, Prentice Hall, Harlow.

Ohmae, K. (1990) *The Borderless World: Power and Strategy in the Interlinked Economy*, Collins, London.

Ohnuki-Tierney, E. (1997) McDonalds in Japan: changing manners and etiquette, in Watson, J.L. (ed.) *Golden Arches East: McDonalds in East Asia*, Stanford University Press, Stanford, CA, 161–82.

O'Loughlin, J. (ed.) (1994) *Dictionary of Geopolitics*, Greenwood Press, Westport, CT.

O'Loughlin, J. and van der Wusten, H. (1990) Political geography and panregions, *Geographical Review*, **80**, 1–20.

O'Loughlin, J., Toal, G. and Kolossov, V. (2005) Russian geopolitical culture and public opinion: the masks of Porteus revisited, *Transactions of the Institute of British Geographers*, **30**(3), 322–35.

Ong, A. (1999) *Flexible Citizenship: The Cultural Logics of Transnationality*, Duke University Press, Durham, USA.

O'Reilly, K. and Crutcher, M.E. (2006) Parallel politics: the spatial power of New Orleans Labor Day parades, *Social and Cultural Geography*, **7**(2), 245–65.

Organization for Economic Co-operation and Development (1999) *Environment in the Transition to a Market Economy*, OECD, Paris.

Organization for Economic Co-operation and Development (2005a) *OECD in Figures 2005*, OECD, Paris.

Organization for Economic Co-operation and Development (2005b) *Potential Offshoring of ICT-Intensive Using Occupations, Working Party on the Information Economy*, DSTA/ICC/IE(2004)19, OECD, Paris.

Organization for Economic Co-operation and Development (2006) *Key World Energy Statistics*, OECD/IEA, available at: www.iea.org

O'Riain, S. (2004) The politics of mobility in technology-driven commodity chains: developmental coalitions in the Irish software industry, *International Journal of Urban and Regional Research*, **28**, 642–63.

O'Riordan, T. (ed.) (2000) *Environmental Science for Environmental Management*, 3rd edition, Prentice Hall, Harlow.

Østergaard, L. (1992) *Gender and Development: A Practical Guide*, Routledge, London.

Oswalt, P. (ed.) (2004) *Schrumpfende Städte. Band 1: Internationale Untersuchung*, Hantje Cantz, Ostfildern-Ruit.

Ó Tuathail, G. (1996a) *Critical Geopolitics: The Politics of Writing Global Space*, Routledge, London.

Ó Tuathail, G. (1996b) An anti-geopolitical eye: Maggie O'Kane in Bosnia, 1992–3, *Gender, Place and Culture*, **3**(2), 177–85.

Ó Tuathail, G. (1998) Postmodern geopolitics? The modern geopolitical imagination and beyond, in Ó Tuathail, G. and Dalby, S. (eds) *Rethinking Geopolitics*, Routledge, London, 16–38.

Ó Tuathail, G. and Dalby, S. (eds) (1998) *Rethinking Geopolitics*, Routledge, London.

Ó Tuathail, G., Dalby, S. and Routledge, P. (eds) (2006) *The Geopolitics Reader*, 2nd edition, Routledge, London.

P

Paasi, A. (1996) *Territories, Boundaries and Consciousness: The Changing Geographies of the Finnish–Russian Frontier*, Wiley, Chichester.

Paasi, A. (2003) Territory, in Agnew, J., Mitchell, K. and Toal, G. (eds) *A Companion to Political Geography*, Blackwell, Malden, 109–22.

Packard, V. (1977) *The Hidden Persuaders*, Penguin, Harmondsworth.

Pahl, J. (1989) *Money and Marriage*, Macmillan, London.

Pain, R.H. (1997) Social geographies of women's fear of crime, *Transactions of the Institute of British Geographers* NS, **22**, 231–44.

Painter, J. (2006) Prosaic geographies of stateness, *Political Geography*, **25**, 752–74.

Pallot, J. (2005) Russia's penal peripheries: space, place and penalty in Soviet and post-Soviet Russia. *Transactions of the Institute of British Geographers*, **30**(1), 98–112.

Palmer, J.A. (ed.) (2001) *Fifty Key Thinkers on the Environment*, London, Routledge.

Panelli, R. (2004) *Social Geographies: From Difference to Action*, Sage, London.

Pantazis, C., Gordon, D. and Townsend, P. (1999) *The Necessities of Life in Britain*, Working Paper, Poverty and Social Exclusion Survey, University of Bristol.

Park, C. (1994) *Sacred Worlds: An Introduction to Geography and Religion*, Routledge, London.

Parker, A. (2004) *Two-Speed Europe: Why 1 Million Jobs will Move Offshore*, Forrester Research.

Parker, J. (2006) Cold War II: the Eisenhower Administration, the Bandung Conference, and the reperiodization of the postwar era, *Diplomatic History*, **30**(5), 867–92.

Parrott, N., Wilson, N. and Murdoch, J. (2002) Spatializing quality: regional protection and the alternative geography of food, *European Urban and Regional Studies*, **9**, 241–61.

Paterson, T.G. (1992) *On Every Front. The Making and Unmaking of the Cold War*, revised edition, W.W. Norton, New York and London.

Peach, W.N. and Constantin, J.A. (1972) *Zimmerman's World Resources and Industries*, 3rd edition, Harper & Row, London.

Pearce, F. (2003) Saving the world, plan B, *The New Scientist*, 13 December, 6–7.

Peck, J.A. and Yeung, H.W. (eds) (2003) *Remaking the Global Economy*, Sage, London.

Peet, R. (1989) The destruction of regional cultures, in Johnston, R.J. and Taylor, P.J. (eds) *A World in Crisis?* 2nd edition, Blackwell, Oxford, 150–72.

Peet, R. with Hartwick, E. (1999) *Theories of Development*, Guilford Press, London.

Pelkmans, M. (2006) *Defending the Border: Identity, Religion and Modernity in the Republic of Georgia*, Cornell University Press, Ithaca and London.

Peluso, N.L. and Watts, M. (eds) (2001) *Violent Environments*, Cornell University Press, Ithaca.

Penrose, J. (2002) Nations, states and homelands: territory and territoriality in nationalist thought, *Nations and Nationalism*, **8**, 277–97.

Pepper, D. (1984) *The Roots of Modern Environmentalism*, Croom Helm, Beckenham.

Perry, S. and Schneck, C. (eds) (2001) *Eye-to-eye: Women Practicing Development Across Cultures*, Zed Books, London.

Peterson Del Mar, D. (2006) *Environmentalism*, Longman, New York.

Petrini, C. (1986) The Slow Food manifesto, *Slow*, **1**, 23–4.

Philip's Great World Atlas (2003) London, George Philip in association with the RGS with the IBG.

Phillips, M. (2002) Distant bodies? Rural studies, political economy and poststructuralism, *Sociologia Ruralis*, **42**(2), 81–105.

Phyne, J. and Mansilla, J. (2003) Forging linkages in the commodity chain: the case of the Chilean salmon farming industry, 1987–2001, *Sociologia Ruralis*, **43**(2), 108–27.

Pickerill, J. (2008) From wilderness to WildCountry: the power of language in environmental campaigns in Australia, *Environmental Politics*, **17**(1), 93–102.

Pickles, J. (2004) A *History of Spaces: Cartographic Reason, Mapping and the Geo-coded World*, Routledge, London and New York.

Pieterse, J.N. and Parekh, B. (eds) (1995) *Decolonization of the Imagination: Culture, Knowledge and Power*, St Martins Press, London.

Pile, S. (1999) The heterogeneity of cities, in Pile, S., Brook, C. and Mooney, G. (eds) *Unruly Cities? Order/Disorder*, Routledge, London, 7–52.

Pinch, S. (1994) Social polarization: a comparison of evidence from Britain and the United States, *Environment and Planning A*, **25**, 779–95.

Pinch, S. and Henry, N. (1999) Paul Krugman's geographical economics, industrial clustering and the British motor sport industry, *Regional Studies*, **33**, 815–27.

Pletsch, C. (1981) The three worlds or the division of social scientific labour 1950–1975, *Comparative Studies in Society and History*, **23**, 565–90.

Poguntke, T. (1993) Goodbye to movement politics? Organisational adaption of the German Green Party, *Environmental Politics*, **2**, 379–404.

Policy Commission on the Future of Farming and Food (2002) *Farming and Food: A Sustainable Future*, HMSO, London.

Political Geography (1998) Space, Place and Politics in Northern Ireland, special issue of *Political Geography*, **17**(2).

Pollard, J.S. (1996) Banking at the margins: a geography of financial exclusion in Los Angeles, *Environment and Planning A*, **28**, 1209–32.

Pollard, J.S. and Sidaway J.D. (2002) Euroland: economic, political and cultural geographies, *Transactions of the Institute of British Geographers*, **27**, 7–10.

Pollard J.S., Oldfield, J., Randalls, S.C. and Thornes, J.E. (2008) Firm finances, weather derivatives and geography, *Geoforum*, **39**(2), 616–24.

Ponte, S. (2002) The 'latte revolution'? Regulation, markets and consumption in the global coffee chain, *World Development*, **30**, 1099–122.

Popke, E.J. (2001) Modernity's abject space: the rise and fall of Durban's Cato Manor, *Environment and Planning A*, **33**, 737–52.

Population Division of the Department of Economic and Social Affairs of the United Nations Secretariat (2007) *World Population Prospects: The 2006 Revision. Highlights*, United Nations, New York.

Population Reference Bureau (2007) *Human Population: Fundamentals of Growth: Future Growth*, Available at: http://www.prb.org/Educators/TeachersGuides/HumanPopulation/FutureGrowth/TeachersGuide.aspx?p=1

Porritt, J. (1984) *Seeing Green*, Blackwell, Oxford.

Porritt, J. and Winner, D. (1988) *The Coming of the Greens*, Fontana, London.

Porter, M.E. (1990) *The Competitive Advantage of Nations*, Free Press, New York.

Porter, P.W. and Sheppard, E.S. (1998) *A World of Difference: Society, Nature, Development*, Guilford Press, New York.

Potter, R.B. (1979) Perception of urban retailing facilities: an analysis of consumer information fields, *Geografiska Annaler*, **61B**, 19–27.

Potter, R., Binns, T., Elliott, J.A. and Smith, D. (1999) *Geographies of Development*, Longman, London.

Potter, R.B., Binns, T., Elliot, J.A. and Smith, D. (2004) *Geographies of Development*, 2nd edition, Prentice Hall, Harlow.

Pounds, N.J.F. (1947) *An Historical and Political Geography of Europe*, Harrap, London.

Pounds, N.J.G. (1990) *An Historical Geography of Europe*, Cambridge University Press, Cambridge.

Power, A. and Tunstall, R. (1997) *Dangerous Disorder: Riots and Violent Disturbances in Thirteen Areas of Britain, 1991–1992*, Joseph Rowntree Foundation, York.

Power, M. (2003) *Rethinking Development Geographies*, Routledge, London.

Power, M. (2007) Digitized virtuosity: video war games and post 9/11 cyber-detterence, *Security Dialogue*, **38** (2), 271–88.

Pratt, G. (2005) Masculinity–feminity, in Cloke, P., Crang, P. and Goodwin, M. (eds) *Introducing Human Geographies*, 2nd edition, Hodder Arnold, London, 91–103.

Pretty, J.N., Ball, A.S., Lang, T. and Morison, J.I.L. (2005) Farm costs and food miles: an assessment of the full cost of the UK weekly food basket, *Food Policy*, **20**, 1–19.

Priestley, J.B. (1937) *English Journey*, William Heinemann, London.

Progress in Human Geography **24**(1) – Classics in Human Geography revisited, 91–9.

R

Raban, J. (1990) *Hunting Mr Heartache*, Collins Harvill, London.

Radcliffe, S. and Westwood, S. (1996) *Remaking the Nation: Place, Identity and Politics in Latin America*, Routledge, London.

Radcliffe, S.A. (1999) Re-thinking development, in Cloke, P., Crang, P. and Goodwin, M. (eds) *Introducing Human Geographies*, Arnold, London, 84–92.

Radcliffe, S.A. (2005) Re-thinking development in Cloke, P., Crang, P. and Goodwin, M. (eds) *Introducing Human Geographies*, 2nd edition, Arnold, London.

Radical Routes (2006) *How to Work Out Your Ecological Footprint*, available at: http://www.radicalroutes.org.uk/pub.html

Rahnema, R. (1997) Introduction, in Rahnema, R. and Bawtree, V. (eds) *The Post-Development Reader*, Zed Books, London.

Rathgeber, E. (1990) WID, WAD, GAD: trends in research and practice, *Journal of Developing Areas*, **24**, 489–502.

Ravenhall, Y. (1999) *Women's Investing*, Part 1, online, available at: www.fool.co.uk/personalfinance

Rawcliffe, P. (1998) *Environmental Pressure Groups in Transition*, Manchester University Press, Manchester.

Reddish, A. and Rand, M. (1996) The environmental effects of present energy policies, in Blunden, J. and Reddish, A. (eds) *Energy, Resources and Environment*, Open University and Hodder & Stoughton, London, 43–91.

Rees, J. (1985) *Natural Resources: Allocation, Economics and Policy*, 2nd edition, Routledge, London.

Rees, J. (1991) Resources and environment: scarcity and sustainability, in Bennett, R. and Estall, R. (eds) *Global Change and Challenge: Geography in the 1990s*, Routledge, London, 5–26.

Reeves, R. (1999) Do we really need gay florists?, *Observer*, 8 August.

Renard, M. (2005) Quality certification, regulation and power in fair trade, *Journal of Rural Studies*, **21**, 419–31.

Renner, M. (2002) The Anatomy of Resource Wars, Worldwatch Paper # 162, Earthwatch Institute, Washington, DC.

Renting, H., Marsden, T.K. and Banks, J. (2003) Understanding alternative food networks: exploring the role of short food supply chains in rural development, *Environment and Planning A*, **35**, 393–411.

Rist, G. (1997) *The History of Development*, Zed, London.

Ritzer, G. (2000) *The McDonaldisation of Society*, 2nd edition, Pine Forge Press, Thousand Oaks, CA.

Ritzer, G. (2004) *The Globalization of Nothing*, Pine Forge/Sage, London.

Robbins, P. (2004) *Political Ecology*, Blackwell, Oxford.

Robinson, G. (2004) *Geographies of Agriculture*, Pearson, London.

Robinson, G.M. and Pobric, A. (2006) Nationalism and identity in post-Dayton Accords: Bosnia-Hercegovina, *Tijdschrift voor Economische en Sociale Geografie*, **97**(3), 237–52.

Robinson, J. (1997) The geopolitics of South African cities: states, citizens, territory, *Political Geography*, **16**, 365–86.

Robinson, J. (1999) Divisive cities, in Pile, S., Brook, C. and Mooney, G. (eds) *Unruly Cities? Order/Disorder*, Routledge, London, 149–200.

Robinson, J. (2002) Global and world cities: a view from off the map, *International Journal of Urban and Regional Research*, **26**(3), 531–54.

Robinson, J. (2005a) *Ordinary Cities, Between Modernity and Development*, Routledge, London.

Robinson, J. (2005b) Urban geography: world cities, or a world of cities, *Progress in Human Geography*, **29**(6), 757–65.

Roche, M. (2002) Rural geography: searching rural geographies, *Progress in Human Geography*, **26**(6), 823–29.

Roche, M. (2005) Rural geography: a borderland revisited, *Progress in Human Geography*, **29**, 299–303.

Rogers, A. and Viles, H. (eds) (2003) *The Student's Companion to Geography*, 2nd edition, Blackwell, Oxford.

Rogers, R. (1999) *Towards an Urban Renaissance, Final Report of the Urban Task Force*, chaired by Lord Rogers of Riverside, E. and F.N. Spon, London.

Rogers, R. and Power, A. (2000) *Cities for a Small Country*, Faber and Faber, London.

Rollins, W.H. (1995) Whose landscape? Technology, Fascism and environmentalism on the National Socialist Autobahn, *Annals of the Association of American Geographers*, **85**, 494–520.

Ronald, L. (2003) The demographic transition: three centuries of fundamental change, *Journal of Economic Perspectives*, **17**(4), 167–90.

Rooum, D. (1995) *What is Anarchism?: An Introduction*, Freedom Press, London.

Rose, N. and Miller, P. (1992) Political power beyond the state: problematics of government, *British Journal of Sociology*, **43**, 173–205.

Ross, M. (2001) *Extractive Sectors and the Poor: An Oxfam America Report*, Oxfam. Available at: www.oxfamamerica.org/cirexport/index.html

Rössler, M. (1989) Applied geography and area research in Nazi society: central place theory and planning, *Environment and Planning D: Society and Space*, **7**, 419–31.

Rostow, W.W. (1960) *The Stages of Economic Growth: A Non-Communist Manifesto*, Cambridge University Press, Cambridge.

Rostow, W.W. (1971) *The Stages of Economic Growth*, 2nd edition, Cambridge University Press, Cambridge.

Rothenberg, T. (1995) 'And she told two friends': lesbians creating urban social space, in Bell, D. and Valentine, G. (eds) *Mapping Desire: Geographies of Sexualities*, Routledge, London, 165–81.

Routledge, P. (2003) Anti-geopolitics, in Agnew, J., Mitchell, K. and Toal, G. (eds) *A Companion to Political Geography*, Blackwell, Oxford, 236–48.

Rowell, A., Marriott, J. and Stockman, L. (2005) *The Next Gulf: London, Washington and Oil Conflict in Nigeria*, Constable, London.

Rubin, B. (1979) Aesthetic ideology and urban design, *Annals of the Association of American Geographers*, **69**(3), 339–61.

Rucht, D. and Roose, J. (1999) The green environmental movement at a crossroads, *Environmental Politics*, **8**(1), 59–80.

Rusten, G., Bryson, J.R. and Aarflot, U. (2007) Places through product and products through places: industrial design and spatial symbols as sources of competitiveness, *Norwegian Journal of Geography*, in press.

Rutledge, I. (2006) *Addicted to Oil: America's Relentless Drive for Energy Security*, I.B. Tarus, London.

S

Sachs, J.D. and Warner, A.M. (2001) Natural resources and economic development: the curse of natural resources, *European Economic Review*, **45**, 827–38.

Sachs, W. (1992) *The Development Dictionary*, ZedBooks, London.

Sack, R. (1986) *Human Territoriality: Its Theory and History*, Cambridge University Press, Cambridge.

Sack, R. (1992) *Place, Modernity and the Consumers World*, Johns Hopkins University Press, Baltimore, MD.

Sage, C. (2003) Social embeddedness and relations of regard: alternative 'good food' networks in south-west Ireland, *Journal of Rural Studies*, **19**, 47–60.

Salway, P. (ed.) (1981) Roman Britain, in *Oxford History of England*, Oxford University Press, Oxford.

Santos, M. (1974) Geography, Marxism and underdevelopment, *Antipode*, **6**, 1–9.

Sassatelli, R. and Scott, A. (2001) Novel food, new markets and trust regimes: responses to the erosion of consumers' confidence in Austria, Italy and the UK, *European Societies*, **3**, 213–44.

Sassen, S. (1991) *The Global City: New York, London, Tokyo*, Princeton University Press, Princeton, NJ.

Sassen, S. (1994) *Cities in a World Economy*, Pine Forge Press, London.

Sassen, S. (2001) *The Global City: New York, London, Tokyo*, Princeton University Press, Princeton, NJ.

Sassen, S. (2006) *Territory, Authority, Rights: From Medieval to Global Assemblages*, Princeton University Press, Princeton.

Savitch, H. and Kantor, P. (2002) *Cities in the International Marketplace: The Political Economy of Urban Development in North America and Western Europe*, Princeton University Press, Princeton.

Saxenian, A. (2006) *The New Argonauts: Regional Advantage in a Global Economy*, Harvard University Press, Cambridge, MA.

Sayer, A. and Walker, R. (1992) *The New Social Economy: Reworking the Division of Labour*, Blackwell, Oxford.

Schama, S. (1995) *Landscape and Memory*, HarperCollins, London.

Schifferes, S. (2007) Chrysler questions climate change, BBC News online, 10 January, http://news.bbc.co.uk/go/pr/fr/-/1/hi/business/6247371.stm

Schlosser, E. (2001) *Fast Food Nation*, Houghton Mifflin, New York.

Schneiders, L. (2006) Is wilderness racist? *Chain Reaction*, Autumn, 25–7.

Schumpeter, J. (1939) *Business Cycles: A Theoretical, Historical and Statistical Analysis of the Capitalist Process*, McGraw-Hill, London.

Schumpeter, J.A. (1942) *Capitalism, Socialism and Democracy*, Harper, New York.

Schuurman, F.J. (2001) Globalization and development studies: introducing the challenges, in Schuurman, F.J. (ed.) *Globalization and Development: Challenges for the 21st Century*, Sage, London, 3–16.

Schuurman, N. (2004) *GIS: A Short Introduction*, Blackwell, Oxford.

Scott, A. (2000) *The Cultural Economy of Cities: Essays on the Geography of Image-Producing Industries*, London, Sage.

Scott, A. (2004) A perspective on economic geography, *Journal of Economic Geography*, **4**, 479–99.

Scott, A.J. and Brown, E.R. (eds) (1993) *South-Central Los Angeles: Anatomy of an Urban Crisis*, The Ralph and Goldy Lewis Center for Regional Policy Studies, Working Paper Series, 06. Available at: http://repositories.cdlib.org/lewis/wps/06

Scott, J. and Simpson, P. (eds) (1991) *Sacred Spaces and Profane Spaces: Essays in the Geographics of Judaism, Christianity and Islam*, Greenwood Press, New York.

Seel, B. and Plows, A. (2000) Coming live and direct: strategies of Earth First! in Seel, B., Patterson, M. and Doherty, B. (eds) *Direct Action in British Environmentalism*, Routledge, London, 112–32.

Service, E.R. (1971) *Primitive Social Organization*, 2nd edition, Random House, New York.

Seton-Watson, H. (1977) *Nations and States: An Enquiry into the Origins of Nations and the Politics of Nationalism*, Westview Press, Boulder, CO.

Seyfang, G. (2002) Tackling social exclusion with community currencies: Learning from LETS to time banks, *International Journal of Community Currency Research*, **6**, http://www.uea.ac.uk/env/ijccr/

Seyfang, G. (2006) Ecological citizenship and sustainable consumption: examining local organic food networks, *Journal of Rural Studies*, **22**(4), 383–95.

Seyfang, G. and Smith, A. (2007) Grassroots innovation for sustainable development: towards a new research and policy agenda, *Environmental Politics*, **16**(4), 584–603.

Shabi, R. (2002) The e-waste land, *The Guardian*, 30 November.

Shakespeare, T. (1994) Cultural representations of disability: dustbins for disavowal, *Disability and Society*, **9**, 283–99.

Sharp, J. (1999) Critical geopolitics, in Cloke, D., Crang, P. and Goodwin, M. (eds) *Introducing Human Geographies*, Arnold, London, 179–88.

Sheehan, J.M. (2000) *The Greening of the World Bank: A Lesson in Bureaucratic Survival*, Cato Institute Foreign Policy Briefing No. 56, April 12. Available from the Cato Institute: http://www.cato.org/

Sheppard, E. and Barnes, T.J. (eds) (2000) *A Companion to Economic Geography*, Blackwell, Oxford.

Sherratt, A. (ed.) (1980) *The Cambridge Encyclopaedia of Archaeology*, Cambridge University Press, Cambridge.

Shields, R. (1991) *Places on the Margin*, London, Routledge.

Shilling, C. (1993) *The Body and Social Theory*, Sage, London.

Shirlow, P. and Murtagh, B. (2006) *Belfast: Segregation, Violence and the City*, Pluto Press, London.

Shiva, V. (1999) We will drive Monsanto out of the country, *Asian Age*, 6 March, 9.

Shiva, V. (2002) *Water Wars: Privatization, Pollution, and Profit*, Pluto Books, London.

Shiva, V. and Moser, I. (1995) *Biopolitics: A Feminist and Ecological Reader on Biotechnology*, ZedBooks, London.

Short, J.R. (1989) Yuppies, yuffies and the new urban order, *Transactions of the Institute of British Geographers*, NS, **14**, 173–88.

Short, J.R. (1996) *The Urban Order: An Introduction to Cities, Culture and Power*, Blackwell, Oxford.

Shurmer-Smith, P. (ed.) (2002) *Doing Cultural Geography*, Sage, London.

Shurmer-Smith, P. and Hannam, K. (1994) *Worlds of Desire, Realms of Power: A Cultural Geography*, Arnold, London.

Sibalis, M. (2004) Urban space and homosexuality: the example of the Marais, Paris 'Gay Ghetto', *Urban Studies*, **41**(9), 1739–58.

Sibley, D. (1995) *Geographies of Exclusion: Society and Difference in the West*, Routledge, London.

Sibley, D. (1999) Creating geographies of difference, in D. Massey, J. Allen and P. Sarre (eds) *Human Geography Today*, Polity, Cambridge, 115–28.

Sidaway, J.D. (1998) What is in a Gulf? From the 'arc of crisis' to the Gulf War, in Ó Tuathail, G. and Dalby, S. (eds) *Rethinking Geopolitics*, Routledge, London and New York, 224–39.

Sidaway, J.D. (1999) American power and the Portuguese Empire, in Slater, D. and Taylor, P.J. (eds) *The American Century*, Blackwell, Oxford, 195–209.

Sidaway, J.D. (2000) Iberian geopolitics, in Dodds, K. and Atkinson, D. (eds) *Geopolitical Traditions: A Century of Geopolitical Thought*, Routledge, London, 118–49.

Sidaway, J.D. (2003a) Sovereign excesses? Portraying postcolonial sovereigntyscapes, *Political Geography*, **22**, 157–78.

Sidaway, J.D. (2003b) Intervention: banal geopolitics resumed, *Antipode*, **35**(4), 645–51.

Sidaway, J.D. and Power, M. (2005) The tears of Portugal: race and destiny in Portuguese geopolitical narratives, *Environment and Planning D: Society and Space*, **23**(4), 527–54.

Sidaway, J.D. and Pryke, M. (2000) The free and the unfree: 'emerging markets', the Heritage Foundation and the 'index of economic freedom', in Bryson, J.R., Daniels, P.W., Henry, N.D. and Pollard, J.S. (eds) *Knowledge, Space, Economy*, Routledge, London.

Sidaway, J.D., Grundy-Warr, C., Park, Bae Gyoon (2005) Asian sovereignty scapes, Editorial, *Political Geography*, **24**, 779–83.

Simmel, G. (1978) *The Philosophy of Money*, Routledge & Kegan Paul, London.

Simmel, G. (1991) Money in modern culture, *Theory, Culture and Society*, **8**, 17–31.

Simmons, I.G. (1996) *Changing the Face of the Earth: Culture, Environment, History*, 2nd edition, Blackwell, Oxford.

Skelton, T. and Valentine, G. (eds) (1998) *Cool Places: Geographies of Youth Cultures*, Routledge, London.

Skidmore, T.E. and Smith, P.H. (1988) *Modern Latin America*, 2nd edition, Oxford University Press, New York.

Slater, C. (1995) Amazonia as Edenic narrative, in Worster, D. (ed.) *Uncommon Ground: Reinventing Nature*, Norton, New York, 114–59.

Slater, D. (1993) The geopolitical imagination and the enframing of development theory, *Transactions of the Institute of British Geographers*, NS, **18**, 419–37.

Slater, D. (1997) *Consumer Culture and Modernity*, Polity Press, Cambridge.

Slater, D. (2003) Beyond Euro-Americanism: democracy and post-colonialism, in Anderson, K., Domosh, M., Pike, S. and Thrift, N. (eds) *Handbook of Cultural Geography*, Sage, London, 420–32.

Slater, N. (2001) Food and eating, *The Observer Food Monthly*, April 2001.

Smith, A. (1776) *An Inquiry into the Nature and Causes of the Wealth of Nations*, edited by Skinner, A. (1974) Penguin, London.

Smith, A. (1977) *The Wealth of Nations*, edited by Skinner, A. Penguin, Harmondsworth.

Smith, A.D. (1988) *The Ethnic Origin of Nations*, Blackwell, Oxford.

Smith, D.M. (1995) *Geography and Social Justice*, Blackwell, Oxford.

Smith, D.P. and Holt, L. (2007) Studentification and 'apprentice' gentrifiers within Britain's provincial towns and cities: extending the meaning of gentrification, *Environment and Planning A*, **39**(1), 142–61.

Smith, M.D. (1996) The empire filters back: consumption, production, and the politics of Starbucks coffee, *Urban Geography*, **17**, 502–24.

Smith, N. (1984) Isiah Bowman: political geography and geopolitics, *Political Geography Quarterly*, **3**, 69–76.

Smith, N. (1996) *The New Urban Frontier: Gentrification and the Revanchist City*, Routledge, London.

Smith, N. (2002) New globalism, new urbanism, gentrification as global urban strategy, *Antipode*, **34**, 434–57.

Smith, N. (2003) *American Empire: Roosevelt's Geographer and the Prelude to Globalization*, University of California Press, Berkeley.

Smith, S. (1989) *The Politics of Race and Residence*, Polity, Cambridge.

Smith, S.J. (2005) Society–space, in Cloke, P., Crang, P. and Goodwin, M. (eds) *Introducing Human Geographies*, 2nd edition, Hodder Arnold, London, 18–33.

Sogge, D. (2002) *Give and Take: What's the Matter with Foreign Aid?*, Zed Books, London.

Soja, E.J. (1989) *Postmodern Geographies: The Reassertion of Space in Critical Social Theory*, Verso, London.

Soja, E. (2000) *Postmetropolis: Critical Studies of Cities and Regions*, Blackwell, Oxford.

Sondheimer, S. (ed.) (1991) *Women and the Environment: A Reader*, Zed Books, London.

Soper, K. (1995) *What is Nature?* Blackwell, Oxford.

Spain, D. (1992) *Gendered Spaces*, University of North Carolina Press, Chapel Hill.

Speer, A. (1995) *Inside the Third Reich*, Phoenix, London.

Spicker, P. (2006) *The Idea of Poverty*, Polity, Bristol.

Spillius, A. (2003) Dialing tone, *Telegraph Magazine*, 19 April, 38–43.

Spirn, A.W. (1997) The authority of nature: conflict and confusion in landscape architecture, in Wolschke-Bulmahn, J. (ed.) *Nature and Ideology: Natural Garden Design in the Twentieth Century*, Dumbarton Oaks Research Library and Collection, Washington, 249–61.

Spivak, G.C. (1985) Three women's texts and a critique of imperialism, *Critical Inquiry*, **12**, 243–61.

Stamp, L.D. and Beaver, S. (1933) *The British Isles: A Geographic and Economic Survey*, Longman, London.

Stamp, L.D. and Beaver, S. (1963) *The British Isles: A Geographic and Economic Survey*, 5th edition, Longman, London.

Stannard, K. (2003) Earth to academia: on the need to reconnect university and school geography, *Area*, **25**, 316–22.

Stempel, J.D. (1981) *Inside the Iranian Revolution*, Indiana University Press, Bloomington.

Stern Review (2006) *Stern Review: The Economics of Climate Change*, HM Treasury, London.

Stevens, P. (2003) *Resource Impact – Curse or Blessing? A Literature Survey*, Centre for Energy, Petroleum and Mineral Law and Policy, electronic journal, available at: www.cepmlp.org

Stienberg, P.E. (2001) *The Social Construction of the Ocean*, Cambridge University Press, Cambridge.

Stobart, J. and Raven, N. (eds) (2004) *Towns, Regions and Industries: Urban and Industrial Change in the Midlands, c. 1700–1840*, Manchester University Press, Manchester.

Stöhr, W.B. and Taylor, D.R.F. (1981) *Development from Above or Below? The Dialectics of Regional Planning in Developing Countries*, John Wiley, Chichester.

Storey, D. (2001) *Territory: The Claiming of Space*, Prentice Hall, Harlow.

Storey, D. (2002) Territory and national identity: examples from the former Yugoslavia, *Geography*, **87**(2), 108–15.

Storey, D. (2003) *Citizen, State and Nation*, Geographical Association, Sheffield.

Storper, M. (1997) *The Regional World: Territorial Development in a Global Economy*, The Guilford Press, New York.

Strange, S. (1996) *Casino Capitalism*, Blackwell, Oxford.

Strange, S. (1996a) *The Retreat of the State: The Diffusions of Power in the World Economy*, Cambridge University Press, Cambridge.

Strange, S. (1999) *Mad Money: When Markets Outgrow Governments*, University of Michigan Press, Ann Arbor, MI.

Stutz, F.P. and de Souza, A.R. (1998) *The World Economy: Resources, Location, Trade and Development*, 3rd edition, Prentice-Hall, Upper Saddle River, NJ.

Sugden, D. (1982) *Arctic and Antarctic*, Basil Blackwell, Oxford.

Sukarno, I. (1955) Modern History Sourcebook: President Sukarno of Indonesia: Speech at the opening of the Bandung conference, 18 April 1955, online, available at: www.fordham.edu/halsall/mod/1955sukarno-bandong.html, accessed 7 April 2003.

Sum, N-L. and Ngai, P. (2005) Globalization and paradoxes of ethical transnational production: code of conduct in a Chinese workplace, *Competition and Change*, **9**, 181–200.

Sumartojo, R. (2004) Contesting place: Antigay and lesbian hate crime in Columbus, Ohio, in Flint, C. (ed.) *Spaces of Hate: Geographies of Discrimination and Intolerance in the USA*, Routledge, New York, 87–107.

Sweetman, C. (ed.) (2000) *Gender in the Twenty-first Century*, Oxfam, Oxford.

Swyngedouw, E. (2000) Authoritarian governance, power, and the politics of rescaling, *Environment and Planning D: Society and Space*, **18**, 63–76.

Sylvester, C. (1996) Picturing the Cold War: an art graft/eye graft, *Alternatives*, **21**, 393–418.

T

Takahashi, L.M. and Dear, M.J. (1997) The changing dynamics of community opposition to human service facilities, *Journal of the American Planning Association*, **63**, 79–93.

Takeuchi, K., Katoh, K., Nan, Y. and Kou, Z. (1995) Vegetation change in desertified Kerqin sandy lands, *Inner Mongolia/Geographical Reports*, Tokyo Metropolitan University, **30**, 1–24.

Talbot, J.M. (2004) *Grounds for Agreement: The Political Economy of the Coffee Commodity Chain*, Rowman and Littlefield, Lanham, MD.

Tan, T.Y. and Kudaisya, G. (2002) *The Aftermath of Partition in South Asia*, Routledge, London.

Taussig, M. (1997) *The Magic of the State*, Routledge, London.

Taylor, I. (1997) Running on empty, *The Guardian*, 14 May, 2–6.

Taylor, P. (2004) *World City Network: A Global Urban Analysis*, Routledge, London.

Taylor, P.J. (1989) *Political Geography*, Longman, London.

Taylor, P.J. (1994) From heartland to hegemony: changing the world in political geography, *Geoforum*, **25**, 403–17.

Taylor, P.J. and Flint, C. (2000) *Political Geography: World-Economy, Nation State and Locality*, Prentice Hall, Harlow.

Taylor, S. and Tyler, M. (2000) Emotional labour and sexual difference in the airline industry, *Work, Employment and Society*, **14**(1), 77–95.

Tesfahuney, M. (1998) Mobility, racism and geopolitics, *Political Geography*, **17**, 499–515.

The Economist (1999), The Next Shock, 4 March, 29. Available electronically from www.Economist.com.

The Independent (2004) Students no longer welcomed, *The Independent*, 21 October, 11.

The Times Archaeology of the World (1999) New edition, Times Books, London.

The Times Atlas of World History (1989) 3rd edition, Guild Publishing, London.

The Times Atlas of World History (1999) New edition, Times Books, London.

The Wilderness Society (2006) A Fringe of Green – Protecting Australia's Forests and Woodlands, available at: http://www.wilderness.org.au/campaigns/forests/fringe/#why-save

Thomas, D. and Middleton, N. (1994) *Desertification: Exploding the Myth*, Wiley, Chichester.

Thompson, E.P. (1963) *The Making of the English Working Class*, Penguin, London and New York.

Thompson, S. (2005) 'Territorialising' the school playground: deconstructing the geography of playtime, *Children's Geographies*, **3**(1), 63–78.

Thoreau, H.D. (1854) *Walden; or Life in the Woods*, Ticknor and Fields, Boston.

Thrift, N. (1996) A hyperactive world, in Johnston, R.J., Taylor, P.J. and Watts, M. (eds) *Geographies of Global Change*, Blackwell, Oxford.

Thrift, N. (1998) Virtual capitalism: the globalisation of reflexive business knowledge, in Carrier, J.G. and Miller, D. (eds) *Virtualism: A New Political Economy*, Berg, Oxford.

Thrift, N. and Kitchin, R. (2008) *The International Encyclopedia of Human Geography*, Elsevier, London.

Thrift, N.J. (2005) *Knowing Capitalism*, Sage, London.

Tickell, A. (2003) Cultures of money, in Anderson, K., Domosh, M., Pile, S. and Thrift, N. (eds) *Handbook of Cultural Geography*, Sage, London, 116–30.

Tiffin, M., Mortimore, M. and Gickuki, F. (1994) *More People, Less Erosion: Environmental Recovery in Kenya*, John Wiley, Chichester.

Time Out (1995) The world on a plate, *Time Out*, 16 August, 31–7.

Tissot, S. and Poupeau, F. (2005) *La spatilisation des problems sociaux*, Actes de la recherché en Sciences socials, **159** (September) 4–9.

Toennies, F. (1887) *Community and Society* (1957 edn), Michigan State University Press, East Lansing, MI.

Tokar, B. (1994) *The Green Alternative*, 2nd edition, R. and E. Miles, San Pedro, CA.

Tomlinson, J. (1991) *Cultural Imperialism: A Critical Introduction*, Pinter, London.

Tønnesson, S. and Antlöv, H. (1996) Asian in theories of nationalism and national identity, in Tønnesson, S. and Antlöv, H. (eds) *Asian Forms of the Nation*, Curzon, Richmond, 1–40.

Tranberg-Hansen, K. (1994) Dealing with used clothing: 'Salaula' and the construction of identity in Zambia's Third Republic, *Public Culture*, **6**, 503–23.

Tratado de Limites (1866) *Tratado de Limites entre Portugal e Espanha assinado em Lisboa aos 29 de Setembro de 1864*, Imprensa Nacional, Lisbon.

Truman, H. (1949) *Public Papers of the President, 20 January*, US Government Printing Offices, Washington, DC.

Tuan, Yi-Fu (1999) *Who am I? An Autobiography of Emotion, Mind and Spirit*, University of Wisconsin Press, Madison, WI.

Tuan, Yi-Fu (2002) *Dear Colleague: Common and Uncommon Observations*, University of Minnesota Press, Minneapolis MI.

Tunstall, J. (1977) *The Media are American*, Constable, London.

Tunstall, R. and Coulter, A. (2006) *Twenty-Five Years on Twenty Estates: Turning the Tide*, Policy Press, Bristol.

Tyler May, E. (1989) Cold War – warm hearth: politics and the family in postwar America, in Fraser, S. and Gerstle, G. (eds) *The Rise and Fall of the New Deal Order, 1930*, Princeton University Press, Princeton, NJ, 153–81.

U

UK Universities (2005) *Studentification: a guide to challenges, opportunities and good practice.* UK Universities, London.

UN *see* United Nations.

UN Millennium Development Goals (2000) Available at: http://www.millenniumgoals.org/

Understanding Cities (1999) DD304, Open University, Milton Keynes.

UNAIDS (2006) *Report on the global AIDS epidemic 2006*, available to download at: http://www.unaids.org/en/HIV_data/2006GlobalReport/default.asp

UNDP *see* United Nations Development Programme.

UNEP *see* United Nations Environment Programme.

UN-HABITAT (2006) The state of the world's cities 2006/7, *The Millennium Development Goals and Urban Sustainability*, Earthscan, London.

UNICEF *see* United Nations Children's Fund.

United Nations (UN) (2001) *The state of the world's cities Istanbul*, Habitat Publications Unit, available at: http://www.unhabitat.org/Istanbul+5/statereport.htm

United Nations (UN) (2002) *International Migration Report 2002*, Department of Economic and Social Affairs, United Nations, New York.

United Nations (UN) (2002a) *Manual on Statistics on International Trade in Services*, United Nations, New York.

United Nations (UN) (2003) *World Population Prospects: The 2002 Revision*, United Nations Population Division, New York.

United Nations (UN) (2006a) *Population Challenges and Development Goals*, available to download at: http://www.un.org/esa/population/publications/pop_challenges/Population_Challenges.pdf

United Nations (UN) (2006b) *World Urbanisation Prospects – the 2005 update*, United Nations, New York.

United Nations Children's Fund (UNICEF) (2000) *The Progress of Nations 2000*, online, available at: www.unicef.org, accessed 22 February 2002.

United Nations Children's Fund (UNICEF) (2006) *The State of the World's Children: Excluded and Invisible*, UNICEF, New York.

United Nations Conference on Trade and Development (UNCTAD) (2006) http://stats.unctad.org/FDI/TableViewer/tableView.aspx?ReportId=5, accessed 31 December 2006.

United Nations Development Programme (UNDP) (1998) *Human Development Report 1998*, Oxford University Press, Oxford.

United Nations Environment Programme (UNEP) (1999) *Global Environmental Outlook 2000*, UNEP-Earthscan Publications, London.

United Nations Population Division (1998) *Briefing Packet*, 1998 Revision of World Population Prospects.

United Nations Population Division (2004) *World Population Prospects: The 2004 Revision*, available at: www.un.org/esa/population/publications/WPP2004/wpp2004.htm

United Nations Population Division (2006) *World Population Prospects: The 2006 Revision*, available at: http://esa.un.org/unpp/index.asp?panel=2

United Nations World Tourism Organization (UNWTO) (2006) *World Tourism Highlights*, UNWTO, Madrid.

Urry, J. (1990) *The Tourist Gaze: Leisure and Travel in Contemporary Societies*, Sage, London.

Urry, J. (2000) *Sociology Beyond Society*, Routledge, London.

V

Valentine, G. (1989) The geography of women's fear, *Area*, **21**(4), 385–90.

Valentine, G. (1995) Out and about: geographies of lesbian landscapes, *International Journal of Urban and Regional Research*, **19**(1), 96–111.

Valentine, G. (2001) *Social Geographies: Space and Society*, Prentice Hall, Harlow.

van Schendel, W. (2005) Spaces of engagement: How borderlands, illegal flows, and territorial states interlock, in Abrahams, I. and van Schendel, W. (eds) *Illicit Flows and Criminal Things*, Indiana University Press, Bloomington and Indianapolis, 38–68.

Van Welsum, D. and Reif, X. (2005) *Potential Offshoring: Evidence from Selected OECD Countries*, OECD, DSTI-ICCP.

Venn, L., Kneafsey, M., Holloway, L., Cox, R., Dowler, E. and Tuomainen, H. (2006) Researching European 'alternative' food networks: some methodological considerations, *Area*, **38**, 248–58.

Visser, M. (1986) *Much Depends on Dinner*, Collier / Macmillan, New York.

Visvanathan, N., Duggan, L. and Nisonoff, L. (eds) (1997) *The Women, Gender and Development Reader*, ZedBooks, London.

W

Wackernagel, M. and Rees, W.E. (1998) *Our Ecological Footprint: Reducing Human Impact on the Earth*, New Society Publishers, Gabriola Island, BC, Canada.

Wacquant, L. (2007) *Urban Outcasts: A Comparative Sociology of Advanced Marginality*, Polity Press, Cambridge.

Wacquant, L.J.D. (1993) Urban outcasts: stigma and division in the black American ghetto and the French urban periphery, *International Journal of Urban and Regional Research*, **17**, 366–83.

Wacquant, L.J.D. (1995) The ghetto, the state and the new capitalist economy, in Kasinitz, P. (ed.) *Metropolis Centre and Symbol of our Times*, Macmillan, London, 418–49.

Wade, R. (2001) Winners and losers, *The Economist*, April, 79–82.

Wait, G., McGuirk, P., Dunn, K., and Hartig, K. (eds) (2000) *Introducing Human Geography: Globalization, Difference and Inequality*, Longman, French's Forest, NSW.

Walker, R. (1996) Another round of globalization in San Francisco, *Urban Geography*, **17**, 60–94.

Wall, D. (1999) *Earth First! and the Anti-roads Movement*, Routledge, London.

Wallace, D. and Wallace, R. (1998) *A Plague on Your Houses: How New York was Burned Down and How National Public Health Crumbled*, Methuen, London.

Wallace, D. and Wallace, R. (2000) Life and death in Upper Manhattan and the Bronx: toward an evolutionary perspective on catastrophic social change, *Environment and Planning A*, **32**, 1245–66.

Wallace, I. (1985) Towards a geography of agribusiness, *Progress in Human Geography*, **9**, 491–514.

Wallerstein, I. (1979) *The Capitalist World Economy*, Cambridge University Press, Cambridge.

Wallerstein, I. (1980) Imperialism and development, in Bergeson, A. (ed.) *Studies of the Modern World System*, Academic Press, New York.

Wallerstein, I. (1991) *Geopolitics and Geoculture: Essays on the Changing World System*, Cambridge University Press, Cambridge.

Wallerstein, I. (1994) Development: lodestar or illusion?, in Sklair, L. (ed.) *Capitalism and Development*, Routledge, London, 3–20.

Wa Ngugi, M. (2005) *New Orleans and the Third World*, http://www.zmag.org/content/showarticle.cfm?ItemID=8694 Accessed 23 May 2007.

Ward, D. (1987) Population growth, migration, and urbanization, 1860–1920, in Mitchell R.D. and Groves, P. A. (eds) *North America: The Historical Geography of a Changing Continent*, Hutchinson, London.

Ward, D. (1989) *Poverty, Ethnicity and the American City, 1840–1925*, Cambridge University Press, Cambridge.

Warhurst, C., Nickson, D., Witz, A. and Cullen, A. (2000) Aesthetic labour in interactive service work: some case study evidence from the 'new' Glasgow, *The Service Industries Journal*, **20**(3), 1–18.

Watney, S. (1987) *Policing Desire: Pornography, AIDS and the Media*, Methuen, London.

Watnick, M. (1952–3) The appeal of Communism to the peoples of underdeveloped areas, *Economic Development and Cultural Change*, **1**, 22–36.

Watson, J.L. (1997a) Introduction: transnationalism, localization, and fast foods in East Asia, in Watson J.L. (ed.) *Golden Arches East: McDonalds in East Asia*, Stanford University Press, Stanford, CA, 1–38.

Watson, J.L. (1997b) McDonalds in Hong Kong: consumerism, dietary change, and the rise of a children's culture, in Watson, J.L. (ed.) *Golden Arches East: McDonalds in East Asia*, Stanford University Press, Stanford, CA, 77–109.

Watts, D.C.H., Ilbery, B. and Maye, D. (2005) Making re-connections in agro-food geography: alternative systems of food provision, *Progress in Human Geography*, **29**(1), 22–40.

Weatherell, C., Tregear, A. and Allinson, J. (2003) In search of the concerned consumer: UK public perceptions of food, farming and buying local, *Journal of Rural Studies*, **19**, 233–44.

Weber, A. (1929) *Alfred Weber's Theory of the Location of Industries*, Chicago University Press, Chicago.

Webster, D. (1988) *Looka Yonder! The Imaginary America of Populist Culture*, Comedia, London.

Weinrich, J. (1982) Is homosexuality biologically natural?, in Paul, W., Weinrich, J., Gonsiorek, J. and Hotvedt, M. (eds) *Homosexuality: Social, Psychological and Biological Issues*, Sage, Beverley Hills, 197–208.

Wellington, C. and Bryson, J.R. (2001) At face value? Image consultancy, emotional labour and professional work, *Sociology*, **35**(4), 933–46.

Werbner, P. (1997) Introduction: The dialectics of cultural hybridity, in Werbner, P. and Modood, T. (eds) *Debating Cultural Hybridity: Multi-Cultural Identities and the Politics of Anti-Racism*, ZedBooks, London, 1–26.

Whatmore, S. (1995) From farming to agribusiness: the global agro-food system, in Johnston, R., Taylor, P. and Watts, M. (eds) *Geographies of Global Change*, Blackwell, Oxford, 36–49.

Whatmore, S. (2002) *Hybrid Geographies*, Routledge, London.

Whatmore, S., Stassart, P. and Renting, H. (2003) What's alternative about alternative food networks?, *Environment and Planning A*, **35**, 389–91.

Whitelegg, J. (1993) *Transport for a Sustainable Future*, Belhaven Press, London.

Whittlesey, D. (1936) Major agricultural regions of the earth, *Annals of the Association of American Geographers*, **26**, 199–240.

Wiener, M.J. (1985) *English Culture and the Decline of the Industrial Spirit, 1850–1980*, Penguin, Harmondsworth.

Williams, C.C. (1996) Local exchange and trading systems: a new source of work and credit for the poor and unemployed?, *Environment and Planning A*, **28**, 1395–415.

Williams, P. and Hubbard, P. (2001) Who is disadvantaged? Retail change and social exclusion, *International Review of Retail, Distribution and Consumer Research*, **11**, 267–86.

Willis, K. (2005) *Theories and Practices of Development*, Routledge, London.

Willis, P. (1990) *Uncommon Culture*, Open University Press, Milton Keynes.

Willis, R.P. (ed.) (2001) New Zealand in the 1990s, special edition of *Asia Pacific Viewpoint*, **42**, 1.

Wilson, A. (1992) Technological utopias, world's fairs and theme parks, in Wilson, A. (ed.) *The Culture of Nature: North American Landscape from Disney to the Exxon Valdez*, Blackwell, Oxford, 157–90.

Wilson, D. (2005) *Inventing Black-on-Black Violence: Discourse, Space, Representation*, Syracuse University Press, Syracuse.

Wilson, G. and Rigg, J. (2003) 'Post-productivist' agricultural regimes and the South: discordant concepts?, *Progress in Human Geography*, **27**(6), 681–707.

Wilson, G.A. (2001) From productivism to post-productivism and back again? Exploring the (un) changed natural and mental landscapes of European agriculture, *Transactions of the Institute of British Geographers*, **26**(1), 77–102.

Wilton, R. (1996) Diminished worlds? The geography of everyday life with HIV/AIDS, *Health and Place*, **2**, 69–83.

Winchester, H. and White, P. (1988) The location of marginalised groups in the inner city, *Environment and Planning D: Society and Space*, **6**, 37–54.

Winchester, H.P.M., Kong, L. and Dunn, K. (2003) *Landscapes: Ways of Imagining the World*, Prentice Hall, Harlow.

Winichakul, T. (1995) *Siam Mapped: A History of the Geo-body of a Nation*, University of Hawaii Press, Hawaii.

Winichakul, T. (1996) Maps and the formation of the geo-body of Siam, in Tønnesson, S. and Antlöv, H. (eds) *Asian Forms of the Nation*, Curzon Press, Richmond, 67–91.

Winter, M. (2003a) Geographies of food: agro-food geographies – making reconnections, *Progress in Human Geography*, **27**, 505–13.

Winter, M. (2003b) Embeddedness, the new food economy and defensive localism, *Journal of Rural Studies*, **19**, 23–32.

Wirth, L. (1938) Urbanism as a way of life, *American Journal of Sociology*, **44**, 1–24, reprinted in Hart, P.K. and Reiss, A.J. (eds) (1957) *Cities and Society*, Free Press, Chicago.

Wiston Spirn, A. (1997) The authority of nature: conflict and confusion in landscape architecture, in Wolschke-Bulmahn, J. (ed.) *Nature and Ideology: Natural Garden Design in the Twentieth Century*, Dumbarton Oaks Research Library and Collection, Washington, DC, 249–61.

Withers, C.W.J. and Mayhew, R.J. (2002) Rethinking 'disciplinary' history: geography in British universities, c.1580–1887, *Transactions of the Institute of British Geographers*, **27**, 11–29.

Wolf, E.R. (1966) *Peasants*, Prentice-Hall, Englewood Cliffs, NJ.

Wolfe, T. (1988) *Bonfire of the Vanities*, New York, Bantam.

Wolfe-Phillips, L. (1987) Why Third World: origins, definitions and usage, *Third World Quarterly*, **9**, 1311–19.

Wolters, O.W. (1982) *History, Culture, and Religion in Southeast Asian Perspectives*, Institute of Southeast Asian Studies, Singapore.

Women and Geography Study Group (1997) *Feminist Geographies: Explorations in Diversity and Difference*, Longman, Harlow.

Wood, A. and Welch, C. (1998) *Policing the Policemen: The Case for an Independent Evaluation Mechanism for the IMF*, Bretton Woods Project, Friends of the Earth US, London.

Wood, G. (1985) *Labelling in Development Policy: Essays in Honour of Bernard Schaffer*, Sage, London.

Woods, C. (1998) *Development Arrested: Race, Power and Blues in the Mississippi Delta*, Verso, London.

Woodward, K. (1997) Concepts of identity and difference, in Woodward, K. (ed) *Identity and Difference*, Sage, London, 8–59.

World Bank (1999) *World Development Indicators*, World Bank, Washington, DC, also available as *World Development Indicators on CD-Rom*, World Bank, Washington, DC.

World Bank (2001a) *World Development Indicators*, World Bank, Washington, DC.

World Bank (2001b) *Attacking Poverty: World Development Report 2000/2001*, Oxford University Press and World Bank, Washington, DC.

World Bank (2002) *World Development Report 2002: Building Institutions for Markets*, World Bank, Washington, DC.

World Bank (2005) *World Development Indicators 2005*, World Bank, Washington, DC.

World Bank (2006) *World Development Report: Equity and Development*, World Bank, Washington, DC.

World Bank (2007) *The Little Green Data Book 2007*, World Bank, Washington DC.

World Federation of Exchanges (WFE) (2006) Data available at: http://www.world-exchanges.org/WFE/home.asp?menu=378&document=3557

Wright Mills, C. (1959) *White Collar Work: The American Middle Classes*, Oxford University Press, New York.

Wright, R. (1995) *The Color Curtain: A Report on the Bandung Conference*, Banner Books, New York.

Wrigley, N. (2002) 'Food deserts' in British cities: policy context and research priorities, *Urban Studies*, **39**, 2029–40.

Wrigley, N. and Lowe, M. (eds) (1996) *Retailing, Consumption and Capital: Towards the New Retail Geography*, Longman, Harlow.

Wrigley, N. and Lowe, M. (2002) *Reading Retail*, Arnold, London.

Wrigley, N., Warm, D. and Margetts, B. (2003) Deprivation, diet and food-retail access: findings from the Leeds 'food deserts' study, *Environment and Planning A*, **35**, 151–88.

Wylie, J. (2007) *Landscape*, Routledge, London.

Y

Yan, Y. (1997) McDonalds in Beijing: the localization of Americana, in Watson, J.L. (ed.) *Golden Arches East: McDonalds in East Asia*, Stanford University Press, Stanford, CA, 39–76.

Yang, Y. (2006) The Taiwanese notebook computer production network in China: implications for upgrading of the Chinese electronics industry, *The Personal Computer Industry Center publication*, UC Irvine (see http://www.pcic.gsm.uci.edu/pubs.asp).

Yawnghwe, C.T. (1987) *The Shan of Burma: Memoirs of a Shan Exile*, Institute of Southeast Asian Studies, Singapore.

Yeates, N. (2004) Global care chains: critical reflections and lines of enquiry, *International Feminist Journal of Politics*, **6**(3), 369–91.

Yergin, D. (2006) Ensuring energy security, *Foreign Affairs*, **85**(2), 69–82.

Yuval-Davis, N. (1997) *Gender and Nation*, Sage, London.

Yuval-Davis, N. (2003) Citizenship, territoriality and the gendered construction of difference, in Brenner, N., Jessop, B., Jones, M. and MacLeod, G. (eds) *State/Space: A Reader*, Blackwell, Oxford, 309–25.

Z

Zelizer, V. (1989) The social meaning of money: 'special monies', *American Journal of Sociology*, **95**, 342–77.

Zouza, L.B. (2002) Vital interests and budget deficits: US Foreign Aid after September 11, *Middle East Insight*, March–April, 25–9.

Zukin, S. (1989) *Loft Living: Culture and Capital in Urban Change*, Rutgers University Press, Newark, NJ.

Zukin, S. (1995) *The Cultures of Cities*, Blackwell, Oxford.

Index

D

G

O

P

Q

R

MANSFIELD COLLEGE
LIBRARY
OXFORD